The Environment
Issues and Choices for Society

The Environment
Issues and Choices for Society

Penelope ReVelle
Essex Community College

Charles ReVelle
The Johns Hopkins University

ΙΛΛ⊑ Willard Grant Press
Boston, Massachusetts

PWS PUBLISHERS

Prindle, Weber & Schmidt · 🐾 · Willard Grant Press · **WG** · Duxbury Press · ♠
Statler Office Building · 20 Providence Street · Boston, Massachusetts 02116

10 9 8 7 6 5 4 3 2

To Parents and Children,
the Two Ends of the Living Arrow of Which We Are a Part

Preface

Students who enroll in courses on the environment often come from a variety of academic backgrounds. This book is designed so that differences in preparation will present as few problems as possible. Specifically, we have organized most of the chapters so that there are three types of material.

In the typical chapter, the general text section provides an overview of each topic and a thorough explanation at a level suitable for all college students. After reading and absorbing this material, students should have a solid foundation in the topic and should be able to take part in class discussions. Graphs, line drawings, and photographs illustrate the material in this section.

The second section is most often imbedded within the first section. This second section includes controversial issues which are set off in boxes from the general text. The controversies are examples of disagreements on environmental policy and contain capsule quotes from opposing parties. In these debates, sometimes surprising and other times predictable conflicts arise. In many of the selections, the identities of the hero and villain are not at all clear. Differing viewpoints are argued, popular and unpopular causes are debated, and the difficulties inherent in making fair, effective laws are illustrated.

In some of these exchanges, even experts may disagree on environmental matters. We hope this exposure will encourage students to sift facts and come to their own conclusions rather than take the word of any one expert. In doing so, students will discover a surprising fact: many questions do not have answers. There are only social agreements, sometimes expressed in laws. The agreements represent society's consensus at the moment and are open to change.

A third section contains supplementary material designed to enrich study of the specific topic. This third section is noted by a shaded asterisk. Students and instructors desiring a more technical treatment and more of the basic scientific information on a particular topic, or further information on an issue, will find such material here.

Use of the supplementary material will vary depending on the particular interests of the students and instructors. In the air pollution chapter, for example, biologists may want to read more about the effects of carbon monoxide on oxygen transport to the cells. Chemistry majors may want to study the chemistry of photochemical air pollution. Physics majors will find a description of the breeder reactor of interest. Although formal coursework will often be of help in pursuing the subject in the supplementary sections, simply an interest in the subject will be sufficient, especially when the supplementary section is devoted to completing the discussion of a particular topic by the addition of related but not central or crucial subjects. For example, the chapter on wildlife resources is followed by a discussion of eskimo survival, which is set against survival of the endangered bowhead whale.

Two additional features of the organization of this text should be mentioned. First, we have provided five chapters on ecology in order to introduce needed ecological concepts. Each begins an appropriate section of the book. Thus, a chapter on the properties of water and the communities found in water environments introduces the section on Water Resource Problems. In a similar fashion, a chapter on weather and climate precedes the chapters on air pollution. In this way we hope the reader will see the importance of ecological concepts to the understanding of environmental issues. Finally, at the end of each chapter, there is a list of books and articles covering the subjects in the chapter. These references are described briefly to help students choose the ones that will be most helpful on a particular topic.

For the instructor, the authors have prepared an instructional package that contains a variety of materials intended to help organize and enrich environment courses. The Instructor's Manual, available separately from the publisher, contains additional references, more controversies, suggestions for class projects, and a set of questions with answers to test comprehension. The Manual also includes names and addresses of distributors of environmental films, environmental film strips, cassettes, and other audiovisual materials. To supplement the text further, the authors have assembled a set of slides on key environmental topics. The slide set is available from the publisher to departments using this book.

Acknowledgments

Many people influenced this book and made it possible. There is, first of all, our good friend John Riina, who recognized in our ideas and prose new possibilities; John steered us with a gentle hand toward creation of a new and better shape for our proposed text. Then there were the patient editors at D. Van Nostrand Company: Ralph DeSoignie and Christina Mikulak helped in many ways, especially as we sheared away corners of the oversize block of marble we had cut out for the text. Katharine Lacher, our good neighbor, aided us in proofreading the endless stream of galleys.

And at Chuck's regular job, Professor Jared Cohon willingly carried heavier burdens to allow Chuck the time to write. His knowledge made possible the careful discussion of the nuclear incident at Three Mile Island. Julie Allen, originally secretary to "the work," became also a valued research assistant in the course of the book's preparation. She typed the manuscripts, all of the drafts, tracked down articles and other references, and penetrated federal agencies in search of photographs. All of this was performed graciously and with good cheer.

We also owe a debt of gratitude to Scott Pyne, whose assistance in the preparation of questions was invaluable.

Our major sources of controversy in environmental issues were articles we found in scientific journals and in the popular press. In this area, two parties contributed particularly. Peter and Linda Rottmann called our attention to the fire at Baxter Park and the sharp debate in the Bangor Daily News. It was an archetype of a problem without an answer. And Donald W. Taube, University of California at Santa Barbara, mentioned his own interest in Palau, the Pacific paradise on the verge of becoming a super-tanker port. His bibliography and articles supplemented our own, enabling us to develop the Palau story more deeply.

Finally, Jeff Wright, an environmentally concerned colleague and fellow educator, helped us form the portion of the final chapter that deals with the opportunities we all have to contribute to the environmental movement. His ideas helped to prevent us from preaching, for which we are all grateful.

Critical Reviewers

The development of *The Environment: Issues and Choices for Society* was accomplished with the suggestions of the following:

Ted L. Hanes, California State University at Fullerton
John Mikulski, Oakton Community College
Clyde W. Hibbs, Ball State University
Jerry Howell, Morehead State University
Arthur C. Borror, University of New Hampshire

Contents

The Environment
Issues and Choices for Society

Introduction: On Solving Environmental Problems

THREE POINTS OF VIEW

The idea is current in the environmental movement today that we must break out of the traditional mold, abandon old methods of problem-solving, and not rely on technological solutions. We must invent a new life-style based on a new environmental ethic. And it is true that technological ways of solving problems may create new problems while solving the old ones. For instance, a farmer can increase production by using chemical pesticides, but this use may also contaminate foods and lead to the poisoning of harmless creatures.

Another feeling, held by parts of the general public, is that problems are best left to the experts, to the investigators who study the science of environmental pollution. For instance, suppose it was necessary to set limits for a certain air pollutant. Some people think it best to leave this sort of job to a group of scientists. In this case, the group might be made up of specialists in the chemistry of the atmosphere, specialists in how winds transport and mix pollutants, and still others who study the frequency of illnesses related to air pollution.

However, such experts cannot actually decide on the appropriate standards for the air pollutant. They can set out the choices, by predicting within fairly wide limits the results of various control actions. But only society as a whole can decide whether particular levels of disease are acceptably small and whether the cost of a certain control measure is a reasonable burden to taxpayers. Only society, and you as a member of society, can make decisions on what to achieve, on what risks to bear and what money to spend.

Here is the way one decision-maker viewed the responsibility of his office:

> I am convinced that if a decision regarding the use of a particular chemical is to have credibility with the public, and with the media who may strongly influence that public judgment, then the decision must be made in the full glare of the public limelight. It no longer suffices for me to call a group of scientists to my office and, when we have finished, to announce that based on their advice I have arrived at a certain decision. Rather, it is necessary for me to lay my scientific evidence and advice on the table where it may be examined and, indeed, cross-examined by other scientists and the public alike before I make a final decision.[1]

Some believe a new life-style is necessary; others say leave the decisions to the experts; still others have given up hope. This last group sees the problems as so overwhelming and the effort to change life-styles so enormous, that they see little future for the world. If we felt that way, we would

---

1 William D. Ruckelshaus, then Administrator of the Environmental Protection Agency, in a speech to the American Chemical Society, 13 September 1971.

not have written this book. We personally believe there is hope for improving the quality of the environment. Humans are a creative lot, as you will see in the book; we only need more people who care. And that number is increasing all the time.

ENVIRONMENTAL DECISIONS

Whether or not you feel that changes in life-style are essential to cure and to prevent environmental ills, you are needed, your ideas and energy are needed, in a decision process that is on-going. You should know that there is an urgency to your becoming involved. That is because actions to resolve environmental problems are being taken right now. In some cases, immediate action is demanded by problems so severe that no thinking person would expect to wait for a change in life-style or for an improved philosophical framework. As an example, when it was discovered that the chemical, vinyl chloride, was almost certainly responsible for cancer deaths among industrial workers, government officials moved swiftly to reduce the amount of this chemical in work places. They also prohibited the use of the chemical in products the general public uses. In the case of vinyl chloride this included hair spray and food wraps.

In a similar case of a present danger, California health officials have ordered a strict reduction in auto emissions by requiring special air-pollution-control devices on automobiles sold in California in order to reduce the level of smog in Los Angeles. The smog, which is caused mainly by automobile emissions, is the worst in the nation because of the low level of mixing of the air over Los Angeles and the unbelievable crush of traffic on the city's free-ways. Almost anyone who has measured or experienced smog in Los Angeles feels the level is too high for good health. It is possible that if people could be educated to walk more, to use carpools, to use public transit, smog could be reduced. But no one wants to wait long enough before taking action to

see if these life-style changes can be accomplished. Instead, control actions, mostly technical and promising quick results, are tried. Public opinion and public health demand this.

These actions, which are taken to deal with clear-cut environmental problems, must be sought and achieved within a political and bureaucratic framework that is unlikely to change in the near future. For a solution to work, it must be acceptable to, and seen as reasonable by, the people who make up this framework.

An example of a solution that has not worked, because it has not been politically acceptable, is one of the air-pollution-control policies of the Environmental Protection Agency (EPA). In the early 1970s, traffic control was seen as a bright hope for reducing air pollution in our major cities. By the end of the decade, however, there had been little if any progress in limiting traffic in those cities. *The New York Times*, in reviewing the national situation, headlined:

Cities' Pollution and Traffic Snarls Are Worse Despite Decade's Effort.

Efforts to Encourage More Use of Mass Transportation Founder on Resistance of Motorists and Officials[2]

One reason for this failure is that public transportation has been inadequate to serve the scattered pattern of housing that has developed since the end of World War II. Plans for limiting city traffic without adequate public transportation have been politically unacceptable.

The need for immediate responses to some problems and the limited range of politically acceptable actions does not mean that we should not search for new ways to solve environmental problems. There is the utmost importance in looking for ways to live so that the environmental ills that beset us do not occur in the first place. Of course, we

---

2 *The New York Times*, 21 May 1979, p. 1.

should continue to look to the long-term, but we must also look to the short-term. We ought to know how environmental decisions are made and who is making them. The important question is: How can we best influence and become part of this decision-making process, *while* we look for the long-term solutions and for environmentally sound ways of living?

HOW DECISIONS ARE BASED AND WHO IS MAKING THEM

Inputs to the Decision Process

Until they examine the situation closely, many people have the comfortable feeling that decisions being made on environmental matters by elected or appointed government officials are taken on the basis of definite information provided by scientists and other experts. Unfortunately, even a brief examination of some environmental issues reveals that in many cases we have conflicting or inadequate scientific information. We are usually not in a position to predict with any precision what the long-term effects on environment and health (or sometimes even the short-term effects) of many actions will be. In short, many decisions have to be made on less than complete evidence. The experts often disagree, even violently.

The scientific aspects of the problem are simply one component of decisions. The social, moral, and economic aspects also carry a great deal of weight. For instance, what if an otherwise reasonable energy conservation measure put a heavier burden on the poor than on the rich? One such measure is a new federal tax on gasoline, over and above that being collected for the Highway Trust Fund. The poor who own cars will be hurt more than the middle income groups by a gasoline tax since more of their money is tied up in purchasing the essentials of living. Is the economist or the scientist alone qualified to judge what amount of tax should be levied? Their concerns may be with tax revenues or

the decrease in oil imports resulting from a lower demand. Yet someone must speak for the urban and rural poor.

Who Makes the Decisions?

Decisions on matters affecting the environment can be divided into two kinds, public decisions such as those made by Congress or federal and state agencies, and private decisions such as those made by business executives or by consumers. All of these decision makers are important. Public decisions are contained primarily in laws passed by elected officials in federal, state, or local governments. These laws are influenced by, among other things, public opinion and industry pressure groups. The laws once passed are then interpreted by appointed officials in the Environmental Protection Agency (EPA) or in the state conservation departments and by elected officials such as state governors. As an example, Congress, by law, directed the Environmental Protection Agency to set air quality standards. After extensive study and public hearings, EPA set these standards.

The courts also are involved in the process of environmental decision-making. When two groups have differing interpretations of the law, a suit forces the courts to resolve the situation. For instance, the Environmental Defense Fund, a citizens conservation group, sued EPA in court because they felt the agency was not acting as quickly as it should to investigate and ban dangerous chemicals, as it was directed to do by the Toxic Substances Control Act. With the help of the court, a compromise schedule of action was agreed on.

Private decisions are no less important than public ones, but here it is less easy to see the decision-making process. As an example, during the early 1950s Hooker Chemical and Plastics Corporation used several pieces of land in the Niagara Falls area as dumping sites for waste chemicals. One area, the Love Canal site, was later sold to the local board of education, and a school was built there. Hooker was informed in 1958 when several children were burned by chemical wastes at the site. In addition, the company apparently knew that chemicals from this and the other dump sites might be draining into the surrounding areas where homes (and also the city's water supply) were located. Company officials notified the school board of the problem but decided not to notify local residents. In 1979, the Love Canal area was declared a disaster area because chemicals were seeping into basements and yards. Hundreds of families were moved out of the area by the state. In other cases, of course, corporate officials have made environmentally sound decisions. Sometimes these are made on moral grounds, sometimes they are made to avoid the sort of publicity that Hooker received after the Love Canal incident.

Consumers, too, make environmental decisions all the time. Should we buy an energy efficient air conditioner, even if it costs more? Should we buy a small car that uses less gas, even though we like the comfort and handling of a larger one? Should we bother to read ingredient listings on labels in order to avoid synthetic chemicals in foods?

As a citizen, you may be asked to take part in solid wastes recycling programs that require the separation of glass and metals in household garbage. You may wonder if you should go to the trouble of carrying the used oil to a collection center after you change the oil in your car, when it is simpler to let it run down the sewer. You might be considering a boycott on Japanese products to help save whales. We could go on with the list. At this point, however, you may see that as a voter and a consumer the individual citizen is a vital link in environmental decision-making.

MAKING SOUND ENVIRONMENTAL DECISIONS

Whether a citizen is being asked to vote on an environmentally related bond issue or to cast a more subtle but equally important vote by buying (or not buying) a particular product, the citizen-consumer needs some background to help make an educated decision. Whether you are working for an environmentally conscientious political candidate, or for a conservation group, or for a responsible corporation, it helps to know what you are talking about.

In part, this text is meant to help you gain the necessary background for understanding environmental problems. But this is only a portion of what is needed. Just as important is the ability to evaluate proposed approaches to environmental problems. Does the approach, in fact, solve the problem? Can it cause new problems? Is there a better way to approach the whole issue?

Whatever the final answers to such questions, one fact is clear. No one has more of a right to exert influence on decisions in environmental matters than you do.

> Crucial value decisions have to be made, and they should not be made only by involved scientists closeted with financially interested industrialists and governmental authorities. They should be made by unbiased and informed members of the general public after hearing all sides of the questions, with balanced input from scientists, humanists, historians, philosophers, theologians and most of all, from ordinary citizens.[3]

Basic Considerations

Before describing how to go about evaluating solutions in a specific way, some basic philosophical issues ought to be considered. Your agreement or disagreement on these issues is very likely to influence how you feel environmental problems should be approached.

Trusting people   The first issue is one of trust—whether people, as individuals, can be trusted to act in an environmentally sound way. There is

---

3 John C. Cobb, Letter to the Editor, *Science* **194**, 674 (12 November 1976).

some vogue, especially in the area of population control, for believing that people only act in their own short-term best interests. The thought is that plans involving long-term benefits for society as a whole, such as limiting population growth, must be accomplished by law.

> Conscience is self eliminating.... It is a mistake to think that we can control the breeding of mankind in the long run by an appeal to conscience....
>
> People vary. Confronted with appeals to limit breeding, some people will undoubtedly respond to the plea more than others. Those who have more children will produce a larger fraction of the next generation than those with more susceptible consciences. The difference will be accentuated, generation by generation.[4]

The opposing opinion, stated perhaps most eloquently by Roger Revelle (no relation), is that an educated people can be trusted to see that their long-term best interests lie in smaller families and slower population growth for their country.

> Some scientists and publicists have seriously advocated a "lifeboat ethic," saying that nations which do not *compel* human fertility control (by what means is never stated) are endangering the survival of our species—hence they should be starved out of the human race by denying them food aid. This obscene doctrine assumes that men and woman will not voluntarily limit their own fertility when they have good reasons and the knowledge and means to do so.[5]

Note that Roger Revelle is talking about people who are not only educated but who also can afford the environmentally sound choice.

The issue of trust comes up often, not only in the area of population control. One can even ask whether corporations, which are, after all, run by

4 Garrett Hardin, *Science* **162**, 1246 (13 December 1968).

5 Roger Revelle, *Science* **186**, 589 (15 November 1974).

individuals, can be trusted to act in environmentally responsible ways. Or, since environmental protection generally costs money, will corporation executives always choose profits at the expense of the environment?

Trusting technology   The second issue also involves trust; in this case, trust in technology. We must ask whether technology can be relied on to solve specific environmental problems and get us out of environmental jams. It is not difficult to find examples of cases in which solutions, especially technological ones, solve one set of problems, but in doing so, create new problems. This was stated more cynically by H. L. Mencken who said, "Every problem has a solution, simple, neat and wrong."

As an example of a technical solution which can cause problems, electric cars are thought by many to be the ideal way to get rid of air pollution in the city. However, cars consuming vast quantities of electricity may not be such a good idea. The number of new power plants needed would be enormous; doubling our current power generating capacity might be necessary if a complete switch to electric autos were made. And the plants would be fired by either coal, a fuel with undesirable effects on air, water, and land, or by nuclear power with all its accompanying problems. A better solution might be excellent public transportation for citizens in order to eliminate a portion of the need for use of the auto. Such transport is far more energy efficient than automobiles.

Weighing risks and benefits   Recognizing that actions involving the environment can cause harm as well as good brings us to the final issue: our willingness to weigh the benefits of environmental actions against the risks involved. For instance, we may be willing to risk some of the long-term effects of a pesticide used to save lives that are threatened by insect-carried diseases such as malaria. We may also justify using pesticides to gain the benefits of increased food production. At the same time, we need to ask whether such uses may significantly increase our chances of dying from chemically-caused cancer. We must also ask whether harmless species are threatened by our use of pesticides and whether the stability of the agricultural ecosystem may be upset. The ability to weigh the benefits of an action with the risks it brings is essential to achieving and preserving a satisfactory environment. Nuclear power can provide electricity to the nation and reduce oil imports, but are the benefits of nuclear power worth the risks, both known and unknown that accompany it? (See Chapter 15.)

This is not to say that such weighing of risks and benefits is easy, or free of conflict. Some individuals see modern society as too concerned about risk:

> No child, no idea is born into this world without the possibility of causing harm as it grows older. Shall we then abort all birth and all innovation for fear of possible environmental damage?[6]

Yet some risks are able to be reduced and are well worth reducing:

> Such spectacular technical failures [as the wreck of the oil tanker Torrey Canyon, or the blowout of an oil well in the Santa Barbara Channel] also brought home to the general public something else:...that environmentalists—those impractical people with their feet in a swamp and their head in a cloud—are not necessarily wrong?

These, then, are the three issues: whether you trust people, whether you trust technology, and your willingness to tolerate risk, known and unknown. These basic issues will arise again and again as you evaluate solutions to environmental

problems, and you will find that your stance on these issues influences to a very large degree how you approach environmental problems.

Examining Specific Solutions

The following questions are designed to help you examine specific proposals for solving environmental problems. They are designed to help you determine the worth of the solution itself. But they are also meant to help us all move towards the definition of new ways of living, ways that do not ignore natural laws and limits (see, for example, Chapter 11 on energy laws). These ways of living may make it possible for us to do more than react to a problem after it occurs. They may make it possible for us to prevent some environmental problems from occurring in the first place.

It is helpful first to classify a given solution as to type, then to go on to ask questions depending on the type of solution.

1. *Is it a "technical fix"?* Is this a technical solution *added on* to a current technology? (For instance, adding scrubbers to coal-fired power plants to remove sulfur oxides from stack gases.)
   Will it solve the problem with reasonable certainty?
   Will the solution cause new problems: technically, economically, socially?
   Could we, instead, substitute another technology that does not cause this problem?
   Could we do without the technology that causes the problem? Or could we do with less of it?

2. *Is the solution an alternate form of technology?* (For instance, substituting nuclear power plants for coal-fired power plants to eliminate sulfur oxide emissions.)
   Will it solve the problem with reasonable certainty?
   Will it cause new problems?
   Could we do without the technology that caused

---

6 Cyrus Adler of the Electric Whale Company, Letter to the Editor, *Science* **178**, 450 (3 November 1972).

7 John B. Oakes, Editor, *The New York Times* Editorial page. Speech given to National Audubon Society, November 1976.

the problem in the first place? Or could we do with less?

3. *Is the solution a non-technological one?* (For instance, a gasoline tax could be levied to increase the cost of gasoline so that people will use their automobiles less, thus saving fuel resources and decreasing air pollution.)

Does this represent a real solution? That is, will it work? Is it politically acceptable?

Is the idea economically justifiable? Is the cost reasonable compared to benefits gained?

Is the idea socially justifiable? Who gains and who loses?

In some cases non-technological solutions are the result of an attempt to go directly to the root of a problem. For instance, we could ask if, rather than attempting to decrease automobile use, there are ways to make the use of automobiles unnecessary. New towns are based partly on the philo-sophy that when shops, recreational areas, and workplaces are located within easy walking distance, people will automatically use their cars less often. For these sorts of ideas we must ask, in addition to the questions above: Will people accept this new idea?

All of the solutions to environmental problems presented in this book can and should be examined in the light of these questions. You may decide that some are good solutions. Some will fail your examination. However, once you've examined a solution and come to a decision, what do you do then? How can you put what you've learned into practice? Some brief thoughts on this are found in the summary material at the end of the book.

We invite you now to read and study the material that follows. We hope it will help you become an environmentally informed citizen, ready to take a needed position on behalf of the environment, ready to enter the debate and influence decisions.

UPI

Part 1

Humans and Other Nations that Inhabit the Earth

It is undeniable that there is too much of many materials in the world today; too much mercury in the water, too much sulfur in the air, too many wrecked automobiles. But surely the most curious excess is people. How has it happened that human populations have grown so rapidly that food and job shortages have gripped some countries, that both wildlife and wilderness are threatened and that cities have grown to unmanageable sizes?

There are many reasons. Throughout the history of humankind, children have been valued. They have provided hands to work on the farm, to help in the blacksmith and carpentry shops and in the kitchen. Children represent security for parents in old age because they provide for parents too aged and infirm to work. And, for different reasons to different people, children are desirable in and of themselves. Children are a form of immortality, both of genes and of values. Through their children, a people's physical characteristics and also their beliefs can survive into the future. In traditional African religion, people are immortal as long as there are descendents to remember them. All societies, from the most primitive to the most cultured, consider the rearing and teaching of children one of their most important tasks.

Children are not only a precious resource; they have also been a most fragile one. In the United States and other countries, before the development of modern medicine and sanitation, five or six children were born, on the average, to each family. Of these, perhaps only two or three lived to adulthood. Because the survival of a child was so uncertain, it was customary to have large families. Among other considerations, this helped to ensure that at least two children would live to care for the parents when they grew old.

In many societies, especially where most of the population is engaged in farming, the tradition of large families has remained. At the same time, however, modern medicine has managed to make child survival much more certain. Within the past 50 years, the *death rate* (or the number of people who die, each year, per 1000 people in the population) has been reduced in most societies by one-half or more. This reduction in the death rate is due only partly to the medical advances which help adults to live longer; it is due in greater part to the survival of infants and young children who now live to become adults. This means that, for many countries, the number of people who die each year is much smaller than the number of people who are born. And so the populations of these countries grow rapidly.

As the population of a country grows, many facets of life undergo stress. Food supplies must grow or shortages, malnutrition, and finally starvation will result. Job opportunities must increase or unemployment spreads. New housing must be constructed, new parks opened, more roads, hospitals and schools built or unbearable crowding will occur. Even if construction needs can be met, stresses of city living are forced on many people.

How, and how well, a country meets the stresses of rapid population growth depends in a large part on how technically advanced that country is. For instance, technically underdeveloped countries may find it difficult or impossible to feed, house, and educate a rapidly growing population. Technically developed countries, on the other hand, can often expand production of food, goods, and services, but they cannot cope with increased pollution, preserve wilderness, or prevent the mushrooming of social problems in growing cities.

Population growth is thus an underlying cause of or contributes to many of the environmental problems we know by other names: food crisis; disappearance of wilderness; energy crisis; air and water pollution; urban sprawl.

We begin this part with a discussion of some of the ecological terms and principles related to populations and their interactions. Next the problem of rapid growth in human populations is explored. How has the population problem come about? What are the effects of rapid population growth in developed and undeveloped nations? What means do we have of controlling population growth? And finally, what is the outlook for the future?

The human population is, of course, not inhabiting the earth alone. Human populations cannot grow without inevitably affecting other populations, both plant and animal, living on the earth. For instance, the Indian tiger, once the most coveted prize of the trophy hunter, is now gravely endangered. Although hunters used to be the most serious danger the tiger faced, now India's rapidly growing population is taking over the little habitat left to the tiger. Before World War II, India literally teemed with great herds of wildlife. Although hunting by wealthy Indian royalty and foreigners made some impact on the numbers of game, it was nothing compared to what happened after independence in 1947. According to Hari Dang, an ecologist: "Postwar exploitation—open season on all resources! Shoot everything, burn what's left, destroy the rest. It was our disaster period. You had the same thing in the American West—in the 1800's everybody started shooting, and the great herds were destroyed." But even worse was to come.

As the human population grew, competition with wildlife increased for food, water, living space, and forest products. John Putman of *National Geographic* tells of a bird-watching walk he took in India's Borivli National Park with Dr. Salim Ali, in 1976: "There was the rustle of palms in a light breeze. We spotted palm swifts, then a rufous-backed shrike, black-headed orioles, racket-tailed drongos. Then out of the forest come a column of tribesmen, each carrying branches. They filed by, looking at us only briefly. 'What is this! What is happening! An army coming out of the forest!' Dr. Salim Ali cried. Part of Borivli's habitat was disappearing before our eyes. 'What can you do? What should you do?' Dr. Salim Ali added. 'They must eat, and need wood for their cooking fires.' He shook his head: 'Population—it is at the root of every problem,'" (Quotes from *National Geographic*, September 1976.)

Chapter 1

Lessons from Ecology: Structure in Ecosystems

A student of the environment uses knowledge from many other fields. Biology, law, sociology, mathematics, anthropology, and physics are only some of the subjects useful in understanding environmental problems. The discipline most closely tied to environmental studies, however, is probably ecology.

The word ecology is made up of the Greek words "oikos" meaning house, and "logos" to study. Thus ecology means, roughly, the study of where organisms live, or their environment. A better way to put it might be that ecology is concerned with the relationship between organisms and their environment.

Many ecological terms are used in environmental studies. The purpose of the ecology chapters, which appear in several main sections of this book, is to explain ecological principles, which you will need to understand environmental problems.

PEREGRINE FALCONS AND THE DEFINITION OF A SPECIES

Birds belonging to the species called peregrine falcons are spectacular hunters (Fig. 1-1). Sailing with wings outspread, peregrines search for the smaller birds which are their prey. Spotting something that looks as if it would make a meal, the falcon plunges downward at speeds up to 200 miles an hour and snatches its unsuspecting dinner in mid-air.

Figure 1-1 *A mature peregrine falcon grows to 15–20 inches in length (37–50 centimeters) and has a wingspread of up to 43 inches (108 centimeters). The birds are noted for their beautiful markings and for the spectacular dives they make in mid-air. (The Peregrine Fund)*

Since the early 1960s there has been no population of peregrine falcons east of the Mississippi. The birds fell victim to poisons in their ecosystem, most likely DDT. But DDT, which was first used in this country in 1945, is now banned and levels in the environment seem to be decreasing. For this reason, Dr. Tom Cade, a Cornell University ornithologist, thinks it is a good time to try to restore the Eastern peregrine falcon population. He breeds the birds in captivity and then trains them to live successfully in the wild areas in which they were once found (Fig. 1–2).

During this project, Dr. Cade and the U.S. Fish and Wildlife Service came into conflict over the definition of a species. Dr. Cade breeds his falcons from European peregrine falcons as well as from native North American birds. Because the European birds live in fairly populated areas, Dr. Cade feels that European falcons have characteristics that would be valuable to falcons attempting to settle along the populated Eastern coast of the United States. The U.S. Fish and Wildlife Service was uneasy about allowing the result of this crossbreeding, which might be called a foreign species, to be established in this country. There is an executive order which prohibits the introduction of "exotic species" into the United States. Many scientists came to Dr. Cade's defense by pointing out that all of the different birds Dr. Cade used were simply subspecies of peregrine falcons. Who was right?

What Is a Species?

(A *species* is most often defined in terms of reproduction. That is, a species consists of a group of individuals who can successfully breed with each other, who share ties of common parentage, and who therefore possess a common pool of genes, or hereditary material. In most cases it is possible to tell species apart on the basis of their different appearance or behavior or physiological makeup. However, these differences in themselves do not

Figure 1–2 *Young falcons compared to a more mature bird held by Tom Cade. Many peregrine falcons flew in North America until about the early 1960s, when they fell victim to pesticides such as DDT in the environment. Now that DDT is no longer used in the U.S., Tom Cade, a Cornell University ornithologist, is breeding the birds in captivity and then reintroducing them to the wild. Over 200 birds have been successfully released. This is equal to as many birds as the whole Eastern population of peregrines once raised in a year, before DDT came into use. (New York State College of Agriculture and Life Sciences at Cornell University.)*

define species.)If two apparently similar groups of organisms cannot breed successfully when given the opportunity, the two groups make up two separate species. Similarly, if two different-seeming groups of organisms are capable of interbreeding, there can be a flow of genes between them. They are thus members of the same species, no matter how different they are in appearance.

(Another important part of the species concept is that, since species are not able to interbreed, they will follow different avenues of evolution as they adapt to their environment. In this way the adaptive zone of different species, which is made up of the resource space they occupy and the parasites and predators[1] they encounter, becomes different from that of any other species.[2]

Populations and Subspecies

Populations, such as the 50 or so birds which make up the entire population of peregrine falcons west of the Rockies, are the members of a species living together in a particular locality. There can be various populations of a particular kind of animal, such as the Grizzly bear, all belonging to the same species (Fig. 1-3).

In general, even though they all belong to the same species, the members of a particular population resemble each other more closely than they resemble members of other populations. This is because pairing is more likely to occur between individuals within a population than between those in differing populations, and because members of a population are all subject to similar environmental influences upon their direction of evolution.

In some cases, geographic conditions in which various populations of a species live are very different or the barriers to travel between the localities are great. For instance, populations of the same

Figure 1-3   *The Grizzly is a prized trophy for hunters, but few are left in the 48 states. Some 1000 are believed to live in small populations in wilderness areas of Idaho, Wyoming, and Montana. The Interior Department has listed Grizzlies as "threatened," a category that indicates an animal needs less protection than an endangered animal. Larger populations still exist in Canada (11,000-18,000) and Alaska (8000-10,000) where hunting is still allowed. (©1977 Jeanne White/Photo Researchers, Inc.)*

species may live on different islands or in different rivers or on opposite sides of mountain ranges. The populations may then be found to have some very different genetic characteristics (although not enough differences to prevent successful interbreeding if they were given the opportunity). (Populations of a species that are unlikely to breed because of geographic factors and that show genetic differences are sometimes defined as *subspecies* or *races*.) Dr. Cade's peregrine falcons seem to breed successfully (although he does use some complicated artificial insemination techniques) and so it is argued that they are all members of one species— or at worst, subspecies—and the executive order

---

1 Predators are generally larger than the organisms they eat, while parasites are smaller than the organism they eat.

2 Plant species are more difficult to define than animal species. Reproductive methods and genetic systems are more diverse in plants, thus clouding efforts to define plant species clearly. The article by D.A. Levin (see Further Reading section) is a thoughtful discussion of the whole problem of defining species.

does not apply. The Fish and Wildlife Service appears to be unconvinced by this argument but has agreed to allow an exception to the rule so the peregrine falcons can still be released.

HOW ORGANISMS LIVE TOGETHER

Communities

(Within a given environment, the various species do not simply act in a random fashion but are organized into a *community.* This community includes all of the living organisms, both plant and animal, interacting in that particular environment. The term "community" can be applied to a rather small group of organisms,) such as those which make up a small pond community. It can also be used in describing a much larger group, such as the Eastern deciduous forest community into which Dr. Cade hopes to reintroduce the peregrine falcon. This community is made up of the populations of small bird species, which are the prey of the peregrine falcon, as well as populations of many other species of large and small animals, of insects, and populations of various species of trees and other plants.

(A minor community is one that is not self-sufficient, but depends on neighboring communities for such inputs as organic food materials. Major communities, on the other hand, are large and complete enough to require little more than energy from the sun in order to function as a unit.)

Because of the interactions among the organisms in a community, you might say that a community adds up to more than the sum of its parts. That is, the community itself has certain properties in addition to the characteristics of the individuals or species making up the community. Examples include the trophic structure of communities (described on p. 18) and the way energy flows through communities (a concept dealt with in Chapter 11).

Ecosystems

(In ecological terms, community refers to a system of living organisms. This living system plus its nonliving components make up an *ecosystem* (Fig. 1–4). The nonliving parts of an ecosystem include such things as the soil type, the amount of rainfall, and the amount of sunlight reaching the system.)

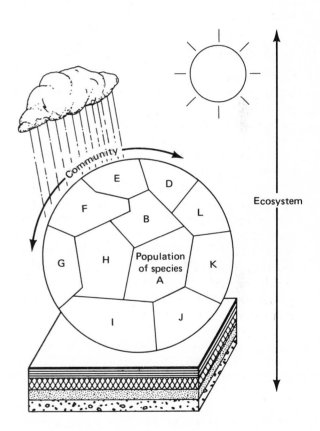

Figure 1–4 *A community is made up of populations of various species that interact in a given physical environment. An ecosystem is the living community plus all its nonliving factors such as sunlight, rainfall, and soil type. The terms "community" and "ecosystem" can refer to small areas such as a pond or a woodlot or to large areas, such as the Eastern deciduous forest.*

Habitats and Niches

Peregrine falcons prefer to build their nest, or eyrie, on a narrow ledge or a steep cliff. This provides protection for the baby falcons from predators such as owls or raccoons. In at least one case, a skyscraper seems to fill a peregrine's requirements. Scarlet, one of the peregrines released by Dr. Cade, has chosen to roost on a 35-story building in Baltimore, to the delight of its inhabitants.

(The physical surroundings in which an organism lives are called its *habitat*.)This is where you would go to find that particular organism. The habitat of bloodworms, or chironimids, for instance, is the mud at the bottom of lakes, while the habitat of deer mice is temperate zone woodlots. These habitats consist, of course, of a variety of physical factors such as temperature, soil type, and moisture but they also contain many other animals as well as plants. (Thus an organism's habitat includes living as well as nonliving factors. Communities also have habitats, but since the community is defined as consisting of all the living organisms, community habitat refers only to the nonliving factors.) By breeding European peregrines, which are used to living near humans, with native American peregrines, Dr. Cade hopes to be able to produce birds that can live in habitats that are becoming more and more urban.

(A somewhat more complicated term is "niche." The *niche* in which an organism lives includes its habitat but also includes its prey (what it eats), what activities it carries out, what its predators or what its parasites are. In short, the niche encompasses how an organism functions in its ecosystem, not just where it lives (Fig. 1–5). By definition, two species cannot occupy the same niche, or they could not be two separate species.)

Trophic Structure

(The *trophic structure* of a community describes how the various organisms in the community obtain their nourishment. In any particular community, the organisms are linked together in a pattern of preying and being preyed upon—of eating and being eaten. This pattern is called a *food chain* or food web.)

Figure 1–5 *An organism's habitat has been called its "address," while its niche is its "profession." Actually, the term "niche" covers even more than this since it includes not only what an organism does in the environment but also what climate it needs, what its predators are, and so forth.*

Food chains   Food chains can be diagrammed in a relatively simple, straight-line fashion. Creatures in the chain generally feed on only one or a few species and are preyed on by one or a few other species. Food chains often exist where conditions are so harsh that few animals or plant species have been able to survive, for instance on the arctic tundra (Fig. 1–6).

Islands also often have simple food chains, made up of relatively few creatures. A hungry animal has few choices about what to eat; organisms are preyed on by only one or a few creatures. Food chains are in very delicate balance.

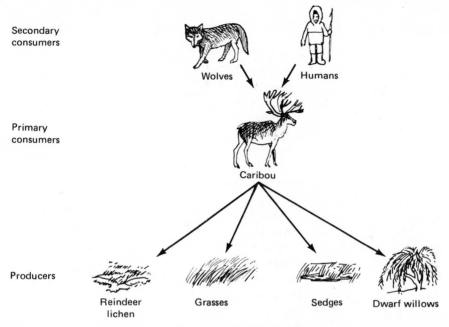

Secondary consumers

Wolves   Humans

Primary consumers

Caribou

Producers

Reindeer   Grasses   Sedges   Dwarf willows
lichen

Figure 1-6 *Arctic food chains are simple and short because few species are able to live in the harsh and unpredictable arctic climate. One arctic food chain is made up of a few arctic tundra plants, the caribou, which eat the plants, and wolves and humans, who hunt the caribou.*

The introduction of new species to islands can have drastic effects. For instance, cats, rats, and mongooses introduced at various times to Puerto Rico, the Hawaiian islands, or the West Indian islands have caused a great loss of native species. The animals on the islands, especially the birds, had no time to evolve defenses against these new predators. Introduced rabbits in Australia and goats in the Hawaiian islands competed with native island species for plant foods.

The food chain described in Figure 1–6 is a grazing food chain. Organisms at the bottom of grazing food chains are the *producers.* These are plant species, which convert sunlight into food energy by photosynthesis[3] (the grasses, sedges, etc., in Fig. 1–6) Creatures that eat producers are herbivores or green-plant eaters (the caribou in Fig. 1–6) and they are called *primary consumers.* Next in line are the meat eaters or carnivores, which eat the primary consumers (men, wolves in Fig. 1–6).

These are the *secondary consumers.* When food chains are longer, as they are in temperate and tropical climates, there are often *tertiary consumers* to eat the secondary consumers and so on.

At another trophic level are the decomposers or detritus feeders. This group of organisms is extremely important since it feeds on dead organic matter, breaking it down into inorganic and organic materials. These materials may be used by plants and animals or may inhibit or stimulate various parts of ecosystems. Fungi and bacteria are commonly decomposers but animal species may also be important (see Fig. 1–7).

Food webs Actually, in temperate and tropical climates, where there are many species interacting with each other, simple food chains are less common than *food webs.* Food webs are made up of a

---

3 See section on Photosynthesis.

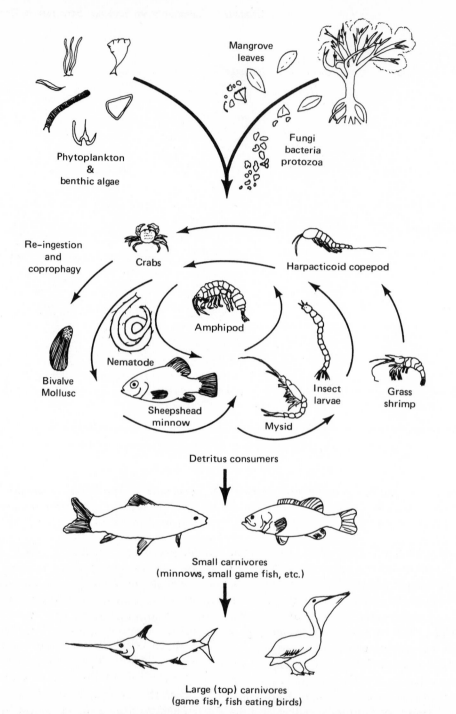

Phytoplankton
&
benthic algae

Mangrove
leaves

Fungi
bacteria
protozoa

Re-ingestion
and
coprophagy

Crabs

Harpacticoid copepod

Amphipod

Nematode

Bivalve
Mollusc

Sheepshead
minnow

Mysid

Insect
larvae

Grass
shrimp

Detritus consumers

Small carnivores
(minnows, small game fish, etc.)

Large (top) carnivores
(game fish, fish eating birds)

Figure 1–7   *A detritus food chain based on mangrove leaves that fall into swamp waters. A variety of detritus consumers eat and excrete the leaf fragments. These detritus feeders are food for the larger creatures in the chain, such as game fish and water birds. (Adapted from W. E. Odum, Dissertation, University of Miami, 1970.)*

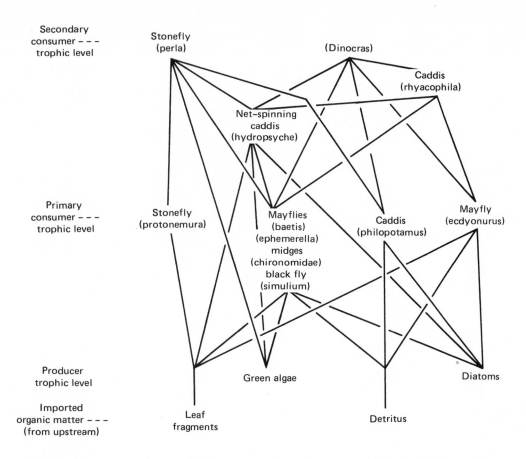

Figure 1-8   *Part of the food web in a small stream community in South Wales. Note that the net-spinning caddis is both a primary consumer (eats green plants) and a secondary consumer (eats primary consumers). [Adapted from J. R. Jones,* Journal of Animal Ecology **18,** *142(1949).]*

number of food chains that are connected to each other. Figure 1–8 shows an example of a food web from a small stream in South Wales. Food webs are more complicated than food chains and also more difficult to sort into the trophic levels of producer, consumer, etc. This is because food webs involve so many species that we cannot always be sure that we know all of the creatures in a particular food web. Also, some species occupy more than one position in a web. That is, an organism may eat both green plants and such primary consumers as insects. That organism would be both a primary consumer and a secondary consumer. The net-spinning caddis in Figure 1–8 illustrates just such a complication.

In this food web, the decomposers feed upon materials washed from upstream, such as leaves, feces, dead organisms, etc. Many food webs, especially in water, are based on detritus. Figure 1–7 shows the detritus-based food web in a mangrove swamp. The leaves fall into the water, are broken down by organisms such as bacteria, and then the leaf fragments enter a cycle of being eaten and excreted and eaten again by the detritus feeders such as crabs, shrimp, and small fish. These detritus feeders are preyed on by small carnivores (such as small game fish), which in turn are eaten by large game fish and water birds.

Biological magnification   One of the reasons why it is important to understand food chains and food webs is that we can then see why a species such as the peregrine falcon is so sensitive to pesticides such as DDT in the environment.

DDT is a chemical that is soluble in fat. When an organism ingests DDT, whether from water or from the detritus of plants that have been sprayed or from insects that have eaten the plants, the DDT becomes concentrated in the organism's fatty parts. DDT is lost very slowly from these fatty tissues. If another creature in the food web eats the first organism, the consumer will be eating a concentrated dose of DDT. The second creature concentrates the DDT still further in its own fatty tissues. A third consumer will receive an even more concentrated dose. In this way, organisms such as the peregrine falcon (or man), which are at the top of their food chain are eating foods that have a much higher level of DDT than is generally present in the environment. The DDT has been biologically magnified. One effect of such levels of DDT on birds is that they lay eggs with shells that are much thinner than normal. These thin shells break easily and so are no protection for the developing chick inside the egg (Fig. 1-9). In this way, the Eastern peregrine falcon failed to reproduce and so became extinct in the 1960s.

LOOKING AHEAD

Dr. Cade's program to reestablish the peregrine falcon population in the United States is an example of wildlife management techniques at their best and most imaginative. Not only has he developed ways to encourage the birds to breed in captivity, but he and his staff also teach the newly fledged falcons to hunt for themselves in the wild. At about four weeks, when the birds can tear and eat their food, they are taken from the laboratory to old falcon nesting sites or to specially made

Figure 1-9   *Baby peregrines bred in captivity. In areas where DDT is found in the environment peregrine falcons lay eggs with thin shells. These eggs are more likely to break before hatching than eggs with normal shells. This is believed to be the reason why there have been no natural populations of peregrine falcons in the Eastern U.S. since the early 1960s. (New York State College of Agriculture and Life Sciences at Cornell University.)*

towers. There they are protected 24 hours a day by attendants who keep away owls and raccoons that would eat the young birds. The peregrines are taught to return to the nest site to eat until they are skilled in catching their own meals.

Still, we do not have anywhere near enough researchers to undertake a similar program for other endangered creatures. There are also many reasons why it would not work in other cases. Clearly we must try to prevent species from becoming extinct rather than try to repair the damage once it is done.

In the following chapters, we explore problems faced by the human species as well as those faced by wildlife populations. For, as the fate of the peregrine falcon was changed by human acts, first for ill and then for good, so are human fates and fortunes tied up with those of the other species on this planet.

Photosynthesis

Green plants use the energy in sunlight to manufacture organic "food material" out of the gas, carbon dioxide, and water. The organic materials formed include a variety of sugars. Carbon dioxide is a chemical compound that can be described as "low energy," while sugars have a great deal of energy stored in their chemical bonds. Thus in photosynthesis the energy from sunlight is turned into, or stored as, chemical energy. In addition to organic materials, oxygen is produced during photosynthesis. Plants themselves use up some of this oxygen but more is produced than they can use. All animal life, including human, depends on this excess oxygen for respiration. We also depend, for food, on the sugars produced during photosynthesis, whether we eat the plants directly or whether we eat animals who eat plants. In summary, the process of photosynthesis can be written as the following equation:

$$\text{carbon dioxide} + \text{water} \xrightarrow[\text{sunlight}]{\text{green plants}} \underset{\text{(sugars)}}{\text{organic material}} + \text{oxygen}$$

Questions

1. How would you define a species? a population? a subspecies?
2. What could be the purpose of an executive order prohibiting the introduction of exotic species?
3. What is the difference between a community and an ecosystem?
4. Why does a community have more characteristics than simply those of the species from which the community is made up?
5. Differentiate between an organism's habitat and its niche.
6. Diagram a simple food chain that you think might exist in an area near your home. Who are the producers? the consumers? the decomposers?
7. What is meant by biological magnification? Why is this an important natural phenomenon?

Further Reading

If you are especially interested in ecology, you may want to read some of the following materials.

Levin, D. A., "The Nature of Plant Species," *Science* 204, 381 (27 April 1979).

In this article, Levin states:

For humans the environment has meaning only when its components can be inter-related in a predictive structure. We try to make sense out of nonsense and put the world into some perspective which has order and harmony.

The species concept has caused a great deal of argument and confusion. This article is a thoughtful consideration of this problem, as well as a thought-provoking discussion of the human need to impose order where it is not always clear that order exists.

Levin, H. L., *The Earth Through Time*, Saunders, Philadelphia, 1978.

Odum, E. P., *Fundamentals of Ecology*, Saunders, Philadelphia, 1971.

Watt, K., *Principles of Environmental Science*, McGraw-Hill, New York, 1973.

Chapter 2

Human Population Problems

GROWING AND CHANGING POPULATIONS

Population Growth Through History

Growth spurts When the beings that we call first humans lived on the earth, life itself must have been a very uncertain proposition. Humans had to hunt and gather their food. They were at the mercy of the weather and victims of fluctuating animal populations. They did not yet grow food or store it for times of drought or for unseasonable cold or when hunting was poor. The average life span is thought to have been perhaps 13 to 20 years.

When people learned to domesticate animals and to grow and store their own food, about 8000 B.C., the human population increased to perhaps five million people. Was this because the death rate decreased as food supplies became more certain? Perhaps not. Some scientists point out that death rates may even have increased as crowded living conditions in stationary communities allowed diseases to spread more easily.

A more important factor was probably the spacing of births. Studies of present day hunter-gatherers such as the Dobe! Kung show that births are spaced in such nomadic societies so that there is at least three years between children. This is necessary because mothers must carry their infants as the tribe moves about. Further, nomadic peoples do not have soft foods, which are necessary to feed young children after they are weaned. Thus chil-dren are usually nursed for a relatively long time (five years, in some cases). Agricultural societies, however, are sedentary, therefore children need not be carried long distances. Also more easily digested foods can be grown. These factors would allow for more closely spaced births and would account for population growth as agricultural societies developed. (You may wonder how primitive people could have spaced births at all. The reference by Dumond has some interesting thoughts about this.)

The next great population growth spurt began much more recently, just before the start of the Industrial Revolution. From the 1800s to the middle 1900s, in the countries we now call the developed or industrialized world, death rates fell because of improvements in sanitation, medicine, nutrition, and agriculture. Food was grown in larger quantities and transported more efficiently. People became healthier because they were better fed.

Demographic transition In these developed nations, (birth rates, or the number of people born each year,) per 1000 people in the population, generally declined along with death rates (Fig. 2–1).

Those (countries which have managed to reduce both their birth and death rates are said to have undergone the *demographic transition.*) (Demographic transition is the name population scientists apply to a distinctive pattern of change in birth and death rates.) First seen in the industrial nations of Europe and America, the transition consists of a decline in the death rate resulting from the many medical or nutritional advances the nation

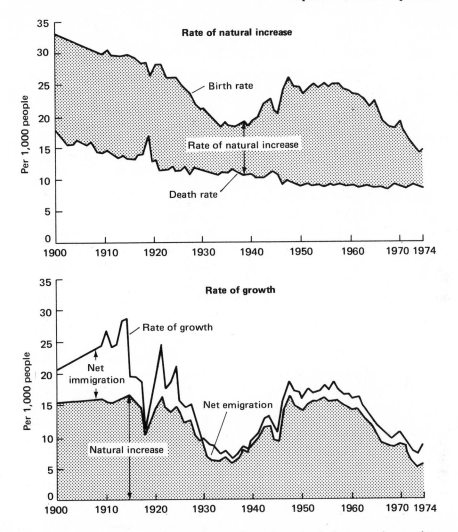

Figure 2-1   *The rate of growth for a country depends on both the natural rate of increase (birth rate minus death rate) and the rate at which people immigrate (move in) or emigrate (move out). These rates are shown in the above figures for the United States from 1900 to 1974. At the present rate of growth the population will double in about 87 years. (Source: Population Reference Bureau, 1975.)*

has come to enjoy. While the decline in death rate is underway, the birth rate, too, falls. It falls to a lower and fairly stable value. It would be wrong to conclude, however, that birth rates in developed nations fell solely because people became aware of the decline in death rates. There is evidence that birth rates actually began to fall before the medical

and sanitary advances that lowered death rates. Most industrial nations have already undergone this demographic transition. Further changes in birth and death rates in these countries in the near future are likely to be small.

The situation is different, however, in the countries labeled developing or undeveloped.

Rapid Population Growth

Why it occurs   In countries all over the world, death rates are being lowered by public health measures. Diseases such as malaria and yellow fever, which are spread by insects, are prevented by the use of insecticide sprays and by draining and filling the swamps where these insects breed. Cholera and typhoid and other intestinal diseases are prevented by chlorinating and filtering public water supplies. The pasteurization of milk can prevent the spread of bacterial diseases in one of the most common foods of childhood. General food sanitation, made possible by refrigeration and a pure water supply, may also contribute to reducing childhood diseases.

Antibiotics save many children who might once have been victims of diseases such as pneumonia and scarlet fever. Drugs that cure tuberculosis and a vaccine that prevents it are available. Vaccinations can now prevent most children from ever having smallpox, diptheria, tetanus, whooping cough, polio, or measles. In addition, better nutrition and care for pregnant women have improved both the health of mothers and the chances that their babies will survive.

All of these medical and public health advances have combined to cause a dramatic drop in the death rate among infants and small children. The death rate of infants and children has fallen not only in countries that are developed, or industrialized. It has also fallen in undeveloped countries where many of the health and sanitation measures have been instituted. Nevertheless, even though the odds on child survival have increased in these countries, even though more of a couple's children survive to adulthood, there has not always been a corresponding drop in the birth rate. That is to say, in many cases, parents still have as many children as they did when it was necessary to have five or six children in order to be sure of raising two to adulthood.

In undeveloped countries, the birth rate is likely to be as high as 30–45 births per 1000 people in the population each year, while the death rate may be 15 per 1000 per year (Fig. 2–2). In contrast, among the developed countries, birth rates have fallen to 10–15 per year for each 1000 people in the population while death rates range from 6 to 13 per year.

If we subtract the death rate from the birth rate, we will have the *rate of natural increase*, a number that tells us how fast a population is growing. In countries classified as economically undeveloped, this rate may range from 15 to 30 people per year for every 1000 people already in the population (Fig. 2–2). The *growth rate* is a com-

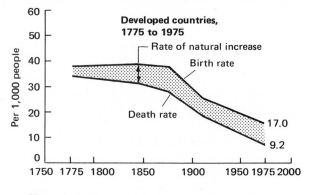

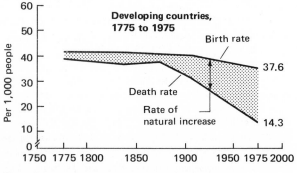

Rate of natural increase = birth rate — death rate

Figure 2-2   *World birth and death rates. The rates of birth and death are shown for both developed and undeveloped countries. Although the populations are growing faster in undeveloped countries, they are growing in developed countries as well. (Source: Population Reference Bureau, 1975.)*

bination of the rate of natural increase and the movement of people into or out of a country (Fig. 2–1).

In developed countries, growth rates vary. In some countries, such as Denmark, Sweden, Federal Republic of Germany, Switzerland, and Yugoslavia, populations are hovering near a rate of zero growth. In others, such as Holland, Luxemborg, and the United States, birth rates still exceed death rates. As a group, the developed nations have a growth rate of about 1% per year, contrasted with an average of 3% per year for the undeveloped nations. The world population as a whole is increasing at the rate of nearly 2% per year. What does this mean? In 1825, the world population was one billion. One hundred and five years later, in 1930, it had reached two billion. Forty-six years later, in 1976, the world population had doubled again to four billion. The time required for the world population to double is becoming shorter and shorter (Table 2–1).

TABLE 2–1   Doubling Time for the World's Population

Year	Approximate population	Time to double
1650	500,000,000	175 years
1825	1,000,000,000	105 years
1930	2,000,000,000	46 years
1976	4,000,000,000	

Why rapid population growth continues   It is in some ways difficult for Westerners to understand why birth rates remain high in undeveloped countries even though death rates are falling. The reasons involve differences in the value of children to African or Asian societies compared to their value in Western society. In Western societies today, most of the benefits of raising children could be called psychological. The pleasures of parenting, the feeling of passing on values to another generation, the fulfilling of other people's expectations (i.e., grandparents)—none of these brings any economic reward to the parents. Most of these psychological rewards are gained by having one or two children, although a third child may be wanted if the first two are of the same sex. This, combined with the high cost of raising children in Western society today and with effective, readily available contraceptive methods, has made the two-child family more and more the accepted norm.

It is otherwise in most developing nations. In many parts of Africa labor is a scarce resource. Land is not in short supply. The family that has many working members can farm more land and become wealthier. As one writer notes of Mali,

> except for recently introduced plows and carts, the physical factors a man needs to produce food are readily available. Land is abundant, "capital goods"—hoes and oxen—are cheap, and the wood and iron for making them are abundant. The crucial factors of production are the family's labor, of course, and "technology."[1]

While in Uganda:

> extensive labor is needed for herding....A moderately prosperous herd owner with say 100 to 150 cattle, 100 sheep and goats and a few donkeys needs about six herd boys ranging from 6 to 25 years of age to maintain a herd by himself. A man with many cattle but few sons must herd together, and share the yield of his stock, with a man who has few cattle and many sons.[2]

Further, in some parts of Africa, for example, the Upper Volta and the Sudan, private ownership of land does not exist. A person owns the crops grown

---

1 W. Jones, "The Food Economy of the Ba Bugu Ijoliba, Mali," in *African Food Production Systems*, P. F. McLoughlin, Ed. Johns Hopkins Press, Baltimore, 1970.
2 R. Dyson-Hudson and N. Dyson-Hudson, "The Food Production System of a Semi-nomadic Society: The Karimojong, Uganda," in *African Food Production Systems*, P. F. McLoughlin, Ed., Johns Hopkins Press, Baltimore, 1970.

on land but no one can own the land itself. Land use priorities are often given out according to the number of people in a family who can work on it.

Also important is the fact that the definition of family differs in many developing countries from that in the West. In parts of Africa, a father and his sons and their wives and children may live together or at least work together under the father's direction. The economic unit, then, of one man, his wife, and their children is often not the usual one. In fact, in parts of Africa, a man may have more than one wife. In situations like this, property rarely belongs to a husband and wife together and inheritance is often lateral, rather than from father to son. For these reasons, certain pressures to limit family do not exist. For instance, peasants in Europe felt the need to limit family size in order that inherited land would not be split among too many heirs. In Africa where land is not held privately or where a father's property is not divided among his sons, factors that help to limit family size are missing.

The problems of inheritance in some Indian states are seen as the problems of the next generation. The father has the right and the duty to work for the maximum benefits for himself and the family he heads. This may very well mean more children now to provide more labor and produce more goods in the present.

African and Asian families are often not as child-centered as those in the West. Parents may be expected to eat more than a proportionate share of food and use more of the family resources for clothing, etc., than are the children. A child's duty is to work for the benefit of the parents. In some African tribes, a son's loyalty and labor belong to his father even after he has a family of his own. Besides providing labor to free the parent from household or field chores so the parent can do other work, working children can allow the parent to have some leisure time. Younger children even allow older children more freedom.

In developing nations children can be useful at as young an age as five years (Fig. 2–3). If they can-

Figure 2–3 *This Venezuelan boy is one of what the International Labour Organization estimates as at least 52 million child laborers in the world today. (UNICEF photo.)*

not do an adult's labor in the fields, they can begin to help take care of younger brothers and sisters and do some of the household chores. In developing countries, such as Bangladesh, marketing and preparing meals for a family may be a full day's work for one person in the family. Children can be trained to help with meal preparation chores. One writer estimated that a man living alone in Bangladesh would spend 9/10 of his working time on household chores[3]—tasks women and children would otherwise do.

Children also provide a sense of security against old age or ill health, especially where there are no government old-age programs. In at least one sense, children are better than pensions or government social security since the benefits they

_____

3 M. Cain, "The Economic Activities of Children in a Village in Bangladesh," Paper presented at the IUSSP International Population Conference, Mexico City, 1977.

promise will not be eaten away by inflation. Even when parents must invest resources in children during their growing up years, they can expect to reap some return when the children are grown. This enables parents to even out their own lifetime consumption. The parents are spending more when they are young and vigorous and able to work and drawing on this investment when they are old or infirm. In Africa, parents who can expect their children to support them when they are old tend to respond to arguments that fewer children are needed as medical care improves. However, African parents living in cities, who seem to feel less certain that the old values will hold, tend to have

A BRAZILIAN COUPLE HAS HAD 32 CHILDREN

BRASILIA (AP)–After 42 years of marriage Mr. and Mrs. Raimundo Carnauba have stopped having babies, barring an unforeseen 33rd. Now the Brazilian superfamily is concentrating on grandchildren.

Mr. Carnauba and his sturdy wife, Maria Madalena, have had more babies than any other living couple in the world, according to The Guiness Book of Records. Twenty-six of the 32 Carnauba offspring are alive, and most have reached the age of parenthood.

The rotund 66-year-old carpenter says his grandchildren have grown in number from four dozen late in 1973 to some six dozen today. Make that yesterday.

"It seems like every day there is a new one," he said. "I don't know for sure, but I think there are 72."

He did some quick calculating: "It was 72. Then two died. But now two more were born."

The most prolific of the second generation so far is Juvencio, 36, with eight of his own.

Mr. Carnauba said his oldest child is 39 and his youngest, a girl, is 8. Seven of the 26 are girls and six are twins.

His wife was 13 and he was 23 when they married.

The Carnauba home is a shabby wooden house with a tin roof on a dirt street in one of several low-income "satellite cities" around Brazil's modern capital. Counting the kitchen but not the adjoining carpentry shop, where Mr. Carnauba works, it has six rooms. "It's pretty tight," he said.

He predicted that ignoring birth control and fathering all those babies will pay off some day. "There has to be one of those who is going to support me."

even more children because it is then more likely that at least a few will provide for their parents' old age.

In developed and in certain undeveloped nations, parents can make choices about how they will educate their children. They may educate all of them, splitting the family's resources; they may educate only the brightest, sending that one child as far up the educational ladder as possible; or they may educate none of them. In countries where there is enough economic development and where social classes are not rigid, parents may see a benefit in producing only a few, well-educated children. These children can be expected to make something of themselves and add to the family's prestige and wealth. In poorer countries such as Bangladesh, parents may not have this choice. Their resources are often so limited that even one child cannot be educated. In such a situation, the only wealth available to a family is a large number of children.

Summary of factors that limit or promote rapid population growth (Thus, in many developing nations children provide parents with wealth, with leisure, and with security in old age. These factors, which promote large families, are largely absent in industrialized societies.)

(Almost all of the nations which have managed to slow their growth rates are also characterized by increased economic development and by what is called social mobility. That is to say, jobs are available to those who are willing and able to work and further, there are no barriers of class or custom to prevent a family from bettering itself. In most of these countries, parents cannot benefit from their children's labor because of laws against child labor and because chilren are often required to be in school. In a climate such as this, parents seem to limit family size so they can take advantage of opportunities offered. Women may marry later and limit family size in order to gain extra income by working. A few well-educated children are seen as a better investment than a large family.)

Rural to urban migration   Although many people are concerned about the rapid growth of the world's population there is another equally far-reaching phenomenon occurring in this century, which affects both the environment and how people interact with one another. It is the way in which society is organizing itself. We are in the midst of a transition from a mainly rural society to a mainly urban society. The transition is occurring in the undeveloped countries as well as in the developed, industrial nations. Although at one time most people lived on farms, soon most will live in cities. In the early 1900s, only one-quarter of the world's population lived in cities. By the year 2000, it is believed that 60% will live in cities.

Cities have certain advantages. They allow for greater division of labor, providing more jobs. There are more specialized occupations in cities where people can find work. Examples are taxicab or rickshaw drivers, policemen and firemen, jobs which may not even exist in the hinterlands. Even crime and begging provide ways to sustain life in the cities. A more concentrated population can also be more easily provided with health and welfare services. These factors are important in drawing people to cities. Another reason for the rural to urban migration is that the land available for farming is limited. That is, as the population in rural areas increases, there will not be enough land to support the entire population in farming or in other traditional rural occupations. In both developed and developing nations, fertilizers, pesticides, and modern farming equipment when available make it possible for fewer people to till the same area of land. Employment opportunities are then fewer and people move to the cities in order to support themselves.

In the developing nations, the poor people who move to cities to find work often find conditions of extreme squalor. In the major cities of India, many people must live on sidewalks, making their home on nothing more than the pavement. (Fig. 2–4). The water and sanitation conditions of

Figure 2-4 *Typical sidewalk squatter dwellings of children in Bombay contrast starkly with modern high-rise apartment buildings that house a fortunate few. (UNICEF photo by UN/UNICEF.)*

these wretchedly poor people are far worse than in the rural areas from which they came.

Cardboard and tar paper shacks seem a step up to people who have so little. The santitation conditions are a modest improvement in communities of shacks but they remain extremely primitive, with no running water and only ground disposal of waste. Intestinal diseases are rampant in these unsanitary conditions, and tuberculosis is spread by such close contact. For most people in developing nations, the move to the city is a misery often

comparable to, and in some ways worse than, the misery left behind.

In the wealthier, developed countries, poor people who move from rural areas to the cities may find noise, dirty air, violence-ridden schools, crime, readily available drugs, and, too often, ugly surroundings.

These problems of cities in the developing and developed world are in part due to the speed of the transition from rural to urban society. Laws and social customs have not evolved quickly enough to cover situations in which enormous numbers of people are crowded together into cities. Perhaps new customs and laws will in time help solve some of these problems. As an example, wise land-use programs may be used to redirect the rural-urban migration. Such programs may preserve open space or limit "urban sprawl" or prevent strip development or confine industrial operations (Fig. 2–5).

EFFECTS OF POPULATION GROWTH

To a large extent, the problems caused by rapid population growth differ depending on whether a country is a developed one, such as the United States and the countries of Western Europe, or whether it is an undeveloped one such as the countries in Latin America.

Consequences of Population Growth in Developing Countries

Food production ⟨In developing countries, the most pressing need is often simply to produce enough food to feed the rapidly growing numbers of people. Government programs must be heavily weighted in this direction to prevent starvation. In fact, the percentage of people in a country involved in growing food gives us two insights. It tells something about how developed a country is, because as a country develops, fewer and fewer

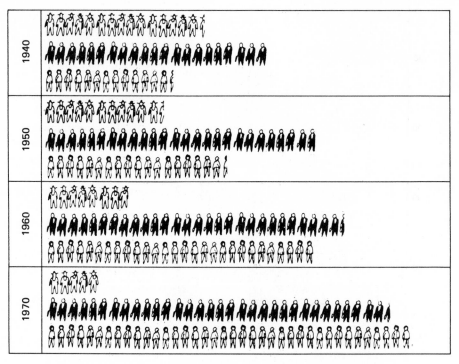

Each symbol = 2 million population

Farm

Central City

Suburb

Figure 2–5   *The transition of the United States population from rural to urban–suburban has been a continuing phenomenon up until 1975. The trend has recently shown a slight reverse in the United States, although it continues worldwide. (Source:* Options: A Study Guide to Population and the American Future, *Population Reference Bureau, 1973.)*

people are needed to grow food for the rest. In addition, this percentage in undeveloped countries also points out how important food production is.) China-watchers estimate that fully 70% of the Chinese people are involved in agriculture. Although, in the United States, one farmer feeds 59 people, on a worldwide basis, one farmer feeds only 5 people. Resources are employed that could be used in many other ways.

(Investments in food production are also important for the stability of the government. A hungry population is a discontented one. This discontent can flare into revolution and overthrow of the government. This possibility sometimes leads governments to act in ways that are not economically wise but will help them to stay in power.)For

instance, food prices are often controlled at a low level in large cities. This keeps the city people happy. However, this policy results in low prices for the farmer, which in turn discourages increases in food production. Farmers will not want to raise more food if they do not receive a reasonable price for their efforts. In fact, low farm prices also mean that farmers cannot invest in modern farming methods that would increase their crop yields. Thus, countries that most need more food may be in some ways hindering greater food production (Figs. 2–6 and 2–7). In this manner, population pressures can force governments to slant their policies in ways that are not in the long-term best interest of that country. The world food situation is covered in more detail in Chapter 33.

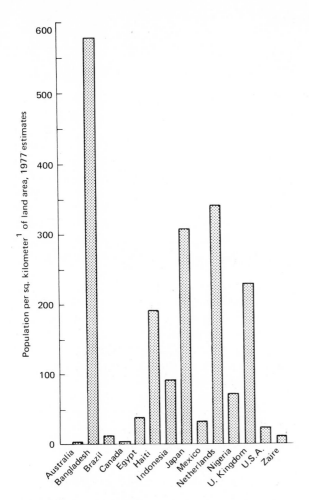

¹1 sq. km. = 0.3861 sq. mi.; 1 sq. mi. = 2.59 sq. km.

Figure 2–6 *Population per square kilometer of land area (1977). Bangladesh leads with the highest density of people per square kilometer of land in the world. Interestingly, the Netherlands is second. (Source: Population Reference Bureau, 1977.)*

Health consequences (Because so much money must be spent on food production in developing nations, little is left over for health programs. The rates of maternal illness and death are higher in undeveloped countries among women who have many children.) The babies born to these women

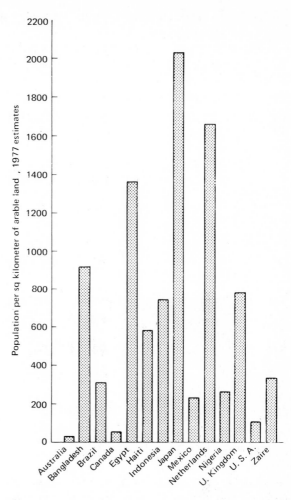

Figure 2–7 *Population per square kilometer of arable land (1977). When population density is measured in people per square kilometer of arable land (land that can be farmed), the picture changes. Now Japan leads the world in terms of population density. The Netherlands is still second but Bangladesh is in a somewhat better position with respect to its ability to grow enough food to feed its people. Actually population density is not always the important measure when population problems are considered. It is the rate of growth of a population that is important. A sparse but rapidly growing population can strain a nation's resources, while a densely populated but nongrowing nation may be able to provide for its citizens relatively well. (Source: Population Reference Bureau, 1977.)*

tend to suffer from malnutrition (Fig. 2–8). The lack of sufficient food in childhood leads to poor health and decreased mental ability.

The developing nation's annual investment in health activities must cover salaries of physicians and nurses, medicines, rents, equipment, etc. Yet the total annual investment in health activities, when divided by the population, comes to the figure of only $1 to $2 per person per year in most developing countries. Obviously, not much health care can be obtained for this sum.

Ill health has its effect on productivity. Burdened by ailments of many sorts, individuals are not likely to be able to pull themselves and their

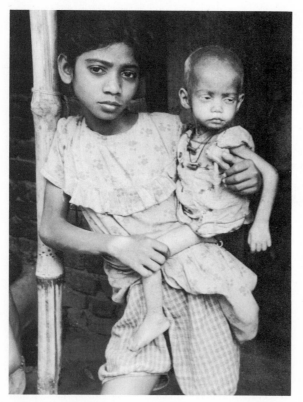

Figure 2–8 *Severe malnutrition affects about 10 million children under age five in developing countries. (UNICEF/Abigail Heyman.)*

families out of poverty by their own labor. In this way, the expanding populations of developing nations diminish the health resources available to each person and so reduce each person's potential to succeed.

Social and economic consequences in developing nations (Not only is there little money for health programs, but education is also likely to be slighted. The government is often unable to increase educational services at the same rate at which the population is growing (Fig. 2–9). Despite actual increases in school enrollment, the number of illiterate people in the world rose between 1950 and 1965 because the population grew so fast. As increasing numbers of young people go out to look for jobs each year, their low educational levels and the scarcity of jobs force them into unskilled positions at best. People who are forced to accept these low-paying jobs will have limited upward social mobility and very little opportunity to improve their economic lot even by very hard work. These factors lead to the development of two groups in a developing nation, one rich and well-fed, one poor and ill-nourished. The rich are likely to remain rich and privileged. The poor and the children of the poor have little hope to escape their fate.

(At a national level, as population grows, the per capita income decreases. That is, the number of people to be divided into the Gross National Product is larger each year, so the result, which is the per capita income, is smaller. There are two ways of looking at this particular fact. Some experts feel that the lower per capita income decreases savings, cuts down on the amount of capital available for investment, and prevents people from buying manufactured goods, thus retarding industrialization. Others point out that in many undeveloped countries, the capital and most of the income is in the hands of a privileged few anyway. They argue that, unless population programs are combined with some form of redistribution of this wealth, population reduction among the poorer people will not help improve their economic status at all.)

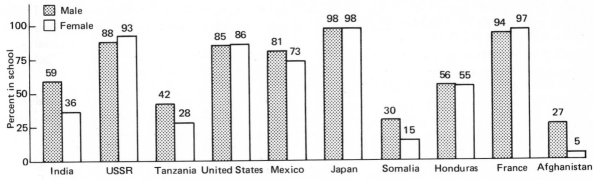

Figure 2-9 *School enrollment in 10 selected countries for 1975. It shows the percent of children, 6-18 years old, in school. (Source: UNESCO Yearbook, 1976.)*

Environmental effects in undeveloped countries
(Increased agricultural development needed to feed a growing population can cause increased soil erosion, contamination of soil and water with fertilizers and pesticides, and the destruction of wildlife habitats and areas of natural beauty. Erosion, pollution, loss of species, and loss of land—all these are the results of unchecked population growth.)

We catalog these events because they matter to us from our perspective. Our perspective is influenced, however, by the mountain of plenty on which we stand. Well fed, well clothed, well educated, we can afford the luxury of seeking quality in our environment. Our concern with environment in the developing countries has a certain hollowness to the people of these nations.

It is not hard to understand why wise land use policies or pollution control or protecting endangered wildlife are very low on the list of priorities in developing nations. Where children and parents are perpetually hungry, often sick, without adequate clothing, where there is little hope for people to better themselves, the immediate future of the environment does not really matter very much. Unless these other, primary needs are met, there will be little enthusiasm about seeking a quality environment.

Nonetheless, certain elements of the environment we seek are among those of long-term importance to people everywhere. Without good agricultural methods, such as proper irrigation techniques, careful use of pesticides, and control of erosion, the productivity of the soil will eventually be lost and food production will fall. Without adequate treatment of water supplies and of waste waters, diseases will continue to ravage and weaken the population and periodic epidemics of chemical poisoning will occur.

(The list could be made longer. The point is that, even in developing nations where the most pressing needs are the most basic human ones, for food and shelter and good health, thought must be taken for the environment. Unless the environment remains healthy it can not long support a human population.)

Resources and technology in undeveloped countries
Since, in undeveloped countries, there is little technological development (which is, after all, why they are called undeveloped), there is little demand for the resources which fuel an industrialized society. Thus, the scarcity of resources is not a problem faced by undeveloped countries. The one serious exception to this is materials used to produce fertilizers. Undeveloped nations do need large amounts of fertilizer to produce food. But fertilizers need nitrates (which are produced from natural gas), phosphates, and energy for their production.

For this reason, rising world oil prices are affecting not only the developed parts of the world, but also undeveloped countries such as India. Except for agricultural development, technology has little to offer undeveloped nations. "Modern technology has increasingly been directed toward doing better or more cheaply or in a greater variety of ways those things that most people in the poorer countries cannot afford to do at all."[4]

Consequences in Developed Countries

Resources and technology (In developed countries, by contrast, technology and the scarcity of resources are subjects of primary concern with respect to population growth.)

The energy crisis has caused many people to be concerned that we are running out of such resources as natural gas and oil. Shortages are also predicted by the turn of the century in many minerals, among them copper, tin, magnesium, and zinc.

It has been pointed out that, looking back in history, it is hard to find instances in which we have actually run out of some critical resource. What happens, commonly, is that scarcity causes a price rise that eventually forces a switch to some other resource. Thus a scarcity of wood in England in centuries past, led to the burning of coal. Whether this mechanism, which economists call "substitution," can be counted on to continue to work in the face of increasing population growth is not clear. For instance, will technology provide us with solar or fusion power before we run out of uranium, oil, and gas? We are at least gambling with the possibility of future resource shortages.

The illustration of the switch from wood to coal fires in England brings up another problem.

By the 1900s, these coal fires were responsible for a level of air pollution that made London almost uninhabitable at certain times of the year. Air pollution incidents, such as the London fog of 1952 that caused 4,000 deaths, were primarily due to pollutants derived from the burning of coal.

(Pollution is one of the most serious problems we in developed countries face.) We need more fertilizer and pesticides to grow ever increasing quantities of food. We need more petroleum to manufacture the gasoline needed to get increasing numbers of people to ever more distant places of work. This results in increasing levels of smog. More people need more housing and more farm products. Housing and farming use up forested lands and lead to increased soil erosion. More people generate more sewage and solid wastes, which either foul natural waters or demand technological solutions for their disposal. And more people consume more products whose manufacture generates chemical waste products. Although technology may be able to solve some of these problems, more and more of our money and effort are consumed in maintaining the quality of life. We run, like the Red Queen in *Through the Looking Glass*, "to stay in the same place."

Limiting choices (As population grows, we limit not only our choices about spending society's money, but we also limit personal choices.) For example, whole areas of the country (such as the Eastern Seaboard) are becoming urban. Recreational areas are threatened by throngs whose presence may even destroy the values of the area (Fig. 2–10). Furthermore, as population increases, available open space decreases or becomes more costly. Wilderness turns into overcrowded parks, and parks degenerate into outdoor slums. (Our lives become increasingly regulated by rules to maintain social order.)

Scientists have studied animals under crowded conditions and some of their results are worth noting. For instance, rats in crowded cages begin to lose their normal patterns of social behavior. They neglect their young and sometimes resort to can-

---

4 *Rapid Population Growth Consequences and Policy Implications*, Volume I, Summary and Recommendations, National Academy of Science, Johns Hopkins Press, Baltimore, 1971, p. 31.

Figure 2–10   *Crowded campground in South Tyrol. In developed countries, rapid population growth leads to overcrowding of recreational and wilderness areas. In a real sense, some of our freedom of choice is lost. (Werner H. Müller/Peter Arnold, Inc.)*

nibalism. Some rats become overly aggressive while others withdraw from the community. Sexual behavior becomes abnormal. Monkeys also show some of the same behavior when they are crowded together.

In the wild, animal populations, too, occasionally reproduce very rapidly. The needs of the expanded population may exceed the food supply or living space available. In such cases, starvation, disease, and reduced fertility will lead to a decline in the animal population. The mass migrations of lemmings to the sea is thought by some to be a response to rapid population growth.

Can the results of these studies be applied to human populations? Probably they cannot be applied directly. Nevertheless, the catastrophes that occur in animal communities stressed by rapid growth and crowding warrant consideration.

It may be that humans can survive under relatively crowded conditions. On the other hand, we may prefer not to see the changes such crowding will cause. If we wish to influence the future, we must make decisions about population growth. Otherwise the choice will have been made for us.

LIMITING POPULATION GROWTH

Problem of Time Scale

Programs designed to reduce population growth have one very serious and built-in problem. It takes some 10 to 20 years before the results of any such program can be seen. This is because fertility reduction programs can have an effect only on the number of babies currently being born. They can obviously have no effect in the group of children, age 0–15 years, already born. These children will grow to adulthood and marry and have children themselves during the next 10 to 20 years.)

In most countries, the size of this group (0–15 years) is larger than that of any other age group in the population. If these people in the 0–15 age category only have enough children to reproduce themselves, they will still swell the size of their country's population. (See p. 50 for more details). (Thus, most countries would continue for a time to grow in population even if birth control programs immediately reduced the birth rate to about 2.3 children per couple. This is the birth rate that provides for each person to reproduce himself. (See Fig. 2–11.))

The fact that population programs take such a long time to make their effects felt is a great handi-cap to effective action. It takes five to six years before decreased school enrollments are seen, 15 years before a reduction in the labor force occurs and 15–19 years before declines begin in the amount of food needs. A government may thus seem to be spending a great deal of money, for many years, with very little obvious effect.

It has been pointed out that it is important to try to separate birth control programs from politics and politicians. Politicians and political movements or ideologies have a way of changing or going in and out of favor in less than 10 or 20 years.

Methods of Limiting Population Growth

Birth control information (When a government wishes to limit its population growth, the first policy to be adopted is usually one of providing people with information and materials for birth control.) In many cases abortions are made legal and thus easier for the poor to obtain (Fig. 2–12). Although there are religious organizations that object to both abortion and to the distribution of birth control information, there is usually also a great deal of popular support for this type of policy. In most situations, enough people will at least tacitly agree that couples should be able to limit their family size if they wish to do so.

(A policy of providing access to information and materials on birth control can be justified on grounds of humanitarianism.) Fewer children will, in general, be better fed and cared for, since a family's resources will not be spread so thinly. These children may even be healthier, since repeated, closely spaced pregnancies have been shown to be harmful both to mothers and to babies. Furthermore, if only wanted pregnancies occur, mothers have no need to resort to abortions, which like any operation carry some risk. For this reason many anti-abortion groups support contraceptive programs. The cost to society of unwanted children who may become wards of the state in one way or another is also reduced.

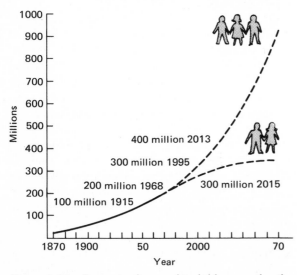

Figure 2–11 *Even at a figure of 2 children per family, the United States population will grow in size until the middle of the 21st century. Of course, at a figure of 3 children per family, growth will be even faster. (Source: Report of the Commission on Population Growth and the American Future, p. 23.)*

Figure 2-12 *A family planning clinic in Dacca, Bangladesh. Men and women wait for advice, medical treatment, or sterilization operations. (Bill Bouders/NYT pictures)*

However, even if the government is willing to provide birth control services, there is no guarantee that people will see a need to use them. As previously mentioned, there are a number of reasons why couples may choose to have more children than is considered ideal for society as a whole.

(If birth control information and services prove ineffective in slowing the expansion of a nation's population, the type of policy to be implemented next is not at all clear. There is a great deal of controversy, in fact, on what types of policies should be adopted. Three general types of policies have been suggested.)

Coercive measures (The first group involves coercive measures, in which the government takes active control of individual reproduction. In such a situation, a license might be issued to allow a couple to have a child. Forced sterilization programs might be carried out) as they were in India in the 1970s. There are a number of valid objections to policies such as these. For instance, who would decide, and on what basis, who was to be privileged to have children? (The tenets of individual freedom would be lost in such programs.) It should also be pointed out that we do not at this time have the technical capability to carry out such a policy. There is no magic chemical that could be put in the water to prevent fertility, without other serious effects. There are not enough doctors and nurses available for forced, large-scale contraceptive or sterilization procedures. (From a purely practical standpoint, family limitation programs, if they are to succeed, must be made to seem desirable to everyone.) (See the Controversy on p. 42 for India's experience.)

Rewards and incentives (The second group of policies involves various economic rewards for not having children or penalties for those who do have children.) One suggestion in the United States was to remove welfare benefits or income tax deductions for dependent children after the second child. Other suggestions have included fines for having more than a specified number of children and educational or other bonuses for families with few children. The main objection to this group of policies is that the extra children, who had no control over whether they were born or not, suffer along with the parents. Other children in the family likewise suffer.

Despite these drawbacks, such policies are, in fact, being followed in several countries and are planned in others, such as Bangladesh. State employees in certain Indian states are not eligible for housing loans and land grants if they have more than three children. In Singapore, there are increasing fines for the birth of a third, fourth, fifth, or sixth child. Maternity leave is not allowed for the birth of a third or subsequent child, and hospital fees rise with the number of children in a family. On the other hand, sterilization after two children is rewarded with education and employment for the children and extra vacation time for the parent. Foreigners who wish to marry Singapore nationals

FAMILY PLANNING IN INDIA– A LESSON ON THE MEANING OF FREEDOM

Among the nation's of the developing world, India has had one of the fastest growing populations. At 300 million just after World War II, India doubled its population by the mid 1970s. Efforts to limit population have been underway there since the early 1950s. In fact, India was the first nation to create a national program to limit population through family planning. Even into the 1970s, however, the results were barely noticeable. In 1976, it was estimated that only 17.5 million couples out of 103 million in the reproductive age groups were using contraceptives.

India's program up to that time was traditional; sterilization or birth control methods and devices were made available to those who asked for them. Although no one was forced to seek sterilization, incentives were offered. For a time, a man could obtain cash or a transistor radio, a coveted article, in return for having a vasectomy.

In April 1976, a new policy was adopted emphasizing sterilization by vasectomy and, in many ways, compelling men to submit to the operation. Mrs. Indira Gandhi, prime minister of India, was quoted as saying, "We must act decisively and bring down the birth rate.... We should not hesitate to take steps which might be described as drastic. Some personal rights have to be kept in abeyance for the human rights of the nation: the right to live, the right to progress." The government's goal: a reduction in the birth rate from 35 per thousand to 25 per thousand by 1984. It is of use to note that India had been declared in a state of emergency in June 1975 and that a number of civil rights, including free speech, had either been suspended or decreased since that time. It was in this political climate that the intensive program of sterilization was begun. Although the government did not mount a national campaign of compulsory sterilization, the individual Indian states were encouraged to do so.

The laws in the State of Maharashtra called for compulsory sterilization for the father of three living children and compulsory abortion of a pregnancy leading to a fourth child. Incentive payments to those who submitted to vasectomies were a part of the law, as were payments to informers. In Delhi, the capital, and in the states of Punjab and Haryana, laws were passed withdrawing vital government benefits and services from married men with two children who did not submit to vasectomy. Loss of subsidized housing, loss of free medical care and loans, and even loss of employment were possible if a man did not comply with the sterilization laws.

Widespread abuses, in which people–especially poor people–were compelled to submit, were reported as the state governments attempted to fill sterilization quotas. Riots broke out in Northern India over the issue. In the final months of Mrs. Gandhi's rule, sterilizations were numbering a million per month. And then in early 1977, Mrs. Gandhi's government fell, as the voters sent her and her Congress Party from office. It is not clear if her government fell because of her near-dictatorial rule and her suspension of civil liberties or because of the highly visible, highly controversial program of sterilization she promoted. Many regarded the sterilization program as the largest factor in her defeat. The government that succeeded hers, led by Prime Minister Desai, abandoned the approach of force and of punishments for failure to be sterilized, presumably because the issue remained highly charged politically.

This episode raises some basic questions about birth control policies. Mr. Kaval Gulhati asks, in an article in Science, *

> Should they stand by and wait for economic development and family planning programs to motivate contraception? Or should they take the destiny of the people in their hands, and force a fertility decline?

Mr. Gulhati asked this question before Mrs. Gandhi's government was rejected. In light of her government's defeat, the question might be expanded to ask,

> Can the government of a free people sucessfully undertake a compulsory program of population control and itself survive?

---

* K. Gulhati, "Compulsory Sterilization: The Change in India's Population Policy," Science **195,** 1300 (25 March 1977).

must agree to be sterilized after the birth of their second child.

Methods which preserve freedom of choice (The third group of policies includes those that do not put burdens on innocent children. These policies also allow for some preservation of an individual's freedom of choice and they often have some other value to society in addition to the reduction in population growth.)

In many undeveloped countries, children may be sent out to work and so represent to the parents the possibility of additional family income. The passage and enforcement of laws prohibiting child labor can make having many children less desirable by removing their potential for income. Parents must then feed and clothe the child for a longer time during which he cannot work. Such laws benefit children by allowing them time for schooling and, depending on what work they would have done, possibly improving their health (Fig. 2–13).

We recall from the earlier discussion that children represent security for parents in old age. If old age pensions, such as the Social Security program of the United States, are created in developing nations, parents may be able to give up the idea that they need many children. In Bangladesh, Social Security benefits are planned, but they are to be awarded only to parents of two or fewer children, in order to further emphasize the desirability of small families.

Another reason many parents have more than two children is to ensure that they will be able to raise at least two children to adulthood. The region with the world's highest birth rate also has the world's highest death rate.[5] Continuing reductions in infant and child diseases should eventually lead to some reduction in births.

Education, besides helping the children themselves and society as a whole, has several benefits related to population control. Educated people tend

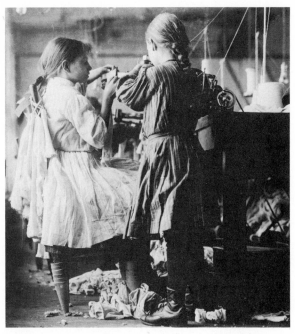

Figure 2–13 *Young, nimble fingered children toiled long hours at low pay in the 19th century factories and mines of today's more developed countries. (U.S. Library of Congress; photo by Lewis Hines.)*

to marry later and have fewer chidren. This is related to the fact that education allows people to move upward in economic and social status. This upward movement would be hindered by large numbers of children to feed, clothe and care for. Educated women tend to limit their families in order to allow themselves the satisfaction of a career and additional family income.

(Improvement in the status of women, in addition to education, may have the same effect of limiting population growth. If the need to care for young children prevents mothers from taking advantage of new or better job openings, they seem to limit their family size to take advantage of those opportunities)(Fig. 2–14).

(Another way to influence population growth is to control housing availability. In addition, the government can require a period of military or other national service for all young people. This ac-

---

5 The region is Africa, which in 1978 had a birth rate of about 45, and a death rate near 20 per 1000 per year.

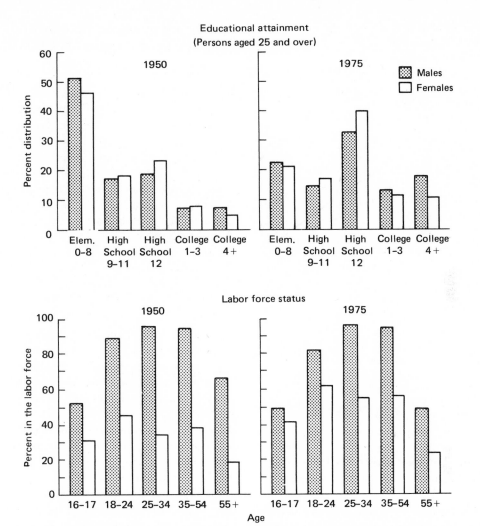

Figure 2-14 *In the United States, the educational achievements of women, as well as men, have been steadily growing. In addition, the percentage of women who work has been increasing. These factors have probably contributed to the downward trend in the U.S. birth rate. On the other hand, in parts of Africa and Asia where there are many family members to help care for young children, or where uneducated and unskilled babysitters are easy to find, an increase in the number of women in the work force does not have much influence on the birth rate. (Source: Population Reference Bureau, 1976.)*

tion will tend to increase the age at which marriage occurs, reducing the span of years in which a couple can conceive.)

These means all allow individuals some freedom of choice on the number of children they plan to have. They also make people aware of the additional costs that children will bring, and they open other avenues of satisfaction. They provide other means of achieving the same benefits that are obtained in large families. Through such mechanisms one would hope to be able to reduce population growth without coercion. There is evidence that, under the right circumstances, people can become aware that having more than two children is not in their own best interests. Countries in which there is the greatest possibility of upward movement in economic and social status have shown the most dramatic decreases in birth rate. Thus, the ideal of bringing population under control may be achieved by means that help people reach, at the same time, a satisfying standard of living.

The main problem in devising ways of reducing population growth is, of course, the fact that the issue is not some abstract scientific problem that can be solved by dollars and technology. Instead, it is a problem that touches people directly. Computers can be programmed to provide predictions of population, but only people's behavior, plans, and aspirations can determine what the future will be. People's needs and feelings affect how they react to incentives and plans. People also exert a degree of control over policies that affect them. In the United States, for instance, population control policies must be seen as desirable to the majority of people before they have any chance of being put into action.

OUTLOOK FOR POPULATION GROWTH IN THE FUTURE

Late in the decade of the 1970s, we became aware that population growth rates in undeveloped countries were decreasing slightly. Is this a sign that undeveloped countries are beginning to undergo a demographic transition in which birth rates will come into line with lowered death rates? In fact, two trends are occurring. First, there is a small but real decrease in world population growth rates. Second, there is a general and important decline in the fertility of women in the less developed countries. Demographers define the *fertility rate* for a nation as the number of live births that a woman would be expected to have during her childbearing years.)

The decline in fertility, which is reflected in the observed slower rates of growth, is a dramatic signal of possible changes on the horizon.)

According to a 1979 publication of the U.S. Bureau of Census, the rate of world population growth fell from 1.98% per year in the period 1965–1970 to 1.88% in the period 1975–1977. While the decline is small, it is significant because it represents a reversal in the pattern of growth seen in recent history; populations now seem to be growing more slowly. It gives hope that future rates of growth will fall still further as family planning becomes an integral part of total planning in the less developed nations. The report noted that exceptional declines in growth rate had occurred in Sri Lanka (down 0.8%) the Philippines (down 0.7%), Republic of Korea (down 0.7%), Thailand (down 0.7%), and Colombia (down 0.6%). Only African countries seemed to be running counter to the trend of declining growth rates.

The second trend, a decline in fertility, is the underlying explanation for the decrease in growth rates. On a worldwide basis, the average total fertility rate fell from 4.6 in 1968 to 4.1 live births per woman in 1975. Fully 80% of the world's population in 1975 was in countries with declining fertility rates.

In a study of 113 less developed countries, 95 of the nations had seen fertility declines from 1968 to 1975. Table 2–2 lists the decline in fertility in a number of the larger less developed nations. Using the new estimates of fertility, the world's population in the year 2000 may be 5.8 billion rather than the 6.3 billion predicted by the United Nations only a few years earlier.

From such statistics as these, there appears to be genuine hope that world population is coming

---

6 "World Population 1977: Recent Demographic Estimates for the Countries and Regions of the World," available from U.S. Bureau of the Census, Population Division, Washington, D.C. 20333, January 1979.

under control (Fig. 2–15). This by no means erases the problem of population growth. On the contrary, it appears to show that rapid population growth is a problem that more effort and money can affect.

(A close look at recent declines in fertility rates in developing countries suggests that there were two equally important factors at work. One was an improvement in economic conditions within some of the countries. In many ways, as we have seen, better economic and social conditions cause parents to feel that it is in their own best interest to have fewer children. The second important factor in many countries was a strong family planning program. The time, effort, and money put into these programs appears to have paid off.)

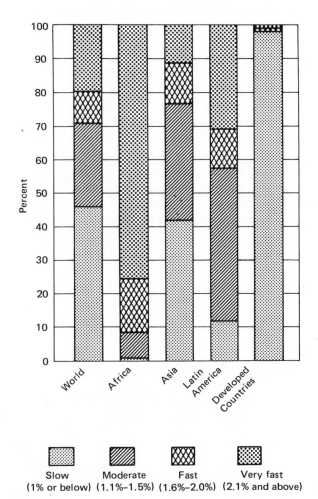

Figure 2–15 *Percentages of world population at various growth rates in 2000: CFSC projections. In 1978 the Community and Family Study Center of the University of Chicago made the above predictions of population growth rates for the year 2000. By this time the Center's predictions indicate that most of the world's population will be experiencing slow to moderate population growth rates. It is believed that the rest of the world will follow suit within 25 years after that. (Source:* Population Bulletin, *Population Reference Bureau, 1978.)*

POPULATION

...Any choice and decision with regard to the size of the family must irrevocably rest with the family itself...

United Nations Universal
Declaration of Human
Rights (1967)

Freedom to breed will bring ruin to all.

Garrett Hardin

There are many difficult decisions involved in the control of population growth. One of the most difficult is whether couples should have the complete freedom to decide how many children they wish to have. That is, should governments set and enforce upper limits of family size?

In 1967, the United Nations passed the Universal Declaration of Human Rights, which stated:

The Universal Declaration of Human Rights describes the family as the natural and fundamental unit of society. It follows that any choice and decision with regard to the size of the family must irrevocably rest with the family itself, and cannot be made by anyone else.

In the past few years, however, a few nations have decided that this concept is unworkable. Thus in Bangladesh, Singapore, and India, the governments have begun to use both economic threats and bonuses designed to limit population size.

Garrett Hardin is a scientist who believes that the decision to limit family size cannot be left up to the individual. He says that such a decision is similar to one which faced cattle owners in 18th century England. At that time, some pastureland was held in common. Hardin argues that there was great advantage to a herdsman to add more cattle to his herd, if he could, since the pastureland was free to all. If everyone did that, however, the pasture would be overgrazed and thus ruined. He argues that, in a sense, our welfare-oriented society is like a commons. Individuals can choose to use up more than a fair share of resources, both social and environmental, by having more than their fair share of children.

Hardin * writes:

Perhaps the simplest summary of this analysis of man's population problems is this: the commons, if justifiable at all, is justifiable only under conditions of low-population density. As the human population has increased, the commons has had to be abandoned in one aspect after another.

First we abandoned the commons in food gathering, enclosing farm land and restricting pastures and hunting and fishing areas. These restrictions are still not complete throughout the world.

Somewhat later we saw that the commons as a place for waste disposal would also have to be abandoned. Restrictions on the disposal of domestic sewage are widely accepted in the Western world; we are still struggling to close the commons to pollution by automobiles, factories, insecticide sprayers, fertilizing operations, and atomic energy installations.

...I believe it was Hegel who said, "Freedom is the recognition of necessity."

The most important aspect of necessity that we must now recognize, is the necessity of abandoning the commons in breeding. No technical solution can rescue us from the misery of overpopulation. Freedom to breed will bring ruin to all. At the moment, to avoid hard decisions many of us are tempted to propagandize for conscience and responsible parenthood. The temptation must be resisted, because an appeal to independently acting consciences selects for the disappearance of all conscience in the long run, and an increase in anxiety in the short.

The only way we can preserve and nurture other and more precious freedoms is by relinquishing the freedom to breed, and that very soon.

Do you agree with Hardin? Should the decision on family size be made by each couple, or should governments have a voice in it? If you feel that population growth should be slowed, should appeals to conscience be tried first? What did Hardin mean when he said "an appeal to independently acting consciences selects for the disappearance of all conscience in the long run"? Suppose such appeals don't work, can you think of economic measures, such as the use of taxes, that could influence family size? Suppose neither appeals nor economic measures are effective, are penalties justifiable?

---

* Garrett Hardin, "The Tragedy of the Commons," <u>Science</u> **162,** 1243 (13 December 1968).

TABLE 2–2 Decline in Fertility in Some Less Developed Nations[a]

Country	Fertility (expected live births per woman) 1968	1975	% Decline	Population in 1975 (millions)
Bangladesh	6.98	6.32	10	79
Colombia	6.54	4.24	35	24
Egypt	6.11	4.72	23	38
India	5.67	5.24	8	600
Indonesia	6.46	4.58	29	130
Mexico	6.59	5.74	13	60
Pakistan	6.84	6.57	4	70
Peoples Republic of China	4.20	3.20	24	896
Thailand	5.86	4.85	17	—

[a] Source: A. Tsui and D. Bogue, "Declining World Fertility," *Population Bulletin* **33** (4), (October 1978).

Population

Demography–the Study of Populations

To understand how we can influence what will happen to a population in the future, we take note of the people in two age ranges, those 0–15 years, who are not yet old enough to have children, and those 15–35 years, who are at the child-bearing age. We can see that knowing the size of the total population is not as important as knowing how many people are in these younger age groups, since people in the older age groups are no longer likely to have children. These older people are thus not able to affect the *future* size of a country's population.

Because the 0–15 age group in most developing nations is larger than any other age group, even if people now in this group only reproduce themselves on a one-to-one basis, the country's total population will grow. In order to have the total population stay at the same level as it is today, parents have to average fewer than two children per couple. This is not a likely situation. Most people feel that they have at least the right to reproduce themselves. Nor is it even necessarily a good idea to have fewer than two children. Birth control efforts in Japan since World War II have lead to a small work force supporting a large aging population, a social and economic burden of great consequence. (See Further Reading for other thoughts about "zero population growth.")

Because changes in the near future are the result of the age structure existing today, population programs that are started right now will have a 40–60 year lag time before they can result in stabilized populations. This is true even if the programs are immediately and successfully implemented. It has been calculated that the United States' population will take 60–70 years to stabilize, even though the fertility rate is now at slightly less than two children per woman. Actual experience has shown that, in most countries, population programs are not immediately sucessful but take a number of years to become accepted.

Figure 2–16 shows the age structure of the U.S. population from 1900 to 2000 as a series of population profiles. Note how the large group of babies born in 1955–1959 acts to form a bulge in the population profile.

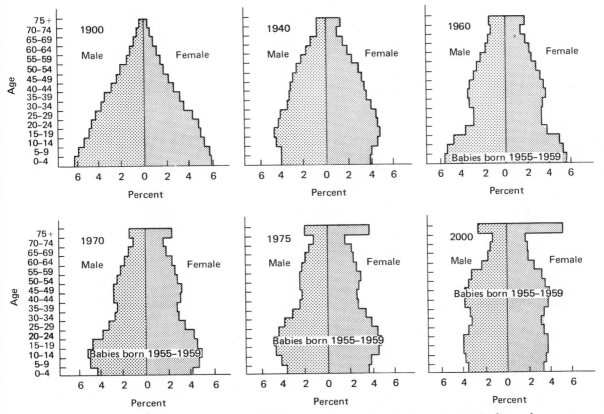

Figure 2–16   *The population of the United States at various times is shown here in the form of bar graphs or population profiles. From these graphs you can see how a large group of babies born during one period (1955–1959) will swell the size of other age group in later years. (Source: Population Reference Bureau, 1976.)*

Certain features of populations show up very well on bar graphs. For instance, many undeveloped countries show a characteristic "pinched profile." A decreased death rate in younger portions of the population, brought about by better health care, has swollen the younger age groups compared to the older ones. The profile has a wide base and a very narrow top (Figs. 2–17 and 2–18). Compare the profile in Figure 2–18 to the one of India in 1951. Note

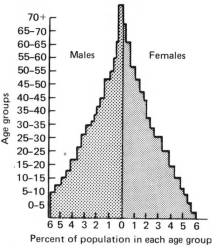

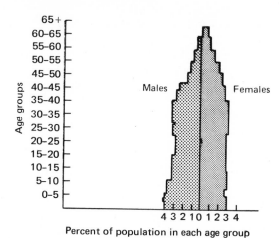

Figure 2–17 *Population of India in 1951, (Source:* W. S. Thompson and D. T. Lewis, Population Problems, *5th ed., McGraw-Hill, New York, 1965.)*

Figure 2–19 *Population of Great Britain in 1959. (Source:* Population Bulletin, *Population Reference Bureau, 1966.)*

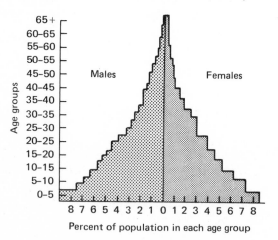

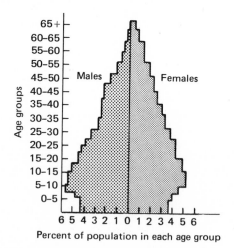

Figure 2–18 *Population of India in 1970 ("pinched" profile). (Source:* Population Bulletin, *Population Reference Bureau, 1970.)*

Figure 2–20 *Population of Japan in 1960. (Source:* W. S. Thompson and D. T. Lewis, Population Problems, *5th ed., McGraw-Hill, New York, 1965.)*

how the younger age groups in 1970 are very large compared to the older groups. This is the result of better infant and child survival as health care improved. Figure 2–19 shows the characteristic obelisk shape of population graphs for many developed countries. In contrast to the undeveloped nations, there is a more uniform distribution among the age classes and not such a broad base.

A successful population reduction program in Japan resulted in the profile shown in Figure 2–20. Note the narrow base of the profile. More recently the trend has reversed in Japan and birth rates are growing.

Questions

1. What is meant by the terms *birth rate, death rate, rate of growth*?
2. Although the United States birth rate for 1978 (15) was a record low, the U.S. population is still growing. At current birth, death, and immigration rates the population would double in 87 years. Explain why.
3. Let us assume that such a doubling would be unacceptable. Outline a program of population growth reduction for the U.S. Explain what methods you consider reasonable and workable in this country.
4. In the latest U.S. census, figures show that migration from rural areas to cities has begun to reverse for the first time in this country's history. However 3 out of 4 Americans still live in a large city or its suburbs. What do you see as the advantages and disadvantages of living in a large city? What are the advantages and disadvantages of living in rural areas? Do you feel that, on the whole, the world trend toward increasing urbanization is good, bad, or inconsequential?
5. What is meant by *demographic transition*? Why do you think this transition has occurred in developed nations but not yet in undeveloped ones?
6. Make a list of the problems caused by rapid population growth in undeveloped nations. Contrast to the problems caused by rapid population growth in developed nations.
7. Can you support the proposition that developing countries should spend some of their desperately scarce funds on protecting the environment?
8. Population reduction policies that penalize families with more than a certain number of children can be viewed as unfair to innocent children. Are policies that reward small families similarly unfair? Explain, using examples such as housing loans, employment and education bonuses, old age social security, etc.
9. How can economic development affect population growth?
10. How many children do you want to have? What advantages do you see to having children in the U.S today? What disadvantages do you envision? If you were a poor farmer living in Bangladesh what advantages and disadvantages might there be? How many children would you want to have if you lived in Bangladesh?

Further Reading

Bachrach, C. A., "Old Age Isolation and Low Fertility," *Population Bulletin* 7 (1), (January 1979).

Once we achieve a nongrowing population, what changes can we expect? Will there be large economic and social changes? This article examines a future that we should, perhaps, prepare for now.

Day, L. H., "What Will A ZPG Society Be Like?," *Population Bulletin* 33 (3), (June 1978).

Dumond, D. E., "The Limitation of Human Population: A Natural History," *Science* 187, 713 (28 February 1975).

This author argues that humans have actually practiced some form of population control all through history. His interesting theory gives new insights to the problems of rapid population growth now affecting the world.

Faucett, J. T. et. al., "The Value of Children in Asia and the United States: Comparative Perspectives," Papers of the East-West Population Institute #38, July 1974.

Green, Ambassador Marshall and Robert Fearey, "World Population: the Silent Explosion," Department of State Bulletin, Fall 1978 (Stock # 044–000–01710–5)

Population growth, and increasing world awareness of the problem over the last 10–15 years, are summarized in this booklet written by the State Department's Coordinator of Population Affairs. Available from the Superintendent of Documents, U.S. Government Printing Office.

Intercom, The International Newsletter in Population, Population Reference Bureau Inc., Washington, D.C.

The Population Reference Bureau, 1337 Connecticut Avenue, N.W., Washington, D.C. 20036, publishes a monthly newsletter as well as many position papers, teaching aids, wall posters, and population data sheets. Memberships, which entitle the owner to almost all the publications (a real wealth of information), are available to teachers and students at reduced rates. The writing is clear and at a level intended for the general public.

Nortman, D., "Changing Contraceptive Patterns: A Global Perspective," *Population Bulletin* 32 (3), (August 1977).

The state of the art in birth control is summarized in this paper.

Rapid Population Growth, Consequences and Policy Implications, Volumes I and II, National Academy of Sciences, Johns Hopkins Press, Baltimore, 1971.

These volumes provide a good background for understanding the results of rapid population growth.

Revelle, R., A. Khosla, and M. Vinovskis, *The Survival Equation*, Houghton Mifflin Company, Boston, 1971.

Revelle presents a thoughtful, humanitarian point of view about rapid population growth and possible solutions to the problems it causes. The book is clearly written and organized. Well worth reading for the arguments against coercion in birth control programs and the arguments that under the right circumstances, all peoples can see that control of rapid population growth is in their best interests.

Teitelbaum, M. S., "Relevance of Demographic Transition Theory for Developing Countries," *Science* **188**, 420 (2 May 1975).

Why did the developed nations undergo a demographic transition? Will this happen in undeveloped nations? These questions are examined.

Ware, Helen, "The Economic Value of Children in Asia and Africa: Comparative Perspective," Papers of the East-West Population Institute #50, April 1978.

This paper and the one by J. T. Faucett, are excellent for students who want to know more about how people in other cultures view children and population growth. This is the kind of background necessary to devise effective birth control programs in developing nations. These papers (and others) are available in single copies from East-West Population Institute, East-West Road, Honolulu, Hawaii 96822.

Chapter 3

Protecting Wildlife Resources

IS A LITTLE FISH WORTH MORE THAN A BIG DAM?

In the free-running waters of the Little Tennessee River, there lived a tiny fish called the snail darter. When this small member of the perch family was first discovered in 1973, all the snail darters in the world lived in this one place. Seven years before anyone even knew the snail darter existed, in 1966, Congress authorized the Tennessee Valley Authority (TVA) to build the Tellico Dam and reservoir across the Little Tennessee River. Builders had half-completed the dam before the snail darter was discovered (Fig. 3–1). Since the snail darter cannot breed in the still waters of a reservoir (it needs free-running water), completion of the 116 million dollar Tellico Dam threatened to destroy the entire newly discovered population at a blow.

In the same year the snail darter was discovered, Congress passed the Endangered Species Act. An endangered species is one that has so few living members that it is in danger of becoming extinct in the near future. The Act states, in part, that actions of federal government agencies may not "jeopardize the continued existence of endangered species and threatened species or result in the destruction or modification of habitat of such species which is determined to be critical."

The Tellico Dam was three-quarters finished in 1975 when the Secretary of the Interior listed the snail darter as an endangered species. Clearly, by destroying the breeding grounds of the snail darter,

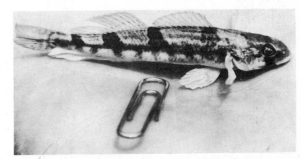

Figure 3–1 *Snail darter. This endangered fish stood in the way of completion of the $116 million Tellico Dam. The controversy over the fish and the dam led the Tennessee Valley Authority to rethink the whole idea of placing a dam on the Little Tennessee River, which provides one of the few remaining stretches of good cold-water fishing in the region. The controversy also caused Congress to rethink the Endangered Species Act, which protects even tiny fish from large construction projects. (For the final results of the controversy see p. 68.) (NYT pictures.)*

the Tellico Dam, if completed, would violate the Endangered Species Act. Several environmental groups sued to have construction stopped and the case went all the way to the Supreme Court. In 1978, with the dam 90% completed, the Court ruled that the project did indeed violate the Endangered Species Act and so must either be stopped or changed.

But was the protection of a small population of 3 inch fish really what Congress had in mind when it passed the Endangered Species Act? As one writer put it: "Undoubtedly many members of Congress were thinking thoughts of brown-eyed creatures and soaring winged things when they cast their vote, and now are finding themselves confronted with a Pandora's box containing infinite numbers of creeping things they never dreamed existed."[1]

What is the value of a species? Why should we attempt to save species from extinction? Can we say that some species are more worth saving than others?

Scientists estimate that probably about 10 million species exist in the whole world. Ecologists have so far discovered and described only some one and one-half million. But the discovery of new species is becoming a race against extinction. In prehistoric times, one species was lost perhaps every 1000 years. Between 1600 and 1950, the rate increased to around one species lost every 10 years. Today one species becomes extinct every year.

THE VALUE OF SPECIES

Practical Value in Medicine and Agriculture

(Lost species can be viewed as lost opportunities. Animals and plants provide us with drugs, with foods, and with raw materials for industry.)As an example, a group of chemical compounds called alkaloids are found in various plants. These chemicals give rise to drugs used to treat leukemia and certain other cancers, heart disease, and high blood pressure. Experiments have been done on only 2% of the world's flowering plant species, yet 1000 different alkaloids have been found in them. Many other important drugs, such as the tranquilizer reserpine and a variety of antibiotics and pain killers, come from plants (Fig. 3–2).

An important technique in farming is the use of biological controls. This involves using one species to prevent another from harming crops. For

Figure 3–2 *Endangered plant—persistent trillium (Trillium persistens). Concern about endangered plant species has come much more slowly than concern about endangered animals. Yet the two are so intimately related that they cannot be conserved separately. There are many examples of animals which are driven to the brink of extinction because a particular plant on which they feed or under which they shelter becomes scarce. Various common and rare plants are also directly useful to man, providing food, medicines, and material for shelter among many other things. The Endangered Species Act does not prohibit taking listed plant species, as long as state laws are followed. It does, however, prohibit the sale of the plants in interstate commerce. (U.S. Department of the Interior, Fish and Wildlife Service; photo taken by Dr. John Freeman, Auburn University)*

instance, certain wasps can successfully prevent the sugarcane borer from destroying whole fields of cane. Another technique of modern agriculture involves the cross-breeding of various plant species to develop crops with higher yields or resistance to disease or drought or heat. Every time we allow a plant or animal to become extinct we run the risk of losing a possibly helpful organism. We also deny future generations the opportunity to benefit from these species in ways that have not yet been thought of.

1 C. Holden, "Endangered Species: Review of Law Triggered by Tellico Impasse," *Science* **196**, 1427 (24 June 1977).

Value as Part of Food Webs and Gene Pool

(In addition, the loss of a particular species or group of species may have far-reaching effects on the community in which the species live. Complex food webs are common in temperate and tropical climates. Only a relatively few webs have been thoroughly studied. Thus, we usually lack the knowledge to predict what the effect will be of the extinction of any particular plant or animal species. In some cases, it might mean the loss of a food supply for which there is no substitute. Many rare insects, snails, and birds depend on a particular kind of plant for food or as a place to live. If the plant becomes extinct, the animal is also likely to become extinct (Fig. 3–3). In another case, a predator that normally keeps some pest under control might be lost.)For instance, ladybugs eat red spider mites. In some areas sprayed with DDT, red spider mites have gone out of control and damaged crops. This occurred because DDT killed the ladybugs, leaving red spider mites, which were not affected by the DDT, free to reproduce in enormous numbers.

Ladybugs are not in any present danger of extinction. Wolves, however, are endangered in part because their role in food webs is not understood. Wolves kill animals, like deer, for food but select the sick, the weak, and the old. They can thus help keep a herd of deer healthy and of a size proper to the available food supply. Humans, hunting competitors of the wolf, pride themselves in taking the finest deer, thus lowering the quality of a herd!

(The loss of species can cause other less obvious problems. DNA, or genes, the hereditary material contained in the nucleus of living cells, determines the characteristics of the organism of which it is a part. Genes are also the means by which characteristics from two parents combine and are passed to their offspring. As environmental conditions change, the existence of a large number or pool of genes can be important, since there is then a greater possibility for the development of organisms able to cope with new conditions. Reducing the number of species in the world reduces the size of the gene pool.)

Intrinsic Value of Species

(Besides all the practical reasons we might give, there are also philosophical arguments in favor of preserving as many species as possible. Any species lost is gone forever. If we fail to do what is in our power to prevent these losses, we make a choice not only for ourselves but also for future generations. We are saying that they will never see the same living creatures that we can see. They will never enjoy the diversity that we enjoy. It may not even be a question of enjoyment. Having evolved in the midst of such diversity, humans may require it to maintain their own mental health.)

All these reasons, of course, consider other species only from the view point of their usefulness to humans. Henry Beston wrote:

> Remote from universal nature, and living by complicated artifice, man in civilization surveys the creature through the glass of his knowledge and sees thereby a feather magnified and the whole image in distortion. We patronize them for their incompleteness, for their tragic fate of having taken form so far below ourselves. And therein we err and greatly err. For the animal shall not be measured by man. In a world older and more complete than ours they move finished and complete, gifted with extensions of the senses we have lost or never attained, living by voices we shall never hear. They are not brethren, they are not underlings; they are other nations, caught with ourselves in the net of life and time, fellow prisoners of the splendour and travail of the earth.[2]

---

2 From *The Outermost House* by Henry Beston. Copyright 1928, 1949, ©1956 by Henry Beston. Copyright ©1977 by Elizabeth C. Beston. Reprinted by permission of Holt, Rinehart and Winston, Publishers.

THE MOST IMPORTANT ENVIRONMENTAL PROBLEM

Despite concern in the U.S. over pollution, it is about the least important aspect of environment.

Lee M. Talbot

Lee M. Talbot, an ecologist at the Council on Environmental Quality, points out that pollution "is about the least important aspect of environment" because it is, in most cases, reversible. But changing land use, such as leveling forests or filling in wetlands, eradicates entire habitats and causes some species to be lost to the world forever. Plants and animals that may now be regarded as dispensable may one day emerge as valuable resources. *
Russell Train, President of the World Wildlife Fund, agrees:

> I have spent most of my time over the past several years working on a variety of pollution problems–air, water, and chemical among others. As I review these efforts, I am struck by the fact that the real "bottom line" is the maintenance of life on this earth. Time is running out rapidly on the natural systems of the earth, and particularly on the survival of species. The loss of genetic diversity which threatens everywhere and the resulting biological impoverishment of the planet have grave implications for our long-term future.
>
> We need nothing less than a comprehensive program worldwide to preserve and protect representative ecosystems....
>
> We human beings are relative newcomers on the face of the earth, but we now possess the power of life or death over our fellow creatures. While the scientific and economic arguments for the maintenance of species are compelling, it seems to me that we have an overriding moral responsibility to help preserve the other forms of life with which we share the earth. †

Do you agree that loss of habitats of wild animals and plants is the most serious environmental problem we face? If not, what do you think is the most serious problem?

---

* Lee M. Talbot, Council on Environmental Quality, in <u>Science</u> **184,** 646 (10 May 1974)

† Quoted from "Letters to the Editor," <u>Science</u> **201,** 324 (28 July 1978).

Figure 3-3   *Otters were almost wiped out by fur trappers in the 18th and 19th centuries. Now, due in part to laws such as the Marine Mammal Protection Act, sea otter populations are recovering. In fact, they are threatening in the process to avenge themselves, if not on man himself, then on species dear to his gastronomic heart: abalone, Pacific lobster, and crabs. From a few individuals discovered near Monterey, California, in 1938, otters have increased to almost 2000 animals, ranging along a 150-mile stretch of coast. Unfortunately, this same stretch of coast is also famous for seafoods such as abalone, a shellfish marketed for $8–$10 a pound. Commercial fishermen are demanding that otter herds be limited in size to prevent further depredation of the profitable fishing industry. On the other hand, ecological studies show that the sea otter is a vital member of the shore community. By feeding on species such as sea urchins, otters help to protect seaweeds such as kelp from overgrazing by sea urchins. In turn, kelp beds are at the bottom of the food webs that sustain species such as harbor seals and bald eagles. (Dr. Daniel Costa, Scripps Institution of Oceanography.)*

HOW SPECIES BECOME ENDANGERED

Habitat Destruction and Hunting

(The reason that comes most quickly to mind for the disappearance of species is probably hunting. And hunting has contributed to the loss of a number of animals, especially vertebrates.[3])The buffalo of the American plains was hunted almost to extinction in the 1800s. Trainloads of hunters came for the sport. They often carried no more than a buffalo head home to mount as a trophy. In Africa, game officials have stopped or limited the hunting of many big game species lest these animals cease to exist except in zoos. (See Figs. 3–4 and 3–5.)

(But hunting is not the main problem faced by most endangered species. The majority of these are threatened with a loss of their habitat, the area in which they breed, seek food, and find shelter.)As human populations grow, they require more houses, roads, and shopping centers. Forests are cut down, marshes, estuaries, and bays are filled in, and land is overturned in the search for coal. All of these processes reduce the land or food supply

---

3 Vertebrates are animals that have a backbone.

Figure 3-4   *The elephant is threatened in some parts of Africa. This is mainly because of loss of its normal range land but is also due to poachers, who are after ivory tusks. However, in other parts of Africa elephants must be purposely killed as they overcrowd the small range areas left to them. In Kenya's Tasavo National Park in 1971, 6000 elephants, as well as many other animals, starved to death due to a combination of drought and over-grazing by the elephant herds. At Wankie National Park in Rhodesia approximately 500 elephants are shot each year to maintain the population at a reasonable size. (Judy Rensberger/NYT pictures.)*

available to various animals and plants. In a sense, humans are increasing their own habitat at the expense of the habitats for other creatures.

In some cases habitat destruction is a result of game management procedures, such as burning or flooding. This is done to make areas more attractive to game species. Populations of elk, pronghorn antelope, white-tailed deer, and mule deer have all increased greatly as a result of such management techniques. In the process, however, the habitat becomes unsuitable for many other, nongame species.

Although destruction of all types of habitats is occurring, the problem is critical in tropical rain forests. These forests have more different types of species than any other ecosystem. Yet the demand for forest products and pasture land causes the clearing of vast areas of tropical rain forest. In Southeast Asia, more than six million acres of the forest are cleared every year. In ten years, all the easily reached virgin forests in Malaysia and the Philippines will have been logged to satisfy demands for veneer and plywood. In the Amazon River basin and Central America, vast forest areas have been cleared for crops and livestock. Two-thirds of the forests in Central America are gone.

Pesticides

The large-scale use of pesticides in agriculture places further stress upon many endangered species. For instance, birds in the Raptor group, which includes hawks and falcons, are affected by the use of DDT. These birds have been laying eggs with very thin shells. The shells are so thin that they crack before they can hatch. Thinning is believed to be due to DDT, a pesticide once used extensively in the U.S. DDT is now banned in the U.S., largely because of the effects it has on certain bird species (Fig. 3-6).

Figure 3-5   *Bones of elephants that starved to death in Kenya's Tsavo National Park in the 1970–1971 drought. Elephants have turned areas that were once dense bush into open grasslands. (Boyce Rensberger/NYT pictures.)*

Figure 3-6   *Brown pelicans. Their population along the California coast decreased steadily in the 1950s and 1960s as the birds accumulated high levels of the pesticide DDT. Since 1970, however, the pelicans seem to be staging a comeback. In that year, a plant manufacturing DDT stopped discharging wastes through a sewage outflow near Los Angeles. At two colonies near Baja, California, reproduction rose from a total of 4 young fledged in 1969 to 1185 fledged in 1974. However, on another front, almost the entire population of brown pelicans in Louisiana died in 1975, apparently from pesticide poisoning. The new problem seems to have occurred because of high tides in the Mississippi. This caused the birds to use body fat, stored for times of stress. Body fat, however, is also where small amounts of pesticide, eaten in fish, are stored. The birds may have gotten a poisonous dose of pesticides from their own body tissues. The brains of all birds tested contained a lethal dose of the pesticide endrin. (Interior—Sport Fisheries & Wildlife/Photo by Luther C. Goldman.)*

The white-faced ibis, a Texas marsh bird, is an example of a bird that has suffered from the use of the pesticides aldrin and mercury. In 1970 all of the ibis chicks died from high levels of these pesticides used on rice seeds. Laws now forbid the use of mercury as a pesticide in water. The use of aldrin eventually stopped because the rice pest against which it was used became immune to it.

(In another kind of pest control program, attempts were made to kill coyotes, foxes, and wolves by using poisoned baits. These methods had a severe effect on populations of endangered species, among them the bald eagle, that also take the bait (Fig. 3-7). The eagle, a symbol of national pride, is in danger not only from pesticides, but also from loss of habitat, and hunting by farmers who believe eagles kill their livestock.)

Pesticides alone probably have not caused the extinction of entire species. There are ex-

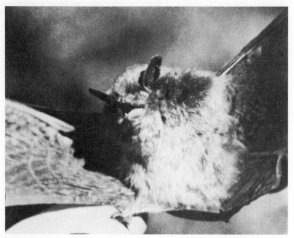

Figure 3-7 *Gray bats, found in 11 states, are endangered partly because of misunderstanding of their role in nature. According to the Office of Endangered Species, the bats have "a totally undeservedly bad public image due to their nocturnal habits and association with witches, hobgoblins, and other Halloween characters." In fact the gray bat, which lives in caves, is harmless to humans and important to our ecosystem. The bats eat insects, thus controlling insect populations. In addition, they create fertilizer necessary for many other cave-dwelling creatures. The bats also appear to suffer from exposure to the pesticide dieldrin. Many young bats have been found dead in caves where they roost, with lethal concentrations of dieldrin in their brains. The pesticide has been used on crops in areas where the bats roost. (Interior—U.S. Fish and Wildlife Service/Photo by Don Rimbach.)*

amples, however, in which an entire population of a species appears to have disappeared as a result of pesticide contamination of the environment. An example is the disappearance of the Eastern U.S. population of peregrine falcons, as a result of DDT contamination.

The Special Case of Islands

(Island ecosystems are especially fragile in terms of loss of species.) One hundred forty nine of the 161 birds that have become extinct since the year 1600 lived on islands. The Hawaiian islands alone have lost 25 bird species. Only 38 native bird species

are left there and 25 of these are now endangered.

(Some island species were lost because their habitat was destroyed, while others suffered when foreign species were introduced. On some islands, introduced predators preyed upon relatively defenseless native creatures. Mongooses were brought to the American Virgin Islands to control rats, but instead attacked other native species. In other cases, new species were able to compete more effectively for food than native species. In Hawaii, introduced mouplan sheep threaten both the mamane tree and the honeycreeper, a bird dependent on the tree as a food source.)

(Living creatures must change as environmental conditions change. Species unable to adapt to new conditions die out and new species evolve to take their place.) There are no longer any dinosaurs or flying reptiles, but there are other creatures that did not exist when the dinosaurs did. Humans, however, are speeding up the rate of change to the point where species cannot evolve quickly enough to replace ones that are lost. One-half of the world's extinct mammals died out within the last 50 years.

PROTECTING WILDLIFE RESOURCES

(The protection of wildlife can be approached in a variety of ways. Many laws have been passed in the U.S. to help preserve native wildlife. The development of specific wildlife management techniques has brought several endangered species back from the brink of extinction and has increased populations of a number of species of game animals. A third approach looks more closely at why species are becoming extinct at such a rapid rate all over the world. From this kind of analysis have come suggestions for economic incentives to help preserve endangered species worldwide.)

Wildlife Protection Laws

Refuge system (In the U.S., during the early 1900s, Congress began to set aside areas of wildlife habitat, or refuges, to help protect endangered wildlife. The development of the National Wildlife Refuge System is explained in the general context of land preservation in Chapter 32.[4])

(Among the first federal laws dealing with wildlife passed in the U.S. were those that taxed hunting and fishing equipment and required permits for these sports. The money collected has been used to buy land for wildlife refuges. Hundreds of millions of dollars have been raised for this purpose (a fact that must be taken into account by those who oppose hunting).)

Plant species, especially, can best be saved by setting aside part of their natural habitat as a preserve. A few individuals of a species in botanic gardens are not enough to ensure reproduction and species survival. A number of animals are also protected in refuges. The trumpeter swan, for instance, flourishes in Red Rocks Lake Refuge in Montana.

Many wildlife experts point out, however, that refuges must be large areas, measured in thousands of square kilometers. Smaller areas may not be able to support certain species, often precisely those most endangered. For instance, large predators such as wolves or the big cats must roam vast areas to find food. In addition, larger reserves are able to buffer species from border pressures, such as pollution or human disturbance. Because the animals and plants living in a particular area are dependent upon each other, we must attempt to save whole communities rather than just single species. With this in mind, (the United Nations Educational Scientific and Cultural Organization (UNESCO) has begun to identify "biosphere reserves" or "ecological reserves." This is meant to be a network of protected samples of the world's major ecosystem types. The reserves must be large enough to support all the species living there, to buffer them from the outside world, and to protect their genetic diversity. In this way, the reserve allows both growth and evolution and acts as a standard against which human effects on the environment can be measured.)

---

4 Pages 646–648 deal specifically with the wildlife refuges.

(Besides laws establishing preserves, stronger laws are necessary to limit the spraying of pesticides near game preserves or near the other habitats of endangered species.)

Engangered Species Act (In 1966, Congress passed an endangered species law. This legislation was designed not only to protect wildlife but to determine just how extensive the problem of disappearing wildlife was. It directed that a list should be made of endangered species. Many private conservation and wildlife organizations labor to provide some of the information needed before endangered species can become listed. Estimates must be obtained on numbers of individuals of a species as well as the range over which the species is found.)

(In 1973 this law was made much stronger by a series of amendments. The new law recognized that we may want to protect species that face extinction in the U.S., though not worldwide. It also established a new category called "threatened species." Threatened species are those not now endangered but whose populations are heading in that direction. By recognizing this fact early, it is possible that more can be done to save them. A further important change was that a new category, endangered plants, was added to the endangered species list. In addition, the new amendments directed that federal agencies could not undertake projects which would threaten endangered species or their habitat. Although this particular provision drew little notice when the amendments were passed, it became the basis for the conflict between the snail darter and Tellico Dam. As written, the law allowed no weighing of benefits versus costs if the extinction of a species was involved (See Controversy, p. 66).)

In 1978, as a result of the snail darter controversy, the Act was again amended to make it more flexible when it conflicted with government projects. Now a committee must first decide whether those in charge of a project have considered all reasonable alternatives. If conflict still exists, another committee, composed of the Secretaries of Interior, Agriculture, and the Army; the chairman

DO WE NEED "FLEXIBILITY" IN LAWS PROTECTING ENDANGERED SPECIES?

To dilute such a law (the Endangered Species Act) merely because some Congressmen see some bugbear, some hobgoblin, some mirage haunting their dam is tragic.

Senator Gaylord Nelson,
letter to the editor,
The New York Times,
21 July 1978.

No compromise is possible when the problem is stated in terms of the question "Do organisms have the right to exist?"

Wayne Grimm, National
Museum of Canada, quoted
in Science **196,**
1428 (24 June 1977)

Are you going to do anything to get the snail darter off our backs?

Unidentified Alabama Congressman
to Interior Secretary Cecil Andrus
quoted in Science **196,**
1427 (24 June 1977)

When the Endangered Species Act halted construction of the Tellico Dam because a completed dam would have wiped out the endangered snail darter, many congressmen began to feel that the Act was "inflexible." That is to say, nowhere in the Act was there a provision for considering the value of a project compared to the value of an endangered species. People began thinking of cases in which, they felt, a project or action could have more value to humans than the continued existence of a species. Senator William Scott of Virginia argued:

> Suppose a bird of some endangered species was in front of an intercontinental ballistic missile.... They could not release that missile. To me that would be a ridiculous offense.... Any commander worth his salt...would go ahead and release the missile, but he would be disobeying the law and would be subject to a fine of $20,000 and imprisonment for up to a year. *

Others argued that the Act was actually working well. They pointed out that the Tellico Dam was the only project ever halted by the law and one of only three to go to the courts at all (some 5000 possible problems were solved by consultation with the Fish and Wildlife Service).

But aside from questions about how well the original Act worked there is the question of whether flexibility is desirable. Do you feel there are instances in which a project could have more significance to humankind than the survival of an endangered species? Can you think of an example? Or do you think that no species should become extinct because of a construction project, however beneficial to humans? Can you justify this view?

---

* Quoted in <u>Science</u> **201**, 427 (4 August 1978).

of the Council of Economic Advisors; and representatives of the state in which the project is located, will rule on whether a disputed project gives benefits that clearly outweigh preserving an endangered species. At its first meeting, in January 1979, this committee ruled against the completion of the Tellico Dam. Although millions of dollars of public money had already been spent on the project, this was ruled not to be sufficient reason to allow a species to be exterminated. This ruling was made even though the species involved had no sport or commercial value. The committee noted further that, the threat to the snail darter was a pointer to other environmental problems that completion of the Tellico Dam would cause. These problems only came to public attention because of the fight over the snail darter. Many acres of fertile farmland would be covered with water when the dam was completed. This would decrease a vital agricultural resource. In addition, a recreational resource would be lost: the last free-flowing stretch of the Little Tennessee River. Third, land of historical value, the ancestral homeland and grave sites of the Cherokee Indians would be flooded. This case illustrates the way in which endangered species, even seemingly insignificant ones such as the snail darter, act as barometers for environmental problems in general.

Jimmie Durham, a Cherokee Indian leader, noted in testimony before a House committee that many people were making fun of the snail darter. "I would like to ask why it is considered so humorously insignificant," he asked. "Because it is little, or because it is a fish?"[5]

Although it seemed that the snail darter and also the stretch of free flowing river in which it lived were now safe, this was not so. In September 1979 supporters of the dam tacked onto an energy development bill an amendment which authorized completion of the dam "notwithstanding the provisions of" the Endangered Species Act. This bill

was signed by the President, to the dismay of many environmentalists.

Besides this type of problem with the Endangered Species Act, the Act does not control the effects of non-governmental projects. Private citizens are not bound by law to consider the effects of projects, such as housing developments, on endangered species.

The 1973 amendments to the Endangered Species Act served one further purpose and that was to ratify the Convention on International Trade in Endangered Species of Wild Fauna and Flora. The treaty sets up a system of permits for both exporting and importing threatened and endangered species, or products made from them. Trade in nearly extinct animals is practically prohibited, while strict controls are set for other endangered or threatened species.

Wildlife Management Techniques

A variety of special techniques have been developed to preserve species in danger of extinction or to increase the range of animals considered very desirable (i.e., those that people like to hunt). Animals may in some cases be transferred from their natural habitat to a similar area where they were not previously found. This has mainly been done with nonendangered game species like Canada geese. The wild turkey, which has been introduced in a number of areas, now occupies more territory than it did during Colonial times.

When it is judged that a species will not survive on its own even if given a fair chance, eggs may be collected and hatched in captivity or breeding programs can be instituted at zoos. The animals can in some, but not all cases, be successfully reintroduced into the wild. Sea turtles, which by instinct run to the sea after hatching and later return to their birthplace to lay eggs, never seem to get their bearings right if hatched in captivity. They swim off into dangerous waters and fail to return to suitable beaches for successful egg laying. On

---

5 *The New York Times*, 26 June 1978.

the other hand, about half of the whooping cranes alive today were hatched and reared in captivity.

(In some cases management procedures on preserves are so successful that limited hunting can again be allowed.)One hundred years ago, the American bison lived in herds so huge it sometimes required several hours to pass by a herd. Fifty years ago there were only a few hundred bison left. Within the past few years, there have been enough bison again to allow some hunting.

Incentives to Preserve Endangered Species Worldwide

(In order to understand the third approach to saving endangered species, we need to consider a little bit about how economics of the marketplace influence decisions people make. Most people would agree that other creatures have a right to survive on the earth. People rarely intentionally set out to wipe out other species. Yet, as a number of experts now point out, our economic system is set up in such a way that we tend to do just that.)

Species, like air and water, are in a sense a common resource. That is, we all stand to benefit from having a wide variety of plants and animals on the planet. None of us, however, "owns" any particular wild species and so no one is directly responsible for the survival of any particular species. Individuals can easily see the benefits to be gained if they hunt tigers or capture apes or build a housing development on some other creature's habitat. Much harder to keep in mind are the benefits, to all of us together, of having a great variety of species, since these benefits (esthetic or moral) are less visible or long term (medical uses, agricultural). (In other words, the short term, visible benefits go directly to the individuals involved, while the losses are mainly long term and are spread over society as a whole. As a result, we are as wasteful of the resource of species as we have been of air and water, other resources which no one owns[6] (Fig. 3–8).)

Figure 3-8 *Pupfish. There are 12 species of pupfish, found in various locations around the U.S. Important to science because of their ability to withstand extremes of temperature and salt concentration, the fish have run afoul of humans in several instances. In 1976 the U.S. Supreme Court ruled that the Endangered Species Act prohibited ranchers from pumping so much underground water for their ranches that the water level was lowered in Devil's Hole, home of the endangered Nevada pupfish. The fish have been marooned there since the last glacier receded, leaving much of Nevada desert country. Not so lucky, California's Tecopa pupfish was recently declared extinct. Thirty years ago builders of a bathhouse diverted waters from the thermal pools and springs in which the fish lived. Unable to adapt to life in the new swift stream, the fish finally died out. The bathhouse is no longer in use. (Tom McHugh/ Photo Reseachers, Inc.)*

(Because these sorts of effects are not an intended result of people's actions, economists call them *externalities,* or spill-over effects. People who undertake the most economic activities are likely to cause the greatest spill-over effects. People in developed countries use the most raw materials, which are often obtained by disturbing the habitat of creatures in less developed countries, and are, in this sense, the most responsible for loss of species.)

---

6 These principles were first clearly applied to environmental problems by Garrett Hardin in his "Tragedy of the Commons," *Science* 162, 1243 (13 December 1968).

ENDANGERED SPECIES VERSUS HUMAN HEALTH BENEFITS

Surely it is not beyond our scientific ingenuity to find alternative methods.

F. B. Orlans

It is not consistent with the genius of the American people to restrict the progress of scientific knowledge.

A. S. Packard, Jr. and
E. D. Cope,
American Naturalist
17, 175 (1883).

The use of animals in scientific research was once opposed mainly on the grounds of possible pain and cruelty. Scientists went to great lengths to assure the general public that the animals used in experiments would never feel pain.

Now, however, new ethical concerns are raised. Are research animals housed in such a way that social and behavioral needs are met? That is to say, normally social animals like chimpanzees should not be kept in individual, isolated cages because this would be a form of mental cruelty.

To go even further, should an animal with a dwindling population be used in research at all, even if man stands to benefit greatly? N. Wade writes:

...production of the (hepatitis) vaccine may well pose a fatal conflict between the interests of mankind and those of chimpanzees. Chimps are the only species, other than man, in which the safety of the vaccine can be tested....if the chimpanzees are protected–the species is already classified as threatened–it may prove impossible to safety test and hence to manufacture the vaccine. Yet even in developed countries, where the disease is comparatively rare, hepatitis B takes a heavy toll. In the United States 15,000 cases were reported in 1976. The true incidence was probably 150,000 according to the Center for Disease Control, of which probably about 1500 cases ended in death....

...officials deny that their chimpanzees would be captured inhumanely. "The method of capture is generally by locating a group of chimpanzees, surrounding them with a number of people and chasing them. The juveniles would usually tire first and these were captured by hand," a Merck official told the Federal Wildlife Permit Office....

"....Totally impossible unless you had big nets," says Jane Goodall. "Utterly fanciful....Given the sort of habitat where wild chimpanzees are found, no human being could keep up with a wild chimpanzee, much less run it to the ground....I can only conclude that someone is seeking to conceal the actual but less humane method of capture used–that is, shooting the mother to recover the young, which is the standard method used in Africa." *

And F. B. Orlans adds:

...a way must be sought to solve this conflict in a manner that is not detrimental to the chimpanzees. In the past, alternative methods of producing other vaccines (notably that for polio) have been found so that animal lives are spared....the ethical concerns for elimination of inhumane killing (in Wade's words, "to capture a chimpanzee: first shoot the mother") and for preservation of this dwindling species of animal are overriding. †

Do you feel that a clear human need should outweigh the need to preserve an animal species?

---

* N. Wade, "New Vaccine May Bring Man and Chimpanzee into Tragic Conflict," <u>Science</u> **200,** 1027 (2 June 1978).

† F. B. Orlans, in "Letters to the Editor" <u>Science</u> **201**, 6 (7 July 1978).

COULD PROTECTIVE LAWS WIPE OUT ENDANGERED SPECIES?

The alligator must still be strictly protected.

> C. Kenneth Dodd, Jr.,
> U.S. Fish and Wildlife Service

It [the Endangered Species Act] represents the most serious threat that the alligator has ever faced.

> S. H. Hanke

Protective laws, such as the Endangered Species Act and the Convention of International Trade in Endangered Species, are believed by almost everyone to be the only hope of saving animals and plants on the brink of extinction. These laws are designed to stop poaching, or illegal hunting, by making it illegal to buy, sell, or own products made from species which are endangered. The idea is that if the species has no commercial value, no one will bother to hunt it (Fig. 3-9).

C. Kenneth Dodd, Jr. writes:

...The alligator must still be strictly protected to prevent illegal trade and other abuses which can rapidly again lead to a precarious status.

Education programs must still inform the public about its conservation, its value, the dangers associated with it, and its role in the maintenance of its ecosystem. For instance, the alligator is known to be extremely important in the existence of many species which dwell in an aquatic environment in the Everglades of Florida. Adult alligators build "gator holes" throughout the grasslands. When the dry season comes, these gator holes serve as refuges for aquatic species until the rains return; thus they are vitally important in the Everglade's ecological balance.

In addition, many individuals now recognize the alligator as a valuable predator and are trying to obtain them to control pest species, such as beavers. The alligator story is an example of how endangered species can be managed successfully if proper time, understanding and care are taken. *

A few economists, however, are presenting a different viewpoint.
Steve Hanke writes:

...Private landowners have fairly clearly defined property rights over alligators residing on their holdings, much as they do over other assets on their land–trees and mineral deposits, for example. As a consequence of the

Endangered Species Act, what were formerly legitimate market values have been done away with, and landowners throughout the South no longer have any incentive to husband their assets in alligators. The net result of this taking without compensation has been to change the attractiveness of various forms of land use. Since alligators are no longer a cash crop, many farmers have begun to drain their lands converting them to dry-land farming. This habitat change, which is by far the greatest threat to many forms of wildlife, has made it increasingly difficult for the American Alligator to survive.

A change in policy is imperative. To save the alligator and to increase the efficiency with which land is used, we must delete the alligator from the Endangered Species List, thereby giving it a market value. The results: alligators will be able to compete effectively with other forms of land use, and landowners will have every incentive to manage, protect, and regularly crop alligators in order to maximize the value of their property.[t]

Is this an example of well-intentioned environmental legislation which will do more harm than good in the long run? Or is it an example of a law that does not go far enough? Can you think of ways to change or supplement the laws so they serve their purpose?

Figure 3-9 *An American alligator from Alachua County, Florida. Alligators were once threatened with extinction in certain parts of the United States because their skins were in demand to make shoes, belts, and other leather goods. (National Park Service.)*

---

* C. Kenneth Dodd, Jr., Staff Herpetologist with the Office of Endangered Species, U.S. Fish and Wildlife Service, <u>Water Spectrum,</u> (Winter 1977-1978).
† Steve H. Hanke, <u>Policy Analysis</u> **1** (1), (Winter 1975).

Several suggestions have been made to remedy this situation.

(In the first place, corporations operating overseas could be required by law to determine the effects their operations have had on animal and plant species in foreign countries. Policies could be worked out to allow tax credits for conservation measures or fines for damaging operations.)

International public organizations, such as U.S. Agency for International Development or the United Nations agencies, could be required to write environmental impact statements for their projects. Even if it was decided, because of great benefits to humans, to go ahead with plans that would damage habitats of other species, the costs would be recognized and weighed. Much unnecessary damage could be avoided.

In some cases, developed countries should help pay the costs of saving species in undeveloped countries, either because the countries in which the endangered species are found are too poor to undertake protection, or because their people are too far from achieving minimal human needs in terms of food, shelter, and health care, to be able to choose wildlife protection over exploitation of natural resources. Tanzania, which has the largest wildlife population in Africa, has asked that other countries contribute to the cost of guarding wildlife from poaching. Needed equipment, such as surveillance helicopters for game wardens, is beyond the reach of many developing countries.

In the final analysis, the fate of other species is a barometer for the fate of the human species. If the outlook is not sunny for the survival of other creatures on this planet, surely human survival is in for a stormy time (Fig. 3–10).

> I heard the song
> Of the world's last whale
> As I rocked in the moonlight
> And reefed the sail
> It'll happen to you
> Also without fail
> If it happens to me

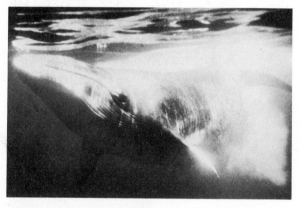

Figure 3–10 *Although this sei whale appears to be smiling, it may have little to be happy about. Along with other species of large whales, sei whales have been hunted until their numbers are alarmingly small. There is more on whales on pp. 75 to 83. (©1974 G. Williamson/Bruce Coleman, Inc.)*

---

7 "Song of the World's Last Whale" by Pete Seeger. Copyright 1970 by Stormking Music, Inc. All rights reserved. Used by permission.

Endangered Ocean Mammals Versus Endangered Native Cultures

Eskimos and the Sale of Products from Sea Mammals

In recent years, the U.S. government has twice been forced to look for a middle ground that would allow the survival of both endangered sea mammals and equally endangered native cultures. The issues involved will come up again in other forms and in other parts of the world. As human populations expand and as their needs for living space and food grow, pressures on wildlife species will increase. In some cases, animals and plants may be endangered because their habitats are turned to human uses. In other cases, hunting for food or other uses may bring species to the point of extinction. In the final analysis, choices will have to be made between human needs and the survival of species. The following examples, drawn from the experiences of Alaskan Eskimos, illustrate how difficult such choices will be.

In 1972, Congress passed the Ocean Mammal Protection Bill. This Act protects arctic fur seals, walruses, and whales, among other ocean mammals, from commerical hunting, which has reduced some species almost to extinction. However, it was soon clear that the Bill itself threatens something else that is endangered; native Eskimo culture. Although the law allows Eskimos to hunt these animals as food for themselves, it prohibits the sale in interstate commerce of products made from the animals. This has created a hardship for the Eskimos. They had established a small trade in skins, carved tusks, and other items derived from the sea mammals (Figs. 3–11 and 3–12). The money they earned from the sale of these items enabled them to buy foodstuffs to supplement their diet of sea mammals. During the hunting season, the Eskimos could sell seals to traders; the money they received bought food during the lean months when hunting was poor. (See Controversy.)

Subsistence Hunting and the Bowhead Whale

In 1976, even the Eskimo right to subsistence hunting (hunting for food) was called into question.

For centuries, humans have hunted whales for their oil and meat. In the 1700s and the 1800s the right and bowhead whales were the main targets. These were the "right" whales because they float after they are killed and because they are slower moving than other species. Now these whales, along with the blue whale, the humpback and the gray whale, are near extinction.

The great blue whale is the largest animal now living on the earth. Although they swim in the oceans, as fish do, whales and their smaller relatives, the dolphins, are mammals. Whales are warmblooded and suckle

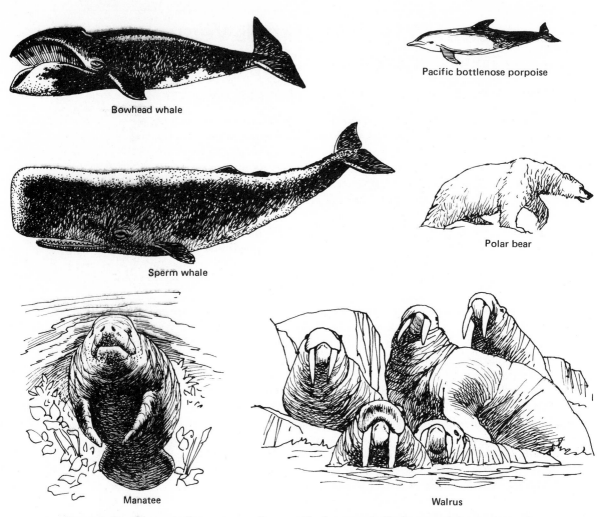

Figure 3–11   *There are four groups of mammals that spend all of most of their time in the ocean: the Cetacea (whales and Porpoises); Pinnipedia (includes seals, walruses, and sea lions); Serenia (sea cows or manatees); and some of the Carnivora (sea otters and polar bears). The whales are further divided into two groups: the baleen whales, such as the bowhead, which strain seawater for the minute plankton that is their food; and toothed whales, such as the sperm whale, which eats fish and squid.*

their young. They are also social animals, living in herds and possibly even in families. Within the past few years, scientists have realized that whales can "talk" to each other. Some beautiful records have been made of the singing sounds they make underwater.[8]

---

8 One record of these sounds is "Songs of the Humpback Whale," Capital Records, Inc., Hollywood and Vine Streets, Hollywood, California.

Figure 3–12   *Eskimo craftsmen carve and sew materials from ocean mammals into useful and beautiful objects. Above is an example of a parka sewn from seal skins. (Dobbs Collection/Alaska Historical Library.)*

Whaling was once a dangerous occupation. The small crew of men with their harpoons and in their fragile boat, and the huge whale were relatively well matched. Mostly the men won, but sometimes the whale did, too. Modern methods, however, have tipped the balance far in favor of men (Fig. 3–13).

Modern non-Eskimo whaling fleets consist of fast killer boats, which kill the whales with harpoon guns, and large factory ships, which are able to process the whales at sea. Whales are located by sonar or small planes, then driven by high frequency sound waves until they are exhausted. A harpoon carrying a grenade, which explodes within the whale, is used to kill them. With these methods, modern whaling fleets have killed more whales in this century than in the 16th, 17th, 18th, and 19th centuries together. Within the past 50 years, the numbers of blue, humpback, and gray whales have been

Figure 3–13 *Whaling, with a hand-thrown harpoon, in a small boat as is still done in the Azores is a difficult and dangerous business. The whale might overturn the boat, with tragic results for the whalers. (American Museum of Natural History) In contrast, gigantic modern whaling ships, such as this Soviet factory ship, ensure there is little danger to the whalers. They are frighteningly efficient at killing whales, however. As a result, a number of whale species are now endangered. (Matt Heron/Black Star)*

reduced from hundreds of thousands to a few thousand. The California gray whale, whose breeding grounds are now protected by the Mexican government, has begun to recover in numbers, perhaps even to reach a population size near that of the 1800s. The bowhead whale, on the other hand, has been protected for 35 years, but does not seem to be increasing in numbers.

Seventeen nations belong to the International Whaling Commission, which sets yearly whaling quotas (the numbers of each type of whale that may be caught in the various oceans). Nations that have whaling fleets, such as Russia and Japan, and those that no longer have fleets, such as the U.S., Canada, and South Africa, send delegates to the Commission meetings.

Over the past few years, quotas were slowly reduced as new scientific data came to light on the population sizes and reproductive ability of whales. However, quotas also declined because, in a number of cases, whalers were simply unable to find enough whales to fill them. Many conservationists would like to see a ten-year moratorium on all whaling, to allow time for the study of whales and their survival needs, before they are unintentionally hunted to extinction. Such a moratorium was proposed several times at IWC meetings in the last few years.

In 1946, the IWC treaty noted that the bowhead whale was the most endangered whale species. The commission at that time forbade all but subsistence hunting.

United States Eskimos have been catching about 10 to 15 bowheads per year since 1946. However, in the past few years, the catch has increased. Further, the number of whales struck but lost has steadily increased (Table 3–1). It is estimated that at least half of these later die. The IWC was concerned about the effect this increased hunting would have on an already severely endangered species. In 1976, the commission called for a total ban on subsistence hunting of the bowhead.

TABLE 3–1   Alaskan Eskimo Bowhead Whale Harvest [a]

	1973	1974	1975	1976	1977*
Landed	37	20	15	48	26
Killed, lost	0	3	2	8	2
Struck, lost	10	28	26	37	77
	47	51	43	93	105

*The 1977 figures represent the spring hunt only.

[a] Source: R. Storro-Patterson, "The Bowhead Issue," *Oceans*, p. 63 (January 1978).

Why did Eskimo hunting of the bowhead whale increase in the last few years? The answer is tied up with economics and the impact of outsiders on native culture. R. Storro-Patterson writes:[9]

Correlated with the increase in the Eskimo harvest of bowhead whales is the growing number of whaling crews:

Number of Whaling Crews—Spring Hunt

1971	1972	1973	1974	1975	1976
25	27	45	46	75	86

The rise in the number of whaling crews may in turn be significantly related to the availability of jobs and money for Eskimos. Traditionally, to be a whaling captain was not only a matter of great prestige, but great difficulty. One had to either inherit the equipment, marry to obtain it, or gain sufficient wealth to purchase it. The latter option was seldom possible. Recently, however, such construction projects as the Alaskan pipeline, and oil drilling operations, as well as the Alaskan Native Claims Settlement have changed this. An ambitious Eskimo can save $9,000 for an "outfit" plus another $2,000 for provisions for a whaling crew.

The Eskimos were outraged by the ban, as were many other people who sympathize with the native-Alaskans' difficult fight to survive and preserve their culture in a relatively hostile natural environment.

"Hunting the bowhead is what keeps our communities together. Our people depend on the bowhead, first for food and, just as importantly, for the survival of our culture," said Dale B. Stotts, an official of Alaska's North Slope Borough, which includes a number of whaling towns and villages along the coast of the Beaufort Seas.[10]

---

9 R. Storro-Patterson, "The Bowhead Issue," *Oceans*, p. 64 (January 1978).
10 Quoted by B. Rensberger in *The New York Times*, 5 October 1977.

CONTROVERSY

THE SEA MAMMAL AND ESKIMO LIFE

The following testimony is excerpted from the hearings before the Sub-committee on Oceans and Atmosphere of the Committee on Commerce of the United States Senate in the spring of 1972 (before the ocean mammals protection bill was passed):

STATEMENT OF JOHN HENRY

Mr. Henry: I don't have no written statements.... So, I'm going to make my own–what people thinks. As people thinks, I'm Eskimo, just like Indian or Aleut, all same. We are from Norton Sound on the coast area, where we could hunt on the sea and on the ground. We have two ways to hunt. To feed ourself from the sea and from the ground. Even we don't have no money. Sometimes we have hard times. We can't purchase from the ground, we can't purchase from the sea. Because we are Eskimo, we can't write a check to purchase something that we could buy. Just sit down and write a check to purchase something for our children....

Sea mammals. It's our living and I born on sea mammals. I born, we didn't have much money. My dad didn't have much money. He hunt and hunt, just the same as like other peoples in the other villages. He try to support us from the hunting, that's all....No other resource we have, just from his hunting. Just like other villages does. And ancestor, from our ancestor I think we born like them. I will be like that, because my ancestors I know. I was born without money. I will be ancestor for my own folks without money....

STATEMENT OF MYRTLE JOHNSON

Mrs. Johnson: I am a resident of Nome, and was born and raised in the village of Golovin....I will draw from people I know personally to paint an imaginary picture based upon the truth.

John and Mary are in their mid-50s. John works part-time seasonally in casual labor, and they care for four foster children plus one grandchild. John is a successful seal hunter. He also catches Beluga. If he had a chance, he would join a crew and hunt black whale as well. This would bring him a share of meat and muktuk–something his family otherwise must buy at $3 per pound or receive as a gift that will obligate him to return in some other form.

John's family depends heavily upon sea mammal meat and oil all year long. The spring greens are stored in fresh seal oil, some of this being sent to family and friends living in the cities. The meat is dried, frozen, or in other ways preserved for the time when it is not so plentiful....

The skins will end up as mukluks, parkas, parka-pants, and vests for the family, as well as surplus skins or garments to be sold or traded to others. Both raw and stretched, skins are a source of income and add to the internal welfare system of their village.

John's wife will sell some articles she makes directly to others, and some at the village store. Some things she will give to the church or school to raise funds. If John carves, he will earn some cash from the tusks he has taken, or those he has gotten from other hunters.

Not all men in a village hunt seal. Some are working or cannot hunt for other reasons. In every village a certain amount of sea-mammal meat, oil, and skins are purchased by nonhunters. This stretches their food dollars. A portion of the village sea-mammal harvest is always provided for the poor or the aged in a fashion that is in keeping with village traditions.

Beef and pork, seldom carried at a village store, must be priced 25 percent to 50 percent higher than at the outlets with jet flight service. Thus, the importance of a local meat harvest cannot be underestimated. Without the meats and oils, many families would be without this basic food in their diet....

Caucasians speak of bread as the staff of life. For coastal Eskimos the seal represents the single basic food staple with all the meanings others may associate with bread....

To take away our Eskimo bread is to deny the men and women of the villages the right to work to meet their needs as they see them and can mean the final end of our way of life before we are fully adapted to the modern world.

W.H. DuBay wrote, in defense of the Eskimo hunters:

> ...People outside Alaska don't seem to realize that the Eskimos are the residents
> of the Arctic and that, were it not for their aggressiveness in protecting their
> Arctic homeland from those who would destroy it, there would be no effective
> environmental safeguards at all operating in the U.S. or Canadian Arctic.

> ...The issue is not just the best way to manage the preservation of the bowhead
> whale, but also the great question of subsistence hunting rights of American
> native peoples and the basic human right to eat what you have to eat in order to
> survive in your own environment.[11]

Environmental groups were torn between their sympathy for the natives
and their need to protect whales. Environmentalists felt that if the U.S. filed a
formal objection to the IWC decision, which would cancel the ban, other
nations, such as Japan or Russia, would feel free to file objections to quotas
the IWC had set on other whale species:

> "For the first time, this has put the United States in the position of being the
> affected party in restricting whaling," said Dr. Roger Payne, a whale specialist
> with the New York Zoological Society. "If the United States files an objection to
> the I.W.C. position this country could lose all its credibility in whale
> conservation."

> "The original exemption on the bowhead was to guarantee what are known
> as the aboriginal rights of the Eskimos," Dr. Payne said. "The Eskimos aren't
> aborigines any more and along with modern hunting weapons there have to go
> some controls."[12]

The U.S. government finally decided to ask that the Eskimos be given a
quota on the number of bowheads they could catch. At an emergency meeting
in December 1977, the IWC voted a quota of 12 whales landed or 18 struck,
whichever came first.

The Eskimos objected that this would mean that some people must go
hungry in the coming year. Nonetheless, they agreed to obey the quota and set
up a self-governing system to do so.

Perhaps the necessary balance has been found for the moment. It seems
unfair that the Eskimos, as they attempt to protect their cultural heritage, must
shoulder one more burden, that of nurturing back to health a whale
population plundered by commercial whalers in the beginning of this century.
Yet, as a New York Times editorial noted,[13] "if Eskimo culture needs the

---

11 W. H. DuBay, Letter to the Editor, *The New York Times,* 15 November 1977.

12 B. Rensberger, *The New York Times,* 29 September 1977.

13 *The New York Times,* 1 October 1977.

bowhead, it must be saved; killing off the few that remain would not long satisfy either stomach or spirit."

Questions

1. How would you explain the value of preserving a wide variety of species, both plant and animal?
2. What do we mean by the terms *extinct, threatened, endangered,* when they are applied to species?
3. Briefly list the main reasons why species become extinct or endangered. Which is the most important reason?
4. Do you believe hunting should be permitted? Give your reasons for or against and any special requirements you would like to see enforced.
5. How does the fact that wildlife species are usually viewed as a common resource, in the same way that water or air are considered common resources, contribute to the rapid loss of species today?
6. Recount some of the ways in which attempts are currently being made to preserve species from extinction. What further measures could be taken?
7. Most of us would agree that it would be terribly wrong to hunt whales to extinction. But to go a step further, should we "harvest" them at all?

   Whale products are used to manufacture such things as fertilizers, transmission fluids, and animal feeds. There are other materials that could be used to make all of these items. Whale meat is eaten in Japan and the Soviet Union. The Japanese claim that although whale meat forms only a small part of the Japanese diet (possibly as little as 1 percent of the protein) it is eaten mainly by the poor who do not have many other choices.

   On the other side is the evidence gathered by scientists, that whales are able to communicate with each other, that at least some whales travel in family groups, and that whales have large brains (the ratio of brain weight to body weight in dolphins—if the blubber layer is ignored—is 2.0 compared to 1.9 for humans). Furthermore, there are even examples of whales caring for injured fellows:

   > "Recently, off Vancouver Island, British Columbia, a ferry captain saw a vivid demonstration of care-giving. He heard a crunch from the stern and, supposing he had struck a partly submerged log, he turned about. To his dismay, a young killer whale, one of a family of four, was wallowing in the sea.

   > "The cow and the bull," he said, "cradled the injured calf between them to prevent it from turning upside down. Occasionally the bull would lose its position and the calf would roll over on its side. When this occurred, the slashes caused by our propeller were quite visible."

   > Fifteen days later, in the same waters, a woman saw "two whales supporting a third one, preventing it from turning over."

   How should we balance these factors, as we deal with this "other nation," with which we inhabit the earth?

**Further
Reading**

Alternatives for Completing the Tellico Project draft, Tennessee Valley Authority and U.S. Department of Interior, 10 August 1978.

The snail darter–Tellico Dam controversy is detailed very clearly in this draft report, for those who wish to follow this issue, which nearly gutted the Endangered Species Act.

Animal Welfare Institute, P.O. Box 3650, Washington, D.C. 20007.

This conservation organization is one of the most active in campaigns to save whales. If you write to them, they'll put you on a mailing list for current information and also send directions for staging your own "Save The Whales" campaigns.

Davis, R. K., et al., "Conventional and Unconventional Approaches to Wildlife Exploitation," Transactions of the 38th North American Wildlife and Natural Resources Conference, 18 March 1973.

This paper and the ones by J. F. Franklin and N. Myers present some of the more current thinking on wildlife conservation. While not downgrading older procedures involving refuges and captive breeding programs for severely endangered animals, these authors try to go one step further in suggesting procedures that could save whole ecosystems or influence the systems that cause animals and plants to become endangered.

The Evolution of National Wildlife Law, Council on Environmental Quality, 1977.

The history of wildlife law and all major U.S. wildlife legislation is covered in this comprehensive report.

Franklin, J. F., "The Biosphere Reserve Program in the United States," Science 195, 262 (21 January 1977).

Graves, William, "The Imperiled Giants," National Geographic, p. 752 (December 1976).

This beautiful pictorial essay and the essay by V. B. Scheffer, explore the life-history of whales as well as their often ill-fated relationship with humans.

Mowat, F., People of the Deer, Pyramid, New York, 1968.

Mowat, F., The Desperate People, McClelland and Stewart Ltd., Toronto, 1975, Volumes I and II.

These three books tell of a vanishing people, the Ihalmiut, who were the Eskimos of Canada's inland region. The story tells of the delicate balance between humans and wildlife, which once enabled the Ihalmiut to live and flourish in a harsh climate. The books will give you an understanding of the plight of native peoples and endangered animals in an increasingly technological world.

Myers, N., "An Expanded Approach to the Problem of Disappearing Species," Science 193, 198 (16 July 1976).

Putnam, John J., "India Struggles To Save Her Wildlife," National Geographic, p. 299 (September 1976).

India provides one of the best examples of how an expanding human population exerts intolerable pressures on wildlife populations in developing countries. The almost unresolvable conflicts between human and animal needs are highlighted in this perceptive article.

Scheffer, V. B., "Exploring the Lives of Whales," *National Geographic*, p. 722 (December 1976).

Grant Heilman Photography

Part 2

Water Resource Problems

Human survival depends upon a number of resources. Water is certainly one example; air is another and energy resources a third. The importance of plants and animals as wildlife resources was discussed in the last chapter, and will come up again when land and food resources are examined in Parts 7 and 8.

The next chapters are concerned with problems involving the resources air, water, and energy. We begin with water, in part because the environmental movement first took shape around efforts to protect water supplies. More than 100 years ago, people began to realize that water could carry disease. Because of that recognition, the profession of Environmental Engineering, or as it was then called Sanitary Engineering, grew up. The environmental movement, as a visible phenomenon, had begun.

Chapter 4 examines water from the ecologist's point of view. Water is a resource, with unique properties, essential to all life on earth. It is a basic factor in the growth of natural communities and human civilizations. The chapter covers the various kinds of water habitats and the organisms that live there.

Chapters 5 to 10 examine water problems from the point of view of human needs. When is water pure and when is it safe? What are the substances that contaminate water supplies and how do we remove them? How do we handle waste waters and what effects do they have on natural waters?

The first of these questions deals with *the water we drink.* They involve basic facts about the safety and quality of drinking water that we should know about our water supply.

The impurities that influence the safety of a water supply for drinking fall into three broad classes. Inorganic chemicals are one class; included in it are the ions arsenate, nitrate, fluoride (at high levels), and other chemicals that can have adverse effects upon our health. Organic chemical compounds, a second category, may also be dissolved in the water; some of these compounds have been linked to cancer. Finally, water may contain microorganisms (microbes) that cause diseases such as typhoid and cholera (Fig. A). Fortunately, these diseases are only a distant memory to most of us in the United States. Once widespread in this country,

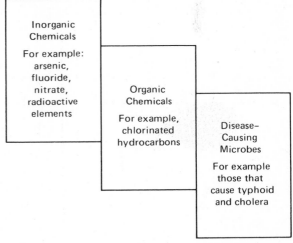

Figure A *The three classes of impurities in drinking water.*

they still occur commonly in nations that have not yet treated their water supplies.

In addition to being concerned with the safety of water supplies, we are concerned with other water characteristics such as clarity, odor, and taste. Water may also be investigated for its "hardness." Hardness is caused by the presence in solution of compounds such as calcium and magnesium carbonate. The "hardness" property decreases the effectiveness of soap and makes it more difficult to wash clothes, dishes, and people. In Chapters 5, 6, and 7, we shall explain the possible hazards in drinking water and then look at water treatment, the methods by which we attempt to make water safe for drinking.

The effectiveness of current water treatment solutions, however, depends in part on the amount and kinds of wastes that we allow to contaminate water we may later wish to drink. The remaining chapters in this part deal with this other kind of water resource problem: problems caused by *the water we waste.* Waste waters flow from cities and industries, from mining operations, from farming and rural homes. These wastes are treated in different ways but are generally disposed of in the same way: into the nearest river or lake or into the ocean.

The problem with waste waters is not only that they might contaminate our drinking water supplies. Waste water can be fairly easily chlorinated so that it will not carry large quantities of the organisms that cause disease. And the water we drink, even if it comes from a highly polluted source, can be treated until it is not only healthy but pleasant to drink. Thus, our insistence that waste water be treated before it is released is not focused only on human health problems. We must also consider how the waste water will affect the natural waters into which it flows.

If the organic pollutants are not removed, they can set up a chain reaction that robs water of the oxygen normally present. Further, certain chemicals, such as pesticides, may be directly poisonous to aquatic organisms. Fish and other aquatic creatures may not be able to live under these conditions, and other less desirable species take over. In addition, certain inorganic elements in waste water, such as phosphorus and nitrogen, cause excessive growths of the microscopic green water plants called algae. These "blooms," as they are called, form unpleasant scums and mats over the surface of lakes.

Water pollution control is the term given to those methods which are meant to clean up waste waters so that they can be released to natural waters without causing problems.

What are the methods by which waste water is treated to prevent the pollution of natural waters? The methods, surprisingly, use biological processes to purify the water.

It should be noted that several waste water pollutants are covered in detail in other places in the book: acid mine drainage is discussed (along with its cause, coal mining) in Chapter 13. Oil pollution is described in Chapter 20. Water pollutants that can cause cancer (i.e. asbestos, chlorinated hydrocarbons, and arsenic) are covered in Chapter 27.

Chapter 4

Lessons from Ecology:
Water and Life

THE HUDSON RIVER

High in the Adirondack Mountains of New York State, the Hudson River begins as a small, clear lake, Lake Tear of the Clouds (Fig. 4-1). Flowing first as a brook and then as the Opalescent River, the waters run south, joined by many other streams. At the town of Newcomb, still in the mountains, the·Opalescent officially becomes the Husdon. Just above Troy, the Mohawk River joins the Husdon, which is now one of the mightiest rivers in the U.S. Three hundred and fifteen miles from its origin, as it passes New York City, the Hudson meets the Atlantic Ocean in New York Bay, which is the Hudson's estuary.

The communities of organisms living in various parts of the Hudson change as conditions in the river change. Small swift streams that feed the river provide homes for different species than those in the slower moving river itself. Organisms living near the point at which the river meets the ocean tides must be able to live in varying concentrations of salt. The ocean itself is a very different environment from the fresh water river, ponds, and lakes.

Organisms living in the Hudson River have many problems with which to contend. For years, a number of cities have dumped raw or poorly treated sewage into the river. Industry has also been guilty of using the Hudson as a sewer (Fig. 4-2).

But before we deal with the problems of polluted waters, let us look at water itself and its importance to life. (Water cycles in the environment. This cycling and the relative availability of

Figure 4-1 *A campground at Flowed Lands in the Adirondack Mountains of New York State. Here, close to the small lake where the Hudson rises, stream waters are so pure that guidebooks assure campers they can drink the water without first boiling it.*

water help to determine the kinds and distribution of life on earth. In addition, water has certain unique properties that have contributed to the development of life as we know it.)

Finally, we will examine the character of the habitats and communities found in natural waters.

WATER—A LIMITING FACTOR FOR LIFE

Availability of Water

(Water is a necessity for all forms of life on earth. In land environments, the abundance of water (a function of rainfall, humidity, and the

Figure 4-2 *View of the Hudson today from near Bear Mountain Bridge, New York. (Peter Arnold, Inc.)*

evaporation rate) determines the kinds of communities that will develop. In water environments, the types of communities also depend on the availability of water. However, in this case, the availability of water means changes in water levels, for instance with tides. It also refers to differences in the salt content of water, which affect the rate at which water enters or leaves organisms. Rainfall, humidity, and the evaporation rate are all factors that help define the climate of a given region.)Climate and how it affects organisms is discussed in Chapter 16, while the different land communities, which develop according to the availability of water, are noted in Chapter 30.

The Watershed

Surface waters such as lakes, rivers, and oceans are highly visible features of the environment. It is easy to see that they provide different habitats for living organisms than do land areas. What is not as easily seen is that(the two kinds of habitats, land and water, are tied together by the cycling of energy, water, and nutrients through the environment.)For instance, the Hudson River is not a self-contained system. There is energy from the sun. Nutrients are washed into the Hudson, by erosion and streamflow, from the river's banks and from land bordering all those streams flowing into

the Hudson. Even the water in the river itself is cycling as some evaporates, and as the total is increased by rainfall and streamflows.(Pollutants, too, reach the river not only directly but also from the land areas surrounding it.)

(We can see, then, that the functioning unit is not simply the river itself, but also the whole land area that drains into the river. This area is the watershed. In terms of understanding and maintaining the quality of natural waters, the whole watershed is the ecosystem that must be studied or managed.)

WATER AS A RESOURCE

The Hydrologic Cycle

(The constant cycling of water in the environment is called the *hydrologic cycle* (Fig. 4-3). It consists of three distinct and continuing events: the evaporation of water, condensation and rainfall, and run-off.)Water evaporates from the surface of lakes, ponds, streams, rivers, and oceans. It also evaporates from soil and vegetation. This water is returned to the earth as rainfall. More water evaporates from the surface of the oceans than returns to the oceans as rain, however. On land the opposite is true. Less water evaporates from soils,

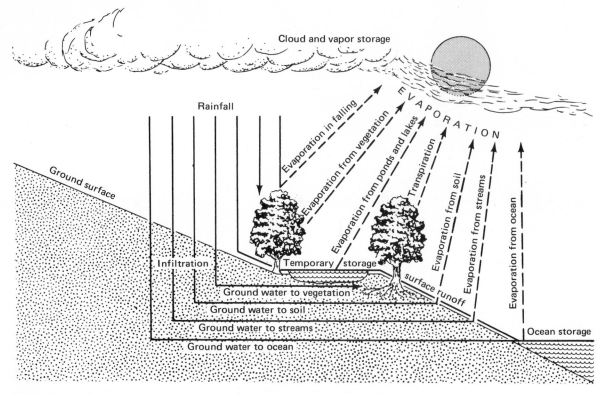

Figure 4–3   *The hydrologic cycle. Water evaporates from soil, vegetation, and surface waters. This water is returned to the earth as rainfall. In addition, water moves from land areas to the ocean as run-off in streams, rivers, and ground-water flows.*

vegetation, and surface waters than is returned by rainfall. The balance is, of course, maintained by run-off, or water flowing from streams, lakes, rivers, and ground waters, to the oceans.

The hydrologic cycle can be compared to the process of distillation, in which water is vaporized by heating it in a flask. The water vapor leaves behind dissolved materials such as salt (Fig. 4–4). (This does not mean that rain is free of contaminants. Rain and snow may become contaminated with gases and particles in the air. However, rain is relatively pure.) In some areas of the world, where there are few rivers and little water in the ground for people to utilize, rain furnishes drinking water.

In such places, rainfall is collected in cisterns and stored for later use.

(The processes in the hydrologic cycle are being modified by human activities; some of the changes are intentional; others are accidental. For instance, rainfall has increased in industrial areas because of minute mineral particles in the air. A water droplet condenses more quickly around such a particle. As another example, run-off increases when vegetation is destroyed. Trees, grasses, and other plant covers capture and hold rainfall, allowing it to percolate down through layers of decaying organic materials and rocks in the soil. Where there is less water percolating through the deep soil layers, there is a

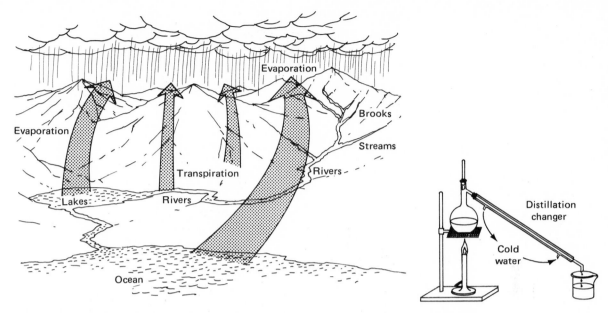

Figure 4–4 *The hydrologic cycle is similar to the distillation of water. Water evaporates from lakes, rivers, and the ocean, leaving behind impurities, such as salt, just as it does when heated in a flask. The water condenses in the atmosphere and falls to the earth as rain in much the same way that steam is turned to water droplets in a condenser.*

decrease in the amount of trace minerals dissolved from buried rocks. These minerals are needed for plant growth, and may have to be added to fertilizer in some areas. Further, the run-off that carries water via streams, rivers, and lakes to the oceans is interrupted at many points and for many purposes. Not only is runoff interrupted, it is also contaminated with chemical and biological wastes.)

Usable Water Supply

How large is it? (Run-off, along with underground fresh waters (ground waters or aquifers), represents all the water available to humans for drinking, growing crops, and manufacturing processes requiring fresh water. Although there is a great deal of water in our environment, 95% is bound up in the rocks and deep crustal regions of the earth. Almost all of the remaining water is in the oceans.

Only 3% of the total available water is fresh water and 3/4 of that is frozen in the polar ice caps.) Salt, of course, can be removed from sea water by the process known as desalinization. In addition, some manufacturers can use salt or brackish (partly salt) waters. However, the cost of desalinization is presently very high, compared to fresh water, in those areas where fresh water is in good supply. In areas where fresh water is scarce, desalinization plants are becoming common. Saudi Arabia, which is chronically short of fresh water, will have spent 15 billion dollars from 1978 to 1981 on desalinization.

(Because of the hydrologic cycle, water is usually considered a renewable resource; water in lakes, rivers, streams, and reservoirs is constantly flowing away and being replenished by rainfall. The water supply is usually measured by determining this rate of replenishment, the annual surface water run-off for an area (Fig. 4–5). The actual

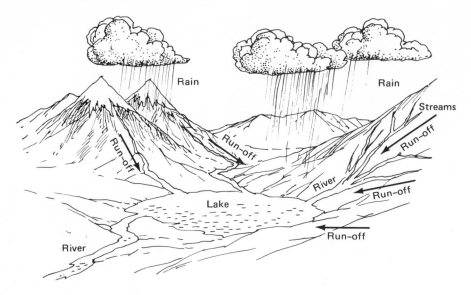

Figure 4-5 *The water supply for an area is usually taken to mean the rate of replenishment by rainfall of the water in lakes, streams, and rivers in an area. This is called the annual surface water run-off. It is the amount of rain that falls minus the amount that evaporates into the atmosphere again. However, the actual available water supply may be larger than this, because water can be reused.*

water supply is probably greater than this because of possible supplies stored underground and because water can be reused.[1] As an example, water users on a river withdraw water, produce a stream of waste water and return this waste stream to the river. Proper waste treatment for the returning water, or water treatment at the next point of use, makes a given amount of water reusable many times over (Fig. 4-6).

Thus water supply estimates, based on water replenishment rates, provide only a lower limit to the actual water supply.

The future   Future water demand is even more difficult to estimate. One of the main reasons is that water has been treated as a common property resource, owned by no one. Except in the Western

U.S., where there is a system of water withdrawal rights, water has generally been available to first comers in any desired amounts. Further, water is cheap. In the U.S., for instance, water rarely costs more than $0.03 per metric ton at the point of use. Compare this to the cost of the cheapest mineral resources, which are rarely less than $30 per metric ton before delivery. These factors combine to produce a relatively careless pattern of water use, which would change greatly if water became scarcer and more expensive. Water demand also depends on population size, and predictions about growth are uncertain. Nonetheless, some brave souls have compared predicted future supply and demand.[2] On a global basis, the predictions are

---

1 In some areas, water is withdrawn from the ground faster than it is being replenished. In these areas, water is a nonrenewable resource. An example is Tucson, Arizona.

2 C A. Doxiadis, "Water and Environment," in Water for Peace, International Conference on Water for Peace, Washington, D.C., May 23–31, 1967.
G. P. Kalinin and V. D. Bykov, "The World's Water Resources, Present and Future," Impact of Science in Society (UNESCO), April–June 1969.

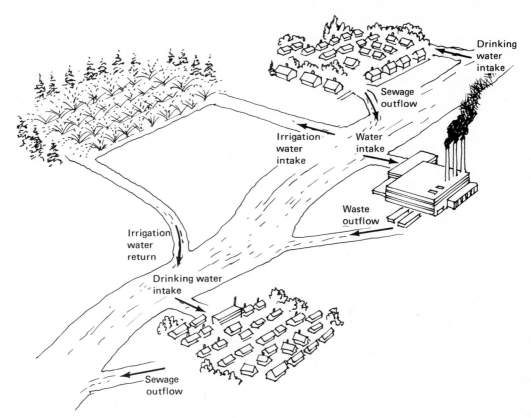

Figure 4-6 *With proper sewage and drinking water treatment, water can be used many times for different purposes.*

comforting. By the year 2000, about 50% of the supply would be withdrawn but only 15% consumed (i.e., not cycled back into the supply for reuse, but lost through evaporation or incorporated into organic material, etc.). This compares to about 3% consumed in 1975. However, global figures are misleading because water is not uniformly available all over the earth.) Even within a country like the U.S. there are dry or desert areas and wet areas. There are also dry years and seasons even in areas with usually good rainfall. There are many areas in which the demand for water is already so great in comparison to the supply, that any increase in water demand will cause a shortage.

Increasing the supply (The supply of usable water can be increased in several ways. In some cases, this

can be done by requiring better waste water treatment by users so that more water can be reused more times.) Price increases or limits on amounts used can also help users share water more evenly and use it more efficiently. There is no incentive to conserve a resource when it is cheap. Yearly or seasonal shortages can be evened out by storing water in reservoirs during good times. (Reservoirs have various effects on the environment, which are discussed in Chapter 22.) Water can also be transported from water-rich areas to water-poor areas, although this is an expensive proposition.

In 1977, a conference was held to discuss the possibility of towing icebergs from the Antarctic to countries short of fresh water. The problems, of course, are stupendous, ranging from how to keep the icebergs from melting or breaking off in chunks (calving) en route, to how to tow them to their destinations.

THE UNIQUE PROPERTIES OF WATER

(The character of water habitats and the kinds of organisms living there are determined by certain properties of water itself.)

Temperature

Water is unusual in that a relatively large amount of heat is needed to change its temperature, or to change solid water (ice) to a liquid or liquid water to a gas (water vapor). For these reasons (temperature changes in water tend to occur slowly and variations in temperature are less than in air. This is important for organisms living in water, since it gives them more time to adapt to changes.)

Water reaches its greatest density at 4° Centigrade. That is, a 1-centimeter cube of water weighs more at 4°C than at any other temperature. Its density decreases as the temperature decreases below 4°C. If you keep in mind that ice forms at 0°C you can see that a given volume of ice (at 0°C) is lighter than the same volume of water at 4°C. This is why ice floats on cold water (Fig. 4–7). This is a very important property, since it prevents lakes from freezing solid. The ice layer floats on the top of the lake and insulates the water beneath it. Many aquatic creatures can winter over in the water below the ice.

Warm water, since it is less dense than cold water, also floats on cold water. This is important in managing reservoirs (Fig. 4–8) and also in determining the effects of pollutants on lakes, such as the phosphorus in detergents (Chapter 9).

Water as a Solvent

(Water is the most common solvent in nature. The amount and kinds of nutrients dissolved in water affect the growth of organisms. In a similar

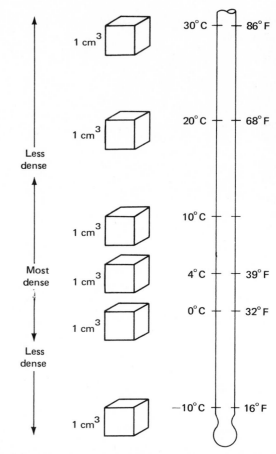

Figure 4–7   *A cube of water 1 centimeter square weighs more at 4°C than at temperatures warmer or cooler than 4°C. Thus ice (at 0°C) is lighter than water at 4°C and will float. This density property of water prevents lakes from freezing solid in the winter and in the summer, allows a stable layer of warm water to float over the cooler bottom waters.*

way, pollutants dissolved in water, even those that are only slightly soluble, affect the organisms living there.)

Plant nutrients   Clear, transparent waters are a sign of an unpolluted lake or stream. Turbid waters, which light hardly penetrates, may be naturally cloudy but may also be polluted with

Figure 4-8 *During the summer in reservoirs such as this one, which is used to generate hydropower, the water forms layers. The warm, oxygen-rich waters float on top of cooler bottom waters, which may be low in oxygen. If, as is common practice, the cooler bottom waters are released through the dam during power generation, stream communities below the reservoir may suffer from lack of oxygen in the water. (Bureau of Reclamation.)*

(nitrates and phosphates. These are nutrients that plants need to grow. Although there are many other necessary nutrients, these two are noteworthy because, together or singly, they are usually the *limiting factors* in natural, unpolluted waters. A limiting factor is the nutrient present in the smallest amount compared to the amount needed for growth.) That is to say, water plants such as algae, will grow until all the available nitrogen or phosphorus is used up (Fig. 4-9). Unnatural amounts of phosphorus (from detergents) or of nitrogen (from sewage) can allow more than the normal amount of algae to grow. This leads in extreme cases to an unsightly, pea-soup-like water condition called eutrophication (Chapter 9). Calcium and other salts may be limiting in certain types of fresh water.

Oxygen and carbon dioxide   Oxygen, which is necessary to animal life, and carbon dioxide, which

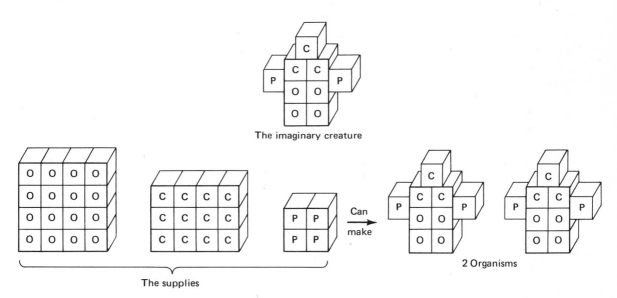

The imaginary creature

The supplies

2 Organisms

Figure 4-9 *The limiting factor concept. Suppose, in order to make up an imaginary, very simplified organism, you needed 4 blocks of O, 2 of P, and 3 of C. If you had 16 blocks of O, 4 of P, and 12 of C you could only make 2 organisms because after that you would have run out of P. In this case P is the limiting factor. No amount of extra O or C blocks will make up for having no more P.*

is needed by plants, are also important factors in water habitats. Either one may limit growth. The supply of oxygen in polluted water is low, which makes life difficult for many organisms, such as fish. On the other hand, the supply of carbon dioxide in polluted water increases. This may help to cause excess growths of unwanted water weeds or algae. (More on this in Chapter 8.)

Salt content of water (Water is an excellent solvent for most salts. Salt waters, such as found in the oceans, are 3.5% salt, or 35 parts of salt for every 1000 parts of water.)(Do you understand what is meant by terms such as parts per thousand, parts per million or parts per billion? See p. 112.)(In contrast, fresh waters are 0.05% salt, or 0.5 parts per thousand. Most of the salt is sodium chloride, but many other salts are present.)

(The salt content of water is one of the major factors determining which organisms will be found there.) Fresh water organisms, both plant and animal, have a salt concentration in their body fluids and inside their cells higher than that of the water in which they live. Because of this concentration difference, water tends to enter and salts tend to leave these organisms. Fresh water organisms have developed mechanisms or structural parts to cope with this problem. In addition, fresh water organisms have evolved so that they contain lower salt concentrations in their bodies than organisms found in salt water.

In salt water, a different situation holds. Some salt water organisms, for instance, marine algae and many marine invertebrates, have a salt concentration in their bodies or cells almost identical to that of salt water. However, many marine organisms have body fluids with a lower salt concentration than the water in which they live. For these organisms, the problem is that water tends to leave their bodies or cells and salts tend to enter. Their regulatory mechanisms must solve a different problem from that of fresh water organisms. The bony fish, for instance, have developed ways of excreting salt and retaining water. The main point is that the two environments, salt water and fresh water, pro-

vide different problems for organisms to solve and thus are inhabited by different kinds of organisms.

In addition to salt and fresh waters, there are brackish waters, with intermediate salt concentrations. Such waters occur wherever salt and fresh waters meet, in estuaries, for instance, or where salt water intrudes on fresh ground waters. Certain organisms are adapted, for all or part of their life cycles, to various intermediate salt concentrations.)

Other Properties that Affect Living Organisms

(The *turbidity* of water, or the depth to which sunlight can penetrate, is important because sunlight is needed for photosynthesis, the process by which green plants convert the sun's energy to food energy. Thus, green plants can only live in the water zone into which sunlight reaches. The presence or absence of a *current* is another important property that helps determine the type of aquatic community. The factor of currents will be examined in the section on river and stream habitats.)

WATER HABITATS

Now that we have noted various properties of water important to life, we will look at the major types of water habitats.(Water habitats are usually differentiated on the basis of salt content (i.e., salt water versus fresh water habitats) and whether a current is present or absent (for instance, streams with swift flowing waters versus lakes or river pools where a current is absent) (Fig. 4–10).)

Fresh Water Habitats

(The most important physical characteristics of fresh water habitats involve: temperature; turbidity; whether the waters are flowing or still; and the amount of dissolved materials, such as nitrate and phosphate salts, and gases, such as oxygen and carbon dioxide.)

Figure 4–10 *The three main types of water habitat. (a) In fresh-water habitats, the water is either still as in lakes and ponds, or moving as in streams and rivers. (Grant Heilman Photography) (b) In marine habitats such as oceans and seas, the water moves continuously as a result of various currents. The salt concentration is, of course, much higher than in fresh water. (U.S. Coast Guard official photo) (c) Estuaries are partly enclosed bodies of water where salt water meets fresh water. An example is the Chesapeake Bay, shown here. (Chesapeake Bay–Bridge Tunnel Authority)*

Lakes Most of the organisms living in lakes and ponds or in the quiet pools of streams are adapted to life in still waters. In the shallow water zone along the shore, light reaches all the way to the bottom. Here live rooted water plants and floating algae. Further from shore, there is an area of open water. This zone is divided into two layers. The upper layer, which light penetrates, is home to minute plants and animals called plankton, as well as to fish.

The plankton are microscopic, floating organisms found in lake waters as far down as light penetrates. Plankton species are, in general, unable to move against currents. They float along with water movements. There are plant species, the phytoplankton, which include many of the species of algae, and animal species, the zooplankton. Phytoplankton are important producers in the lake ecosystem. They capture the sun's energy and turn it into organic nutrients that form the basis of many of the lake's food chains. The zooplankton feed on phytoplankton and so are primary consumers in the lake ecosystem. Along with the phytoplankton in the open waters, plants in the shore zone are producers in lakes and ponds. However, despite their small size, phytoplankton species are more important producers than rooted plants.

Living in the deeper water layer and on the bottom, where there is not enough light for photosynthesis, are organisms which can live on dead organic matter. Bacteria, fungi, small clams, and blood worms all "re-process" organic matter, which is then carried by currents or swimming creatures back to the other lake zones. Thus, the deep zone houses the detritus feeders in lake and pond communities.

Oxygen is sometimes in short supply in deep lake waters. For this reason many organisms living there are adapted to low oxygen concentrations or even able to live with no oxygen at all.

While consumers such as frogs and snakes live along the shore line, other consumers such as fish may range over all three zones depending on the season and availability of food. Figure 4–11 summarizes these lake communities.

Ponds Ponds differ from lakes in that the shore zone is relatively large and the open water zone is comparatively small. Ponds are often too shallow to have a layer of water that light does not reach. Thus, photosynthesis takes place at all depths. In addition, lakes often stratify, or have layers of different temperatures. In summer, a warm upper layer floats on top of a cold layer. Nutrients produced by photosynthesis in the upper layers filter down into the lower layer where decomposition reactions take place. This sometimes lowers the oxygen content of the lower layer drastically. (See Chapter 9, Eutrophication.) Ponds usually have no temperture stratification because they are too shallow to prevent thermal currents from mixing the waters. Some ponds dry up during part of the year and this creates particular stress on their communities. Organisms living there must have a dormant stage to survive the dry period. For instance, fairy shrimp lay eggs capable of surviving for

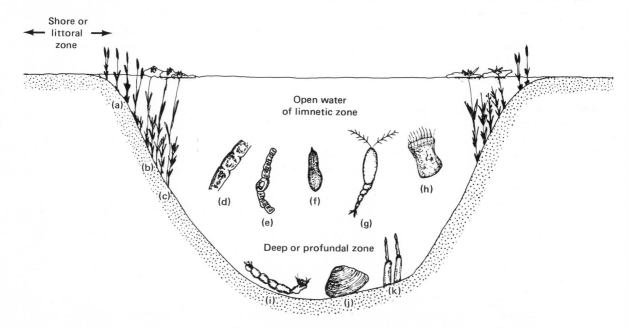

Figure 4–11 *In the shallow waters along lake shores grow rooted water plants such as bulrushes, cattails and muskgrass. In the top layer of open water, the limnetic zone, a variety of plankton, or drifting organisms, find a habitat. Examples include algae such as spirogyra, navicula, or anabaena and zooplankton such as copepods and rotifers. In the deep or profundal zone live bacteria, fungi, and other creatures that live in or on the bottom muds. (a) cattails, (b) waterlillies, (c) muskgrass (chara), (d) scenedesmus, (e) anabaena, (f) navicula, (g) senecella, (h) asplanchna, (i) bloodworm (chironomid larva), (j) peashell clam (musculium), and (k) tubifex. (d) through (k) are not drawn to scale, we need a microscope to see the limnetic zone organisms shown here.*

months in dry soil. Some organisms are able to live both on land or in water, as can amphibians such as frogs.

Rivers and streams (Three features of the environment in rapidly flowing waters are very important to understanding the types of organisms that can live there: the presence of a current, the high oxygen concentration, and the source of nutrients.) A current is one of the main factors making life in a stream or river different from life in lakes and ponds. However, the difference is not found in all parts of these environments. Streams have pools or areas of quiet flow where organisms find similar habitats to those in lakes. In addition, a lake shore, where waves keep water moving, provide organisms with a habitat similar to a rapidly flowing stream or river. (There are thus two types of stream or river communities, those in flowing water and those in quiet water.)

(A major, and very understandable, feature of organisms living in moving water is that they usually have some way of hanging on to surfaces such as rocks or stream bottoms.) Some are attached firmly to stones or other objects in the flowing waters. Others have hooks, suckers or sticky undersides. Stream creatures also have streamlined bodies to reduce resistance to flowing water and are often flattened so they can crawl under rocks to escape the pull of the current (Fig. 4–12). Because of the current, rapidly flowing streams or rivers usually have a high oxygen content. The waters, moving and tumbling over rocks, are kept well mixed with air and so absorb a great deal of oxygen. Organisms living in rapidly flowing water are used to these high concentrations of oxygen. When pollutants that use up oxygen in water are added to streams, the clean stream organisms cannot survive the low oxygen levels. The stream communities found in polluted and clean water are contrasted in Chapter 8.

(A large part of the nutrients in streams and rivers either wash or fall into the water from the banks and surrounding watershed. Plant nutrients, such as nitrate and phosphate, and organic material, such as leaves on which detritus feeders

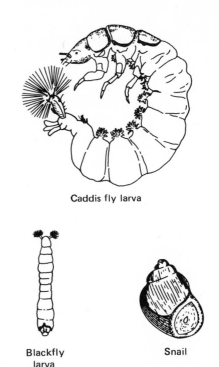

Caddis fly larva

Blackfly larva

Snail

Figure 4–12 *Organisms living in flowing water have special adaptations which help them stay in one area instead of being swept downstream. (a) Some organisms are permanently cemented or rooted in place. Caddisfly larvae cement themselves to stones, while various plants are rooted in the bottom mud or attached to stones or logs. (b) Other species have developed hooks or suckers. Semulium, the blackfly larva, has a sucker and also attaches itself to a surface by a thread. If the sucker loses hold, the larva can pull itself back by the thread. (c) Organisms like snails have sticky undersides to help them cling to smooth rocks.*

live, all enter the stream from its watershed. Stream organisms are adapted to this constant flow of fresh nutrients and also to the removal of their waste products by the current. Waters with currents thus provide an environment fundamentally different from the still waters of ponds and lakes.)

Plant nutrients are not the only materials washed into rivers and streams from watersheds. Pesticides or industrial wastes in ground waters may also wash in. For this reason, when river and stream pollution problems arise, we must, in examining possible solutions, take into account not only the stream itself but also the land surrounding the stream.

The Hudson as a Sewer

Partly because currents usually carry wastes downstream, rivers and streams have always seemed especially handy and inexpensive ways to dispose of wastes. The Hudson River was used for years as a sewer for many communities along its banks. This practice resulted in a series of typhoid epidemics around 1890 in communities which not only dumped sewage but also drew drinking water from the Hudson (Chapter 5). The latest story about pollution in the Hudson River, however, only goes back to 1975 when it was discovered that as a result of industrial sewage, river fish contained dangerous levels of the toxic chemicals called PCBs. PCBs are lethal in very small amounts (as little as 10 parts per billion) to many insect larvae and to some small fish.[3] Large fish, at the top of their food chains, were found to have PCB concentrations well over the legal limit for human foods (Fig. 4–13).

As a result of laws such as the Toxic Substances Control Act of 1976, the manufacture and disposal of PCBs are now controlled much more closely. For the Hudson River, however, this comes too late. (See Chapter 6.)

---

3 For some comparisons that may help you understand this number better, see tables on pp. 112–114.

Figure 4–13   *Pavel Petrovitch Svinin (1787–1839), Shad Fishermen on the Shores of the Hudson River. Henry Hudson anchored his ship Half Moon off this point as he returned from exploring the river that was later given his name. But Indians fished for shad in the Hudson long before Hudson arrived in 1609. Fishing for certain species of fish is no longer allowed in the Hudson. Unhealthy levels of the chemicals known as PCB's are now found in the river fish. (The Metropolitan Museum of Art, Rogers Fund, 1942.)*

Estuarine Habitats

What they are   As the Hudson winds its way to the Atlantic Ocean, it reaches at last an area where its fresh waters mix with the salt waters brought by ocean tides. In the Hudson, salt and fresh waters meet in New York Bay and some salt water moves as far upriver as Troy, New York.

River mouths, salt marshes, bodies of water behind barrier islands and coastal bays such as New York Bay into which the Hudson empties, are all estuaries (Fig. 4–14). *Estuaries are coastal bodies of*

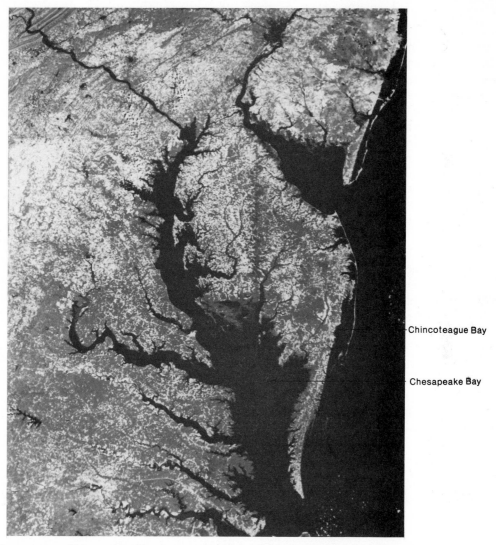

Chincoteague Bay

Chesapeake Bay

Figure 4–14   *Chincoteague Bay is an estuary formed by the mainland on one side and the barrier islands, Chincoteague and Assateague, islands on the other. The Cheasapeake Bay is a huge estuary where fresh water from many large rivers such as the James and the Susquehanna and the Potomac mix with salt waters from the Atlantic Ocean. (NASA)*

water partly surrounded by land but still having an open connection with the ocean. In areas such as these, fresh water drains from the land and mixes with tidal currents of salt water. Estuaries, especially marshes, are often looked upon as wasteland, best dredged or filled. But this is a real misunderstanding of the role estuaries play in their ecosystem.

Productivity Estuaries are very productive systems. They are generally more fertile than either the neighboring ocean or the fresh waters that flow into them. This is because nutrients are easily trapped in estuaries. The nutrients are trapped in a physical sense by the action of tides and fresh water flow (Fig. 4-15) and also in a biological sense because nutrients are recycled rapidly by a network of producers, consumers, and detritus feeders. Unfortunately, pollutants are also recycled in estuaries, so the effect of toxic materials such as DDT can be more serious than in a river or the ocean.

Another factor that contributes to the fertility of estuaries is tidal action. Tides, which cause the water to flow back and forth in estuaries, make it possible for organisms like the oyster, which feeds by filtering sea water, to sit and have their food brought to them. In the same way, their wastes are removed.

Furthermore, estuaries provide good habitats for a variety of producer organisms, from the large rooted grasses like eel grass, turtle grass, or salt marsh grass, to the tiny floating plants or bottom dwelling algae. In fact, more organic material is often produced than can be recycled in the estuary

Sea — River

Figure 4-15 *Fresh water, which is lighter, tends to float on the heavier seawater. As the one rolls over the other, mixing currents are set up which tend to recirculate nutrients. (Adapted from E. P. Odum,* Fundamentals of Ecology, *Saunders, Philadelphia, 1971, p. 354.)*

itself. The excess nutrients flow out into the ocean and fertilize these waters. Good fishing is the result.

Estuaries serve yet another important purpose and that is to provide a nursery for many ocean species, such as shrimp. The larvae, or immature stages, of these species find protection and food in the estuary. Fish, such as salmon or Hudson River shad, which live in salt water but return to fresh water to breed, require estuaries as places to rest during their journey. Thus, when estuaries are unthinkingly filled in, the effects fall not only upon the creatures that spend their whole lives there, but also upon many ocean species that use estuaries or the food produced there.

Marine Habitats

Geographic features of the sea bottom Actually, the Hudson River does not disappear into the Atlantic Ocean without a trace at New York Bay (Fig. 4-16). Three miles southeast of Ambrose Lightship an underwater channel begins and runs 130 miles along the ocean bottom. Geologists believe that at the end of the last ice age, before humans existed, the shore was 150 miles farther into the ocean than it is now. They speculate that the roaring Hudson, fed by melting ice, cut a channel though this land on its way to the sea. At some points, it created a gorge 36,000 feet deep, deeper than the Royal Gorge of the Colorado River. But the sea has returned and now lies thousands of feet above the place where the waters of the Hudson once sparkled in the sunlight between sheer cliffs almost two miles high. A natural wonder, Hudson Canyon lies thousands of feet below the surface of the ocean.

Besides Hudson Canyon, there are many interesting features below ocean waters. Looking at the sea in profile, as if it were cut in half from top to bottom, several well defined areas (Fig. 4-17) can be seen. For some distance, the ocean floor slopes gradually away from the land. This area is called the continental shelf. The floor then drops off sharply (continental slope) and again levels off, into the continental rise. Finally the floor drops off once

Figure 4-16 *(a) Staten Island and the Narrows, Thomas Chambers. View looking southeast to lower New York Bay and the Atlantic, about 1833. (The Brooklyn Museum: Dick S. Ramsay Fund.) (b) Contrast this view of New York Bay, the Hudson's estuary, today (New York Convention and Visitors Bureau.)*

more to a level plain, the abyssal plain, 2000 to 5000 meters below the surface. Towards the middle of the ocean, a series of ridges are scattered across this abyssal plain.

Currents   Unlike lakes, where layers of water may be still for long periods, the water in the sea moves continually. Currents are caused by a variety of forces such as temperature differences, differences in salt content, the rotation of the earth, and winds. Because of these mixing currents, even the deep parts of the oceans have a constant supply of oxygen in the water. In some areas, along steep coastal slopes, winds continually blow the surface water away from the shore. This allows cold bottom waters, rich in nutrients to rise to the surface. This is called *upwelling*. The areas where this occurs, for instance along the coast of Peru, are the most fertile in the seas. In general, although life is found in all areas of the sea, the major commerical fisheries are all located on or near the continental shelf. This is because many ocean food chains are based on the microscopic green plants, which grow best in areas of coastal upwelling.

Communities   In Figure 4-18, some of the species found in the coastal zone are shown. The communities that live in the intertidal zone are composed of organisms specially adapted to the periodic absence of water when the tide goes out. There is concern that these organisms might be endangered by the development of tidal power, the use of the tides to generate electricity. We shall examine this more fully in Chapter 22. In the open ocean, live species adapted to life far from shore. In the top layer of water, where light penetrates, floating microscopic plants and animals live. The sunlit zone in the open oceans does not support as much life per square meter as the light zone in coastal areas does. However, the oceans cover 70% of the earth's surface and much of this is open ocean. For this reason the photosynthetic organisms in the open ocean are very important in world oxygen and carbon dioxide balances. Figures 4-19 and 4-20 show some of the organisms of the open ocean.

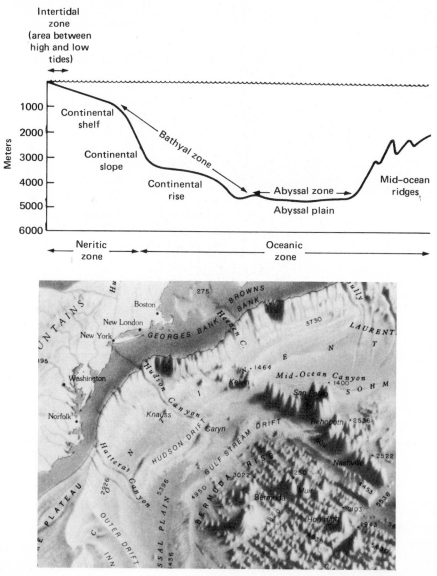

Figure 4–17   *The sea in cross-section. The shallow water of the continental shelf is called the neritic zone. This includes the area where the tides cover and uncover the shore twice a day. The rest of the open sea is called the oceanic zone. The bottom of the sea along the continental slope and rise is called the bathyal zone, while along the deeper plain it is called the abyssal zone. As in fresh water, light penetrates only the top layer of water. All below this zone is in darkness. (Bottom photo) Major deep sea currents are believed to have piled ocean sediments into cliffs and ridges along the bottom. Note the Hudson Canyon shown in the upper left-hand corner of the picture. (Panorama of ocean bottom showing Hudson Canyon is from World Ocean Floor Panorama by Bruce C. Heezen and Marie Tharp 1977. ©Marie Tharp)*

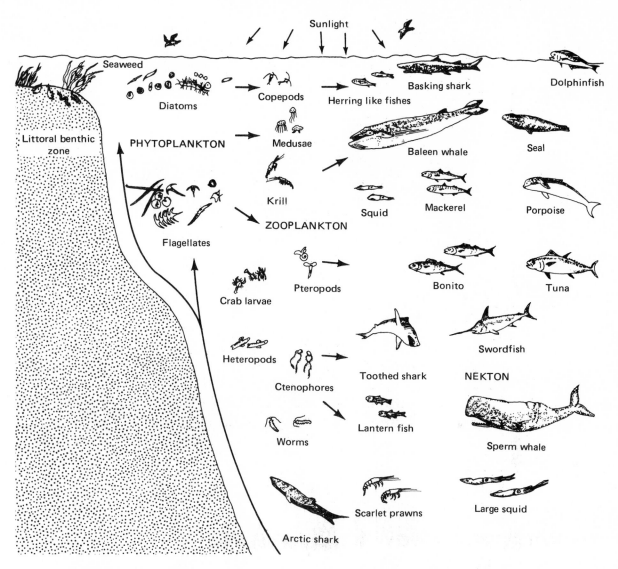

Figure 4-18 *Communities in the region of the continental shelf. Diatoms, dinoflagellates, and microflagellates are the producers, which form the base of the food chains in the neritic or coastal zone. Sea weeds, algae adapted to hold on to rocky shores as the tide washes over them, are also producers in this zone. Tiny zooplankton such as copepods and shrimplike krill feed on the producers and are fed on by fish, crustaceans such as lobsters and crabs, and sea mammals such as whales. Sea birds, large fish, such as tuna and swordfish, and humans are top consumers in these food chains. Also found in this region are the larvae of immature stages of many marine species. Buried in the bottom muds or clinging to rocks are the worms, clams, snails, crabs, and bacteria that form the benthos or bottom dwellers, which feed on detritus. (Adapted from John D. Isaacs, "The Nature of Oceanic Life," Scientific American, September 1969.)*

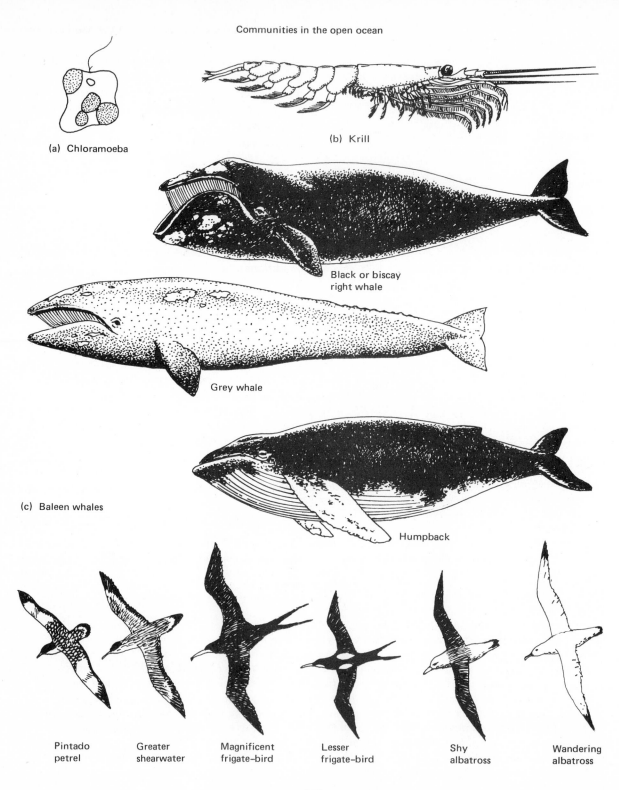

Communities in the open ocean

(a) Chloramoeba

(b) Krill

Black or biscay
right whale

Grey whale

(c) Baleen whales

Humpback

Pintado
petrel

Greater
shearwater

Magnificent
frigate-bird

Lesser
frigate-bird

Shy
albatross

Wandering
albatross

Figure 4–19 *(Facing page). Communities in the open ocean. Microplankton species such as chloramoeba (a) are the main producers in the open oceans. The shrimplike krill (b) as well as small zooplankton species feed on microplankton. Larger fish and sea mammals such as the baleen whales (c) range over both the open ocean and the shore areas in search of the krill and zooplankton, on which they feed. Oceanic birds such as petrels, albatrosses, and frigate birds feed on the open oceans except during breeding time when they fly to land.*

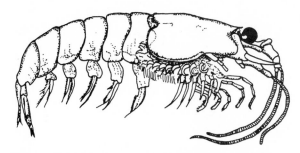

Figure 4–20 *Shrimplike krill (Euphausia superba) are a vital link in ocean food chains between the microscopic producers and larger creatures such as whales. However, nations hungry for protein are now looking at the 5-cm long krill as a possible food source for their human populations. Krill can be made into a sort of shrimp paste. What effect massive harvesting of krill would have on ocean communities is unknown.*

We know relatively little about communities in the deep zones of the ocean. Only recently have a number of facts come to light, quite literally, since one of the main physical characteristics of the deep sea regions is relative darkness. Sunlight does not reach these regions (although they are not completely dark). Since there is not enough light for photosynthesis, organisms are dependent on the producers in the top layer of water for organic nutrients. The major portion of the organic material reaching the deep ocean zones is probably composed of fecal pellets from zooplankton on the surface.

Diversity and deep sea communities   Much of the deep sea bottom is covered with thick layers of mud. In some areas, this appears to be pushed by

continuous currents into topographic features such as ridges and cliffs tens to hundreds of kilometers high. In and on the bottom muds of the plains live many species of worms, clams, and crustaceans (Fig. 4–21). These organisms live in an area that maintains a very stable temperature and supply of energy. Although the organisms living here are small, there is an enormous variety of them. In fact, the variety is comparable to that found in tropical rain forests and in coral reefs, two other areas noted for their great variety of species. These habitats are similar in that they maintain physical stability (that is, little or no changes in physical conditions take place) and have a long evolutionary history. Some scientists believe that these are the very conditions that favor the development of a *diversity* of species. A high diversity of species means that there are many different species in an area compared to the total number of individuals there. This diversity is characteristic of ecosystems that have not been disturbed for long periods of time (mature systems).

In ecosystems having a high diversity of species, there are many relationships between species. Organisms are bound together not only by complex patterns of predators and prey but also because some organisms produce materials that other organisms need. Because there are more "options,"

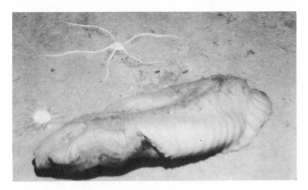

Figure 4–21 *Although a few large species are visible on the surface, most of the diversity of life in the deep sea is found in the mud. Here, a holothurian, an urchin, and brittle stars share the sediment surface. (Dr. Fred Grassle, Woods Hole Oceanographic Institute.)*

Abyssal Stars, Urchins, Cucumbers Crustaceans, Spiders, and Fish

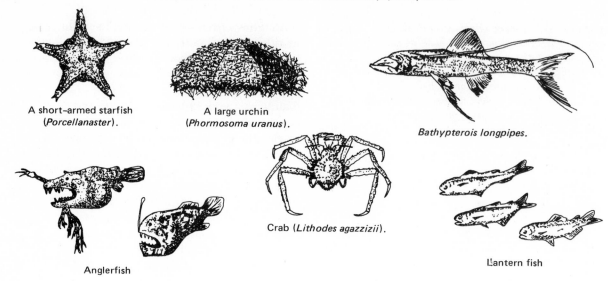

A short–armed starfish
(*Porcellanaster*).

A large urchin
(*Phormosoma uranus*).

Bathypterois longpipes.

Anglerfish

Crab (*Lithodes agazzizii*).

Lantern fish

Animal Sponges and Flowers of the Deep Sea

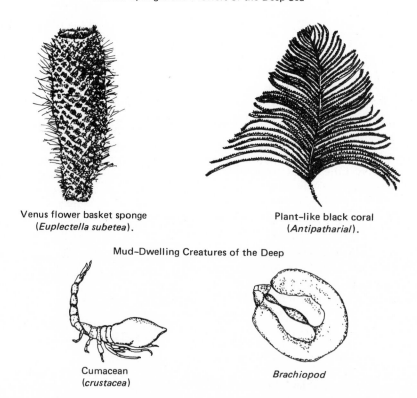

Venus flower basket sponge
(*Euplectella subetea*).

Plant–like black coral
(*Antipatharial*).

Mud–Dwelling Creatures of the Deep

Cumacean
(*crustacea*)

Brachiopod

Figure 4-22 *Common deep sea dwellers. Some species have interesting adaptation such as enlarged eyes, to gather in the little available light, or even luminescent organs to generate their own light (angler fish, lantern fish).*

such systems seem to be more stable, less subject to rapid changes in populations.) Some common organisms found in the deep waters are illustrated in Figure 4–22.

Phosphorus Cycle

(The sea is a major part of a number of important cycles. One example is the hydrologic cycle; another is the *phosphorus cycle* shown in Figure 4–23. Phosphorus enters the environment mainly from rocks or deposits laid down in past ages. Erosion gradually releases the phosphorus. Some is used by biological systems but much of it is washed into the ocean, where it settles in the sediments. Phosphorus in shallow sediments is cycled by bottom dwellers through fish and then to birds, which deposit the phosphorus back on land as droppings or guano. But a portion of the phosphorus is not recycled. This is the amount that is buried in deep sediments.)

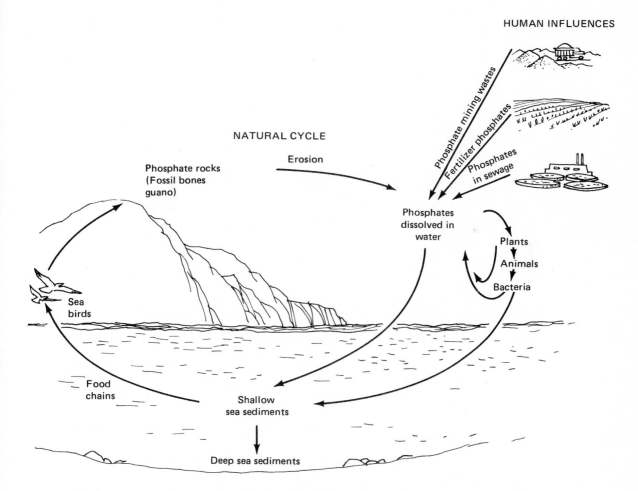

Figure 4–23 *The phosphorus cycle. Erosion and mining release phosphates, from rocks and other deposits, to water environments. Phosphates are eventually deposited in shallow or deep sea sediments. From shallow sediments phosphate is recycled to land environments.*

Large quantitites of phosphorus are used as fertilizer and most of this is eventually lost to the deep sea sediments. At some time, though not in the near future, shortages could result. Phosphate, which enters natural waters from sewage, fertilizer run-off, and mining wastes, contributes to the problem of eutrophication (Chapter 9).

In this chapter we have examined the Hudson River from its source in the Adirondack Mountains, high above sea level, to its final disappearance into the ocean floor, thousands of feet below the surface of the ocean. Like other large rivers the Hudson supports many different kinds of aquatic communities along its length and, in addition, suffers from a variety of pollution problems.

In the following chapters various factors are discussed which can disrupt the functioning of aquatic communities and make waters unfit or unpleasant for human uses.

How Much is a Part per Trillion Anyway?

A number of environmental problems, especially those involving water pollutants, are defined in terms of parts per million or parts per billion or even parts per trillion of pollutant chemicals or particles in a given quantity of the resource (in this case, water). These are very small numbers, indeed. But are the numbers so small they can be safely ignored? This, of course, depends on the frame of reference. The following tables attempt, in two very different ways, to put these small numbers into proper perspective.

The December 1976 issue of <u>ChemEcology</u> carried the following table, by Dr. Warren B. Crumett of the Dow Chemical Company, which was designed to show just how small trace concentrations are.

Trace Concentration Units

Unit	1 part per million (ppm)	1 part per billion (ppb)	1 part per trillion (ppt)
Length	1 inch/16 miles	1 inch/16,000 miles	1 inch/16,000,000 miles (A 6 inch leap on a journey to the sun.)
Time	1 minute/2 years	1 second/32 years	1 second/320 centuries
Money	1¢/$10,000	1¢/$10,000,000	1¢/$10,000,000,000
Weight	1 oz. salt/31 tons potato chips	1 pinch salt/10 tons potato chips	1 pinch salt/10,000 tons potato chips

Trace Concentration Units

Unit	1 part per million (ppm)	1 part per billion (ppb)	1 part per trillion (ppt)
Volume	1 drop vermouth/ 80 "fifths" gin	1 drop vermouth/ 500 barrels gin	1 drop vermouth/ 250,000 hogheads of gin
Area	1 ft²/23 acres	1 ft²/36 mi²	1 in²/250 sq. mi.
Quality	1 bad apple/2000 barrels	1 bad apple/ 2,000,000 barrels	1 bad apple/ 2,000,000,000 barrels

Feeling that Dr. Crumett may have overstated his case, Colin Chriswell of Iowa State University wrote (ChemEcology, November 1977):

> Even a minute concentration can add up to a sizable total amount of material when contained in a large sample. For example, consider the amount of material that would be involved in the amount of water used in the Los Angeles area (3,000,000 acre-feet/year) if some selected contaminants were present at a part per billion level.

Substance	one ppb in 3,000,000 acre-feet/year would amount to enough to:
lead	cast 1,000,000 bullets
chromium	plate 50,000 car bumpers
mercury	fill 4,000,000 rectal thermometers
phenols	produce 250,000 bottles of Lysol
herbicide	kill all the dandelions in 100,000 lawns
insecticides	fill 5,000,000 aerosol cans of bug killer
gold	run the federal government for nearly 20 minutes or support 50 average families for eternity.

A few numbers from the biological world might also be worth comparing to those in the table:

parts per million	parts per billion	parts per trillion
if there is 1 part per million oil in the water, 1/2 of the exposed Dungeness crab larvae will be killed	at levels of 20 parts per billion in their blood, humans show symptoms of mercury poisoning	brook trout cannot grow properly or reproduce at levels of toxaphene over 39 parts per trillion

Questions

1. Why is the watershed of a lake or river considered the important ecosystem rather than just the river or lake itself?
2. Describe what is meant by the hydrologic cycle. Give some examples of ways in which humans are affecting its balance.
3. What is meant by the usable water supply?
4. What major factors define water habitats?
5. Why are estuaries important ecosystems?
6. Do you think the oceans are a vast untapped resource of food to feed the hungry nations of the world? Why or why not?
7. Describe how phosphorus cycles in the environment.
8. What is the importance of plankton to fresh and salt water communities?

Chapter 5

Waterborne Diseases

LEARNING FROM PAST MISTAKES

How It Was Discovered That Water Can Carry Disease

As, it seems with all science,(the first notions that water could carry disease occurred to the ancient Greeks. The physician Hippocrates, the ancient innovator of medical ethics, advised that polluted water be boiled or filtered before being consumed.)Despite Hippocrates' recommendations, only 150 years ago most people were still not aware that human diseases could be spread by water. For this reason, many communities dumped their raw sewage into the same lake or river that they, or other communities, used for drinking water.

To some extent, this type of system (or nonsystem) can work for a time. If an area is sparsely populated, if the next town on the river is some distance off, then the disease organisms added to water by human sewage will die off before the next person drinks the water. This is because bacteria, viruses, and protozoans, the microscopic organisms (microorganisms) that cause human diseases, can only survive for a limited time outside the human body. As areas become more crowded, however, there is not enough time for water to "purify itself." That is, between the time raw sewage is put into the water and drinking water is taken out, the organisms do not die. For many years, epidemics were spread through water because of this fact and few people realized what was happening.

Typhoid and Asiatic cholera are two diseases that are spread by water polluted with human wastes. These diseases attack and affect the intestinal tract of humans. The bowel discharges of individuals infected with these diseases contain the pathogens (disease-producing microorganisms) that spread the diseases. If these discharges enter a water source, there is a high probability that new infections will occur among those who drink the water. Typhoid, besides being communicated by water polluted with the specific pathogens, may also be spread by carriers of the disease.(In the past, typhoid was spread principally by water polluted by the specific pathogen, *Salmonella typhi*.)

In the United States today, most cases of typhoid are due to contaminated food. The food was probably contaminated in its preparation by an asymptomatic carrier of typhoid (an individual who carries the pathogens but exhibits no symptoms of the disease). One famous case involved a woman who was a cook for a family on Long Island in the summer of 1906. Typhoid infections occurred among the members of this family and an investigation followed. It was found that typhoid had occurred in all the families for whom the woman had cooked in the past several years; Mary, the cook was a chronic carrier of typhoid. The state of New York eventually found employment for "Typhoid Mary" in a laundry during the remainder of her life to ensure that she would not endanger others.

Chronic carriers of cholera, unlike typhoid, are not common. However, in the midst of an epidemic, individuals may temporarily carry the pathogen without showing signs of the disease. Such people are capable of transmitting the disease through contact and food-handling. Cholera, in a number of countries, is still spread via the water route.

During the period from 1850 to 1900, people gradually became aware that such diseases as typhoid and cholera were associated with polluted water. Although it is common knowledge today, it was not common knowledge in 1854. In that year, an epidemic of Asiatic cholera struck hundreds of residents in the London parish of St. James. The pump on Broad Street in St. James served a large number of families and industries in the area. Its water was regarded, in fact, by many as being even better than that of other wells. Nevertheless, in the 17-week epidemic of Asiatic cholera that occurred in that year, over 700 of the 36,000 residents of the St. James district died. Although the rate of death from cholera was over 200 deaths per 10,000 population in St. James Parish, in two districts adjacent to St. James, Charing Cross and Hanover Square, the cholera death rates during the 17-week period were much lower, only 9 and 33 per 10,000 (Fig. 5–1). Clearly, some special circumstance was responsible for the greater number of cholera deaths in St. James.

In the hope of discovering the cause of the outbreak, an inquiry committee was appointed to study conditions in the parish. The committee looked at the density of population in the parish, the weather conditions, and the cleanliness of the houses. They also studied the cesspools, the sewers, and the water supply. The committee, it must be concluded, was fundamentally unaware of the cause of the epidemic. Picture, if you will, a gentleman in white wig and frock, delivering the following pronouncement, in the name of the committee:

> ...a previous long-continued absence of rain...; a high state of temperature both of the air and of the Thames...; an unusual stagnation of the lower strata of the atmosphere, highly favorable to its acquisition of impurity...;

> ...their combined operation, either by favoring a general impurity in the air or in some other way, concurred in a decided manner, last summer and autumn (1854), to give temporary activity to the special cause of that disease...."[1]

Though still unaware of the cause, however, they recognized an unusual circumstance.

> But, as previously shown in the history of this local outbreak, the resulting mortality was so disproportioned to that in the rest of the metropolis, and more particularly to that in the immediately surrounding districts, that we must seek more narrowly and locally for some peculiar conditions which may help to explain this serious visitation.

One of the members of the inquiry committee, John Snow, began an investigation of the deaths in the area near the Broad Street well. His report of the epidemic in St. James Parish, revealed that 73 of

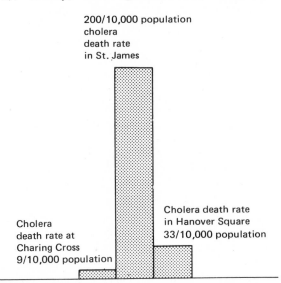

200/10,000 population
cholera
death rate
in St. James

Cholera death rate
in Hanover Square
33/10,000 population

Cholera
death rate at
Charing Cross
9/10,000 population

Figure 5–1 *Cholera death rate in St. James and neighboring districts, London 1854.*

---

1 These citations are from Report on the Cholera Outbreak in the Parish of St. James, Westminster, during the Autumn of 1854, presented to the Vestry by the Cholera Inquiry Committee, July 1855, London, J. Churchill, 1855.

the first 83 deaths in the epidemic occurred among individuals whose closest source of water was the Broad Street pump. Interviews indicated that about 90% of these 73 deaths had occurred among individuals who had been known to drink from the well either constantly or occasionally. The ten additional deaths in the first stage of the epidemic occurred in households nearer to some well other than the Broad Street well (Fig. 5–2). However, it was found that five of the 10 households were in the habit of sending to the Broad Street pump for their water. Dr. Snow writes in the Report of the Inquiry Committee:

I had an interview with the Board of Guardians of St. James's Parish on the evening of Thursday, 7th of September, and represented the above circumstances to them. In consequence of what I said the handle of the pump was removed on the following day....[2]

In the spring of 1855, the Reverend Mr. Whitehead, a member of the Inquiry Committee and Curate of St. Lukes in Berwick Street, began an independent investigation of the circumstances of the epidemic. He discovered that residents of the

2 Report of Cholera Inquiry Committee.

Figure 5–2 *Asiatic cholera and the Broad Street pump, London 1854. (Source: Sedgwick's* Principles of Sanitary Science and Public Health, *revised and enlarged by S. Prescott and M. Norwood, Macmillan, New York, 1948.)*

house at No. 40 Broad Street had had an unidentified disease prior to the epidemic, and further that their dejecta had been disposed of in a cesspool which was not far from the well of the Broad Street pump. The brickwork of the cesspool was found to be defective in the sense that

> the bricks were easily lifted from their beds without the least force; so that any fluid could pass through the work...into the earth...,[3]

and the main drain leading from the house was in a similar state. The cause of the epidemic was clear.

Such were the circumstances surrounding the 1854 epidemic at St. James. It was now clear that cholera infections could be transmitted through a water supply. It was to be 70 years more, however, before engineers had perfected ways to purify water to guard citizens from disease.

In the spring of 1885, the mining town of Plymouth, Pennsylvania, was drawing its water from two sources. One was a mountain stream of seemingly high quality, and the other was the Susquehanna River into which Wilkes Barre, upstream from Plymouth, was discharging its wastes. In April of that year, an epidemic of typhoid struck the town, affecting 1100 of the town's 8000 inhabitants; 114 people died.

Several explanations were advanced for the epidemic. Some people blamed the generally poor sanitary condition of the town for the epidemic. The sudden onset was attributed to the warm rays of the April sun falling upon the dirt in the town and thereby releasing poisonous vapors. Dr. L. H. Taylor, however, who investigated the outbreak, noted that Plymouth was no dirtier than neighboring towns and no dirtier in that year than in previous years. "Thoughtful minds," he tells us, "turned to the water supply as furnishing the true cause of the invasion."[4]

It was the mountain stream, in fact, which carried the typhoid pathogens and brought the epidemic to the town. A house had been located on the stream banks and an inhabitant of that house had contracted typhoid in late December of 1884. His illness persisted for three months, and during this period his dejecta were thrown or deposited on snowy or frozen ground. From there, a rain or thaw had carried them down the slopes into the stream that fed the town's reservoir. An apparently pure mountain stream, trusted for its quality, had caused the epidemic.

Dr. Taylor drew two conclusions from the epidemic: first, that the dejecta of typhoid fever patients must be carefully disposed of and disinfected; second, that water companies "should be compelled to remove from the banks of their streams and reservoirs not only all probable, but all possible sources of pollution,"[5] a warning that water managers carefully heed to the present time.

Many areas of the United States experienced typhoid in this era. In Massachusetts, in 1890 an epidemic was traced to the Merrimac River, which supplied water to the numerous manufacturing towns of the region. In that same year, in New York State, an epidemic of typhoid occurred, and again the path of infection followed the river, this time the Mohawk and Hudson Rivers. The cities of Schenectady, Cohoes, West Troy, and Albany all experienced typhoid epidemics. Eighty years later, we find the same stretch of the Hudson River as the route of contamination of the chemicals known as PCBs.

Early Efforts to Make Water Safe

By 1872, filtration through sand beds was being recognized as a way to make water clear and relatively safe for drinking. In that year Poughkeepsie, New York, became the first city in the

---

3 Report of J. York, Secretary and Surveyor to the Cholera Inquiry Committee.
4 First Annual Report, State Board of Health and Vital Statistics of Pennsylvania, 1886.

---

5 First Annual Report, State Board of Health and Vital Statistics of Pennsylvania, 1886.

SHOULD WATER SUPPLY RESERVOIRS BE USED FOR RECREATION?

Without strict controls over recreation, waterborne diseases could become a serious problem.

American Water Works Association, 1971

No amount of properly planned recreation can generate enough contamination of reservoir water to pose any technical difficulties in treatment.

President's Council on
Environmental Quality
1975

The American Water Works Association issued the following statement on the use of water supply reservoirs for recreation: *

The American Water Works Association supports the principal that water of highest quality be used as source of supply for public water systems. Since each water utility is responsible for its product, determination of type and extent of recreational use of impounding reservoirs shall be vested in the water utility.

Water utilities of the United States and Canada provide water of the highest quality in the world. To improve further upon this record, each water utility must continue to recognize its responsibility to deliver a safe and appealing product to its consumers. A growing demand for use of reservoirs for recreational purposes, however, may make this responsibility more difficult to carry out in the future. In some areas, uncontrolled use of reservoirs already has resulted in deterioration of water quality. Without strict controls over recreation, water-borne diseases could become a serious problem....

Research on the effects of recreation on water quality should be expanded. Reservoirs used for recreation must receive close supervision. In all cases, water from reservoirs used for recreation must be treated, with the degree of treatment to be determined according to state or provincial and local utility laws, regulations, and policies....

Public water supply reservoirs should not be used for recreation if other surface waters are available. Bodily-contact sports such as swimming should not be allowed in public water supply reservoirs. For the purposes of this policy statement, reservoir means an impoundment reservoir subsequent to which water receives treatment before consumption. Distribution reservoirs from which water is supplied directly to the public require the strictest of controls and under no condition should be used for recreation.

The recreation referred to as permissible under certain circumstances is presumed to include boating, fishing, picnicking on shores, etc. The Association takes the position that water supply reservoirs should not be used for recreation activities unless there are no other possibilities in the area. Essentially this is a weak statement. It admits that recreation on reservoirs is not so serious a threat to water quality that it must be prevented at all costs. It recognizes that such practices already exist. The only strong statement is the directive against swimming in public water supply reservoirs. In fact, however, a number of reservoirs in the Midwest and West are used for swimming.

The President's Council on Environmental Quality disagrees with the American Water Works Association.[†] It advocates (and attempts to justify) more, not less, recreation, including swimming in larger reservoirs. The council argues that modern water treatment processes are fully adequate to safeguard water supplies.

If your community is served by a reservoir, find out whether recreational activities are permitted in the watershed area or on the reservoir. Do you think it wise to prevent any public use of the reservoir and watershed area? Can you support your opinion? If recreational uses other than swimming are permitted, what precautions would you recommend to go along with such use? That is, what design features would you incorporate in the recreational area? And what activities would you restrict?

---

[*] Recreational Use of Domestic Water Supply Reservoirs. A statement adopted by the Board of Directors on 26 January 1958, and revised 25 January 1965, and 13 June 1971.

[†] President's Council on Environmental Quality, Recreation on Water Supply Reservoirs, September 1975. This pamphlet can be obtained from the U.S. Government Printing Office, Stock No. 041-011-00027-1.

United States to install sand filters to purify their water. (It was 1892, however, before the effectiveness of sand filters was fully recognized.) The filters in Poughkeepsie were still in operation over 100 years later, in 1977.

The effectiveness of the sand filter at removing disease causing organisms was demonstrated dramatically in 1892. Altona and Hamburg, sister cities on the Elbe River in Germany, shared a common boundary, but did not have the same water supply. Hamburg's water was drawn from the Elbe above the two cities. In contrast, Altona drew its water from a site downstream from *both*. Thus, its supply was potentially polluted by the sewage of both cities. To produce water of sufficient quality, Altona began to filter its water supply through beds of sand. Hamburg, however, drew its water directly from the river without any treatment.

The cholera epidemic that struck in Hamburg in the fall of 1892 killed more than 1 citizen in 100 and made nearly twice that number ill. In Altona, the cholera death rate and case rate was less than 1/7 of Hamburg's, and the difference was correctly attributed to the filtration of its water. The cases of cholera that did occur in Altona are explained by noting that many residents of Altona worked in Hamburg. The protection provided by sand filtration was clear, and cities moved quickly in the coming years to install filters in an effort to prevent such epidemics. Hamburg was among the first to take the step.

During the early part of the 20th century, the city of Chicago was discharging its municipal wastes into Lake Michigan. It was also drawing its water supply from the same source via an intake pipe extending into the lake, just as it does today. Not surprisingly, typhoid fever was endemic in Chicago in this era. The city recognized the connection between its high rate of typhoid fever and its water supply, but rather than purify its water supply mechanically by filtration, Chicago chose to divert its sewage from the lake. To accomplish this, the Chicago River, which had emptied naturally into Lake Michigan, was diverted into a canal. The Chicago Drainage Canal, as it is called, was created for the purpose of carrying the river's flow and the sewage of Chicago all the way to the Mississippi River, over 350 miles distant. This practice, of course, polluted the Mississippi, and eventually Chicago was forced to treat its sewage before it could be discharged into the Drainage Canal.

Typhoid and cholera are not the only bacterial diseases transmitted through polluted water (Table 5-1). Paratyphoid and dysentery can also be spread by water. The failure of the water treatment system of Detroit, Michigan led to an epidemic involving 45,000 cases of bacterial dysentery in 1926. Nor are bacteria the only microorganisms that spread disease. A form of dysentery referred to as amoebic dysentery is caused by an amoeba, a one-celled animal living in water. This disease is accompanied by a diarrhea that brings severe weakness. Chicago suffered an epidemic of amoebic dysentery caused by *Entamoeba histolytica*. Visitors to the Chicago World's Fair of 1933 who stayed at two of the city's first-class hotels were the principal victims. The water supply mains of the hotels were discovered to have cross connections with the hotel sewer lines.

Why Drinking Water Is Chlorinated

(Filtration was not the only process devised to purify water. A chemical "sterilization" using chlorine and chlorine compounds was also being investigated in the early part of the century. Experiments with chlorination using calcium hypochlorite (also called bleaching powder) were conducted in the United States in 1896 by George Fuller in Kentucky. In the following year, sodium hypochlorite was successfully introduced to the water supply of Maidstone, England, to arrest a then-occurring epidemic of typhoid. The process was first used on a continuous basis in Middlekirke, Belgium in 1902. In England, the Metropolitan Water Board of London began continuous chlorination of a

TABLE 5-1   Diseases Caused by Water Problems[a]

Waterborne diseases[b]	Water-related diseases	
	Carriers living in or on water	Insufficent supply leading to uncleanliness
Cholera	Clonorchiasis	Ancylostomiasis (hookworm)
Diarrhea, enteritis	Dengue	Ascariasis (roundworm)
Gastroenteritis	Encephalitis	Mycoses
Infectious hepatitis	Filariasis	Scabies
Paratyphoid fever	Malaria	Trachoma
Schistosomiasis	Onchocerciasis	Trichomoniasis
Typhoid fever	Paragonimiasis	Typhus fever
Poliomyelitis	Yellow fever	Dysentery

[a] F. E. McJunkin, "Community Water Supply in Developing Countries: A Quarter-Century of United States Assistance" (a report for the Office of the War on Hunger, U.S. Agency for International Development, under agreement with the Office of International Health, U.S. Public Health Service), Chapel Hill, North Carolina, 1969.

[b] Diseases are also caused by chemically-poisoned water.

portion of its supply two years later, using sodium hypochlorite.)

In 1908, the East Jersey Water Company, a private firm that furnished the water supply of Jersey City, was ordered by the courts to make its water pure and safe. The East Jersey Water Company drew its water supply from the Rockaway River, storing it at a reservoir at Boonton before delivery to the city. The polluted character of the water caused legal action against the company, and the court ordered that the water be made pure. Calcium hypochlorite was introduced in order to eliminate the bacterial contamination, and the court agreed that the bacteria had indeed been destroyed and issued the following opinion:

> From the proofs before me of the constant observations of the effects of this device, I am of the opinion and find that it is an effective process, which destroys in the water the germs, the presence of which is deemed to indicate danger, including the pathogenic germs, so that the water after treatment attains a purity much beyond that attained in water supplies of other municipalities. The reduction and practical elimination of such germs from the water was shown to be substantially continuous....I do therefore find and report that this device is capable of rendering the water delivered to Jersey City pure and wholesome for the purposes for which it is intended, and is effective in removing from the water those dangerous germs which were deemed by the decree to possibly exist therein at certain times.[6]

The adoption of chlorination, now sanctioned by law as a means to purify water, proceeded rapidly.

---

6 George Johnson, "The Purification of Public Water Supplies," U.S. Geological Survey Paper No. 315, 1913.

Today, nearly all municipal water supplies in the United States utilize chlorination to disinfect their water supplies, although liquid chlorine has replaced calcium and sodium hypochlorite. Calcium hypochlorite still remains a valuable tool to employ on a temporary basis for disinfection when disasters such as floods or earthquakes strike, or in times of war. (The United States Public Health Service still tells travellers how to disinfect water by using bleach[7]: four drops of a bleach mixed into each quart or liter of water is sufficient to disinfect water within 30 minutes.)

The use of liquid chlorine instead of calcium and sodium hypochlorite is based upon a variety of reasons. First, it is more effective in destroying bacteria. In addition, it is easily stored in large quantities and its introduction into the water is easily regulated. We will have more to say about the specific process of chlorination when we discuss the current methods of water treatment in more detail.

(Since those early times, chlorination has been properly regarded as one of the most important tools that scientists have to ensure the safety of drinking water. It has prevented illnesses and deaths in untold numbers around the world. Nonetheless, there is growing reason to question its side effects on human health, and we will explore this issue in Chapter 27.)

Virus Diseases Spread by Water

(Viruses, too, can cause waterborne epidemics. Viruses, the simplest of living organisms, cannot reproduce themselves as bacteria can. Instead they invade plant, animal, or bacterial cells and use the reproductive capability of these cells to produce more of their kind. Smaller than even the bacteria, viruses are very difficult to detect in water samples and require highly specialized equipment for detection.)

Our knowledge of the potential for waterborne viral epidemics has come only recently. It is believed that certain polio epidemics have resulted from sewage contaminated water supplies. There are also reports of cases in which viral respiratory diseases were spread by swimming in contaminated pools. (The only proven waterborne epidemics caused by a virus, however, have involved the hepatitis virus. In hepatitis, the victim suffers an inflammation of the liver. The symptoms include a feeling of weakness, nausea, and fever. The liver is unable to perform its normal function of breaking down certain pigments in the blood; these pigments build up in the body and turn the skin and the whites of the eyes a yellow color. This condition is called jaundice.)

Most hepatitis epidemics have occurred in small private water supplies or resulted from the consumption of raw clams or oysters grown in sewage-polluted water. One large scale epidemic, however, is recorded in New Delhi, India, in 1956. In this outbreak, 50,000 people were infected with hepatitis from the city water supply, which had become contaminated with sewage. Interestingly, the water supply had been chlorinated before distribution to the public. During the hepatitis epidemic, no increase in waterborne bacterial illnesses such as cholera or typhoid was observed. (This points up a difference between bacteria and viruses. Bacteria are much more easily killed by chlorine than viruses are, although chlorination is effective against viruses if it is properly applied.)

Waterborne diseases in epidemic proportions have continued to occur in the U.S. through this century, although with a decreasing frequency. The cities struck by these epidemics are typically small and have insufficient treatment of their water supplies. Although such epidemics are becoming more rare in the United States, they continue to occur frequently where water supplies are not treated or where treatment systems fail or where the water distribution systems may become contaminated.

---

7 Health Information for International Travellers, 1979, USDHEW, Public Health Service, Center for Disease Control.

HOW CAN YOU TELL IF WATER IS UNSAFE TO DRINK?

(There are three major classes of disease-causing organisms found in water: bacteria, viruses, and protozoans. The maladies that they spread range from severe diseases, such as typhoid and dysentery, to minor respiratory and skin diseases.)

To examine water in order to isolate all the possible disease-causing organisms in it would be a difficult, costly task. To avoid such exhaustive testing, special tests have been devised. These tests detect the presence of a group of bacteria, which if present, is very likely to be accompanied by disease-producing microorganisms. If this group of bacteria is absent or if the level of these bacteria in a water sample is sufficiently small, it is doubtful that the less numerous disease-causing organisms will be present.

(The bacteria for which scientists test are non-pathogenic; that is, with rare exception, they do not cause disease. These bacteria are found naturally in the intestines of warm-blooded animals, including humans and they are known as "coliforms." Although coliforms rarely cause disease, their presence in water indicates that the water may have been contaminated with raw (untreated) sewage. To drink such water is, obviously, unwise.)

Thus, the procedure of examining water for the presence of disease-causing organisms does not involve looking for the pathogens themselves;(it is sufficient to show that water has in some way been contaminated with sewage or not treated enough to kill all the coliform bacteria in it.) Now let us look again at the careful wording of the court's opinion on chlorination (p. 123). We can see that the judge was referring to the coliforms when he referred to organisms "the presence of which indicate danger."(The coliforms are often called "indicator organisms.")

The federal Environmental Protection Agency has recently drawn up drinking water regulations, setting forth limits to coliform bacteria. Although the standards have actually existed for nearly half a century, they could not be enforced by the federal government unless the water moved in interstate commerce (for example, bottled water).(According to these new regulations, if more than four coliform bacteria are present in a 100 milliliter sample taken from water destined for human consumption, the supplier must notify the state; further, the supplier must take action that will decrease the coliform count to less than one in each 100 milliliters of water.)

(Four coliforms per 100 milliliter sample is the point at which corrective action for drinking water is required. In contrast, water with more than 2300 coliforms per 100 milliliters is considered unsafe to swim in. Boating is allowed in water with up to 10,000 coliforms per 100 milliliters (but don't fall in!).)

Coliform Level	Activity Permitted
1 coliform or fewer per 100 milliliters of water	Water safe for drinking
4 coliforms or more per 100 milliliters of water	State must be notified and corrective action taken
2300 coliforms or fewer per 100 milliliters of water	Swimming is allowed
10,000 coliforms or fewer per 100 milliliters of water	Boating is allowed

In 1969, 969 public water supplies were examined to see how pure the tap water really is in the United States. Somewhat surprisingly, 12% of the water supplies were found to exceed the permissible limits of coliform bacteria. Figure 5-3 shows that the water supply sytems which violated coliform

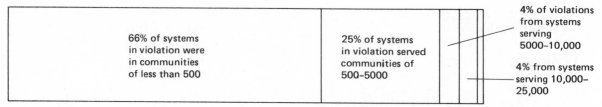

Figure 5-3   *Distribution of violations of coliform standards among communities of various sizes. [Source: J. McCabe et al., "Survey of Community Water Supply System," Journal of the American Water Works Association 62, 670(1970).]*

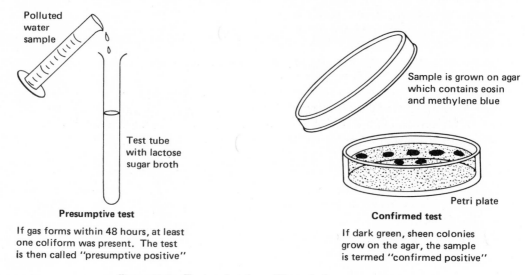

Figure 5-4   *Testing for the coliform indicator organisms.*

standards were generally in smaller communities. For instance, 79 of the 120 systems found in violation were in communities of less than 500 people, and 108 of the 120 systems which failed to meet coliform standards served communities of 5000 people or fewer.

Testing for "Indicator" Organisms

Two methods are in use to test water for the presence of the coliform indicator organisms. In the older tests, water samples are added to a special broth containing lactose sugar (Fig. 5-4). A number of test tubes prepared in this way are set up. Relatively few bacteria besides coliforms are able to use lactose sugar as their energy source. Therefore, if gas forms within 48 hours in the broth (an indication of bacterial activity), it is considered evidence that at

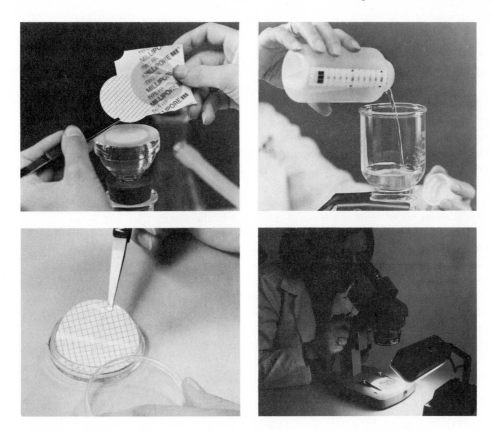

Figure 5-5 *Using the Millipore filter method to identify and count coliform bacteria. (a) A sterile Millipore membrane is placed on a filter which rests in the base of a funnel. (b) A glass top has been added to the funnel base, and the sample of water to be tested is poured into the funnel. The water is drawn through the funnel by applying vacuum suction. Coliform bacteria are too small to pass through the filter and are retained on its surface. (c) The membrane filter is removed from the funnel and placed in a petri dish. The petri dish contains a medium rich in the nutrients which favor the growth of the coliform bacteria. The dish is covered and placed in a constant temperature "incubator" at 35°C for 24 hours. If coliforms were captured by the filtering step, "colonies" of the bacteria will grow up around each bacterium that lodged on the filter. (d) After the 24 hours of incubation, the coliform colonies are counted under a low power microscope. In this photo, the cover of the petri dish has been removed and the bottom of the dish with the medium and colonies has been placed in the microscope field. (Photographs courtesy of the Millipore Corporation, Bedford, Massachusetts.)*

least one coliform organism is present. Such samples are called "presumptive positive." These samples are then grown on a special agar (which is a gelatinlike substance), to which the chemicals eosin and methylene blue have been added. These chemicals cause colonies of coliform bacteria to take on a special appearance: they are dark green and have a "sheen" to them in addition to having a dark center. If coliform colonies appear on the agar, the sample is called "confirmed positive."

There is a simpler method now in use, which gives information more directly and more quickly than the test tube procedure. A measured sample of water is filterd through a special filter membrane. The membrane is made by the Millipore Company and is called a millipore filter. This membrane is manufactured with pores or holes so minute in size that bacteria cannot pass through them, although water can. The filter itself, on which the bacteria are retained, is then added to a medium containing the foods that coliform bacteria need to grow. On that medium (a nutrient mixture), each coliform bacterium captured on the membrane filter divides and forms a colony of bacteria. These colonies are visible to the naked eye; they can be identified and counted in order to estimate the level of pollution in the water. (See Fig. 5-5.)

Determining the presence of viruses in water is not as simple as detecting the presence of bacteria. Most viruses are too small to be trapped by a filter and cannot be grown in a simple broth. However, just as water is not routinely examined for the disease-causing bacteria themselves, water is not commonly checked for disease-causing viruses either. Again, the presence of coliform bacteria in a water sample is taken as evidence that disease-causing viruses as well as bacteria are likely to be present in the water.

Bacteria and viruses are not the only kinds of pollutants that occur in water, however. Poisonous metals and a variety of organic chemicals can also pollute water supplies. These problems are the subjects of the next sections.

Questions

1. Name three classes of organisms that can affect the safety of a water supply for human consumption. Give examples of each class.

2. The spread of certain diseases to epidemic levels can be described in terms of a "cycle." Describe the role of water in the spread of diseases such as typhoid and cholera.

3. The engineering solution to the spread of infectious disease has been to break this cycle. List several ways in which this can be accomplished.

4. What is the function of sand filtration in supplying water for human consumption? Does this process alone ensure safety against infectious disease?

5. What is the function of chlorination in supplying water for human consumption?

6. Explain what is meant by "indicator organism." Discuss the role of coliform bacteria in monitoring water quality.

7. Describe briefly two methods used in testing a water sample for the presence of coliform indicator organisms. Are these methods suitable tests for virus contamination? Explain your answer.

Further Reading

Drinking Water and Health, by the Safe Drinking Water Committee, National Academy of Sciences, Washington, D.C. 1977.

As part of the Safe Drinking Water Act of 1974, Congress directed that a study be made of the possible contaminants in drinking water and their effects of human health. This publication is the result of the definitive and comprehensive study which was undertaken by the National Academy of Sciences at the request of Congress.

National Interim Primary Drinking Water Regulations, from the Office of Water Supply, Environmental Protection Agency, EPA 570/9–76–003, 1976.

This document, which replaces the Public Health Service Drinking Water Standards of 1962, contains the standards for all contaminants that influence the safety of water supplies. The publication contains, in addition to the standards, background material which explains the basis for the values specified for the standards. Contaminants that do not affect safety but that influence appearance and palatability are not listed here. A separate document, Secondary Drinking Water Regulations, will eventually be issued to address standards for these contaminants. The standards for contaminants that are not safety related are, for the time being, those listed in the Public Health Service Drinking Water Standards. The drinking water standards provided in this document became effective in 1977, but will probably be modified, in time.

Sedgwick's Principles of Sanitary Science and Public Health, rewritten and enlarged by S. Prescott and M. Horwood, Macmillan Company, New York, 1948.

The classic public health text. It is easy reading, and especially absorbing for those interested in the historical roots of the enviromental/public health movement.

Standard Methods for the Examination of Water and Waste Water, prepared and published jointly by the American Public Health Association, American Water Works Association, Water Pollution Control Federation, 1975, 13th ed. Available from the American Public Health Association, 1790 Broadway, New York, N.Y. 10019.

This book provides step-by-step descriptions of all the laboratory tests which are used to assess the quality of water and wastewater. The tests "are the best available and the generally accepted procedures for the analysis of water and wastewater."

Chapter 6

Chemicals in Drinking Water

DRINKING WATER: CHEMICAL STANDARDS

Unwanted chemicals find their way into natural waters from many sources. To a young nation, with an apparently limitless supply of fresh water, it seemed reasonable to dispose of industrial waste chemicals into waterways. Mining wastes have been allowed to leach or flow unchecked into nearby rivers. Pesticides and other agricultural wastes are washed by rain into nearby rivers and lakes. Rain also washes a variety of contaminants from the air into water. And in the winter, the salting of icy roads leaves chemicals that will wash into the nearest watercourse at the first thaw (Fig. 6–1).

Inevitably, some of these chemicals find their way into lakes or rivers or wells from which drinking water is taken. In order to protect consumers, the Environmental Protection Agency sets standards limiting the levels of hazardous chemicals in drinking water (Table 6–1). Any water source exceeding these limits is not supposed to be used for drinking water. The standards, and how well this system works, are explained on p. 154.

First, however, we shall discuss some of the chemical water contaminants for which EPA has set standards. *Mercury, cadmium,* and *nitrates* are examples of *inorganic chemicals* that contaminate water supplies. Mercury and cadmium are metals that are poisonous at relatively low levels to all humans and animals. Nitrates on the other hand, are poisonous only to infants at levels found in contaminated drinking waters. *Arsenic* and *lead* are also listed in the table as inorganic chemicals found in drinking water. Because it appears to cause cancer, arsenic is discussed in Chapter 27. Lead, which is a common air pollutant from leaded gasoline, is covered in Chapter 18. *Iron* and *manganese* may also contaminate waters that serve as community water sources. In most cases, high levels of these metals are a result of *acid mine drainage* (see p. 267).

Phthalates, PCBs, and *pesticides,* which are discussed next, are typical of the *organic chemicals* that find their way into drinking waters. All are, or were, widely used chemicals. Water contamination, as well as contamination of air and food, is the result of this widespread use. Another important category of organic chemicals in drinking water is known as *chlorinated hydrocarbons.*[1] Evidence is accumulating that some of these possibly carcinogenic chemicals are actually formed by a water purification procedure, chlorination. This especially difficult problem is covered in Chapter 27. The following sections detail a few typical chemical drinking water problems.

---

1 There are many different chlorinated hydrocarbons. Some have been widely used as pesticides; DDT is one of these.

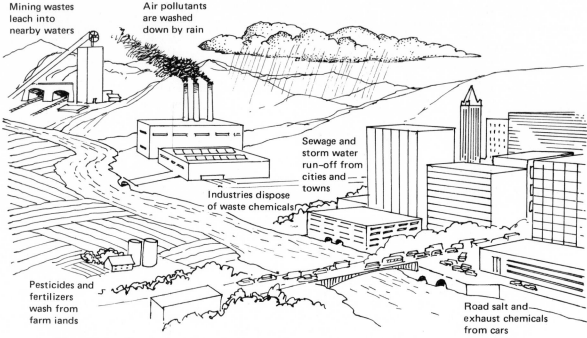

Figure 6–1   *Chemical water pollutants come from a variety of sources. In fact, chemical pollutants usually end up in water, regardless of where they are disposed of first. Air pollutants are washed down by rain. Chemicals disposed of in landfills leach into nearby waters. And agricultural chemicals run off farm land into lakes and rivers.*

TABLE 6–1   **National Interim Primary Drinking Water Standards (1977)** [a]

Chemical	Maximum allowable level (milligrams per liter)	
Inorganic Chemicals		
Arsenic	0.05	
Barium	1.00	(mainly a problem in agriculture, it is toxic to plants)
Cadmium	0.010	
Chromium	0.05	
Lead	0.05	(see p. 389 for a discussion of this contaminant)
Mercury	0.002	(no standard had been set previously)
Nitrate	10.00	(this standard is reduced from 45 mg per liter)
Selenium	0.01	
Silver	0.05	

TABLE 6–1   (Cont.)

Chemical	Maximum allowable level (milligrams per liter)	
Fluorides	1.4–2.4	(depending on the average daily maximum air temperature)
Organic Chemicals[b]		
Chlorinated hydrocarbon pesticides		
Endrin	0.0002	
Lindane	0.004	
Methoxychlor	0.1	
Toxaphene	0.005	
Chlorophenoxy pesticides		
2, 4-D	0.1	
2,4,5–TD	0.01	
Radioactivity (see p. 566)		
Turbidity		the presence of particles in the water, or turbidity, is both esthetically undesirable and also can interfere with disinfection since particles can combine with chlorine and use it up. The standard is set at a maximum of one turbidity unit.[c]
Microbiological contaminants		no more than four coliform bacteria per 100 milliliters may be found when the water is sampled (see p. 125 for a discussion of this standard and what it means).

[a] National Interim Primary Drinking Water Standards, Environmental Protection Agency, Office of Water Supply, EPA 570/9–76–003.

[b] Although the Safe Drinking Water Act was passed in large part due to public concern about organic chemicals like chloroform in drinking water, it was finally decided that there was no good, simple method available to test for them. For this reason, no standard was set for many organic chemicals. Rather the EPA will do expensive and sophisticated tests on samples from certain water supplies to keep track of levels found, in a general way, and will also require certain water supplies to treat their water to remove organic chemicals. At the same time research is being carried out to find a simple test for the chemicals about which there is the most concern. More on this in Chapter 27.

[c] For an explanation of this term see Standard Methods for the Examination of Water and Waste Water, American Public Health Association, American Water Works Association and the Water Pollution Control Federation, Thirteenth Edition. Available for the American Public Health Association, 1790 Broadway, New York, N.Y. 10019.

THE MERCURY PROBLEM

A Strange Disease

In 1953, people who lived in the region around Minimata Bay in Japan began to suffer from a mysterious nervous disease. The symptoms included a narrowing of their field of vision and a lack of coordination. One unusual aspect of the epidemic was that animals and birds seemed affected as well as people. Because of this, public health officials were led to suspect that an environmental poison was causing the epidemic. The symptoms were, in fact, those of mercury poisoning.

Mercury has long been known to be a poison. The expression "mad as a hatter" originated in the times when many workers in the felt-hat industry, who were exposed to high mercury levels in their work, suffered from mental problems. In mild cases, mercury causes symptoms that mimic mental and emotional disorders: insomnia, inability to accept criticism, fear, headache, depression, and generally exaggerated emotional responses.

In Minimata, some 120 people were affected and 46 died before it was realized that people and animals were being poisoned by eating mercury-contaminated fish and shellfish from the bay. The origin of the mercury was a plastics factory located on a stream leading into Minimata Bay. Although mercury is toxic to fish as well as people, mercury levels in the waters were not high enough to prevent fish and shellfish from living there. Two natural processes changed the factory's waste product from a trace chemical in the environment to an epidemic-causing pollutant.

In the first place, biological magnification increased mercury concentrations in fish and shellfish to levels many times greater than those found in the bay waters. In the second place, a transformation of the mercury itself was taking place. There are several forms of mercury. Elemental mercury (used in thermometers) and inorganic salts of mercury (for instance, mercuric chloride) are excreted relatively quickly from the body. Much more poisonous are the alkyl mercury compounds, like methyl mercury and ethyl mercury. These compounds are excreted very slowly from the body, perhaps only 1% of the total amount present is removed each day. Although a great deal of the mercury that finds its way into natural waters is in the form of inorganic mercury, the mercury found in fish is almost always the more toxic methyl mercury. Studies have shown that bacteria in the bottom muds of lakes and rivers and in the slime on fish themselves are capable of transforming inorganic mercury to methyl mercury.

Some of the mercury released into Minimata Bay was in the form of methyl mercury, but a good deal more methyl mercury was apparently formed by bacteria in the bottom muds. Fish and shellfish in the bay concentrated the methyl mercury to levels that were toxic to humans who ate the seafood.

The possibility of these two processes, transformation of a substance in the environment and concentration or magnification by living organisms, must always be taken into account when a decision is made about the hazard of a particular chemical.

Sources of Mercury in the Environment

Mercury in water   Mercury enters natural waters from many sources. The English River in Ontario, Canada, has been contaminated with mercury discharged from a chlorine-caustic soda plant. It has been reported that mercury levels in fish caught in the river are as high as in fish from Minimata Bay. Furthermore, cats fed on fish from the river, have been found to have symptoms of mercury poisoning. Since fish from the river form a major portion of the diet of Indians living in two nearby reservations another Minimata-type epidemic may soon be revealed. (See Controversy.)

In the U.S., it is estimated that chlorine-caustic soda plants released 1/4 to 1/2 pound of mercury for each ton of caustic soda (sodium hydroxide)

MINIMATA REVISITED?

In Dryden, Ontario, there is a chlorine-caustic soda plant that discharges 3.3 g of mercury per metric ton of chlorine produced. Between 1962 and 1970, 9000 to 11,000 kilograms of mercury were discharged into the Wabigoon River on which the plant is located.

Fish caught in this river system show mercury levels comparable to those found in Minimata Bay. Further, two cats fed on fish from the river were found to have symptoms of Minimata Disease, or mercury poisoning. Concern has been voiced because there are two Ojibway reservations nearby, the Grassy Narrows and the White Dog Reserves on the lower English River, which is part of the same river system. The Indians eat a diet high in fish. In fact, government studies have found that many of the Indians have excessive mercury levels in their blood. In addition, a Japanese group that came to study the Indians found signs of Minimata-like mercury poisoning, such as visual problems and lack of coordination.

One group of scientists has written:

(V)ery little has been published even though 26% of the 110 individuals tested at Grassy Narrows had mercury blood concentrations in excess of 100 ppb. * One individual's mercury blood level was reported to be 385 ppb, while the mean value for the group was about 70 ppb.... Furthermore, since the highly contaminated fish were discovered in 1969, no provincial or federal government agency has attempted to establish, by conducting autopsies on Indians who died of unknown causes at either Grassy Narrows or White Dog Reserves, whether either chemical or pathological evidence of Minimata Disease existed. Since cats develop the symptoms of methylmercury poisoning more readily than human beings, they serve as indicator organisms for this disease. Thus, the autopsy evidence of Minimata Disease in two domestic cats fed mercury-contaminated fish from the lower English River in the vicinity of the Grassy Narrows and White Dog Reserves is a significant indication of the hazard to Indians and others who also eat fish from the same waters. [†]

Why do you think so little has been published or done about this problem? What steps could the government take? What effects would these steps have on native culture?

---

* Humans generally begin to show signs of mercury poisoning when blood levels reach 20 parts per billion.

† T. Takeuchi, F. M. D'Itri, P. V. Fischer, C. S. Annett, and M. Okabe, "The Outbreak of Minimata Disease (Methyl Mercury Poisoning) in Cats on Northwestern Ontario Reserves," Environmental Research **13,** 215 (1977).

produced until the early 1970s. Today, strict U.S. laws prohibit the discharge of mercury by industry. However, in areas where mercury was formerly discharged, for instance near pulp and paper mills or chlorine-caustic soda plants, mercury in the bottom mud often still contaminates the water and the organisms living there (Table 6–2).

Lakes and rivers may take from 10 to 100 years to cleanse themselves of mercury after all additions of mercury have stopped. Fishing restrictions have been set in many states because fish are concentrating mercury dumped into the water many years ago. The following story is in many ways typical of the manner in which mercury contaminates water sources and aquatic life.

From the 1930s to about 1950 DuPont manufactured acetate, using mercuric sulfate, in a plant at Waynesboro, Virginia. The plant was located on the South Fork of the Shenandoah River. It appears that mercury, spilled from the plant over 25 years ago in the 1930s and 1940s, is contaminating the river. Writing in *Science*, Luther J. Carter noted:

> Not all of the Shenandoah River is contaminated, only the South Fork, which many regard as the best of it....
>
> For the canoeist or the float fisherman (the South Fork is famed for its small-mouth bass fishing), the scene is ever-changing but is always good and sometimes spectacular, especially when the winding river turns toward the steeply rising slopes of the Massanutten. Along with the rest of the Shenandoah, the South Fork has long been a prime candidate for consideration as part of the national system of wild and scenic rivers...
>
> Although the Shenandoah was known to have some water quality problems, especially overfertilization from the runoff from farmland and other sources, it was not until this spring that state officials got word that part of the river might be heavily polluted with mercury.[2]

Seeping from the abandoned plant site, the mercury seems to have settled into crevices and nooks on the limestone bottom of the river, and is slowly dissolving into the water. Bass caught more than 77 miles downstream from the plant contain over twice the legal limit for mercury.

Mercury in air Mercury is not only a water pollutant, but also an air pollutant from sources such as coal-fired power plants and mercury ore refining plants. The fossil fuels, coal and oil, contain mercury, which is released to the air when they are burned. It has been estimated that some 5000 tons of mercury may be added to the air each year from fossil fuels. A significant amount of this airborne mercury is washed out of the air by rainwater.

The mercury level in the air in cities tends to be much higher than in rural areas, perhaps due to the greater use of fossil fuels in cities. For instance, in rural areas, mercury levels of 0.003 to 0.009 micrograms per cubic meter of air have been measured.

2 Luther J. Carter, *Science* **198**, 1015 (9 December 1977).

TABLE 6–2 **Mercury Found in Organisms Living Upstream and Downstream from a Paper Mill** [a]

	Living in stream above the paper mill	Living in stream below the paper mill
Sowbug	65 parts per million	1900 parts per million
Burrowing alderfly	49 parts per million	5500 parts per million
Caddis fly	—	1700 parts per million

[a] A. G. Johnels *et al.*, *Oikos* **18** (2), 323 (1967).

In Chicago, levels average 0.01 micrograms per cubic meter and in New York City values of 7 to 14 micrograms per cubic meter have been found. Mercury ore processing plants and chloralkalai plants have added significant amounts of mercury to the air in the past. However, new standards are reducing air pollution from these sources to much lower levels.

Mercury in industrial and consumer products
(About six million pounds of mercury are used each year in the U.S. Mercury is used in electrical devices, thermometers, fungicides, dental fillings, drugs, and paints. Although three quarters of this mercury could be recycled, at least half is not recycled. That is, it finds its way into the environment and eventually into natural waters.)

(Mercury could be called a "permanent" pollutant. That is to say, once released to the environment, it appears to be cycled from the air to water, to organisms living in the water, to human food supplies, and perhaps to humans themselves, in seemingly endless cycles. Many years go by before environmental mercury becomes covered with such thick layers of mud at the bottom of lakes or the ocean that it becomes harmless (Fig. 6-2).)

Closer to home, mercury used as a preservative in latex paints may be a hazard. Tests in houses recently painted with latex paints show that mercury levels in the air can remain for at least a week at five times the level considered safe (Table 6-3). The same study shows that broken thermometers can also be a local source of high mercury levels if the mercury is not cleaned up properly. Spilled mercury should be covered with powdered sulfur and then vacuumed up. The area can then also be coated with commercial hairspray.

(Because of its toxicity and tendency to accumulate in living organisms, the standard for mercury in drinking water is set at 0.002 milligrams per liter. Fish with more than 0.5 parts per million mercury are considered unsafe for human consumption)

TABLE 6-3   Mercury Concentrations Inside Buildings[a]

Location	Date	Mercury concentration (ng/m³)[b]	Remarks
Home 1, study	9 Dec. 71	68.2	Home 1 painted with latex-base paint, March 68
Home 1, bedroom 1	9 Dec. 71	66.5	
Home 1, living room	9 Dec. 71	69.0	
Home 1, den	9 Dec. 71	70.0	
Home 1, bedroom 2	9 Nov. 71	139	Repainted with latex-base paint, July 71
Home 1, bedroom 2	9 Dec. 71	130	
Home 1, bedroom 3	9 Dec. 71	92	
Home 1, bedroom 4	9 Nov. 71	78	
Home 2, living room	9 Dec. 71	164	Painted with latex-base paint, March 71
Home 3, bedroom	9 Dec. 71	159	Painted with latex-base paint, July 71
Home 4, living room	9 Dec. 71	148	
Home 4, bathroom	9 Dec. 71	141	
Home 5, bedroom	10 Dec. 71	103	

TABLE 6–3   (Cont.)

Location	Date	Mercury concentration (ng/m^8) [b]	Remarks
Home 6, bedroom	10 Dec. 71	262	New home, painted with latex-base paint 30 days before
Home 7, living room	29 Dec. 71	1560	New home, painted with latex-paint 7 days before
Home 7, bedroom 1	29 Dec. 71	1560	New home, painted with latex-base paint 7 days before
Home 7, bedroom 2	29 Dec. 71	3070	
Home 8, bedroom	9 Nov. 71	12.9	Home paneled
Home 8, kitchen	9 Nov. 71	5.0	
Office building 1, room 1	9 Dec. 71	172	Painted with latex-base paint, June 70
Office building 1, room 2	9 Dec. 71	245	Painted with latex-base paint, June 70
Office building 1, room 3	9 Dec. 71	203	Painted with latex-base paint, June 71
Office building 1, room 4	9 Dec. 71	116	Paneled office
Office building 2, room 1	17 Dec. 71	183	
Doctor's room 1	13 Dec. 71	4950	Hg thermometer broken in the past
Doctor's room 2	13 Dec. 71	5680	Hg thermometer broken in the past
Doctor's room 3	13 Dec. 71	4550	Hg thermometer broken in the past
Dentist 1, room 1	23 Dec. 71	5550	Mixing area for Ag amalgam
Dentist 1, room 2	23 Dec. 71	5030	
Dentist 1, room 3	23 Dec. 71	4770	
Dentist 2, room 1	28 Dec. 71	1295	Inactive for previous 4 days
Dentist 2, room 2	28 Dec. 71	1135	
Dentist 2, room 3	28 Dec. 71	1160	
Hospital laboratory	13 Dec. 71	307	
Hospital ward	13 Dec. 71	336	
Chemical building, laboratory 1	14 Dec. 71	930	At laboratory sink
Chemical building, laboratory 2	14 Dec. 71	592	At laboratory desks
Chemical building, office	14 Dec. 71	398	Office away from laboratory
Washington, D.C.	2 Feb. 72	3.25	Potomac River at Key Bridge
San Francisco	19 Jan. 72	3.14	8 km south of San Francisco on the beach
Dallas	11 Jan. 72	3.38	16 km southwest of Dallas

[a] From R. S. Foote, *Science* **177** 513 (11 August 1972).

[b] Some recommended maximum levels for exposure to mercury in the air are: 40 hour work week, 50 micrograms per cubic meter; 24 hour exposure, 300 nanograms per cubic meter. A nanogram (ng) is 1/1000 of a microgram.

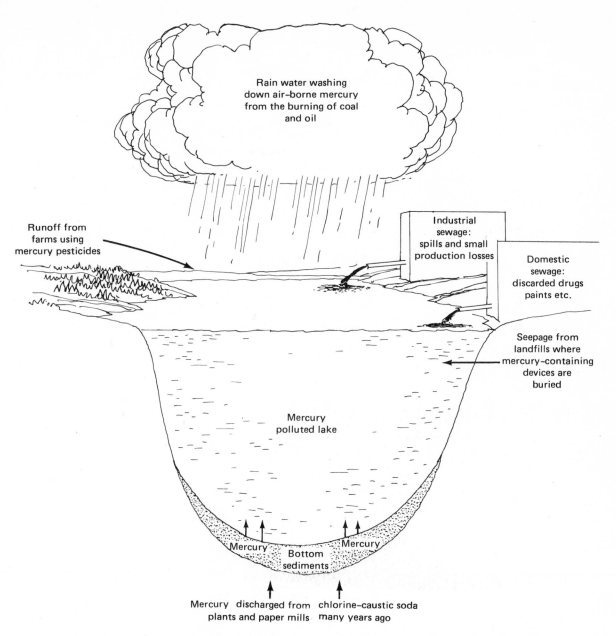

Figure 6-2 *Mercury pollution of a lake. In this imaginary lake, mercury enters the water from industrial and domestic sewage and in seepage from landfills where mercury batteries, switches, and other devices are buried. Rainwater washes mercury-containing pesticides into the lake and also washes down mercury in the air from the burning of fossil fuels. Furthermore, mercury washed into the lake many years ago has settled into the bottom muds where it is slowly being converted to toxic methyl mercury by bacteria and is then entering food chains. Unfortunately, although this particular lake is imaginary, the processes shown are not. Examples of all of them can be found in many areas of the country.*

139

CADMIUM POLLUTION

Environmental Sources

In the Fuchu area of Japan, an unusual disease is found. (It is called *itai-itai*, or "ouch-ouch" disease, because it is so painful.) The victims' bones become brittle and break easily. Even walking can be painful. The cause of the disease has been traced to contamination of the Jintsu River, the area's water supply, with waste from the Kamioka mine. (Cadmium is believed to be the contaminant responsible for the disease.) Cadmium levels in the rice grown in the district are at least three times higher than in unpolluted areas. In several other prefectures in Japan, cadmium from mines or refineries pollutes water supplies.

The natural occurrence of cadmium in water in more than minute amounts is almost unknown. In the past, detectable levels were usually the result of contamination from mining or industrial wastes. Cadmium is used in the manufacture of such products as paints, alloys, light bulbs, pesticides, and nuclear reactor parts. Wastes from electroplating plants have contaminated ground water in the United States. Experts now worry that cadmium-containing products burned at dumps release cadmium to the air, while those buried in landfills are contaminating ground waters.

Cadmium's Toxic Effects

(The main effect of cadmium in the body appears to be on the kidneys, where it accumulates. In addition to kidney disease, cadmium may contribute to high blood pressure.)

In fish, cadmium accumulates in the liver, kidneys, and other organs. The flesh is usually low in cadmium. Products made from whole fish, however, can have quite high levels of cadmium (fish paste has been found to contain up to 5.3 parts per million). Shellfish may also accumulate cadmium. So do wheat and rice.

Smokers seem to face a special risk from cadmium, which is found in tobacco products (1.5 to 2

micrograms per cigarette). Autopsies on smokers have shown twice the amount of cadmium in the kidneys, liver, and lungs of smokers as compared to nonsmokers. The topic of chemicals in cigarette smoke is covered more thoroughly in Chapter 27.

Once a dose of cadmium is absorbed by the body, either from the digestive tract or from the lungs, an extremely long time may pass before it is excreted again. (It may take 10 to 30 years before one-half of the cadmium absorbed is excreted by the body.) Some scientists feel that cadmium levels found in the kidneys of the general population in the U.S. already approach half of the level known to be toxic. They express concern that increasing environmental levels will endanger health.

(Cadmium may be removed by water-softening treatments used at drinking water plants. On the other hand, drinking waters that are relatively acidic and high in oxygen content may corrode pipes easily and pick up cadmium after leaving the treatment plant. The Environmental Protection Agency has set a limit of 0.010 milligrams per liter for cadmium in drinking water.)

NITRATES IN DRINKING WATER

A Special Hazard to Infants

In the late 1940s it was noticed that certain infants, rushed to the hospital because they were listless and suffering from a blueish coloration of the skin, would recover spontaneously while they were in the hospital. Strangely, they would turn blue again when they returned home. These babies lived in rural areas and were fed on formulas made with well water. Doctors became suspicious that something in well water was making the babies ill.

For instance, an eight-week-old baby who lived on a farm in northwestern Iowa was brought to the hospital because he was listless and a blueish-gray color. After a few days in the hospital, he seemed well and was released. Nine days at home caused the return of the blue color. This time after a three day hospital stay, the parents were told not to use their own well water to feed the child. After six weeks, during which time the baby was normal, the parents decided to try the well water again. In one week the blue color returned. The parents, as well as the doctors, were now convinced that the well water was the cause of the baby's illness.

(Thousands of similar cases have been reported and are now known to be caused by nitrates. Only infants seem to be endangered. Parents and older children drinking the same water are unaffected. Nitrates contain the chemical grouping NO_3^-. Water that has more than 10 milligrams per liter of nitrate is considered unsafe for drinking, mainly because it is likely to be poisonous to infants.)

Why are only infants affected? It appears to be due to the fact that in some infants the stomach does not yet produce enough acid to prevent the growth of bacteria that convert nitrates (NO_3^-) to the highly poisonous nitrites (NO_2^-). (If you want to know more about why nitrites are poisonous, see below.)In older children and adults, who produce the proper amount of acid, bacteria that convert nitrates into nitrites cannot grow in the stomach.

Sources of Nitrate Contamination

In rural areas, poorly built wells can be contaminated by nitrates from nearby septic systems.

Wells have also been contaminated by fertilizers that have back-siphoned into the wells during spraying or diluting of the fertilizers. In some areas, well water is naturally high in nitrates. In other places, ground water has become contaminated with nitrates from industrial or agricultural sources. In parts of Israel, for instance, some wells are increasing in nitrate concentration by 2 milligrams or more per liter per year. Plans for recharging ground water supplies with treated waste water, which is high in nitrates, will increase this problem.

There is some evidence that Vitamin C, from orange or tomato juice, can prevent nitrite poisoning in infants, perhaps by combining with the nitrites. Actually, however, only a part of the nitrates we receive come from water. Nitrates (NO_3^-) are also found in many vegetables and both nitrates (NO_3^-) and nitrites (NO_2^-) are added to cured meats, such as hot dogs and bologna. This is discussed in Chapter 28.

PCBs: ORGANIC CHEMICAL WATER POLLUTANTS

What are PCBs?

In 1968, health officials became aware of the spread of yet another curious disease in Japan. Vic-

Why Nitrites Are Poisonous

Nitrite (NO_2^-) is an oxidizing agent. It can change the iron atom in the blood protein, hemoglobin, from Fe^{2+} to Fe^{3+}. The changed hemoglobin is called methemoglobin. Hemoglobin gives blood its red color and carries oxygen from the lungs to all other parts of the body. Hemoglobin can only do this if the iron atoms tucked into the center of the heme groups are Fe^{2+} rather than Fe^{3+}. If enough hemoglobin is changed from hemoglobin (Fe^{2+}) to methemoglobin (Fe^{3+}), the victim will suffocate.

tims developed acne and brown blotches on their skin. In addition, they often complained of severe weakness and eye discharges. The disease was at first named Yusho, or rice-oil disease, because it seemed to result from eating a certain brand of rice oil. Japanese public health detectives finally discovered that the rice oil was contaminated with PCBs (polychlorinated biphenyls). But before the cause was found, over 2000 people were affected and 16 died.

The abbreviation, PCBs, actually refers to a whole family of similar chemicals. Because they do not burn readily, they are widely used to transfer heat from one material to another, as in rice oil manufacture. The Japanese tragedy was caused by leaking pipes, which allowed PCBs to flow into the rice oil.

PCBs are used in electrical devices, such as transformers and capacitors, where a spark could ignite a flammable material. PCBs have also been used in a wide variety of other ways: as solvents for paint and ink, in pesticide sprays, and in plastics. In fact, so much PCB has been manufactured and used, we now find small amounts of contaminants in the water all over the world (Fig. 6–3). Like mercury, PCBs are concentrated by fish and other water-dwellers (see Fig. 6–4). (Even people retain PCBs. Some 40% to 45% of the American public has 1 part per million or more PCBs in their fatty tissues.[3])

Although the effect of large amounts of PCBs are known (in the Japanese epidemic, people ate an average of two grams of PCBs in a short time), the effect small amounts of PCBs may have over a long period of time, is not as clear. People exposed to PCBs at work have developed nerve, skin, and liver ailments. In the laboratory, PCBs cause birth

---

3 This is mainly a result of PCBs in the food we eat. More on PCBs in food on p. 148.

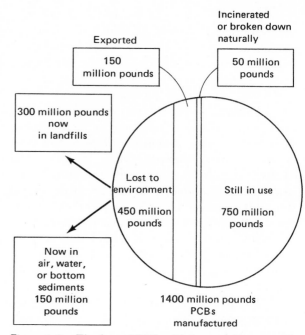

Figure 6–3 *The fate of PCBs. Of the 1.4 billion pounds of PCBs manufactured since 1929, some 450 million pounds are now in the environment. Another 750 million pounds are still in use. Some of what is in use can be expected to enter the environment. Since PCBs are broken down very slowly, environmental levels of PCBs will probably decrease very slowly.*

defects and cancers in animals (possibly at levels as low as 100 to 300 parts per million in their diet).

PCBs in the Environment

By 1970, there was enough concern about the contamination of the environment with PCBs that the only U.S. manufacturer, Monsanto Chemical Company, announced that it would sell PCBs only for use in closed systems, that is, systems in which no PCBs would be released to the environment. Allowable uses included the manufacture of devices such as capacitors and transformers but did not include PCBs in paints, inks, or plastics.

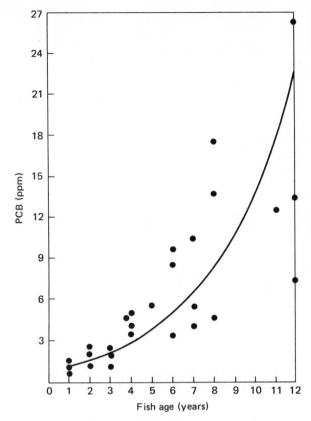

Figure 6-4   *The concentration of PCBs in Cayuga Lake trout as a function of age. [Source: C. A. Bache et al., Science **177**, 1192 (29 September 1972) Copyright 1972 by the American Association for the Advancement of Science.]*

Studies soon after these restrictions went into effect showed PCB concentrations in some rivers to be dropping. For instance, in rivers flowing into Green Bay, Wisconsin, levels dropped from between 0.15 and 0.45 parts per billion to almost nothing. Many people heaved a sigh of relief—unfortunately, prematurely. In August 1975, the New York State Commissioner of Environmental Conservation issued a warning about eating salmon and bass from the Hudson River or Lake Ontario, because the fish had been found to contain well over the five parts per million of PCBs that is considered safe. Headlines collected from September 1975 through January 1977 provide a chronicle of an environmental disaster.

Chronicle of an Environmental Disaster

STATE ISSUES A WARNING ON TOXIC CHEMICALS IN STRIPED BASS AND SALMON

Federal researchers sampling fish for toxic materials notified Mr. Reid, the New York State Commissioner of Environmental Conservation, that striped bass and salmon taken from the Hudson River and Lake Ontario had as much as 37 parts per million PCB. He issued a warning to consumers not to eat these fish and began a state study of the contamination.

The New York Times, 8 September 1975

WARNING IGNORED ON STRIPED BASS–FISH IS STILL BEING ORDERED HERE DESPITE REPORT OF A TOXIC PERIL

"I had striped bass for lunch yesterday," said Andrew Garzel co-owner of Sweets (a Manhattan restaurant). "It was delicious. I'm going to have more. Our bass come from Provincetown, Massachusetts and what happens in the Hudson doesn't affect the fish we serve here."

<u>The New York Times</u>, 9 September 1975

PCB IN FISH STIRS NEW STATE STUDY–REID REPORTS HIGH LEVELS OF THE TOXIC CHEMICAL IN 3 MORE WATERWAYS

Further study showed more rivers, including the St. Lawrence, to be contaminated with PCBs. A wide variety of fish from these waterways contained over the allowable limit of 5 parts per million PCBs. However, fish from the Hudson River still had the distinction of having higher PCB levels than fish from anywhere else in the country.

<u>The New York Times</u>, 16 September 1975

Should the diners have paid more attention to the warning? In fact bass tagged in the Hudson River have been caught off the Massachusetts coast. The Hudson River is a major spawning ground for the entire East coast from Massachusetts to Delaware. Thus, pollution of the Hudson has widespread effects.

PCBs at diet levels as low as 2.5 ppm have been shown to cause reproductive problems and loss of hair in monkeys. They are extremely toxic to fish eggs as well as other aquatic creatures. Water containing 10 to 100 parts per billion PCBs is fatal to some species. It is also suspected that PCBs may be carcinogenic.

Where were the PCBs coming from? A check of manufacturing plants on the river showed that two General Electric Plants, making capacitors, had permits from the federal government to discharge PCBs into the river. Further investigation showed that General Electric had been discharging PCBs into the Hudson for almost 20 years, possibly as much as 48 pounds per day.

Meanwhile, the U.S. Environmental Protection Agency announced that at least 12 other plants in the United States were also dumping PCBs into waterways or municipal sewage systems. Russel

G.E. TESTIFIES ON PCB–DISCHARGE CUTS

The New York State Department of Environmental Conservation ordered G.E. to stop discharging any PCBs into the Hudson by the following September. However, G.E. officials testified at hearings held by the Department of Environmental Conservation that they could not reduce their discharge, now estimated to be down to two pounds per day, below three and one half ounces a day. The Department charged G.E. with violating New York State water quality standards with these discharges.

The New York Times, 10 December 1975

Train, then EPA Administrator said that PCBs "are present in our environment to a far greater degree and at higher levels than we have previously thought." He said that PCBs were detected in the drinking waters of cities all over the country.

As a result of the hearings held by the New York State Department of Environmental Conservation, G.E. was judged guilty of violating state water quality laws. The Chairman of the hearings called the situation a regulatory agency failure as well as a corporate abuse, since G.E. made no secret of what it was doing from the responsible Federal and State agencies.

Some 500,000 pounds of PCBs are still in the Hudson River, 80% within 10 miles of the two G.E. plants. Between 5000 and 10,000 pounds of PCBs move down the river at 4 to 10 miles per year. In February of 1976, the Hudson was closed to most commercial fishing (Fig. 6–5) and in 1971, PCBs found in Lake Michigan fish closed that lake to commercial fishing. The Wisconsin Department of Natural Resources is still waiting for the PCBs to degrade so the fisheries can resume, but the department admits it may take decades.

In New York, General Electric had clearly violated state water quality laws. In October 1976, the state and G.E. agreed to a compromise settlement in which the company contributed three million dollars to a six-million-dollar Hudson reclamation program and another million to a research and testing progam for PCB substitutes.

The Aftermath

(The United States Congress passed an amendment to the 1976 Toxic Substances Control Act,[4] which banned the manufacture of PCBs in the U.S. after January 1979. The Environmental Protection Agency followed through by forbidding any discharge of PCBs into U.S. waters. In addition, rules were made to phase out the use of PCBs, even in totally closed systems.)

But is this really the final note? PCBs have proved to be long-lived in the environment. Early proposals to clean up the Hudson, by dredging up PCB contaminated mud and burning off the PCBs, have been abandoned as "totally impractical." No one, at the moment, seems to know what to do about the Hudson. The then administrator of the EPA, Russel Train, said on January 19, 1976: "The present problem of PCB contamination in the environment is so severe that in many waters

---

4 The act is discussed on p. 613.

DO ENVIRONMENTAL IMPROVEMENTS ALWAYS COST MORE?

For the state to require zero discharge would put us on the "same old self-destruct course" of creating an unfavorable climate for industry here.

A top aide to Governor Carey quoted in
<u>The New York Times</u>, 30 January 1976

tommyrot and balderdash!

Ogden Reid, quoted in
<u>The New York Times</u>,
18 February 1976

Efforts to improve the environment often appear to run head-on against economic interests. For instance, when the New York State Commissioner of Environmental Protection, Ogden Reid, ordered General Electric to stop all discharges of PCBs into the Hudson River immediately, he was quickly over-ruled by the governor of New York, who gave G.E. a whole year to reduce its PCB discharge. Governor Carey thought that the more strict ruling might cause General Electric to move to a more lenient state. New York would lose tax money and jobs.

"For the state to require zero discharge of PCB's," one of the Governor's top aides said..., would "put us on the same old self-destruct course" of creating an unfavorable climate for industry here.

"Furthermore," this aide said, "after all those years of discharging 30 pounds a day, no one can tell me there's one bit of difference between zero and 3.5 ounces."

"It's a totally phony, symbolic issue," he added. *

Mr. Reid said it was "tommyrot and balderdash" that New Yorkers had to choose between jobs and a healthy environment. "There are alternatives to PCB's and we believe it is possible to work this out so that G.E. can continue," Mr. Reid said.

He also criticized those who say G.E.'s proposal to discharge no more than 3.5 ounces a day by 1977 is reasonable. "That amount would impair water quality and we estimate that minnows would carry over four parts per million of PCB's as a result," Mr. Reid said. "Striped bass that eat the minnows would bioaccumulate those PCB to perhaps 11 parts per million. Saying that a little of PCB's don't hurt is like saying you're a little bit pregnant." †

It will undoubtedly cost G.E. more money to discharge absolutely no PCBs. However, this is not the only cost involved in the case. How many other costs can you think of that result from the polluting of the Hudson. Who pays these costs?

Suppose G.E., unable or unwilling to reduce its discharge to zero, planned to move out of state instead. Do you feel that New York State should then compromise its standards? Why?

---

* The New York Times, 30 January 1976

† The New York Times, 18 February 1976

Figure 6-5 *Closing a river to commercial fishing can mean the loss of thousands or even millions of dollars to the industries depending on those fishing grounds. It also means a loss of livelihood to the fishermen involved. (Edward Hausner/NYT pictures.)*

throughout the United States PCB loads are already in excess of the criterion in our proposals."[5]

Yet to be discovered is the effect on the environment, and inevitably, on people, of long term exposure to low levels of PCBs in water and in food.

TABLE 6-4   Average Values for PCBs in Foods

	Average level PCBs in parts per million
Fish [a]	1.87
Cheese [a]	0.27
Milk [a] (cow, fat portion)	2.25
Eggs [a]	0.55

[a] Data from A. Kolbye, "Food Exposures to PCBs," Environmental Health Perspectives #1, NIH 72–218, U.S. Department of Health, Education and Welfare, April 1972.

5 *The New York Times*, 20 January 1977.

PCBs in Food

Because they are so widespread in the air and water, PCBs have found their way into foods, mainly fish and dairy products, and also into packaging materials used for foods. The FDA at present considers unsafe for human consumption any meat, fish, or poultry containing more than 5 ppm PCBs, milk with more than 5 ppm in the milk fat, and eggs with more than 0.5 parts per million in the fat. Table 6–4 presents some average values for PCBs in food.

The Food and Drug Administration is considering lowering the allowed level of PCBs in fish from 5 parts per million to 2 parts per million. Canada adopted the lower limit in 1975. However, the FDA may be restrained by the knowledge that if the 2 ppm rule were passed, much of the fish now eaten in the U.S. would be classed as unfit for human consumption. A Wisconsin official estimated that 75% of the 90 million pounds of fish caught there each year could not meet the 2 ppm PCB limit.

PHTHALATES: AN ENVIRONMENTAL PROBLEM IN THE MAKING?

In the past, environmentally harmful chemicals have been identified because they have caused a problem: people became sick, fish were killed, trees died. However, we are becoming more sensitive to the possibility of harm. We look more suspiciously at chemicals that are likely to become widespread in the environment, whether or not we have evidence that they are dangerous, because these are the chemicals in a position to cause widespread damage.

A case in point is that of phthalates.[6] Phthalates are chemicals used in polyvinyl chloride plastics. Phthalates change polyvinyl chloride (PVC) from a hard, glass-like material into a flexible plastic. Phthalate containing plastics have been incorporated into automobile interiors, clothing, food wraps, and medical products such as the plastic bags that hold blood for transfusion (see Fig. 6–6). Phthalates are not actually attached to the plastic. They fill spaces between the plastic molecules and act, in a way, as lubricants. Because they are not bound to the plastic, they can and do migrate out into materials with which the plastic comes in contact. Because phthalates have a tendency to migrate and because of the widespread use

of plastics, we might have predicted that phthalates would become a widespread environmental pollutant. In fact, that is just what has happened. Plastics burned in incinerators contaminate air, plastics buried in landfills seep into ground water and run off into lakes and rivers. Very low levels of phthalates are even detectable in water samples taken from the open oceans and the Gulf of Mexico (0.005 to 0.090 micrograms per liter). For organisms living in water, phthalates appear to be concentrated up to 13,000 times over the water concentrations.

The next logical question is, what effects do phthalates have on living creatures? Phthalates do not appear to be highly toxic chemicals, in large doses, over the short term. There is no evidence that they cause cancers. We do not know, however, what the effect will be of small doses, such as are now found in water, over long periods of time. Some experiments have shown that phthalates interfere with reproduction in water creatures, for example, waterfleas, guppies, and zebra fish, even at levels as low as 3 micrograms per liter (a level found in some rivers). Higher doses of phthalates cause reproductive problems in mice.

Does this group of chemicals pose a long-term threat to humans perhaps through direct exposure via water, food (food wraps), or blood transfusions? Or are we faced with a situation in which subtle changes in food webs will occur as certain aquatic creatures accumulate phthalates and/or suffer reproductive problems? The questions are serious ones but the answers may be a while in coming.

---

6 (thal´ātes) For information on the chemical structure, see the Environmental Health Perspectives volume mentioned in Further Reading Section.

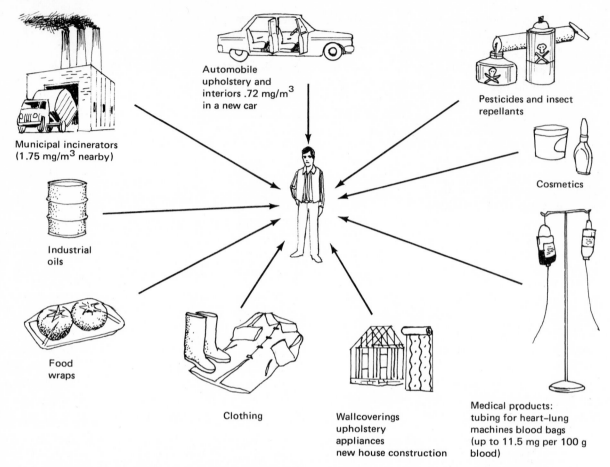

Automobile
upholstery and
interiors .72 mg/m^3
in a new car

Municipal incinerators
(1.75 mg/m^3 nearby)

Industrial
oils

Food
wraps

Clothing

Wallcoverings
upholstery
appliances
new house construction

Pesticides and insect
repellants

Cosmetics

Medical products:
tubing for heart-lung
machines blood bags
(up to 11.5 mg per 100 g
blood)

Figure 6-6 *Exposure to phthalates. Phthalates have been used in a variety of products for home and industry. Although phthalates are relatively nontoxic (humans could probably survive a daily dose of 3-480 milligrams of phthalates; workers in plastics industries begin to show symptoms of poisoning at about 2-70 milligrams phthalate per cubic meter of air), long-term effects are unknown.*

PESTICIDES IN DRINKING WATER

Sources

Pesticides are chemicals designed to kill insects and other pests.[7] These chemicals are widely used.

7 More on the topic of pesticides can be found in Chapters 28 and 35.

In fact they have become part of our daily lives. A study by the U.S. Environmental Protection Agency showed that DDT, a pesticide once very popular in this country, can be found at levels of a few hundred parts per trillion in many U.S. waters, including drinking water supplies. Other studies have shown that Americans have an average of five to 10 parts per million DDT in their fatty tissues.

How do pesticides find their way into waters

we drink? There are a number of ways. Pesticides run off farm lands during rains, either dissolved in the rain water or attached to dirt particles. Rainfall itself may be contaminated with pesticides that have evaporated or remain in the air after the spraying of crops or woodlands. Industrial plants manufacturing pesticides may accidently spill or intentionally discharge pesticides into nearby rivers and lakes. An example of several of these processes can be seen in the story of how two related pesticides, Kepone and Mirex, have come to contaminate such major fisheries and water sources as the James River and Lake Ontario.

Kepone and Mirex: Cases in Point

An occupational hazard becomes an environmental problem   In June 1975, an ailing employee of the Life Sciences Products Corporation was finally persuaded by his wife to see a doctor. The doctor thought that the dizziness and trembling the man suffered might be due to a chemical to which he was exposed at work. When a blood sample was sent to the Public Health Service Center for Disease Control in Atlanta, it was indeed found to contain a high concentration of the pesticide Kepone. Thus began a series of inquiries into an incident that a United States attorney later called "the environmental disaster of the decade."

A state official who visited the Kepone plant in Hopewell, Virginia, found piles of Kepone dust and contaminated water. He also found seven workers ill enough to require immediate hospitalization. Over 100 people, including wives and children of plant employees, were found to have Kepone in their blood. Thirty people were hospitalized and found to be suffering from tremors and visual problems. Further, some of these people may be permanently sterile and all are threatened with the possible development of liver cancer, since Kepone is known to cause such tumors in experimental animals.

But this was only the tip of the iceberg. Studies showed Kepone in the air and water as far as 40

miles away from the site of the factory. In addition, Kepone was found first in fish and shellfish in the James River, on which Hopewell is located, and then in fish and oysters in the Chesapeake Bay into which the James River empties. Eventually traces of Kepone were found in bluefish along hundreds of miles of the Atlantic coast. Commerical fishing and oyster harvesting was banned in the James River and parts of the Chesapeake Bay for a time and are still forbidden in the lower James River (Fig. 6–7). Millions of dollars were lost by the men who usually fish these waters for a living.

Is it possible to fix the blame for what happened? Life Science Products was a small company started by two former employees of Allied Chemical Corporation. The smaller company manufactured Kepone from raw materials supplied by Allied and then sold the finished product exclusively to Allied. Officials of Allied Chemical said the smaller company knew that Kepone was a hazardous substance. Workers at Life Science said they were never told it was dangerous. It has been calculated that those workers received daily doses over a period of 15 months that were within a factor of five of the amount producing cancers in experimental rats and mice exposed for about one and a half to two years.

Wastes from the Life Science plant found their way into Hopewell's sewage treatment plant and caused sewage digesters there to fail within a few months after the plant began operations (Fig. 6–8). Kepone, which was marketed as a control for ants, roaches, and the banana-root borer, had killed the bacteria in the digesters. These bacteria normally oxidize waste materials (this process is explained on p. 199). The city of Hopewell asked the U.S. Environmental Protection Agency for information on how to handle the wastes. State agencies advised that the wastes should be specially treated before being run through the plant. This recommendation was apparently ignored.

Oysters, shrimp, and fish accumulate Kepone 425 to 20,000 times over the level found in the water in which they live. Kepone is toxic to some

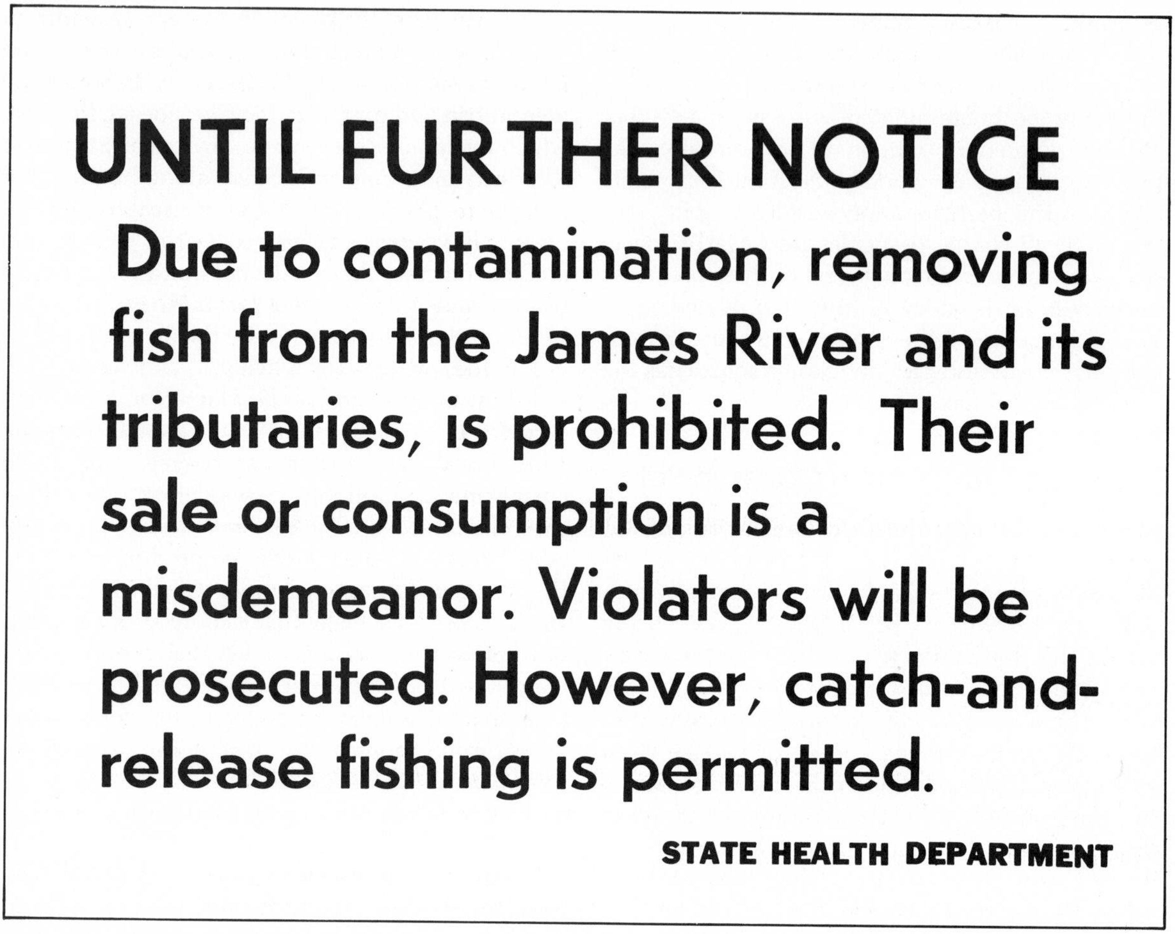

UNTIL FURTHER NOTICE
Due to contamination, removing fish from the James River and its tributaries, is prohibited. Their sale or consumption is a misdemeanor. Violators will be prosecuted. However, catch-and-release fishing is permitted.

STATE HEALTH DEPARTMENT

Figure 6-7   *Signs such as this one are posted along the lower James River due to Kepone contamination of the water and fish.*

water creatures at very low levels. For instance, shrimp fail to reproduce properly at 0.2 ppb. Kepone is also slowly degraded in the environment. As a result of the discharges from the Hopewell plant, fish and shellfish taken from the James River in 1976 were found to contain from 0.1 to 1 parts per million Kepone (safe levels were considered to be 0.1 parts per million at that time but have since been raised to 0.3 parts per million).

Allied Chemical Corporation was eventually fined eight million dollars for polluting the nations waterways under the Water Pollution Control Act of 1972. Another $5.3 million was paid to the city of Hopewell and the State of Virginia for repairs on the sewage treatment plant and patrols on the river to ensure the fishing ban. Cleaning Kepone out of the river doesn't seem to be technically reasonable at the present time. The river may take 50 to 100 years to bury the Kepone in natural sediments or move contaminated muds out to sea.

Mirex   Meanwhile in several southern states, an almost identical pesticide, Mirex, was being used to control the fire ant. This species of ant first entered

Holding tanks

Allied

Figure 6–8   *(a) The former Life Science Products Company in Hopewell Virginia. Nothing is left on the site today. The plant has been razed to the ground, by order of the Virginia State Health Department, and pieces of it were buried in a special lined pit. (b) View from the Life Science Products catwalk down Route 10 shows the proximity to Allied. (Photos by R. S. Jackson, M. D. Virginia State Department of Health.)*

the United States in the late 1950s from Argentina. It resembles ants found here, except that it can deliver a painful sting similar to that of a bee or wasp. The sting is not fatal except to certain people who develop an allergy to it. Fire ants are obviously a serious nuisance in residential areas. They also build large hills of earth, which can interfere with farming.

Mirex was used against fire ants starting in 1962. It was mixed in small amounts with ground corncobs and soybean oil to form a bait, which the ants carried back to their nest, thus killing the whole colony. Unlike aldrin and dieldrin, which had been used previously on fire ants, Mirex at first seemed to be harmless to other creatures in the area. The Department of Agriculture began an eradication campaign in which the bait was dropped from airplanes in an attempt to wipe out the fire ant.

Evidence began accumulating, however, that Mirex was not as harmless as was originally thought. It was found to cause cancer in mice, to be toxic to crabs and shrimp, and to last longer in the environment than DDT. The Environmental Protection Agency joined the fight against Mirex in 1973 when it began hearings on the pesticide. Despite mounting evidence of hazard, the Department of Agriculture and the State of Mississippi were reluctant to give up their goal of eradication of the fire ant. When Allied Chemical Corporation, as a result of increasing pressures brought to bear by environmental groups, decided to discontinue Mirex production, the State of Mississippi took over the Allied plant and bought its inventory. Possibly as a result of the Kepone disaster and its publicity, however, a compromise was reached between the Environmental Protection Agency and the State of Mississippi in which the use of Mirex bait to control fire ants was to be phased out by June 1978.[8] Before that happened, a study was carried out which showed Mirex in the fatty tissue of about 25% of the people living in the area sprayed with Mirex. Thus, food and water must have become contaminated either directly, by spraying, or indirectly. An example of the latter might be the

8 The Mississippi Fire Ant Authority received permssion to use a related pesticide, Ferriamicide. This is a mixture of Mirex and a few other chemicals, supposed to help degrade Mirex more quickly in the environment. It is probable that this was, in large part, a political solution. In any case, use of Ferriamicide was prevented, pending further tests, by a suit brought by environmental groups.

contamination of milk from cows grazing in pastures onto which Mirex had drifted.

This was not the whole story, however. Mirex was also found in fish in Lake Ontario, which is certainly a long way from the southern states where the fire ant program was carried out. This time the source was found to be the Hooker Chemical plant, located on the Niagara River, which flows into Lake Ontario. Hooker was the supplier for technical grade Mirex to the Mississippi plant.

(Thus, one pesticide spray program resulted in the absorption and concentration of the substance by organisms, including humans, in the sprayed area and also in the industrial contamination of waters where the pesticide was being manufactured, far from the sprayed area. In cases where a possibly hazardous chemical is to be used or manufactured, only extreme care can prevent the material from contaminating the environment and concentrating in food chains. The quantities of Kepone and Mirex released from Life Science at Hopewell and by the fire ant eradication program will remain in the environment and as part of food chains for many years to come.)

ARE SAFE DRINKING WATER LAWS THE ANSWER?

Why We Need Them

In December 1974, the President signed into law the Safe Drinking Water Act. Several unsuccessful attempts had been made in other years to pass such a law, but it required some hint of an environmental disaster to provide the final push that would propel the law through Congress. In this case, the push was the publicized discovery of a number of possibly carcinogenic substances in the drinking water of the city of New Orleans. (For more on this, see p. 545).

The Safe Drinking Water Act directs the Environmental Protection Agency to establish regula-

tions ensuring the purity and safety of United States drinking waters. Although each state is expected to be responsible for setting and enforcing its own standards, the federal government, under the new law, can step in if the states' regulations are not at least as stringent as federal regulations, or if the state doesn't enforce its regulations. This is a major change. In previous years, there were only suggested guidelines set by the U.S. Public Health Service. The federal government had no power to regulate any water supply except that used by interstate carriers such as buses or trains.

That the old system was simply not working is illustrated by the following facts, gathered by the Senate Committee on Interstate and Foreign Commerce:

> During the ten-year period 1961–1970, there were 130 outbreaks of disease or poisoning attributed to drinking water. These outbreaks resulted in 46,374 illnesses and 20 deaths. On the average, this represents one reported waterborne outbreak per month with something over 350 persons becoming ill.[9]

A 1970 study by the Department of Health, Education, and Welfare, sampling 1000 water systems, found that:

> Thirty-six percent of 2,600 individual tap water samples contained one or more bacteriological or chemical constituents exceeding the limits in the Public Health Service Drinking Water Standards.

---

9 It should be noted, however, that these figures represent only those incidents: (1) that have been reported; (2) that involve at least two cases of communicable disease; and (3) for which an epidemiological investigation was performed and the waterborne route established as the cause. From: Safe Drinking Water Act, Report from the Committee on Interstate and Foreign Commerce, House of Representatives, 93rd Congress, Second Session, #93, 185, 1974.

Fifty-six percent of the systems evidenced physical deficiencies including poorly protected groundwater sources, inadequate disinfection capacity, inadequate clarification capacity, and/or inadequate system pressure.

Seventy-seven percent of the plant operators were inadequately trained in fundamental water microbiology; and 46 percent were deficient in chemistry relating to their plant operation.

Seventy-nine percent of the systems were not inspected by State or county authorities in 1968, the last full calendar year prior to study. In 50 percent of the cases, plant officials did not remember when, if ever, a State or local health department had last surveyed the supply.

A 1970 survey by the Conference of State Sanitary Engineers indicated that most state sanitary engineers judged their own surveillance program to be deficient.

The New Standards

Standards dealing with the microbiological contamination of drinking water are fairly easy to set. We have methods to detect such contamination and know approximately what levels of contamination will cause outbreaks of disease. It is not an easy task to set standards for chemicals in drinking water. In some cases there is not sufficient evidence about a chemical to know if there are safe levels. For instance, a number of possibly carcinogenic chlorine-containing organic chemicals can be found in water. Their long-term effects at low levels are unknown.

In other cases, chemicals may be harmless, or even necessary, to life in trace amounts, but toxic in larger quantities. This is true of copper, for example, which is an essential in trace amounts in human and animal diets, but which, in larger amounts, has been responsible for epidemics of sheep poisoning in the Netherlands. Large amounts of fluoride are poisonous. However, fluoride is added to water in trace amounts to reduce tooth decay (Fig. 6–9).

Furthermore, not all chemicals can be detected by a simple, inexpensive and reliable test. The chlorine-containing organic chemicals fall into this category. In addition, standards are meant to ensure that drinking water is safe when it leaves the water treatment plant. However, water may pick up metal contaminants from the pipes in which it is transported. Soft waters with low acidity, are especially corrosive. Studies have shown that, depending upon the type of pipes it travels through, water will pick up cadmium, chromium, cobalt, copper, iron, lead, manganese, nickel, silver, and zinc between the time it leaves the treatment plant and the time it reaches the consumer.

The possible dangers that are associated with drinking chemically contaminated water, especially by such groups as children and pregnant women, are such that Congress felt it necessary to direct that standards should be set on the basis of the information and technology that is available. The standards that have been set by the Environmental Protection Agency are interim standards, that is, they will be revised as new information becomes available. The EPA has also begun research in areas where information is needed.

CONTROVERSY

SHAPING UP UTILITY COMPANIES

No one knows at this point how much time, money and effort, or blood will have to be devoted to...this provision..., and certainly no one knows how many utility personnel it will drive to an early retirement or to another profession.

John M. Gaston, California State Department of Health

One provision of the Safe Drinking Water Act is of special interest to both consumers and to the people who run the water treatment plant. This is the requirement that after June 1977, water treatment agencies must notify their customers if the water they supply does not meet all of the standards.

There has been some concern that the public will become unnecessarily alarmed if they are told their drinking water does not meet standards. However, John Gaston, a senior sanitary engineer at the California Department of Health, sees the provision as a way of forcing utilities to "shape-up." He provides the following data from the California water supply system (only water supplies with at least 200 connections are included).(Assuming the notification provision had been in effect in 1976, this is what would have happened.)

In the first eight months of 1976, 72 water utilities would have notified the public a total of 80 times. One unfortunate utility would have notified four out of the eight months of record in 1976. This notification would have been because of bacteriological quality failure, and would have included television, radio, newspaper, and direct mail notification. Sixty-six utilities would have notified the public a total of 102 times because they failed to take enough bacteriological samples. One utility didn't take enough samples five out of the eight months of record.

In the "best" month only four utilities would have failed to meet the bacteriological standard and have had to notify; the worst month, August, would have seen 21 utilities notifying. If this year is any indication of the activity to come we can expect that each month about ten utilities will have to notify via television, radio, newspaper, and direct mail that they failed to meet the bacteriological quality standards, and about 13 utilities will have to notify via direct mail that they failed to take enough samples. *

Certainly the public should be notified of emergency health hazards such as the 70 ton spill of chloroform, which travelled down the Ohio River in February 1977 (the city of Cincinnati takes its drinking water from the Ohio). But do you think that consumers really want to know every time their water suppliers "forget" to take a sample to check for bacteria? Do you think the new provision will improve the record in terms of water supplies meeting the drinking water standards? Have you ever been notified by your utility that your water did not meet the standards?

---

* John Gaston, "Consumer Notification-Public Awareness or the Smoking Pistol," Journal of the American Water Works Association, p. 574 (November 1977).

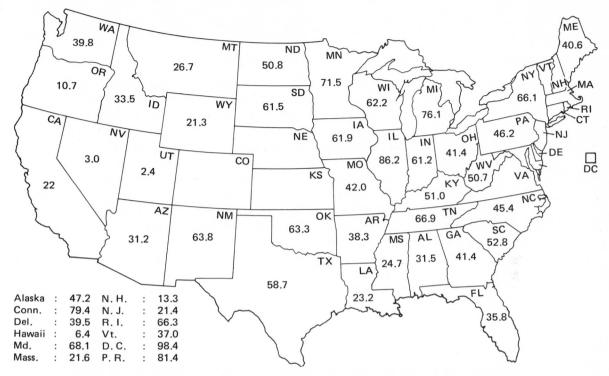

Figure 6–9   *The percentage of state population using fluoridated water (1975). One trace chemical that is added to water is fluoride, which is intended to help prevent tooth decay. Although the American Dental Association is in favor of the addition of fluoride to drinking water supplies, there are still many people who feel that fluoride may be harmful. (Source: Morbidity and Mortality Weekly Report, U.S. Department of Health, Education, and Welfare, 1977.)*

Questions

1. Explain why the mercury concentration found in the waters of Minimata Bay was much lower than those found in the fish and shellfish of the Bay. What does this imply for people living in that region?

2. Suppose you are a public health official in charge of deciding whether to allow a manufacturer to sell a new chemical, which is supposed to kill nuisance plants in lakes. List three questions you would want answered before you decided whether the chemical seemed safe to add to the water environment.

3. What does it mean to recycle mercury? Name some uses of mercury in which it recycles. What are some uses of mercury in which it does not recycle? Why can't the mercury be recycled?

4. Explain how the cadmium concentration of municipal drinking water can depend, in part, upon the characteristics of the water itself.

5. In order to discover how much of a particular pollutant people are exposed to, all the possible sources of exposure must be examined. List all the ways you can think of that people in Fuchu were exposed to cadmium once it contaminated their water supply.

6. If a septic system is located too close to a well or uphill from a well, the wastes from the septic system may contaminate the well. Even if there are no disease-causing bacteria or viruses in the contaminated well waters, problems may still occur if a young infant drinks the water. Explain. Why are high nitrate levels in drinking water only a problem with respect to young infants?

7. PCBs have been found in water all over the country but only in trace amounts. Nowhere have sampling studies found enough PCBs to be immediately poisonous to humans. Give three reasons why PCBs at these low levels are still a serious problem.

8. Should General Electric be legally responsible for cleaning the PCBs from the Hudson River in addition to paying fines for exceeding allowed discharge levels? In a broader sense, should industry be responsible for cleaning up environmental pollution caused by illegal discharges? by accidents? by employee negligence? by "acts of God"? (Oil companies are legally obligated to clean up oil spills that occur during drilling for oil in waters off the coast *whatever the cause of the spill*.) Is this an acceptable burden on a particular industry? Are there other ways to achieve the clean-up?

9. The fishermen on the Hudson River have effectively been deprived of their jobs because G.E. dumped PCBs into the river. Should they be recompensed? Who should help them and why?

10. PCBs are found in almost every flourescent light fixture as well as in some 895 million capacitors scattered all over the country (it is estimated that there are 250 million pounds of PCBs in capacitors now in service.) If special plans are not made and followed for the disposal of these items, the PCBs they contain will almost certainly find their way into dumps or landfills or incinerators and then into the nation's waters.

   a. Diagram how the PCBs in a flourescent light fixture might end up in a large lake.

   b. The EPA is looking for ideas on how to interrupt this flow of PCBs into the environment. High temperature incineration destroys PCBs. Can you think of ways to encourage people to dispose of PCB-containing items appropriately? *Note:* Because of the smallness of the items and their wide distribution, a simple law requiring proper disposal probably would not be very effective.

11. Compared to chemicals like DDT, which is both very poisonous and a cancer causing substance, phthalates are not very toxic. Why, then, should we worry if they become widespread in the environment?

12. Trace a possible route of the phthalates in a piece of food wrap into the ocean environment.

13. Although there was at one time enough Kepone or other toxic material in the Life Science plant effluent to actually overcome a worker at the Hopewell sewage treatment plant, the levels of Mirex that leached or were sprayed into local ponds and

rivers during the fire ant program were generally low, not enough to kill fish directly. Why then was the Mirex still a problem in water? (You will probably be able to think of more than one reason.)

14. Briefly diagram three routes by which Mirex could have found its way into the fatty tissues of Mississippians.

15. How could wives and children of workers at the Life Sciences plant have had Kepone in their blood (assume they didn't visit their husbands or fathers at the plant)?

16. What is the source of your drinking water? Is it a protected source?

17. Has your state set drinking water standards that the EPA has approved?

18. Not all trace minerals are undesirable. Some give water the flavor to which we are accustomed. Try the taste of distilled water, which has none of these minerals: catch the steam from a teakettle on a piece of aluminum foil (be careful, steam burns are painful) and let the drops run into a glass. Compare this to a glass of water straight from the tap.

Further Reading

Cadmium

Friberg, L., M. Piscator, and G. Nordberg, *Cadmium in the Environment*, Chemical Rubber Company Press, Cleveland, 1971.

Friberg, L. *et. al.*, Cadmium in the Environment II, 1973; prepared for the U.S. Environmental Protection Agency; available from National Technical Information Service, Report #EPA–R2–73–190.

These two reports, although somewhat technically written, provide a comprehensive picture of the effects of environmental sources of cadmium on human health.

Mercury

Evans, R. J., J. D. Bails, and F. M. D'Itu, "Mercury Levels in Muscle Tissues of Preserved Museum Fish," *Environmental Science and Technology* 6 (10), 901 (1972).

One of the fascinating aspects of environmental mercury contamination is that fish near the top of salt-water food chains, like tuna or swordfish, seem to have dangerous levels of mercury even in areas where no human sources are found. This paper gives an introduction to the literature on the subject of mecury in food fish.

Mercury and the Environment: Studies of Mercury Use, Emission, Biological Impact and Control; Organization for Economic Cooperation and Development, 1974.

This report is a general survey of the impact of mercury on the environment. It also mentions economic considerations involved in mercury emission control.

Wallace, R. A., W. Fulkerson, W. Shults, and W. Lyon, Mercury in the Environment: The Human Element, Oak Ridge National Laboratory Report ORNL–NSF–EP–1, Oak Ridge, Tennessee, 1971.

A lively, well written description of the mercury problem—especially the sources of mercury.

Mirex and Kepone

Holden, C., "Mirex: Persistent Pesticide on Its Way Out," *Science* **194,** 301 (15 October 1976).

Kaiser, K., "Mirex: An Unrecognized Contaminant of Fishes from Lake Ontario," *Science* **185** 523 (9 August 1975).

Kaiser, K. L., "The Rise and Fall of Mirex," *Environmental Science and Technology* **12** 520 (May 1978).

These three articles span much of the Mirex controversy. Other references to Mirex and Kepone can be found through *The New York Times* index for this period.

Zim, M. H. "Allied Chemical's $20-Million Ordeal With Kepone," *Fortune,* p. 82 (11 September 1978).

It is fascinating to read Allied's view of the Kepone incident. The effect of this disaster on corporate policy has been far-reaching.

Nitrates

Gruener, N. and H. Shuval, Health Aspects of Nitrates in Drinking Water, H. Shuval, ed., Developments in Water Quality Research, Ann Arbor, 1970.

Shuval, H. and N. Gruener, "Epidemological and Toxicological Aspects of Nitrates and Nitrites in the Environment," *American Journal of Public Health* **62**(8), 1045 (1972).

These two articles summarize current knowledge about the effects of nitrates in drinking water and also the current status of the problem in many countries.

PCBs

Environmental Health Perspectives #1, NIH 72–218, U.S. Department of Health, Education and Welfare, April 1972.

This entire volume is dedicated to the problem of PCB contamination of the environment. Most of the information on PCBs gathered up to 1972 is presented or summarized here. For an update, see the following:

PCBs—A Review, Great Lakes Focus on Water Quality, Vol. 2 #1, Winter 1976, International Joint Commission, Regional Office, 100 Ouellette Avenue, Windsor, Canada N9A 6T3.

Phthalates

Environmental Health Perspectives experimental issue #3, U.S. Department of Health Education and Welfare, Public Health Service, National Institute of Health, January 1973.

> This entire issue is devoted to phthalates. While some of the papers are rather technical, others give good summaries of the problems and facts about phthalates.

Giam, C. S. *et. al.*, "Phthalate Ester Plasticizers: A New Class of Marine Pollutant, *Science* **199** 419 (27 January 1978).

> Some of the more recent work on phthalates and their effects on aquatic organisms is mentioned and referenced in this paper.

Safe Drinking Water Laws

Draft Analytical Report, New Orleans Water Supply, Lower Mississippi River Facility, Surveillance and Analysis Division, Region IV, The U.S. Environmental Protection Agency, Dallas, November 1974.

The Implication of Cancer Causing Substances in Mississippi River Water, The Environmental Defense Fund, 6 November 1974.

> These two reports provided a severe blow to the confidence Americans had in their drinking water. The result was the Safe Drinking Water Act of 1974.

McCabe, L. J., *et al.*, Survey of Community Water Supply Sytems, *Journal of the American Water Works Association* **62**, 670 (1970).

> This paper caused a great furor when it first came out (including articles written for the popular press by Ralph Nader). It was, in a sense, the first time the American public was notified that its drinking water was not as pure and safe as it had come to believe.

Safe Drinking Water Act, Report from the Committee on Interstate and Foreign Commerce, House of Representatives, 93rd Congress, Second Session, #931185, 1974.

> The report summarizes and explains the provisions of the Safe Drinking Water Act. In addition it provides some background about why the Act was needed.

Chapter 7

Purifying Water

THE DIFFERENCE BETWEEN WATER TREATMENT AND WATER POLLUTION CONTROL

(The term "water treatment" should be distinguished from the term "water pollution control." These are very different activities. Water treatment and water pollution control are the two sides of the coin of water quality. Taken from the lake, river, or well, water must be made both safe and desirable to drink. Water treatment purifies the water for the consumer whether in the home, in businesses, or in factories. Safety is primarily achieved by destroying disease-causing microorganisms. Palatability is achieved by removing tastes and odors and by making the water clear.)

(Water pollution control, on the other hand, is directed toward restoring the quality of the water that the consumer has used. Most of the organic wastes and bacteria that have been added by use must be removed. The organic wastes are removed primarily to prevent the depletion of oxygen in the waters that will receive the waste. Oxygen is vital to the normal aquatic life of a stream. The relation of these wastes to the level of dissolved oxygen is discussed in Chapter 8. In addition to the organic wastes, disease-causing microorganisms are also destroyed to prevent disease from occurring among those who later use the water.)

The relative positions of water treatment and water pollution control are indicated in Figure 7–1.

WATER TREATMENT

To remove disease-causing organisms and harmful chemicals and to make the water pleasant to drink, cities treat their water. Not all cities treat their water in the same way because of the presence of different substances in their basic supply. We will describe a set of processes that are fairly typical among cities and the common order in which these processes take place.

Taste and Odor Removal

(The tastes and odors that exist in some city supplies may be offensive to the consumers of drinking water and require removal. The causes of such odors and tastes include the following conditions:

1. Hydrogen sulfide gas may be dissolved in the water; the condition is most common in water from wells.
2. Dead or decaying organic matter stemming from algae or aquatic weeds may be present. Water from reservoirs may contain such materials.
3. Chemicals used in agriculture to control weed and insect pests and chemicals from industrial wastes may enter the water supply.)

(The methods to control tastes and odors may be classified in two groups. There are methods to

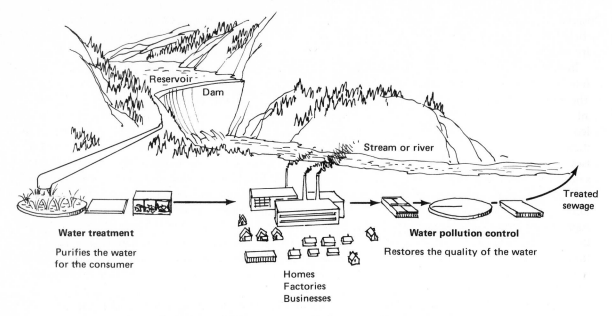

Figure 7-1 *Water treatment and water pollution control.*

prevent algae growth and the growth of aquatic plants. And there are *treatment methods* to remove tastes and odors once they occur. To prevent algae and other aquatic plants from growing, chemicals may be added to the water while it is in reservoirs. Herbicides were once considered for this task, but have now been rejected. The one herbicide that was thought to be unobjectionable (but never used) is now known to cause cancer. This is the compound Lindane.

Copper sulfate is a commonly used substance for algae control. It may be slowly dumped from boats crisscrossing the reservoir.

Aeration (In the treatment plant itself, tastes and odors may be removed by the use of aeration or by chemicals. Water taken from a river or lake or from community wells may often be aerated. The water is commonly sprayed into the air from banks of fountains or allowed to flow over racks in a thin film. You may have seen the spraying step at a local water distribution reservoir. (The aeration (exposure to the air) accomplished by this spraying or by the rack method removes dissolved gases (e.g., hydrogen sulfide) from the water. With the removal of hydrogen sulfide, a substantial control of odor may be achieved. (It is hydrogen sulfide gas that we smell when an egg becomes rotten.))

First Chlorination

(After aeration, chlorine gas is added to the water to kill disease-causing organisms. The step, which is called chlorination, occurs again after all other treatment steps are completed.)

Removal of Particles and Colors

Coagulation (The minute suspended particles that give color to water will neither dissolve nor settle rapidly from the water; particles behaving in this way are termed *colloidal*.) A teaspoonful of rich earth (humic soil) mixed into a gallon jug of clear water illustrates the concept of a colloid. Note how the passage of light is restricted through such water.

Observe the yellow-brown color imparted by the particles. (The particles can be removed by a process called coagulation.)

In the first step of coagulation, alum or ferrous sulfate is continuously added and rapidly mixed into the stream of water which flows through the plant. When the chemicals are added, a precipitate, called a "floc," forms; the floc consists of mineral particles that are insoluble in water. The flow then passes through a flocculation basin where the floc is mixed into contact with the suspended particles.

The water then is pumped slowly through a settling tank. Here the water is detained long enough to allow most of the floc or precipitate to settle to the bottom of the tank. In the process of settling, the "blanket" of floc will capture and drag down a large share of the particles that were suspended in the water. The precipitate is scraped away from the bottom by a scraping bar to prevent build-up.

The water leaving the settling tank has had much of its turbidity (cloudiness) removed. In part, these preceding steps protect the effectiveness of the next step, filtration through beds of sand, by removing materials that could clog the filter.

Carbon treatment   In many water treatment plants, a small amount of activated carbon particles may be added along with the alum or ferrous sulfate in the first step of the coagulation process.

Colloidal particles will be bound to the carbon. When the carbon is removed in a later step, much of the colloidal material responsible for color is carried with it. Not only is color improved when the carbon particles are removed, tastes and odors are decreased as well because many of the dissolved organics have been bound to the carbon particles and removed from the water.

Contacting the water with powdered activated carbon particles is a very popular method to remove tastes and odors, dating to the 1930s. In that era, filters using activated carbon as the filter medium were used in Bay City, Michigan. In contrast, the practice in the recent era has been to add a small amount of carbon in the first step of the coagulation process. Interestingly, the filtration of water through beds of granular carbon is now being recommended as a part of water treatment by the Environmental Protection Agency as a means of removing low-level but important contaminants.

(The ability of carbon particles to remove organic compounds makes carbon filtration a potentially attractive method to remove not only tastes and odors but also pesticides and other resistant organics. Some of the compounds are now viewed as potential causes of cancer. Carbon filtration, therefore, is suggested as a means to remove possible cancer-causing substances which may occur in water.[1] Thus, we may see carbon filtration reintroduced as a common water treatment method.)

Filtration to Remove Microbes

The water from the settling tanks is passed downward through beds of sand. From top to bottom, the sand placed in the filter bed gets larger and larger. The construction of a typical sand filter is shown in Figure 7–2.

The removal of the larger solids, which was achieved by the coagulation step, prevents the sand filter from clogging rapidly. The sand filter does cause a further removal of particles. However, the primary purpose of the filter is to capture and retain bacteria and viruses. Most disease-causing microorganisms are thus prevented from entering the system that distributes clean water to the city.

Periodically, the sand beds must be washed if they are to retain their effectiveness at filtering out microorganisms. Some of the filtered water is saved in a "clear well" for this purpose. The washing proceeds by passing the clean water up through the filter to dislodge particles, which have been retained in the spaces between the particles of sand.

---

1 More on this in Chapter 27.

IS THERE SUCH A THING AS A ZERO-RISK ENVIRONMENT?

We are dealing with a time bomb with a 25 year fuse.

Dr. Robert H. Harris
Environmental Defense Fund

Crash programs . . . to meet "yesterday" deadlines are not understandable at all.

A. E. Gubrud
American Petroleum Institute

Several environmental groups were very unhappy with the way the Environmental Protection Agency set about fulfilling (or not fulfilling) the toxic pollutants portion of the 1972 Federal Water Pollution Control Act–so unhappy, in fact, that they sued the EPA in federal court. The petition charged the Environmental Protection Agency with failure to require that water supplies be checked for the presence of heavy metals such as cadmium and with failure to require proper removal of pesticides, viruses, and asbestos.

"We are dealing with a time bomb with a 25-year fuse," said Dr. Robert H. Harris, at a news conference at which the petition was discussed. He said Americans could not afford to wait to see what would happen, as they had done with cigarette smoking.

Dr. Harris noted that it often took many years for a cancer to develop. Therefore, he argues that exposure of a population to cancer-causing substances might not be reflected in an obvious rise in the cancer rate until 25 years or so after the exposure began....

Dr. Harris, who is associate director of the Environmental Defense Fund's toxic chemicals program, said traces of known and suspected cancer-causing chemicals had been found in water samples from several major cities....

Dr. Harris said federal regulatory agencies had done "virtually nothing" to protect the public from cancer-producing chemicals in water supplies. *

The suit was settled by an agreement between the EPA and the environmental groups, which was approved by a judge of the District Court of Appeals. The settlement agreement required the EPA to start a huge program of experiments on 65 specified toxic pollutants. The agency also agreed to begin studies of the available methods of control and economic effects of controlling these pollutants. Many industry groups were upset by the agreement. As one industry spokesman put it:

> …(A) disturbing trend is emerging, ever more clearly. In the past, the scientific and technical communities have experienced difficulty enough in responding to "technology-forcing" legislation and regulation. Now, however, as a result of the Toxics Effluent Guidelines Settlement Agreement, the scientific and technical communities are being asked not merely to accelerate development of toxic pollutant controls, but to improve the whole state of toxic pollutant knowledge by several orders of magnitude. It is extremely doubtful whether all of the combined talents in government, industry, and academia could achieve that end in the time allowed by the Settlement.
>
> The occasional acceleration of a government program in response to a clear and present environmental danger certainly is understandable. Crash programs in the absence of such danger to meet "yesterday" deadlines are not understandable at all. They arise not in response to specific, scientifically demonstrated needs, but from the belief in some quarters that all people can be protected at all times from all real or suspected environmental dangers–the belief, in short, that a "zero-risk" environment is attainable.
>
> Zero-risk is not attainable in the environmental area, any more than it is attainable in any other area of human endeavor. I know of no health specialist who would maintain that there are absolute thresholds, other than zero, below which no health risks will exist for anyone.
>
> We must continue to work hard and systematically to improve the quality of our environment and eliminate true threats to the public health. But crash programs arising out of a kind of national hypochondria not only can yield dubious and harmful results, but also are extremely wasteful. †

Do you feel that this is the type of question that can be settled in the courts, in the full glare of public view? Are we requiring too much of industry in this case? Can we depend on their good faith attempts to achieve pollution control without these kinds of legal threats? Can we achieve a "zero-risk" environment? Should we try?

---

* Dr. Robert H. Harris, quoted in The New York Times, 18 December 1975.

† A. E. Gubrud, Environmental Affairs Director, American Petroleum Institute at the Federal Water Quality Association Conference on Toxic Substances in the Water Environment, Washington, D.C., 28 April 1977.

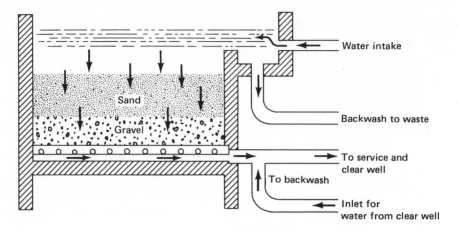

Figure 7-2   *A sand filter.*

The dislodged particles are now suspended in the upward flowing wash water. The water carrying the particles is then dumped back in the river. The filter is now clean and ready for continued use. The dirtier the water entering the filter, the more frequently is backwashing needed. All of these steps are shown schematically in Figure 7-3.

Final Chlorination

(Even though the sand filter is very effective at removing bacteria and viruses, the removal of these microorganisms is not certain. An additional step, a second step of chlorine addition, finally destroys any microorganisms remaining after sand filtration. Chlorine also reacts with any ammonia that may be present in the water. Chlorine is added beyond the level required to kill all microorganisms and beyond the amount required to react with ammonia (Fig. 7-4). This results in "free" or unreacted chlorine in solution. If free chlorine is found to be present, the water may be regarded as "safe" because any new contamination by bacteria will be prevented.)

There is a simple, quick and inexpensive test to determine whether there is unreacted chlorine in water. One reason that chlorination is so favored as a means to disinfect public water supplies is that there is this "left-over" or residual chlorine and a quick, simple way to test for it. When the test shows that "free" chlorine is present, one can be confident that any newly entering microorganisms will be killed. (The simple test for free chlorine is described on p. 169.)

(One of two alternatives to chlorination is disinfection with ozone (the advantages of ozonation are described on p. 171) and disinfection with ultraviolet light (irradiation). Neither however, leave any substance or by-product remaining in solution that can be measured quickly to check whether disinfection has been successful. In one sense, the lack of any residual or by-product is good, but from the standpoint of judging the water's safety, ozonation and ultraviolet irradiation are at a disadvantage. To check the safety of these disinfection methods requires the millipore direct count of coliforms (18 hours). To someone interested in discovering if the water is safe, a day's delay seems long.)

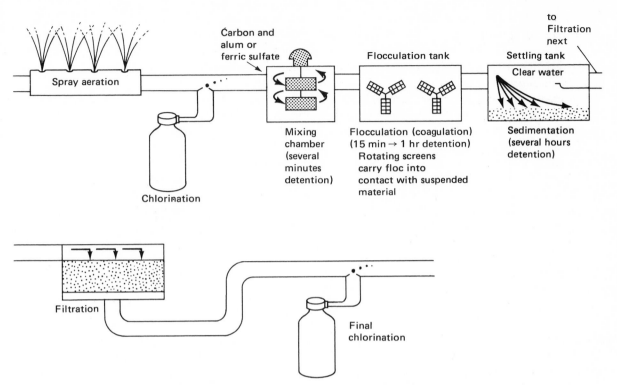

Figure 7–3 *A typical water treatment plant scheme.*

(Chlorination has recently come under attack. We mention here only the negative impacts of chlorination *on humans,* since the water we are discussing is that destined for human use. Chlorination acting on water in which hydrocarbon compounds are dissolved has been found to produce chlorinated hydrocarbons; some of the compounds which may be produced are potential causes of cancer.[3])

Treatment of water with ultraviolet light is probably not practical for municipal water supplies because the light cannot penetrate solid particles and hence cannot kill bacteria inside such particles.) (See Table 7–1.)

---

3 This aspect of chlorination is discussed in Chapter 27.

A Simple Test for Free Chlorine

Chlorine dissolves in water to produce hydrochloric and hypochlorous acids (both very dilute) and free chlorine. Both free chlorine and hypochlorous acid act as disinfectants. That is, they are capable of destroying disease-producing microorganisms. However, they will also react with substances dissolved in the water.

Ammonia and hydrogen sulfide are two common chemicals dissolved in water that react with hypochlorous acid and chlorine. These reactions use up the disinfectant potential of chlorine. Thus, when these two substances are present, more chlorine may be required to achieve disinfection.

A simple test for the presence of free chlorine involves the use of starch and iodine. To water from a swimming pool or to chlorinated tap water, add a small amount of starch plus a few crystals of potassium iodide. Free chlorine oxidizes the iodide to free iodine (chlorine is simultaneously converted to the chloride ion). Free iodine in the presence of starch produces a characteristic blue-purple color. Presence of the color indicates that free chlorine is available to oxidize the iodide ions. The starch-iodine test is used to indicate whether free chlorine is present in sufficient concentration to destroy the pathogenic bacteria. It gives a very swift answer to the question: Is the disinfection process operating properly; that is, is the water safe to drink?

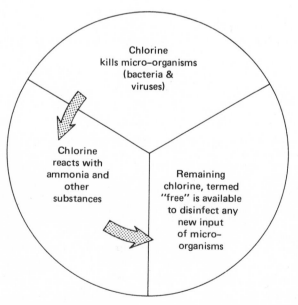

Figure 7-4 *Chlorine is the favored disinfectant for water supplies. Microorganisms are killed by chlorine and ammonia reacts with chlorine. Any chlorine in excess of the amount needed to disinfect and react with ammonia will remain "free" in solution, protecting the water supply from any new sources of pollution en route to the consumer.*

TABLE 7-1   A Comparison of Chlorination and Ozonation

Process	Substance Used	Effect	Objectionable Reactions, By-Products	Check on Water Safety
Chlorination	Chlorine gas	Kills microorganisms	Chlorinated hydrocarbons, taste and odor substances	Immediate check: safe if free chlorine is present
Ozonation	Ozone gas	Kills microorganisms	None	Must run standard tests for coliforms (minimum 18 hours)

Ozonation To Purify Water

The ozone process, like the chlorination process, is a simple contacting of water with gas. Ozone gas is a powerful oxidizing agent, which destroys bacteria and viruses. In contrast to chlorination in which chlorine may combine with hydrocarbons, ozone may actually destroy some of the hydrocarbon compounds by oxidation. Also, potential taste and odor problems created by chlorination are avoided when ozone is used. The combination of chlorine reacting with phenol in water produces an undesirable taste and odor, but ozone does not. The creation of chlorinated hydrocarbons is also avoided with ozone. Furthermore, ozone is effective in removing color.

Ozonation of water supplies is practiced in a number of European cities, but chlorination is still the process most water supply engineers choose. In the United States, ozonation is in use by a number of companies that produce bottled water. They choose this method of disinfection to avoid the taste chlorination gives to water, a taste which the consumer may associate with a municipal water supply. The city of Whiting, Indiana, does use ozone for purifying its water supply. Their water supply, drawn from Lake Michigan, contains phenols, which would react with chlorine to give the undesirable taste and odor mentioned earlier.

Questions

1. Complete the following table, which summarizes the procedures or steps in treating water to make it safe to drink and the reason for these steps.

Water Treatment

Procedure	Reason for this Procedure
Spray aeration	Removes dissolved gases like hydrogen sulfide
Addition of activated carbon	
Coagulation with alum or ferrous sulfate	
Sand filtration	
Chlorination	

Further Reading

Miller, G. W. and R. G. Rice, "European Water Treatment Practices," *Civil Engineering, ASCE,* p. 76 (January 1979).

The arguments for using ozone rather than chlorine are given in this paper, along with the experiences of European water supply managers, who use ozone regularly.

Water Quality and Treatment, 3rd ed., American Water Works Association, Inc., New York, 1971.

Water Treatment Plant Design, American Water Works Association, Inc., New York, 1969.

Both of these sources are somewhat technical in nature, intended principally for environmental engineers and chemists.

Chapter 8

Organic Wastes and Dissolved Oxygen

WHY ORGANIC WASTES ARE POLLUTANTS

Water pollution means different things to different people. It means the presence of toxic chemicals, to some; to others, the presence of disease-causing bacteria. To still others, pollution is seen in floating weeds and algae. To many engineers and scientists, however, water pollution is indicated by organic wastes. Such wastes can come from industry, from farming, and from cities. Organic wastes, composed mainly of carbon, hydrogen, and nitrogen, are responsible for many of the unpleasant conditions that exist in polluted rivers and lakes.

A Simple Experiment

To understand what organic water pollution is, we can consider the following experiment. If we mix sugar and gelatin in an open bottle of water, the solution that results is clear. We would not label the solution as "polluted." A fish should be able to live in the water (after any chlorine has been allowed to diffuse out), at least for a time. Yet, the contents of the bottle will, in less than a week, become clouded. Dissolved oxygen, which fish need to live, will have been used up, and the fish will die from lack of oxygen.

The organic wastes, in this case, were sugar and gelatin. Together, they caused a condition in which fish and many other water creatures cannot live. That condition is the lack of oxygen, resulting from the breakdown of the organic materials, sugar and gelatin, by bacteria. These materials were converted in the presence of oxygen to simpler substances. This process is a biochemical oxidation. If enough species of bacteria are present, the carbon (C) in the sugar and gelatin is oxidized to carbon dioxide (CO_2). Furthermore, the hydrogen (H) in the sugar and gelatin is oxidized to water (H_2O). Last, the nitrogen (N) in the gelatin would eventually be converted to nitrate ions (NO_3^-). All of these reactions utilize oxygen. The net effect is to reduce the amount of oxygen dissolved in the water. (For the measurement of effects of pollutants on dissolved oxygen levels see p. 179.)

Oxygen in Natural Waters

When water is well-mixed in the presence of air, the water will be able to dissolve a maximum amount of oxygen, called the saturation concentration. The saturation concentration is the milligrams of oxygen that can be dissolved in one liter of water before the water can hold no more. The saturation concentration, achieved by thorough mixing, is between 8 and 9 milligrams of oxygen per liter of water, depending on the temperature. In summer it approaches 8 milligrams per liter at higher water temperatures. In winter, at colder water temperatures, it is nearer 9 milligrams per liter.

If the concentration of organics in a sample of water is sufficiently large, oxidation of the organics by bacteria and protozoa may use up all the oxygen in the sample. That is, the concentration of oxygen in water may be reduced to zero. When there is no oxygen in natural waters, we say the water is in an *anaerobic condition*. Under anaerobic or very low oxygen conditions, normal aquatic life such as fish species die away. The few species that do survive and thrive in any numbers are those especially adapted to the low-oxygen condition.

In low-oxygen conditions, the normal bacterial population, which makes use of oxygen, will die away. In its place a population of bacteria grows up which lives on sulfur instead. The sulfur atom occurs in organic wastes and is structurally similar to oxygen, except that it has one more ring of electrons. Sulfur then takes the place of oxygen in the "oxidation" reaction, producing, for instance, hydrogen sulfide (H_2S) instead of water (H_2O). The well known smell of rotten eggs is the smell of hydrogen sulfide.

To summarize, the breakdown of organic wastes by bacteria leads to a loss of oxygen from water. An environment in which oxygen is absent possesses no fish and only a few specially adapted animal species. In this environment, specialized bacteria produce, among other end products, the foul smelling gas, hydrogen sulfide.

STREAM HEALTH AND DISSOLVED OXYGEN

What Happens at a Sewage Discharge

Now that we know how an organic pollution source can alter the dissolved oxygen in a body of water, we can investigate how aquatic life responds to such pollution. When organic substances from the waste outfall of a community or industry enter a river or stream, the concentration of dissolved oxygen in the stream will decrease. This is due to the oxidation by bacteria and protozoa of the organic materials. Natural mixing of the stream with air does tend to replace the oxygen which has been removed, but not immediately. There exists instead a competition between the oxygen-depleting forces and the oxygen-restoring forces. This produces a typical picture of oxygen concentrations along the stream. The profile is shown in the top portion of Figure 8–1.

The four basic sections of the stream are (1) a clean water zone (high dissolved oxygen) upstream from the pollution source, (2) a zone of decline (falling level of dissolved oxygen), (3) a zone of damage (relatively constant and low level of dissolved oxygen), and (4) a zone of recovery (rising dissolved oxygen). Of course, if more than one sewage outfall occurs along the stream or river, the zone of damage may extend for many miles. Such conditions may occur along rivers in major urban and industrial areas.

Clean Water Zone

The clean water zone upstream from the waste discharge may sustain fish, mayflies, clams, stoneflies, and other species. These species require oxygen in water in order to survive (Fig. 8–2). If the oxygen concentration falls, these sensitive species will be among the first to disappear. Such fish as trout, bass, salmon, and minnows are among the sensitive species. While the needs of various species of fish vary, it is generally thought that dissolved oxygen levels of five milligrams per liter or better are necessary to ensure survival of fish.

Zone of Decline

A zone of decline follows the introduction of an organic waste. Species that can survive at these somewhat lower levels of dissolved oxygen are referred to as "intermediately tolerant." They are pictured in Figure 8–3. Solids from the waste water outfall may cloud the water.

Damage Zone

The damage zone, which follows the zone of decline, is one in which dissolved oxygen is almost

	Zones of pollution				
	Clean water	Decline	Damage	Recovery	Clean water
9 milligrams per liter Dissolved oxygen level 1 milligram per liter	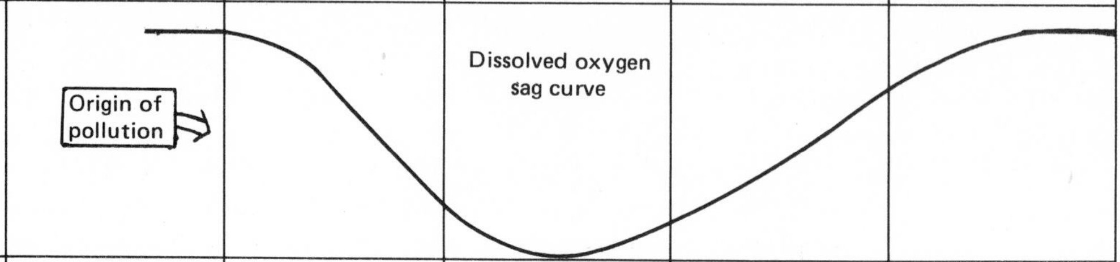 Origin of pollution ⇨		Dissolved oxygen sag curve		
Physical indices	Clear, no bottom sludge	Floating solids, bottom sludge	Turbid, foul gas, bottom sludge	Turbid, bottom sludge	Clear, no bottom sludge
Fish present	Game, pan, food and forage fish	Tolerant fishes—carp, buffalo, gars	None	Tolerant fishes—carp, buffalo, gars	Game, pan, food and forage fish
Bottom animals present	Caddisfly Stonefly	Bloodworm Snail	Sewage mosquito larvae Sewage fly larvae Rat–tailed maggot	Bloodworm snail	Caddisfly Mayfly naiads
Algae & protozoa present	Dinobryon Cladophora Ulothrix Navicula	Paramecium Vorticella	Phormidium Stigeocloneium Oscillatoria	Euglena Spyrogyra Pandorina	Dinobryon Cladophora Ulothrix Navicula

Figure 8–1 *The zones in a polluted water course. (Adapted from K. Mackenthun,* Toward a Cleaner Aquatic Environment, *U.S. Government Printing Office, Washington D.C., 1973.)*

gone. When the dissolved oxygen has fallen to very low levels, only a few species are capable of survival, and the numerous species that characterized the clean stream have disappeared. In their place are a group of organisms we refer to as "pollution tolerant" because of their ability to survive under low dissolved oxygen conditions. One such organism is the sludgeworm, which consumes sludge and thrives in water with as little as 1/2 milligram of oxygen in a liter. Another is the rat-tailed maggot, a resident in the sludge of stream bottoms; the maggot breathes by a long tube which reaches to the water surface. The maggot is the larva of the drone fly. Its numbers may so increase that they cover the bottom of the stream bed in a waving red sheet. Another resident of this zone is the bloodworm. This organism, like the sludgeworm, consumes sludge as its diet. These species are illustrated in Figure 8–4.

In the clean water zone, many species exist side by side, and each is moderately represented in terms of numbers of organisms. In the damage zone, there are very few species, and those are present in enormous numbers. Occasionally, however, one fails to find large numbers of the organisms of the pollution-tolerant species. In this case, some chemical poison in the sewage may be preventing their increase.

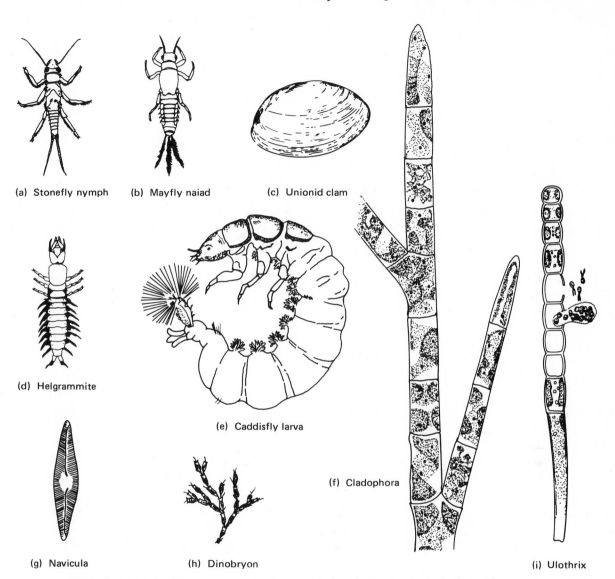

(a) Stonefly nymph   (b) Mayfly naiad   (c) Unionid clam

(d) Helgrammite

(e) Caddisfly larva

(f) Cladophora

(g) Navicula   (h) Dinobryon   (i) Ulothrix

Figure 8-2   *(a)-(e) Clean water animals associated with the stream bed. (f)-(i) Clean water (sensitive) algae.*

Zone of Recovery

A zone of recovery follows the damage zone. Here the water is likely to be clear, allowing the penetration of sunlight. The oxygen is on its way up to more reasonable levels. With the clearing of the water and recovery of dissolved oxygen, algae may begin to grow. Their presence can result in a fluctuation in the oxygen content in the water. During the hours of daylight, the algae produce oxygen as a by-product of photosynthesis. But in the hours of night, the respiration and decomposition of algae

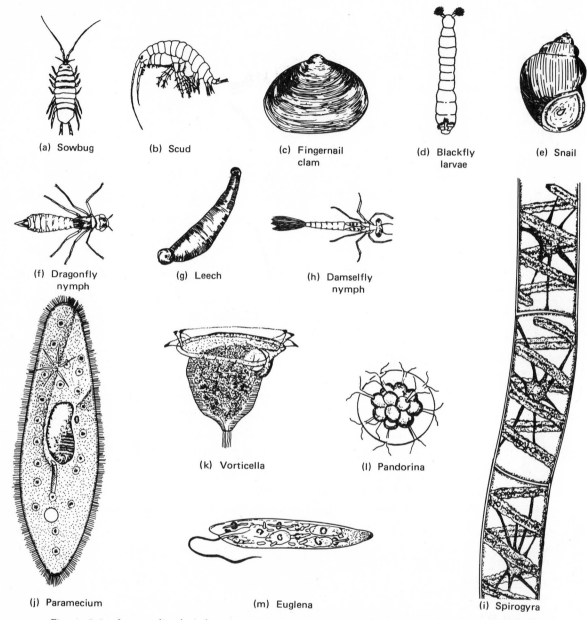

(a) Sowbug

(b) Scud

(c) Fingernail clam

(d) Blackfly larvae

(e) Snail

(f) Dragonfly nymph

(g) Leech

(h) Damselfly nymph

(k) Vorticella

(l) Pandorina

(j) Paramecium

(m) Euglena

(i) Spirogyra

Figure 8–3 *Intermediately tolerant aquatic species. (a)-(h) Intermediately tolerant animals associated with stream bed; (i)-(m) intermediately tolerant algae and protozoa.*

removes oxygen from the water, yielding a low oxygen concentration. It is possible for the algae-caused swings in dissolved oxygen to deplete the stream's oxygen to such an extent that the normal aquatic life, which requires oxygen, may not be re-established. Beyond the zone of recovery, the species of the clean-water zone are found again.)

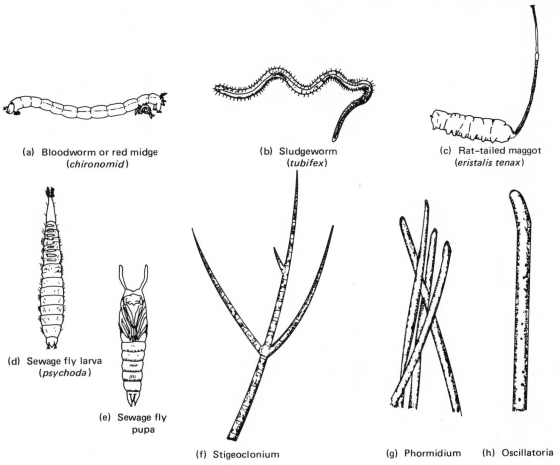

(a) Bloodworm or red midge
(chironomid)

(b) Sludgeworm
(tubifex)

(c) Rat–tailed maggot
(eristalis tenax)

(d) Sewage fly larva
(psychoda)

(e) Sewage fly
pupa

(f) Stigeoclonium

(g) Phormidium

(h) Oscillatoria

Figure 8–4 *Very tolerant aquatic species. (a)-(e) Very tolerant animals associated with the stream bed; (f)-(h) very tolerant algae.*

Organic Wastes: How to State Their Levels and How to Measure Them

How We Summarize the Concentration of Organics in a Single Number

It may occur to you to ask how we determine the amount of organic wastes in a stream. This knowledge would help you to predict the amount of oxygen that would be used up in oxidizing the organic substances. It would also help you to predict the species that will be present in the water.

Unfortunately, to measure the quantity of every organic substance in a waste stream is an exhausting and perhaps impossible job. Each substance that is present would first have to be identified, already an enormous task. Then, once a substance is shown to be present, the quantity of that substance has to be measured. Then, if you also knew the oxygen consumption when bacteria oxidize each gram of each substance present, you could predict how much oxygen could be removed by all the wastes in a liter of the polluted water.

There is a much easier way, however, to arrive at the amount of oxygen that would be consumed. An ingenious and simple measure of the concentration of organic wastes in water was proposed in England near the turn of the century. The method determines not only the concentration of organics, it also reveals the amount of oxygen that all of the organic wastes, acting together, will eventually remove from the water. The number reported is the Biochemical Oxygen Demand (BOD). It is the amount of oxygen that would be consumed if all the organics in one liter of the polluted water were oxidized by the bacteria and protozoa. It is reported in milligrams of oxygen.

Suppose we are studying a sample of polluted water discharged as a waste stream from a community. The sample is reported to have a BOD of 120 milligrams per liter. This means that the bacteria and protozoa oxidizing all the organic substances in one liter of the water would consume 120 milligrams of oxygen. Now suppose that 50 milliliters (0.05 liter) of the polluted water were mixed with 950 milliliters of clean water. The sample has been diluted to 50 parts in 1000 parts or to 1/20th of the mixture. The BOD of the mixture will be (1/20) × (120) or 6 milligrams per liter.

This dilution is similar to what happens when the flow of sewage from a community enters a watercourse that has a substantial water flow. Suppose the BOD concentration in the discharge is the 120 milligrams per liter we were discussing. Assume further that the community empties 1 million gallons of wastes a day into the stream. Assume the clean stream is flowing at 19 million gallons per day past the outfall. The BOD of the mixture is diluted to 1/20th of the discharge value or 6 milligrams/liter. This number, 6 milligrams per liter, is the amount of oxygen that can be removed from every liter in the combined flow of stream and waste water. Given that clean water can only possess up to about 9 milligrams of oxygen per liter, the oxygen of the mixture can be reduced to a very low level. How low it will actually go depends on many factors, but especially on the rate of oxidation by bacteria and on the rate of restoration of oxygen to the stream.

Thus, the BOD tells a biologist the ability of the polluted water to deplete the oxygen resources in the stream. This is a valuable indicator of pollution since the lack of oxygen is responsible for fish kills; it is the cause also of foul odors and of undesirable pollution-tolerant organisms. The BOD does not explain what organic substances are present in the water or in what quantity they are present. Nonetheless, it provides a rapid and important insight into what pollution damage the waste water could cause.

Measuring the Concentration of Organic Wastes

The specific procedure to determine the value of Biochemical Oxygen Demand involves a number of steps. First, a carefully measured sample of the polluted water is diluted in a much larger, measured volume of unpolluted water. A 300-milliliter bottle is specially made for this test. The unpolluted water has previously been well shaken in the presence of air so that the water has absorbed as much oxygen as it can. This water is said to be saturated with oxygen. The water may also have been "seeded" with the microorganisms known to break down organic wastes in the presence of oxygen. If the BOD of polluted river water is to be measured, however, the needed microorganisms are likely to be present in the sample of polluted water already. The mixture of polluted water and clean water completely fills the bottle. The bottle is then closed to the air by inserting a glass stopper. This closure prevents the entry into the bottle of any new oxygen from the air. Two bottles are commonly prepared in this way from samples of the same polluted water.

One of the bottles is set aside in the dark at 20°C, to be taken out at the end of five full days. The darkness is necessary to prevent the growth of algae. Algae may contribute oxygen to the water as a by-product of photosynthesis. Such oxygen would interfere with the measurement of BOD and so is avoided. One bottle is not set aside. The quantity of oxygen dissolved in it is determined immediately (Fig. 8–5).

There are several methods to measure dissolved oxygen. There is a chemical test and a test done by electrodes. The chemical test requires special solutions and procedures. The electrode test only requires the insertion of a probe into the water. However, one must first ascertain that the electrode is calibrated, i.e., that it reads the appropriate value, and this requires prior testing.

Suppose the quantity of dissolved oxygen in the bottle not set aside was 7.5 milligrams. After five days, the quantity of oxygen in the bottle that was stored in the dark was 6.0 milligrams. No oxygen has been added to the second bottle from any other source. Therefore, the difference between the initial quantity of oxygen and the quantity present after five days must be the oxygen removed by microorganisms. The oxygen removed in five days is 1.5 milligrams.

Suppose a 10 milliliter sample of polluted water was diluted initially. Since 10 milliliters of this polluted water caused the removal of 1.5 milligrams of oxygen, 150 milligrams of oxygen could be removed by a 1 liter sample (1000 milliliters) of the polluted water. The BOD (Biochemical Oxygen Demand) of the polluted water is 150 milligrams per liter.

By the end of five days, the rate of oxygen removal has become very slow. Only a little more oxygen will be removed each day after that. The five-day BOD then is the number commonly used to measure the oxygen-consuming ability of polluted water.

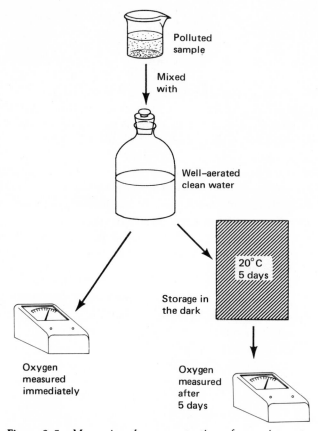

Figure 8–5   *Measuring the concentration of organic wastes.*

Questions

1. When organic wastes are added to natural waters, they may cause fish to die even though the wastes themselves aren't directly poisonous to fish. Can you explain this?

2. Describe how sewage changes the stream communities found downstream from the outfall.

Further Reading

Hart, C., Jr., and S. Fuller, *Pollution Ecology of Freshwater Invertebrates*, Academic Press, New York, 1974.

A research volume with papers by a number of investigators, each dealing with a particular group of invertebrates. Includes methods of using organisms as indicators of water pollution.

Kemp, L., W. Ingram, and K. Mackenthun, eds., *Biology of Water Pollution*, Federal Water Pollution Control Administration*, U.S. Department of Interior, 1967.

This is a collection of biologically oriented papers from journals treating stream pollution, biological waste treatment, and public health as influenced by stream pollution. The classic paper "Stream Life and the Pollution Environment" by Bartsch and Ingram is reprinted here. This article alone makes a copy worth obtaining from the library.

Mackenthun, K., *Toward a Cleaner Aquatic Environment*, U.S. Environmental Protection Agency, 1973, available from U.S. Government Printing Office, Stock No. 5501-00573.

This is virtually a textbook on the biology of water pollution. It is interesting, well-written, well-illustrated and is rarely too technical for the novice to follow. Directions for identifying algae associated with water pollution are provided.

Mackenthun, K., *The Practice of Water Pollution Biology*, Federal Water Pollution Control Administration*, U.S. Government Printing Office, 1969.

Another of Mackenthun's well-written and well-illustrated texts.

Mitchell, R., *Water Pollution Microbiology*, Wiley-Interscience, New York, 1972.

A text with 17 papers for advanced students in pollution microbiology and environmental engineering.

---

* The organizations marked by an asterisk are not separate entities but evolved through adminstrative change to become part of the Environmental Protection Agency.

Chapter 9

Eutrophication

FEEDING LAKES

In 1974, Russell Train, Director of the U.S. Environmental Protection Agency, warned a meeting of the New Hampshire Lakes Region Clean Water Association that nutrients which cause lakes to age rapidly are increasing in U.S. waters. What did he mean by this? How can a lake age? What nutrients are overfeeding lakes?

Carbon, oxygen, hydrogen, nitrogen in nitrates, and phosphorous in the form of phosphates are some of the elements, or "foods," needed by plants. Smaller amounts of many other substances such as iron, calcium, and copper are also required.

Lakes that have large amounts of the nutrients plants need are called *eutrophic* (*eu*, well and *trophos*, nourishment). Relatively high concentrations of plant nutrients in eutrophic lakes allow huge amounts of water plants, such as algae, to grow. This over-population of algae makes a lake unpleasant to swim in. Some kinds of algae, which grow in long strands, wind themselves around boat propellors, making boating impossible. The water in such a lake tends to be scummy, cloudy, or even soupy green. A rapidly growing population of algae is called a bloom. The algae in a bloom may wash onto the lake shores in storms and high winds, and die and decay there, producing bad smells. Sport fish have trouble living in a lake with blooms. The algae can use up most or all of the oxygen in the lake at night, leaving none for fish. Such a lake is clearly not the kind people enjoy using for recreation or drinking water. Cities or industries that want to use eutrophic lake water find they must filter it first to remove the algae. This increases costs. Lake Mendota in Wisconsin is an example of a eutrophic lake (Fig. 9–1).

An *oligotrophic* lake (*oligo*, poorly and *trophos*, nourishment) is one with low levels of nutrients. Such lakes, because they lack plant nutrients, have only a small community of algae. These lakes are clear and sparkling. They are fun to swim in or boat on, delightful to look at, and should provide good drinking water. Lake Tahoe in Nevada is an example.

Figure 9–1 *A eutrophic lake supports the growth of many kinds of plants, both rooted and floating. (Grant Heilman Photography.)*

HOW WATER BECOMES EUTROPHIC

Natural Versus Artificial Eutrophication

Many lakes are oligotrophic when they are first formed. However, over the centuries, silt, debris, and dissolved nutrients accumulate in these lakes, filling in their originally deep basins and turning the lake waters into a nutrient-rich soup. In this way, after a few thousand years, a lake can change, naturally, from an oligotrophic lake to a eutrophic lake. This process is called aging. Not all lakes age, naturally. Some lakes, in the normal course of events, maintain their oligotrophic characteristics. However, any beautiful, clear lake can be made eutrophic if large amounts of plant foods are added. The most common way this happens is when city or industrial sewage is dumped into lakes. The sewage from a city or town is rich in nutrients, such as nitrates and phosphates, that plants need to grow.

There is some controversy about which plant nutrients are most responsible for causing the premature aging of U.S. lakes, rivers, and coastal waters. In general it is agreed that phosphates and nitrates are the most likely culprits.[1]

In his lecture to the Clean Water Association, Russell Train said: "Nitrogen shows an increase during the past decade in about 74 percent of those sections of the [U.S.] waterways that were measured, and phosphorus increased in about 84 percent.... More than half of the waterways measured by the Environmental Protection Agency, in a nationwide survey, are exceeding guidelines for phosphorus and about one-fourth of them show nitrogen nutrients exceeding recommended levels."[2]

Nitrates and Eutrophication

In most U.S. coastal waters and in lakes such as Tahoe, which are fed by steep mountain streams, a low concentration of nitrates seems to be the factor which normally limits plant growth.

Nitrates can enter natural waters from a number of sources. Sewage from cities or animal feedlots are two of the main sources, since human and animal wastes are about one-half nitrates. Fertilizer, which runs off crop lands or suburban lawns during a rainstorm, also contains a large amount of nitrates. There are, in addition, several natural sources. Volcanic eruptions and lightning can change the nitrogen gas in the air into nitrates. Some of the blue-green algae can also turn nitrogen from the air into nitrates. The process is called nitrogen-fixation. These algae, which are usually considered the worst nuisance plants, are thus not dependent on an outside source of nitrate.[3] Finally, waterfowl, which feed on the shores of lakes and ponds and then fly over the water, are the source of what one wildlife expert termed "bombed in" nitrates.

Phosphates and Eutrophication

In contrast, phosphorus in the form of phosphates, is found naturally in waters in only trace amounts. Most pollutant phosphorus comes from human activities. Phosphate mines pollute certain areas of the country, such as central Florida. Fertilizer run-off contains large amounts of phosphate. Domestic sewage is high in phosphates, too, with one half coming from human wastes and the other half from detergents.

Animal feedlots, where cattle are grouped together in large pens to be "finished" or fattened for

---

1 This is the limiting nutrient concept, which was explained on p. 97.

2 "Train Warns of Increasing Nutrients in Nations Lakes," *Environmental News*, 17 July 1974, U.S. Environmental Protection Agency, Washington, D.C., 20460.

3 There is some evidence that decreasing the amount of nitrates in water, without decreasing the amount of phosphate as well, may even favor the growth of the undesirable blue-green algae, because they can fix nitrogen while other water plants cannot.

market, are sources of both nitrates and phosphates. Although growth of algae in coastal waters and some wilderness lakes is limited by low concentrations of nitrates in the water, growth in most U.S. lakes and rivers is believed to be limited by the amount of phosphate present (Fig. 9–2).

HOW CAN EUTROPHICATION BE CONTROLLED?

The Detergent Phosphate Problem

Since phosphate is the nutrient that is limiting plant growth in most U.S. waters, and because detergents are a major source of phosphates, it is natural to ask whether we wouldn't be better off banning the use of phosphates in detergents in an effort to stop eutrophication.

Figure 9-2 *The role of phosphorus in lake eutrophication is dramatically illustrated in this photograph of Lake 226. The lake has two basins. Phosphorus, carbon and nitrogen were added to the far basin (upper part of photo). An algal bloom covered the basin in two months. The lower basin was treated at the same time with the same amounts of nitrogen and carbon, but no phosphorus. This basin remained clear and sparkling. (Photograph courtesy of D. W. Schindler.)*

As you can see, from Table 9–1, the chemical present in the largest amount in phosphate detergents is the phosphate builder. Dry detergents usually contain STP or Sodium Tripoly Phosphate. Liquid detergents may contain sodium or potassium phosphates. Builders are necessary because water often contains calcium and magnesium ions. (Water that has more than 75 milligrams per liter of calcium and magnesium[4] is called hard water.) The calcium and magnesium ions combine with soap, forming a hard precipitate. This precipitate does not make suds and doesn't dissolve grease or dirt. In a similar fashion, calcium and magnesium ions can tie up the surfactant, or dirt-dissolving, molecules in detergents. Without builders, manufacturers would have to include a great deal of relatively expensive surfactant in their detergents to be sure that their product performed well in hardwater areas. In addition, builders help to keep dirt from reattaching itself to clothes.

TABLE 9-1   **A Typical Heavy-Duty Laundry Detergent**

Surfactant (dissolves the dirt)	20%
Phosphate builder	50%
Carboxymethyl cellulose (CMC)—(keeps dirt from redepositing on clothes)	0.5%
Sodium silicate—(prevents caking of detergent)	0.6%
Optical brighteners	0.3%
Perfume	
Sodium perborate—(bleach)	
Sodium carbonate and sodium sulfite—(fillers, water softeners, or generally inert ingredients.)	

Many scientists feel that a complete ban on phosphates in detergents would remove about half of the phosphates in sewage. In a number of areas, a ban on or a reduction in the amount of phosphate allowed in detergents has been tried. One such

4 Measured as calcium carbonate.

locality is the area around Onondaga Lake in New York. The lake had, for years, been the dumping ground for sewage from the city of Syracuse. As a result, blooms of scum-forming algae were covering the lake every summer. In July 1971, areas around Onondaga Lake passed laws limiting the amount of phosphates allowed in detergents. Two years later it was obvious that the blooms were not as bad as in previous years. There is now more oxygen in the lake water than there was before.

This example brings up a serious problem, however. If manufacturers can not put phosphates in detergents, what can they use instead? In some areas water is naturally soft enough (that is, it has little or no calcium or magnesium) for soap to be used. Water can also be made soft by passing it through a water softener or ion exchanger. However, in most houses the water is hard enough so that some sort of builder is necessary.

The choice of a chemical builder to replace phosphates must be carefully made. Just as people did not realize a few years ago that phosphates could harm the environment, there is no way of knowing whether a substitute for phosphates might not be harmful. If a single substitute chemical were chosen, there is the possibility that we would be adding as much as two million pounds of some new material to the waters, every year.

At the present time, phosphate-free detergents contain:

1. carboxymethyl cellulose, an organic carbon compound that may present a problem because it may not be biodegradable.[5]

2. carbonates and silicates, in the form of sodium metasilicate and sodium carbonate. Both combine with calcium and magnesium ions to give insoluble precipitates. However, there is more of a hazard of chemical burns with these chemicals than with phosphates. Also, large amounts of carbonates

and silicates can make phosphates more soluble. There are vast amounts of phosphates already captured in the muds of lake bottoms. Any substance that helps mud phosphates dissolve into lake waters is not a good ingredient in a phosphate-free detergent.

3. borates, in the form of sodium tetraborate decahydrate or borax. However, waters having more than one part per million of borax are toxic to most plants.

Although many states and localities either enacted or considered bans on phosphates in the early 1970s, at least 20 of these bans have now been revoked. Indiana and New York and several other states and cities still do not allow phosphates in detergents, but many other areas have settled for a limit on the amount of phosphate (8.7% phosphorus, which is equivalent to 50% phosphate). Phosphate-free detergents, which once accounted for 14% of the market, now have about 3% or 4% of sales. The main reason for the failure of the anti-phosphate campaign has been the lack of a suitable substitute.

Sewage Treatment as a Solution

There is, however, another way to attack the phosphate problem. Phosphates can be removed from sewage by a method called *tertiary treatment*. In this method, phosphates are precipitated out of sewage before the treated water is released into lakes or rivers. (See Chapter 10.) Eighty to 90% or even more of the phosphates in sewage can be removed. According to studies done in Sweden, the cost is not very great, perhaps one Swedish crown per person per year. Why then shouldn't we allow the use of phosphate detergents and remove the phosphate from sewage by tertiary treatment? One reason is that at present not everyone in this country is served by a sewage treatment plant. Some have septic tanks, others have only treatment ponds. Less than 1% of the people in this country are served by sewage treatment plants that have the

---

5 "Biodegradable" is another term, along with "phosphate-free", that you often see on detergent labels. What this means is explained on p. 194.

facilities for phosphate removal. This particular solution then, is tied up in the economics and politics of building and improving sewage treatment plants. It is a solution that would seem ideal and should be worked for. But it also must be viewed as still in the future, while eutrophication of lakes is occurring right now.

The Environmental Protection Agency is carrying out a study that should help to discover exactly which U.S. lakes and rivers are in danger of becoming eutrophic due to the addition of phosphates. Meanwhile, a combination of solutions is called for. People who live in areas served by a sewage treatment plant that removes phosphates should use a phosphate detergent. Those of use who live in soft water areas or have water softeners could use soap. The rest of us can use a detergent that has no phosphates or the very least amount of phosphates that will get clothes clean. And we should very strongly support the building or equipping of sewage treatment plants to remove phosphates in our area.

Controlling Other Phosphate Sources

Treatment of sewage to remove phosphates will not only help to solve the detergent phosphate problem but will also remove phosphates from other sources, such as human wastes. What then remains? The main problem is nutrients which enter natural waters in small amounts but from many sources. This is called non-point source pollution because it is usually hard to pinpoint exactly where the phosphates or nitrates are coming from. The total amount of nutrients from non-point sources can be very high. In some areas it is equal to or greater than the amount from sewage treatment facilities.

The run-off of fertilizers from agricultural land is one example of non-point source pollution. Poorly working septic systems, drainage from cattle feedlots, and city storm drains, which contain nutrients from lawn fertilizers, pet wastes, and other sources, are more examples.

Nutrients from manure and chemical fertilizers, contained in agricultural runoff, can be controlled by good farming practices, such as strip cropping, terracing, and careful timing of fertilizer application so that fertilizer is not immediately washed away by rains.[6] Feedlot operations may need to install sewage treatment equipment to prevent animal wastes from enriching nearby streams.

The municipalities near Lake Minnetonka, Wisconsin, have tried to solve the problem of storm waters, which often have too much volume for sewage treatment plants to handle, by routing storm waters through nearby wetlands. Organisms in the marshes remove phosphates and nitrates from the storm waters before they enter the lake.

Some communities have adopted the solution of piping municipal sewage and storm waters around threatened lakes and into some other body of water. For instance, sewage that once flowed into Lake Washington in Seattle now is routed around the lake and into the Puget Sound. Although Lake Washington is slowly recovering from its gradual slide into a eutrophic state, we need only think about Puget Sound to realize that this is a temporary "solution". Other solutions to non-point source pollution are discussed in Chapter 10.

Helping Eutrophic Lakes Recover

Once lakes or ponds are eutrophic, simply preventing the addition of any more phosphates is often not enough to help them become clear and sparkling again. This is because phosphates that have settled into the bottom muds can, in many cases, dissolve back into the waters. Thus phosphate levels may remain high for years even with no new additions. A number of ways of combatting

---

6 Strip cropping and terracing also reduce the damage done to farmland by erosion. These techniques are explained in Chapter 34.

WHEN SHOULD WE SOUND THE ALARM?

We must be sure that a full-sized, hungry, four footed wolf with teeth is coming before we cry out about it....

B. J. Kilby and G. Zetterberg

I believe that it is my social responsibility to tell my fellow citizens....

B. Gillberg

In 1971, Swedish scientist Bjorn Gillberg wrote a paper about a minor ingredient in modern detergents, optical brighteners. He had found evidence that optical brighteners caused mutations, or changes in the hereditary material, in yeast. There is evidence that chemicals which cause mutations are also likely to cause cancer. * However, when scientists in another laboratory tried to repeat his experiment, they did not find the optical brighteners to be mutagenic. In a letter to <u>Science</u>, B. J. Kilbey and G. Zetterberg wrote:

> We do not think that our experiments indicate unequivocally that no danger exists from optical brighteners. The data are insufficient at present for this conclusion to be drawn....We also do not wish to imply that public watchdogs, such as Gillberg, do not perform a useful function. However, we must be sure that a full-sized, hungry, four-footed wolf, with teeth, is coming before we start crying out about it. For environmental biologists, this means doing all in our power to be sure that the right experiments are done, positive results are reproducible, and any artifacts of method are excluded....If we startle the public too many times with sensational claims that are later retracted, we run a real risk of losing our most valuable ally if and when a real crisis comes. †

Gillberg replied that he thought the optical brightener manufacturers must have either changed their formulation since his original experiments were run or that there are contaminants which appear now and then in the brighteners and the contaminants had caused the mutations. He wrote:

> However, I have not made any sensational claims about brighteners; the only thing I say in my paper is that I consider it of importance to carry on with genetic studies of brighteners against the background of my results. Research has now begun in other laboratories that should have been undertaken before the brighteners were released on the market. The benefits of a product must of course always be weighed against the risks it may create. In such a situation I prefer not to give the product the benefit of the doubt if there are some questions raised. Questions have been raised about these compounds, and I believe that it is my social responsibility to tell my fellow citizens about them. ‡

Who is right? At what point should the public be told that some doubt has been cast on the safety of a product that is in common use? Must we be careful not to "cry wolf" too many times to avoid producing boredom about environmental dangers, or is it important to give people all available information as soon as it comes to light?

---

* More on this in Chapter 26.

† B. J. Kilbey and G. Zetterberg, "Optical Brighteners," <u>Science</u> **183,** 798 (1 March 1974).

‡ B. Gillberg, "Optical Brighteners and Social Responsibility," <u>Science</u> **184,** 901 (13 September 1974).

this problem have been tried, with varying success. In some cases the flow of water through a lake can be increased. This will help to wash nutrients and algae out of the lake. Low phosphate water from Seattle's sewage treatment plant is used to increase the flow through Green Lake in western Washington.

Chemicals such as aluminum, iron and calcium have been used to precipitate the phosphate out of lake waters. The cost of this sort of treatment is about $150 to $300 per hectare (depending on how polluted the lake is) and results last about three years. Fly ash from power plants has also been used to precipitate phosphates. In addition, the fly ash can seal off the lake sediments, preventing phosphates from dissolving back into the waters. This provides a way for the power plants to dispose of fly ash. However, the ash is known to contain heavy metals, which could be toxic to aquatic life.

Oxygen can be added directly to the oxygen-poor lower layers of eutrophic lakes. This allows fish to take up residence there again. It also helps prevent phosphate release from lake muds. (Release is more likely to occur under conditions of little or no oxygen.) The method has been used in Europe for years and has been tried successfully in several lakes and reservoirs in the U.S.

In some cases, there seems to be no alternative but to actually dredge the phosphate rich sediments out of a lake (Fig. 9–3). This is, however, the most expensive method of lake restoration. Long Lake, a 59 hectare lake in Michigan, was dredged at a cost of $185,000 in 1961–1965. One and one quarter meters of sediment were removed from the lake bottom. The lake recovered and has remained in good condition since then. A similar attempt at Buckingham Lake in New York failed. The failure illustrates the fact that it is not yet clear what techniques will work with any particular lake.

The Environmental Protection Agency is authorized to give grants to communities wishing to restore eutrophic lakes. In order to be eligible for a grant, the lake must be available to the public, it must be classed as fresh water, and the applicants

Figure 9–3   *The town of Växjö, Sweden, stopped pouring sewage into overexploited Lake Trummen in 1958. (a) By 1969 the unrecovered lake contained no oxygen, no fish, no underwater vegetation, and was of little use to humans. The main problem was the rapidly increasing black muddy sediment fed by decaying plankton. Starting in 1970, the sediment was sucked out and put into settling ponds; runoff water was cleansed of phosphorus and returned to the lake. (b) This is how Lake Trummen looks now, a revitalized recreational asset. (Photographs by S. Björk.)*

must show that their plan for restoration of the lake is likely to produce long-lasting benefits. Some 68 lakes were helped under the EPA's Clean Lakes Program between 1976 and 1978 (Fig. 9–4).

LAKE ERIE: A CASE HISTORY

The Great Lakes: Superior, Michigan, Huron, Erie, and Ontario, lie between the United States and Canada. Combined, they represent the largest reservoir of fresh water in the world. However,

Figure 9-4 *Volunteers used this truck to pull stumps out of the lake bed as part of an EPA sponsored clean lakes project. (Photo by Harold Woodworth.)*

because they are the site of large population centers in both the U.S. and Canada (Duluth, Green Bay, Milwaukee, Chicago, Gary, Toledo, Cleveland, Erie, Buffalo, and Toronto among others), the lakes are becoming more and more polluted. In fact Lake Erie, the smallest and shallowest of the lakes, has been described as "dead." This really is not the right term to use. The lake still yields a large commercial fish catch each year. What has changed, however, is the kinds of fish that can be caught in the lake. When the lake area was first settled in the 1700s, Lake Erie was home to large numbers of small-mouth and large-mouth bass, muskellunge, white-fish, sturgeon, walleye pike, lake herring, northern pike, blue pike, sauger, freshwater drum, white bass, and lake trout. There are now no longer any blue pike, sauger or native lake trout. Only a few lake herring, sturgeon, white fish, and muskellunge are left, along with some walleye and northern pike. Instead the fishery is composed of yellow perch, white bass, channel catfish, freshwater drum, carp, goldfish, and rainbow smelt.

Intense commercial fishing of sturgeon, lake trout, whitefish and herring were responsible in part for the change in species. The introduction of new species into the lakes also had unwanted effects. Sea lamprey found their way into Lake Erie some-

time before 1921 via the Welland Canal. Although they feed upon desirable food and game fish, the lampreys are probably not the most serious pest species in the lake. This is because breeding grounds for the lamprey are not in good supply. This limits the possible size of the lamprey population. More serious is the invasion of rainbow smelt, which occurred about 1931. The smelt have become very abundant, preying on the young of desirable fish such as lake trout, blue pike, and lake white-fish.

However, eutrophication of the lake has also played a large part. The amount of algal growth in the lake has increased 20 times since 1919. When these algae die and decay, the level of oxygen in the water plummets, especially in the cooler bottom waters. Not only does this destroy the summer habitat of many preferred fish species, but it also destroys important fish food species, such as the burrowing mayfly. The kinds of algae in the lake have also changed. Blooms of blue-green algae are now common. In contrast to green algae, many species of blue-green algae are not eaten by fish. Some blue-green algae produce toxins capable of poisoning livestock, waterfowl, fish, and people.

In addition to providing a large part of the ni-trates and phosphates that have led to the eutrophi-cation of Lake Erie, sewage inflows have added toxic chemicals such as mercury, PCBs, and Mirex. These chemicals accumulate in fish and have led to severe restrictions on the kinds and sizes of fish that can be kept or eaten when caught in the Great Lakes.

In 1972, the U.S. and Canada agreed on a plan to clean up the Great Lakes. A central feature of the plan was building sewage treatment plants to reduce the amount of phosphate entering the Lakes. Canada met the deadline (31 December 1975) for phosphorus reductions but the U.S., for political and economic reasons, did not. In 1973, 1974, and 1975, the administration impounded funds voted by Congress for construction of needed sewage treatment plants. In fact, most major U.S.

cities added almost three times as much phosphorus to the lake in 1976 than they did in 1975. As one Canadian official, annoyed at the lack of U.S. progress, put it, the present situation is "like mixing a glass of clean water with a glass of dirty water, you end up with dirty water."

In 1978, the agreements were revised, with a new deadline of 1983 for completion of all industrial and municipal sewage treatment construction.

In summary one might say that reports of the death of Lake Erie have been exaggerated. However, if we do not make progress in the reduction of sewage inflows to all the Great Lakes, both because of the nutrients and the poisons they contain, the lakes may become effectively dead to man. We will no longer be able to use the lake waters for recreation nor will we be able any longer to catch and eat those creatures still able to survive in the water.

Other Problems with Detergents: Foaming Waters and Biodegradability

In the early 1950s, many people noticed a strange sight. Brooks and streams, rivers, and even lakes were beginning to foam. Wherever water tumbled over stones or waterfalls, wherever winds rippled the surface, accumulations of bubbly froth built up. The explanation for this odd circumstance was not hard to find. Huge amounts of detergents were being used to wash clothes (close to two billion pounds per year). The dirty wash water ran down millions of drains and into nearby streams, lakes, and rivers. Sometimes it passed first through sewage treatment plants and sometimes it did not. In fact, it made no difference whether it did or not. The chemical in detergents that made them foam was not removed by sewage treatment plants. The material, called alkyl benzene sulfonate (ABS) has several special properties. The property that makes it useful in washing clothes is that it can help greasy dirt dissolve in water. That is the good part. The not-so-good part is that its long, branched shape makes the molecule hard for bacteria to break down. Treatment plants depend upon the use of bacteria. Because bacteria do not have the necessary catalysts (called enzymes) to destroy ABS compounds, these molecules began to accumulate in the early 1950s wherever wash water ran into natural waters. When the concentrations of ABS were high enough, foam formed, just as it did in washing machines and dishpans.

This problem, fortunately, could be solved. Detergent manufacturers now use surfactant molecules having straight rather than branched side chains. Such surfactants are known as linear alkyl sulfonates or LAS. These compounds are called biodegradable because bacteria are able to digest them, and so the foam has almost disappeared from our rivers, lakes, and streams.

Questions

1. The amounts of nitrates that lead to algal blooms are smaller by a factor of at least 10 than the amounts that cause nitrate poisoning in infants. Contrast the sources and effects of nitrates at low and high levels.
2. What is hard water? Why is it a problem? Why are phosphates added to detergents?
3. What problems are caused by too much phosphate in natural waters? Would you call phosphate a poisonous chemical?
4. Why must we be careful about what kind of substitute chemicals are used in detergents instead of phosphates?
5. Does your sewage treatment plant remove phosphates? (You could call and find out.) Does your home have soft water? What would you recommend that your family use to wash clothes? Why?

Further Reading

Eutrophication of Surface Waters: Lake Tahoe's Indian Creek Reservoir, U.S. Environmental Protection Agency, Ecological Research Series, 1975, EP 1.23.

One community's answer to the problem of lake eutrophication was to build a whole new reservoir, Indian Creek, to hold the area's waste water so that it would not have to be dumped into Lake Tahoe. This report details some of the problems they encountered.

Lee, G. F., "Eutrophication," *Supplement to the Encyclopedia of Chemical Technology*, Wiley, New York, 1970.

A good basic reference to the understanding of the process of eutrophication and lake aging.

Phosphates in Detergents and the Eutrophication of America's Waters, 23rd report by the Committee on Government Operations, U.S. Congress, Washington, D.C. (1970).

This report contains much of the controversy about phosphates in detergents. Comments by manufacturers and scientists are interesting and also illustrate how decisions must often be made on evidence that can be interpreted in several ways.

Regier, H.A. and W.L. Hartman, "Lake Erie's Fish Community: 150 Years of Cultural Stress," *Science* 180, 1248 (22 June 1973).

Historical perspective on the effects of civilization on the Great lakes.

Schindler, D.W., "Eutrophication and Recovery in Experimental Lakes: Implications for Lake Management," *Science* 184, 897 (24 May 1974).

This article provides an entry into the literature on the limiting-nutrient controversy.

Welch, Eugene B., Nutrient Diversion: Resulting Lake Trophic State and Phosphorus Dynamics, U.S. Environmental Protection Agency, Ecological Research Series, EPA 600/3-770003 (1977).

The results of piping sewage around instead of into Lake Shammamish have been different than the results for Lake Tahoe. This report attempts to explain why.

Chapter 10

Water Pollution Control

In Köln,[1] a town of monks and bones
And pavements fang'd with murderous stones
And rags, and hags, and hideous wenches;
I counted two and seventy stenches;
All well defined, and several stinks!
Ye nymphs that reign o'er sewers and sinks,
The river Rhine, it is well known,
Doth wash your city of Cologne;
But tell me, Nymphs! What power divine
Shall henceforth wash the river Rhine?

Samuel Taylor Coleridge

CONTROLLING WATER POLLUTION

Point Sources Versus Non-Point Sources

There are a number of ways to combat water pollution, but the methods chosen depend on the origin of the pollution. We can distinguish two origins of pollutants. There are those pollutants that arise in urban areas and enter water from a single pipe; we say such pollutants come from a *point source.* Domestic wastes and wastes from industries are collected by sewers; the sewers carry the wastes to a treatment plant. After it has been treated, the waste water is discharged to a waterway; we call such pollution "point source" because it occurs at a distinct and identifiable point.

1 The city of Cologne.

In contrast, there are pollutants that come from relatively rural areas; these pollutants enter waters in run-off from the land. The run-off occurs in numerous rivulets and streams, and it will reach the river at many, many points. We can think of the run-off as being spread out along the waterway, and we call such run-off with its burden of pollution a "non-point source."

The methods of controlling pollution from point sources and non-point sources are very different. To point source pollution, we can apply the "technical fix"; the technical fix consists of chemical and biological processes that remove contaminants from the waste water. Non-point source pollution, however, is not so easy to control, for it does not lend itself to technical methods. To control non-point source pollution, changes in land use and in agricultural practices are needed. Changes are also needed in mining practices and in tree growing and harvesting. We discuss in this chapter the technical solutions that apply to point source pollution; we also discuss some of the changes in land management needed to control pollution from non-point sources.

WATER POLLUTION CONTROL AT POINT SOURCES

Combined Sewer Systems

When waste water enters a treatment plant (some call it a water pollution control plant), it passes through a rack and a coarse screen that prevent large objects from passing into the plant. The objects could include boots, cloth, branches, etc. Such objects would seem to be unexpected in municipal waste water, but they are quite common. Their presence is an indication of a very large problem which many communities in the United States continue to face: the storm water drain system and the system of domestic sewers are combined. Thus, storm-water flows, which may carry large objects such as those we mentioned, are mixed with municipal waste water during and after a rainfall. The combined flow will overwhelm a treatment plant and must be diverted past it. The concept of combined sewers is illustrated in Figure 10–1. The problem posed by combined sewers is one of continuing concern and is not open to easy, low-cost solutions. (This widespread problem is discussed on p. 214.) To separate the storm sewer system from the waste water sewer system would cost billions of dollars in a number of U.S. cities.

Primary Treatment

After the waste water has passed the rack and screen, the flow moves into a settling tank. Settling is a purely physical process; no chemical or biological reactions take place. Settling removes organic *solids* from waste water by slowing the velocity of the waste stream. As the velocity decreases, the tendency of the water to keep particles in suspension will decrease and such organic particles will slowly settle to the bottom of the tank. The organic solids removed by a "settler" (Fig. 10–2) may account for 35% of the organic matter in the waste water from a typical city. The initial settling is referred to as "primary treatment." The settling tank is also commonly called a "clarifier."

Secondary Treatment

The processes that follow the settling tank are designed to remove those organic materials dis-

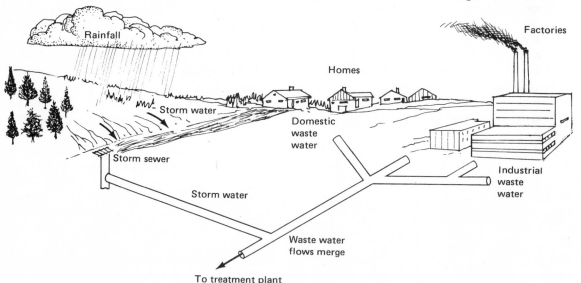

Figure 10–1 *The combined sewer system of most U.S. cities. The runoff from a rainstorm enters a city's sewers and there mixes with the wastes from homes and industries. A treatment plant built for the normal waste water flow will not be able to handle the combined flow. Much of the combined flow will be diverted past the treatment plant and will enter a water body without treatment.*

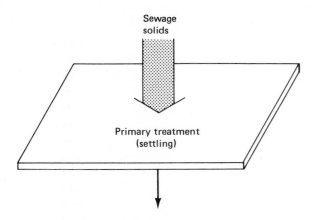

Figure 10-2   *Primary treatment.*

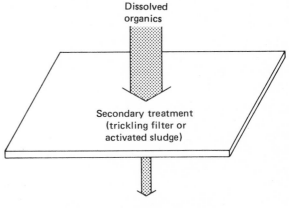

Figure 10-3   *Secondary treatment.*

solved (as opposed to suspended) in the waste water. This process step is referred to as "secondary treatment." The process known as "activated sludge" uses microorganisms, in a large well-aerated tank, to break down dissolved organic substances. A steady stream of waste water from the settler enters one end of the activated sludge tank and exits at the other end. The flow, on exit, is richer in microbes and poorer in dissolved organics.)

The microbial population that grows steadily in the tank uses the organic substances in solution for growth and energy. Thus the organics are removed from solution and are converted within the tank either to more microbes or to the end products of biological oxidation (Fig. 10-3). Among the end products in the presence of oxygen are carbon dioxide and water.

The microbial solids produced in the activated sludge tank are suspended in the waste flow that leaves the tank. This waste flow is passed through another settler, which removes the solids. These solids from the second settler flow in a slurry which is 98% to 99% water; this flow is combined with the solids from the primary settling tank. The combined solids are then treated in a device called a digester. We shall describe the digester shortly.

A portion of these solids, which are primarily microbes, is returned to the activated sludge tank as "seed." Here the returned solids mix with the waste stream from primary treatment and provide an initial bacterial population to start the mixture fermenting in the tank (Figs. 10-4 and 10-5).

Another common technology used in secondary treatment to remove dissolved organics is the trickling filter. Here the waste water is distributed over and passed down through a bed of rocks in a large concrete basin. A slime of microbes has grown up on the surface of the rocks during long operation of the filter. The microbes in the slime remove dissolved organics for growth and energy just as the mircrobes in the activated sludge process do. The slime slides off the stones a little at a time and into the waste water. The result is a waste stream with solid materials in suspension and much of the dissolved organics removed, just as occurred in the activated sludge tank. The solids are removed in a settling tank, and these solids are combined with those from the primary settling step and are pumped to the digester. The trickling filter may be as large as 200 feet in diameter at large installations (Fig. 10-6).

Digestion

(The process of waste digestion takes place in a large, heated tank appropriately called a digester.)

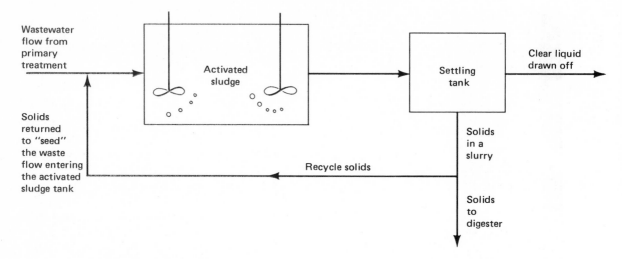

Figure 10-4 *Activated sludge secondary treatment. The activated sludge plant uses the natural activity of microorganisms to remove organics from the waste flow.*

Figure 10-5 *The Package plant. Often used for subdivisions, these small activated-sludge plants are sold as prefabricated steel package plants. The process flow diagram is essentially the same as in the larger system, except that these small plants typically have no primary treatment and aerate the raw wastewater for a 24-hour period rather than the 6-8 hours used in conventional plants.*

The tank is closed so that oxygen does not reach the wastes. Specialized microbes, which live at high temperatures and do not require oxygen, grow in the tank. These microbes convert the wastes to stable end products, which include the gases

methane and hydrogen sulfide. The methane gas is often burned to provide the heat needed to keep the digester at the proper temperature. A stable, non-degrading sludge is the result. This stable sludge is then dried.

In smaller communities, the sludge may be spread in drying beds in greenhouses. Often, the dried sludge is available to citizens willing to come and dig it; its use as a fertilizer is common. In larger cities, the solids are placed on a rotating drum. A vacuum is applied to the drum and moisture drawn from the solids. The dry solids are scraped from the drum surface with a sharp blade. After a further heat drying stage, the solids are then ready for disposal to a landfill. An alternative to drying is burning of the sludge. Air pollution results unless proper control devices are installed on the incinerator. The ash that is left is buried at a landfill site.

The primary treatment (settling) and secondary treatment (activated sludge or the trickling filter) remove organic substances. Together they may remove up to 90% of the organic wastes in the water (Fig. 10-7). Primary treatment is common at many communities in the United States. Secondary treatment is coming into wider use.

Figure 10–6   *(a) Trickling filter. Wastewater is distributed over rocks. A slime growing on the rocks removes organic material from the wastewater. (b) Trickling filter schematic.*

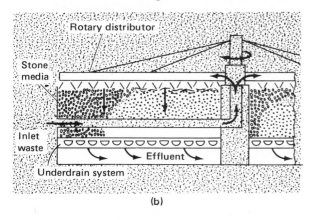

(b)

Tertiary Treatment

(The processes that follow the secondary treatment step are designed to remove the plant nutrients responsible for "over nourishing" lakes and rivers. Taken together, these processes, which follow secondary treatment, are known as tertiary treatment. Tertiary treatment, in contrast to primary and secondary treatment, is quite uncommon at present.)

The serious effects of over-enrichment (eutrophication) of lakes and rivers with plant nutrients have only been recognized in the last few decades. Thus, the methods to control eutrophication have not yet been widely applied. At present, there are probably only a few dozen such installations in the U.S. Nevertheless, there is promise that the processes in tertiary treatment will be added to sewage treatment plants in many areas.

(The main purpose of tertiary treatment is to remove compounds containing the elements nitrogen and phosphorous. Thus, nitrogen compounds such as ammonia, nitrate ions, and nitrite ions are removed by one set of processes. The phosphate ions, found both in human wastes and detergents are removed by another process (Fig. 10–8). Tertiary treatment may also include removal by carbon absorption of those resistant organic substances not removed by secondary treatment. (If you would like

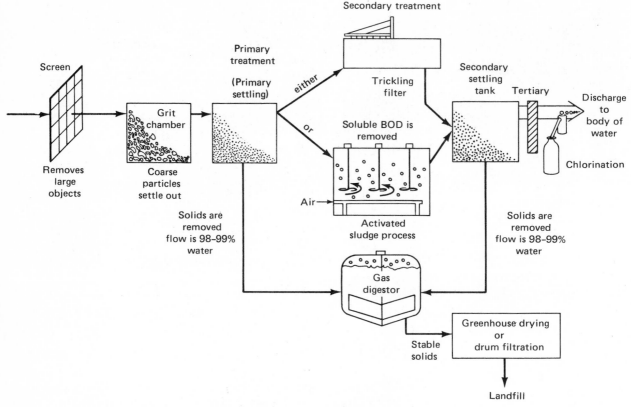

Figure 10-7 *Water pollution control processes.*

to know more about tertiary treatment processes, read p. 210.) These processes are very expensive.)

Chlorination

Bacteria that inhabit the human intestine find the sewage treatment plant cold comfort indeed. Their natural habitat is considerably warmer, so that a gradual reduction in these bacteria is noted in the plant. Nonetheless, considerable numbers of bacteria still exist in the waste stream before it is chlorinated. Where the sewage treatment plant was merely a torturer to these bacteria, chlorination is the executioner.

(In contrast, viruses from the human intestine die away more slowly in the plant. Primary treatment reduces their numbers little, if at all. Secondary treatment via the trickling filter may reduce their numbers by 40%, but secondary treatment via the activated sludge process can reduce their numbers by up to 98%. The final step of chlorination reduces their numbers once again. Yet even after chlorination, one can still find live viruses from the human intestine in the waste stream.)

In the U.S., the waste water leaving the sewage treatment plant is commonly chlorinated in order to destroy disease-causing microorganisms (pathogens). Chlorination of waste water is uncommon in Britain. The health of people who may swim in or boat on the water downstream from the treatment plant is protected by destroying the pathogens in the waste water. The process of chlorination is inexpensive compared to other methods of disinfecting waste water. Furthermore, the effectiveness of

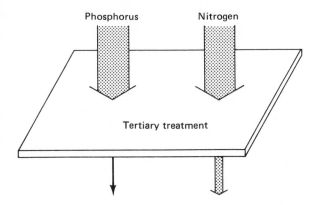

Figure 10-8 *Tertiary treatment.*

the disinfection step is easily checked by using the test for free chlorine remaining in the waste water. (Although chlorination is in wide use at waste treatment plants, people do recognize that its presence in water is harmful to fish, even at low levels. Chlorine is also known to react with hydrocarbons to produce some substances which are suspected to cause cancer. (See Chapter 27.) There are several ways to remove chlorine from waste water once it has been added. These include treatment of the waste water with sulfur dioxide gas and filtration through carbon towers.)

The approximate reduction in pollutants as well as the reduction in bacteria and viruses by each step in the total treatment scheme is indicated in Figure 10-9.

The devices and processes we have discussed so far are the waste water treatment methods used most commonly in American cities today. There are, however, other processes and concepts being studied or used on a small scale. These include tertiary treatment, land application of waste water, the biodisc process, the racetrack pond, and others.

Control of Water Pollution in Rural Areas

Septic systems In rural areas and small subdivisions where there are no waste water treatment

plants, or where there are no sewers to connect to treatment plants, households most often treat their wastes by using septic tanks. The septic tank is the successor to the cesspool. The older device, the cesspool, was a structure that resembled a well; it was lined with stones or concrete blocks. No mortar was placed in the joints between the stones or blocks. Raw, unsettled sewage was disposed of directly into the cesspool (Fig. 10-10). Ground water contamination is a distinct possibility.

The septic tank, in contrast to the cesspool, is a water-tight tank. It is constructed either of metal or of concrete (Fig. 10-11). Depending on the tank size and the size of the home it serves, the tank may hold sewage flows for anywhere from half a day to three days. During this time, solids settle to the bottom of the tank where they are degraded by bacterial action.

The overflow from the septic tank is a liquid from which some of the solids have been removed. The overflow passes to a distribution box which "distributes" the water to a buried leaching system. The leaching system consists of clay pipe pieces, and the waste water seeps into the soil from the joints between the pieces (Fig. 10-12). Microorganisms in the soil will decompose the wastes. Lush grass may often be observed above the "arms" of the leaching system. The leaching system helps to disperse the wastes and decreases the possibility of ground water contamination.

If a family uses both a septic sytem and a well, it is good practice to keep the water supply far away (several hundred feet), preferably uphill from the waste disposal system. Furthermore, the soil into which a leaching field is placed should be a fairly porous one. It should not have fractured or creviced rock beneath the surface, for such features could potentially channel the wastes to the ground water. Lots in suburbia are often required to be very large in order to handle the required leaching field for septic tank wastes.

Lagoons and ponds (One device that finds fairly wide use in smaller communities and also in industry is the "waste stabilization lagoon" or "oxidation

CAN YOU SAVE A RIVER WITHOUT REMOVING THE POLLUTANTS?

There are two additional ways to improve the quality of rivers. These are options that deal with the rivers themselves and that are not treatment processes in the usual sense. The first is called "low flow augmentation" because additional water from an upstream reservoir is released to dilute the downstream concentrations of dissolved organics. The dilution is intended to add oxygen resources to the stream and thereby reduce the effect of the oxygen-demanding waste water. For example, the waste water from a particular community might be able to reduce the dissolved oxygen in the average river flow during August to three milligrams per liter. If the flow is augmented sufficiently, the dissolved oxygen in the increased flow may only fall to five milligrams per liter.

Among environmental engineers, low-flow augmentation is humorously called "the dilution solution to pollution." It is not really pollution control, however; it is a method of water quality improvement. The same quantity of organic wastes are still entering the watercourse. Only their effect is different. Until the early 1970s, the water quality improvement brought about by flow augmentation could be counted as a benefit when the Corps of Engineers proposed a reservoir project to Congress, but this is no longer the case.

In the same category as flow augmentation is the process known as "instream aeration." This, too, is a means to improve the dissolved oxygen content of a river without removing dissolved organics. We know that dissolved organics, through the action of bacteria, will remove dissolved oxygen from a stream. We also know that the flowing water of the stream will capture and dissolve oxygen from the air. Why not help the natural process by bubbling in compressed air? The process is known to work, and aquatic life can be restored to the river by raising the dissolved oxygen level.

The City Council of Paris, France, plans to offer the Seine River, which flows through the city, such a "face-lift." The chairman of the City Council remarked, "We can easily and without spending huge sums of money give back to the river the oxygen which it lacks in the summer due to pollution and a lower water flow." The chairman did not mention that the pollution will still be present. Is in-stream aeration or flow augmentation a better way than pollution control to restore the quality of a river? Is this the kind of process that should be considered in planning the clean-up of a river? Or should clean-up mean only the removal of wastes?

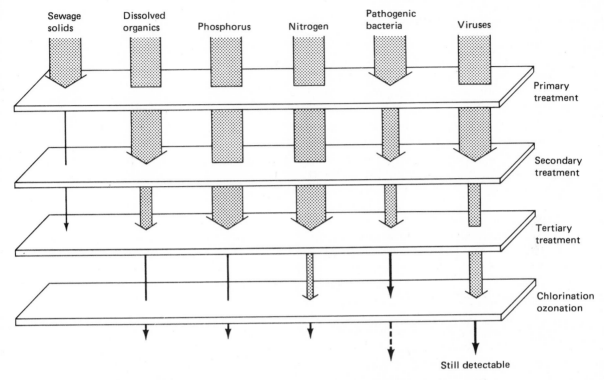

Primary treatment

Secondary treatment

Tertiary treatment

Chlorination ozonation

Sewage solids · Dissolved organics · Phosphorus · Nitrogen · Pathogenic bacteria · Viruses

Still detectable

Figure 10-9 *Removal of contaminants by sewage treatment.*

pond." The lagoon is essentially a wide shallow pond. It is typically two to four feet deep and raw waste water is discharged directly into it. Its use depends on the availability of low-cost land and the character of the surrounding area. Odors may make its use a nuisance in populated areas.)

Most ponds are designed to retain about two weeks worth of flow from the community. Larger ponds can hold nearly two months of flow. Solids settle out as the waste water flows slowly through the pond. The solids decompose in the absence of oxygen at the bottom of the pond. Near the surface, however, bacteria oxidize organic materials in the presence of oxygen. Weeds and other aquatic plants, including algae, may grow in the pond, making use of the nutrients from the wastes. In addition, the algae supply oxygen to the system as a byproduct of the photosynthesis they carry on. The oxygen helps keep the system operating with minimal odors. Sometimes the lagoon is aerated; that is, air is bubbled into the water, much as it is in the activated sludge process. The lagoon thus can act as a kind of combination of primary and secondary treatment. In fact, no primary treatment is usually given to the waste entering a lagoon.

In 1978, the contents of a 36-acre (14 hectares) sewage lagoon in West Plains, Missouri, leaked into the ground. Numerous subterranean streams carried the sewage over many miles and contaminated the ground water in the region. Over 750 cases of diarrheal disease were associated with the leak. The incident illustrates a hazard associated with waste lagoons and septic systems: the possibility of leakage to the ground water through channeled rock.

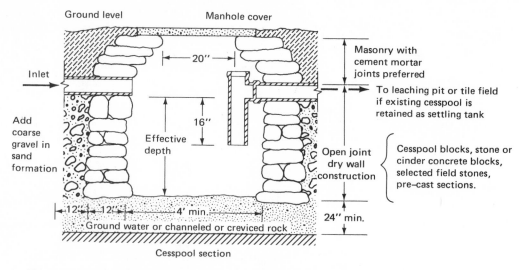

Figure 10–10   *Leaching pit and cesspool details. One of the first devices used to dispose of wastes, the cesspool is still a familiar concept. It is now recognized, however, that the cesspool is inadequate and dangerous in disposing of wastes. The potential contamination of ground water is the principal objection to use of the cesspool. (Adapted from J. Salvato,* Environmental Sanitation, *Copyright ©1958, New York. Reprinted by permission of John Wiley & Sons, Inc.)*

CONTROL OF WATER POLLUTION FROM NON-POINT SOURCES

By non-point sources of water pollution, we mean those many streams and rivulets which may carry pollutants off the land into major bodies of water, such as rivers or lakes. Non-point source substances include organic wastes, nitrogen compounds, phosphorus compounds, and sediment or soil particles.

Dairies and livestock farms produce animal manure which, unless disposed of properly, can end up in streams and rivers. The use of feedlots, as opposed to the grazing of steers on pasture land, can result in a very concentrated source of pollution. The average steer generates about 15 times more wastes than an average person, the hog twice as much. In total, farm animals may be producing 10 times the wastes of the U.S. population. To control pollution from feedlots and dairy farms, waste treatment devices such as lagoons may be installed at the larger operations.

Growing crops can lead to water pollution because of the fertilizer applied to them. Fertilizer is likely to include compounds containing both phosphorus and nitrogen, the two elements most likely to cause eutrophication. In addition to fertilizer, decaying vegetation and soil particles from erosion may enter streams in run-off. Erosion occurs from cropland, from grazing land, from logged areas, unreclaimed mining areas, and construction sites. Soil particles may be suspended in the flowing water, resulting in a murky stream. Or they may settle to the stream bottom. Such a stream is less capable of carrying large water flows and hence is more prone to flooding.

To control the pollution arising from farm land requires more careful application of fertilizers and pesticides. Contour plowing on small hills and terracing on steeper hills will also help decrease run-off. (See the discussion on Erosion in Chapter 34.) In the last decade, many farmers have gone to "no-till farming" as a labor saving device. This method involves the use of a herbicide, which kills weeds

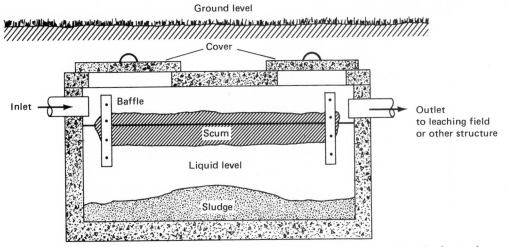

Figure 10–11 *Cross section of a typical concrete septic tank. The septic tank is the device that replaced the cesspool in rural areas. Closed to the soil by its concrete or metal sides, the device provides a modest amount of "treatment" to the waste water from a home. Solids will settle out in the tank and be degraded there, while a relatively clear overflow passes out to the leaching field. (Adapted from "Cleaning Up the Water," Maine Department of Environmental Protection, 1974.)*

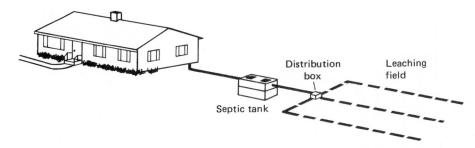

Figure 10–12 *Typical layout of a septic tank and leaching field. The outflow from the septic tank is distributed to a buried leaching field consisting of clay pipes with spaces between them. The waste water enters (leaches into) the soil from these spaces, and soil bacteria degrade the wastes further. (Adapted from "Cleaning Up the Water," Maine Department of Environmental Protection, 1974)*

before they can sprout from the ground. Such ground does not need to be turned over in the spring to eliminate weeds; hence much less erosion occurs. However, these herbicides may be washed into the stream along with other pest control chemicals that have been applied to the crop. Many of the chemicals used to control insects and plants have been found to be cancer-causing substances. Therefore,

it is important that the herbicides selected for use have as little effect on humans as possible.

Septic systems can give rise to non-point source pollution. Urban run-off that does not enter storm sewers but flows from gulleys or culverts into streams is a contributor as well. The run-off from storm drains on highways carries metals such as the lead from leaded gasoline. Roads and shopping centers with their vast impermeable areas create conditions in which storm water run-off can occur very rapidly and hence with considerable erosive power. Because of the rapid build-up of flow from highways and shopping centers, stream channels can be severely eroded.

The magnitude of non-point source pollution is now thought to be larger than we once suspected. The National Commission on Water Quality has projected that when all industrial and municipal sources have installed secondary treatment, the amounts of nitrogen and phosphorus from non-point sources will then match or exceed the quantities from the treated waste water of industry and cities.

At the same time, even though pollution that originates from non-point sources is large in quantity, it is, by its very nature, more diffuse, more spread out, than the point source pollution from cities and industries. In our discussion of organic wastes, we noted that a stream has the ability to recover from the addition of a certain amount of pollution. A stream that receives a small steady amount of pollution recovers to a satisfactory condition more quickly than a stream that has a large, concentrated pollution load dumped into it.

OTHER WAYS OF TREATING WASTE WATER

Physical-Chemical Treatment and New Biological Processes

All the secondary treatment processes we have discussed have utilized microorganisms to remove dissolved organics from waste water. There are defects in this biological treatment of wastes, however. Probably the most important problem is the sensitivity of microorganisms to toxic pollutants. An industry may discharge a waste that will poison the microorganisms. As the waste flows through the treatment plant, the special populations of microbes, which had been thriving in ordinary waste water, may be partially or wholly killed.

When this happens, the effectiveness of the entire plant is reduced to the removal levels achieved by primary treatment alone. A whole new population of microbes must grow up to renew the effectiveness of secondary treatment.

In the last decade, much progress has been made on waste water treatment plants that use no biological steps. The processes, taken together, have been named "physical-chemical treatment." They remove the same substances as are removed by a plant that has conventional primary, secondary, and tertiary processes.

Physical-chemical treatment begins as conventional treatment does, with screening and grit removal. The next step is the addition of a chemical such as lime or alum to create a mineral phosphate precipitate. The resulting solid particles as well as the usual suspended solids are then settled out together in a settling basin, much as in the usual primary treatment step. An optional step of filtration through sand or crushed anthracite coal will remove the few solids left in suspension.

The waste water is then passed through an activated carbon column where dissolved organics are adsorbed onto the surface of the carbon particles. The organic removal achieved in this step matches that of the best biological treatment processes. Eventually, the carbon particles become saturated with organics; then the column is removed for cleaning and replaced by another. Careful heat treatment of the saturated carbon particles burns off only the organics and restores the ability of carbon to adsorb organics. The column can be placed in service again.

Physical-chemical treatment plants may soon be in service in a number of areas including the city of Niagara Falls, New York State and Cleveland, Ohio.

A new biological treatment process planned for a number of cities in the U.S. is the rotating biological contactor or "biodisc." The device, which was developed in Europe, removes dissolved organic substances. Another waste water treatment device from Europe is the "racetrack" pond, so named for its unusual shape. Originated in the Netherlands, the process makes use of an oval loop around which waste water circulates.

Land Application of Waste Water

One concept for waste treatment being studied at government supported projects is land application of waste water. The concept of land application has some appeal. Waste water contains phosphorus, nitrogen, and organic substances. It is a reasonable fertilizer for crops, and, of course, it is a source of moisture. Though such use seems attractive, there are several issues to be considered.

Before describing the three basic methods of land application, it is important to discuss the level of treatment of the waste water before it is applied to the land. First of all, we know that primary treatment ought to precede any of the three methods. If it did not, the system that distributes the water could be clogged by the solids in the waste water. Furthermore, without primary treatment, a very large portion of the bacteria and viruses might be applied to the land. Some researchers feel that secondary treatment is not needed before applying the waste water; they assert that the land treatment will remove dissolved organics. Others feel that applying waste water to the land is an alternative method only for tertiary treatment. The tertiary processes, you will recall, are designed for phosphorus and nitrogen control and for removal of resistant organics.

Each of the three methods of land application is designed to accomplish a different objective. In the process referred to as "overland flow," waste water is applied to sloping land that has been stabilized by grasses. Gravity carries the water down the slope, across the soil and through the grasses to a run-off collection ditch. The water does not penetrate the soil deeply, and at the base of the slope it joins other surface run-off. Because the removal processes are biological in nature, areas where freezing temperatures occur regularly are not good candidates for the overland flow treatment.

The process does not remove much phosphorus and hence, may not be a good choice for tertiary treatment.

Irrigation, in contrast to the overland flow process, is designed to fertilize crops. It provides a high degree of treatment. Irrigation could be thought of as an example of water re-use. Probably irrigation with waste water is best suited for the water-short western states.

When waste water is applied at a slow rate, the removal of organic matter is more efficient. Also, more water will evaporate from the soil and transpire from the plants. This, unfortunately, sometimes creates greater build-up of salts in the soil. When waste water is applied more rapidly, relatively less evaporation and transpiration occur, but the removal of dissolved organics is decreased, as is the potential value of the crop. Muskegon County in Michigan and the city of Lubbock, Texas, have large-scale spray irrigation installations.

The third method of land application, infiltration-percolation, is primarily a method for recharging ground water (Fig. 10–13); a crop need not be involved. Treated waste water is applied to the soil by pumping it into basins from which the water percolates into the soil. Its use is most appropriate where wells are depleting fresh (low salt content) ground water and where waters with high salt content are replacing them. Treatment is primarily by filtration through the soil.

Nassau County on Long Island plans to use such a system. After its sewage is "polished" to drinking water quality, it will be recharged to the ground instead of discharged into the ocean. Nassau has been facing a decrease in its reserves of fresh gound water. Ground water recharge is practiced also in Israel and in the Netherlands. In the Netherlands, the recharge is used to filter the dirty water of the Rhine River before it is used by the city of Amsterdam.

To apply waste water to the land requires a number of conditions. First, land with no better uses (such as housing, industry, etc.) must be available at a distance not far from the treatment plant.

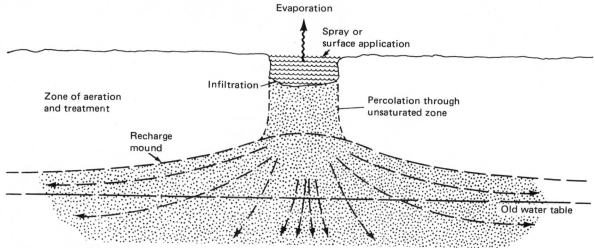

Figure 10–13   *Infiltration-percolation—one of the three methods of land application.*

If the land is too distant, the electrical energy needed to pump the waste water will make the project too costly. Second, citizens must be convinced of the value and safety of the project. It would be untruthful to suggest that applying waste water to the land is known to be safe in all circumstances. The Water Pollution Control Federation issued a policy statement in 1974 on land application of waste water. It felt that "present knowledge of the effects of large scale use of land disposal is too meager to justify widespread use for man-made wastes."

In 1976, a report from Israel studied the frequency of infectious diseases in agricultural communities (kibbutzim) that spray-irrigated their crops with waste water. The investigators found that the infectious disease rate was about four times higher in those areas than in communities that did not irrigate crops with waste water. The waste water used for irrigation had been partially treated, but it had not been chlorinated. Droplets containing bacteria and viruses were probably carried in the air by the wind and caused the increase in disease in the communities.

There are other concerns in applying waste water to the soil; long-term irrigation with river water or with waste water can cause salts to build up and eventually make the land unfit for agriculture. The Imperial and Central Valleys of California are experiencing a build-up of salts in the soil. A much earlier example occurred in the region of Iraq once known as the Fertile Crescent. Watered by the Tigris and Euphrates Rivers, the land was irrigated to stimulate agricultural production. The term "Fertile Crescent" is no longer used to describe this relatively barren area.

Attention must also be given to the possibility that nitrates may appear in the ground water unless removed from the waste water before it is applied. The presence of nitrate ions in high enough concentrations in drinking water can bring about the disease methemoglobinemia. (See Chapter 6.)

Tertiary Treatment–How It Works

Primary and secondary treatment are effective methods of removing sewage solids and dissolved organics. Chemical processes are needed, however, to remove several undesirable chemicals in waste water. These substances are nitrogen as ammonia or as ammonium ion, nitrate or nitrite ions, and phosphorus in the form of phosphate ion.

$$NH_3 \qquad NH_4^+ \qquad NO_3^- \qquad NO_2^- \qquad PO_4^{3-}$$

ammonia gas   ammonium ion   nitrate ion   nitrite ion   phosphate ion

These substances contribute to the eutrophication of natural water, characterized by excessive algae growth. The biological basis of eutrophication is described in Chapter 9.

The processes used to remove these chemicals are often referred to as tertiary treatment when they follow the secondary stage of treatment. Sometimes the processes for chemical removal can actually be included within the primary and secondary treatment stages. When this is done, the added processes may be referred to as "advanced waste treatment."

Phosphate removal is accomplished by chemical precipitation and settling. Chemicals such as ferrous and ferric salts, or aluminum salts or lime are added to the waste stream. When one of these chemicals is mixed well with the waste stream flowing from secondary treatment, a precipitate of solid matter forms. For example, the calcium ion from lime (calcium oxide) will combine with the phosphate ion in solution to produce calcium phosphate as solid particles. After the addition of the lime or other chemical, the waste water flows into a well-stirred tank where the calcium phosphate will clump into large particles. The clumping process is called "flocculation," just as in water treatment. The mixture then passes to a settling tank where the precipitate is collected. A clear flow much reduced in phosphorous content leaves the settling tank.

As we have seen, settling tanks are used for primary treatment, and they are used also to collect solids after secondary treatment. Hence, engineers have suggested that the addition of chemicals to precipitate phosphate should take place upstream of either of these settling tanks. This would allow collection of both the organic solids and the phosphate precipitate to take place at the same time.

After the phosphate is precipitated and settled out, any remaining suspended matter may be removed by filtration. Beds of sand, crushed coal, garnet, and gravel are used in a number of layers as the filter materials for this step. As in filtration of drinking water, backwashing to unclog the filter is needed from time to time. Because of the several components of the filter bed, the process is referred to as "multi-media filtration."

The removal of nitrogen is much more difficult. There are no common insoluble nitrate or nitrite compounds that can be settled out; nor is there any common insoluble ammonium compounds. Hence, there is no way that nitrogen can be removed by precipitation in the treatment plant.

It is important to remove nitrogen that occurs as ammonia (NH_3) for a number of reasons. First, ammonia is harmful to fish. Second, ammonia combines with the chlorine added to destroy harmful bacteria and viruses.

Reactions with ammonia tie up chlorine in substances known as chloramines and hence decrease chlorine's effectiveness. Third, ammonia is oxidized by bacteria in the stream, and this oxidation process may remove large amounts of oxygen from the water. Thus, we have many reasons to remove ammonia.

The ammonium ion may be removed as ammonia gas by a process known as gas "stripping." In the process, the waste water is exposed to air in a metal tower which is packed with small inert plastic shapes. The water is allowed to run down over the plastic surfaces while air passes upward through the column. Ammonia moves from the water into the passing air stream.

Ammonia may also be removed by biological treatment by passing the waste water through an aerated tank with a specialized population of microbes. These microbes convert the nitrogen in ammonia and nitrites to the nitrate ion. The process is somewhat unstable because of the need to maintain the proper type of bacteria growing in the tank at all times.

It is also desirable to remove nitrate ions, the most oxidized form of nitrogen. Although nitrate ions may have been present initially in the waste water, they could also have been produced by the oxidation of ammonia and nitrite ions. Nitrate ions should be removed because they increase eutrophication, but there is an even more compelling reason to remove nitrate and nitrite ions. If high nitrate waters should be used for drinking, a blood disorder in infants, methemeglobinemia, could result. (See Chapter 6.)

Nitrate removal may be achieved by specialized populations of microbes that convert nitrate to nitrogen gas and water. Methyl alcohol must be present for this reaction which produces carbon dioxide, water, and nitrogen gas. Again, keeping the proper population of microbes growing in the tank is a problem.

A final procedure in tertiary treatment is to pass the waste stream through a tower which contains particles of activated carbon. The activated carbon adsorbs dissolved organic substances which were not removed in any of the previous processes. The organic materials "stick" to the surfaces of the particles. This "carbon polishing" step restores the waste water to such an extent that people have considered re-using this water, after it has been filtered and disinfected, in public water supplies.

One of the earliest successful applications of tertiary treatment was at Lake Tahoe, Nevada, in the late 1960s. There a resort community realized that its waste water was destroying one of the world's deepest (1600 feet) and most beautiful lakes. The clarity of this shining lake was threatened by algae growths caused by phosphorus and nitrogen compounds in the community's waste water. Scientists recognized that the usual primary and secondary treatment steps could not stop the degradation of the lake.

A tertiary treatment plant was opened at Lake Tahoe in 1965. Phosphate is precipitated at the plant and the particles removed in a settling tank. Ammonia is stripped from the waste water by air. Then the waste water is "polished" by passing it through carbon columns to remove the last and resistant dissolved organics. The waste water—now clear, colorless, and odorless—is chlorinated before it leaves the plant. Over 95% of the dissolved organics are removed by the combination of primary, secondary, and tertiary processes at Tahoe, and the same portion of the phosphorus has disappeared as well. Nitrogen removal at Tahoe ranges from 50% to 95%.

The waste water from the plant is never discharged to the lake. Instead, the water is pumped to Indian Creek Reservoir, some 27 miles distant, a reservoir created expressly to receive the plant's waste water.

Problems from Chlorination of Sewage and Options Available

Though chlorination is commonly used at most treatment plants, there are other ways to disinfect waste water. And there are reasons to consider these ways. At the same time, environmental engineers are reluctant to abandon chlorination. They understand how well chlorination has served in the past to purify water and prevent disease.

Nevertheless, ecological problems do arise as a result of chlorination. Chlorine and the compounds it forms with ammonia are toxic to fish. Fish can be poisoned even at concentrations of chlorine as low as 0.002 milligrams per liter (two parts per billion). Suppose a waste water were being discharged into a river and the available chlorine (free and reacted with ammonia) were two milligrams per liter. This is a fairly common level for available chlorine in such discharges. Such a waste water would need to be diluted 1000 times to decrease the concentration in the receiving body to less than 0.002 milligrams per liter.

We mentioned in our discussion of drinking water that chlorinated hydrocarbons could form from the reaction of chlorine with hydrocarbons. These persistent and toxic substances are suspected of being carcinogenic (cancer-causing) agents. In Chapter 27 these substances, which have been detected in public water supplies, are discussed in greater detail.

The chlorine can be removed from waste water by contacting the waste stream with sulfur dioxide gas. Sulfate and chloride ions are the principal remainders in solution after the process is complete. The presence of the sulfate ion, which commonly occurs in natural waters, has so far not worried investigators. The sulfur dioxide will even strip chlorine away from the compounds it forms with ammonia. It will not, however, remove chlorine from the chlorinated hydrocarbons which may have been formed. Removal of chlorine by passing the waste water through towers containing activated carbon is also being investigated. Such towers would also be useful in removing chlorinated hydrocarbons.

Among the alternatives to chlorination of municipal wastes are irradiation with ultraviolet (uv) light and ozonation. The use of ultraviolet light leaves no harmful by-products, but the solids suspended in waste water are likely to prevent the uv light from penetrating deeply enough to be fully effective. Furthermore, when viruses and bacteria are inside larger particles, the uv light is unable to reach and kill them.

Ozone, though a powerful disinfectant, disappears quickly in contaminated water. No protection is left behind to destroy new micro-organisms that may enter the water later. Chlorine does provide that continued protection, but it is just that continued presence which also makes it ecologically unwise. In contrast to chlorine, which reacts with ammonia and hydrocarbons, ozone does not seem to form toxic by-products when it reacts with substances in water.

In the United States only about twenty small communities use ozone to disinfect waste water but its popularity is increasing. Paris, France, however, does disinfect its vast quantities of waste water with ozone. Ozone also seems to be effective at destroying some of the resistant organic compounds that persist through the last stages of treatment. Ozone is manufactured at the treatment plant in most applications. This eliminates the risk of transporting chlorine, which is a dangerous substance. Nevertheless, high costs for ozone disinfection continue to make chlorination popular.

Control of Storm Waters

Run-off that enters storm drains is simply water which was not absorbed into the ground. Its volume may at times be as much as 100 times the rate of domestic waste water flow (called dry-weather flow) which averages 100 gallons per day per person. Since the volume so greatly exceeds the flow of domestic waste water, the sewage treatment plant is designed only for the domestic flow (as opposed to 101 times larger). When a storm does occur and run-off is generated, the two flows mix. However, the treatment plant can handle only its usual volume, the dry-weather flow. The remainder must overflow to the stream or river and is left untreated.

Storm water carries many contaminants, including substances washed from the streets by the rains. Such substances as lead (from the lead in gasoline), particles of earth, leaf litter, and unburnt fuels may be captured in the flow before it enters the sewers. Once the flow enters the sewers, it may scour up organic solids (which include bacteria) deposited there by the slower moving dry-weather flow. Thus, storm water is not at all clean; it is carrying metals, organics, and possibly pathogenic bacteria into the watercourse.

There are four ways to alleviate this problem: (1) The sewer systems can be separated (at a cost of billions of dollars). (2) The streets may be washed and/or swept on a regular basis. This action reduces the contaminants washing off from streets but does not alter the scouring up of sediments in the sewer. (3) Very large storage tanks can be inserted in the system to hold the large flows; the tanks will release these volumes gradually to the treatment plant. The tanks may also serve as settling basins so that organic solids settle out of suspension while the water waits for treatment. (4) Facilities may be built to treat combined sewer overflows. Of course, an alternative may include a mixture of these basic options.

While such options exist, in most cities little has been done to deal with the problem because of the expense involved.

Questions

1. Complete the following chart, which lists the steps or procedures in treating sewage before it is released to natural waters, and the reasons for those steps.

Water Pollution Control

Procedure	What Is Done	Reason
Primary treatment	Settling, removes solid materials	Removes 35% of organic materials, which would otherwise use up O_2 in natural waters
Secondary treatment	either: or:	
Tertiary treatment or advanced waste treatment	Phosphate removal: Nitrogen removal:	
Digestion of sludge from primary and secondary treatment		

2. What is the difference between water treatment and water pollution control?
3. How is the sewage from your home treated? Does it go to a sewage treatment plant or into a septic tank or cesspool? If it goes into a septic tank or cesspool, summarize what happens in a few sentences. If it goes to a sewage treatment plant, does the plant give only primary treatment or does it have secondary and/or tertiary treatment equipment too?

Further Reading

Environmental Pollution Control Alternatives: Municipal Wastewater, U.S. Environmental Protection Agency, Technology Transfer, EPA–625/5–76–012.

This is a profusely illustrated, well-written document which deals with most primary, secondary, and tertiary processes for waste treatment, including European practice and emerging technologies. Since it does not go into process design, the reading remains nontechnical at all times.

Fair, G., J. Geyer, and D. Okum, *Water and Wastewater Engineering*, 2 Volumes, Wiley, New York, 1968.

These two volumes are basic texts in environmental engineering programs and so are very complete and explanatory. They do, however, become technical in the sense of giving mathematical treatment to many subjects.

Berger, B. and L. Dworsky, "Water Pollution Control," *EOS* **58** (1), 16 (1977).

The legislative and administrative history of water pollution control efforts by the federal government are covered. It is written by two very knowledgeable individuals who have served in the government in this area. Nontechnical.

The following three paperbacks, all with the same basic title, total about 200 pages. While there is some technical-engineering material in them, on the whole, they are quite readable since they were designed to educate a general audience.

Land Treatment of Municipal Wastewater Effluents, Environmental Protection Agency, Technology Transfer Seminar Publication, 1976: (1) Design Factors—I; (2) Design Factors—II; (3) Case Histories.

Crites, R. and C. Pound, "Land Treatment of Municipal Wastewater," *Environmental Science and Technology* **10**, 549 (June 1976).

This article describes the three land application methods and reviews several examples of land application in the U.S.

Bureau of Land Management

Part 3

Conventional Sources of Energy: Resources and Issues

In the fall of 1973, the United States came abruptly to the end of an era. This era was marked by the growth of cities, sprawling across the landscape, by record levels of auto ownership and use, and by the spread of interstate highways, slicing the nation. Transit and train travel dropped dramatically during this era. It was an era in which it was fashionable to erect glass buildings with windows that did not open. Insulation was largely ignored. It was, in short, an era of cheap energy.

The sudden increase in oil prices, which occurred in 1974 after the Arab oil embargo, was a clear notice to the United States that the era of cheap energy was coming to an end. Inflation, combined with a decreased energy resource base in the United States, contributed to the death of the era. A more fundamental reason, however, was the realization on the part of a cartel of oil producers that the industrial world was feeding at their trough.

Western Europe, the United States, and Japan were consuming far more energy than they themselves were producing. The industries of these nations require oil in vast quantities. Without oil, the entire industrial structure of the Western world could be threatened. As a result of this realization, a new price for oil was set by the cartel. The new price was high enough so that the industrial world began to awake to the fact that it had unwittingly become dependent on a group of supplier nations capable of draining off their money in huge gulps.

The embargo in 1973-74, followed by the six-fold price increase in the spring of 1974, was the first signal. By 1979, prices had risen from the pre-embargo level of $2.00 per barrel to nearly $30 per barrel. These are signals that we cannot ignore. The United States and the rest of the industrialized world must learn and are learning to conserve energy and need to discover and utilize new energy sources.

With the embargo and following price rise, we entered a new era—the era of energy conflict, in which energy users are squeezed by energy producers. Such conflict logically precedes an era of scarcity. Since our energy resources such as coal, oil, natural gas, and uranium are limited, energy use will eventually strike up against the barrier of limited resources; this will be the era of scarcity. Although it has not happened yet, we now have warnings of the scarcity to come, warnings communicated by the higher prices we are paying for energy of all kinds. What has happened can be seen as fortunate in the sense that we have received an early warning of problems still only glimmering on the horizon.

The events that occurred, the embargo and meteoric price rise, were the results of a reduced energy resource base in the industrial world and an enormous energy base in certain less developed countries. The reality of the global resource situation has now brought us squarely up against the issue of energy resources, and it has happened a great deal sooner than most people would have predicted.

The reason the cartel could raise the price of oil so high so quickly is that the cost of finding and producing new oil in the industrial world is much higher than the price the producing nations had been charging for their oil. There was room for a price increase. Furthermore, in the short term, the industrial nations had no option but to pay the price, since new oil of their own could not be discovered and produced quickly enough to take the place of oil from the cartel.

In a way, the drastic price rise is an illustration of a classic argument of resource economists. These economists have said that resources are never actually used up, but that they simply become so costly that substitute resources begin to be found. Resource economists envision that, as more of the resource is consumed, additional quantities become more difficult for suppliers to find. To continue to sell the resource at a profit when it costs more to find and produce, the supplier must charge a higher price. As the price of the resource is pushed higher by continued use and gradual depletion, users become interested in seeking substitutes that are less costly. Substitutes can even replace the original resource and become more widely used than the resource they displace.

Whale oil, shale oil, and coal oil are examples of energy resources from an earlier era that were eventually displaced by a cheaper substitute. These oils were used to light lamps in colonial days and into the middle 1800s. It was in the last half of the 19th century that kerosene, a liquid derived from petroleum, replaced these oils because of its lower cost.

The harder it is to obtain a resource, the higher its price. This notion is basic. It is so basic that the very statement of reserves depends crucially on the price of the resource in the market place. Oil in the North Sea, which cost $5–$6 per barrel in the middle 1970s, was not even counted as a reserve in the early 1970s when the world price of oil was $2.00 per barrel.

The main point to be learned from events taking place from 1974 to the present is that the energy crisis is the result of the geographic and political separation of producers and suppliers. The crisis is not yet one of global demand butting up against global supply. Such a crisis may one day come to pass, but the likelihood of its happening is now far less, for we have had a clear warning notice nailed on our collective door.

In Part 3, we shall discuss conventional ways of generating power and the conventional energy resources we regularly employ. We shall describe the fundamental limit in the efficiency of the conversion of heat energy to work or to electricity. This limit, imposed by the second law of thermodynamics, explains the source of one of our basic inefficiencies in energy use. Ways to improve the efficiency of electric generation are also noted.

With this background in our use of energy, we go on, in Chapter 13, to an investigation of conventional sources of energy. Coal, the troublesome, dirty fuel we are so rich in, is explored first. Its abundance and uses are contrasted to the host of environmental problems it creates. From the impact on the land of strip mining to the impact on humans of deep mining and power generation, there are numerous points at which coal can degrade the environment.

Oil and gas, two fuels cleaner than coal but in shorter supply, are discussed in Chapter 14. More than most other resources, oil and gas are surrounded by economic, political, and conservation issues that must be aired. What are our current sources of these fuels and what are the patterns by which we consume them? Strong clues on how we should conserve oil and gas will emerge from the answers to these questions. New sources of oil and natural gas are also being sought. For instance, oil-bearing shale exists in abundance in the Rocky Mountains, but the environmental consequences of developing this resource are very great. Natural gas, too, is available from many alternate sources.

In Chapter 15, we shall discuss uranium resources which, upon enrichment in the component known as uranium-235, can serve to power nuclear electric plants. Nuclear energy, however, poses basic questions to society about safety, risk, and the spread of nuclear weapons. It is a resource which, if it is fully developed, will demand the most rigid control and security. The "nuclear option" raises grave issues, to say the least.

In the next major part of the book, Part 4, the effects of these conventional energy sources on air and water resources will be discussed in greater detail. For example, we shall explore how our pattern of consumption of the various fossil fuels is at the root of most of our air pollution difficulties. Our energy use also produces global water pollution problems. While the transport and production of oil pollute the oceans, the disposal of waste oil provides an unnoticed but very large source of the oil that reaches our waters. Not only does oil pollu-

tion threaten aquatic species, the waste heat from power plants, dumped into rivers and oceans, upsets the natural ecological balance, upon which a diversity of species depends.

We seek new sources of fossil fuel to extend our resource base, but we can also seek new sources of energy not linked to the fossil fuels—energy from tides, winds, the sun, and the earth's heat. These new sources, which we refer to as natural sources, not only extend our resource base, they also have far less impact on the environment. Many of the air and water pollution problems are simply swept aside by the use of these natural sources.

In Part 5 we shall explore alternatives to conventional energy resources. Do they provide a solution to our energy crisis, or do they themselves have undesirable environmental effects? Finally, we look at methods of energy conservation, since here, at least, we seem to have a solution guaranteed to do no harm to the environment.

Chapter 11

Lessons From Ecology: Energy Laws and the Environment

THE FIRST LAW

Coal and the Conservation of Energy

When the ancient forests of prehistoric earth were buried beneath sediments, energy from the sun was locked within the earth in the form of dead organic matter. Over the eons, some of these deposits of early plant life were changed by biological and physical processes into coal. The sun's energy, which plants transformed by photosynthesis from radiant energy to chemical energy, is stored in this fossil fuel. That energy is liberated when coal is burned (Figs. 11–1 and 11–2).

This is an example of one of the most basic scientific laws, sometimes called the *first law of thermodynamics* or the *law of conservation of energy:* energy can be changed from one form to another but it cannot be created and it cannot be destroyed.[1] In forming fossil fuels, energy in the form of light is changed into energy in the form of chemical bonds. When the coal is burned the stored chemical bond energy is turned into heat energy.

Energy in Living Systems

The flow of energy through systems is of major concern to ecologists and should be to everyone else too. This is because an understanding of energy flow, and the basic laws that govern it, leads to a better understanding of why natural systems work the way they do. Equally as important, such an understanding helps us see the limits on our ability to make changes in our environment without destroying it altogether.

Living systems must be considered with their surroundings. Energy and materials flow into natural systems from their surroundings and flow back out again into the surroundings. They are thus called open systems.

In a stream ecosystem, for instance, energy flows into the system as sunlight, while materials enter in the form of dead organic material, or inorganic nutrients washed in from the watershed. Energy and materials leave the system as heat and as organic material and other nutrients washed downstream (Fig. 11–3).

Energy Budgets

Figure 11–4 illustrates another way of talking about the first law of thermodynamics: all of the energy in a system can be accounted for. That is, an

---

1 Energy can, of course be produced from matter according to Einstein's famous equation,

$$E = mc^2$$
[(energy) = (matter) × (speed of light)^2].

This involves splitting atoms, however. Here we are concerned with ordinary chemical and physical processes. In these processes, we can say that energy is neither created nor destroyed.

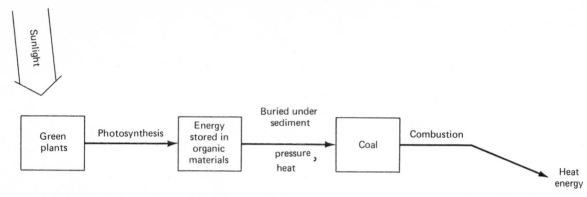

Figure 11-1   *Green plants capture part of the energy in the sunlight by manufacturing high energy organic materials during photosynthesis. In past centuries, some of the plants and the animals that ate them were buried beneath layers of sand and mud. Intense pressure and heat over vast stretches of time turned these fossil deposits into coal. By burning coal, we free the energy from ancient sunlight, in the form of heat.*

Figure 11-2   *Carboniferous swamp forest. Much of the coal we use today comes from the remains of plants that grew in forests that flourished during the carboniferous period, some 345 million years ago. This is a reconstruction of what such a forest might have looked like (Vince Abromitis/Carnegie Museum of Natural History).*

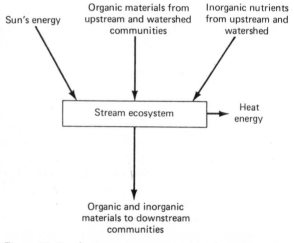

Figure 11-3   *A stream ecosystem viewed as an energy system. Living systems are open systems. Materials and energy flow into the system from its surroundings and out again into the environment.*

"energy budget" can be set up that balances the energy entering an ecosystem, the energy leaving the system, and the energy remaining in the system. In Figure 11-4, the energy budget for the whole earth's ecosystem is presented in broad outlines. But the flow of energy can also be shown in much smaller systems: streams or bogs or cornfields (Figure 11-3).

Limits Explained by the First Law

The first law of thermodynamics explains certain facts about the environment. For instance, it can help us understand limits on food production. Food is a form of stored chemical energy used by living organisms. When we eat food, it is broken

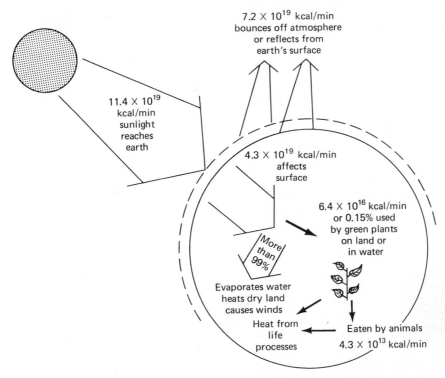

7.2 × 10¹⁹ kcal/min
bounces off atmosphere
or reflects from
earth's surface

11.4 × 10¹⁹
kcal/min
sunlight
reaches
earth

4.3 × 10¹⁹ kcal/min
affects
surface

6.4 × 10¹⁶ kcal/min
or 0.15% used
by green plants
on land or
in water

More
than
99%

Evaporates water
heats dry land
causes winds

Heat from
life
processes

Eaten by animals
4.3 × 10¹³ kcal/min

Figure 11–4 *An energy budget for the whole earth would start with the amount of sunlight reaching the earth, about 11.4×10¹⁹ kilocalories per minute. (114 million trillion kilocalories per minute. For comparison, a person needs about 2000 kilocalories a day as energy from food.) Of this only 3/8 affects the earth's surface, because 5/8 is reflected back by the atmosphere (clouds, dust, gases) and bright surfaces such as the ocean. Most of the nonreflected sunlight evaporates water in the hydrologic cycle, heats dry land and air, causing winds. Only a tiny percentage is absorbed by green plants and used in photosynthesis. The energy which plants capture is used in plant life processes or, when the plants are eaten, used in animal life processes.*

down in our bodies with the release of energy. This energy powers the reactions essential to life.

(In order to increase food production, especially in the developed world, we use great quantities of synthetic fertilizer. However, this fertilizer is produced by using fossil fuels, both as raw materials and to run the manufacturing process itself. That is, we are increasing the production of one kind of energy, the food energy in plant crops and the animals that eat them, by using up stores of high energy fossil fuels. In a similar way, synthetic foods cannot be manufactured without using some other energy resource to run the process. Energy cannot be created.)

THE SECOND LAW

Coal and the Transformation of Energy

Coal and other fossil fuels are burned to free energy to do useful work: to run a steam engine or to generate electricity. But when coal is burned, all of the energy that is released cannot be used to do work. This is an example of the *second law of thermodynamics*, which can be stated several ways. One way is as follows: when energy is transformed from one type into another, some is always lost as heat energy. For instance, when coal is burned in the boiler of a coal-fired power plant not all of the heat goes into steam. Some heat is lost into the environment. Furthermore, when the steam generates electricity, only some of the energy in the steam can be transformed into electricity. Again, part of the energy in the steam becomes waste heat. All of the energy in the coal can be accounted for, as the first law of thermodynamics states, but only part can do useful work. The rest is lost into the environment as heat. In the case of a power plant, either coal or nuclear, this lost heat can cause thermal pollution problems (Chapter 21).

Randomness and Order

A much broader and sometimes more useful statement of the second law of thermodynamics is that *all systems tend to become random, or disordered,* on their own, that is *spontaneously.* For instance, if you dump marbles onto the floor from a bag, they will spontaneously spread out in a disorderly fashion. The release of air pollutants, such as those formed when coal is burned (Chapters 17 and 18), is a demonstration of the second law in action. The pollutants, released from the power plant smokestack, disperse spontaneously in the atmosphere. We actually depend on this phenomenon. Because the pollutants disperse and thus become more dilute in the atmosphere, they are less dangerous. Of course, this sort of system only works up to a point. As populations grow, power demands and pollutant levels increase. Dispersal or dilution of pollutants becomes a less and less effective solution.

On their own, the marbles we spilled will not regroup and hop back into the bag to restore order in the room. We will need to expend some energy

hunting for them and picking them up and putting them back in the bag. The formation of order is a nonspontaneous process requiring energy whether someone is cleaning up a floor or whether an organism is growing by the orderly arrangement of molecules.

When one first looks at living organisms, however, it may seem that here is a system that is becoming more orderly spontaneously. A growing living organism, if it is considered alone, might seem to fit this description. However, at the same time that it is growing in an orderly fashion, biochemical reactions going on in the organism produce heat. This heat, which is given off into the environment, increases the random motion of molecules in the organism's environment. Thus the organism plus its environment is still tending spontaneously towards disorder.

From the second law, it is possible to project some of the effects of population growth. Increasing numbers of people, who themselves live and grow by the production of order, will cause increasing levels of disorder in their environment. Air pollution caused by increased power demands of growing populations is one example of this. Another example is the erosion of soil, which was once bound into tightly knit forest ecosystems, after the forest has been cleared to plant crops or after the land has been turned over in the search for minerals.

Efficiency of Energy Transfer in Living and Non-Living Systems

We can see, when we consider the efficiency of energy transfer when coal is burned, that this system, too, is tending spontaneously toward disorder. All of the energy released by burning coal cannot be used to do useful work (e.g., bring order into a system). Waste heat is also produced and released. This creates disorder in the environment by increasing the random motion of air and water molecules to which the heat is transferred. The system as a whole, then, tends toward disorder. In this case the disorder involves the release of heat into the environment.

(In a similar fashion, the second law operates in food chains. The amount of energy existing, as orderly arrangements of molecules in living organisms, decreases at each higher level of the food chain. At the same time, heat is released into the environment from biochemical reactions when energy is transfered from one level in the food chain to the next. If we look at a food chain in terms of how much energy flows through each level, we can see that there is less and less energy moving up the chain (Fig. 11-5). The energy transfers are never 100% efficient, as the second law of thermodynamics predicts.)

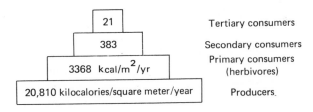

21	Tertiary consumers
383	Secondary consumers
3368 kcal/m^2/yr	Primary consumers (herbivores)
20,810 kilocalories/square meter/year	Producers

Figure 11-5 *An energy pyramid for Silver Springs, Florida. The numbers represent the energy flowing through each level of the food chain. To calculate these numbers, a researcher must determine how much energy is being used to synthesize biological materials and carry on the life processes at each level. Note that only about 10% of the energy is transferred at each level. [Source: H. T. Odum,* Ecological Monographs **27**, *55 (1957).]*

This is one of the reasons why there are few individuals (such as the large carnivores) at the top of food chains compared to the vast numbers of individuals at the bottom. The loss of energy at each step in the food chain makes it impossible for more than a few individuals to be supported. And this is also why (it is not energy-efficient to eat from the top of food chains as humans often do. When we eat meat or fish, we eat food that contains only a small part of the energy with which the food chain began.)

(All ecosystems must be supplied with large amounts of energy from the outside (sunlight, detritus) if they are to keep going.) The large amount of energy lost to the system as heat during energy transfers in the food web must be made up from the outside, or the whole system would run out of energy and collapse. Materials, such as minerals or water, cycle in the environment, but energy does not. (Energy flows into ecosystems and is lost as heat.) In Figure 11-6, some of the energy transfers in Cedar Bog Lake are shown. You can see how energy is lost at each step in the food chain as heat from the life processes of plants and animals.

Although these laws of thermodynamics are important to understanding relationships between organisms and the effects of organisms, including humans, on their environment, most people fail to take them into account. A study of how these natural laws operate, with respect to power generation, for example, can help us find ways to minimize our impact on the environment.

Questions

1. Explain the first law of thermodynamics in terms of the energy in coal.
2. The example is given in the text of using synthetic fertilizers to increase food production. The synthetic fertilizer is often needed to replace nitrogen used up by food crops. This method of producing more foods uses energy from fossil fuels. Suppose instead of synthetic fertilizers a system of crop rotation is used in which a nitrogen-fixing crop is planted to refertilize the ground between crops of the food species. Does this method of increasing food production violate the first law of thermodynamics? Hint: What energy source is being used?

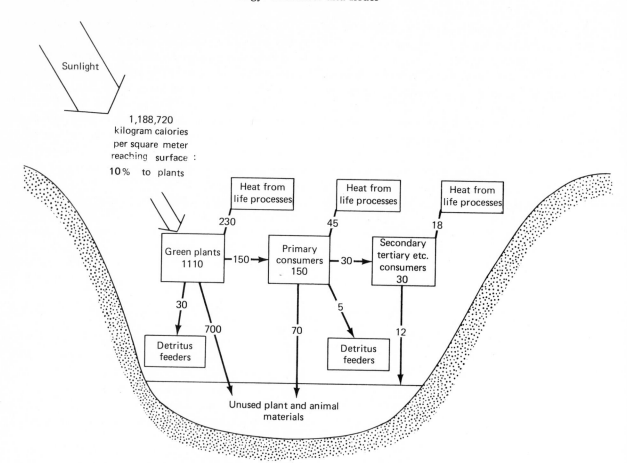

Figure 11-6 *Energy flow diagram for Cedar Bog Lake, Minnesota. Numbers are kilocalories per square meter per year. Note how the amount of energy at each level of the food web decreases. A large amount of energy is lost at each level in the form of heat from the life processes of each organism. About 1,188,720 kilogram calories per square meter reach the surface of the bog each year. Ninety percent of this amount is absorbed by the water and about 10% reaches plants in the water. By photosynthesis, the plants produce 1110 kg·cal/m² per year of chemical energy, of which part is used by the plants themselves, part is decomposed, part goes to herbivores which eat the plants, and part accumulates in the bog as unused plant material. (This is one reason why the bog fills up year after year.) [Source: R. Lindeman, Ecology 23, 399 (1942).]*

3. What is meant by the second law of thermodynamics? Explain, using a simple example other than a bag of marbles. Are there exceptions to this law?

4. Make up an energy budget (without numbers, just general relationships) for a grassy meadow, bordering a small stream.

5. How does the second law explain the fact that there are many fewer organisms at the top of food chains than at the bottom?

6. Give some examples, other than those in the text, of how growing human populations, which must be supported at high levels of order, cause increasing disorder in their environment.

7. How would you explain that dumping sewage wastes into a river is an example of depending on the second law of thermodynamics to solve a disposal problem. Does it work?

Chapter 12

How We Use Energy Resources

ENERGY CONVERSIONS AND THE SECOND LAW

We commonly utilize energy resources in three fundamental ways. We may release heat energy by burning fossil fuels and use this energy directly as heat: heat for homes, schools, factories, and shops. Or we may convert this heat energy into work, using refined oil to drive machinery and to power automobiles, trucks, trains, planes, and ships. Last, we may convert the heat energy from burning fossil fuels or from the fission of uranium into electricity and then use this transformed energy either for heat or to do work. Falling water is also commonly used to generate electricity. Electricity, in effect, is acting as the middle-man between energy resources and the final use. Just as the presence of a middle-man in a market creates higher prices, the use of energy in the form of electricity brings a mark-up as well, in this case, an energy loss or penalty (Fig. 12–1).

The practice of converting resources into electrical energy arose for many reasons. In some cases, it was not possible to make effective use of these direct sources of energy except to convert them to electricity. The energy from falling water, before the era of electricity, could only be used to drive machinery at the waterfall itself. Textile mills, grain mills, and lumber mills all operated at waterfalls for a time in early manufacturing history. Hydropower could be used in no other way until it could be used to generate electricity, which then made it possible to operate machines at distant sites. Uranium, likewise, is not generally used to fulfill energy requirements except in the form of electric energy. Then, however, like hydropower, it can be used not only to drive machinery but also to produce heat for homes, hot water, etc.

The fossil fuels, in contrast to falling water, were first recognized as sources of light and heat and not as sources of energy to power machinery. Wood and coal, and often dried peat, were burned for heat in homes and commercial buildings, and coal was used to provide the heat needed to manufacture iron. "Carbon oil" from coal was used in lamps. It was not until Watts' invention of the steam engine in the 1700s, however, that the potential of fossil fuels to drive machinery was discovered. By the early decades of the 1800s, railroad locomotives were being built in the U.S. that burned coal to generate the steam for turning power. And by the early decades of the 20th century, coal was burned to provide steam to generate electricity, although the production was very inefficient at the time.

Thus, fossil fuels began to assume a new role in the industrial world. From merely providing heat and light, they came, through the use of engines and electricity, to be sources of power for industrial machinery and for transport, as well.

These historical events color our present attitudes. The use of coal and other fossil fuels to generate electric power for machines is most appealing. Since electric power can be transmitted from place to place, the machinery can be located

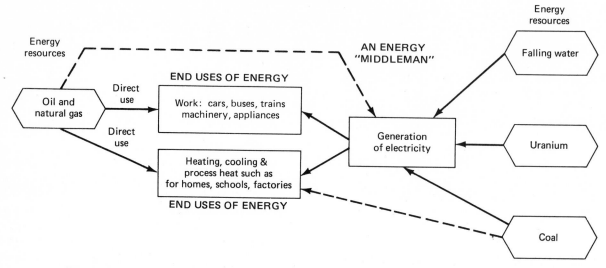

Figure 12–1   *Common routes for the use of energy resources. Electricity is an energy middleman that takes the basic energy resource, processes it, and transfers it to the final user. Like any middleman, it demands that a toll be paid.*

far from the site of generation. Electric heating of homes is also appealing; no fuel is consumed in the home; no ashes and smoke are produced. But there is a penalty for using this form of power, far beyond the extensive pollution we so easily ignore at the point of electric generation. The penalty is an enormous loss of efficiency—an enormous waste of the heat energy in the fossil fuels.

As a consequence of the second law of thermodynamics, there is an inherent limit to the amount of heat energy that can be converted into work and hence into electricity. The reverse is not true; that is, other forms of energy can be converted completely into heat. This limit in the efficiency in the conversion of heat into work was not an obvious concept to the early designers of engines which used heat for power. An engineering text[1] tell us that

    .... the low efficiencies encountered in the conversion of... heat to work were at first attributed to incorrect design of the mechanism

employed, or to practical difficulties in operation. Even after careful improvement, however, low efficiencies prevailed. When extensive efforts at improvement proved fruitless, it was finally inferred that there existed some underlying natural limit to the conversion of heat to work.

In fact, most of the possible design improvements in the efficiency of the generation of electric power from steam have now been made. The modern coal-fired power plant, which produces steam to turn the turbine for power generation, has reached 40% efficiency. That is, 40% of the heat in the coal burned is converted into electric energy. Oil-fired power plants push up into this range as well. Conventional approaches to electric power generation cannot be expected to increase this efficiency by very much (Fig. 12–2). Nuclear electric power plants also produce steam to turn a turbine and these have attained efficiences of only 30%–32%. That is, only 30%–32% of the heat of fission is converted to electrical energy. Attempts to increase this conversion percentage have yet to succeed.

---

1 H. C. Weber and H. Meissner, *Thermodynamics for Chemical Engineers,* 2nd ed., Wiley, New York, 1957.

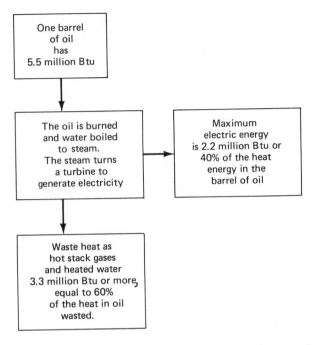

Figure 12-2   *The inefficiency of generating electricity: The example of oil.*

(Electric generation then, which is a convenience and a very great convenience, to be sure, is wasteful of the heat energy of the fossil fuels, coal, oil, and natural gas. It is doubtful that we would go back to heating homes with coal, so the use of coal for electric generation, though wasteful from a heat standpoint, may be one of the few uses which we can now make of coal.) Because so much of the energy we use is in the form of electric energy, we need to discuss more completely the sources and uses of electric power, as well as the impact upon the environment that accompanies this use.

ELECTRIC POWER

A flick of the finger and we have

light,
sound,
visual communication,
heat,
hot water,

refrigeration,
air conditioning,

all from electric energy.

A flick of the finger and we also have

river valleys buried in sediment;
sulfur oxides, acid rain, acid mine drainage;
oil pollution;
particulate matter;
nitrogen oxides and smog;
strip mining;
thermal pollution;
radioactive wastes;
plutonium for bombs;

all from electric energy.

Generation, Use, and Impacts

(Electric energy is one of the most easily used forms of energy.) One electric line enters a house and the energy to light the home, cook and bake, warm water, and heat the home is all at our fingertips. The first list above gives us an idea of how widely useful electric energy is. The second list reminds us that the clean fuel at the point of use is anything but clean at the points of mining and generation.

The use of electricity for household tasks is steadily increasing. Electrical appliances are becoming more widely available as time goes on, as the graph in Figure 12–3 shows.

Because electricity is such a convenient form in which to use energy, it is expected to capture more and more of the energy market. According to the Congressional Research Service, the quantity of electric energy used in the U.S. will increase by almost 100% from 1976 to 1990, even though total energy use will increase by only about 50% in that same period (Fig. 12–4). Looked at another way, the projected increase in the demand for electrical energy is about three-fifths of the total projected increase in all forms of energy use.

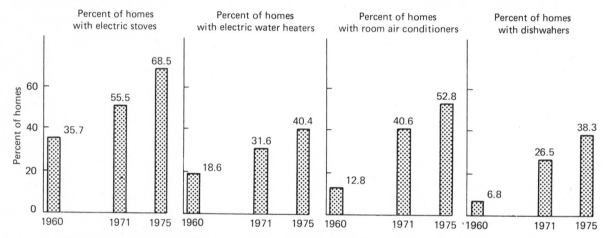

Figure 12–3   *Growth in percentage of homes with certain electric appliances. (Note that the number of homes increased by about 25% from 1960 to 1971.) (Sources: "Energy and the Environment: Electric Power," The President's Council on Environmental Quality, August 1973; and* EPRI Journal, *December 1977.)*

BEWARE THE METER

(Following is an excerpt from an editorial that appeared in The New York Times or, as it was spelled then, The New-York Times on June 11, 1881.)

Now, there is not the least reason to believe that electricity exerts any better moral influence than gas, and we know that when a number of reasonably Christian men form themselves into a gas company they immediately become pirates of the most merciless and extortionate character.

Why should we look for better things from the electric light companies? They espect to have us at their mercy, and they will be merciless as the gas men.

We shall have electric meters in our cellars that will be as mendacious and unprincipled as the gas meters, and the moment we refuse to pay for ten thousand feet of electricity which we have not used our lights will be cut off and we shall be left to candles and kerosene.

To understand how fuels are used to generate electricity we need to describe the operation of a typical "thermal" electric plant (Figure 12-5). By thermal electric plant, we mean a power plant that utilizes heat to produce steam to turn a turbine. A coal-fired plant is a thermal electric plant; so is one that burns oil or gas. A plant that utilizes the heat from nuclear fission is also a thermal electric plant.

The principle of operation in all three is the same (Fig. 12-5).

Heat is produced by combustion (or fission). That heat is used to boil water to steam. The steam, at a high temperature and pressure, is used to turn the turbine and generate electricity. The turbine's rotation of an armature in a magnetic field induces the flow of electric current. The steam that leaves the turbine has its pressure and temperature much reduced. This "spent" steam is converted back to water by a condenser,[2] through which cooling water flows.

The water used to condense the steam may be returned to the body of water from which it was taken. Alternatively, it may be passed through towers to be cooled and used again in the condenser. The water which was condensed re-enters the boiler to be again converted to steam and circulated to the turbine. The water or steam is referred to descriptively as the "working fluid."

The coal station wastes 60 units of heat to the environment for every 40 units converted to electricity; it is working at 40% efficiency. That is, 1.5 units of heat are wasted per unit of energy pro-

---

2 A condenser is a device which brings two streams of fluid into close contact with a metal barrier between them. Heat is exchanged between the two fluids, the cooler stream accepting heat, the warmer stream donating it. The condenser in a power plant consists of parallel tubes inside a large cylinder through which water runs. The steam circulates in the cylinder around the outside of the tubes, losing heat to the water flowing in the tubes and condensing.

1976: 74.0 quadrillion Btu, 52.5, 21.5 quadrillion Btu
1985: 94.8 quadrillion Btu, 70.6, 34.2 quadrillion Btu
1990: 108.9 quadrillion Btu, 67.4, 41.5 quadrillion Btu electrical

(1 quadrillion = 1 million billion = 10^{15})

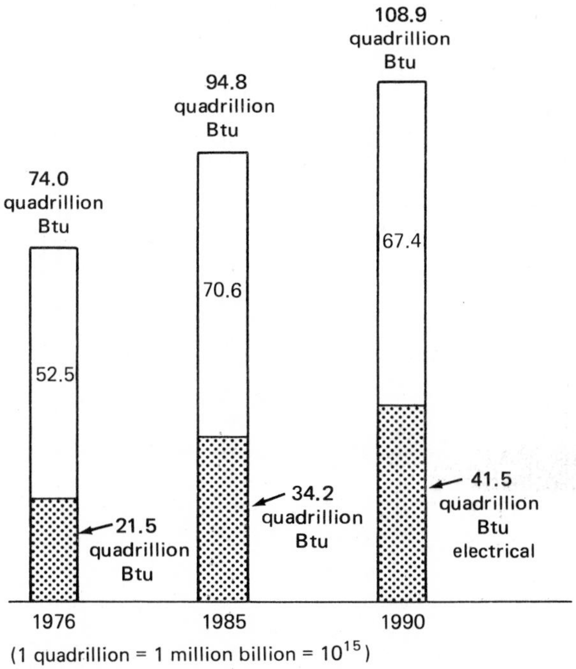

Figure 12-4 *Estimated total energy use and energy for generating electricity through 1990. (Source:* Project Interdependence, *Committee Print, 95-33, Congressional Research Service, U.S. Senate, 1977.)*

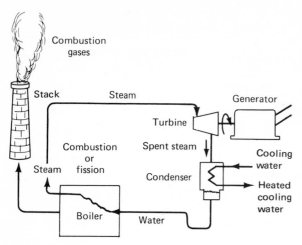

Figure 12–5   *Elements of a thermal electric power plant.*

duced. The nuclear plant, at 30% efficiency, wastes 70 units of heat for every 30 units converted to electrical energy. That is, 2.33 units of heat are wasted per unit of energy produced at the nuclear power plant, or about 55% more waste heat than the coal plant per unit of electric energy produced.)

(The problem is compounded for the nuclear plant, because all its waste heat is transferred to the condenser cooling water. The coal-fired power plant, in contrast, directs only 75% of its waste heat to the cooling water. The remainder of the heat goes up the smokestack.)

Electric energy is generated at fossil-fuel power plants, at nuclear power plants, and at hydroelectric plants. The combination of plants from which we draw electricity has been constantly changing for a variety of reasons. In the late 1960s and into the late 1970s, new nuclear plants were being added to the stock of generating stations. During the late 1960s, coal-fired power plants were being converted to residual oil in an effort to reduce sulfur dioxide levels in urban areas. With the onset of the 1973 oil embargo, and a massive oil price rise, coal plants, with better pollution control or burning lower sulfur coal, began to replace the oil-fired plants. A program to convert to this more abundant resource was carried out in an effort to reduce oil imports. And as resistance to nuclear power grew, the number of orders for new nuclear stations dropped to only several plants in the latter years of the decade of the 1970s. Thus, the combination of plants, as we show it for 1978 (Fig. 12–6), is actually a snapshot of an evolving system. By the time you read this, oil's share of electric generation will have fallen, reflecting the fact that no new oil-fired units will be added to meet growing demands. Gas use will also fall as natural gas is diverted to use in heating homes. Coal use will increase not only because of conversions from oil to coal, but because much of the new generating capacity will be in coal-fired units. Nuclear use will have grown because the nuclear units already under construction will have come "on line" in this period. Hydropower will lose a portion of its market share as the market expands because few new large hydroprojects are possible.

(That electric generation is not a clean source of power is vividly shown in Table 12–1. Here it can

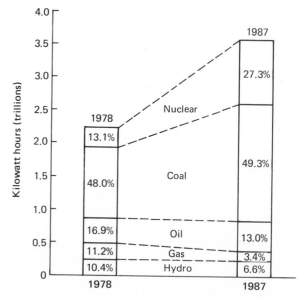

Figure 12–6 *Fuels for electric generation: electric energy generation by principal sources (contiguous U.S.). (Source: National Electric Reliability Council, 8th Annual Review, August 1978. Copyright by NERC, Research Park, Terhune Road, Princeton, N.J. 08540.)*

TABLE 12-1   Sources of Man-Made Pollution[a]

	Particles million tons per year	Sulfur Oxides million tons per year	Nitrogen Oxides million tons per year	Hydrocarbons million tons per year	Carbon Monoxide million tons per year
Electric Generation	4.5[b]	16.0	7.1	0.1	0.4
Residential and Commercial Combustion	0.6	2.9	1.1	0.1	0.6
Industrial Combustion	2.0	3.2	2.7	0.1	0.2
Other Industrial Processes	14.4[c]	7.5	0.2	5.5	12.0
Transportation	0.8	1.1	11.2	19.8	111.5
Miscellaneous	12.8	0.4	2.4	11.2	26.1
TOTAL	35.1	31.1	24.7	36.8	150.8

[a]From Energy / Environment Fact Book, March 1978; USEPA and USDOE EPA-600 / 9-77-041.

[b]If not under control, this amount could be multiplied by a factor of 10.

[c]Grinding, spraying, demolition, construction, mining, cement plants, grain handling, etc.

be seen that the emissions from electric generation account for a high proportion of the annual emissions of three of the five major air pollutants. The table also illustrates the other leading culprit in air pollution activities, transportation. The wide range of environmental insults caused by electric power generation will become clearer in the chapters that follow.

New Developments in Conventional Electrical Generation

The efficiency of conventional fossil-fuel electric power plants has improved dramatically in the last half century, but the improvements have almost come to a halt in the last decade. The technological problems are becoming much more difficult to solve. Yet the promise of improvement still exists, so research is going forward to solve the basic problems. The solutions may be quite novel as researchers veer off the conventional path to add new features and processes to the basic electric generation technology. The thrust of the research is to put processes together to capture the heat from combustion, which used to be wasted. We will discuss three promising ideas. The first is called "district heating" and uses spent steam for heating. The second, known as "combined cycle," puts together the turbine engine and the ordinary boiler/turbine system. The last is magnetohydrodynamics (MHD for short) and is designed to be used with a coal-fired boiler/turbine system or possibly a nuclear system.

District heating   One option to make electric generation more efficient is to make use of the waste heat, the heat from combustion or fission not con-

SHOULD LOCAL ELECTRIC COMPANIES BECOME PUBLICLY AS OPPOSED TO PRIVATELY OWNED?

[The] impetus toward municipal ownership now is the higher rates of the private utilities and the feeling that local ownership would be more responsive to public needs.

> Alex Radin, General Manager,
> American Public Power Association,
> Wall Street Journal, 23 October 1974

Higher electric bills are simply a reflection of the higher cost of everything a utility must buy to provide service.

> W. Donham Crawford, President,
> Edison Electric Institute
> Wall Street Journal, 23 October 1974

About 80% of the power generated in the U.S. comes from investor-owned utilities. A number of electric companies, however, are not investor-owned but owned by local governments. Public ownership has some benefits. These include generally lower electric rates than investor-owned utilities and a better reception for public opinion. The responsive attitude follows directly from public ownership, but lower rates come about for other reasons. A municipally owned utility does not have to pay dividends to stockholders or federal taxes on its profits; nor does it pay local property taxes. Finally it can finance new construction by selling tax-exempt bonds.

The interest on the bonds issued by an investor-owned utility are liable for taxation as income for the owner of the bond, as is the interest on all privately issued industrial bonds. In contrast, the interest on the bonds issued by state and local governments is not liable for federal taxation. If bond purchasers who are in the 50% income-tax bracket can choose between an industrial bond paying 12% and a municipal bond paying 7%, they are likely to choose the municipal bond. It pays 7% free of taxes; the industrial bond after 50% taxation pays only 6%. Because the interest on the municipal bond is not subject to taxes, purchasers are willing to accept a lower interest rate. Thus, the municipal utility is able to raise money for construction at a lower cost than the investor-owned utility. As a consequence of this and the lack of taxation on its property or on its income, the municipally owned utility is generally able to offer lower electric rates to its customers.

The immunity to property taxes is a two-edge sword, however. Since a municipal utility can escape taxes, the local government must obtain its revenues by increasing taxes on its remaining property. The lower electric rates that stem from the fact that the utility need not pay property taxes may then be an illusory saving for consumers. Lower electric rates due to the tax-free bonding power of communities and lower electric rates because no stockholder dividends need be paid are very real.

In addition, public utilities, because they have no reason to expand their investment base, should be more willing to save the public money by using peak-load pricing. Peak-load pricing places higher charges on electricity generated at the hours of peak use and, hence, decreases peak demands. Lower peak demands mean less investment required for new capacity and, thus, lower electric rates. Privately owned utilities, in contrast, like to expand their investment because they are allowed to earn a rate of return on their investment.

Finally, public utilities are expected to be more accessible to public opinion on issues such as location of its facilities and whether to use nuclear power. On the negative side, one must face the issue of efficiency–whether public enterprises are as efficient as private ones. It is this issue that gives the argument for public power two sides.

verted to electrical energy. That waste heat carried by the spent steam can be transported via underground pipes to homes, offices, and factories to provide space heating during the winter months. (District heating, as this process is called in Europe, has been achieved in Finland. Its prospects are diminished, however, by the large quantity of pipes that must be placed in order to deliver the heat to customers. In addition, if the pipes must go long distances, heat losses through the walls of the pipe become so great that the service is not economical. Thus, the concept requires small power stations with customers for the heating service located very nearby. Since smaller plants are less economical than larger ones and since most Americans do not like to live near power plants, the prospects for district heating are not bright at the present time.)

Combined cycle To describe the combined cycle method requires a brief description of the turbine engine, the component that makes it different (Fig. 12–7). Since World War II, there have been steady improvements in the turbine engines used to power jet aircraft. These improved turbine engines have been adapted to power generation and have become common equipment in electric generating stations.

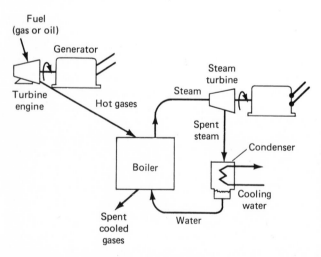

Figure 12–7 *Combined cycle concept for electric power generation.*

Burning kerosene or natural gas, the engines themselves produce the rotation power necessary for electric generation rather than boiling water to make steam. Their efficiency is low; on the order of 25% of the heat energy produced is converted to electric power, and the fuel is expensive compared to coal. The equipment, however, compared to the boiler/turbine system of conventional plants, is relatively inexpensive. Electric utilities buy these turbine generators and hold them on standby for times when power loads reach levels beyond the capacity of their base-load equipment. At those times, the turbine engines are pressed into service. For these peak loads, a slightly more expensive fuel does not drive costs too high.

(The combined cycle idea is really quite simple. The turbine engine and the boiler/turbine system are put together. The burning gases in the turbine engine generate power from the rotation of the engine shaft. The very hot gases discharged from the turbine engine are used to boil water to steam. The steam, in turn, is used in the ordinary way in electric power plants to turn the turbine.)

Magnetohydrodynamics (MHD) (The last major extension of the modern generating plants is known as magnetohydrodynamics; it makes use of the hot gases from combustion to generate electricity, after which the gases can be used to produce steam.)The hot gases from the combustion of coal or other fossil fuel are seeded with potassium, which ionizes into charged particles in the very high temperature flow. The hot gases are directed through a channel surrounded by coils to induce a magnetic field. The movement of charged particles through the magnetic field sets up a current flow that is drawn off by electrodes spaced along the channel. Upon exiting the channel, the hot gases are used to produce steam, which is then used to turn a turbine in an ordinary way.

(The combined MHD-conventional system could reach efficiencies of up to 60%. That is, 60% of the heat in coal would be converted into electrical energy. If such plants are built, they will be far less polluting than conventional plants. For every kilowatt-hour of electrical energy produced, the 40% efficient plant will consume 50% more fuel than the 60% efficient plant. That means 50% more thermal pollution, particles, sulfur oxides, and nitrogen oxides.)

(One of the problems still to be overcome is obtaining materials that will not corrode in the high temperature (2000°C) in the MHD equipment. No large generators have yet been built, although a good-sized developmental unit has been operated in Russia. MHD research in the U.S. is apparently well-funded at the experimental level.)

Fusion–Electric Energy in the Future?

In the midst of recurring energy shortages, the question occurs again and again, "What about fusion?" Fusion has become not only a specific atomic event which scientists hope to harness to produce electric power; it has to many become the name for the "dream of limitless energy." Is fusion one piece of the answer to energy supply problems of the future? How soon could it contribute to our basic needs for energy? The answers to such questions have a philosophical content and their meaning is only clarified by a description of fusion, the atomic event, and the concept of power from fusion.

(While nuclear fission refers to the *splitting apart* of the nuclei of large *heavy* atoms with a release of energy, nuclear fusion refers to the *combining* of the nuclei of certain *light* atoms such as deuterium and tritium, two isotopes of hydrogen. This combination, which takes place constantly in the sun and stars, also releases large quantities of energy. Just as the modern light water nuclear reactors are designed to capture the heat in nuclear fission and use it for electric generation, the fusion reactor of the future would capture fusion heat and convert it to electric power.)

The fuel for the future fusion reactors, the substance which is so abundant and hence makes fusion seem so attractive, is deuterium, an isotope of hydrogen with one proton and one neutron in its nucleus. Ordinary hydrogen has only one proton and no neutrons in its nucleus. The fusion of a deuterium atom with another deuterium atom or with a tritium atom (two protons and one neutron in the nucleus) produces vast quantities of energy. Deuterium is so abundant in the surface waters of the earth that these sources could contain up to 10 million tons of this element—even though there is barely 1 pound of deuterium for every 30,000 pounds of water. Still, if the energy from deuterium reactions were fully extracted, only eight pounds of water could supply the energy equivalent of 300 gallons of gasoline. (Such an abundant fuel suggests a low cost for electric power generated from fusion, but, in fact, containment of the high temperature fusion reaction cannot be had cheaply. In fact, at this point in time, it is not clear that fusion can ever be exploited because of our inability to somehow "house" this reaction. Thus cost estimates really have no basis, for we are not at a stage where we can say that fusion will work.)

(Containment is needed for the fusion reaction (say the combination of a deuterium and tritium atom to produce helium) because the fusion (combination) will take place only at the star temperature, 100 million degrees Celsius.) Since no substance is known which remains solid at this temperature, scientists have devised several novel concepts to "hold" the fusion reaction. These include the famous *tokamak* concept in which the fusion reaction is held in place by powerful magnetic fields. Experiments with tokamak containment are underway at Princeton University, the Oak Ridge National Laboratory in the U.S. and in the Soviet Union. Other magnetic shapes than the tokamak "bottle" are possible and are being pursued as well. A concept which does not use magnetic confinement is laser fusion in which converging laser beams focus on a near microscopic pellet of deuterium-tritium fuel. The heat from the laser beams causes the pellet to explode within the minute region defined by the beams, producing the temperatures needed for fusion. Many nations are now conducting fusion experiments.

The environmental safety of the future fusion reactor is an often mentioned advantage for fusion. In fact, there may be a significant quantity of radioactive waste to dispose of. Most of the fusion reactions being considered now produce high speed neutrons, and the neutrons bombarding the walls of the reaction vessel will make the wall radioactive and a waste to be disposed of somehow.

To this time, no experimental fusion reactor has been able to achieve a sustained reaction. It follows that power production, even at a net loss of energy, is still some time away. When active research on fusion began in the late 1950s, the first fusion power was estimated at 20 years away. Today, nearly a quarter of a century later, scientists are still saying "20 years away." It makes sense to pursue dreams; dream and ideas are the forerunner of reality. It makes sense to pursue fusion, but we should understand that it cannot contribute to our needs "here and now."

In a certain way though we are using fusion already. The energy reaching us from our nearest star, the sun, is being captured by green plants in the process of photosynthesis. Thus, our food is linked to the fusion process taking place in the sun; likewise our oil, coal and natural gas, which are derived from plant materials, are also linked to fusion. And solar energy for heating water and buildings is the result of our utilizing the heat from fusion being released from the sun.

Fuel Cells and the Hydrogen Economy

The fuel cell is the name given to the concept of producing electricity from the chemical reaction of hydrogen and oxygen. The aerospace program pushed this conept far enough along that fuel cell power plants are now being built, though the first plants are small in size. At the site of electric generation fuel cells produce very little in the way of air pollution, thermal pollution or noise pollution. Hence, as urban neighbors they are expected to be more welcome than the smoky, sooty coal-fired plant or the nuclear power plant with its built-in hazard for populations.

Whereas the oxygen needed for the fuel cell can be supplied from the air, the hydrogen gas will be obtained by chemical processing of a liquid hydrocarbon fuel such as naptha or of a hydrocarbon gas. Hydrogen may also be obtained by processing natural gas or petroleum gas. Gas produced from coal gasification may also eventually be used. The first generation fuel cell uses phosphoric acid as the electrolyte or conducting liquid. Such a cell, about five percent of the power capacity of the large modern coal or nuclear plants, is being installed in New York City to demonstrate the feasibility of fuel cells. Since direct current electricity would be produced by the fuel cell, an invertor will be needed to convert the electricity to the alternating form that we use throughout the country.

In a second generation concept, in which the fuel cell electrolyte is molten carbonate, the two fuels which combine with oxygen are hydrogen and carbon monoxide. Both of these fuels are produced in the initial stage of coal gasification (see enrichment section of Chapter 14). At present this cell is the subject of research; full scale cells are at least a decade away. Whether the coal gasification stage might be carried out in the same geographic area in which the fuel cell is located is an open question. The fuel cell's advantage of almost pollution-free power generation would disappear if the coal gasifier were nearby—because the coal gasifier is a generous producer of air pollutants. It is this second generation of fuel cells that electric utilities see as important contributors to power for cities.

The hydrogen-using fuel cell is one example of how hydrogen may be used in the future. Hydrogen is not a new energy resource, however. Instead it is a new way to convey energy. Hydrogen from the electrolysis of sea water might be produced by first generating electrical energy from the wind; it is the wind that is the primary energy resource.

Hydrogen might also be produced from sea water using electricity from conventional power plants; the fuel which fired the power plants is the base energy resource.

In addition to its use in fuel cells, hydrogen is being explored as a fuel for the internal combustion engines of automobiles. It is also being considered to power turbine engines such as that of the jet plane, and since a similar turbine engine is used to produce extra electrical power at times of peak demand, hydrogen might be used here as well. Such future uses of hydrogen are not at all certain. Some see such uses in their crystal ball; others do not.

THE USES OF ENERGY AND THE FUELS THAT SUPPLY IT

To save energy, we need an idea first of the ways in which we presently use energy. We also need to know what energy resources we rely on. This information will help us to determine the ways we can conserve energy and the fuels that can be saved.

Figure 12-8 shows how we have historically consumed energy. In general, the pattern of consumption has not changed drastically over the past quarter century. Roughly the same proportions exist in each economic category from 1950 to the present. Actual energy consumption[3] more than doubled between 1950 and 1970. Although the population increased by 40%, accounting for a portion of the increase in consumption, the actual increase in average annual energy consumption per person was about 55%, from 210 million Btu per person annually in 1950 to 329 million Btu per person in 1975.

Transportation has steadily accounted for about 25% of the energy we consume each year.

Nearly all of this energy is consumed as some form of petroleum.

The industrial use of energy and the use of energy in the household/commercial sector are both rivals to transportation in terms of their percentage consumption of energy. In contrast to the transportation sector, however, both of these categories are "mixed bags" of energy sources. Large quantities of coal and natural gas in addition to oil are used in these two categories, and the electrical middle-man plays an important role.

Although nearly all of the U.S. transportation energy comes from oil, in Europe this is not the case. In Europe, where trains still play an important role, some of this energy is electrical energy, as most rail systems in Western Europe have been electrified. Of the energy consumed for transportation in the U.S. about 52% to 55% is burned in automobiles. Another 22% is consumed in the transportation of goods by truck. Airplane operations account for another 10% and railroads for only about 4%. Buses, pipelines, ships and barges, military air and ground vehicles, and other minor consumers utilize the remainder of transport energy (Fig. 12-9).

We can break down the household/commercial category of energy use as well. Nearly three quarters of the energy consumed in this category goes into keeping people either warm or cool [see Fig. 12-10(a)]. Heating used 63% of all energy in this category when figures were last tabulated. Cooling consumed 8% of the energy in the household/commercial sector.[4] Water heating accounted for 14% of the category; refrigeration and freezing, 10%, and cooking, 4%. Air conditioning and refrigeration together make up 18% of the use in this category. Both of these uses draw on electricity as virtually their sole source of energy.[5]

---

3 Total energy consumption is the sum of oil, natural gas, and coal when burned plus hydroelectric and nuclear electric energy, the latter converted to the energy required to create the electric energy.

---

4 The figures are for 1970, but a hotter summer or colder winter could change the numbers, especially as air conditioning spreads to more of the population.
5 There has been a very small amount of cooling and refrigeration with natural gas.

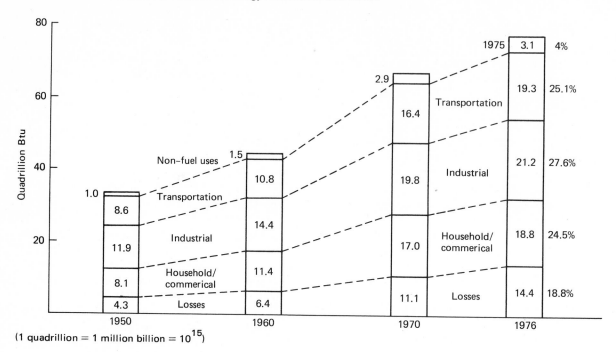

(1 quadrillion = 1 million billion = 10^15)

Figure 12-8   *The pattern of growth in U.S. energy consumption by economic sector. (Sources: U.S. Department of the Interior, 1976 and 1977; Project Interdependence, Committee Print, 1977.)*

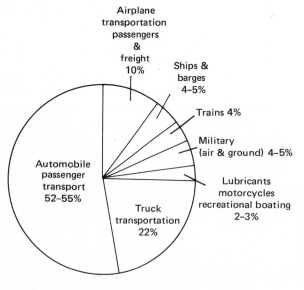

Figure 12-9   *How transport energy is consumed.*

If we include the electrical energy consumed in water heating, and in space heating and cooking, another 10% of this category's energy derives from electricity, for a total of 28%.

In Figure 12-10(b), the space heating category is broken down by fuel type to show the contributions of the various fossil fuels.

As the energy uses in various sectors expand, so also must our use of fuels. Figure 12-11 displays our energy growth in the last quarter century in terms of the fuels we have consumed. From this chart, we can see that petroleum and natural gas have expanded their portion of the energy market since 1950, while the portion using coal has declined. From 58% of the energy used in 1950, petroleum and natural gas expanded to 75% of the energy market in the 1970s. While coal appears to have lost ground in percentage terms, it has not lost ground in actual amounts used. The 1976 produc-

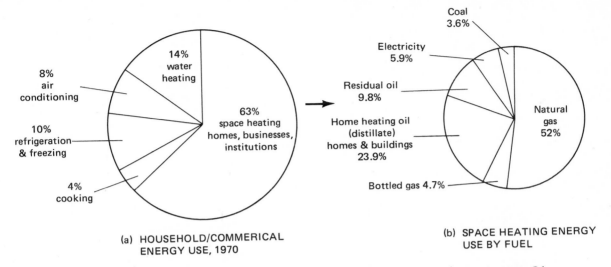

Figure 12–10   *(a) Pattern of household/commercial energy consumption by use in 1970. Of the energy consumed in this sector, 28% was electrical. (b) Pattern of space heating energy consumption by fuel type in 1970. The 9.8% residual oil was used principally to heat office buildings, schools, hospitals, etc. (Adapted from D. Kash et al.,* Our Energy Future, *University of Oklahoma Press, Norman, Oklahoma, 1976.)*

tion of this fossil fuel was 20% more than in 1950. What did happen was that coal lost the growth increase in two important markets, the railroads where coal gave way to petroleum, and home heating where coal was displaced by oil and natural gas.

From the 1950s onward, coal as a means to heat homes lost ground steadily to natural gas and oil. Coal was dirty and produced large quantities of ash, which the homeowner added to his trash. It also required a large storage area in the home (the coal bin), and the coal furnace needed frequent shovel feeding. Natural gas heating systems were compact, automatic, and needed no storage area in the home. Oil heat, although it required a tank for storage, was also compact and automatic. These fuels, still inexpensive, when they began to become popular, displaced coal in most space heating appli-

cations and still hold most of this market today.

Since the oil embargo of 1973-74, with the following sharp increase in oil prices and the shortages and price increases in natural gas, the nation has begun to re-evaluate the importance of our coal resource. It is the fossil fuel in which we are richest. Perhaps 20% of the world's supply of coal lies within our borders. All plans to fill the growing U.S. demand for energy focus upon coal. Our coal, however, is a mixed blessing. It has serious effects upon the land and upon our health. Because coal is now seen as so important to the nation's economic health, and because it also has a very large negative impact we need to explore fully the consequences of our plans for its use. As we enter onto a new highway, we should be aware, to the largest extent possible, of its condition. It may be a rocky road indeed.

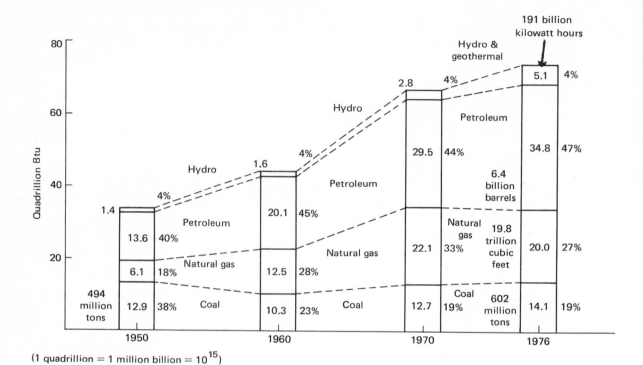

(1 quadrillion = 1 million billion = 10^{15})

Figure 12–11   *Pattern of growth in U.S. energy consumption by fuel type. The 1985 projection by the Congressional Research Service is 91.2 to 98.4 quadrillion Btu total energy consumption; 1990 projection is 104.5 to 113.4 quadrillion Btu. (Adapted from D. Kash et al.,* Our Energy Future, *University of Oklahoma Press, Norman, Oklahoma, 1976.)*

Questions

1. From what you have read about electric power, do you think that the use of electric autos will help to solve the pollution problems of the United States? Explain as fully as possible your position. Will it help one kind of geographical area and hurt another?

2. The loss of energy in the United States in 1976 (waste energy that heated up the environment, lost in friction, etc.) was 14.4 Quads (Fig. 12–8). According to Figure 12–4, 21.4 Quads of energy were consumed in generating electricity in 1976. With what you know about electric generation and its efficiency in converting heat to electricity, can you identify these losses? For the sake of discussion, assume the average efficiency of all electric generation is about 30%, including transmission losses; that is, about 70% of the heat energy consumed will be wasted to the air or water.

**Further
Reading**

Council on Environmental Quality, "Energy and the Environment: Electric Power," August 1973, available from the U.S. Government Printing Office, Washington, D.C. (stock number 4111-00019).

This well written document provides an excellent overview of the major issues surrounding the generation of electric power. Although now aging in its analysis, the document's discussion of the issues remains current even if numbers have changed slightly.

Nagel, T., "Operating a Major Electric Utility Today," *Science* **201**, 985 (15 September 1978).

An interesting general reference by a power company executive.

Also, see almost any issue of the EPRI Journal, published by the American Electric Power Research Institute, Palo Alto, California. Fascinating insights into the utility industry are provided by an industry newspaper "Public Utilities Fortnightly."

Phillip, Owen, *The Last Chance Energy Book*, John Hopkins University Press, Baltimore, Md., 1979.

This short (140 pages) book is an eminently readable, entertaining and thoughtful treatment of our energy dilemma. Its discussion of alternative energy sources is very well done as is the treatment of the theory of resource exploitation.

Energy: The Next Twenty Years, Report by a Study Group sponsored by the Ford Foundation and Administered by Resources for the Future, Ballinger Publishing Company, Cambridge, Mass., 1979.

Oil, coal, nuclear, solar, projections, economics—all are here, thoroughly documented and comprehensive; one of the most complete studies of the decade of the 70s.

Hayes, E. T., "Energy Resources Available to the United States," *Science* **203**, 233 (19 January 1979).

Although this work is a nontechnical treatment of conventional sources, the reader will find very little discussion of natural sources of power (see Part 5 of this book).

Kash, Don, *et al.*, *Our Energy Future*, University of Oklahoma Press, Norman, OK, 1976. (Originally published as Energy Alternatives: A Report to the President's Council on Environmental Quality, 1975).

Slightly dated, but still a comprehensive source of information on most conventional energy alternatives, resources, and use.

Project Interdependence, U.S. and World Energy Outlook through 1990, U.S. Government Printing Office, Washington, D.C., 1977.

Jam-packed with information but, as with jam, parts run together, the going gets sticky, and the flow is uneven.

Tokamak and Magnetic Confinement Fusion

Steiner, D. and J. F. Clarke, "The Tokamak: Model T Fusion Reactor," *Science* **199**(4336), 1395 (31 March 1978).

Metz, W. D., "Magnetic Containment Fusion: What Are the Prospects?", *Science* **178**, 291 (20 October 1972).

Laser Fusion

Metz, W. D., "Laser Fusion: One Milepost Passed—Millions More to Go," *Science* **186**, 1193 (27 December 1974).

Metz, W. D., "Laser Fusion: A New Approach to Thermonuclear Power," *Science* **177**, 1180 (29 September 1972).

General Discussions of Fusion

Rose, D. J. and M. Feirtag, "The Prospect for Fusion," *Technology Review,* pp. 21–43 (December 1976).

Holdren, J. P., "Fusion Energy in Context: Its Fitness for the Long Term," *Science* **200**, 168 (14 April 1978).

Fuel Cells and The Hydrogen Economy

Maugh II, T. H., "Hydrogen: Synthetic Fuel of the Future," *Science,* **178**, 849–852 (24 November 1972).

Gregory, D. P., "The Hydrogen Economy," *Scientific American* **228**(1), 13–21 (January 1973).

Fickett, A. P., "Fuel-Cell Power Plants," *Scientific American* **219**(6), 70–76 (December (December 1978).

Chapter 13

Coal:
A Mixed Blessing

INTRODUCTION

It was coal that took us out of the era in which only human muscle, wind, and water power were available to manufacture goods. Coal took us into the era of machinery powered by combustion. Watt's steam engine, powered by coal, made it possible for us to have machinery that was not linked to rivers. Coal was also the fossil fuel that first changed our methods of transportation; it was coal that powered the steam locomotive.

Although the major use of coal in the U.S. today is as a fuel for steam electric power plants, it was once an important fuel for home heating and for transportation. After World War II, homeowners began to switch from coal to oil and the newly available natural gas. The railroads also converted, from coal-fired locomotives to the more dependable and cleaner engine, the diesel.

(There are four basic kinds of coal: peat, lignite, bituminous, and anthracite. All are in use as fuels. These basic forms reflect differing levels of carbon content.)

Peat (Peat is the most primitive form of coal. Still highly woody in nature, there is a great deal of moisture in the fuel; this moisture must be removed before peat can be used. Peat is found in bogs where wood, immersed in water, has decayed in the absence of oxygen. Historically, peat has been used only for heating.)

Lignite (The heat and carbon content of lignite or brown coal is greater than that of peat, but less than the higher forms of coal. Most of the lignite resources of the country can be strip-mined. Sub-bituminous coal is next up on the scale from lignite in terms of heating values.)

Bituminous coal (Characterized by a higher heat and carbon content than lignite, bituminous coal is used mainly for the production of steam to generate electric power. Lower sulfur bituminous coal, however, is generally used in iron and steel industries. Sub-bituminous coal is a grade of coal whose heat content is just below that of bituminous coal, but above that of lignite. Of the four types of coal, we rely most heavily on bituminous coal because of its relatively high carbon content and wide availability.)

Anthracite coal (Although anthracite has the highest carbon content of all coals, its current production in the United States is very small because its reserve base is so low.)

(Coal is not a pure substance. It contains an inorganic material, which remains after coal has been burned; we know the material simply as ash. Sulfur occurs in coal, sometimes as iron sulfide, and sometimes combined with organic compounds. There is also arsenic in coal, and radioactive elements. In fact, coal is the dirtiest of the fossil fuels)

(Dirty though it is, it is a magnificent source of heat energy. Burning one pound (0.454 kilograms) of bituminous coal releases 13,000 Btu (13,700 kilojoules) of heat energy. Furthermore, coal is the most plentiful fossil fuel we possess in the United States.)

COAL AND ITS USE

How Much Coal Is There?

It is potentially misleading to give a simple figure in tonnage for the coal reserves of the United States. Coal, unlike most oil and natural gas, has, as we have seen, a range of heating values depending on what form the coal is in. Peat has the lowest heating value; lignite is next followed by sub-bituminous and bituminous coal and finally by anthracite which has the highest heating value. To give a more accurate impression of our coal reserves, we need to provide reserve estimates for each kind of coal. We also need to know how much of the reserves can be surface mined and how much must be mined underground.

(Of the 397 billion metric[1] tons of discovered coal reserves in the United States, 54% occurs in the western states (see Fig. 13-1). Of these western coal reserves, 44% can be obtained by surface mining methods while the remaining 56% requires underground mines. Eastern coal, in contrast, is concentrated as underground reserves, which constitute about 83% of total eastern coal. The coal reserves in the U.S. amount to about 20% of the world's coal reserves.)

These reserves are discovered resources, resources we are reasonably confident exist. They are, however, not fully recoverable by mining technology. The coal that we obtain from underground mining is only about 50% of the actual coal in the deposits, since supporting structures or pillars must

_____

1 One metric ton = 1000 kilograms = 2200 pounds = 1.10 English ton.

be left in the mine. In contrast, 80%–95% of the coal in surface deposits may be recovered by the principal surface mining techniques, leaving only minor amounts of residue. Some feel that these recovery factors, which are used by the Bureau of Mines of the Department of Interior, are too optimistic. Nonetheless, these reserves are sufficient for several centuries at the present rate of use.

The coal reserves must also be described by rank (type) as well. The rank of coal tells us its heating value per pound and how efficient it is in such applications as steam production. About 92% of the coal reserves in the United States is of either the bituminous or sub-bituminous grade. Lignite is the other major coal resource (about 6% of the total) and virtually all this low heating value coal is available by surface mining. The anthracite coal reserves, which all require underground mining, make up only about 2% of U.S. coal reserves, and are not considered further.

The heating value of bituminous coal is 13,000 Btu per pound (30,200 kilojoules per kilogram) and that of sub-bituminous coal 10,000 Btu per pound (23,200 kilojoules per kilogram). Lignite provides only about 7000 Btu per pound (16,250 kilojoules per kilogram). Using the heating values of the different coals, we can restate U.S. reserves of coal in Btu as well as tons (see Table 13-1).

The coals of the various ranks are not evenly distributed in the United States. Lignite and sub-bituminous coal, for instance, are found only in small quantities east of the Mississippi River. Bituminous coal, in contrast, is concentrated in the eastern portion of the United States; only about 17% of this resource lies west of the Mississippi (Fig. 13-2).

Certain states have much more coal than others. In the East, coal resources are concentrated in Ohio, Illinois, Pennsylvania, West Virginia, and Kentucky. In the West, North Dakota, Montana, Wyoming, and Colorado have rich coal resources (Fig. 13-3). About 45% of the land area of North

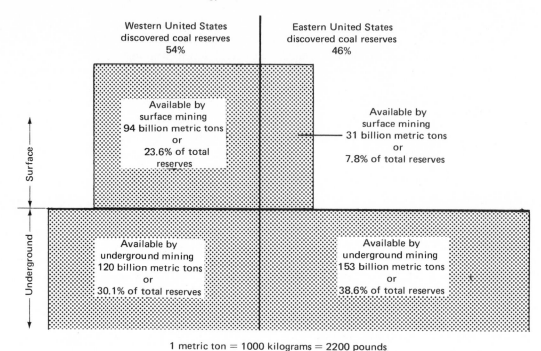

Figure 13-1   *Discovered coal reserves of the United States (398 billion metric tons).*

Dakota and 41% of the area of Wyoming contain substrata of coal bearing rock. About 35% of Montana and 28% of Colorado sit atop coal resources.

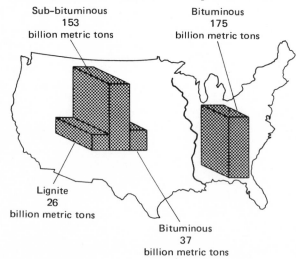

Figure 13-2   *Distribution of grades of coal.*

Nearly 85% of the coal that is less than 1% sulfur by weight is found west of the Mississippi (Fig. 13-4). Thus, to obtain low-sulfur coal in the eastern United States, where a larger percentage of the coal is used, the East must bring its coal from the western United States. These imports may come in by train or barge, or if the newest technology takes hold, by coal slurry pipeline. However, the lower heating values of most western coals requires greater quantities to be shipped.

How is Coal Being Used?

(The use of coal is growing after a period in which it had fallen steadily (see Fig. 13-5). The long decline in coal consumption came about as the railroads switched from coal to diesel fuel and homeowners switched to gas and oil. The upward trend is the result of the growing demand for elec-

TABLE 13-1   Reserves of Coal in U.S. in Energy Units and Tons

	Billions of metric tons	Quads of energy[a]	Quadrillion kilojoules	Percent of coal resource based on energy
Bituminous	211.7	6032	6390	62%
Sub-bituminous	152.9	3364	3550	34%
Lignite	25.6	395	416	4%

[a] 1 quad = 1 quadrillion Btu = 1000 trillion Btu.

tricity, a demand which increasingly is being filled by coal-fired electric generating stations.)

In 1976, about 600 million metric tons of coal were produced in the United States. With discov-ered reserves of nearly 400 billion metric tons, coal appears to be a secure energy source for the next several centuries. Of the 600 million metric tons of coal produced in 1975, nearly 55% was obtained by

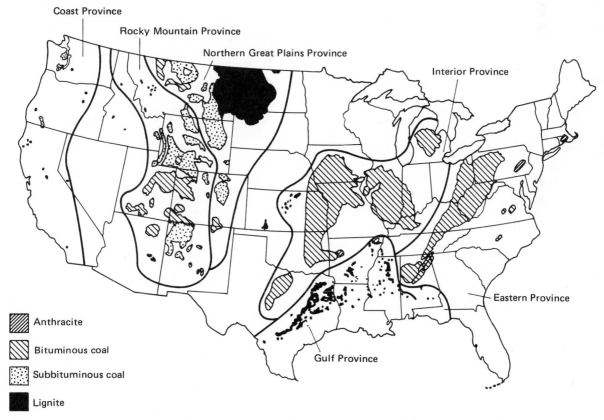

Figure 13-3   *Distribution of United States coal resources. (Source: Bureau of Land Management, 1974.)*

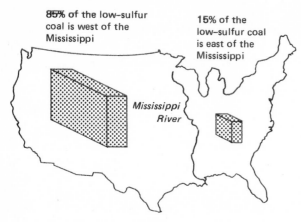

85% of the low–sulfur coal is west of the Mississippi

15% of the low-sulfur coal is east of the Mississippi

Mississippi River

Figure 13-4 *Low-sulfur coal distribution.*

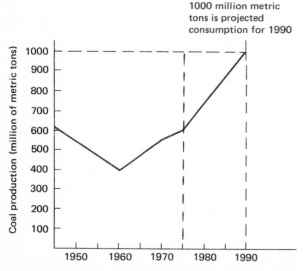

1000 million metric tons is projected consumption for 1990

Figure 13-5 *Plot of coal production versus time. (Data from Congressional Research Service.) Note that 1 metric ton = 1000 kilograms = 2200 pounds.*

surface mining methods. Only 5 years before in 1970, surface mining had accounted for only 44% of annual coal production.

(Coal is currently finding its greatest use in the electric utility industry.)In 1976, coal consumed in the generation of electricity amounted to a fraction over two-thirds of the annual coal production in the United States. Other uses of coal include coking

in the iron and steel industry (13%), general industrial use (9%), and export (9%) (Fig. 13-6). A small amount of coal is still used in home heating, although this use has fallen dramatically since the end of World War II. The foreign countries that purchase our coal generally use it in their iron and steel industries.

ENVIRONMENTAL AND SOCIAL IMPACTS OF COAL

With so much coal, it appears to make good sense to get the maximum use from this fuel. It can replace oil and gas in electric generating stations that now use these costly or scarce fuels. And it can be converted from solid form to a liquid fuel and from solid form to a gaseous fuel, although there are costs to the conversion.

(The National Coal Association, an industry organization, calls coal "America's Ace in the Hole" in the energy crisis.)Although they urge us to let the mining industry go in and get it, we must remember that tons of coal do not rise like magic from the mine. Both the natural environment and human be-

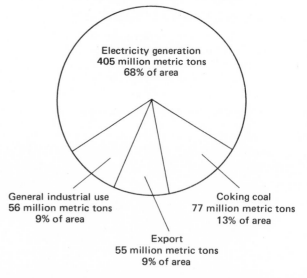

Electricity generation 405 million metric tons 68% of area

General industrial use 56 million metric tons 9% of area

Coking coal 77 million metric tons 13% of area

Export 55 million metric tons 9% of area

Figure 13-6 *Coal consumption in 1975 by sector.*

SHOULD WE ALLOW COAL TO BE EXPORTED?

We've got long-term contracts to honor with the countries as far as metallurgical coal is concerned, and there should be no regulation (on trade)....between this country and others.

Nicholas Camicia,
President of the
Pittston Coal Company *

I'm very strongly against the exporting of coal at a time when we are in an energy crisis.

Kenneth Heckler (M.C.), *
U.S. House of Representatives

We are currently exporting about 10% of our annual coal production. This coal, known as metallurgical grade coal, is destined for use in steel production in Japan and other nations. Mr. Camicia notes that "coal is needed to help in our balance of trade, and it is one of the few things we can afford to export." *

Nonetheless, Representative Heckler feels that, "The exports are unjustified and inflate the price of coal domestically." Even the American Electric Power Company, the largest electric utility in the country, agreed with Representative Heckler's position. Donald Cook, the company's chairman observed, "I fully agree on the value of foreign trade..., but in a crisis of domestic power supply the relief (by not exporting)...would greatly outweigh the harmful effects." *

How do you feel on the issue of coal export? Explain your position.

---

* All quotes from <u>The New York Times</u>, 26 September 1974.

ings may be sacrificed along the way. Nor does the coal reach its destination without impact on people and on the environment. And at its destination the by-products from the burning of coal pollute the air, injuring vegetation and affecting human health. To understand why no route to the future is perfect, we need to examine carefully the environmental and social impacts of coal mining.

How Coal Is Mined

The impact of coal mining on the environment is enormous. Where coal is mined underground, human safety is threatened by mine explosions and collapse. Underground miners are subject to black lung disease, a crippling shortness of breath. Acid drainage from underground mines pollutes streams and rivers.

Where coal is surface mined without adequate reclamation, erosion will choke streams with particles of earth. Mountains of rubble degrade the landscape and highwalls ring the scalped hillsides. To understand the origins of these environmental insults we need to explore the ways that coal is mined.

Coal is mined in two principal fashions, although variations exist. Coal may be surface mined, a practice which has grown dramatically in the last decade. And coal may be drawn from underground mines, the practice which is giving way to surface mining.

Why is surface mining capturing such a large share of the coal mining market? It is a matter mainly of economics: the price of surface mined coal at the mine is about half the cost of coal from underground mines. This price difference is not the same in all sections of the country, and transport costs will influence whether a consumer chooses surface mined coal or not. The price difference is so large, though, that more and more companies are turning to surface mined coal (Fig. 13-7).

The low price of this coal is due to the technology of surface mining, which has eliminated a great deal of human labor. The surface "miner" (he

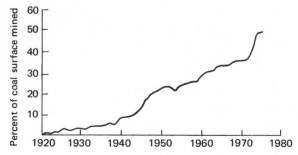

Figure 13-7 *Increase in coal production by surface mining. (Source: D. Kash et al.,* Energy Alternatives, *U.S. Environmental Protection Agency, EPA 430/9-73-011, 1973.)*

may operate a bulldozer or power shovel or truck) is about twice as productive as the underground miner. That is, the tonnage of coal hauled per miner each day in surface mining is about twice the tonnage hauled by the underground miner. In addition, the special safety precautions to avoid cave-ins and explosions in underground mines are not needed. Furthermore, the health precautions relative to particles in the air in mines become unnecessary in surface mining.

The price of surface-mined coal has also been low because, until 1977, very few states required the reclamation of land scarred by the surface mining. The federal Surface Mining Control and Reclamation Act of 1977 changed the rules for reclamation, so that the difference in price due to a lack of adequate environmental protection will be diminished. A gap in price will still remain, however. Surface-mined coal, where transport costs do not alter the cost structure, will still be cheaper than coal mined from underground mines.

Surface Mining

How the coal is removed   Whether a modern coal company decides on surface or underground mining of a coal seam depends on the depth of the seam beneath the surface of the earth. The more layers of soil and rock (called overburden) which must be removed, the greater the cost to mine from the surface. The decision also depends on the thickness of the seam, since the amount of coal recovered will influence profits.

Much of modern surface mining depends on giant equipment (Fig. 13–8). To bring this costly equipment to a site, a very large deposit of coal must be present. It follows that a small number of surface mines account for a large fraction of the coal produced each year in the United States. The largest surface coal mine in North America is the West Decker Mine of the Decker Coal Company. Situated in southeastern Big Horn County, Montana, the mine produced about 8.4 million metric tons of coal in 1975, nearly 1.5% of the entire U.S.

coal production in that year. Extensions of the mine have been planned in order to ship coal to Illinois, Michigan, and Texas for coal-fired electric plants.

(There are two basic kinds of strip mining; their difference stems from the different land forms on which they operate. Contour mining is used in hilly terrain where coal seams are beneath hills. A power shovel cuts a groove into a hillside to expose the coal seam. The overburden may be discarded to lower levels of the hill, but reclamation is more difficult if this is done (Fig. 13–9).)

When the overburden becomes too thick to remove, giant auger drills up to 7 feet in diameter may bore into the coal seam to bring out more coal.

Figure 13–8   *The equipment used in strip mining is massive. (a) This is the bucket of a "dragline" machine which operates at the Medicine Bow Coal Mine in Hanna, Wyoming. The bucket holds 78 cubic yards, a volume equal to that of a 16 foot square room with an 8 foot ceiling. (b) This dragline operating at a surface mine at Marrisa, Illinois is so large (20 stories tall) that it had to be assembled at the mining site. (Photos courtesy of U.S. Department of Energy.)*

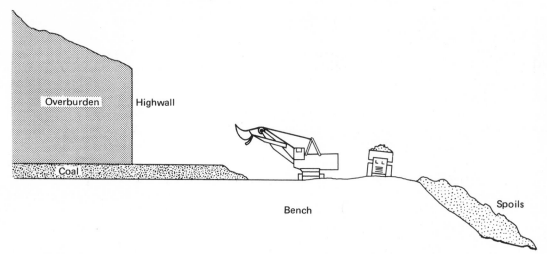

Figure 13-9   *Contour mining using a power shovel. (Adapted from D. Kash et al.,* Energy Alternatives, *U.S. Environmental Protection Agency, EPA 430/9-73-011, 1973.)*

When no attempt is made to restore the area, a highwall remains (Fig. 13-10). This vertical cut into the side of the hill may have vegetation at the top followed by a layer of soil and rock. Then comes a dense black stripe, the coal left behind when further removal of the overburden becomes too costly. Such a stretch of highwall may run for miles. Over 20,000 miles (32,000 kilometers) of highwalls ring the hills of the Appalachian region, scars from an era when profit came before environment.

Area mining, in contrast to contour mining, operates on flat terrain (Fig. 13-11). Area mining lays up the soil in rows to expose the coal in a long path. When the coal is removed from that open path, the overburden on the coal next to the path is excavated and dumped into the mined area, exposing another long path of coal. The process is repeated for row after row leaving hills of rubble resembling miniature mountain ranges (Fig. 13-12).

Strip mining begins with surface preparation. Bulldozers, scrapers, loaders, and trucks remove vegetation and overburden. The trucks will transport topsoil to a stock pile area for later use in restoring the mined area. Explosive charges are detonated to fracture the rock overburden and the coal into sizes that can be removed.

Once the overburden has been fractured, power equipment excavates it from the mine area. In small operations, a bulldozer, a scraper, or a front end loader will be sufficient. In larger area mine operations, a power shovel may be utilized along with a dragline. The power shovel is used to scoop overburden and load coal into trucks. The dragline is used primarily to move overburden from place to place. The bucket of the largest dragline can carry up to 220 cubic yards (168 cubic meters) of earth. This is equivalent to a cube of earth 18 feet (5.5 meters) on a side or a room 26 feet (7.9 meters) wide in both directions with an 8 foot (2.4 meter) ceiling. The weight of the earth in the bucket of the dragline may run to 200 tons (182 metric tons).

Reclamation of surface mined land   Even into the 1970s the strip miners were allowed to withdraw the coal and depart. The result was a surrealist picture of huge mounds of rubble, sometimes arranged in rows, sometimes strewn haphazardly on the landscape. The land from which the coal was torn was often left useless and without vegetation. Black-striped highwalls ringed the tops of hills, cutting off any plant and animal life from the nearby surroundings.

Figure 13–10 *The highwall is left after a vertical cut into a hillside has been made to expose the coal seam for mining. The dense black stripe is coal, coal that was too costly to recover. Highwalls, that may run for miles, isolate the hills from many animal species and so destroy valuable wildlife habitat. It is illegal now to leave highwalls, but the hills of Appalachia are scarred from pre-1977 operations.*

Without vegetation to hold them, soil particles can wash off the barren slopes of the hills. The eroded soil particles can fill stream beds. Because the stream channels are decreased, the flows from large rain storms can not be carried in the channel and the stream will flood more frequently. Furthermore, the barren hillsides with loose earth are unstable. A soaking rain can increase the weight of a soil mass and decrease its hold to the stable earth. The result is landslides occurring off the sides of strip mined hills.

Eventually vegetation will return to many of these areas, but the process is slow. The rubble left

on the land is acidic because of the presence of impurities in coal, and the acid condition (see Acid Mine Drainage) hinders the regrowth of grasses and trees. In addition, acid rivulets run off these lands and enter nearby streams. Acid waters are typically unfit for aquatic life and undrinkable for human beings.

Such land areas as these have been all too common. States began to pass laws to deal with the devastation, but these laws, with a few exceptions, were not enforced. The federal government, after six years of great difficulty, finally passed a reclamation law in 1977. The concept of reclamation is simple. Restore the original contour of the land if possible and re-establish vegetation to "anchor" the soil. The reality of reclamation, however, may range from simplicity to impossibility.

The two kinds of surface mining are very different in their requirements for reclamation. In gently sloping or flat areas, where area mining was utilized, the topsoil is saved so that it can be replaced on top of the regraded rubble from the mining operation. Once the topsoil has been replaced, fertilizer and grass seed are applied. With adequate rainfall and 6 inches (15 centimeters) or more of topsoil, the vegetation has a good chance of being established. The vegetation, which binds the soil, also prevents rapid run-off of water and checks the erosion process. The land can be returned to grazing or crops if these were the original uses. To restore forest land is, of course, impossible in a short time.

Where contour mining has occurred, special methods have been designed to reclaim the land. In one of the reclamation procedures, full backfilling, the spoils are pushed back up the bank and fully cover the highwall (Fig. 13–13). The restoration matches the surface of the filled area to the surface at the top of the highwall. This insures that water from run-off will not fall rapidly over a sharp slope; the hillside is thus made more stable. Because the older vegetation with its established root system has been removed, the slope, even with grasses, may still be somewhat unstable. To prevent erosion

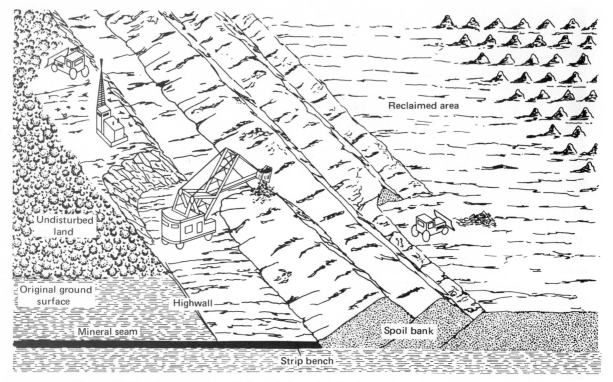

Labels in figure: Reclaimed area; Undisturbed land; Original ground surface; Highwall; Mineral seam; Strip bench; Spoil bank

Figure 13-11   *Area mining. (Adapted from D. Kash et al.,* Energy Alternatives, U.S. *Environmental Protection Agency, EPA 430/9-73-011, 1973.)*

gullies from forming, terraces may also be cut around the restored hillside to catch and slow the water that runs off the slope.

Backfilling is an afterthought, though, an effort to repair damage after it has been done. In contrast, the "modified block cut" is an economically attractive procedure designed to minimize damage in the process of mining (Fig. 13–14). The first step in the modified block cut is the same as in conventional contour cuts. Overburden is excavated from above the coal seam and is dumped downhill. This cut, however, is confined to a narrow area. The second cut is smaller in size and is taken at the same elevation as the first, but to one side of the initial cut. The overburden removed from the second cut is dumped into the adjacent space made by the first cut. The third cut is taken to the opposite side of the first cut, and the spoils are discarded into the remaining space of the first cut. The spoils from the fourth cut are dumped into the second cut to which it is adjacent. The spoils from the fifth cut fill the void created by cut number three, and so on.

These are only a few of the many methods that have been used to restore contour-mined land, but they give an impression of the range of possibilities and their relative advantages. Experience in restoring steep slopes is limited, although the modified block cut is reported to have been successful on slopes of 20° and above. In addition, the modified block cut appears to be no more expensive than ordinary contour mining, suggesting that it may become an important procedure in the future.

Replanting a strip mined area   The best of revegetation, however, cannot soon bring back forested

Figure 13-12   *Area strip mining left unreclaimed. This land near Nucla, Colorado was area mined for coal by the Peabody Coal Company and left. Rapid erosion of soil occurs from such areas clogging streams and making them subject to flooding. Acid conditions in the rubble and soil hinder the regrowth of trees or other vegetation. The land is unfit for human or wildlife habitation. (EPA-Documerica, Bill Gillette.)*

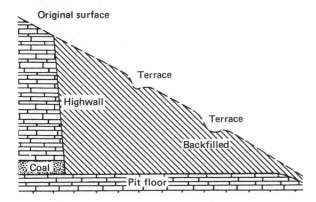

Figure 13-13   *Full backfilling of contour-mined land. (Adapted from D. Kash et al.,* Energy Alternatives, *U.S. Environmental Protection Agency, EPA 430/9-73-011, 1973.)*

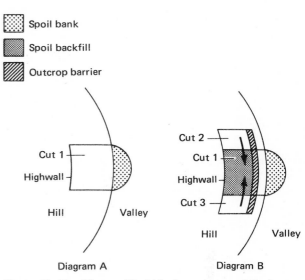

Figure 13-14   *The modified block cut is one of the best methods to reclaim contour-mined land, and appears to cost no more than ordinary mining. (Adapted from D. Kash et al.,* Energy Alternatives, *U.S. Environmental Protection Agency, EPA 430/9-73-011, 1973.)*

area. Wildlife characteristic of forests cannot survive in grassland; their food supply and nesting places have been destroyed. Thus the ecological balance is unavoidably altered by strip mining, even with the best of reclamation.

The whole effort of reclamation, be it of contour mined land or area mined land, appears to be very dependent on rainfall. In regions where rain is sparse, it may be that the best of methods are inadequate to restore vegetation to the land. The concern here is mainly with the vast coalfields of the West where huge strip mining operations are getting underway, and the consequences are difficult to predict.

A committee of the National Academy of Sciences studied the issue of western surface mining from the standpoint of the potential for reclamation of the lands.[2] The panel indicated that sites with less than 10 inches annual rainfall were probably not good candidates for reclamation.

The committee felt that sandy soil supporting only desert shrubs probably can not be restored to its original form since no topsoil is available to recover the earth after surface mining. Because so few seeds would be present in the sterile, exposed

---

2 National Academy of Sciences *Rehabilitation Potential of Western Coal Lands,* Ballinger Publishing, Cambridge, MA, 1974, 198 pages.

soil, the growth of new plants would be extremely slow. The natural succession of plant species, in which one diverse community of plants replaces another, would be delayed in this adverse environment, so that 100 years or more might pass before the native vegetation returned.

Figure 13–15 summarizes the views of the committee on these probabilities for the four vegetative zones encountered in the Western coal states.

Surface Mining Law Established   In 1977, after years of debate and disagreement, Congress passed a law establishing partial control over strip mining. The Surface Mining Control and Reclamation Act essentially forbids leaving highwalls after mining. It also forbids dumping overburden on slopes of over 20°. The legislation demands that strip mined lands be returned to approximately original contours. Prime farm lands, according to the legislation, are not to be mined unless they can be returned to their original productivity. A tax is to be collected on the coal currently being produced. This tax is to be used to pay for restoration of land that was marred in the era before reclamation.

While the enforcement of the standards are left to the states, the Department of the Interior may step in when the states fail to act.

(Although reclamation costs could run from $1000 to $5000/acre ($2500–$12500/hectare) depending on the slopes involved, it will still be profitable to surface mine coal. Further, the relative position of strip mined coal as cheaper than deep mined coal will generally not change.)

Surface mining: Is the issue settled?   During the decade of the 1970s, federal laws to control strip mining were proposed again and again only to be defeated either by Congress or by presidential veto. One group in Congress favored a complete ban on all strip mining. On the opposite side were the coal companies and their associations. The arguments were fierce. Each side had its truth to tell. Strip mining was ravaging counties in Kentucky, Virginia, and West Virginia, where irresponsible operators would simply devour a coal seam and

depart. At the same time, responsible companies were restoring the land they had mined. In general, however, the state laws on surface mining were weak[3] and the conscience of the operator was all people had to rely on. In such cases, pangs of conscience meant lower profits for operators.

Now, as the strip miners move into the West under a new law, it seems wise to review the local impact of strip mining as it took place in the 1970s. Perhaps the most eloquent spokesman against strip mining was the Honorable Kenneth Heckler, representative to Congress from West Virginia. He gave a speech in the House of Representatives which was entitled "The Hills of Appalachia are Bleeding,"[4] and the statement below is from that speech:

> The human suffering of those who live near strip mining sites is pitiful. The blasting and the bulldozers have frequently sent boulders onto the property and even into the homes of those on the fringes of strip mining. I have called at homes where embarrassed owners have almost cried because they cannot draw me a glass of water, for their water supply has turned black or brackish. The entire Appalachian area is honeycombed with moon-scapes. Foundations of homes have been destroyed. The giant mounds of earth which are pushed over the hillsides are unstable. They swell with the rains, get heavy, crack and slide. They erode away, washing rock and soil down the hillsides into the streams below.... Do you want more areas of the country to become instant Appalachias? Do you want to allow these strip-and-run exploitation artists to rip up your land the way they have ripped up ours?

Now surface mining is increasing dramatically in the West. In 1975, a portion of Death Valley Na-

---

3 A notable exception was Pennsylvania where a tough law was being enforced by a colorful and tough administrator, William Guckert.

4 Honorable Kenneth Heckler, Member of Congress, *Regulation of Surface Mining,* Part I, Hearings before Subcommittees of the Committee on Interior and Insular Affairs, House of Representatives, Serial 93–11, 1973.

(a) Ponderosa Pine and Mountain Brush

"...where precipitation is favorable for plant growth and where rather deep, fertile soils have developed, [such sites] do not present a revegetation problem; it must be remembered, however, that many years are required to grow mature pines."

(b) Mixed Grass Plains

"...a rather high probability for satisfactory rehabilitation... predicting such results assumes the best technology will be applied, including the addition of top soil,...."

(c) Sagebrush Foothills

"...a delicate reconstruction process in the handling of substrata and top soil is demanded for favorable rehabilitation... in some places disturbed ground may have to be repeatedly reseeded...."

(d) Desert

"Disturbing such areas for the surface mining of coal amounts to sacrificing such values permanently for economic reward."

Figure 13–15 *Reclaiming surface mined lands in the West. (Quotes are from National Academy of Sciences,* Rehabilitation Potential of Western Coal Lands, *Ballinger Publishing, Cambridge, MA, 1974.) (Photos a-c, Grant Heilman Photography; photo d, National Park Service.)*

tional Monument was being surface mined for coal by the Tenneco Company. Montana, Wyoming, North Dakota, and other western states are being surface mined. The law that now governs surface mining is a federal law, The Surface Mining Control and Reclamation Act of 1977, that we described earlier. In terms of western mining, there are several flaws in that legislation. When a state fails to enforce the law, and illegal and destructive stripping takes place, much damage could occur before the federal government discovers the operation and is able to take action. The act dictated that prime farm lands could not be stripped unless their productivity could be restored, but who is to judge what has not yet been determined even by scientists? And there is no prohibition on surface mining the desert, where we are told, that restoration of the land is extremely unlikely.

Thus, it appears that the issue of coal surface mining is not yet laid to rest. As the 1980s open, you may wish to observe the effect of vastly increased surface mining on the states of the western plains.

Underground Mining

How coal is removed In the underground coal mine, men and machines cut the mineral from the coal face and load it into cars or onto conveyors for removal to the surface. The "room and pillar" method of mining is the traditional means to extract the coal (Fig. 13–16). From a central passage, parallel rooms or tunnels up to 20 feet (6 meters) in width may penetrate the coal seam for 200 to 300 feet (60–90 meters). Another set of parallel tunnels driven perpendicular to the first result in a tic-tac-toe pattern of rooms. Pillars of coal between the rooms hold the earth up; large timbers and roof bolts provide added support. Room and pillar mining has an efficiency on the order of 45% to 50%. That is, about 45% to 50% of the coal in the seam is actually brought out.

Nearly three-quarters of the cutting and loading is now done by machine while the remaining one-quarter is still accomplished by the more

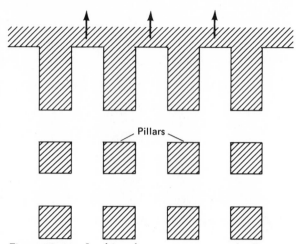

Figure 13–16   *Looking down on a room-and-pillar mining operation.*

labor-intensive operations involving either hand cutting or hand loading.

Of growing interest is the highly efficient method imported from Europe known as "longwall mining." Although longwall mining has been in use in Europe for a number of years, it is only in the last decade though that the procedure has been used in the United States. The aim of longwall mining is to remove the coal without disturbing the layers of earth above the coal seam and without creating permanent deep mines. It resembles a cross between surface and underground mining (Fig. 13–17).

Movable jacks hold the roof of the mine in place, while mining is going on, but when the coal has been removed the jacks are moved forward to the new mining "room." When the jacks have been relocated, the earth collapses above the mined space because no columns support it. The surface remains untouched and most of the mining area has been sealed off from water or oxygen, preventing the formation of acid mine drainage.

Coal waste piles or refuse banks Once the coal has been extracted from the mine, it may be sent directly to customers such as the electric utilities. Two thirds or more of the current bituminous coal production moves directly to use in this way. The

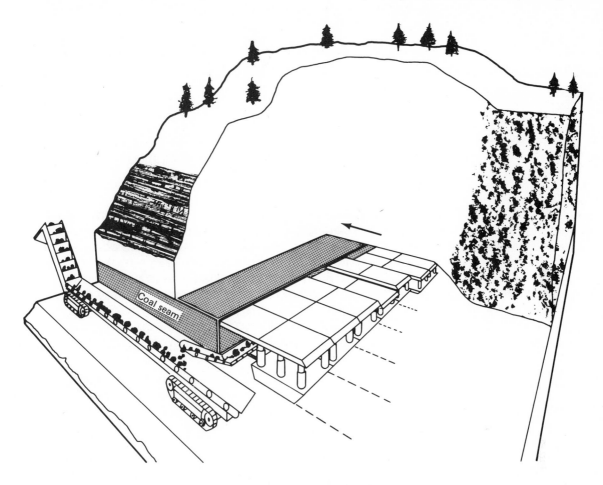

Figure 13–17   *A longwall mining operation. (Adapted from D. Kash et al.,* Energy Alternatives, *U.S. Environmental Protection Agency, EPA 430/9-73-011, 1973.)*

remainder is sent to a preparation plant where it is crushed, washed and graded in size.

The wastes from the preparation plant consist of low grade coal, shale, slate, and coal dust. These wastes are stored in huge hills known as *refuse banks* or "gob" piles. Water contaminated with coal dust from the washing process may also be produced. This water may be settled in ponds to remove the coal particles, but in some cases the water is discharged directly into streams. One practice is to dam streams with the refuse bank and thus create a settling pond behind the dam. The practice is hazardous as you can see from the excerpts of news stories on the next page.

The refuse banks that are not used to dam streams may also be a source of acid waters as rainfall drains through them. The rainwater extracts the iron pyrites, which cause the formation of acid. Acid waters are not the only pollution from refuse banks. These mounds of coal debris often catch fire and may smolder for years. Air pollution, including carbon monoxide from the incomplete combustion of coal, results from these fires. One bank in the northern Appalachian region discovered burn-

WEST VIRGINIA FLOOD TOLL AT 60 WITH HUNDREDS LOST

Bodies Found in River 24 Miles Away After Water Behind Coal-Waste Pile Rushes Down Over 14 Communities

Man, W. Va., Feb. 27– Up on Buffalo Creek, where people used to live, only the bulldozers and the helicopters belonged today. The people were gone, scattered by yesterday morning's flash flood that was unleashed by the collapse of a coal waste pile accumulated over the last 15 years....

The tragedy occurred around 8 A.M. yesterday when a huge pile of waste material from the operations of the Buffalo Mining Company gave way on a mountain outside Man. The wall of water, estimated by some residents as ranging from 20 to 50 feet high at the beginning, rushed downhill through 14 mining communities, wrecking most of the houses in its path and displacing as many as 5,000 persons....

George Vecsey,
Special to The New York Times, 28 February 1972

HOMELESS THOUSANDS AWAIT FUTURE IN FLOODED VALLEY

Willis Jacobs stood in the doorway and watched the other refugees wander around....

"I saw the whole Gunnels family go under," Mr. Jacobs said. "I could see their heads and hear their voices before they went under the water and the timbers. I can still hear them."...

At the Pittston office this morning, company officials began to assemble. A company lawyer, Robert Reineke of Abingdon, Va. answered questions about Pittston's role...

Mr. Reineke said the pile of waste material had been formed for "functional" purposes–"to filter acid and fine coal and to back up the water."...

A vice president of Pittston, who asked not to be identified, said that the company considered the damage to be caused by "the flood, which we believe, of course, to be an act of God."

George Vecsey
Special to <u>The New York Times</u>, 29 February 1972

---

ing in a 1963 study had been burning despite efforts at control since 1884. To control the fire in a refuse bank may require actual burial of the mound under an earth cover. There are known to be 800 refuse banks of varying composition in the anthracite region of Pennsylvania alone.

Subsidence   The removal of coal from underground mines leaves another legacy, *subsidence.* The mine space, as the roof supports give way, may fill with overburden. At the surface, sink holes and cracks may appear as evidence of the events beneath the earth, and tremors may shake the area.

Subsidence is a nuisance when it occurs in farm land or forest or recreational land. It is a serious problem, however, when human dwellings are near the subsiding area. Structures such as roads and foundations may crack, buckle, and crumble as the earth settles into a firmer position below. Railroad tracks may be bent and lose their support; gas mains may break and threaten explosion; sewer lines may crack. In some areas of Pennsylvania, homes have had to be abandoned as tremors and subsidence have shaken and swallowed the dwellings.

In new workings, subsidence can be controlled by leaving enough supporting pillars and structures. The practice of removing some of the pillars for their coal at end of the operation increases the risk of subsidence.

Acid mine drainage   Still another effect of underground coal mining is acid mine drainage. From underground mines and from strip mines flows a substance called "yellow boy," which eventually settles and coats stream bottoms. Yellow boy,

which is derived from one of the impurities in coal, makes the water into which it flows acidic. Waters that carry yellow boy are called *acid mine drainage.* Acid mine drainage results because the mining of a coal seam is rarely complete. Some coal is unavoidably left behind in mining operations.

An area that has been strip mined and has not been restored will have numerous ditches and small hillocks covering the landscape. Rain leaves pools of surface water in the hollows created by mining. Surface water will wind its way through a maze-like course, sometimes pooling, sometimes rushing downhill, until it reaches a stream. On its route, the flowing water comes in contact with the coal remaining in the soil and leaches from the coal the substance known as iron pyrites. This substance is the impurity which gives rise to the acid waters.

The process which creates acid mine drainage in surface mining is much the same for the underground mine. Water may enter the mine directly through the mine shaft, or groundwater may seep through the layers of soil into the mine. Sometimes the earth above the mine has subsided (sunk into the mine). In this common situation, the water may easily penetrate the mine through the "sink hole" which has been formed. It is even possible that entire surface streams may go underground through such sink holes. Iron pyrites from the coal remaining in the mine dissolve in these waters.

Iron pyrite is known chemically as ferrous sulfide, a compound composed of iron and sulfur. It is also well known as the shiny golden mineral, "fools' gold," which has raised false hopes for so many prospectors. As ferrous sulfide dissolves in water, the oxygen in the water begins to oxidize both the sulfide and the ferrous ion. The oxidation, in which bacteria play a role, converts the sulfide to sulfate ion and the ferrous ion to ferric ion. Oxygen from the water is used up in the conversion, resulting in such low levels of oxygen that the normal aquatic life of the stream is threatened. (The effects of low levels of dissolved oxygen are discussed in Chapter 8.))

(Other reactions make matters worse. The ferric ion combines with hydroxyl ions in the water to form the yellow-brown precipitate, ferric hydroxide, the substance we referred to as "yellow boy." Yellow boy may sink and coat the bottom of the stream. Creatures such as the fresh-water crayfish, which feed on the stream bottom, will thus be deprived of their food source. A visitor to the coal mining areas of Appalachia will often remark on the yellow-orange color of the streambeds in the region.)

Although yellow boy coats the stream bed and the low levels of dissolved oxygen harm aquatic life, the worst consequence is the acidity of the water. The precipitation of ferric hydroxide leaves an excess of hydrogen ions in solution. The water is now rich in both hydrogen ions and sulfate ions. This situation is equivalent to adding sulfuric acid to water. The highly acidic water with low levels of dissolved oxygen and a stream bottom coated with yellow boy together make a rather unpleasant environment for aquatic species.

Acid mine drainage has historically been a regional problem. Pennsylvania, West Virginia, and neighboring coal states have seen the greatest share of the problem. An estimate in the 1970s by the Environmental Protection Agency placed the length of streams affected by acid mine drainage at more than 11,000 miles, mostly east of the Mississippi River. However, the surface mining of coal in the western states was only beginning in this period. Because so much of the nation's coal now originates in the West, the regional aspect of the problem of acid mine drainage is disappearing. Its control is a problem of national scope.

Solutions to the problem of acid mine drainage
Even though coal mining operations will undoubtedly continue and expand, there are a number of possible ways to prevent or reduce acid mine drainage. The methods are applied in different situations and are not equally successful.

(Where an area is to be deep mined, it is helpful if the mining company uses "down-dip" as opposed to "up-dip" recovery (Fig. 13–18). The shaft of the down-dip mine slopes downward to the coal seam; the shaft of the up-dip mine angles upward to the coal seam. In down-dip mining it is difficult for the pooled waters in the down-dip mine to escape to surface waters.)

At long-abandoned mines, however, there is no one left to run a treatment plant or to pump pooled waters out of the mine. Such abandoned mines are responsible for a very high portion of the acid mine drainage that occurs. This fact is largely due to the traditional methods of coal discovery and mining. In years gone by, mine shafts would be cut into hillsides at the point at which the prospector saw the coal seam. The shaft would be carved into the hillside at an upward angle so that coal could be loaded on carts at the coal face and allowed to roll down to the outside of the mine. The water that accumulated in the mines from groundwater seepage or that "wept" off the walls due to condensation would simply run out of the mine by gravity.

How then to control the pollution from long abandoned mines? You might think one way would be to seal the entrance to the mine, and by doing so to seal surface water and oxygen out. But when the mine shaft rises upward at an angle from the entry, mine sealing efforts have been unsuccessful. When water accumulates in the mine behind the seal, tremendous pressures can cause the seal to leak or give way. On the other hand, where the mine shaft is level or where it falls away at an angle from the entrance, mine sealing has been effective.

Another option to control acid drainage from abandoned mines is to prevent surface water from entering the mines. This can be done by filling sink

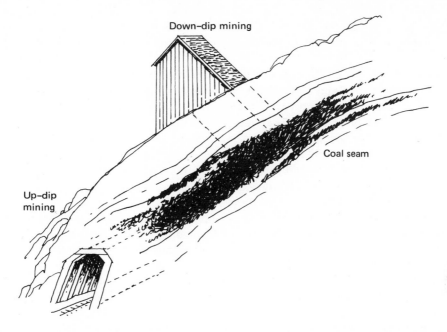

Figure 13–18 *Two methods of reaching a coal seam. In down-dip mining, surface water may enter through the downward sloping shaft or via groundwater flow. It may also condense on the walls of the mine because of humid conditions there. Such waters are likely to be trapped in the mine. During the active operation of the mine, such water must be pumped out and, if necessary, treated to reduce its acidity. When the mine is abandoned, the water may accumulate, having nowhere to go but into deeper groundwater. In contrast, water that enters the up-dip mine via the groundwater flow and through condensation will flow out naturally down through the mine shaft itself. When the mine is in active operation, little or no pumping is needed to get rid of water in the work area, although discharged acid waters must be chemically treated. When the mine is abandoned, however, acid waters continue to pour out of the mine, and it is extremely difficult to seal the mine entrance.*

holes which were formed when the earth above a mine subsided into the cavity. Gulleys formed by erosion may sometimes flow above an abandoned mine; the channels for this storm water can be changed via ditches or dikes or conduits. If the mine site can be avoided, surface water will be unable to seep into the mine. The acid waters from an abandoned or active mining operation can also be transported via pipeline to a site further downstream where larger flows occur.

HEALTH AND SAFETY OF THE COAL MINER

Mine Safety

Coal mining, more than any other process of fuel resource extraction, has an impact on human life itself. We read and hear often of mine disasters where rescue operations have failed, of cave-ins or coal mine explosions. Several hundred coal mining deaths occur each year. It is a grim, grimy, and dangerous business. Although the statistics tell us the mines are getting safer, coal mining is still one of the most hazardous jobs in modern times. The dangers are immediate; a loading car may break loose and careen down a slope and strike a miner. And the dangers are delayed; the sooty air, laced with coal particles, will implant in a miner's lung and eventually destroy his ability to breathe.

About 60% of the injuries in underground coal-mining are related to roof falls and face falls (the coal face landslides in miniature). About half of the deaths in underground coal mining stem from roof falls alone. Hauling coal underground is also extremely hazardous. Accidents occurring in haulage are often fatal. About 15%–20% of underground accidents occur in hauling operations. Explosions and fires account for another 15%–20%.

Explosions in coal mines are a result of the ignition of methane gas (though coal dust, itself, can explode). Pockets of methane commonly occur in underground coal deposits. If these are undetected and the gas builds up in the mine, a simple spark can detonate an explosion. While the explosion itself can injure and kill, the rock falls that result may be even more hazardous, even to the extent of sealing men inside a portion of a mine. Such explosions from methane gas can be prevented by adequate ventilation of the mine, but to be sure the mine air is safe, methane detection equipment must be used.

The high accident rate in the coal industry leads to a great number of cases of disability. Unable to work, these former miners may also require extensive medical attention. Not only do the accidents contribute to disability, the working conditions do as well. Performing heavy labor in a stooped position or crawling on hands and knees lead to inflammations of the joints, conditions known as "beat hand" and "beat knee."

Black lung disease (Accidents, fatalities, crippling deformities, and finally *black lung*. A wheezing and shortness of breath, an inability to climb stairs or perform labor, a disease so disabling that some vic-

tims are only able to sit in chairs, their lungs destroyed by the particles they inhaled.)

The public and the coal miners call this condition "black lung disease," because the lungs of miners who die from the disease are black (Fig. 13–19). The physician knows the disease as "coal workers pneumoconiosis" (pronounced new-mō-cōn-io´–sis) and abbreviates the name as CWP.)

The disease is caused by the inhalation of dusts composed of minute particles of carbon and rock.) The carbon particles account for the black color of the lungs. A growth of fibers occurs within the lungs around the sites at which particles have deposited. These fibers can, if they grow extensively within the lung, destroy the normal elasticity of the lung. The elasticity of the lung, its ability "to bounce back," is what makes breathing easy for most of us. A miner's lung, overgrown with fibers due to coal dust, lacks the ability to bounce back. For him, breathing is torture. There appears to be no treatment for black lung other than for the bacterial infections that are side effects of the disease. That is, the symptoms cannot be relieved, the disease cannot be cured.

The risk of the disease appears to increase with quantity of coal dust inhaled. Two factors influence this quantity significantly. First, the greater the number of years spent underground, the greater the quantity of dust inhaled. The specific mining job, whether it is a dusty one or relatively clean one, also influences the risk of contracting the disease. Cutting machine operators show a very large risk compared to workers who are stationed outside the mine. To bring the number of new cases down, the levels of dust in the mine must be brought under control. Although a federal standard for dust in coal mines has been set, the extent of compliance with the standard is not clear because of the absence of checks on company data.)

How many individuals are afflicted with black lung disease? The question requires answers at a number of levels. First, there were about 230,000 working coal miners in the late 1970s and this number was increasing. In 1947, however, before

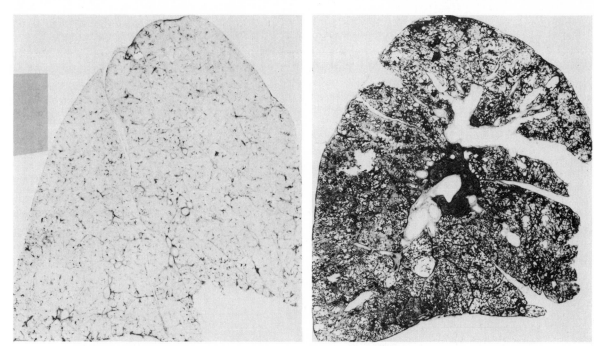

Figure 13–19 *A comparison of the lungs of a non-miner and a coal miner. Photo left shows a section from the lung of a man who died at age 86 and who never was a coal miner. Photo right shows a section from the lung of a man who died at age 78 and who was an underground coal miner for 36 years. Although the man did not smoke, he suffered from severe emphysema, a lung disease in which breathing is badly impaired.(Photos courtesy of Dr. Frank Green, M.D., Chief, Pathology Section, Appalachian Laboratory for Occupational Safety and Health.)*

the era of mechanization began in a serious way, there were about 400,000 coal miners. At one point in the 1920s there were just over 700,000 active coal miners. That means that many men have retired from the mines in the past 30 years or so. We need to know this fact to understand why over 250,000 claims of black lung have been approved by the Social Security Administration. These claims were submitted either by miners or their widows and more claims are continually being approved. More than a billion dollars per year were being paid in 1978 as benefits to coal miners or their widows. These are tax dollars, not dollars from the coal industry where the miners contracted their disease.

In 1978, a bill to amend the first Black Lung Law (passed in 1969) was enacted by Congress. The original bill had called for the coal industry to pay black lung compensation, but the coal industry successfully fought this law in court, and the law was never implemented. The 1978 Black Lung Benefits Reform Act established a Black Lung Disability Trust Fund which comes from a tax on each ton of coal produced. Money from this fund would be used to pay any newly established claims of black lung disease.

COAL AT THE POINT OF USE

The Coal-Fired Electric Power Plant and Its Impact

Coal, as we pointed out earlier, finds its main use in the generation of electricity. That is, coal is

STRIP MINING: WHAT ARE THE IMPLICATIONS OF EXPANSION?

...Superscale western mining means—between twenty thousand and forty thousand jobs will be lost in the East.

Arnold Miller

Which resources are we most concerned about conserving—human lives, the terrain, the vegetation, or the mineral?

David L. Kuck

As surface mining expands its share of the coal market, the number of mining injuries and fatalities declines. As David Kuck pointed out in a letter to Science, *

> Having spent the last 23 years working in and around both surface and underground mines, I have a different view of strip mines...
>
> Black lung and silicosis are rare diseases among strip miners. Very few men are injured by rockfalls in surface mines. In an underground mine, if something goes wrong, there is nowhere to go; the miner is surrounded by rock....
>
> ...Which resources are we most concerned about conserving–human lives, the terrain, the vegetation, or the mineral?

From the viewpoint of worker health, it appears that strip mining holds an edge over deep mining. Also, strip-mined coal costs less to produce than deep-mined coal. Yet Arnold Miller, former President of the United Mine Workers, sounds like an environmentalist when he commented on strip mining:

> Fifty years ago they promised to develop Appalachia; and they left it a wreckage. Now they promise to develop the Great Plains. They will leave it a ruins. †

Miller knows Appalachia, so he knows whereof he speaks. As environmentalists, we hear him warning one section of a country of what the coal industry did elsewhere, but he has another motive, which he does not hide:

> We must not rush into development of the West at the expense of the East. A headlong commitment to superscale Western mining means that over the next five years between twenty-five thousand and forty thousand jobs will be lost in the East. Of course, that concerns us as a union of miners. It concerns us also because we have lived through an unending depression in Appalachia, and we simply cannot sit silently and watch another one come rolling in on us. Finally it concerns us because you cannot turn underground coal production on and off like a light switch. If we arrive at a rational fuels policy (finally) and decide to strengthen our emphasis on Eastern mining, the mines will not be there, and neither will the miners.

Do you agree with Miller now that you understand his motivation? Defend your position.

If you agree, can you think of ways to keep coal mining in the East alive?

* Science **183** (4120), 28 (11 January 1974).

† Center Magazine, November-December 1973.

burned to produce the heat needed to create steam; the steam is used to turn the turbine to generate electricity. The burning of coal produces more than heat; many air pollutants including gases and particles are produced as well. Coal has now become our mainstay in electric generation as a means of preventing our dependence on foreign energy sources.

With any thermal electric power plant such as the coal-fired power plant, the conversion of heat to electric energy is incomplete. The modern coal-fired power plant, however, is more efficient than its nuclear competition. About 40% of the heat in coal may be converted to electrical energy, as against conversion of only about 30%–32% of the heat from nuclear fission.

The burning of a carbon-containing fuel produces carbon dioxide. (Only a very small amount of carbon monoxide is produced from burning coal.) The carbon dioxide is of concern because of its hypothesized effect on the earth's climate. (See Chapter 17 for a further discussion of carbon dioxide and climate.)

(The three most important pollutants from coal-fired power plants are oxides of nitrogen (NO_x), oxides of sulfur (SO_x) and particulate matter. Nitrogen oxides have been very difficult to control in electric power generation. Although these gases have direct effects upon our health, they are noted primarily for their interaction with hydrocarbons to produce ozone, which aggravates diseases of the lung and respiratory tract. Sulfur oxides and particulate matter are the current prime targets for control at coal-fired power plants.)

Sulfur oxides and particles (Sulfur oxides injure plants, materials, and people.) (See Chapter 18.) Sulfur dioxide was present in the air in great quantities during the air pollution disasters at Donora, Pennsylvania, and the Meuse Valley in Belgium. It was also present in large quantities in the infamous London Fog and in the episodes in New York City in the 1950s and 1960s.

(Particles or particulate matter are "the partners in crime" of sulfur oxides. Some particles in the atmosphere may in fact be tiny droplets of sulfuric acid dissolved in water. Others may be calcium or magnesium sulfate compounds, formed when the metal oxides in fly ash react with sulfur oxides in the air.)

Sulfur oxides are present in the stack gases of the coal-fired generating stations because sulfur is present in coal. When carbon is burned, the sulfur is oxidized almost entirely to sulfur dioxide (SO_2). A ton of coal at 2.5% sulfur by weight (about 50 pounds or 23 kilograms of sulfur) will produce about 100 pounds (46 kilograms) of sulfur dioxide when the ton of coal is burned; the weight increase is due to the addition of oxygen. Since the sulfur content of coal may range from 0.5% to 6%, a level of 2.5% is not unusual. A single 1000 megawatt coal-fired electric plant burns four to five million tons of coal a year. If the coal is 2.5% sulfur, the plant will discharge 200–250 thousand *tons* of sulfur dioxide per year.

As the nation swings more and more to coal for reasons of national security, the sulfur dioxide problem will multiply unless something is done. That something consists of two steps, which may be used in combination: ridding the coal of its sulfur by cleansing it and "scrubbing" the stack gases. (See Chapter 17.)

Particles or particulate matter are also discharged into the atmosphere when coal is burned; these particles are known as fly ash. If the emission of fly ash were not controlled, the annual production of particles from a 1000 megawatt plant could be as high as 200–250 thousand tons, depending on how much of the year the plant was in operation at its capacity. Instead, because particles are so well controlled, only about 10% or less of this quantity actually reaches the atmosphere.

Coal also contains arsenic, which is released when the coal is burned. On the average about 2.5 g of arsenic are released per ton of coal. Because arsenic is released in the fly ash or particulate mat-

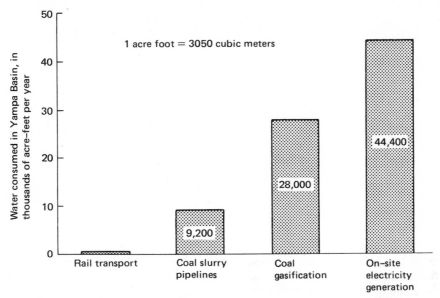

Figure 13–20 *Comparison of water consumed assuming 12.5 million tons per year are hauled or consumed. Although rail transport of coal consumes the least of the West's precious water supply, coal slurry pipelines take far less water than other major efforts to bring the energy East. Gasifying the coal and transporting the gas will consume three times the water of coal slurry pipelining. Producing electricity that would be transmitted to the East consumes even more water than gasification. (Source: U.S. Geological Survey, Open File Report 77-698, August 1977.)*

ter, 70% to 90% of it will be removed if power plants install scrubbing, or an electrostatic precipitator. In the case of very high arsenic coals, however, this may not be enough protection for people living near the power plant. In an area of Czechoslovakia where very high arsenic coals are burned at a power plant, children have been found to suffer hearing losses, apparently due to inhalation of arsenic in the air.

Selenium is also present in small amounts in the fly ash discharged by coal-fired power plants. Much of this selenium seems to appear in the smallest fly ash particles, which often elude capture by the electrostatic precipitator. Fly ash has been used to enrich soil, but high doses of selenium to cattle from the grasses would be injurious.

The last category of emissions from coal-fired power plants is unexpected. There are radioactive elements and their decay products, which are found naturally in fly ash. Oil-fired power plants produce these elements as well, but since there is so little ash from burning oil, the quantities are very much smaller. The quantity of radiation from a coal-fired electric plant is very low, less than 1% of natural background radiation, about the quantity of radiation from a nuclear plant operating under the stringent standards set in the mid-1970s.[5] Nonetheless, this aspect bears watching to illustrate again the remarkable list of environmental consequences that stem from energy production.

---

5 See J. McBride, R. Moore, J. Witherspoon, and R. Blanco, "Radiological Impacts of Airborne Effluents of Coal and Nuclear Plants," *Science* **202**, 1045 (8 December 1978).

WHAT SHOULD BE DONE WITH THE COAL IN THE WEST?

Assuming the coal is to be mined, and there is every indication that it will be, the options are as follows:

1. ship it east and west to population centers via train or coal slurry pipeline;
2. burn the coal in generating stations in the West and transmit the electricity to the population centers;
3. gasify the coal, removing sulfur, and send the gas via pipeline to the population centers.

If the coal is shipped to the population centers, it will be burned there. Even if scrubbers are installed, the total pollution burden on individuals living in the cities will be increased. Not only will pollution levels rise, the number of individuals exposed increases as well. The pollution impact of burning the coal is thus more serious. However, even when electric utilities build their plants in the remote areas of western states, they are being forced by federal government regulation to install such devices as scrubbers and precipitators.

On the other hand, a task force, not connected with the government, composed of industrial executives and environmentalists appears to be recommending "that new coal-fired power plants should be built in the regions where the power will be consumed–which is to say, the people who get the electricity should also have to live with the environmental effects of generating it." * The issue is whether to keep clean areas clean or allow clean areas to be polluted in order to prevent polluted areas from being degraded still further.

There are other issues as well. The panel of the National Academy of Sciences which studied strip mining in the West (reference on p. 263 <u>Rehabilitation of Western Coal Lands)</u> referred to the huge water demands that would be placed on the arid West if energy development takes place in that region (Fig. 13-20).

If water resources are to be committed to large electrical generation and coal gasification plants, huge supplies of water will be necessary. Based on current technologies and projected rates of consumption many power plants are planned for an operating life of less than fifty years. What will be done with the proposed aqueducts, dams, and water when the coal economy no longer needs them? Does society wish to commit water resources to an ephemeral industry at a particular place?

And water demands aside, it has also been pointed out that if the coal is to be gasified (made into a synthetic natural gas) and the sulfur is to be removed in the process, there is no reason why gasification should not take place in the East. Not only could the coal be gasified in the East where water is more abundant, eastern coal could be used rather than western coal. Thus, water is not taken from a place where it is scarce and neither gas nor coal needs to be transported long distances. Unfortunately, coal gasification plants also produce air pollution, again placing the pollution burden on the East.

These questions of policy are not easily settled; the debate will continue. Where do you stand on the issue of where energy development should go and why? Do you think the plants should go in remote areas of the West? Should they have pollution control equipment? Why did the Task Force feel that people who live in population centers should bear the pollution burden?

* Science **198,** 276 (21 October 1977).

TRANSPORTATION OF COAL

Coal does not reach its destination without having an impact on the environment. Coal may move from mine to point of use in a number of ways: by train, by barge, by slurry pipeline, as synthetic natural gas, or as electricity. In recent years, the "unit train," a train devoted exclusively to moving the huge quantities of coal needed by coal-fired electric plants, has come into use. Coal also may be moved by barge on a portion of the trip from mine to plant. The most recent development in coal transportation is the coal slurry pipeline; in this system, a mixture of water and crushed coal is pumped from mine to point of use. Finally, the energy in coal may be transported by converting coal to a synthetic natural gas (SNG) or directly into electricity at a coal-fired generating station.

Train transportation of coal seems a harmless enough activity at first glance; after all, the trains will move on existing lines along rail networks long established and accepted. In fact, however, the vastly increased movement of coal will have an impact. The coal car is an open car so that the dust from coal is scattered along the right of way. Where trains pass through communities, the dust will affect people as it settles from the air. A coal-fired electric plant that produces power at the rate of 1000 megawatts has a coal requirement equal to about 12,000 tons (10,900 metric tons) of coal per day. This is the quantity delivered daily by a single unit train of about 100 cars.

The coal trains may be moving enormous distances. Contracts have already been completed for the sale of Montana coal to a power plant in Paducah, Kentucky, on the Ohio River and to Detroit, Michigan, on Lake Huron. Coal from Wyoming has been sold to power companies in Illinois and Indiana, where coal production is already large. These sales were probably based on the low sulfur content of western coal. A geographical portion of these long-distance trips may be undertaken by barge. Coal trains reaching the Mississippi River may transfer their loads to barges, which complete the trip to market. Coal trains may also unload at Lake Superior ports, leaving barges to finish the journey to the midwestern cities on the Great Lakes.

In contrast to barging, the coal slurry pipeline is viewed as a serious competitor to the train as a means of coal shipment. First attempted successfully near the turn of the century, coal slurry pipelines are now attracting close and favorable attention from the coal companies. Although a coal slurry pipeline was delivering coal to London in 1914, it was not until 1957 that a 100-mile (160-kilometer) pipeline for coal was established in Ohio because of that state's high rail rates for coal. Reaching from Cadiz, Ohio, to Cleveland, the 10-inch (25-centimeter) pipeline was capable of moving 1.3 million tons (1.2 million metric tons) of coal each year. The pipeline closed in 1963 because local rail rates had finally become competitive.

The pipeline had been well publicized, however, and its success encouraged another installation in Arizona. There, the Peabody Coal Company was to supply a power plant in the southern tip of Nevada some 270 miles (432 kilometers) distant from the mine. Because of the rugged terrain, direct rail haul was considered infeasible. A roundabout routing by railroad, avoiding the mountains, would have been very costly. Completed in 1970 by the Black Mesa Pipeline Company, the pipeline now carries a river of water and finely ground coal through the deserts and mountains of the rugged Southwest. The mixture is about 50% water and 50% coal.

With the Black Mesa experience in mind, companies have been planning more coal slurry pipelines. The longest pipeline currently under consideration would reach 1036 miles (1658 kilometers) from Gillette, Wyoming, to White Bluff, Arkansas. The 38-inch (1-meter) pipeline is capable of delivering 25 million tons (22.7 million metric tons) of coal each year to the power plants in Arkansas.

Unfortunately, surface water is scarce in the West. The most important uses for water are irrigation, community water supply, and recreation. Water destined for these purposes should not be diverted for coal slurry pipelines. The Black Mesa pipeline as well as the pipeline proposed from Wyoming to Arkansas draw water from deep wells rather than consume precious surface water.

In the face of these problems, the slurry pipeline remains attractive because of its comparatively low cost. When the coal must move long distances where the railroad bed does not yet exist, rail haul will not compete. Neither will rail/barge shipment, or the transmission of electricity (Fig. 13–21).

Questions

1. What are the four types of coal, ranked by heating value? Which is most abundant in the United States?

2. Why are lignite and sub-bituminous coal being shipped from western states to the Midwest and further east? Give two reasons. What portion of lignite is to be mined by underground mining?

3. Where are the majority of coal reserves which can be mined by surface mining, in the eastern U.S. or the western U.S.? Where are the reserves of low sulfur coal concentrated, in the eastern U.S. or the western U.S.? What does this imply for the percentage of future coal production which might take place in the East or West?

4. What single end-use consumes two-thirds of all the coal mined each year in the United States? Why was coal-use falling until about 1960? What fuels were replacing it? Why has coal use increased since that time?

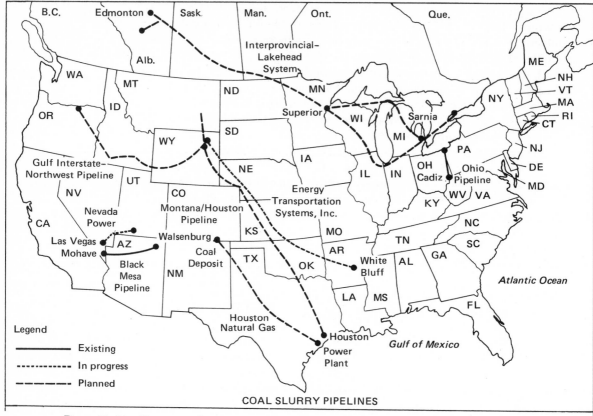

Figure 13–21 *The status of coal slurry pipelines. (Source: Energy Transportation Systems, Inc.)*

5. Why is surface-mined coal gradually replacing coal from deep mines? You may wish to give several reasons, but one is most important.

6. What are the two basic kinds of strip mining and how does a mine operator decide between them?

7. What is a highwall and how does it come about?

8. When strip mined land is not "reclaimed," i.e., left "as is" after mining, a number of problems occur. List and briefly explain three of the problems.

9. What is reclamation? What is the best way to reclaim a hillside that has already been strip mined? is about to be strip mined? Why were coal companies generally unwilling to reclaim the land they surface mined until laws were established?

10. Strip mining is on the increase in the West. What are the two vegetative areas with the least probability of being successfully reclaimed?

11. The Surface Mining Law passed in 1977 calls for reclamation of lands previously stripped. Where will the funds come from to do this?

12. Why are people concerned about coal waste piles (coal refuse banks) created from the wastes from underground mining? What is subsidence?

13. What is acid mine drainage?

14. What environmental problems does acid mine drainage cause?

15. Why could you say that acid mine drainage is one of the environmental costs of generating electric power?

16. Who should pay for controlling acid drainage from long abandoned inactive mines and from mines only recently in operation? Should it be the coal company, local or state governments, or federal agencies? Why?

17. Coal miners are at risk from explosions in underground mines. What substance(s) is (are) exploding?

18. How does coal dust affect a miner's health? Why do the miner's lungs lose their elasticity?

19. What air pollutants not including carbon dioxide come from burning coal? Include three little known pollutants among the six you list.

Further Reading

Acid Mine Drainage

Ackenhail, Alfred, "Pennsylvania Erases its Mining Scars," *Civil Engineering Magazine*, p. 54 (October 1970).

This is a short but extremely well-written article which covers causes, effects and simple control approaches.

Goldberg, Everett and Garrett Power, "Legal Problems of Coal Mine Reclamation," University of Maryland School of Law, prepared for the Environmental Protection Agency, 1972. For sale by the Superintendent of Documents, U.S. Government Printing Office, Washington, D.C., 20402.

While aimed at legal problems, this publication provides clear, well-written, and extensive descriptions of the pollution problems which stem from the mining of coal.

Lackey, James, "Aquatic Life in Waters Polluted by Acid Mine Waste," *Public Health Reports* 54, 740–746 (1939).

This is a classic article identifying the nature of the problem of acid waters. It is nicely written and well illustrated.

Coal Slurry Pipelines

Gray, W. and P. Mason, "Slurry Pipelines: What the Coal Man Should Know in the Planning Stage," *Coal Age* (August 1975).

Clearly written, surprisingly environmental, description of the status, promise and problems of these pipelines.

General References

Kash, D., *et al.*, "Energy Alternatives," prepared for the Council on Environmental Quality, 1975. Published as *Our Energy Future*, University of Oklahoma Press, Norman, OK, 1976.

Nephew, E., "The Challenge and Promise of Coal," *Technology Review*, p. 11, (December 1973).

Energy: The Next Twenty Years, Report by a Study Group Sponsored by the Ford Foundation and Administered by Resources for the Future, Ballinger Publishing, Cambridge, MA, 1979.

Oil, coal, nuclear power, solar, projections, economics—all are here, thoroughly documented and comprehensible; one of the most complete studies of the 1970s.

Project Interdependence: U.S. and World Energy Outlook through 1990, Committee Print 95-33, by the Congressional Research Service, November 1977, pp. 208–262. Available from the Government Printing Office.

Strip Mining

Alexander, Tom, "A Promising Try at Environmental Detente for Coal," *Fortune*, p. 94 (13 February 1978).

This is the story of the National Coal Policy Project, a fascinating experiment in which environmentalists and coal industry people met head to head to resolve their differences. This is the group referred to and quoted in Controversy—What Should be done with the Coal in the West?

"Processes, Procedures, and Methods to Control Pollution from Mining Activities," U.S. Environmental Protection Agency, Washington, D.C., Document Number EPA-430/9-73-011, 1973.

This publication describes approaches to controlling mining pollution rather than the causes and effects of mine drainage. As such it is probably most interesting to consult when studying strip mining.

Surface Mining and Our Environment, U.S. Department of Interior, 1968.

Vivid color pictures of strip mining and resulting erosion and acid mine drainage.

Chapter 14

Oil and Natural Gas

A BRIEF HISTORY OF OIL IN THE UNITED STATES

In the decade preceding the Civil War, a black oily liquid was collected from salt wells near Pittsburgh, Pennsylvania. Although originally bottled as a medicine, some of the liquid was processed by a small "refinery." Five barrels of "carbon oil" for use in oil lamps could be produced each day by the refinery, thereby depriving whale oil and coal oil of a portion of their traditional markets. Oil from elsewhere in Pennsylvania found use as a grease for the machinery in textile mills.

A remarkable invention, the automobile, was to account for much of the growing demand for oil. Rich oil strikes in Oklahoma, Texas, and Louisiana brought competition to the oil fields of western Pennsylvania and surrounding states. Oil use continued to grow as the automobile became part of America's way of life.

In the 1950s, the United States found itself importing cheap foreign oil in larger and larger quantities. By the late 1970s the proportion of U.S. demands met by imports was approaching 40%.

Because of this growing dependence on foreign oil, we went for the oil discovered in northern Alaska as a means to establish greater independence from foreign sources. The 1968 discovery of oil on the North Slope of the Brooks Range, just south of the Beaufort Sea, brought a dozen oil companies to explore for the riches buried in that frozen land. Alaskan oil extended our proved reserves by

nearly 25%, or 10 billion barrels. In the mid 1970s, the Trans-Alaska pipeline was cut across the Brooks Range from Prudhoe Bay, Alaska, bringing oil to the port of Valdez for shipment via tanker to West Coast ports. When the first Alaska oil finally arrived, the West Coast, as predicted by those who advocated a pipeline through Canada to the Midwest, did not need the oil. A new pipeline to carry the Alaskan oil inland may be built.

In 1978, drilling began offshore in the eastern United States. Though populated areas and ecologically important wetlands were not far away, national needs apparently dictated that the search for petroleum go on.

FINDING AND PRODUCING OIL AND NATURAL GAS

We link oil and gas together not only because they are often found together geologically, but because these two preferred fossil fuels are both in short supply in this country, and require imports to fill demands. We also link these two fuels because conservation efforts can help make these fuels more secure.

Petroleum is a mixture of hydrocarbon compounds which include simple methane gas, liquids, and the very complex asphalt materials (solid). All of these substances may occasionally be found together in underground reservoirs within sedimentary rock. Their presence in sedimentary rock is never certain, even with knowledge of surrounding deposits. This uncertainty makes predictions of recoverable oil in the world a subject of controversy.

Because the oil reservoir is deep underground, precise knowledge about the limits of the field and the volume of oil-in-place is not available. A large number of wells, many of them dry holes, would have to be drilled at considerable expense in order to define the field size precisely. Thus, many errors necessarily creep into estimating the volume of oil in a field. Quite naturally, petroleum engineers tend to estimate on the low side so that their efforts in developing the field are sure to be economical.

The quantity of oil that flows out of the well by natural pressure alone is referred to as *primary recovery* and averages about 20% of the oil-in-place. In fact, however, primary recovery may range from one-fourth of this average up to four times this value. Since about 1960, petroleum engineers have developed methods to force up more of the petroleum that remains behind after the initial pressure has subsided. Known as *secondary recovery* methods, these techniques involve either injection of water beneath the oil or injection of gas above the reservoir; both methods place increased pressure on the oil-in-place.

Secondary methods are applied when the expense involved in producing another barrel of oil by these methods is less than the average cost of finding a barrel of new oil. Secondary methods can increase the yield of oil reservoirs to 90% of the oil-in-place.

With so much uncertainty about petroleum in a given field, we can infer that statements of oil reserves on a national or international basis are generally subject to a tremendous lack of precision. Petroleum scientists and engineers attempt to classify reserves by the degree of their uncertainty and this leads to a language of oil reserves, which we will discuss in a moment.

WHAT ARE OIL AND GAS?

It is believed that petroleum was derived from the plankton in the ocean. Algae which drift with the currents are a large component of plankton. In much the same fashion that coal was derived from woody land plants, bacteria break down buried plankton in the absence of oxygen. Chemical and physical processes, operating over many thousands of years, further convert the organic material until at last petroleum compounds become abundant.

Oil consists mainly of liquid hydrocarbons; hydrocarbons are compounds whose only elements are carbon and hydrogen. Ninety to 95% by weight of oil is usually hydrogen and carbon. Carbon makes up about 80%–85% of the weight of oil. Hydrogen adds another 11%–15% of the weight. Nitrogen contributes 1% or less of the weight of oil, but sulfur and oxygen each may account for up to 5% of the weight of the oil. Oil with a sulfur content of less than 1% is referred to as "sweet" crude oil. High-sulfur oil, because of the odor of hydrogen sulfide, is called "sour" crude oil.

To separate oil into its components, it is distilled or "refined." A distillation column with trays is used to separate the components of oil (Fig. 14–1). Petroleum gases exit the top of the tower; liquids collect at various levels in the tower. The components that boil first condense at the topmost plate. Those that boil last, the tars and pitches, collect at the bottom of the column. On the trays between are the various components separated in an order reflecting the ease with which they boil.

Natural gas, in contrast to oil, has a bit less carbon; perhaps as low as 65% carbon by weight and the hydrogen content is variable. Sulfur is usually very low but nitrogen levels may be high, up to 15%. Natural gas is a close relative of petroleum; the same kinds of sediments and geological conditions that give rise to oil also give rise to natural gas. It is not unusual, therefore, to find oil and gas together. When they are found associated, the gas is often dissolved in the liquid hydrocarbons. In this situation, the gas is separated from the oil when it is brought to the surface. Gas is also often found trapped above the liquid petroleum; the industry calls this "gas cap" gas. And, of course, gas is most often found alone, but about 25% of natural gas is found in the search for oil.

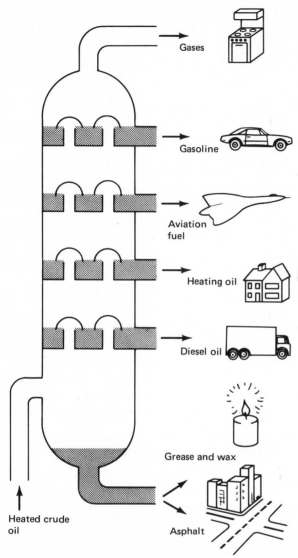

Figure 14–1 *The refining of crude oil into its components. (Adapted from* The Open University, The Earth's Physical Resources, Block 2, Energy Resources, *The Open University Press, Walton Hall, Milton Keynes, United Kingdom, 1973.)*

OIL AND GAS RESOURCES

In our discussion of coal, we spoke of "reserves" and did not explain the term. In fact, we were referring to coal known to exist in quantities and positions that are economic (profitable) to mine. In our discussion of oil and natural gas, we need to introduce new terms since our knowledge of how much of these fuels is present is much less certain. These new terms are used to indicate how certain we are of the presence and quantity of the fuel resource.

Oil Resources

If one were to ask the question, "How much oil is there in the United States?" the reply would come in at least two parts. The first part of the reply would probably take this form: "The *Oil and Gas Journal* estimated *proved* oil reserves in the United States as 29.5 billion barrels, as of January 1, 1978." The meaning of the term "proved" is that this oil is known to be contained in the portions of oil fields that have already been drilled. Furthermore, we are assured that this oil will be profitable to recover. These reserves have been falling since the early 1960s as new discoveries have failed to keep up with our rate of consumption. As a consequence, oil imports rose steadily.

The second part of the reply would be: "There is actually more oil than this, but for one reason or another we cannot count this oil at the present time." Some small portion we can estimate as quite likely to be recoverable on the basis of experience when new but proven methods of recovery are utilized. These are called "indicated" reserves. Another larger portion can be "inferred" on the basis of drilling to date; that is, we already have limited evidence of its existence. Another much larger quantity for which we have much less assurance is, as yet, "undiscovered *resources*." One estimate of these quantities is arrayed in Tables 14–1 and 14–2.

The quantity of "undiscovered" oil is estimated on the basis of geologic evidence, by the use of statistical methods, and by extrapolation of past trends. Unfortunately, these procedures can lead to very different estimates of the oil remaining to be

A FISH STORY

M. King Hubbert, who provided the lowest recent estimate of U.S. oil reserves remaining to be discovered, is one of the most renowned energy analysts in the world. In a chapter of a Congressional Committee Print, Project Interdependence, Hubbert * explains the philosophical basis of his calculations in an amusing and concise way.

> The problem of estimating the ultimate amount of oil or gas that will be produced in any given region has been aptly likened to that of estimating the abundance of fish in a lake by the success in fishing. Although the fish in the lake are not seen by the fisherman, if he catches a fish by almost every cast the inference is justified that the lake is teeming with fish. If, after such a lake has been heavily fished by vacationers for a number of seasons, the same angler is able to catch only a fish or two per day, he is justified in the inference that the lake has been about fished out.

> Similarly, in a virgin petroleum-bearing region, if most of the initial exploratory wells succeed in discovering oil fields, the inference is justified that the given region is rich in petroleum resources. When, at a more mature stage of exploratory drilling, it is found that a steadily decreasing fraction of the exploratory wells drilled succeed in finding oil, the inference is unavoidable that most of the oil in the region has already been discovered, or that the lake is about fished out.

> However, in the case of the lake, if fishing is discontinued for a few years, the lake will become restocked, whereas in the oil-field analogy this cannot happen because oil fields do not breed. There was only a fixed and finite number of oil accumulations in the sediments of the region initially and every time one of these is discovered, the number remaining is reduced by one. It is inevitable that as the remaining fields become fewer in number, and probably also deeper and smaller in size, the difficulty of making a discovery must accordingly increase. Conversely, this record of fewer discoveries with increased exploratory drilling affords one of our more reliable means of estimating how far we have progressed toward the ultimate discoveries likely to be made in the region.

To many of us, a few fish per day sounds like a great deal, so Hubbert must have some very special spots to fish. Of course, Hubbert is right about how we might respond to a lake that once yielded fish in plenty and now provides only a few per day. Suppose, however, that we had a cabin on the lake and a canoe with which to explore. How quickly would you as a fisherman abandon your lake? Would you take other actions first in terms of tackle and bait? What

---

* Chapter XIX, World Oil and Natural Gas Reserves and Resources, of Project Interdependence.

actions? How about in terms of exploration? What advanced methods might you think of to locate the fish? Might you consider catching fish that were less desirable than those in your earlier catches? In the context of oil, a less desirable fish might be oil shale. Can you think of others? Perhaps a different way of cooking would make them better to eat. You begin to understand now how the oil companies respond to their fewer successes on the U.S. mainland.

Hubbert's story gives us another insight on oil. Because oil exploration is likened to fishing, we are drawn to a comparison of fish to oil. Fish is a renewable resource; unless we remove nearly all the fish, the fish will breed and repopulate the lake. Oil is a nonrenewable resource; each removal is final and irrevocable. A much better energy resource on which to rely is one that is renewable, such as solar energy.

TABLE 14–1   Unproven Oil Resources of the United States[a]

	Billions of barrels of indicated & inferred reserves	Billions of barrels of undiscovered resources, lowest estimate
"Lower 48" onshore	17.0	110
Alaska onshore	5.0	25
Offshore (Alaska and "lower 48")	3.0	65
	25.0	200

[a] Source: *U.S. Geological Survey*, March 1974.

TABLE 14–2   Undiscovered Offshore Oil Resources of the United States[a]

	Billions of barrels undiscovered oil (lowest estimate)
Atlantic offshore	10
Gulf of Mexico	20
Pacific offshore	5
Alaska offshore	30
Total offshore	65

[a] Source: U.S. Geological Survey.

discovered. The most recent, widely cited studies provide estimates of undiscovered liquid hydrocarbons that range from 72 billion barrels to 200 billion barrels.[1] The lowest estimate (72 billion barrels) of unproven U.S. oil reserves was provided by Hubbert of the U.S. Geological Survey.

There is one further category in oil resources— that oil not presently profitable to produce. Some is already discovered; the remainder is yet to be discovered. We mentioned earlier that North Sea oil became profitable after the price of oil jumped from $2.00 per barrel to $12.00 per barrel.

United States and World Use of Oil

(The United States is currently using over one-third of all oil consumed in the world each year.)It was not always this way, however; we used to consume even a greater fraction. We are gradually losing our dominance as a consumer of oil, not because we are consuming any less, but because other nations have increased their use so dramatically.

1 *Energy: The Next Twenty Years,* Report by a Study Group Sponsored by the Ford Foundation and Administered by Resources for the Future, Ballinger Publishing Company, Cambridge, MA, 1979.

In 1976, U.S. consumption of 17.0 million barrels of oil per day is equivalent to 6.21 billion barrels of oil per year. This number is broken down in the following way (Fig. 14–2). Gasolines account for 7.42 million barrels of oil per day or 44%; use of middle distillate, which consists of home heating oil, diesel fuel, and kerosene (jet fuel), is 4.24 million barrels per day or 25%, and residual fuel oil consumption averages 2.67 million barrels per day or 16%. Residual oil is used for heating commercial and industrial buildings and as a fuel in a decreasing number of steam electric plants. The total of all other uses including asphalt, plastics, chemical manufacture, etc. gives us the remaining 15%. These shares have remained relatively constant over the last several decades. The portion of oil that goes for gasoline in western Europe is only about 21%, but this figure is growing as car ownership continues to grow in the western European nations.

With the U.S. using over a third of the world's oil production, and producing only 9.1 out of the 17.0 million barrels it consumes each day (1976), it makes sense to examine where the U.S. is getting its 7.9 million barrels per day of imported oil (Fig. 14–3).

Figure 14–4 shows us where proved reserves of oil are located worldwide. As we can see, the Middle East has by far the greatest proved reserves in the World. It is not hard to see why our imports from Middle East nations had increased to 36% of all our oil imports by 1976.

The future statistics on oil production and oil importation are uncertain for a number of reasons. If imported oil is taxed or OPEC raises its prices steeply or if major oil discoveries occur in the U.S. offshore, domestic production could climb. A shale oil industry is a distinct possibility.

If the economy is not healthy and many people are out of work, the use of the automobile will diminish as people try to save money. This was seen in the period following the 1973–74 embargo, when the country went through an economic slump due to combined inflation in the price of oil and other goods. Any combination of these events could occur, leading to an enormous uncertainty in predicting the U.S. oil future.

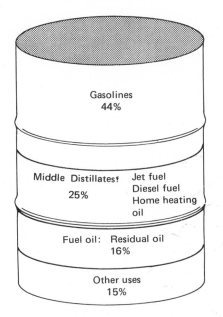

Figure 14–2 *Oil consumption in the United States, by use category, in 1976.*

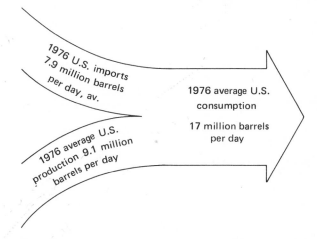

Figure 14–3 *Production, imports, and consumption of oil in the United States in 1976.*

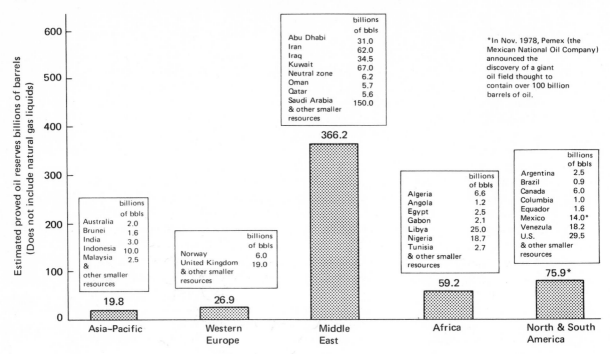

Figure 14-4   *Proved oil reserves as of 1 January 1978. (Source:* Oil & Gas Journal, *26 December 1977.)*

Natural Gas Resources, Production, and Use

We can describe the resources of natural gas in the United States in the same way that we described oil resources. There are proved reserves; we know they are in place and profitable to produce. There are indicated and inferred reserves for which we have some degree of evidence. Finally, undiscovered resources are yet to be found. Table 14-3 provides the estimates of natural gas resources in these various categories and indicates the potential in Alaska and the offshore for natural gas.

TABLE 14-3   **Natural Gas Resources of the United States**[a]

Trillions of cubic feet[b]	Proved reserves	Indicated and inferred	Undiscovered
		(lowest estimate)	(lowest estimate)
Lower 48	189.8 (5.30)	93 (2.63)	500 (14.2)
Alaska onshore	28.5 (0.81)	14 (0.40)	105 (2.97)
Offshore (lower 48 and Alaska)	47.8 (1.35)	23 (0.65)	395 (11.2)
Total proved reserves	266.1 (7.54)	130 (3.68)	1,000 (28.3)

[a] Source: *U.S. Geological Survey,* March 1974.

[b] Trillions of cubic meters in parentheses.

PARTICIPATION

Participation seems an unusual word to use in the context of oil and gas. It implies a partnership, and that is precisely what has evolved in the North Sea where oil companies have searched for and are now producing large quantities of oil. The partnership does not involve one industry working with another to develop an offshore field such as the Brent field where Shell and Exxon have a joint venture. The partnership is between industry and government. The British government "participates" in the revenues of the oil companies operating in the British Sector of the North Sea.

The British National Oil Company (BNOC) came into existence on January 1, 1976. It was not the first such government oil company. In fact, it is only part of a long trend toward government oil companies. At the present time, nearly all oil producing nations, except for the United States, have government-owned oil companies. The British National Oil Company was formed to return some of the profits from the sale of North Sea oil to the Exchequer (Treasury) and hence to the British public. The rationale was that the oil belonged to the British people rather than to the companies that found and produced it. The rate of exploitation, too, was a factor the government wished to control. The oil companies are, after all, international enterprises rather than firms of British origin, and many have their origins in the United States.

As the Labour Party put it, *

> (The BNOC participation in the oil industry) will give this nation a direct right to the oil extracted from our own seabed. Oil underground belongs to the nation, but once it is extracted it becomes the property of the licensee. Under participation, we shall gain a right in the property and a say in the disposal of our own oil ....

The Labour government also viewed the BNOC as a means to obtain the most accurate information possible on actual reserves.

Whether BNOC will explore or refine or market oil as the independent oil companies do is still an open question. For the time, it merely brings revenues to the British government.

Should the United States have a state-owned oil company? Should it explore, refine, market? Or should it simply participate in revenues? Or are our private companies efficient enough and honest enough that the tasks should remain theirs? Is there the possibility that a state-owned oil corporation will be inefficient in the way in which it spends money? Do independent oil companies already return sufficient benefits to the American consumer?

---

* Statement by John Smith, Parliamentary Under Secretary for Energy, House of Commons, April 1975. Quoted in Controlling Oil: British Oil Policy and the British National Oil Corporation, by E. Krapels, Committee Print, prepared at the request of Sen. Henry M. Jackson, October 1977, Public No. 95-59.

SHOULD THE PRICE OF NATURAL GAS BE HELD LOW?

Let the Bastards Freeze in the Dark!

> A popular bumper sticker in
> oil and gas rich states.

"I stayed in bed to keep warm and to keep from being sick," she said. "My brother used to sit by the oven."

> The New York Times, 7 February 1977,
> an unidentified elderly woman in
> Queens, New York, describing how
> her brother had gotten frostbite
> and gangrene.

The conflict on natural gas pricing probes deeply into one of the controversies in our free enterprise society. On the one hand, we have a society that believes in free enterprise as a practical and efficient means to provide the goods and services people demand. On the other, we have chosen to provide social security and medical care to the elderly. We attempt to be at once a society that both supports a free market system and cares for those who have little.

The issue of natural gas pricing illustrates the conflict between these objectives. The notion of a free marketplace involves people who need a product and producers who can provide it. When many producers can provide the product, let us say bread, the producers are assumed to compete; that is, they are assumed to lower their prices to similar levels so that they can continue to capture a portion of the demand. Unfortunately, perfectly free markets do not exist for every product.

A free market for the sale of natural gas, which moves in interstate commerce, did not exist for some years. The market was controlled by the government because it was thought to be open to abuse. The rationale was that monopolies could occur when a single pipeline supplied an urban area. A single gas supply firm would charge very high prices for natural gas because it was the only supplier and because the gas was essential. To prevent excessive prices in this monopoly situation, the government controlled the price of natural gas.

The government control of the price of gas sold interstate, however, appears to have been too tight. The low price of this gas led many millions of families in urban areas of the eastern U.S. and the Midwest to heat their homes with natural gas. Meanwhile, gas sold within the state in which it was produced become a valued good. In addition, its price was not regulated. The gas was used in the producer states for fertilizer and chemical manufacture and electric power generation as well as for heating homes and hot water. The price of gas sold within the state in which it was produced soared to four and five times the regulated price for interstate gas. With such a market in the producing states, it was no wonder that producers did not want to sell gas on the interstate market.

Supplies to the East and Midwest did not keep pace with demand; new customers for natural gas had to be turned away by the local utilities. Some industries had their supplies of natural gas diverted to homes and schools as severe winter weather set in. The producers told us that the government should not regulate prices. Mobil Oil Company put it this way *

> What is needed is decontrol of prices of new supplies of natural gas....This approach would give producers the incentive to step up their already active search for gas...

On another occasion, Mobil added, †

> The government's focus on low prices to the consumer has ignored his need for secure and adequate supplies.

As we pointed out in the text, Congress has ordered the regulation of natural gas prices to end by 1985. It would be a mistake, however, to suppose that the decision of 1978 will no longer be debated. When Congress originally ordered the regulation of natural gas in 1954, the gas industry lobbied and buttonholed Congress for deregulation the next year, the year following that, the year after, and so on until 1978, almost a quarter of a century later. And the vote to end regulation in 1978 was very close. In 1985, we are scheduled to have our first taste of a "free" market in natural gas. Some senators and representatives, as well as ordinary citizens, do not trust the oil and gas industry and can be expected to work to re-establish price regulation to prevent sky rocketting gas prices. And, in a sense, the issue of regulation does boil down to whether you trust the oil and gas industry.

It is clear that both sides have something of value in their argument. Low interstate prices have caused shortages as new supplies have been sold within their state of origin, but low prices have also protected the consumer. An unregulated natural gas industry has the potential to exploit the consumer. Furthermore, if prices rise sharply, reserves may come up, but the poorer people among us will be less able to afford heat for their homes. Do you have suggestions for what we should do? Do you favor the government's removing or restoring controls on natural gas? Why or why not?

---

* Mobil Oil Company advertisement, The New York Times, 18 January 1976 and 26 January 1977.

† Mobil Oil Company advertisement, The New York Times, 18 December 1976 and 26 January 1977.

Figure 14-5 indicates the pattern that U.S. proved reserves of natural gas have followed in the past quarter century. The "spike" in additions to natural gas which occured in 1970 was due to the gas finds in Alaska in that year. Without that spike, the trend would have been steadily downward since 1969. The downward trend is the result of annual consumption exceeding annual additions to reserves since that time.

The observant reader will note that the graph of proved reserves through time does not give the same number at the same time as Table 14-3, which shows proved, indicated-inferred, and undiscovered gas resources. The reason is that the two estimates were prepared by two different organizations, the American Gas Association, an industry group, and the U.S. Geological Survey, an office of the Department of the Interior. Since the American Gas Association represents the natural gas companies (which by and large are the oil companies), low statements of reserves tend to support the point that we are running out of gas. These statements support the industry's demand that the federal government end its control of the price of natural gas. The industry, in 1978, finally convinced Congress that an end to control was necessary. In that year, Congress passed

and the President approved a bill that would phase out all price controls on natural gas by 1985.

Criticisms of the estimates of reserves by the American Gas Association have been going on for a number of years. These criticisms led the U.S. Geological Survey to make its own estimates. The Survey, too, was criticized and a study of reserves by the Federal Energy Administration[2] was ordered by Congress. This study, as well, was roundly criticized.

The future of natural gas production is as cloudy as the present resource base. The rate at which new gas in the traditional areas of search onshore is found has been declining, according to the American Gas Association. By that we mean that more exploratory wells have been needed on average for each new discovery. Unless offshore drilling becomes a bonanza or subeconomic resources become economic by virtue of price increases, future additions to the resource base will not compare with the past.

Of late, we have begun to import natural gas by tanker, although we have in the past been sup-

---

2 The Federal Energy Administration (FEA) is now part of the Department of Energy.

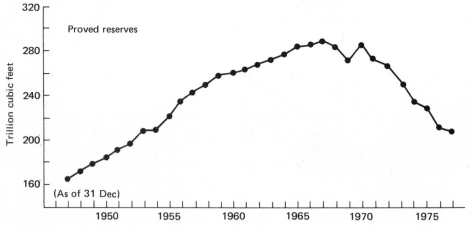

Figure 14-5   *Proved reserves of natural gas in the United States, 1947-1977. [Source: E. T. Hayes, "Energy Resources Available to the United States, 1985-2000." Science* **203**, *236 (19 January 1979). Copyright 1979 by the American Association for the Advancement of Science. Data from the American Gas Association.]*

plied with gas from Canada by pipeline (Fig. 14-6). The tanker gas has been imported principally from Algeria. In order to save space in shipment, the gas has been made liquid by cooling it to very low temperatures and putting it under pressure.

The liquified natural gas (LNG) has been received at coastal terminals to supplement gas received via pipeline. At the coastal terminals, the liquid is converted back to gas, and injected into pipelines that already carry U.S.-produced natural gas. The LNG has been considerably more expensive (two to three times) than domestically produced natural gas.

Since the era of LNG transport has barely begun, it is difficult to predict the safety of the procedures used. Thus we cannot predict the results if an LNG tanker ran aground or was involved in a collision and lost its cargo. The concerns about the safety of LNG stem from two major accidents. The first occurred in Cleveland in 1944 when an LNG tank leaked millions of gallons into the sewer system. The spilled material caught fire and 128 people were killed. In 1973, an LNG tank on Staten Island was being repaired when it exploded. Forty workmen were killed. The tank had been believed to be empty. A 1976 study by the Rand Corporation warned that LNG might "spill in port or at sea, vaporize into a cold gas cloud, be carried inland by the wind, and ignite with tremendous explosive force and heat over a heavily populated area." "To avoid serious accidents," the study went on, "... . it would only be prudent to locate all facilities for handling LNG at remote sites—not major port facilities..."

NEW SOURCES OF OIL

The discussion of oil resources has dealt only with the oil we are used to producing, the oil trapped in geologic structures beneath the earth. There are, however, three other sources of the liquid hydrocarbons we need. One is the oil shale, principally in Colorado. The technology to extract oil from shale is ready, but the environmental impact may be very great. Another is the Tar Sands in Canada, a reliable but hard-bargaining trading partner. The third source of liquid hydrocarbons is coal. Coal, the most plentiful of U.S. fossil fuels, can be converted to a liquid. We call the process coal liquefaction.

The Tar Sands of Athabasca

In the northern portion of the province of Alberta, Canada, is a region where deposits of a tarry black sand are being exploited to produce oil;

Figure 14-6 *LNG tanker. A number of these specially designed tankers already are in service transporting natural gas in liquefied form at minus 260 degrees Fahrenheit (−162 degrees Centigrade). Concerns over the hazard of this method of transport have been expressed because of the explosive nature of the cargo. (American Gas Association.)*

the sands are coated with pitch or tar. These petroleum deposits, instead of being trapped in an underground reservoir, have migrated up into porous sands near the surface of the earth. The tar, a hydrocarbon known as bitumen, may account for up to 16% of the weight of sandstone. Three tons of a rich tar sand, one which is 14% bitumen by weight or more, is sufficient to produce two barrels of hydrocarbon liquid. The efforts in Alberta have had a difficult history; the oil has proven difficult to extract in an economical fashion. Efforts continue and a commercially profitable process is now thought possible, especially given the high cost of oil on the world market. Geologists estimate that there may be 300 billion barrels of oil in the tar sands deposits, about 10 times the proved oil reserves in the United States.

Tar sands are also found in the United States, but most deposits are small. By far the largest deposits are in Utah and on federal land; 19–29 million barrels of liquid hydrocarbon might be produced from these deposits.

Oil Shale

Another source of liquid hydrocarbons is the shale rock that contains the organic material known as kerogen. A liquid much like oil can be extracted from the kerogen by distillation. The organic matter locked in the shale is the result of geologic processes acting on the ancient sediments accumulated in inland lakes. The world's largest deposits of this oil-bearing rock are apparently in Wyoming and Colorado, where the equivalent of 600 billion barrels of oil may be present. Compared to U.S. proved reserves of 30–40 billion barrels of oil in ordinary reservoirs, this resource is enormous.

If it's so enormous, then, why aren't we producing oil from this source? There are several reasons. First, to remove the oil requires the mining of rock. About 1/2 to two barrels of oil can be extracted from each ton of rock, leaving 1700 pounds of waste rock to dispose of. If we were to produce from this shale one billion barrels of oil, or about

one-sixth of our annual (1978) consumption of oil, we would produce about 900 million tons of waste rock. This waste rock will have a huge impact on the land; the mining operation could be massive. Caution in pursuing this resource is appropriate (Fig. 14–7).

The Institute of Ecology provided a review of the government's Environmental Impact Statement on oil shale leasing. That report indicated that there are other problems besides disposal of the waste rock, itself a considerable obstacle. Oil shale processing is expected to cause the release of mercury, cadmium, and lead to the air. Water used in the mining operation will become rich in dissolved salts; the brine must be disposed of in a way that will not contaminate ground water. Revegetation of mined areas is probably difficult because of the saline content of the waste rock (Fig. 14–8).

The second reason we are not producing our oil from shale has to do with costs. Estimates of the cost to produce a barrel of oil from shale are very uncertain. In 1977, some said shale oil could be produced for as little as $8 per barrel; others thought the cost of production (not the sale price) might be as much as $28 per barrel. With such uncertainty, energy companies have been reluctant to invest in oil shale production.

OTHER SOURCES OF GAS

Geopressured methane If there ever was a phantom resource, the natural gas known to be dissolved in deeply buried caverns of salt water is that phantom. The trapped subterranean waters are at very high temperatures and under enormous pressure. In this environment, natural gas may be formed from the breakdown of oil deposits and dissolve in the salt water.

If the salt water caverns themselves can be tapped, a very large resource of natural gas may await us. The theory is that if the pressure on the reservoirs can be released, perhaps by allowing some of the water to escape as steam, the gas will come out of solution forcibly and can be captured

Oil shale
conventional recovery

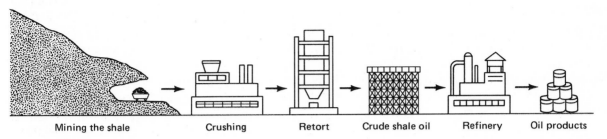

Mining the shale   Crushing   Retort   Crude shale oil   Refinery   Oil products

Oil shale
in–situ process (underground)

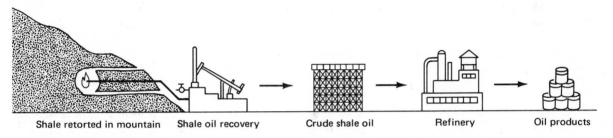

Shale retorted in mountain   Shale oil recovery   Crude shale oil   Refinery   Oil products

Figure 14-7   *There are two basic ways shale oil can be produced. The mined material may be retorted ("cooked") above ground or the rock may be cooked in place (underground). Of the two, the second is more acceptable from an environmental standpoint, as the surface is not modified and waste rock is not produced. The difficulty or unknown is how efficient the underground process can become. At the moment, it seems that surface mining and surface re-torting are the most economical. The oil produced from shale must, in any case, be cleansed of sulfur and nitrogen before it is comparable to oil used in refinery operations.*

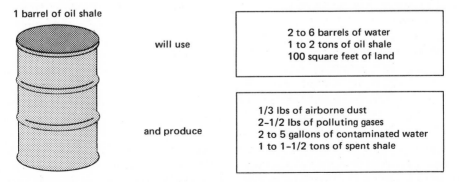

1 barrel of oil shale

will use

2 to 6 barrels of water
1 to 2 tons of oil shale
100 square feet of land

and produce

1/3 lbs of airborne dust
2-1/2 lbs of polluting gases
2 to 5 gallons of contaminated water
1 to 1-1/2 tons of spent shale

Figure 14-8   *Physical impacts of oil shale.*

at a well. There is controversy over whether or not this gas can, in practice, be recovered, but agreement that it is there. A Chevron well driven 21,000 feet (6300 meters) into the earth in Louisiana is said to have produced geopressured gas in 1977, but the company asserts it was an ordinary gas deposit. There are potential environmental problems if we use this resource, such as disposal of the hot brine and concern over subsidence above the cavern if geopressured gas is tapped. The resource, may, however, be worth thinking about.

(The U.S. Geological Survey estimates that there may be 24,000 trillion cubic feet (680 trillion cubic meters) of gas in the geopressured zone of the continental U.S. The estimate should be compared to current annual use in the U.S. of about 20 trillion cubic feet (0.57 trillion cubic meters) of gas; this is a 1200-year supply at the current rate of use. How much of the geopressured gas could be recovered is not known.)

Methane associated with coal (There are other ways to produce a methane gas of the quality presently being produced from drilling. One such method is to remove the methane produced in coal deposits before opening the mine. This procedure both produces gas and makes the mine safer to work. In fact, the Equitable Gas Company of Pittsburgh has been quietly producing natural gas from a coal field since 1949.)

Methane in Devonian shale (Still another source of natural gas is the shale created in the geologic era known as the Devonian. This shale lies beneath the surface of about 250,000 square miles (640,000 square kilometers) in the eastern and central U.S. Each ton of Devonian shale may contain between 20 and 30 cubic feet (0.57 to 0.85 cubic meters) of natural gas; so that there may be as much as 500 trillion cubic feet of gas in these deposits. To produce sufficient gas from these deposits requires fracturing the rock either with explosives or with water injected under high pressure; such fracturing makes it possible for the numerous, but tightly trapped gas deposits, to connect and thus for the gas to flow.)

Methane from landfills (Another way to obtain gas is to capture the methane that is produced by sanitary landfills, the areas where trash and garbage are buried. The decomposition of the buried garbage in the absence of oxygen is known to produce methane, sometimes contaminated with the foul smelling gas, hydrogen sulfide. The methane level, however, is good and the gas can be burned.)

Methane from wastewater treatment (Methane has been produced for some years in the sewage treatment plant in the device known as the anaerobic digestor. In this process, sewage solids are degraded by microbes in the absence of oxygen (anaerobic). The methane is consumed (burned) in the plant to provide the heat needed for the digestor and elsewhere in the plant.)

Methane from animal wastes (Methane may also be produced from animal wastes. When cattle are crowded into feedlots for beef production, animal wastes may build up. The decomposition of these wastes in the absence of oxygen produces methane. Plans to use gas from cattle manure have been formulated by the Peoples Gas Company, a Midwestern firm; they would use the gas as an addition to their basic supplies.

Methane in tight sands (One unusual method of gas production from wells is quietly slipping into oblivion. Known as the Plowshare Program,[3] the undertaking began in the 1960s under the leadership of the Atomic Energy Commission (now part of the Department of Energy). Underground nuclear explosions were to be used to free gas which was in "tight" formations. The nuclear explosions, referred to as "nuclear stimulation," were to fracture the rock walls between the small deposits in order to create a single reservoir cavern into which the gas could collect.)

Three experiments were conducted in New Mexico and Colorado. These were Project Gas-

---

3 A hopeful name derived from the biblical phrase "... They shall beat their swords into plowshares, ..." (Isaiah ii, 4).

buggy (New Mexico), Project Rulison (Colorado), and Project Rio Blanco (Colorado). Two factors probably account for the end of the program. First, the residents of the states where the atomic explosions were to be set off were concerned about the effects. Second, the gas itself had a low level of radioactivity. The gas in these tight sands in the Western U.S., however, could amount to as much as 600 trillion cubic feet, and engineers will probably attempt to release the gas by hydraulic fracturing of the rock (the injection of water and sand at high pressures).

Coal gasification (The last method we shall mention to produce pipeline quality gas is coal gasification (see pp. 303–307). In this process, the carbon in coal is converted by the addition of hydrogen into methane. Government financial support for coal gasification is large, and there are a number of competing methods.)

The Power of the Cartel and Its Impact on Oil Companies (Economic and Political Aspects of the Global Oil Situation)

The oil cartel has had enormous influence on the quality of life in the United States. It could be argued that the cartel's pricing policy will have long-run benefits in stimulating our use of renewable energy resources. It could also be argued that the high prices will eventually lead us to develop shale oil processes, a step which could have adverse effects on the states in which the shale is located. In the short run, the pricing policy will have marked effects on our ability as a nation to afford other improvements to our environment. Constantly increasing oil prices increase inflation and cause citizens to ask if pollution control can be afforded. To understand the urgent need for oil conservation, we need a closer look at the organization that sets oil prices.

To explain the power of the cartel, we must first discuss the notion of a monopoly. A single firm monopoly is an enterprise that is alone in its field. No other firm exists that supplies to customers the same product or service.

When one company produces such a very large portion of the total demand for the good, the other firms which supply the good are in a happy but risky situation. If the giant firm has the financial resources or its production costs are very low in comparison to competitors, it can decrease its price to a level at which competitors cannot make a profit. If the competitors are driven from the scene, the giant becomes a monopolist. On the other hand, if the giant firm does not care to increase its share of the market for some reason, competitors are in an excellent financial position. They can share the monopoly power of the giant firm to sell at an inflated price.

Now in the international oil market, the giant firm is represented by OPEC (the Organization of Petroleum Exporting Countries). The competitors are the international oil companies, many of which are based in the United States. Even though OPEC consists of a number of countries, they are still able to dominate the marketplace as though they were a single firm; that is, they set the price at which they will sell their oil. Such an organization, which attempts to dominate the international market, is referred to as a cartel. In order to ensure that each member of the cartel is able to obtain the price set by OPEC, each country participating in OPEC agrees to produce so many million barrels of oil per day and no more.

Since OPEC both sets a price and allocates production to the member nations, each member is, in effect, agreeing to a particular revenue from oil sales; each will earn the product of price and production. The stability of the cartel is anchored on the agreement that each member will continue to accept its stated share of production so that it can obtain the price set by the organization. The cartel, any cartel, may not be stable if one of its members needs increased revenues or if new suppliers appear on the scene.

We mentioned in our brief discussion of monopolies that the other producers in a near-monopoly situation would be in a "happy but risky" situation. The international oil market illustrates this point well. Just before the 1973–74 embargo, oil produced within the U.S. was selling on the U.S. market for $3.50 per barrel. Just after the oil embargo of 1973–74, OPEC began to demonstrate its strength; it set the price of its oil at $11 per barrel. By 1980, the price was set at $28 per barrel. Oil now being produced in the United States by the oil companies can be sold at the OPEC price.

This new high price for oil is the "happy" situation of producers who operate in the shadow of a monopolistic firm. The price of their goods rises to the giant firm's price. U.S. and international oil companies reap a handsome benefit from the cartel's control of the market.

Their situation is not without risk, however. In their search for greater profits, the firms do invest heavily in exploration, production, and in research in methods of enhanced recovery. To see the origin of the risk, we use the example of the North Sea, where the pace of exploration and development has quickened since OPEC's rise to power. In the middle 1970s, an offshore oil platform, built to withstand the lashing winds and towering waves of the North Sea, cost about 200 million dollars. This investment had to be made before the first drop of oil could be produced from a field. (See Figure 14–9.) It was estimated that oil from the North Sea would cost on the order of $5–$6 per barrel to produce, but so long as the cartel keeps its price high, the investment is well spent. It is of interest that the oil in the North Sea was not even counted as a reserve in the early 1970s. The world price for oil was then about $2.00 per barrel, so that North Sea oil was not profitable to produce. If the cartel were to decide to lower its price to less than about $7 per barrel, the North Sea investments would again no longer be profitable.

Could the cartel lower its prices so drastically? The plentiful oil in the Middle East costs less than 25 cents per barrel to produce, as opposed to $5 in the North Sea. And to transport it via tanker to market costs only about $1 per barrel. OPEC has awesome power to price cut and destroy the competition's investments. This is the "risky" situation to which we referred earlier.

Figure 14-9 *Production platform nearing completion. It is destined for the Brent Field in the North Sea. The investment to produce this oil is evident. (A Shell photograph.)*

Oil Conservation–More Than One Reason to Save

Conservation of oil is important to future generations. The basic goal of oil conservation is to postpone the days of oil scarcity until new sources of energy become available. It is an unarguable goal, worthy in and of itself. Nonetheless, there are two other important reasons to conserve oil. These other reasons make our interest and the interest of future generations one. First is the hazard of dependence by the United States on foreign oil; an embargo like the one undertaken in 1973–74 could stall the engine of our industrial society and could influence the freedom we have in choosing our policies relative to other nations. And second is the major influence of oil imports on the economic health of the nation. Our massive oil imports may go in part to heat our homes, but they also fuel the fires of inflation. How can this be? It is quite simple.

In 1976, the U.S. imported about 2.9 billion barrels of oil; more than 35 billion dollars were paid for this oil. Those 35 billion dollars left the country, entering the coffers of foreign treasuries and foreign companies. When dollars leave the country, the companies, the individuals, or the nations that obtain them may be willing to sell their dollars. They are interested in exchanging them for either their own currencies or the money of other nations. As an example, if Saudi Arabia were to purchase a plant to desalt sea water from West Germany, it would pay West Germany in German marks. If Saudi Arabia did not hold at the time enough German marks, but had many U.S. dollars from the sale of oil, it would offer to sell dollars in return for marks. It would seek a nation or buyer who was willing to accept dollars and give marks in return (Fig. 14–10).

Of course, there is a competing process in which U.S. goods such as computers are purchased by foreign buyers. Dollars are needed by the nations or companies that purchase these goods. To obtain the needed dollars, these firms or nations enter the international money market to purchase U.S. currency; they sell the currencies they hold in return for dollars. On any given day, there are those willing to sell dollars so they can have other currencies for their transactions, and those seeking to buy dollars so that they can purchase U.S. goods.

On a particular day, the average price at which the dollar is exchanged for the currency of some other nation may be viewed as the value of the dollar relative to that currency. There is an exchange rate of the dollar with the

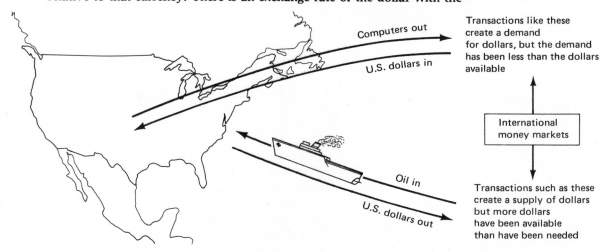

Figure 14–10   *The flow of U.S. dollars into and out of the country.*

British pound, the dollar with the German mark, the Swiss franc, the Dutch guilder, the Japanese yen, and so on. All this has to do with oil, with energy conservation, and with inflation. We trace the effect in the following way.

Each year, economists add the dollars paid for U.S. goods by foreign firms and nations and compare it with the dollars leaving the country for the purchase and importation of foreign goods. Each year they find that more dollars leave the country than enter it, and this can largely be traced to oil purchases. The balance of trade for the U.S. is said to be negative. That is, more dollars have fled the country than were needed by foreign interests to purchase U.S. goods; hence there must be a supply of U.S. dollars on the international money market that exceeds the demand. Whenever supply exceeds demand at a given price, the price will be decreased, causing demand to increase and the excess supply to be absorbed. Or more simply, when supply exceeds demand, the price falls. At the root of the fall of the dollar's value is the outflow of dollars for foreign goods and most particularly for oil.

The decline in value of the dollar seems a remote event as reported in the newspapers, but the daily changes in the dollar's value relative to other currencies makes a considerable difference when added over a period of time.

The impact of this fall in value becomes clear when the purchase of a foreign-made product is considered. If a consumer is looking at a Japanese car or German camera, he will find that the price of these items has risen considerably in the past decade. The price increase results from a decline in the value of the dollar. Suppose a German camera sold in Germany for 1000 marks in the early 1970s and because of manufacturing efficiencies still sold for about that amount in the late 1970s. It took $330 to buy this camera in the early 1970s, because each mark cost 33¢. In the late 1970s, however, the same camera cost $500, because each mark now costs 50¢.

This kind of price increase on goods imported from the industrial nations has been a contributing factor to the excessive inflation in the price of goods that the United States has experienced. Many of the goods and raw materials we utilize may have been produced abroad. Examples are autos, cameras, radios, tape recorders, watches, clothing, iron, steel, copper, cobalt, and so on. As the dollar falls in value, the cost in dollars of these imported goods and raw materials goes up in response. The impact of these price increases caused by the fall in value of the dollar leads to a rise in the cost of living and contributes importantly to the annual inflation rate. The fall in the value stems from the enormous quantities of oil we import from foreign nations.

Gas from Coal (Coal Gasification)

Gas from coal is not a new idea, although the quantity and quality of the product now being considered is quite different. Gas was being manufactured on a wide scale from coal in the early 1800s, first in Britain then in the United

States. The product of the local "gas works" was known as "town gas" or "illuminating gas"; this low heating value gas was used for community lighting, for heating homes and for cooking. By the end of World War II, natural gas had displaced town gas in the United States.

Now, as our supplies of natural gas appear to dwindle, the gasification of coal again may become economically attractive. As a consequence, a great deal of research is going into new processes to convert coal to a high quality gas.

There are two kinds of gasification, based on the quality of the product. The product of one process is a gas of relatively low heating value, consisting primarily of carbon monoxide and hydrogen; the process is referred to as low-Btu gasification. [The Btu (or British thermal unit) is the heat energy needed to raise the temperature of 1 pound of water by 1 °F.] The product of the second process is a gas of a heating value nearly that of natural gas, and the process is known as high-Btu gasification. The gas itself may be referred to as SNG (for Synthetic Natural Gas), and consists primarily of methane.

It is reasonable to substitute this high-Btu gas for the natural gas transported by pipeline to many sections of the country. The low-Btu gas, however, can be used in industries at the place it is produced or it may be used to generate electric power at the site of production.

The processes currently being studied are known by such trade names as Lurgi, HYGAS, BI-GAS, and SYNTHANE. The only commercially proven process for coal gasification, however, is the Lurgi process, which originated in Germany before World War II.

Although the processes differ in their details, the broad picture is very nearly the same.

Low Btu Gas.   Coal is burned in the presence of either air or pure oxygen. Then hydrogen is produced by the reaction of steam (water vapor, H_2O) with the burning coal (carbon, C). The oxygen in the water is captured by the carbon, resulting in the production of carbon monoxide gas (CO) and hydrogen gas (H_2).

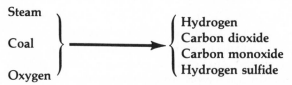

Steam
Coal
Oxygen
→
Hydrogen
Carbon dioxide
Carbon monoxide
Hydrogen sulfide

The gas mixture that results from this single step has a relatively low Btu rating, perhaps 200 Btu per cubic foot of gas (7400 kilojoules per cubic meter). That is, when burned, the heat that results is low. However, if pure oxygen, rather than air, was used in the first step of the process, the heating value may be doubled or more. To produce a gas suitable to inject into pipelines for transport, the gas mixture must undergo further steps to clean it and to increase its heating value.

High-Btu Gas. The reaction used to increase the heating value of the gas is called "methanation"; it means simply that methane is formed. In the presence of a catalyst, carbon monoxide reacts with hydrogen to produce methane (CH_4) and water

$$CO \quad + \quad 3H_2 \quad \xrightarrow{\text{catalyst}} \quad CH_4 \quad + \quad H_2O$$

carbon monoxide   hydrogen   methane   water
(gas)   (gas)   (gas)   (as vapor)

With upgrading by cleaning and by methanation (and if pure oxygen is used rather than air in the first step), the product will be a gas of heating value nearly that of natural gas, about 1000 Btu per cubic foot of gas (37,200 kilojoules per cubic meter) (See Fig. 14–11.)

Gasification Plant Requirements

The typical coal gasification plant is expected to be large. To produce 250 million cubic feet (7.08 million cubic meters) of high-Btu gas each day will require 25,000 tons (22,700 metric tons) of coal, or about one ton of coal for each 10,000 cubic feet of gas (Fig. 14–12). If the plant is in operation for 90% of the year, 8.5 million tons (7.71 million metric tons) of coal can be converted to gas. This quantity of coal for a year's operation is about that of the largest coal output in the early 1970s of any mine in the United States. In 1976, the

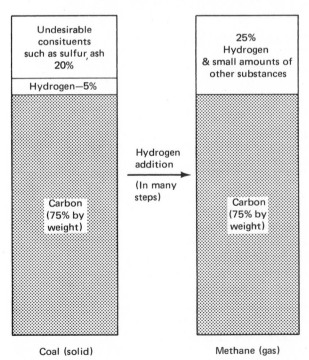

Figure 14-11 *Coal gasification replaces undesirable constituents of coal with hydrogen.*

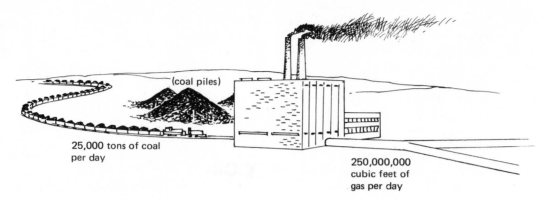

(coal piles)

25,000 tons of coal
per day

250,000,000
cubic feet of
gas per day

Figure 14–12   *Inputs and outputs of a coal gasification plant.*

cost of a coal gasification plant producing 250 million cubic feet per day of
pipeline quality gas was estimated at $1 billion.

Each coal gasification plant of this size is expected to consume 10–15,000
acre-feet (1235–1850 hectare-meters) of water per year in the process of
producing the gas. An acre-foot of water equals a volume one acre in area and
one foot in height, so that 10,000 acre-feet would be a 10,000 acre area (about
16 square miles) one foot deep with water.

All plants presently planned would use the Lurgi gasification process, a
process presently being used in Sasol, South Africa, for gas production. All
plants also plan to convert the carbon monoxide and hydrogen produced by
Lurgi gasification to methane to produce a pipeline-quality gas. Nearly a dozen
such plants were in the planning stages as the decade of the 1970s was drawing
to a close.

There is another method of coal gasification, but it is not carried on in a
factory. Designed to exploit coal seams that cannot be mined profitably, the
process takes place underground and should not disrupt the land extensively.
Underground coal gasification produces gas of low heating value, which may
be upgraded to a high-Btu pipeline quality gas by chemical processes in a plant
on the surface.

All of the processes being studied for underground gasification inject air
into the coal seam. Two methods appear particularly promising. These are the
methods which are known as linked vertical wells and the packed bed process.
Although neither have been implemented commercially at this time, their
potential for profitable production of gas appears to be excellent.

The production of gas by underground gasification has the potential to
multiply our coal reserves, possibly by a factor of three, according to the
Department of Energy. Since these reserves were uneconomic to exploit for
mining, their use for gas production gives us a new resource base. No miners
are needed, so that none of the health and safety problems associated with

mining will be encountered. No solid wastes are produced, and the surface disturbance is likely to be less than would occur with strip mining. Investments are lower as well. But surface water and underground waters may be polluted in the process and the surface will be expected to subside. The extent of subsidence is unknown, but it is thought, since the burns will occur several hundred feet below the surface, that a foot of subsidence might occur for each ten feet of coal seam removed.

Oil From Coal (Coal Liquefaction)

There are three basic methods to manufacture liquid hydrocarbons from coal. The catalytic method uses various catalysts to convert the gases from the initial step of coal gasification into liquids. That is, instead of converting carbon monoxide and hydrogen to methane, the gases are treated in such a way as to produce liquid hydrocarbons. The method of pyrolysis, in contrast, involves heating the coal to drive off hydrocarbon liquids, leaving a solid of nearly pure carbon called char. The last method utilizes hydrogen donation to enrich the carbon in coal with hydrogen.

The catalytic method is the oldest of the processes. A commercial coal-to-oil plant in Sasol, South Africa, using the catalytic method has been in operation for more than 20 years. The plant converts 3500 tons (3175 metric tons) per day of coal to about 30 million cubic feet (0.85 million cubic meters) of gas. The gas is then converted to a liquid. The principal products of the plant are gasoline and other motor fuels along with a high-Btu gas suitable as a substitute for natural gas. The first step in the production of the oil is the step we referred to earlier as the Lurgi process for making synthetic natural gas from coal.

Unfortunately, the process is quite costly. It is likely to produce a synthetic oil that costs perhaps 50% to 100% more than the 1980 cost of crude oil. An industry that involves itself in the manufacture of oil from coal by one of these processes must be betting that OPEC will raise the price of oil quite steadily through the next decade.

Pyrolysis of coal is accomplished by heating coal in the absence of oxygen. Coal is generally about 1 part hydrogen to 16 parts of carbon by weight. To produce a liquid hydrocarbon, such as fuel oil, requires a shift in the hydrogen-carbon ratio to about 1 part hydrogen to 6 parts carbon. The intense heat of pyrolysis drives off hydrocarbon gases, leaving behind a hydrocarbon liquid and char. The char is basically all carbon, the hydrogen in the coal having been shifted to the liquid and to the gases that were produced. Several pyrolysis processes have been developed, but none appears competitive with the leading coal-to-liquid process called <u>solvent refining of coal.</u>

Solvent refining of coal (abbreviated SRC) is an example of a process in which hydrogen is donated to coal to increase the hydrogen-to-carbon ratio. The product can be liquid or solid; it is a solid at room temperature and is low in sulfur and ash. The material, which melts at 350 to 375 °F (177 to 191 °C), may find use as the boiler fuel for electric power plants. In July 1978, the United States government announced that Morgantown, West Virginia, had been picked for the first commerical coal-to-oil plant in the U.S., a joint venture of the U.S. Department of Energy and Gulf Oil that uses Solvent Refining of Coal.

Questions

1. What is *primary recovery?* What are two *secondary recovery* methods? What level of recovery can be achieved by secondary recovery methods?
2. What are the two basic elements in oil? What other elements or substances may be present in the oil?
3. Why is crude oil distilled or refined? What are some of the substances into which it is separated?
4. In terms of oil, what are "proved reserves"? How do these differ from "undiscovered reserves"? To what extent do experts, agencies, and companies agree on the oil remaining to be discovered?
5. Hubbert likens a lake with intensive fishing to a region with petroleum deposits and intensive exploration. Finish the analogy.
6. What portion of the oil that we use is consumed in gasoline combustion? What are some other uses of oil?
7. What portion of the oil used worldwide is consumed in the United States? Why is this portion falling?
8. What part of our oil consumption was imported in 1976? What portion of this was from the Middle East?
9. What is the meaning of the term "participation" as applied to North Sea oil?
10. Why was the price of natural gas controlled by federal law in the first place? What were the effects of price control as seen by the oil companies?
11. How is natural gas imported from other continents to the U.S.? Is the procedure safe? What precautions should be taken?
12. What two unusual sources of oil exist in North America and in what estimated quantities? Where are these deposits located? How will the price of OPEC oil affect whether they will be recovered?
13. What are some of the untapped resources of natural gas or synthetic natural gas? What two factors generally are responsible for the fact that we have not used these resources in quantity?
14. Cartels set prices. How do they manage to get their prices? Use the case of oil.
15. How do competitors with the cartel benefit? What risk do they face when they develop a new and costly oil field?

16. Explain how the large quantity of oil we import each year leads to a fall in the value of the dollar and hence to inflation.
17. The idea to make gas from coal is not new. How old is it? What was one of the names given to gas from coal in those days?
18. How is low-Btu gas produced?
19. How many tons of coal would be required each year by a coal gasification plant which produced 250 million cubic feet of gas per day? Compare this to the output from the largest coal mine operating in the 1970s. Was this largest mine a surface mine or a deep mine?
20. Underground coal gasification is not expected to decrease our estimated reserves of coal. Why?
21. Coal gasification and liquefaction processes have been operating overseas for many years. Where? Is this the first place these processes were used? If not, where were they used before?

Further Reading

General

Energy: The Next Twenty Years, Report by a Study Group Sponsored by the Ford Foundation and Administered by Resources for the Future; Ballinger Publishing, Cambridge, MA, 1979.

Oil, coal, nuclear power, solar, projections, economics—all are here. Thoroughly documented and comprehensive; one of the most complete studies of the 1970s.

International Petroleum Encyclopedia, see latest.

For the facts as the industry sees them, on oil production, reserves, locations around the world, this is the document. Many tables and charts.

Kash, D., *et al.*, *Energy Alternatives: A Comparative Analysis*, for the President's Council on Environmental Quality. Published as *Our Energy Future*, University of Oklahoma Press, Norman, OK, 1976.

As thorough a description of energy technology and resources as can be found. Language is largely non-technical and new words applying to the technologies are defined as they are used.

Metz, W., "Mexico: The Premier Oil Discovery in the Western Hemisphere", *Science* **202**, 1262 (22 December 1978).

"Mexico's Reluctant Oil Boom," *Business Week*, p. 64 (15 January 1979).

These two articles are both non-technical.

"The Natural Gas Shortage", *Business Week*, p. 66 (27 September 1976).

A very readable and relatively unbiased article describing the background to the natural gas shortage.

Project Interdependence: U.S. and World Energy Outlook through 1990, a report by the Congressional Research Service, November 1977, 939 pages.

This authoritative report contains more information on the energy subject than most of us have time to consume. If you want to become an instant expert on some energy subject, say fusion or oil, the material is here, *and* it is not terribly complicated by equations and technical jargon.

Oil Shale and Tar Sands

Fletcher, K. and M. Baldwin, eds., *A Scientific and Policy Review of the Final Environmental Impact Statement for the Prototype Oil Shale Leasing Program of the Department of the Interior,* The Institute of Ecology, 1973.

Understandable and worth reading for the environmentalist viewpoint.

Maugh, T., "Tar Sands: A New Fuels Industry Takes Shape," *Science* **199,** 756 (17 February 1978).

Oriented toward the science, technology, policy aspects of tar sands exploitation.

"Oil from Shale Is Still a Distant Hope," *Business Week,* p. 126 (23 April 1979).

Non-technical, oriented toward problems of oil companies who are interested in developing the shale oil resource.

"Oil Shale and The Environment," Office of Research and Development, U.S. EPA, EPA 600/9-77-033, October 1977. Available from National Technical Information Service, Springfield, Virginia.

Recommended: non-technical and profusely illustrated.

Tippee, R., "Tar Sands, Heavy Oil Push Building Rapidly in Canada," *Oil and Gas Journal* (30 January 1978).

Oriented toward the commercial, economic aspects of the tar sands effort.

Chapter 15

Nuclear Power

INTRODUCTION

It began under a cloud—a mushroom cloud rising over a Japanese city. The reputation acquired by that beginning has proved very difficult to change. Nuclear power remains suspect in the minds of many despite the continuing assurances of its safety.

President Eisenhower foresaw the peaceful use of atomic energy to generate electric power. Under his program of "Atoms for Peace," private corporations were given the privilege of owning nuclear reactors. Industries in joint projects with the Atomic Energy Commission then began the development of nuclear power reactors.

In 1956, a boiling water nuclear reactor began experimental production of electricity at the Argonne National Laboratory. In the following year a pressurized water reactor at Shippingport, Pennsylvania, began delivering 60 megawatts of electrical power. The size of new plants increased as operating experience grew. By 1963, several nuclear plants were delivering 200 megawatts of electric power and commitments had been made for the larger Oyster Creek plant in New Jersey and the Nine Mile Point plant in New York. These new plants, when they went on line in 1969, delivered up to 600 megawatts.

A tide of commitments to nuclear plants began in 1965. There were seven orders that year, 20 the following year, and 30 in 1967. The rate then began to fluctuate but continued strong into the 1970s. By mid 1974, more than 200 nuclear plants had been ordered, and most of the new plants on order were to provide 1000 megawatts or more of electric power.

Between 1974 and 1979, however, only 11 new orders were placed, and in 1978, no electric utility in the U.S. ordered a nuclear reactor. The tide was receding (Fig. 15-1). The course of development of nuclear power has been slowed by debate in the scientific community and in Congress about the possible levels of radiation released from the plants. Scientists have also argued over the reliability of the structure in which the fission process is allowed to proceed. Several states have had referendums on the further development of nuclear electric power in their jurisdictions. Debate continues on the use of plutonium as a nuclear fuel to extend fuel supplies. Is it the cloud at the beginning that has limited the acceptance of nuclear electric power? Is it a problem of public misunderstanding of its actual safety and its potential for effectiveness? Or is there more substance to the debate than poor public relations?

To understand the debate and controversy that surrounds nuclear power will require discussion of the nuclear power system as it has evolved to this time. The principles of operation and the reasons for engineering design features require explanation. In addition, the fuel cycle, which begins with the mining of uranium and may end with the disposal of radioactive wastes, must be fully explored. With such principles and facts in mind, it becomes possible to examine the statements of opponents to

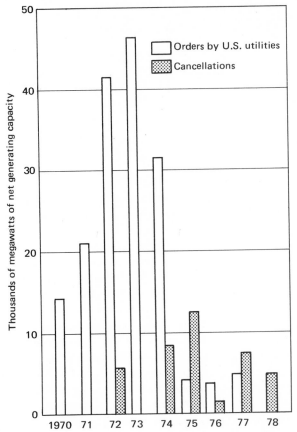

Figure 15-1 *Orders and cancellations for nuclear generating facilities in the United States between 1970 and 1978. In the last half of the 1970s, new orders for nuclear reactors from U.S. utilities fell sharply. No new orders were placed in 1978 and 1979. At the same time, a number of utilities began to cancel orders previously made. (*Source: Business Week, *25 December 1978.)*

nuclear power to gauge their validity. To discuss these statements in any other way would deny readers the right to draw their own conclusions.

LIGHT WATER NUCLEAR REACTORS

How They Operate

There are two fundamental reactor types in common use in the United States and elsewhere around the world, although others exist or are in the design stage. These two are the boiling water reactor (BWR) and the pressurized water reactor (PWR). Their technology is relatively proven and reliable, and they promise to be the work horses of nuclear electric generation for the next several decades. Other reactors are being developed and investigated, but their contributions to our electric energy supply are likely to be small in the near future.

The boiling water reactor (BWR) and pressurized water reactor (PWR) are actually very similar, the PWR having one additional cycle of operation. Both reactor types generate heat energy in a "core" where uranium–235 undergoes controlled fission.

Long cylindrical metal rods containing enriched uranium dioxide are arranged vertically in the interior of the core (Fig. 15–2). From the fission which takes place within the fuel rods comes the energy to heat the water flowing through the core (Fig. 15–3). Since neutrons and gamma rays are produced by the fission process, the core must be heavily shielded to prevent worker exposure to these particles and rays. Layers of concrete and iron or steel are used to encase the core and prevent the escape of these radiations. About 3% of the uranium in the rod is uranium–235. The remainder of the uranium is uranium–238, a substance which does not undergo fission.

The core of the nuclear reactor functions in much the same way as the boiler of a fossil fuel power plant. In the boiler, coal or oil is burned and the heat released is transferred to the water, which is boiled to produce steam. The steam under high pressure is then used to turn the turbine and generate electricity. In the core of the nuclear reactor, uranium–235 atoms are struck by neutrons and break apart (fission) into fragments (known as fission products) with the release of heat energy and more neutrons. The heat energy from fission is taken up by the water which circulates through the core.

As the reactor operates over a period of time, the amount of uranium–235 within a rod falls and

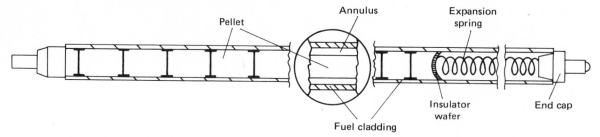

Figure 15-2   *Cutaway of oxide fuel rod for a commercial LWR reactor. (Adapted from Final Environmental Statement, Energy Research and Development Administration, April 1976.)*

the fragments or fission products build up. The fission products, however, capture some of the neutrons that sustain the chain reaction, and the fission process slows down. After about three years, when the rod is no longer an efficient source of heat because of a low rate of fission, the rod is replaced by a fresh fuel rod. The spent fuel rod is now relatively low in uranium-235 and rich in the fragments or fission products.

One of the two reactor types we are discussing, the boiling water reactor (BWR), functions in precisely the same way as the boiler in the fossil fuel power plant; that is, the water in the core is converted directly to steam which will be used to turn a turbine (Fig. 15-4).

The other reactor type, the pressurized water reactor (PWR), does not convert the water in the core to steam. An enormous pressure is maintained on the water and prevents it from boiling. All the heat energy from fission goes into raising the temperature of the liquid water. The heated water that leaves the core passes through a steam generator where it is cooled; it is then returned to the core to be heated again. In the steam generator/heat exchanger, the heat from the water in the first loop is transferred to water flowing in a secondary loop. The water in the secondary loop is boiled to steam as it leaves the steam generator under pressure. It is this steam in the second loop that is used to turn a turbine and to generate electricity (Fig. 15-5).

Thus, in the BWR, steam is generated in the core itself and is used directly to turn a turbine. In

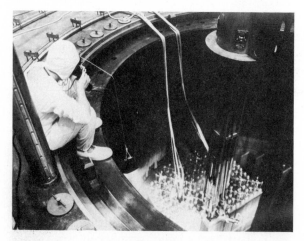

Figure 15-3 *Fuel loading operation. A technician monitors the fuel loading operation at Unit No. 1 of the Calvert Cliffs Nuclear Power Plant, owned by the Baltimore Gas & Electric Company. Each fuel bundle— of a total 217 bundles—is being positioned by an automatic fuel handling machine, shown extending down to the upper core level. (Combustion Engineering, Inc.)*

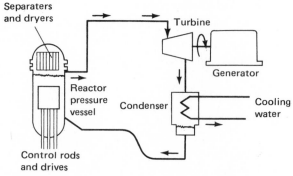

Figure 15-4 *Boiling water reactor power plant.*

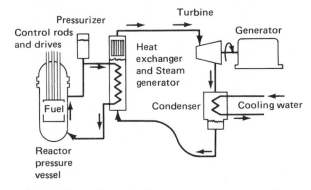

Figure 15-5 *Pressurized water reactor power plant.*

contrast, in the PWR, steam is produced by heat exchange in a steam generator and flows only in the secondary loop where it turns a turbine.

The steam that leaves the turbine lacks the energy to generate more electric power. But it is still quite hot. The heat from this hot steam is transferred to a flow of cold water (called cooling water) in a heat exchanger. The water that left the turbine as steam is condensed to liquid water in this heat exchanger. The condensed water circulates back to be converted to steam once again. In the PWR, the conversion to steam takes place in the steam generator. In the BWR, it occurs in the core itself.

The use of the cooling water and its effects on aquatic life are discussed in the section on thermal pollution (see Chapter 21).

Radioactive Wastes and Radiation Releases from Light Water Reactors

Just as the fossil fuel power plant emits pollutants, so also does the nuclear power plant. Whereas the emissions from fossil fuel power plants include the commonly known chemical air pollutants, the emissions from the nuclear power plant are radioactive elements. These radioactive elements stem almost entirely from the fission process.

The elements which are products of the fission process include the radioactive forms of krypton, xenon, cesium, strontium, and iodine. These are among the fragments that came from the fission of uranium–235. Although the fission process is supposed to be confined to the interior of the fuel rod, defects in the rod's metal shell allow fission products to escape into the water in the core. Fission products, then, appear in water in the primary loop.

Other radioelements show up in the water circulating through the core. These radioelements are created when neutrons are captured in the nuclei of the metals that make up the shell of the fuel rod. When the neutron is captured, the atom becomes another form (an isotope) of the same element. This new form is likely to be radioactive.

Radioactive forms of iron, cobalt, manganese, and zinc may all be created as a result of neutrons being captured in the nuclei of the metals in the shell of the fuel rod. These radioelements are all found in the water that circulates through the core. Corrosion of the metals carries them into solution.

In summary, the water that cools the core contains in it not only fission products but also radioactive metals derived from the metals in the shell of the fuel rod. If this water should leak from a seal or joint in the core, it must be considered to be a radioactive waste, and it must be cautiously handled and disposed of.

Solid, liquid, and gaseous wastes are produced by the BWR and the PWR. Gaseous wastes from the PWR include radioactive forms of the gases krypton and xenon, both of which were produced in the fission of the uranium. Gaseous wastes from the BWR include a radioactive form of nitrogen. Both of the two reactor types discharge their gaseous wastes through tall stacks, which allow the substances to be widely dispersed. The total amount of radioactivity discharged from the stacks of the PWR is far less than that from the BWR.

Liquid wastes stem from leaks and from drains in the system; the wastes of concern are those from the primary loop, where water circulates through the core. Although the liquid wastes are, for all practical purposes, water, radioactive substances are dissolved or suspended in the water. When liquid wastes are treated to remove these substances, solid wastes are created.

The liquid wastes are treated by evaporation followed by condensation or by passing the water through filters. The water once treated may be diluted in the condenser cooling water and discharged to the body of water which furnishes the cooling water. To avoid such discharges, the water can be further treated by passing it through ion-exchange resins; it is made so pure by this treatment that it can be reused in the primary loop that passes through the core.

Solid radioactive wastes are created by these treatment steps. The slurry of solids left after evaporation, filtering, and passing through the ion-exchange resins are all radioactive. These items are encased in concrete within 55-gallon metal drums and shipped to special sites for burial.

The burial of the low level solid wastes is conducted at federally licensed burial grounds. The grounds, though they are commercially operated, are supervised by the states. There have been six such sites in operation, located in Illinois, Kentucky, Nevada, New York, South Carolina, and Washington.

Although originally six sites had been established, only three sites were in operation through 1979. The Sheffield site in Illinois and the West Valley site in New York had long been closed. The Maxey Flats site in Kentucky had been closed in 1977 after the discovery that groundwater nearby had been contaminated by radioactive substances. In 1979, the state of Washington closed its Richland site and Nevada ordered its burial ground at Beatty shut as well, both closings resulting from improper practices at the sites. Whether the closings are final, even if practices improve, depends to a degree on political pressures.

How well do nuclear plants do in preventing routine radioactive releases to the environment? When one considers the potential for such releases, the answer is: "Exceptionally well." Most plants are meeting the rules set by the federal government on releases, and these rules in the last decade have become very stringent. The rules call for routine releases of less than 5% of the natural background radiation that most people receive. It should be added that the guidelines were not always so strict and that criticism by scientists helped to bring the guidelines down.

In addition to the spent fuel there is one other type of waste from the nuclear power plant; it is the plant itself. No major plant has yet been "decommissioned." If a plant is to be dismantled and hauled away, the cost could run to $100 million for the modern 1000 megawatt plant or about 10% of its initial cost, a cost not yet taken into account by utilities. The alternative is to seal and guard the site and reactor for 100 years or more.

Safety of Light Water Reactors

Routine releases of radioactivity, then, are negligible. That is, on a day-to-day basis, a resident nearby a nuclear power plant is exposed to radiation that is barely more than natural background levels. Nonetheless, criticism is still directed at the safety of nuclear power plants. The day-to-day emissions, however, are not the target of this criticism. It is the radiation release that could occur if a serious accident or sabotage took place within the plant. One particular form of accident is of more concern than all others.

We recall that water circulates through the reactor core to withdraw the heat of fission. If the circulating water is suddenly lost, the heat would accumulate rapidly, and the contents of the core could melt. In the jargon of the nuclear industry, this is called a "loss-of-coolant accident." This kind of accident might occur if a steam or water line that carried water into or out of the core should break. The high pressures in the line would probably cause most of the water or steam to be rapidly ejected from the core. The temperature in the core would rise rapidly, melting the fuel rods. Radioactive xenon and krypton gases, products of uranium fission, would be released from the melted rods and enter the building through the break in the pipe.

The molten mass would fall to the floor of the reactor vessel, perhaps melt through the floor and fall into the containment room that surrounds the reactor core. More radioactive gases would be released into the containment area, but the containment room is designed so that the leakage of radioactivity to the outside will be very small. A water spray system will condense the steam in the containment room and reduce the temperature there.

The reactor core is also equipped with a spray system. The Emergency Core Cooling System is designed to inject water automatically into the core if a break in the coolant line should occur; its purpose is to cool the reactor core and prevent melting. However, very intense heat would probably develop within the core after a loss-of-coolant accident, and the heat would produce steam at a very high pressure. The pressure could be so high that water sprayed into the core would be blocked from even entering. The reliability of the emergency core cooling system was finally tested in 1979 at a test reactor, more than 20 years after the first nuclear power plants went on line. The initial test did prove successful, but it was run at less than full-power operation.

(The loss of coolant is regarded as the most serious accident that could occur at a light water reactor.) The severity of the accident, in terms if its impact on the population, would depend on how effectively the emergency core cooling system functions. If a meltdown was not prevented, the impact would then depend on how effectively the containment room prevented radioactivity from entering the environment. And if radiation escaped, the impact would depend upon the wind speed and direction and whether a populated area was downwind from the plant.

Because reactor safety is so prominent an issue in the nuclear power debate, the Atomic Energy Commission[1] directed that an extensive study (WASH–1400) of the problem be undertaken. Professor Norman Rasmussen of M.I.T. conducted the study which cost 4 million dollars, involved more than 100 people, and took 3 years to complete. The study concluded that the odds against the worst-case accident occurring were astronomically large. The worst-case accident projected an estimate of about 3000 early deaths and 14 billion dollars in property damage due to contamination. Cancers occurring later due to the event might number 1500 per year. The odds against such an event, however, were estimated at 10 million to one, given that 100 reactors are in operation.

At the same time, the probability of a meltdown is not so small. With 100 light water reactors in operation (a likely figure, by the time you read this, Fig. 15–7), the probability of a meltdown is 1 in 200 per year. Thus, although the worst-case accident is estimated only as a remote possibility, the meltdown alone is not such an unlikely event. The Rasmussen study essentially attempts to evaluate the probability that a meltdown will turn into a serious, life-threatening accident. The study concludes that the safety features engineered into the plant are very likely to prevent such serious consequences.

The Rasmussen study has had serious criticisms, however. One of the most scathing was that of a panel of 21 scientists, economists, and political scientists assembled for a study of nuclear power by the Ford Foundation. Their 400-page report, issued as a book, *Nuclear Power, Issues and Choices*, looks at many parts of the nuclear electric system. The panel felt that the Rasmussen study seriously underestimated the uncertainties of reactor safety. Flaws in methodology, they asserted, could make some of the estimates of the probability of accidents low by a factor of 500.[2]

Another panel, this one composed only of nuclear scientists and created in 1977 by the Nuclear Regulatory Commission itself, also reviewed the

---

1 Now split into an office within the Department of Energy and an office known as the Nuclear Regulatory Commission.

2 *Nuclear Power, Issues and Choices*, the Nuclear Energy Policy Study Group, S. Keeney, Jr., Chairman, Ballinger Publishing, Cambridge, MA, 1977.

Middletown, Pennsylvania, the Three Mile Island Nuclear Power Station: On March 28, 1979 at 36 seconds past the hour of 4:00 am, two pumps in the secondary cooling water loop of Reactor 2 stopped functioning. This was the first in a series of events that eventually resulted in a partial meltdown of the reactor core and an accident generally accepted to be the most serious at a commercial nuclear power reactor in the United States.

No one has died as a direct result of the Three Mile Island (TMI) accident, and though the eventual effects cannot be known at this time, the expected number of additional cancers has been placed between 0 and 10. Though there have been no deaths, the accident at Three Mile Island came very close to being much worse.

The accident at Three Mile Island hardly resembles the many scenarios in the Rasmussen report. In that report, the accidents of interest were the major loss of cooling accidents, resulting from large equipment failures. In contrast, the events which led to this accident were (1) the malfunctioning of two relatively insignificant pumps, a common enough occurrence and (2) a sticking automatic valve in the cooling system, which could have been easily closed if it had been noticed in time. The "accident" at Three Mile Island was really a collection of small events–small equipment failures and human errors, and it led nearly to a complete core meltdown. Releases of significant and potentially lethal amounts of radioactive material could have occurred.

The role of human operators at Three Mile Island was terribly important. Even though the operators behaved in a manner that was largely consistent with their training, their actions nevertheless made the accident far worse than it needed to be. It is clear that operators must be better trained in how the plant operates, as opposed to learning rules of thumb for responses to certain situations. Control room design and information display can also be improved substantially. During the first eight minutes of the accident, the operators were subjected to 100 different alarms for bells, buzzers, and lights, and one of the critical indicator lights was virtually out of sight, being on the back of a control room panel, 7½ feet above the floor.

The response of local, state and federal authorities to the incident revealed a most disturbing lack of preparation for accidents at nuclear plants. Optimistic reports issued by the utility and the accounts of the representatives of the Nuclear Regulatory Commission·were in conflict. The public was confused by this contradictory information and not informed. No usable plan for evacuating the public existed, although many people left voluntarily and children and pregnant women were advised to leave. We have also learned from the event that a federally maintained team, which could respond to a nuclear accident may be required.

There remains still the question of clean-up and decontamination of the crippled plant. As of December 1979, the containment building was continuing to fill with highly radioactive water issuing at a rate of 400 gallons per day from a leak in the primary loop. Tanks in the auxilliary building hold another 400,000 gallons of less contaminated, but still radioactive water. The containment building is filled with radioactive krypton-85 gas that is threatening to breach the seals that hold the gas in.

Figure 15–6 *The Cooling Towers and Auxiliary Buildings at the Three Mile Island Nuclear Plant. (United Press International photo.)*

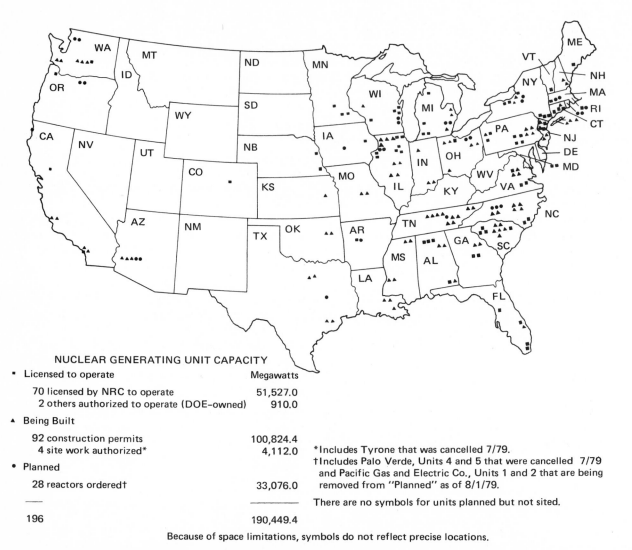

NUCLEAR GENERATING UNIT CAPACITY

▪ Licensed to operate   Megawatts

   70 licensed by NRC to operate   51,527.0
   2 others authorized to operate (DOE–owned)   910.0

▲ Being Built

   92 construction permits   100,824.4
   4 site work authorized*   4,112.0

● Planned

   28 reactors ordered†   33,076.0

   ———   ————————

   196   190,449.4

*Includes Tyrone that was cancelled 7/79.
†Includes Palo Verde, Units 4 and 5 that were cancelled 7/79 and Pacific Gas and Electric Co., Units 1 and 2 that are being removed from "Planned" as of 8/1/79.

There are no symbols for units planned but not sited.

Because of space limitations, symbols do not reflect precise locations.

Figure 15-7 *Nuclear power reactors in the United States. (Source: Department of Energy, 31 October 1979.)*

Rasmussen report and found it lacking in a number of respects. In particular, the estimates of risk set forth in the report were criticized as being subject to much larger errors than the report implied, errors that could not be reduced because of a fundamental lack of data. That is, whereas the report provided estimates of the risk of various accidents, the panel felt the risks could not be stated nearly so precisely.[3] The report, once the cornerstone of arguments for

3 "Risk Assessment Review Group Report to the Nuclear Regulatory Commission," Prepared for the U.S. Nuclear Regulatory Commission, NUREG/CR-0400, September 1978.

the safety of nuclear plants, has been reduced in status to merely a piece of evidence, to be read very carefully for flaws.

The Rasmussen study when it appeared had silenced many doubters of the safety of nuclear power plants, and until 1979, the issue of nuclear safety lay buried in the jumble of questions which had piled up about nuclear power. Accompanying the safety question were such concerns as thermal pollution, radioactive waste disposal, fuel reprocessing, spread of atomic weapons, and transport of nuclear materials. In 1979, however, after a nuclear reactor near Harrisburg, Pennsylvania, experienced an accident of a character not thought possible, nuclear safety came bubbling to the top of the nation's and the world's nuclear concerns.

PROBLEMS IN THE NUCLEAR FUEL CYCLE

From Mine to the Power Plant

There is more to a system of nuclear electric generation than a single power plant. There is a cycle that begins with uranium mining and proceeds through use of uranium at the plant to the recycling and/or disposal of the fuel rods. We find serious safety questions focussed at a number of points in the nuclear fuel cycle. Figure 15–8 illustrates the operations in the nuclear fuel cycle along with the transportation of products from one operation to another.

The first operation in the cycle is the actual mining and milling of uranium ore followed by the extraction of an oxide of uranium called "yellowcake." In the past the tailings from the processing of uranium ore, although a potential hazard, have been simply dumped on the ground at older mining operations. (See Chapter 27.)

The oxide of uranium consists of two types of uranium. There is uranium–235 and uranium–238, both forms of uranium except that an atom of uranium–238 has three more neutrons in its nucleus. Only about 0.7% of the uranium is present as uranium–235. The remainder is uranium–238. It is the fission (splitting) of uranium–235 that provides the heat energy in the core of the BWR and PWR. Since the amount of uranium–235 is so low, a step is needed in which the concentration of uranium–235 is increased. This step is called "enrichment," and the process is known as "Gaseous Diffusion." In this step, the level of uranium–235 is increased to 3% of the total uranium present.

The enriched uranium is then formed into fuel rods at a fuel fabrication plant. The fuel rods are shipped in turn to the power plant where, as we

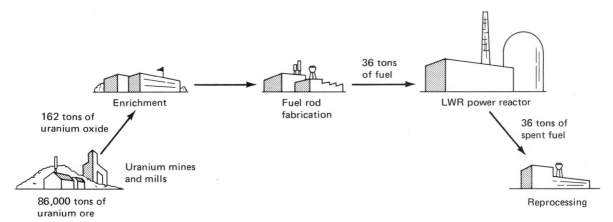

Figure 15–8   *The nuclear fuel cycle. The numbers on the diagram are the approximate annual requirements for one 1000 megawatt (electrical) pressurized water reactor.*

discussed earlier, they produce heat in the core of the reactor. When fission products have built up in the rods to a level that slows the chain reaction, the rods are removed from the core of the nuclear power plant. Such spent rods were originally intended to go to a fuel reprocessing plant where remaining fuel resources could be recovered. At the moment, however, the rods are simply piling up.

Fuel Reprocessing and Spent Fuel Rod Storage

The most serious problems in the fuel cycle begin after the spent fuel rods have been removed from the core of the power plant. Because there are still quantities of reusable fuel in the rods, the advocates of nuclear power have viewed the step of fuel reprocessing as a crucial one. However, there are also hazardous radioactive elements, the fission products, in the rods.

Of the reusable fuels, there is first the uranium–235 that has not fissioned and can still be used if recovered. Another substance present in the spent rod can also be used as a nuclear fuel. Plutonium, which was not present in the rod initially, was "bred" from uranium–238 during the process of fission. Uranium–238, you recall, does not participate in the fission process even though it is the most abundant element in the rod. A small but significant portion of the uranium–238 is converted to plutonium, however, during the fission process. The fuel reprocessing plant is supposed to separate the useful uranium–235 and the plutonium from the hazardous and unneeded fission products (Fig. 15–9).

Through most of the 1970s, no reprocessing plant was operating in the United States. The plant at West Valley, New York, was closed for repairs at the beginning of the decade and never reopened. The G.E. reprocessing plant at Morris Plains, Illinois, scheduled to open in the early 1970s, never opened its doors because of design problems. And the General Nuclear Services Plant at Barnwell,

South Carolina, was unable to meet its licensing requirements.

Storage at the power plant   Although the fuel rod when it entered the core was not particularly hazardous, the spent fuel rod contains the products from uranium fission. One might refer to the intensely radioactive products of fission from the "burning" of uranium as the "ashes" of the nuclear fuel cycle. Unlike the ashes from coal, these substances emit lethal quantities of radiation. Because they emit dangerous radiation and are at high temperatures due to the radiation, the rods must be unloaded from the reactor core through an underwater canal. Exposure of plant personnel to radiation is minimized by handling the rods underwater. About one-third of the rods in the core are removed and replaced each year.

The spent rods are moved in the underwater canal to a storage basin also filled with water. In

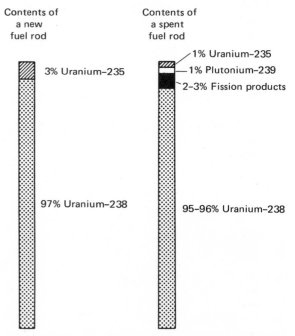

Figure 15–9   *New and spent fuel rods. Plutonium-239 is created when neutrons are captured in the nucleus of uranium-238. This capture process creates uranium-239, which decays in two steps to plutonium-239.*

the first few months of storage, much of the fast, initial radioactive decay within the rods takes place. By the end of these several months of storage, most of the dangerous iodine–131 has disappeared. The rods can then be placed more safely into heavily shielded canisters for transport to a fuel reprocessing plant (Fig. 15–10).

As we pointed out a moment ago, there were no operating fuel reprocessing plants in the United States through most of the 1970s. Thus, spent fuel rods have been accumulating at power plants for some time. By 1976, some reactors were already in need of additional storage capacity to proceed with refueling on schedule. Recognizing the need for a method to handle spent fuel, utilities are now building at their plants very large water-filled basins which can store as much as 15 years' accumulation of spent fuel rods.

In 1978, the Department of Energy finally began to come to grips with the issue of the spent fuel rods that were piling up at the power plants. The pile-up of spent fuel rods at power plants

Figure 15–10 *Crash test of a spent fuel cask. A rocket propelled railcar carrying a 74-ton spent fuel cask smashes into a concrete wall at a speed of 81.4 mph. The impact demolished the front of the railcar, turning it into a mass of twisted metal, but the cask was essentially undamaged. Such crashes are meant to duplicate the worst possible accident conditions that could occur in order to predict how well containers will survive. (Sandia Laboratories.)*

because of the lack of reprocessing requires some action if the reactors are to continue to operate. In the near future, it seems likely that the U.S. Department of Energy will rely on *Away From Reactor* (AFR) storage as the means to relieve the space squeeze on nuclear power plants. The concept is to transport the spent rods to several centrally located sites where they will again be stored intact in swimming pool-like vaults. There they would remain until the government can decide what to do next in the way of final disposal.

At the fuel reprocessing plant (West Valley days) In 1966, the first fuel reprocessing plant began operation at West Valley, New York, a town just south of Buffalo (Fig. 15–11). Privately owned by Nuclear Fuel Services, the plant processed spent fuel rods from light water nuclear power plants, separating the fission products from the reusable fuels, uranium and plutonium. When the plant closed in 1972 to modernize and expand, it had already processed 600 metric tons of spent fuel[4] (a metric ton is 1000 kilograms or 2200 pounds).

In 1976, Nuclear Fuel Services announced that it could not reopen its West Valley plant. It offered as reason for its action the additional expense of meeting new earthquake protection requirements. In fact, the motive for the decision to remain closed may have been the presence of a carbon-steel tank holding about 600,000 gallons (2.25 million liters) of high-level radioactive wastes. The tank is expected to leak. A tank of similar material in Hanford, Washington, eventually leaked its radioactive wastes because its steel corroded (Fig. 15–12).

No one is certain how to prevent the tank at West Valley from leaking. An estimate in the mid-1970s for the cost to transfer the wastes safely to a stainless steel tank was 600 million dollars. In contrast to the tank that leaked at Hanford, the

---

4 About 60% of the material received was waste from nuclear weapons manufacture. The material was supplied to the plant as an incentive to open the plant, since spent fuel from power reactors was not available in sufficient quantity at that time to make the operation profitable.

Figure 15–11   *The Nuclear Fuel Services, Inc. plant at West Valley, NY before it closed in 1972. (NYT pictures)*

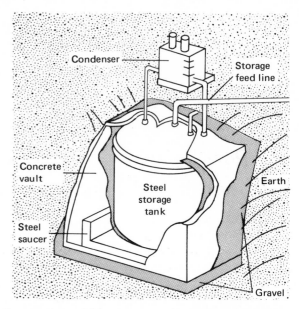

Figure 15–12   *The wastes stored at West Valley. When Nuclear Fuel Services closed its doors at its Fuel Reprocessing Plant at West Valley, it left behind a carbon steel tank containing high level radioactive wastes. At Hanford, Washington, several carbon steel tanks holding high level radioactive wastes have already corroded and leaked. Although the tanks at Hanford were single-walled and the tank at West Valley is double-walled, the potential for leakage at West Valley is still great. No one wants the responsibility for the wastes, not Nuclear Fuel Services, the Federal Government or New York State.*

West Valley tank does have double walls. Nevertheless, corrosion is still seen as certain.

The leak of high-level wastes "waiting to happen" at West Valley is one result of its six years of operation. Other activities at West Valley do not bolster our confidence in industry's ability to handle nuclear wastes. In 1969, a group of 42 fuel rods at West Valley broke open and could not be processed. The rods were sealed in concrete only and buried in a hole 50 feet deep. If the West Valley plant is ever to be decontaminated, the wastes will probably have to be dug up.

Although surveys of the environment around the plant found radiation doses to people to be very small, the control of radiation received by employees inside the plant was less successful. Apparently to save money in building the plant, the company planned repair by hand of broken equipment even though the broken machines were likely to be contaminated by radioactivity. The technology for remote repair of broken machines was not used

even though the methods used earlier at Hanford could have been employed.

This resulted in high doses of radiation to the regular employees of the plant. The "dirty jobs," the ones with the worst exposures, were given to unskilled labor hired off the streets of Buffalo. The unskilled workers were sent in to do a job, allowed to go to a maximum level of radiation exposure, and then taken off the job. For their daring or ignorance, they received half day's pay for as little as a few minutes of work.

Two other reprocessing plants have been built. The General Electric plant at Morris Plains, Illinois, never worked and so never opened. The Allied–General Nuclear Services[5] Plant at Barnwell, South Carolina, was never licensed to operate by the Department of Energy, though it is substantially complete.

_____

5 A partnership of General Atomic and Allied Chemical.

'REACTIONARY' SEABROOK DEMONSTRATORS

To the Editor:

The recent anti-nuclear-energy demonstration in Seabrook, N.H., should be placed in perspective. The animating rationale of the protest movement is based upon a rejection of progress and technology....

The anti-nuclear movement is doomed to failure, for it seeks to undermine the scientific and technological revolution which has created the modern world. Such a protest is unable to provide the guidelines necessary to confront the economic challenges of a thriving, growing America. The Seabrook demonstrators, unconcerned for economic and social progress, exhibit a reactionary mentality. They are temporary and inconsequential impedimenta to the tide of history and are destined to become a forgotten footnote to the constructive advancements of science in this century. For without scientific and technological progress, backed by economic power, the world's problems are insoluble.

Appeared in The New York Times, 6 July 1978

Alex A. Vardamis, Research Fellow, Program for Science and International Affairs Harvard University Cambridge, Mass., June 29, 1978

Who was Ned Ludd? And who were the Luddites?

Is Vardamis suggesting the demonstrators are modern-day Luddites?

Is the comparison a valid or invalid one? Defend your position. The following letter also deals with this topic.

NUCLEAR DEVELOPMENT PREREQUISITE

To the Editor:

Questioned as to the possibility of another blackout in New York this summer, Con Ed Chairman Charles Luce sensibly pointed out that in a complex system of machines, operated by fallible human beings, there can be no 100 percent guarantee.

His words should be carved in stone for the benefit of those who see nuclear power as the answer to our future energy needs....

The catalogue of hazards ranges from an admitted inability to guarantee that storage of nuclear wastes will not fatally pollute our land, water and atmosphere, to the very real possibility that proliferation of weapons-grade fuel will make possible the production of explosive devices...

If man is to survive on this earth, it will not be through his genius for technical innovation. It will be through his wisdom in choosing between those innovations that he can control and those that he cannot.

Appeared in The New York Times,
7 July 1978

George Dryfoos
Hastings-on-Hudson, June 18, 1978

The closing at West Valley, the failure at Morris Plains, and the delays at Barnwell have contributed to an uncertainty in the utility industry about the future of nuclear power. In the late 1970s, it was apparent that the electric industry's enthusiasm for nuclear power was decreasing; the number of orders for new reactors fell dramatically. No reactors were ordered in 1978. This decline in orders will mean a declining rate of new nuclear units going into service in the early to mid-1980s.

Arguments for and against reprocessing Three reasons have been put forward for the reprocessing of spent fuel. First it is thought that there "may be"

reduction of costs for nuclear fuel when the uranium and plutonium remaining in the spent fuel are purified and recycled. About 1% of the uranium in the spent fuel is uranium–235; there is about 3% in fresh fuel. Further, plutonium "bred" from uranium–238 during the fission process may have increased from virtually nothing to a bit less than 1% of the contents of the spent fuel rod.)

The reuse of the fissionable uranium–235 and plutonium in new fuel rods can decrease the need for fresh fuel, but there is controversy over how much savings this will make possible. Those who favor reprocessing estimate up to a 20% savings in electric power cost if spent fuel is reprocessed.

Other responsible scientists[6] estimate savings in electric costs on only 1% or 2% if reprocessing of spent fuel is carried out.

(The second reason cited for fuel reprocessing in the U.S. is to decrease dependence on foreign sources of supply.) If foreign suppliers have a large share of a nation's market, they could raise prices without worry or could put an embargo into effect in order to influence the nation's foreign policy. France and Germany, two industrial nations that have decided to reprocess nuclear fuel, have little uranium resources of their own. They also have relatively low amounts of other energy resources. Reprocessing makes these nations less dependent on uranium fuel imports.

(The third reason given for reprocessing is to reduce the difficulty of disposing of radioactive wastes. This reason, however, is vigorously debated, since the high level wastes will remain neatly packaged inside the fuel rods only so long as no reprocessing takes place.)

In the United States, the plans to proceed with reprocessing were brought to a standstill in the late 1970s. Probably one of the reasons for the delay was the need for road shipment of plutonium from the reprocessing plant to the plant manufacturing the fuel rods. If plutonium, the raw material for atomic bombs, were hihacked, the risks would be great. If spent fuel were not reprocessed there would be no plutonium recovery nor shipment.

Concerned that plutonium could be hijacked for nuclear weapons, U.S. diplomats have asked Britain, France, and Germany to reconsider their decisions to reprocess nuclear fuel. Yet Britain is building a reprocessing plant at Windscale in Cumbria. It is to have a capacity twice that of Barnwell. The British public is aroused, however, because of a nuclear accident at an earlier plant at Windscale in 1972. The British people have also protested a

decision to use the plant at Windscale to reprocess the spent nuclear fuel of Japan. The nation's energy authority was accused of making Britain the world's nuclear "dustbin," a British term for trash can. A number of delays in reprocessing have occurred as the British ponder their decision.

In Marcoule, France, an existing plant offers a reprocessing capacity of 1000 metric tons (MTU) per year. At La Hague, France, an 800-MTU plant has been built, which is scheduled to be expanded several times over. In Germany, a plant with a capacity of 1500 MTU per year has been planned for operation in 1984. The plant is to be located at Gorleben near the border with East Germany.

Unfortunately, France has contracted to sell reprocessing technology to Pakistan. In addition, West Germany has offered to sell reprocessing technology to Brazil. France and Germany have agreed under pressure not to export reprocessing technology in the future. Their agreements with Pakistan and Brazil may still be honored, but pressure from the U.S. is being applied to prevent the actual completion of these sales.

Why did the United States oppose these sales and work against the ownership of reprocessing plants by other countries? The answer lies in what reprocessing, specifically, the PUREX process does. (PUREX stands for Plutonium and Uranium Extraction; it was developed during World War II.)

(The process begins with the mechanical chopping of the fuel rods into fragments; the fuel is then dissolved in nitric acid (Fig. 15–13). Uranium and plutonium are extracted from the solution by mixing the solution with a solvent that selectively dissolves these elements. The reprocessing plant produces three separate products: (1) uranium, a portion of which is uranium–235, (2) plutonium, and (3) wastes (the fission products). The amount of plutonium that could be recovered annually from the spent fuel of a 1000 megawatt light water reactor would be about 250 kilograms (about 550 pounds) or enough to produce about 15 atom bombs per year.)

---

6 See *Nuclear Power, Issues and Choices*, Ballinger Publishing, Cambridge, MA, 1977.

DOES REPROCESSING PRODUCE WEAPONS-GRADE PLUTONIUM?

Conflicting views exist on the potential for plutonium recovered from spent fuel rods to be used in bombs. Bebbington * tells us,

> It is not generally appreciated that during the long exposure of fuel in a power reactor, there is an accumulation of plutonium isotopes other than plutonium-239, particularly plutonium-240, which makes it much more difficult to assemble a supercritical mass of plutonium without an inefficient premature explosion. Weapons-grade plutonium is made with much shorter exposure in the reactor.

In contrast, an anonymous official in Washington is quoted in <u>Science</u> † as saying, "For proliferation, you can't find anything worse than Purex. It was developed for bombs."

A member of the Nuclear Regulatory Commission also addressed the issue of the suitability for bombs of plutonium recovered from the spent fuel of power reactors. ‡

> There is an old notion, recently revived in certain quarters, that so-called "reactor-grade" plutonium is not suitable to the manufacture of nuclear weapons....

> The obvious intention here is to create the impression there is nothing to fear from separated plutonium from commercial power plants. This is not true...

> The fact is that reactor grade plutonium may be used for nuclear warheads at all levels of technical sophistication....we now know that even simple designs, albeit with some uncertainties in yield, can serve as effective, highly powerful weapons–reliably in the kiloton range.

In support of this latter view, we note that of the seven nations that had tested nuclear weapons (by 1977), six had obtained the materials for their bombs from reprocessing. It should be pointed out that the material reprocessed was not necessarily produced in nuclear power reactors.

From these excerpts, it is clear that experts disagree on the usefulness of reactor plutonium for atomic weapons. What would you tell people if you were asked whether plutonium recovered from the rods from nuclear power plants was useful for bombs? Given that the experts disagree, suppose the issue of plutonium recovery were on a national election ballot as a yes or no question, how would you vote and why? Why might you like more information on who these "experts" are in order to make your decision? Does where these statements were published influence your opinion?

---

* "The Reprocessing of Nuclear Fuels," <u>Scientific American,</u> **235(6),** 30 (December 1976).
† <u>Science,</u> **196,** (1 April 1977).
‡ Victor Gilinsky, "Plutonium, Proliferation and Policy," <u>Technology Review,</u> **79(4),** (February 1977).

Figure 15-13 *A technician at NFS West Valley, NY, plant watches operations inside the chemical process cell where short lengths of nuclear fuel are put into nitric acid to dissolve the fuel material and thus separate it from its metal cladding. (Photograph courtesy of Nuclear Fuel Services, Inc.)*

Mixed Oxide Fuel and Plutonium Transport

The reprocessing of spent fuel produces separate streams of uranium, plutonium, and waste material, the last consisting principally of fission products. The uranium is transported elsewhere to be enriched in uranium–235. The plutonium can be transported to a special fuel fabrication plant. There plutonium is mixed with uranium that has been enriched in uranium–235; the mixture is then encased in fuel rods (Fig. 15-14). These "mixed oxide" fuel rods can be used in today's light water reactors. They are called mixed oxide fuel because

both uranium and plutonium will be present in the oxide form.

If spent fuel is reprocessed, should plutonium be recovered for use as a supplemental fuel in the light water reactors? One issue is most important to us here; it is the transportation of the plutonium. Plutonium is likely to move by truck along ordinary highways from the reprocessing plant to a plant where the mixed oxide fuel will be made. Although the truck's contents will be clearly labeled as "radioactive" to warn the public, there will occasionally be collisions, overturns, flat tires, etc.

The standards that the specially designed packing must meet are strict so that the contents of the shipment cannot escape should an accident occur. Furthermore, the shielding that prevents radiation exposure must stay intact so that no radiation escapes. In addition, the contents of the shipment cannot come in close enough contact to allow a chain reaction.

The older design package for plutonium transport was built up with three layers of steel and insulation, all surrounding about 10 pounds of plutonium (4.5 kilograms). Improved designs call for even more insulation, and may allow transport of up to 15 pounds (6.4 kilograms) of plutonium in a package. At 40 drums per truck, about 600 pounds (0.26 metric tons) of plutonium can be moved in a given shipment. An estimate for 1990 places the number of such shipments at 360 annually with a total quantity of plutonium of 108 tons (91 metric tons) moved each year (Fig. 15-15). The total annual travel of plutonium shipments could reach 108,000 miles per year by 1990, according to one conservative projection. Centers of population could be avoided by the trucks and the individual shipments could be made smaller, but the added mileage of the extra trips would increase the probability of an accident.

We are concerned about the movement of plutonium on the highways for two reasons. First, it is a dangerously poisonous substance; inhaling the most minute quantities can cause cancer. Second, forceful redirection of the shipment (hijacking)

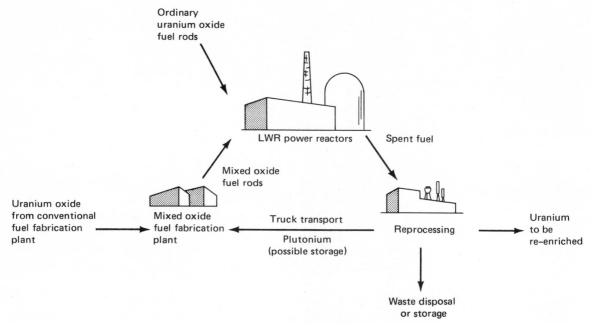

Figure 15-14   *Uranium and plutonium recycle system.*

could occur. The plutonium could be used by terrorists for either atomic bombs or dispersal weapons. Escorts for the trucks carrying plutonium were temporarily rejected in the mid 1970s as too costly, but armed guards may eventually be needed to escort armored convoys of plutonium; such convoys would be equipped with special radio links to police. Knockout gases also may be carried to foil sabotage attempts. Finally, the plutonium might be "spiked" with a radioactive element that emits high levels of radiation, exposing hijackers to an additional risk.

Perhaps the most effective way to reduce the possibility of the hijacking of plutonium is to place

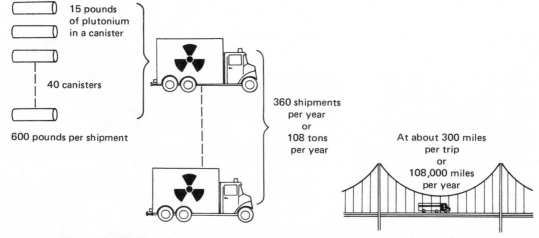

Figure 15-15   *Plutonium shipments (estimate) in 1990 if spent fuel is reprocessed.*

the reprocessing plant and mixed oxide fuel fabrication plant at the same site. The plutonium produced at the reprocessing plant would move only to an adjacent building to be combined into fuel rods. Nuclear electric power plants might be located at this same site.

International Safeguards

The term "safeguards" refers to the procedures and methods of preventing both the misuse of nuclear materials and the spread of nuclear weapons. A nation that receives a nuclear power plant or a reprocessing plant may be subject to international treaties which force it to accept safeguards.

When the United States first began to export nuclear technology, it insisted on treaty guarantees from the buying nations that neither the material nor technology would be used for atomic weapons. The United States reserved the right to inspect all the records and reports from the buyers as well as the nuclear power plants themselves. Over a thousand inspections were conducted by the United States through 1974, but an international agency has now picked up this function. This agency is the International Atomic Energy Agency (IAEA). The IAEA's purpose is to:

> ensure that special fissionable and other materials...equipment...and information made available by the Agency...are not used in such a way as to further any military purpose.

The IAEA does not itself protect fissionable material; this function is left to the nation that owns the material. Nor can the IAEA prevent theft or sabotage even if it is aware of them. All "police type" responsibilities are reserved to the owning nation. If a prohibited act is detected, the IAEA may notify the nation that supplies the violator so that it can end its cooperation agreement with the country, but that is virtually all the agency can do.

The Nuclear Non-Proliferation Treaty, which gives the IAEA its power to monitor and inspect, is designed to prevent new nations from obtaining nuclear weapons or from obtaining nuclear explosives for "peaceful uses."[7] A summary of the treaty is given in Table 15–1. The Treaty was established in 1970.

Since 1970, more than 100 nations have signed the treaty and most have approved it internally as well (see Table 15–2). Nevertheless, many nations have not signed, among them Argentina, Brazil, Chile, China, France, India, Israel, Pakistan, South Africa, and Spain.

Of the nations which did not sign the NPT, India was able to explode an atomic bomb in 1974, violating a treaty with Canada, which was supplying it with nuclear technology. By diverting nuclear material from a power plant to the manufacture of explosives, India was able to manufacture a nuclear explosive, illustrating the hazard of giving nuclear technology to a nation that has not signed and ratified the NPT. Sales of full nuclear technology, including reprocessing, have also been made to Brazil (by West Germany) and to Pakistan (by France). Although these nations deny intentions to obtain

TABLE 15–1   The Non-Proliferation Treaty [a]

Article I prohibits the transfer of nuclear weapons or other nuclear explosive devices (including devices for peaceful nuclear explosions) to any state....Nuclear-weapon states are also forbidden to assist non-nuclear-weapon states to acquire nuclear weapons or explosive devices.

Article II prohibits non-nuclear-weapon signatories from manufacturing or otherwise acquiring nuclear weapons or devices, including peaceful nuclear explosives.

Article III obligates the non-nuclear-weapon parties to accept international safeguards...to ensure that there is no diversion of nuclear material to the manufacture of nuclear explosives.

[a] Stockholm International Peace Research Institute, "Preventing Nuclear Weapons Proliferation," Stockholm, Sweden, January 1975.

7 A party to the Treaty can legally make all the preparations needed to manufacture a nuclear weapon so long as it does not actually assemble the warhead.

DOES THE PUBLIC HAVE THE RIGHT TO KNOW
(HOW EASY IT IS TO STEAL NUCLEAR MATERIAL AND BUILD AN ATOMIC WEAPON)?

The choice [on nuclear safeguards] is ours to make as a nation, and we believe it should be made on broad economic and social grounds after full public discussion.

> M. Willrich and T. Taylor, in
> Nuclear Theft: Risks and
> Safeguards,
> Ballinger Publishing,
> Cambridge, MA, 1974

As seasoned professionals know only too well, truly constructive contributions to an effective, balanced safeguards system...require competent, dedicated and sustained hard work—nearly always at low profile.

> G. R. Keepin, in
> "Nuclear Materials Safeguards:
> A Professional Speaks Out,"
> Nuclear News, p. 76, September 1974

A book published in 1974 by Ballinger Publishing Company set off a controversy in the nuclear community. The book, Nuclear Theft: Risks and Safeguards, was written by Mason Willrich and Theodore Taylor. Willrich is a professor of law who has been involved with the Arms Control and Disarmament Agency of the United States Government. Taylor is a nuclear physicist, now a corporation executive, who has been involved with the design of atomic weapons.

The second chapter of Nuclear Theft contained a rather general discussion of bomb building, with extensive quotations from an encyclopedia entry by J. S. Foster, a nuclear weapons expert. The authors chose to omit material, though the references were not classified, on the specifics of bomb building.

Willrich and Taylor conclude that:

> Under conceivable circumstances, a few persons, possibly even one person working alone who possessed about ten kilograms of plutonium oxide and a substantial amount of chemical high explosive (the implication is that the material will be squeezed by the explosive to decrease the needed critical mass) could, within several weeks, design and build a crude fission bomb... This could be done using materials and equipment that could be purchased at a hardware store and from commercial suppliers of scientific equipment for student laboratories.

In their opening chapter, Willrich and Taylor explain their reasons for writing the book and for the level of detail they provide:

> ...how much does the public need to know about these matters? This question haunts us, and we believe it merits discussion before proceeding further.

> To us, the most compelling argument against informing the public about the risks of nuclear theft is that such an effort might inspire warped or evil minds....However, a large amount of information in much greater detail than we present here is already in the public domain....
>
> ...when security risks are inherent in a long-term activity, which is clearly the case with nuclear theft, the public in a democratic society has a right to know, and those with knowledge have a duty to inform....
>
> ...The years just ahead provide the last chance to develop long-term safeguards that will deal effectively with the risks of nuclear theft. Once the material flows in the nuclear power industry are as enormous as expected a few years from now, it will be too late. *

Willrich and Taylor closed their introductory chapter with an appeal to readers to arrive at their own conclusions on risks and safeguards, saying that the choice belonged to the people and (by implication) not to the technical experts.

A number of nuclear professionals, however, felt that Willrich and Taylor had provided aid to would-be terrorists by collecting so much information in one source on ways to steal and use nuclear materials. G. R. Keepin, who reviewed the book in a magazine read by nuclear professionals, disagreed with Willrich and Taylor. His position is that the public would be better off if the book had not been printed, that the nuclear industry was moving along satisfactorily with safeguards methods.

> Without elucidation, this reviewer would simply record here the opinion that it is both unseemly and counterproductive for a former professional in the weapons field to speculate–however hypothetically–on how a would-be diverter might proceed with design and fabrication of an illicit atomic bomb. Genuine concern with safeguarding...nuclear materials could surely take more constructive forms than indulging in what could turn out to be self-fulfilling prophecy.
>
> The dictates of reason and prudence would appear to reject Taylor's assertion that "it seems necessary to be quite specific" in order to make the risks of nuclear terrorism credible, and to convince the public of the gravity and urgency of the nuclear materials diversion problem....
>
> The important over-all point to be made here is that–notwithstanding certain glaring shortcomings and sins of the past–the AEC and much of the nuclear industry are in fact making great strides toward effective, stringent control of the nuclear materials which are the lifeblood of that industry. †

Does the public have the right to know that a technology is unsafe if publishing this information could increase the risks? Would you rather leave the management of nuclear material to the experts as Keepin suggests or did Willrich and Taylor make the right decision? Is the discussion here in itself a disservice because it brings the controversy to light?

References to assist your understanding of this controversy are provided at the end of the chapter on pp. 344-345.

---

* M. Willrich and T. Taylor, <u>Nuclear Theft: Risks and Safeguards,</u> Ballinger Publishing, Cambridge, MA, 1974, p. 2-4.

† G. R. Keepin, in <u>Nuclear News</u> **17(12)**, 76 (September 1974).

TABLE 15-2   Status of the Non-Proliferation Treaty (as of December 1975)

Nations that have both signed and ratified [a]

Afghanistan	Gabon	Lesotho	San Marino
Australia	German Democratic	Liberia	Senegal
Austria	Republic	Luxembourg	Sierra Leone
Belgium	Germany, Federal	Malagasy	Somalia
Bolivia	Republic of	Republic	Sudan
Botswana	Ghana	Malaysia	Swaziland
Bulgaria	Greece	Maldive	Sweden
Burundi	Guatemala	Republic	Syrian Arab
Cameroon	Haiti	Mali	Republic
Canada	Holy See	Malta	Thailand
Central African	Honduras	Mauritius	Togo
Republic	Hungary	Mexico	Tonga
Chad	Iceland	Mongolia	Tunisia
China, Republic of	Iran	Morocco	Union of Soviet
Costa Rica	Iraq	Nepal	Socialist Republics
Cyprus	Ireland	Netherlands	United Kingdom
Czechoslovakia	Italy	New Zealand	United States of
Dahomey	Ivory Coast	Nicaragua	America
Denmark	Jamaica	Nigeria	Upper Volta
Dominican Republic	Jordan	Norway	Uruguay
Ecuador	Kenya	Paraguay	Viet-Nam (South)
El Salvador	Korea, Rep. of	Peru	Western Samoa
Ethiopia	Laos	Philippines	Yugoslavia (S,R)
Fiji	Lebanon	Poland	Zaire (S, R)
Finland		Romania	

Nations that have signed but not ratified [a]

Barbados	Japan	Singapore	Yemen (Arab
Colombia	Kuwait	Sri Lanka	Republic)
Egypt	Libya	Switzerland	Yemen (People's
Gambia	Panama	Trinidad and Tobago	Democratic
Indonesia	Rwanda	Turkey	Republic)

Nations that have neither signed nor ratified [b]

Albania	China	Khmer Republic	Saudi Arabia
Algeria	Comos	Malawi	South Africa
Argentina	Congo	Mauritania	Spain
Bahrain	Cuba	Niger	Taiwan
Bangladesh	Equatorial Guinea	North Korea	Uganda
Bhutan	France	Oman	Ukraine
Brazil	Guinea	Pakistan	United Arab Emirates

Nations that have neither signed nor ratified[b]

Burma	Guinea–Bissau	Papua New Guinea	United Republic of
Byelorussia	Guyana	Portugal	Tanzia
Cape Verde	India	Qatar	Zambia
Chile	Israel	Sao Tome and Principe	

[a] U.S. Arms Control and Disarmament Agency, Publication No. 70, April 1975.

[b] Compiled by the Congressional Reference Service from U.N. and IAEA listings of membership.

nuclear weapons, neither has signed the Non-Proliferation Treaty, and hence cannot be inspected by IAEA. There is a great deal of insecurity in providing such nations with nuclear technology.

Besides the Non-Proliferation Treaty, a number of European nations belong to the European Atomic Energy Community,[8] known as EURATOM. In the 1950s this organization created a safeguards system that includes inspection and audit. The IAEA accepts the safeguard findings of EURATOM, but the IAEA reserves the right to conduct independent studies. The IAEA cannot, however, investigate procedures in France since France has not signed the NPT.

Unfortunately from the safeguards standpoint, it appears from recently published information that a reprocessing plant could be secretly built in as little as four to six months. The plant could, once in production, process spent fuel rods from light water reactors and produce enough plutonium for half a dozen bombs per month. The time for the discovery of violations and sanctions against the offending nation is terribly small.

Storage or Disposal of the Wastes From Fuel Reprocessing

The safe storage of the high-level wastes from a fuel reprocessing plant is an issue of great importance to the nuclear industry. The plans for storing

these wastes are drawn from experience in the manufacture of nuclear weapons, where wastes similar to those from the fuel reprocessing plants also require disposal.

At the Hanford site in Washington, at the Savannah River site in South Carolina, and at the National Reactor Testing Station in Idaho, experience has accumulated in handling these high-level wastes. At Hanford, a concrete tank with an exterior carbon-steel liner was used initially. At Savannah River the carbon steel tank was inside a concrete tank and steel was used to line the outside. Leaks have occurred from both these tanks because of corrosion of the steel. Exposure of the public has been negligible but the underground storage sites of the tanks now must be controlled for several hundreds of years.

The corroding or rusting evidently took place at points of stress in the tanks; new tanks are now being heat-treated to relieve such stresses. Although no leaks have yet occurred from tanks so treated, it was a long time before the leaks were discovered in the original types of tanks. The tanks used at the Idaho site, on the other hand, were of stainless steel, and leaks from these tanks have not been reported.

The era of liquid wastes from reprocessing is ending, however. At the Idaho site as well as at Hanford, methods to solidify the liquid wastes have been developed. Solidification makes it much more difficult for these wastes to contaminate the environment. Concrete and clay are viewed as useful substances into which the wastes can be physically mixed and fixed in solid form. Fixing the wastes

---

8 France, Belgium, Luxemburg, West Germany, Netherlands, Italy were initial members. The United Kingdom, Ireland, and Denmark joined later.

into glass may provide even greater safety because of the continuous character of the solid form of glass. Studies of this method are underway.

The new regulations for high-level wastes, if reprocessing should ever resume in the U.S., call for the wastes to be solidified before shipment from the fuel reprocessing plant. A number of proposals have been studied for safe storage alternatives or for disposal of high level wastes. One of the most unusual suggestions is to rocket the wastes into deep space, but the costs to dispose of the many tons of high level wastes in this way are very high. Furthermore, one cannot rule out an accident at the launching pad or exit from the earth's gravity due to a failure of the boosting rockets.

Other unusual suggestions have been put forward, all involving burial in the deep earth. One concept would see the wastes placed in the geologic trench in the ocean where the ocean's tectonic plates are gradually sliding beneath those of the continent. Another idea is to dispose of the wastes in a deep shaft drilled into the earth's core or in a cavity carved out by a nuclear explosion. In none of these processes can the wastes be retrieved at some future time. Nor can the results of these plans be carefully evaluated with present knowledge.

As early as 1955, however, nuclear scientists and engineers were considering the possibility of storing high-level wastes in deep geologic formations, such as salt beds. Natural underground salt beds exist in a number of places in the United States and apparently remain stable in both their physical and chemical characters for thousands of years. The judgment that salt beds are a safe place to store nuclear wastes has been reaffirmed many times, but recent criticism suggests that the heat generated by the wastes could alter the salt formations.

In 1972, the Atomic Energy Commission[9] selected a site near Lyons, Kansas, for Project Salt Vault. The project at Lyons was designed to show that storage in salt beds would work, but the project was abandoned when it was discovered that salt mining (by dissolving salt in water) within the formation was being conducted not far away. The unknown effect of the mining on the stability of the formation was the key factor in ending the project at Lyons. The idea of storing high level wastes in bedded salt, however, was not abandoned because of this setback. A site in New Mexico is still being considered.

A plan to store the wastes on a temporary basis was still needed, however, and the concept of a retrievable surface storage facility (RSSF) has been studied to solve the temporary storage problem for high level wastes. Retrievable storage is designed to hold high-level solid wastes accumulating from reprocessing while a permanent solution is sought.

Opponents of the surface storage plan have argued that perpetual security would be needed for the site. Advocates counter that it is only a temporary measure, but their argument is weakened because it is not known what plan of disposal will replace surface storage. The high-level wastes to be placed in the surface storage repository would be fully retrievable. Thus, if and when a permanent solution to waste disposal is worked out, the accumulated wastes can be included in the permanent solution.

Two concepts are being considered for retrievable surface storage. In one, the wastes would be in canisters stored underwater. In the other, the canisters would be open to the air, either indoors or outside on concrete pads (Fig. 15–16).

---

9 The AEC's functions in nuclear waste management now belong to the Department of Energy.

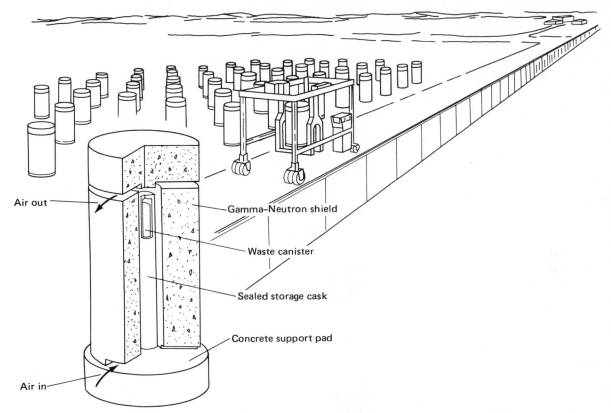

Figure 15-16 *Retrievable surface storage facility, sealed storage cask concept, storage area.*

Alternatives to Reprocessing

It is possible to modify the PUREX process so that uranium and plutonium are recovered together as a mixture. The mixture of uranium and plutonium would be mainly uranium–238, a substance that cannot be used for bombs. Only about 1% would be uranium–235 and 1% plutonium. Since this material could not be used for bomb building, it would be much less valuable to would-be hijackers. The uranium–plutonium mixture can be used in fuel rods if it is first mixed with uranium extra enriched in uranium–235. The final mixture would then be about 3% fissionable material. The modification of the PUREX process is called "coprocessing" and lately has been called CIVEX (for civilian extraction) (Fig. 15–17). The CIVEX process would, in addition, leave some small quantity of fission products in the recovered uranium–plutonium

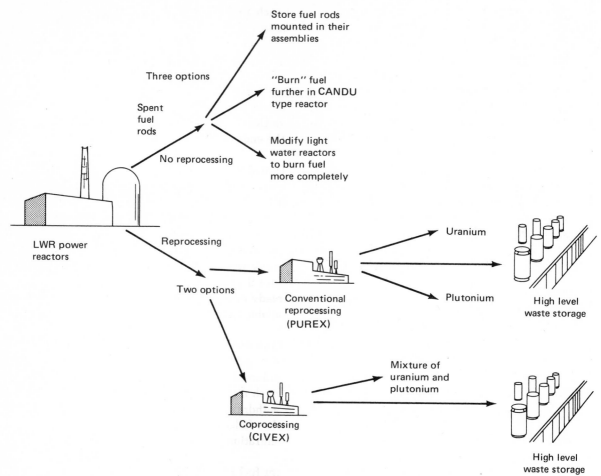

Figure 15–17 *Reprocessing and Alternatives.*

mixture. The radioactivity from the fission products would make the mixture physically unapproachable for humans. Only remote handling would be possible.

Because purified plutonium will not be carried between the reprocessing plant and the plant making fuel rods, the risk of hijacking will be decreased. A nation given only this reprocessing technology would not have the pure plutonium needed for bombs, but the separation of plutonium from the mixture would require only one additional chemical step. Coprocessing would also continue to produce the fission products as high-level wastes from reprocessing just as the PUREX process does.

We pointed out earlier, but it bears repeating, that reprocessing is based on the concept of recovering "unburned" uranium–235 and plutonium, which are viewed as usable energy resources. It is also possible that the light water reactors could operate in a "throw-away" cycle. In such a cycle, spent fuel rods would never be opened. They would simply be stored until the levels of radiation and radioactive elements are decreased enough to allow ordinary disposal.

The throw-away cycle, while simpler, may be wasteful of resources. Thus, if there were a way to recover the energy in the spent fuel without reprocessing, the methodology would be very attractive, because plutonium would not become available. Hence, the spread of nuclear weapons would be delayed. In fact, there are several ways to recover the energy remaining in spent fuel without reprocessing.

The light water reactors, which dominate the United States and European markets, are not the only ways that nuclear electric power can be generated. There is another economically competitive and relatively proven technology that originated in Canada. This technology is more efficient at capturing the energy of uranium fission than is either the BWR or PWR. It is the heavy water reactor built by Atomic Energy of Canada, Ltd. and it is known as the CANDU reactor. The CANDU reactor circulates "heavy water" through the core to capture the heat from the fission process. More than a dozen commercial scale units of this type are operating already or are under construction. By 1983, CANDU reactors will be available to deliver 12,500 megawatts of electric power.

The CANDU reactor does not require enriched uranium fuel as the light water reactors do. While the light water reactors use a uranium fuel enriched to about 3% uranium-235 (the rest of the fuel is uranium-238), the CANDU reactor can operate with unenriched uranium fuel which may be only about 0.7% uranium-235. When the fuel rods are finally withdrawn from the CANDU for replacement, the level of uranium-235 has fallen to 0.25% or slightly less. This level is far below the natural concentration of uranium-235 in uranium ore.

In contrast, the level of uranium-235 in the spent fuel rods withdrawn from the light water reactors is nearly 1%. This level in the fuel rods discarded from the light water reactors is actually higher than the level in fresh fuel rods used by the CANDU. It follows that what is "spent" to the light water reactors can potentially still serve as fuel for the CANDU. Thus, it has been suggested that the discarded rods from the light water reactors be transferred to CANDU reactors where more energy could be derived from the rod without reprocessing.

Another suggestion that makes use of the Canadian design is to modify the pressurized water reactors by using heavy water in the coolant in addition to light water so that more complete fission of uranium-235 would be possible. However, heavy water would have to be manufactured in the U.S. in order to accomplish this change. These alternatives are summarized in Figure 15–17.

Liquid Metal Fast Breeder Reactor

We have discussed so far the light water reactors that utilize the fission of uranium–235 as a source of heat, but the light water reactors have surprisingly been viewed by nuclear advocates as a short-term source of energy. These people see light water reactors as a means to supplement and replace fossil fuels until the time when more secure sources of nuclear power, such as the breeder reactor, become available.

Viewed as the successor to the light water reactor, the LMFBR (liquid metal fast breeder reactor) is under development in the U.S.S.R., Japan, France, the United States, and possibly other countries as well. The light water reactor uses uranium–235 for fission, but uranium–235 is not as plentiful in nature as uranium–238, the dominant component of uranium ore. It is the abundant uranium–238 that the breeder reactor utilizes to produce power. The idea of using uranium–238, even though it does not naturally undergo fission, is ingenious.

In our discussion of the light water reactors, we pointed out that the fission of uranium–235 "bred" plutonium–239 from uranium–238. Plutonium–239 will undergo fission just as uranium–235 will when properly struck by a neutron. In much the same fashion as uranium–235, plutonium breaks into two fragments with a release of neutrons and energy. The neutrons released will sustain the fission of more plutonium–239, producing heat. The neutrons will also "breed" more plutonium from uranium–238 atoms. The process, illustrated in Figure 15–18, both produces heat from fission and "breeds" more fuel to undergo fission. This explains the word "breeder" in the name of the reactor.

Instead of a water coolant, this reactor utilizes a liquid metal (pure, melted sodium) to withdraw the heat of fission from the reactor core. Liquid sodium holds a good quantity of heat and is a good conductor of heat. In addition, it will not slow down the neutrons released in the fission process, and the high energy neutrons are essential to breeding more plutonium. Liquid sodium has other advantages. Sodium melts at 210 ° F but does not boil until it reaches 1640 ° F. Because its boiling point is so distant from its melting point, it is unlikely to "flash" to a vapor if an accidental rupture of a line occurs. The loss-of-coolant accident, which is a concern in light water reactors, is therefore thought to be much less probable in the breeder.

The design of the LMFBR calls for a primary loop containing liquid sodium, which circulates through the core of the reactor. This sodium, which may become radioactive as it circulates through the core, will pick up the heat produced by the fission process. A second loop also contains circulating liquid sodium. This sodium picks up the heat from the first loop in a heat exchanger. A third loop contains water which is boiled to steam by the heat from the second loop. The steam turns the turbine, is condensed, and returned to the boiler (Fig. 15–19).

The fuel rods in the reactor will contain a mixture of plutonium oxide and uranium oxide (called "mixed oxide"). In time, as fission proceeds, the fission products build up in the rod and, just as in the light water reactors have the effect of "poisoning" the reaction by capturing a portion of the neutrons. When neutron capture slows heat production too greatly, the rods are withdrawn and sent to a reprocessing facility. There, plutonium and uranium are separated from the fission products. The plutonium content will have increased during the breeding cycle in the reactor. The fission products are

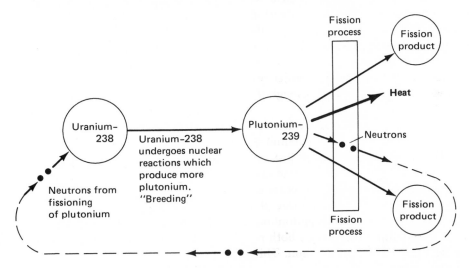

Figure 15-18 *Nuclear reactions in the breeder reactor. Plutonium is "bred" by bombarding uranium-238 with neutrons. These neutrons are furnished by the fission of plutonium-239. The fission process produces heat, which will eventually be used to make steam and the neutrons needed to breed more plutonium-239 from uranium-238.*

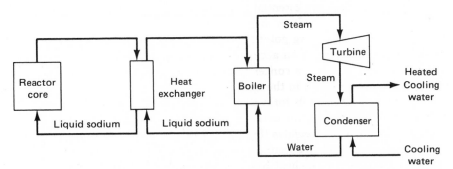

Figure 15-19 *Schematic diagram of the liquid metal fast breeder reactor (LMFBR).*

wastes and require disposal. New mixed oxide fuel rods are fabricated from the plutonium and from uranium.

The familiar problems of reprocessing, plutonium movement, and disposal or storage of fission product wastes are all present in the breeder reactor. Furthermore, the quantities of plutonium will be enormous, even in comparison to the quantities potentially produced by the light water reactors. A core meltdown after a loss of the sodium coolant would pose an even greater problem than a meltdown in a light water reactor. This is because the core contains so much more of the highly toxic plutonium. Thus, although the breeder reactor is an idea of cunning ingenuity for extending fuel supplies, the problems and the issues that surround its development must be critically examined.

Questions

The Light Water Nuclear Reactors

How they operate

1. What are the two basic types of light water reactors?
2. Describe the basic difference between the two reactor types.
3. Which process is more like that of a fossil fuel power plant?
4. As the reactor operates, what eventually happens to the level of uranium–235 in the fuel rods? To the level of fission products?
5. What is the function of the condenser, as shown in Figures 15–4 and 15–5, in the electric generation process?

Radioactive wastes and radiation releases from light water reactors

6. What accounts for the presence of fission products and other radioelements in the water that circulates through the core of a nuclear reactor? Why should we be concerned about the ultimate disposal of water in which these fission products are dissolved?
7. Identify as many sources of solid radioactive waste resulting from the nuclear power production process as you can. What is done with these wastes?

Safety of light water reactors

8. Explain what is meant by core melt or "meltdown." What would cause a meltdown? What safeguards are designed into nuclear power plants to prevent this from happening?

Problems in the Nuclear Fuel Cycle

From mine to power plant

9. What is accomplished by "enrichment" in the processing of nuclear fuels? Is uranium–235 or uranium–238 used for the fission process?

Fuel reprocessing/spent fuel rod storage

10. What is initially done with "spent" fuel rods from nuclear power plants?
11. What is meant by fuel reprocessing? What substances are separated out?
12. Which of these substances can be used again? Where did plutonium come from?
13. Discuss the pros and cons of reprocessing as an alternative to indefinite underground storage of spent fuel rods.
14. Was the reprocessing plant at West Valley a technological success even though it went out of business?
15. Did the West Valley Plant use remote handling of the rods? Did it use remote repair of broken equipment? What abuses resulted?
16. Why did France and Germany proceed with reprocessing in the 1970s?
17. Why did the U.S. work against the sale of reprocessing technology to Brazil and Pakistan?

International Safeguards

18. What rights does the Nuclear Non-Proliferation Treaty give to the International Atomic Energy Agency (IAEA)?
19. Does the IAEA have the right to seize fissionable material to prevent manufacture of nuclear weapons?

Storage or disposal of the wastes from fuel reprocessing

20. Are carbon-steel tanks adequate for storing reprocessing wastes?
21. List and discuss briefly at least three methods that have been proposed for the storage and/or disposal of high-level radioactive wastes.

Further Reading

General References

Bupp, I. and J. Derian, *Light Water: How the Nuclear Dream Dissolved*, Basic Books, New York, 1978.

Hammond, A., "Fission: The Pro's and Con's of Nuclear Power," *Science* **178**, 147–149 (13 October 1972).

Hohenemser, C., R. Kasperson, and R. Kates, "The Distrust of Nuclear Power," *Science* **196**, 25–34 (1 April 1977).

"Nuclear Dilemma: The Atom's Fizzle in an Energy-Short World," *Business Week*, p. 54 (25 December 1978).

"Nuclear Energy: How Bright a Future," *Environmental Science and Technology* **11(2)**, 128–130 (February 1977).

"Nuclear Power: Can We Live With It?," a panel discussion including D. Kleitman, N. Rasmussen, R. Stewart, and J. Yellin, *Technology Review* **81(7)**, 32 (June/July 1979).

Nuclear Power, Issues and Choices, the Nuclear Energy Policy Study Group, S. Keeny, Jr., Chairman, Ballinger Publishing Company, Cambridge, MA, 1977.

Rose, D. J., "Nuclear Electric Power," *Science* **184**, 351–359 (19 April 1974).

Weinberg, A., "Social Institutions and Nuclear Energy," *Science* **177**, 27–34 (7 July 1972).

Weinberg, A., "Salvaging the Atomic Age," *The Wilson Quarterly*, pp. 88–112 (Summer 1979).

Nuclear Plant Reactor Safety

"NRC Panel Renders Mixed Verdict on Rasmussen Reactor Safety Study," *Science* **201**, 1196 (29 September 1978).

"Reactor Safety Study: An Assessment of Accident Risks in U.S. Commerical Nuclear Power Plants," Nuclear Regulatory Commission, October 1975, Wash–1400 (the Rasmussen Report).

"Risk Assessment Review Group Report to the U.S. Nuclear Regulatory Commission," prepared for the U.S. Nuclear Regulatory Commission, NUREG/CR–0400, September 1978 (review of the Rasmussen Report).

Shapely, D., "Reactor Safety: Independence of Rasmussen Study Doubted," *Science* **197**, 29–31 (1 July 1977)

The Accident at Three Mile Island (The Need for Change: The Legacy of Three Mile Island), Report of the President's Commission, J. Kemeny, Chairman, October 1979.

Reprocessing, Wastes, and Nuclear Weapons Spread

Lester, R. K. and D. J. Rose, " The Nuclear Wastes at West Valley, New York," *Technology Review*, pp. 20–29 (May 1977).

Rose, D. and R. Lester, "Nuclear Power, Nuclear Weapons, and International Stability," *Scientific American* **238(4)**, 45–57 (April 1978).

Bebbington, W. P., "The Reprocessing of Nuclear Fuels," *Scientific American*, 30–41 (December 1976).

References to "Controversy: Does the Public Have the Right to Know?"

Keepin, G. R., "Nuclear Materials Safeguards: A Professional Speaks Out," *Nuclear News*, p. 76 (September 1974).

Willrich, M. and T. Taylor, *Nuclear Theft: Risks and Safeguards,* Ballinger Publishing, Cambridge, MA, 1974.

Other publications that stir the pot of controversy are:

Gillette, R., "Nuclear Safeguards: Holes in the Fence," *Science* **182,** 1112 (14 December 1973).

Ingram, T. H., "Nuclear Hijacking: Now Within Grasp of Any Bright Lunatic," *Washington Monthly,* pp. 20–28 (December 1972).

Lapp, R. E., "The Ultimate Blackmail," *The New York Times Magazine,* p. 13 (4 February 1973).

McPhee, J., "The Curve of Binding Energy," *The New Yorker Magazine* (3, 10, and 17 December 1973).

McPhee, J., *The Curve of Binding Energy: A Journey into the Awesome and Alarming World of Theodore B. Taylor,* Farrar, Straus, & Giroux, New York, 1974.

Shapely, D., "Plutonium: Reactor Proliferation Threatens a Nuclear Black Market," *Science* **172,** 143–146 (9 April 1971).

(Hugh Rogers/Monkmeyer)

How Conventional Fuels Affect Environmental Quality

When we burn the fossil fuels, coal, oil and gas, we furnish heat for homes, energy for mobility, and power for the production of goods. Fossil fuels have become our servants, and our skill at capturing their energy has transformed civilization. This mastery of energy has brought abundance and comfort, but energy has turned out to be a dangerous servant.

Our misuse of these fuels has resulted in air pollution episodes so severe that humans have died. Ships carrrying oil to fuel power plants have foundered, spilling cargoes of black oil which foul beaches and bring death to countless sea birds. And while we fear such events, they are little more than the "tip of the iceberg," visible to us because of their nightmare drama. Beneath the episodes is the history of chronic and accumulated effects on humans and other animal species and on plants, which taken together may be found to be even more dangerous. Some of the environmental effects of power generation were mentioned in Part 3—acid mine drainage, strip mining damage, and nuclear waste contamination of the environment.

Part 4 begins with a chapter on atmosphere and climate. The concepts in this chapter are essential to understanding how air pollutants, produced when energy is generated, affect climate and the natural cycles of materials such as carbon, sulfur, and nitrogen. Such disturbances eventually threaten the health and survival of human and wildlife populations.

Numerous pollutants foul the air we breathe. Solid particles such as lead, asbestos, and soot are dispersed in the air. Liquid droplets of hydrocarbons and sulfuric acid float in the atmosphere. Gases such as carbon monoxide, nitrogen oxides, and sulfur dioxide are dissolved in the atmosphere that surrounds us. The contaminants in the air have direct biological effects upon us. Our breathing may be impaired; diseases of the heart and lung are complicated and made worse. In addition, the landscape is influenced because the vegetation is also sensitive to air pollutants. Construction materials such as mortar and metal are corroded by certain pollutants, and the fibers of our fabrics are damaged by others.

Near Los Angeles, the most sprawling city in the country and the one most dependent upon the automobile, smog is destroying the pine trees in the hills that surround the city. And almost 200 miles away, about the distance between New York and Boston, the desert holly, a plant that grows in the isolation of Death Valley, is threatened by the creeping smog from Los Angeles. The smog results from reactions between hydrocarbons and nitrogen oxides. The predominant source of these substances in the Los Angeles basin is the automobile. Sunlight stimulates the reactions between hydrocarbons and nitrogen oxides.

In the northeastern United States as well as in Scandinavia an "acid rain" appears to be stunting the growth of forests. The acidity is caused by the solution of sulfur oxides in the atmosphere into rainwater. Sulfur oxides are pollutants produced from the burning of coal and oil, two sulfur-containing fuels. These fuels are burned to produce heat, to generate electric power and to manufacture steel. Chapters 17, 18, and 19 examine the air pollutants produced by the generation of power and the possible solutions to these problems (Fig. A).

However, the impact of energy on the environment extends far beyond its effects on the air we breathe. Energy activities reach out as well to influence the oceans and rivers of the globe. The effects begin at the moment the energy is extracted from the earth. Where oil is drawn from offshore wells, blowouts may occur that threaten both marine species and beaches. The transport of oil in tankers from the producing nations of the world to the industrial consumers is also hazardous. Tankers may be shipwrecked or may spill oil during normal operations.

Other major problems occur when coal and oil are burned in the steam boilers of electric power plants. About 2½ times more heat is produced than we can use in the production of electricity. This is roughly true at nuclear power plants as well except that even a smaller fraction of the heat can be used. Heat not converted to electricity is wasted; it goes into the cooling water of the power plant.

Environmental problems can arise in three areas as a consequence of the steam power plant cycle. First, drawing the cooling water at high velocities into the narrow tubes of a condenser subjects organisms living in the water to an extraordinary shock. We might compare it to the trauma we would suffer if we were to attempt to ride on the outside of an airplane. The second area of concern is the effect on aquatic life of the increase in temperature of a body of water into which the heated condenser water is discharged. The third area of concern is the effect of the numerous chemicals added to the cooling water to prevent the cooling apparatus from clogging. If most future power plants were designed to use the method known as once-through cooling, by the year 2000, over two-thirds of the daily run-off could be required to cool power plants. Although such an extremity is unlikely, there is certainly the potential for waste heat to have a serious effect on the water environment. Thus, it is necessary to examine the effects hot water can have on aquatic organisms and also to consider what alternative cooling methods are available. The final two chapters in this section describe these water pollution problems related to the production of energy, and suggest some solutions and alternatives.

Solid wastes 3%	Other 3%	Other —2%		Other —3%
Residential/com. 3%		Res/com —3%	Other 9%	Solid wastes —3%
Other 6%				Forest fires —4%
Transportation 7%	Industrial 33%	Transportation 44%	Oil & gas production and marketing 14%	
Industrial 62%			Organic solvents 27%	Transportation 80%
	Electric utilities 64%	Industrial 23%	38% Transportation	
Electric utilities 19%		Electric utilities 28%		
			Industrial 11%	Industrial 10%
Suspended particles	**Sulfur oxides**	**Nitrogen oxides**	**Hydro-carbons**	**Carbon monoxide**
16.4	29.9	22.0	28.1	87.4

1975 emissions nation–wide (millions of metric tons*)

*1 metric ton = 1000 kilograms = 2200 pounds

Figure A *Sources and quantities of pollution—1975. (Source: Energy/Environment Fact Book Decision Series, U.S. Environmental Protection Agency, March 1978.)*

Chapter 16

Lessons from Ecology: The Atmosphere and Climate on Earth

BIRTH OF THE SOLAR SYSTEM

One Theory

What was the earth like three billion years ago, before life began? Does the answer lie buried in time, lost because there was no intelligence to record it? Perhaps not. Most scientists believe we can reason backward from looking at the rocks that were formed when the solar system was born, by observing the other planets in our solar system, and from experiments. Observations and experiments have given rise to several theories about the origin of the solar system.

Many experts think our solar system condensed from a cold cloud of dust and gases four and one-half to five billion years ago. In their picture of the birth of our solar system, the cloud of dust and gases, swirling in a counterclockwise direction, began to shrink and flatten. Masses of material concentrated in the center of the turning disk while the shrinking generated enormous amounts of heat. Eventually pressure and temperature rose so high in this central core that nuclear fusion began and the core, now the sun, started to give off radiant energy. Meanwhile, other centers of condensed matter in the outer part of the rotating dust clouds formed planets, including the earth (Fig. 16–1). As the earth took form, gravity pulled heavier elements such as iron and nickel towards the central core.

Because it was condensing from a loose mass of dust and rocks into a dense body, the earth, too, began to heat up.

The atmosphere (To start with, the earth's atmosphere was probably made up of gases from the original cosmic cloud, held by the earth's gravitational forces. However, this first atmosphere was swept away by solar radiation from the newly formed sun. Later, a second atmosphere formed of gases from the center of the earth, brought up when volcanoes erupted. Water vapor, carbon dioxide (CO_2), and carbon monoxide (CO), hydrogen gas (H_2), and sulfur compounds (H_2S, SO_2) were all part of this primitive atmosphere. Oxygen gas, which we need to live, was not present.)

The seas form (At first, any rain that fell must have sizzled immediately back into steamy vapor. But slowly the ground cooled and the heavy rains that fell ran in rivulets and streams to fill depressions in the ground. The rains washed gases from the atmosphere as well as salts from the earth's crust, and chemically rich seas were formed (Fig. 16–2).)

Life begins (In this nutrient soup, the first life probably began. In fact, it is possible to restage some of the early events, today, in a laboratory. If a mixture of gases similar to those believed to be in the early atmosphere is shot through with electric sparks, (used to imitate lightening), some of the building

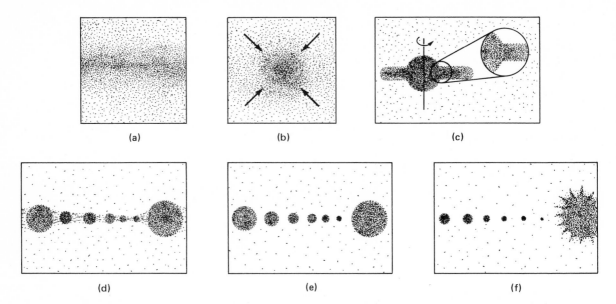

Figure 16-1 *How the solar system was formed. According to one theory, the solar system formed from a rotating dust cloud in space which shrank and flattened (a and b). The central bulge became the sun (c), while concentrations of material in the outer part of the cloud became the planets (d, e, and f). (Adapted from J.M. Pasachoff,* Contemporary Astronomy, *Saunders, Philadelphia, 1977.)*

blocks of life will form, among them, amino acids.) Under certain conditions, such as those found at the vents of erupting volcanoes, scientists have demonstrated that amino acids will form and combine to make protein-like materials. Short nucleic acid pieces also form, under conditions believed similar to those existing on earth before life began. Tiny globs of such materials come together and even begin to hint at some of the properties that are characteristic of living matter. It is reasonable to think that in the prehistoric oceans such globules led eventually to true life forms.

One of the reasons why we believe life began in water is that the sun gives off ultraviolet radiation, which can destroy the delicate materials that make up living creatures. Today, ozone in our atmosphere absorbs enough ultraviolet radiation to make life on land possible but in ancient times life needed water as a protective shield against ultraviolet rays.

Free oxygen appears in the atomosphere (At some point, more than three billion years ago, the simple cells, which were nourished by chemicals floating in the water in which they lived, evolved into organisms capable of photosynthesis. While using the sun's energy and carbon dioxide to make their own organic compounds, they also produced oxygen. This oxygen began to build up in the atmosphere, and a new era began. Part of the oxygen (O_2) was changed by sunlight into *ozone* (O_3). Protected from harmful ultraviolet rays by ozone in the atmosphere, life could evolve on land. Then, provided with a source of oxygen, complex life forms could evolve (Fig. 16-3).)

As the next few chapters will show, humans, who arrived fairly late on the scene (about 10 million years ago), now threaten to make noticeable changes in this atmosphere which has evolved over billions of years. One example is the use of fluorocarbons as spray-can propellants. These chemicals

Water
CO(Carbon monoxide)
CO_2 (Carbon dioxide)
SO_2(Sulfur dioxide)
N_2(Nitrogen)
HCl(Hydrogen chloride)

Rain clouds

Hydrogen(H_2)
Hydrogen sulfide(H_2S)
Methane(CH_4)
Ammonia(NH_3)

Runoff

Salts

Volcanoes

Runoff

Salts

Seas

Figure 16–2   *Before there was life on earth, the seas became a rich chemical soup, as volcanic gases and salts from the earths crust were washed in by rainfall. Today large amounts of these compounds cannot accumulate in the oceans because living creatures use them up. Billions of years ago, lightning flashes may have turned such chemicals into the building blocks of life.*

destroy ozone in the atmosphere and so threaten the protective ozone layer (Chapter 27).

THE CARBON CYCLE

(Carbon dioxide was added to the early atmosphere by erupting volcanoes. Carbon dioxide is used by green plants during photosynthesis and given off by both plants (at night) and animals (especially the decomposers). This is only a small part of the *carbon cycle*, however. Most carbon involved in the cycle is in the oceans. This carbon (in the form of carbonates) controls, to a large extent, the amount of carbon dioxide in the air. Excess car-

bon dioxide in the atmosphere can dissolve in the oceans. In the opposite direction, the oceans can release carbon dioxide to the atmosphere.)Scientists believe this mechanism kept the amount of carbon dioxide in the atmosphere relatively constant until industrialization began. Human activities since then appear to be unbalancing the carbon cycle in several ways. An extra six billion tons of carbon dioxide are added to the atmosphere each year when fuels are burned. A further two billion tons per year enter the atmosphere when fields are plowed. Plowing releases carbon dioxide that would otherwise have stayed in the soil. Some scientists believe that the rapid cutting of forests to satisfy human needs for wood and wood products over the past

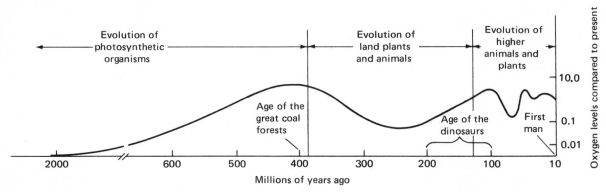

Figure 16-3   *About 2 billion years ago, the level of free oxygen in the earth's atmosphere began to rise. After a protective layer of ozone was formed from part of the oxygen, land plants and animals evolved. Oxygen levels have varied a great deal as levels of production and use have changed. (Source: E.P. Odum,* Fundamentals of Ecology, *3rd ed., Saunders, Philadelphia, 1971.)*

few years may be decreasing the amount of carbon dioxide used by plants.

Although some of the excess carbon dioxide from agriculture and the burning of fuels dissolves in the ocean, half or more remains in the atmosphere. Measurements show that the level of carbon dioxide has increased steadily since the 1950s. The possible effects of an increase in atmospheric carbon dioxide are explored in Chapter 17.

THE IMPORTANCE OF CLIMATE TO LIFE

We are concerned about carbon dioxide because the level of carbon dioxide in the atmosphere may affect the earth's climate. By *climate* we mean a whole complex of factors including the temperature of an area and its humidity; how much rainfall it receives, the amount of snow and hail that fall, and how fast these last three evaporate. In addition, climate includes the winds that blow and the amount of sunlight. All of these factors and the way they vary from day to day and year to year affect the creatures living in the area.

The earth's climate is determined by a number of influences, some of which are not well under-stood and some of which undoubtedly have not even been discovered yet. Variations in the tilt of the earth's axis and in its orbit cause temperature changes in 20,000- to 40,000-year cycles. Changes in the amount of radiation given off by the sun also affect climate but in much shorter cycles. The 11-year sunspot cycle, for example, seems to be related to these changes in the amount of solar radiation.

Another factor affecting climate is the amount of dust in the air. In the past, this has been due mainly to volcanic eruptions. Volcanoes spew into the air huge clouds of dust, which travel around the world and decrease the amount of sunlight reaching the earth. (The dust is also responsible for beautiful sunsets.) A change in the amount of sunlight reaching the earth causes changes in all climatic factors. Some of these effects are diagrammed in Figure 16-4.

By a sort of domino effect, humans feel economic and social effects and possibly even political effects from changes in the amount of sunlight. For instance, Figure 16-5 shows how a volcanic eruption in 1815, by cooling and worsening weather conditions probably caused the price of flour in London to double eighteen months later.

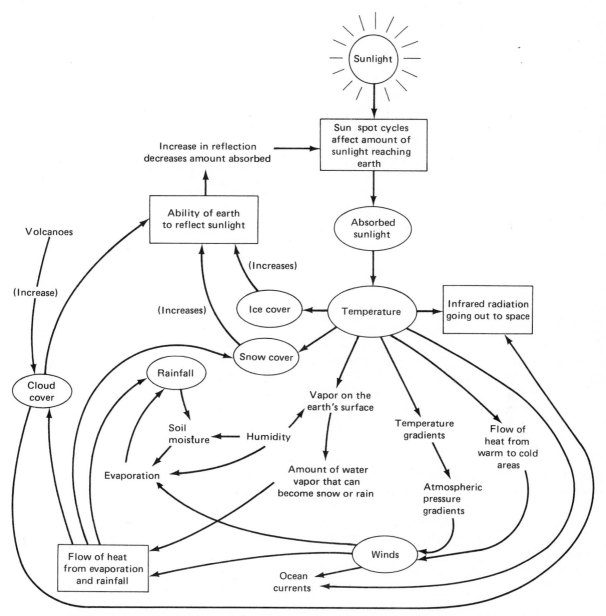

Figure 16–4   *Radiation from the sun and the earth's climate. In this diagram are some of the cause and effect relationships between radiation from the sun and the factors making up the earth's climate. [Adapted from W. W. Kellogg and S. H. Schneider,* Science *186, 1163 (1974) Copyright 1974 by the American Association for the Advancement of Science.]*

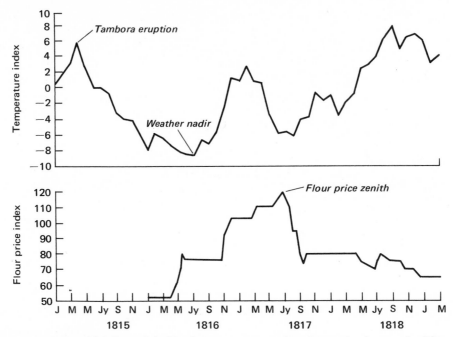

Figure 16-5 *Possible effect of the Tambora eruption on the asking price for a sack of flour in the London, England, commodities market. The temperature indices are 5-month running sums of departures from the long-term mean for the corresponding month, using data from historical reconstruction of the central English climate by Manley (1959). Flour prices are from Duffy's Farmers Journal. This is not a fortuitous relationship, in all likelihood, because newspapers of the time describe the effect of inclement weather conditions on harvesting operations and crop growth following the eruption. Also, we expect a lag in the effect of weather on crop prices, because most countries have about a one-year carry-over of grain in storage. (Source: K.E. Watt,* Principles of Environmental Science, *McGraw-Hill, New York, 1973.)*

Think of the social and political events that might follow this sort of climatic event. Humans are in a position to affect climate in a similar fashion by increasing the amount of particulate matter in the atmosphere. Human sources of particulates and their effects are detailed in Chapter 18.

Changes in climate affect agriculture because of the possibility of crop failures. Temperatures may be too low, there may be more or less rainfall than normal, or more severe storms. Wildlife is also affected. For example, the survival of fish eggs or of young hares depends on certain temperatures. Furthermore, effects of climate on one species may in turn affect other organisms in food webs. An interesting example of this was pointed out by K. E. Watt (Fig. 16-6).

THE SULFUR CYCLE

Two other natural cycles that humans affect are the sulfur cycle and the nitrogen cycle. In the *sulfur cycle*, sulfur is changed from one form to another by a variety of microorganisms in the soil and sediments (Fig. 16-7). Plants and animals require sulfur as sulfate (SO_4) in order to build organic materials such as amino acids and proteins.

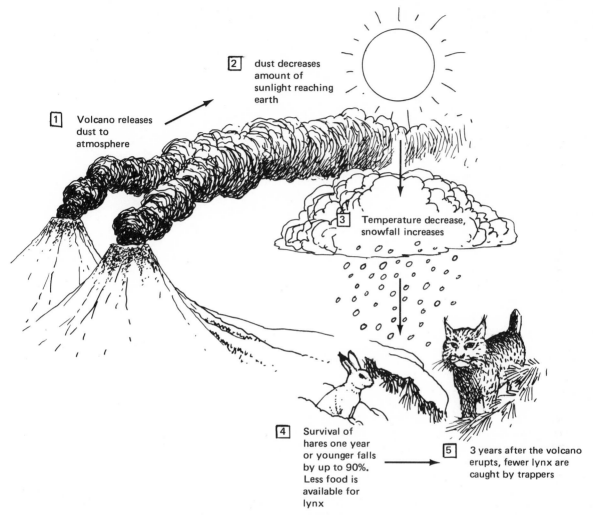

2 dust decreases amount of sunlight reaching earth

1 Volcano releases dust to atmosphere

3 Temperature decrease, snowfall increases

4 Survival of hares one year or younger falls by up to 90%. Less food is available for lynx

5 3 years after the volcano erupts, fewer lynx are caught by trappers

Figure 16–6   *The volcano–temperature–snow–hare–lynx–trapper effect. Note that the effect of the volcano was not seen in the lynx population until 3 years after the eruption. In a similar way, the effects of human caused climate changes may not be seen immediately.*

In the deep sediments and soils, sulfur is changed to iron sulfides at the same time that phosphorus is changed from an insoluble to a soluble form. In this way, cycles of sulfur and phosphorus interact.)

(Human activities, notably the burning of high-sulfur coal to generate electric power, are affecting the sulfur cycle by overloading the atmospheric pool of sulfur with sulfur dioxide, a compound nor-mally present at very low levels (see the effects this has in Chapter 18).)

THE NITROGEN CYCLE

(The *nitrogen cycle* (Fig. 16–8) involves changes in the form of nitrogen by microorganisms. And,

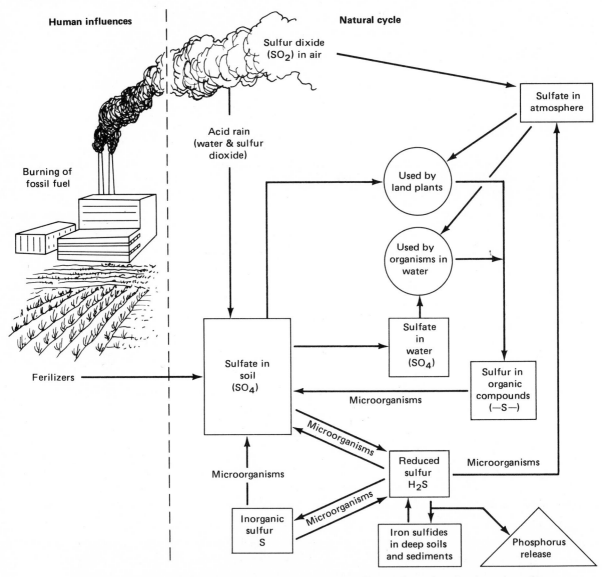

Figure 16-7   *The sulfur cycle. There is a large pool of sulfur in the soil and sediments and a smaller pool of sulfur in the atmosphere. Plants and animals use sulfur, as sulfates, from soil, water, and air. This sulfur is then incorporated into organic sulfur compounds such as sulfur containing proteins. Sulfur is changed from sulfates to reduced sulfur compounds and back again by a variety of microorganisms in the soil and sediments.*

like sulfur, nitrogen is an essential ingredient in biological materials.)

(Nitrogen is present in a very large pool in the atmosphere, as nitrogen gas (N_2). The air we breathe is about 80% nitrogen. This nitrogen becomes part of biological materials almost entirely through the bacteria and algae that can "fix" atmospheric nitrogen into organic compounds and nitrates. Legumes,

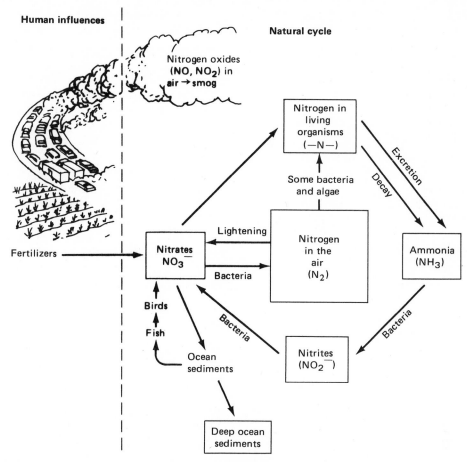

Figure 16–8   *The nitrogen cycle. Plants and animals generally require nitrogen in the form of nitrates, except for some bacteria and algae that can "fix" nitrogen gas from the air into organic materials. Biological materials are broken down into ammonia, which is cycled by bacteria through nitrites and back to nitrates.*

such as clover, alfalfa, soy beans, walnut and locust trees, form little nodules in their roots in which such nitrogen fixing bacteria live. Farmers often fertilize their land naturally by growing these types of crops and then plowing them under.)

(Nitrogen is lost from the cycle to deep sea sediments but this is just about balanced by additions from volcanoes. Humans interfere with the nitrogen cycle by producing large quantities of nitrogen oxide gases which, like sulfur dioxide, are normally present in very small amounts. These gases contribute to the formation of smog (Chapter

19). In addition, excess nitrate from agricultural fertilizers can lead to overgrowths of algae, or eutrophication (Chapter 9).)

INVERSIONS

(Before leaving the subject of atmosphere and climate, a special weather condition should be explained. This is the condition known as an *inversion*. Inversions are weather phenomena that tend to trap air pollutants in the lower layer of air. To understand how this happens it is helpful to examine the way the air pollutants in a city are normally carried off into the upper atmosphere. Upward currents known as *thermals* normally serve this purpose. When the sun warms the earth, the layer of air closest to the earth is also warmed. This layer is normally the warmest of all levels of the atmosphere. (The temperature of the air falls at a fairly steady rate the further out one moves in space.))

(A warm gas is less dense than a cooler one and will rise up through the cool one. The pollutants contained in the warm air are carried along in the upward currents and are diluted high up in the atmosphere. At the same time, clean air from the surrounding countryside may replace the rising city air. These natural currents along with strong surface winds are the means by which pollutants are usually dispersed from their original sources. When such currents and winds exist, the buildup of

dangerous substances will not tend to occur. Sometimes, however, the pattern is reversed: Cooler air lies next to the earth and a warm air mass sits on top of it. This phenomenon is termed an inversion (Fig. 16–9). When currents and winds are slight, and when cool air huddles close to the earth, the stage is set for an air pollution episode. All that is needed are the players: sulfur oxides, nitrogen oxides, particulate matter, hydrocarbons, and carbon monoxide.)

(Inversions commonly build up in the following way. When night falls, the warmth of the day, which has been accumulating in the earth, in buildings, and in vegetation, radiates away from the earth into space. Cold air, which is denser than warm air, naturally settles towards the earth's surface. Now, in the hours just before dawn, the temperature in the layer nearest the surface reaches its

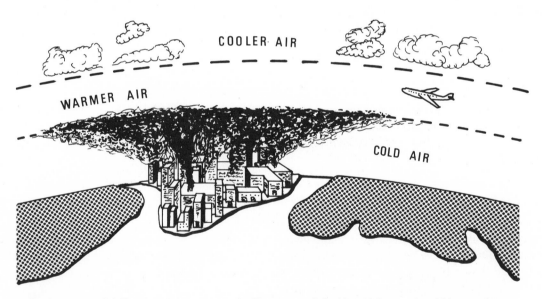

Figure 16–9   *A fall morning in a city on the East coast of the United States. A cold layer of air lies next to the earth extending to about 2000 feet. Warmer air lies above the cold air mass. Smoke plumes from chimneys will cool as they rise in the cold air. They become so dense that they are heavier than the air in the layer of warm air above. Because the plumes become heavier they will not escape into the upper atmosphere. Pollutant levels will increase dramatically.*

lowest level. A layer of cold air surrounds the surface of the earth. Pollutants have accumulated in the lower atmosphere through the night, because there is no warmth from the sun to set warm air currents in motion. (We have assumed also that surface winds are light.))

(As the sun begins to rise, the pollutants trapped during the night are joined by new emissions. The autos carrying people to work and trucks carrying goods discharge carbon monoxide, nitrogen oxide, and hydrocarbons to the atmosphere. Industrial activities discharge these contaminants plus sulfur oxides and particulate matter. Concentrations in the cold lower air begin to increase even further. The sun attempts to work. Vertical thermal currents are set in motion and these will rise until their temperature has reached that of the surrounding air. However, if the cold layer is thick and extends far above the surface of the earth, the currents may be unable to escape into the upper air. The concentrations of contaminants grow through the day as little pollution escapes. Sometimes, these conditions are repeated for several days. The level of the contaminants becomes increasingly dangerous. Strong winds are needed desperately to break up the cool lower layer. Sometimes they arrive in time; sometimes they do not.)

The following chapters explain the effects air pollutants have on living organisms. Also explained are the ways in which the pollutants are generated and how they can be controlled.

Questions

1. Why do most scientists believe life began in water rather than on land?
2. In what ways do human influences threaten to unbalance the carbon cycle?
3. Climate is important in determining how living communities are distributed over the earth's surface. What is meant by climate?
4. Describe the sulfur cycle. In what ways do humans influence this cycle?
5. In what ways do human activities influence the nitrogen cycle?
6. What is an inversion? How can inversions increase the effects of air pollution emissions? Outline the process that results in the formation of an air pollution inversion. Begin with a description of the "normal" temperature stratification of the earth's atmosphere.

Further Reading

Kerr, R. A., "Climate Control: How Large a Role for Orbital Variations," *Science* **201**, 144 (14 July 1978).

Levin, H. L., *The Earth Through Time*, Saunders, Philadelphia, 1978.

Siegenthaler, U. and H. Oeschger, "Predicting Future Atmospheric Carbon Dioxide Levels," *Science* **199**, 388 (27 January 1978).

Watt, K. E., *Principles of Environmental Science*, McGraw-Hill, New York, 1973.

Chapter 17

Air Pollution: Air Pollution Episodes, Carbon Dioxide, and Carbon Monoxide

AIR POLLUTION EPISODES: THE AWAKENING

Strangely, our recognition on a national scale of the permanent threat of air pollution comes long after serious events in the past momentarily rivetted our attention on these hazards. These events serve, today, as constant reminders of the consequences of doing nothing. As such, they are the touchstone of the clean air movement.

The earliest recorded incident of air pollution in the United States occurred in October 1948. A blanket of cold air enveloped the town of Donora, Pennsylvania, and refused to rise. Pollutants were trapped in the stable air mass and began to accumulate. No air pollution sensors were present to signal the dramatic rise in pollution in the small valley town. Most of the residents carried on their normal occupations, but as evening approached, the evidence could no longer be ignored; people were dying. Twenty deaths that day and afterward were attributed to the fog, and America suddenly entered an era of awareness of air pollution.

Four years later, in December 1952, an air pollution episode gripped the city of London for five successive days. In the days following the onset of the famous London Fog, over 4000 deaths in excess of the number expected for that time of year were recorded. The elderly and those with chronic respiratory diseases were often victims and suffered most. Episodes occurred again in London in the years following, but fortunately the tragic extent of the 1952 fog was never repeated.

Episodes have occurred in New York City, and alerts have been announced in other major cities in recent years. The mathematical science of statistics and the monitoring of air pollution have, however, provided new information. When our senses have failed to tell us of the decline in the quality of our air, chemical sensors have detected pollutants. When the effects on health have not been visible to the untrained eye, statisticians have shown deterioration of health at times of high pollution levels.

These episodes or incidents are not the result of sudden and vast emissions of pollution from many sources. That is, we cannot contend that they are caused by certain individuals or industries that have taken a momentary irresponsible action. In fact, industries have probably been operating in a "business as usual" fashion. The pollutants present during an episode are typically only a portion of the normal output from domestic, industrial, and transportation activities.

The episode represents a massive accumulation, a gathering together of these pollutants on a huge scale. The massive accumulation is the result of weather conditions (inversions) that hinder the natural mixing of the atmosphere.[1] Because the emissions leading to and continuing during an

---

1 Inversions are described more fully in Chapter 16, p. 359.

THE DONORA STORY, OCTOBER 1948

During the last week of October 1948 a heavy smog settled down over the area surrounding Donora, Pa. Weather men described it as a temperature inversion and anticyclonic conditions characterized by little or no air movement, prevailing over a wide area encompassing western Pennsylvania, eastern Ohio, and parts of Maryland and Virginia. This prolonged stable atmospheric condition was accompanied by fog and permitted the accumulation of atmospheric contaminants resulting in dense smog, particularly in highly industrialized areas. Smogs of short duration are not unusual and except for discomfort due to irritation and nuisance of the dirt and poor visibility, no unusual significance is attached to such occurrences.

This particular smog encompassed the Donora area on the morning of Wednesday, October 27. It was even then of sufficient density to evoke comments by the residents. It was reported that streamers of carbon appeared to hang motionless in the air and that visibility was so poor that even natives of the area became lost.

The smog continued through Thursday, but still no more attention was attracted than that of conversational comment.

On Friday, however, a marked increase in illness began to take place in the area. By Friday evening the physicians' telephone exchange was flooded with calls for medical aid, and the doctors were making calls unceasingly to care for their patients. Many persons were sent to nearby hospitals, and the Donora Fire Department, the local chapter of the American Red Cross, and other organizations were asked to help with the many ill persons.

There was, nevertheless, no general alarm about the smog's effects even then. On Friday evening the annual Donora Hallowe'en parade was well attended, and on Saturday afternoon a football game between Donora and Monongahela high schools was played on the gridiron of Donora High School before a large crowd.

The first death during the smog had already occurred, however, early Saturday morning–at 2 a.m., to be precise. More followed in quick succession during the day and by nightfall word of these deaths was racing through the town. By 11:30 that night 17 persons were dead. Two more were to follow on Sunday, and still another who fell ill during the smog was to die a week later on November 8.

On Sunday afternoon rain came to clear away the smog. But hundreds were still ill, and the rest of the residents were still stunned by the number of deaths that had taken place during the preceding 36 hours. That night the

town council held a meeting to consider action, and followed with another on Monday night. By this time emergency aid was on its way to do whatever possible for the stricken town....

From "Air Pollution in Donora, PA," Public Health Bulletin No. 306,
U.S. Public Health Service 1949

---

episode are no more than normal, if an episode occurs once in a particular city, the potential exists for an episode to occur again. That potential will only be decreased if some positive action is taken to reduce the output of pollutants.

Sadly, the weather conditions that now threaten to bring on air pollution episodes were regarded in the past as among the finest weather of the year (Fig. 17-1). In rural areas, the fine clear days of fall that follow cool cloudless nights are still enjoyable. In polluted urban areas, however, such weather is a danger signal. This weather may lead to the inversions which act to trap air pollutants in the lower layers of the air. Weather conditions such as these subjected the town of Donora to a deadly concentration of pollutants over a quarter of a century ago. Only the control or removal of air pollution sources can prevent it from happening again.

What are the air pollutants which so injure our health? Where do they come from? How do they harm us? How can we control them? In the sections that follow we shall describe the various air pollutants, their origins, their effects, and methods to decrease them.

CARBON DIOXIDE

Carbon Cycle

The oxidation of fossil fuels produces two oxides of carbon. One, carbon dioxide, is not poisonous. It could, however, change the climate of the earth. The other, carbon monoxide, has known harmful effects on humans.

Figure 17-1 *The sunny days of Fall are beautiful to behold in rural areas. In urban areas, however, these same days can threaten air pollution episodes. If the air over a city remains windless for too many days, the pollutants from autos, industries, homes, and ships may become dangerously concentrated to the point where human health can be harmed. (Grant Heilman Photography.)*

As we have said before, plants utilize carbon dioxide and animals and plants produce it. Plants *need* carbon dioxide for photosynthesis. Photosynthesis is the manufacture of carbohydrates from carbon dioxide and water in the presence of light. Oxygen is produced during photosynthesis. Most plant and animal species consume oxygen in respiration. They also produce carbon dioxide as a waste product of that respiration. Hence, carbon dioxide is naturally present in the atmosphere in reasonable quantities. A normal sample of air contains about 0.05% carbon dioxide by weight.

Not only is carbon dioxide consumed in photosynthesis; it also may dissolve in the vast oceans. When carbon dioxide dissolves in the ocean, carbonic acid is produced at a *very low concentration*. The carbonic acid dissociates in part to bicarbonate and carbonate ions. These ions combine with calcium and magnesium from the natural weathering of rocks which have been carried into the ocean. The reaction of the calcium ion with carbonate ion produces calcium carbonate. We know this relatively insoluble substance as limestone. Magnesium and calcium may jointly react with the carbonate ion to produce dolomite.

These precipitation reactions remove carbonate from the water, and make room for more carbon dioxide to dissolve. Thus, there are two natural mechanisms for carbon dioxide removal from the atmosphere; (1) solution in the ocean followed by precipitation and (2) the utilization of carbon dioxide by green plants in photosynthesis (Fig. 17–2).

Carbon Dioxide from Fossil Fuels

Carbon dioxide is now being produced in massive quantities from the burning of fossil fuels. The complete oxidation of carbon produces the gas carbon dioxide:

$$C + O_2 \longrightarrow CO_2$$

In the late 1970s, the United States alone was burn-

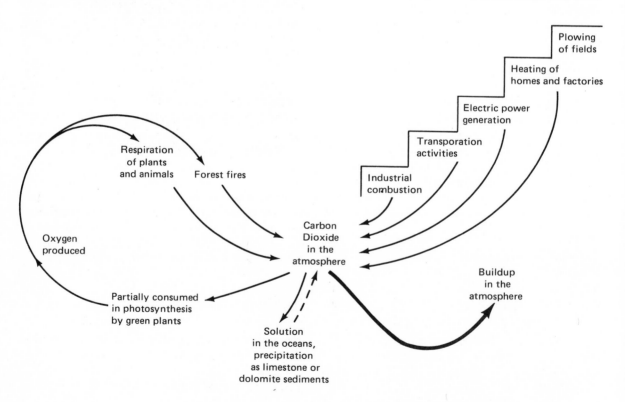

Figure 17–2   *The carbon dioxide cycle as modified by human activity.*

ing annually about 400 million tons of coal in electric power plants. We also consumed annually a comparable tonnage of gasoline in that period. This worldwide increase in combustion has increased the rate at which carbon dioxide enters the atmosphere. Carbon dioxide from fossil fuel combustion now adds about 5%–7% of the carbon dioxide production from green plants. About half of this additional quantity is accumulating. That is, it is not being consumed in photosynthesis; nor is it being dissolved in the ocean.

As time goes on, the amount of carbon dioxide produced annually from combustion is expected to grow enormously. As an example, one projection of the use of coal for electric power generation in the United States shows consumption doubling from 1980 to the year 2000.

The extent of the increase in fossil fuel combustion depends on energy conservation measures taken. It also depends on the extent to which nuclear and solar energy replace fossil fuels. Total fossil fuel combustion might be expected to increase by 3% or 4% annually through the year 2000. On that basis, the amount of carbon dioxide in the atmosphere could potentially increase by as much as 25% from today's level by the year 2000. (Another factor that could hasten the increase in carbon dioxide levels is the destruction of forests for agriculture, as is happening in countries like Brazil. Carbon dioxide intake by plants would then be decreased.)

The Greenhouse Effect

For nearly three quarters of a century, scientists have been aware of this buildup of carbon dioxide. It is now being carefully watched. The concern of thoughtful scientists is that carbon dioxide will trap heat in the earth's atmosphere. The effect of carbon dioxide has been compared to the effect of the glass panes of greenhouses. The panes let sunlight into the greenhouse, which is warmed by the solar radiation. At night, the heat radiates away from the buildings. However, the glass panes decrease the rate of heat radiation out of the structure.

Carbon dioxide is expected to act in a similar way. Solar energy would continue to reach the earth without being affected by the gas, and the earth would be warmed. But the radiation of heat away from the earth would be slowed by the carbon dioxide. The earth would then heat up (Fig. 17-3). Could the ice caps melt? Some scientists think it entirely possible.

From the turn of the century until World War II, a small warming trend was noted. Since that time, however, the trend has reversed, and a modest cooling has occurred although the trend now appears to be only in the Northern Hemisphere. The theory was that the earth would warm as combustion went on. That theory, however, would not predict a cooling trend. From appearances alone, the theory has not stood up.

Another factor, though, could have prevented the expected warming. From combustion comes not only carbon dioxide and water but masses of tiny unburned particles. These particles, suspended in the atmosphere, are reflecting sunlight away from the earth and robbing the earth of the heat the sun's radiation would have provided. Is this the cause of the cooling trend? Scientists do not know the answer. Nor are they sure of the effects of carbon dioxide on climate. There may be built-in mechanisms that soften the heating trend. One likely possibility is that a warming trend would increase evaporation from the oceans. This, in turn, would increase cloud cover and decrease the quantity of sunlight reaching the earth. For the moment, however, we cannot predict the consequences of the carbon dioxide buildup we have observed and which we expect to continue (Fig. 17-4).

CARBON MONOXIDE

Motor Vehicles Are the Major Source

(When carbon is not oxidized completely, the colorless and odorless gas, carbon monoxide, results.) Carbon monoxide surrounds the city dweller

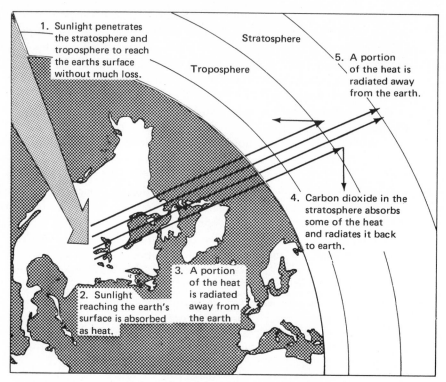

1. Sunlight penetrates the stratosphere and troposphere to reach the earths surface without much loss.

Stratosphere

Troposphere

5. A portion of the heat is radiated away from the earth.

4. Carbon dioxide in the stratosphere absorbs some of the heat and radiates it back to earth.

2. Sunlight reaching the earth's surface is absorbed as heat.

3. A portion of the heat is radiated away from the earth

Figure 17–3   *The effect of carbon dioxide on the heat balance of the earth.*

in concentrations greater than any other air pollutant. Because the gas is colorless and odorless, however, we cannot detect it with our senses. It cannot be seen; it cannot be smelled, but it is present nevertheless (Fig. 17–5).

The greatest source of carbon monoxide in our cities is the motor vehicle. Over 120 million vehicles are on the road in the United States and are discharging the gas into the atmosphere. In most cities, over 90% of the carbon monoxide in the air comes from the incomplete combustion of carbon in motor fuels.

$$C + \tfrac{1}{2} O_2 \longrightarrow CO$$
carbon in
motor fuel

There is another source of carbon monoxide to which people are exposed. Only smokers and their neighbors, however, have this special privilege. We can compare the individual who smokes moderately and lives in a clean environment with an individual who does not smoke and lives in a highly polluted environment. The smoker will absorb daily twice as much carbon monoxide as his nonsmoking counterpart.

Carbon Monoxide and Hemoglobin

Carbon monoxide competes with oxygen for the molecule that carries oxygen to the cells. For this reason, at high enough concentrations, it is a deadly poison. Hemoglobin, a complex protein present in the blood, transports oxygen from the lungs to the cells and carries carbon dioxide from the cells back to the lungs. Carbon monoxide, however, at-

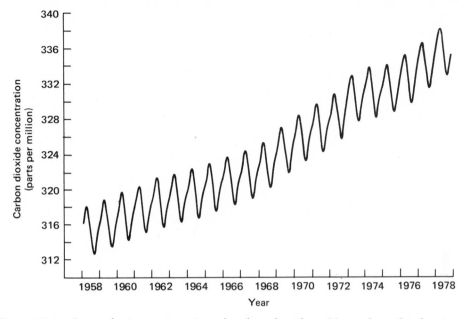

Figure 17-4   *Atmospheric concentration of carbon dioxide at Mauna Loa. Combustion of fossil fuels is producing carbon dioxide faster than green plants can use it and faster than the ocean can dissolve it. As a consequence, levels of carbon dioxide in the atmosphere are building up and will build up faster as combustion rates increase. Scientists suggest that carbon dioxide may prevent the earth from radiating heat back to space and so cause a warming trend. Whether the trend will occur and the ice caps melt are still large unknowns. (Source: R.B. Bacastow and C.D. Keeling, "Models to Predict Future Atmospheric CO$_2$ Concentrations," in* Workshop on the Global Effects of Carbon Dioxide from Fossil Fuels, *edited by Elliott and Machta, NTIS, U.S. Department of Commerce.)*

taches more strongly to hemoglobin than oxygen does. The more carbon monoxide present in the air, the more hemoglobin is "tied up" and the less oxygen can reach the cells. (What happens to you when your hemoglobin is tied up? See p. 372.)

In work places such as tunnels and loading platforms, carbon monoxide may reach concentrations of 70 milligrams per cubic meter. After 8 to 12 hours at such a concentration, a workman may lose the service of 10% of his hemoglobin. If an individual were exposed to an average of 16 milligrams per cubic meter for an 8 hour period, about 3.0% of his hemoglobin would be unavailable for oxygen transport. This level is not uncommon on city streets. At this point, the impact of smoking on the avail-

ability of hemoglobin can be seen. Even the *light* smoker will tie up 3.0% of his hemoglobin *by smoking alone.* This is equivalent to being in a room with a carbon monoxide concentration of 16 milligrams per cubic meter. The moderate smoker achieves nearly a 6% loss of useful hemoglobin.

Recently proposed air quality standards for carbon monoxide are 10 milligrams per cubic meter, averaged over 8 hours. This level is not to be exceeded more than once a year. We can see that smoking alone produces a personal environment worse than that required by the air quality standards. From 1973 to 1976 a Los Angeles resident would have experienced 90 or more days per year on which carbon monoxide levels exceeded the stand-

Figure 17–5   *Carbon monoxide emmissions for 1975 by Air Quality Control Region (AQCR), in thousands of tons per year. (Source: U.S. Department of Energy, December 1977.)*

ard. In 1976, 120 days exhibited levels above the standard in Los Angeles. New York City, specifically Manhattan, is even worse.

The effects of carbon monoxide at low levels are subtle and difficult to determine. When 3% of the hemoglobin is bound by carbon monoxide, tests have shown that the ability of individuals to judge differences in light intensities has decreased. In experimental situations that produced levels of 10% unusable hemoglobin, the skills needed to drive a car were impaired, responses to brake lights and to the speed of the auto ahead were poorer. The possible influence on safety is obvious.

The effects of low levels of carbon monoxide on health are not well established, because when carbon monoxide concentrations are high, the concentrations of other pollutants are typically also high.

Emission Controls–A Solution?

The initial control of carbon monoxide emissions was achieved in part by increases in the ratio of air to fuel in the gasoline engine. Additional air is provided to burn the gasoline more completely, resulting in lower carbon monoxide in the exhaust. Some vehicles also utilized "air injection." Air was mixed with the exhaust stream for a final combustion of the carbon monoxide to carbon dioxide.

$$CO + (½) O_2 \longrightarrow CO_2$$

The typical motor vehicle of the mid-1960s would exhaust an average of 73 grams of carbon monoxide in every mile of travel. A standard of 23 grams of carbon monoxide per mile was set for 1971 vehicles, and tests by the Environmental Protection Agency established that new passenger cars met the standard easily. Carbon monoxide emissions of 3.4 grams per mile are to be achieved by autos by 1981 (Fig. 17–6). To achieve this standard, the exhaust gases are mixed with a stream of air in the presence of a catalyst. Further oxidation of the remaining carbon monoxide occurs in this "catalytic converter." The catalyst system appears for the present to be the chosen method to reduce carbon monoxide emissions. Nevertheless, auto manufacturers are probably studying the potential of the stratified-charge, dual carburetor engine system. This system has been used in the Honda, a vehicle of Japanese manufacture, and has achieved the required reduction in carbon monoxide.

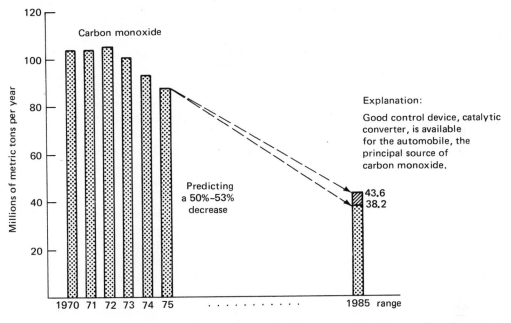

Figure 17–6   *Trends in air pollution emissions of carbon monoxide. (Source: U.S. Environmental Protection Agency, SEAS model, REPS model.)*

The Effects of Carbon Monoxide on the Body

In an atmosphere rich in carbon monoxide, death results from asphyxiation. This is another way of saying that the body tissues become starved for oxygen. Scientists have suspected for more than a decade that the concentrations of carbon monoxide found in our cities were harmful. Nevertheless, it has only been in the last few years that the needed data have been obtained to brand carbon monoxide in the air as a serious health hazard.

Carbon Monoxide Disrupts the Transport of Oxygen in the Body

To understand the danger of small concentrations of carbon monoxide, we need to review the process that supplies oxygen to the tissues of the body. Oxygen, a component of the air, is brought into the lungs with each breath. At the alveoli, the small sacs at the end of the tree-like branches of the lungs, the oxygen gas is transferred into the bloodstream. In the blood, the oxygen attaches to hemoglobin, a complex protein molecule carried in the red blood cells (the erythrocytes). The red cells transport the oxygenated hemoglobin through the arteries of the body and finally to the capillaries, the narrowest tubes of the arterial circulatory system. Here, oxygen is transferred across the walls of the capillaries into the cells. (See Fig. 27–7.)

Carbon dioxide, one of the waste products of cell activity, flows in the opposite direction, from the cell into the bloodstream. Some of the carbon dioxide takes the place of the oxygen that was attached to hemoglobin, and some of the gas dissolves in the blood fluid as bicarbonate ion. The blood, now rich in carbon dioxide, returns via the veins to the lungs. There, carbon dioxide diffuses from the blood into the alveoli, while oxygen from the air in the sacs moves into the blood. The carbon dioxide is then exhaled in the breath.

This normal pattern of transportation is disturbed when carbon monoxide is present in the air we breathe. Even minute quantities can disrupt oxygen transport, because carbon monoxide is about 200 times more attractive to hemoglobin than is oxygen. The carbon monoxide attaches tightly to the hemoglobin and deprives oxygen of its carrier to the cells. The greater the amount of carbon monoxide present in the air, the more hemoglobin is "tied up" and unavailable for oxygen transport. When hemoglobin binds carbon monoxide to itself, it is called carboxyhemoglobin. In contrast, when hemoglobin has oxygen attached to it, it is called oxyhemoglobin.

Table 17–1 shows how even small quantities of carbon monoxide gas in the air produce high levels of carboxyhemoglobin in the blood. Note that the table lists carboxyhemoglobin percentages after eight to ten hours of breathing the contaminated air. This level is termed the "equilibrium value." A longer exposure at that concentration will not increase the percent of carboxyhemoglobin any further. Note that even when the air is free of carbon monoxide, some small amount of hemoglobin is tied up. This carbon monoxide stems from natural body processes.

Low Levels of Carbon Monoxide Hinder Judgment and Increase the Tendency Toward Heart Attacks

Individuals with increased levels of carbon monoxide-bound hemoglobin are subject to two important kinds of effects. One effect is a decreased ability

TABLE 17–1   Small Amounts of Carbon Monoxide in the Air Can Remove a Great Deal of Hemoglobin from Service

Level of carbon monoxide per cubic meter of air (milligrams)	Estimated percent of hemoglobin tied up as carboxyhemoglobin after 8 to 10 hours[a]
0	0.4
5	1.0
10	1.6 (1972 air quality standard and resulting carboxyhemoglobin)
20	2.9
30	4.1
40	5.4
50	6.6
60	7.8

[a] This is the percentage of hemoglobin to which carbon monoxide is attached. This hemoglobin is not available to carry oxygen to the cells.

to perceive one's personal environment. This perception has been measured by a number of tests. For instance, individuals have been asked to report on sound signals. At levels of carboxyhemoglobin in the range of 2% to 5% of total hemoglobin, the signals were often missed. The ability to tell which of two tones was longer in duration decreased when carboxyhemoglobin was in the range of 2.5% to 4%. The processes of the mind are also interferred with. Simple tests, such as the additions of columns of numbers took longer to complete as carboxyhemoglobin levels increased. The ability to distinguish when a light becomes brighter also decreased. Tests of perception of brightness showed fewer correct answers even when carboxyhemoglobin levels were as low as 3%.

No formal link has yet been established between elevated carbon monoxide in the outdoor atmosphere and driving performance. Nevertheless, carbon monoxide levels may rise to 60 milligrams or more per cubic meter of air in freeway traffic, producing carboxyhemoglobin levels close to those at which skills are impaired. One scientist found higher carboxyhemoglobin levels in individuals involved in auto accidents, but we are still unable to blame carbon monoxide as a cause of auto accidents.

The past decade has brought us much new information on carbon monoxide's relation to heart attacks. Earlier, it was suspected that carbon monoxide might be a cause of heart attacks because it was found that more heart attacks occurred during periods of high carbon monoxide concentrations. Because other pollutants were also at high concentrations, no firm conclusions could be drawn. Now the suspicion is being substantiated in new data from individuals with angina pectoris. Angina pectoris is a chronic form of heart disease characterized by chest pain; it is less severe than the acute heart attack which is life threatening when it occurs. Individuals who suffer from angina pain have been tested for their susceptibility to carbon monoxide. They first were asked to breathe air with carbon monoxide at levels sufficient to raise their carboxyhemoglobin to 3%; then they were asked to exercise. At this blood level of carboxyhemoglobin, when the subjects were stressed the onset of the pain of angina arrived sooner than expected under normal conditions; furthermore the pain continued longer than expected.

Angina is only one form of heart disease. In total, 35% of the annual deaths in the United States are attributed to some form of heart disease. Carbon monoxide is known to decrease the supply of oxygen to the tissues. One tissue of enormous fraility if deprived of oxygen is the myocardium (muscle tissue of the heart). The experiments with angina patients tend to buttress the contention that carbon monoxide may be an agent in causing heart attacks. (We do not say that carbon monoxide causes heart disease itself, which is most frequently defined as a narrowing of coronary blood vessels.) Another factor supports the idea that carbon monoxide is one of the villains in bringing on heart attacks. The inhaled smoke of a cigarette may contain as much as 4% carbon monoxide. Heavy smokers may have carboxyhemoglobin levels as high as 10% to 15%. One study showed an average carboxyhemoglobin level of 4.4% in smokers of all types who were not exposed to carbon monoxide in their job. We know from statistical studies that when people quit smoking, their risk of heart attack is quickly reduced. It appears that a causative agent has been removed from their personal environment. That causative agent may be carbon monoxide, but other substances are absent as well when they stop smoking, so that we are still uncertain of the relationship.

Questions

1. Regions A and B experience air temperature inversions with the same frequency. Describe at least three differences between the two regions that could cause region A to experience air pollution episodes much more often than region B.

2. Name and describe the two natural processes by which carbon dioxide is removed from the atmosphere. What factors account for the fact that the total quantity of carbon dioxide in the atmosphere is increasing? Discuss the possible effects of this accumulation.

3. What is the major source of carbon monoxide in urban areas?

4. What process in the human body is hampered by carbon dioxide?

5. What health effects does carbon monoxide have?

6. What measures have been taken to reduce carbon monoxide emissions by motor vehicles? Are these measures working?

Further Reading

General References

"Air Quality and Automobile Emission Control," a report by the Co-ordinating Committee on Air Quality Standards prepared for the Committee on Public Works of the U.S. Senate (Vol. 2, Health Effects of Air Pollutants), 1974 (Serial No. 93-24).

Restricted to Carbon Monoxide, Nitrogen Oxides, and Photochemical Pollutants (all auto emissions). Contains the reports of three committees. Very thorough on these three pollutants.

"Controlling Air Pollution," American Lung Association, 1974.

A very well-done short book on methods to control air pollution from stationary and mobile sources. Drawings and text are excellent. Writing level is for the layperson.

"Proceedings of the Conference on Health Effects of Air Pollutants," prepared for the Committee on Public Works of the U.S. Senate by the Assembly of Life Sciences, November 1973 (Serial Number 93-15).

A massive collection of papers on health effects, some very technical, others quite readable; a wealth of information.

Waldbott, George, *Health Effects of Environmental Pollutants*, 2nd ed., C. V. Mosby Company, St. Louis, 1978.

This well-illustrated (136 pages of illustrations) book has a wealth of information and documentation on all of the major air pollutants. It is a technical work, but clearly written. The book could serve as an excellent reference for the health aspects of air pollutants.

Carbon Monoxide References

Air Quality Criteria for Carbon Monoxide, U.S. Environmental Protection Agency AP-62, 1970.

Control Techniques for Carbon Monoxide, Nitrogen Oxide and Hydrocarbon Emissions from Mobile Sources, U.S. Environmental Protection Agency AP-66, 1970.

While technical, these documents are very clearly written and very complete.

Carbon Dioxide Reference

Woodwell, G., "The Carbon Dioxide Question," *Scientific American* **238**(1), 34–43 (January 1978).

A clear exposition by a research scientist in the field of the global carbon cycle.

Chapter 18

Sulfur Oxide
and Particulate
Air Pollutants

SULFUR OXIDES

Sulfur Comes From Coal and Oil

(Compounds of sulfur, mainly derived from the burning of sulfur-rich fuels, pollute the air but these fuels are needed to produce heat, to generate electricity and to power machinery.)

Not all fuels contain significant quantities of sulfur but some coals may contain as much as 6% sulfur. Other coals may have as little as 0.5% sulfur. Coal is used in steel-making and is burned to generate steam to produce electricity. While the average sulfur content of the coals used in electric power generation is on the order of 2.5%, the coal used in steel-making must be much lower in sulfur. In fact, long-term contracts by the steel industry tie up a large proportion of the low sulfur coal in the United States.

For every million metric tons of coal burned in electric power generation, about 25,000 metric tons of sulfur are released. Of course, the sulfur is not released in the elemental form. It is released mainly as sulfur dioxide gas. In 1976, about 405 million metric tons of coal were burned for the sole purpose of generating steam to produce electricity. How many tons of sulfur were released?

Sulfur is also present in crude petroleum (raw or unrefined), but most crude oils contain less than 1% sulfur. Refining will coax much of the sulfur out of such petroleum products, as gasoline and kerosene. The waste sulfur compounds are "flamed" at the refinery; they are burned to sulfur oxides at the top of a tall metal stack. Gasoline and kerosene, then, are responsible for only a small portion of the sulfur compounds in the atmosphere. In the United States, probably less than 5% of the yearly release of sulfur compounds comes from the burning of gasoline. Home heating oil is also relatively low in sulfur, having an average of about 0.25%.

In the refining process much of the sulfur is shifted to residual oil, one of the most dense of the

TABLE 18–1   Sulfur Oxides Emissions, 1975: The Worst Regions and the Cleanest Regions[a]

The Five Air Quality Control Regions with the Worst Sulfur Oxides Emissions (in Tons per Square Mile)

Louisville, KY	352
Steubenville, OH	276
St. Louis	182
Chicago	171
Cleveland	162

Air Quality Control Regions with the Least Sulfur Oxides Emissions (in Tons per Square Mile)

S.E. Oklahoma	0.067
N.W. Oklahoma	0.066
N.E. New Mexico	0.032
Comanche, CO	0.029
N.W. Maine	0.001

[a] From U.S. Department of Energy, December 1977.

products of refining. Anywhere from 0.5% to 5% of residual oil may be sulfur, although special refining steps may be used to reduce the sulfur content of residual oil. Alternatively, low-sulfur crude oil may be used at the outset so that a low-sulfur residual oil is produced. Residual oil is used to heat apartments and institutions such as schools and hospitals. In the past it had been widely used in boilers to produce steam for electric generation primarily because it is lower in sulfur content than many coals.

Natural gas, in contrast to coal and oil, is almost free of sulfur. From this standpoint, it is an environmentally useful fuel. In 1976, burning coal and oil in electric generating plants produced 65% of the sulfur oxides emitted that year. Other coal and oil burning accounted for half of the remainder. Sulfur compounds also enter the air from the smelting and refining industries as well as from other sources, such as burning municipal wastes.

Sulfur is Oxidized to a Corrosive Mist

When coal or oil is burned, the sulfur in the fuels is oxidized. Two compounds are formed— sulfur dioxide and sulfur trioxide. Less than 3% of the sulfur is oxidized to the trioxide form during the initial burning. Although sulfur dioxide is the prin-

cipal sulfur compound formed in the combustion process, sulfur dioxide in the air is oxidized gradually to the trioxide. Any sulfur trioxide formed will immediately react with water vapor to yield sulfuric acid. The sulfuric acid is present in the air as a fine mist of liquid droplets. This very corrosive substance will damage many materials, including building materials such as marble and mortar.

Metal oxides are also formed in the process of combustion. Oxides of calcium and iron are typical metal oxides that enter the air in great quantities from the burning of coal. Particles from the burning of oil are less numerous. These oxides may react with sulfuric acid to yield the corresponding metal sulfate in particle form. As as example, calcium oxide may react with sulfuric acid to produce calcium sulfate and water. Iron oxides may react in a similar fashion. The reactions are summarized in Figure 18–1. Metal sulfate particles plus sulfuric acid droplets may account for between 5% and 20% of the particulate matter in urban air.

Effects on Materials, Plants, and People

Many materials may be attacked by the mists of sulfuric acid, among them carbonate building

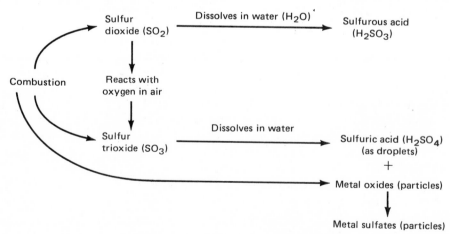

Figure 18–1   *The reactions of sulfur in air.*

materials such as marble and mortar. Calcium sulfate, water, and carbon dioxide are produced in the reaction. Metals such as steel, copper, and aluminum are corroded, and common cloth fabrics are damaged as well.)

(Many plants may be injured and stunted by sulfur oxide pollution.) Elevated levels of sulfur oxides common in our major industrial cities may cause a yellowing of plant leaves, an effect we shall discuss in more detail. People, however, are of greatest concern. Elevated levels of sulfur oxide have been linked to human illnesses and even to deaths. In air pollution episodes in New York, Osaka, and London, investigators have noted an increase in the normal death rate following periods of high concentrations of sulfur oxides. (Effects on human health of sulfur oxides and particulates are difficult to separate because the two kinds of pollutants tend to interact. The effects of the two together are discussed on p. 391.) In the famous London Fog of 1952, the concentration of sulfur oxides reached 4000 micrograms per cubic meter. In the 1962 episode in New York, the average concentration during one day was 2500 micrograms per cubic meter. Respiratory illnesses such as bronchitis are seen to increase with sulfur oxide levels. One study found the illness rate increased in an area where the average annual concentration was only 100 micrograms per cubic meter.

Rural areas have background concentrations at about 0.5 micrograms per cubic meter but urban areas have levels 50–1000 times greater. The people of Chicago, for example, lived at an *average* concentration of 470 micrograms per cubic meter in 1964. This is, however, an extreme, and levels in Chicago have probably decreased under the stimulus of the air quality standards issued in 1971 (Fig. 18–2).

The daily standard for sulfur oxides is 365 micrograms per cubic meter (an average over 24 hours), and this level is not to be exceeded more than once per year. An air pollution "alert" is to be called when the 24-hour average exceeds 800 micrograms per cubic meter.

Acid Rain

(High concentrations of sulfur dioxide and its derivatives may cause acute injury to plants.) After such acute injury, where sulfur dioxide concentrations may reach nearly 3000 micrograms per cubic meter, leaves, needles, and other plant tissues may appear as though bleached. Eventually, leaves or needles may take on a scorched, reddish-brown look and the damaged foliage may fall from the plant. Smelters that refine copper and lead ores that contain sulfur may pollute extensive areas with their gaseous emissions. A near absence of vegetation may be observed in the amazing scene around the smelting operation at Sudbury, Ontario, for example.

Even where sulfur dioxide levels average only on the order of 100 micrograms per cubic meter, plants may have a yellow discoloration. Such fruit trees as apple and pear and such forest trees as ponderosa pine and tamarack (larch) are susceptible to damage from sulfur oxides. The cotton plant is also susceptible, as are alfalfa and barley.

The visible damage to plants is only part of the effect of sulfur oxides. There is another and more subtle damage, which scientists are only beginning to observe and understand. When, for instance, sulfur dioxide is emitted from electric power plants, some of the gas exerts its effect and is consumed in reactions in the locality of the plant. We have already described damage that occurs in areas near the source of emissions. The higher the smoke stack, however, the more of the gas is dispersed into the moving air stream of weather and climate patterns. That is, the gas becomes more diluted and is spread far from its original source.

(People who speak for much of the electric power industry argue that such tall stacks are the correct way to deal with sulfur oxides. Their philosophy is one of "dilute and disperse." In this way, they can avoid the high concentrations of sulfur dioxide at ground level, which have been shown to have effects on human health.) In contrast, the Environmental Protection Agency of the United States

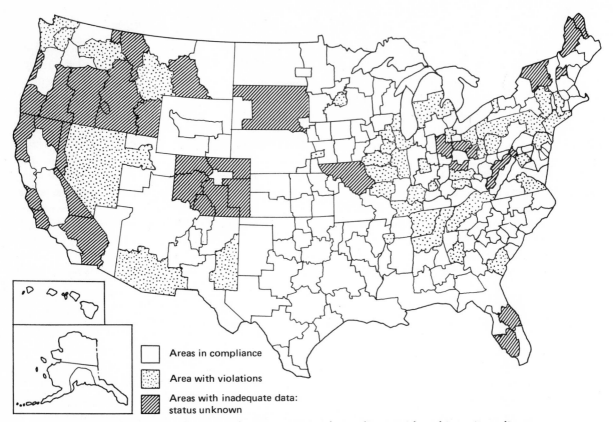

Figure 18-2 *Air quality control regions: status of compliance with ambient air quality standards for sulfur dioxide. (Source:* Energy/Environment Fact Book, *U.S. Environmental Protection Agency, 1978.)*

advocates actual *removal* of the sulfur oxides from stack gases by use of scrubbing devices. The limestone "scrubber" captures sulfur oxides by reacting them with lime. The argument and debate over tall stacks versus scrubbers is still going on, although it appears that scrubbers will finally be required. The following discussion should begin to explain why some scientists believe that "tall stacks" are not the answer.

First in the Scandinavian countries and now in the northeastern United States, we are discovering that rainwater has become highly acidic due to solution of sulfur dioxide in the water in the atmosphere. The more sulfur dioxide dissolves in water, the more acidic the water becomes. The

measure of acidity is the number of hydrogen ions per liter of water. Water molecules (H_2O) are normally dissociated into hydrogen ions (H^+) and hydroxyl ions (OH^-). In a sample of pure water, one would expect to find about 0.0000001 (one ten-millionth) of the water molecules dissociated into hydrogen ions and hydroxyl ions; in pure water the two ions are present in equal numbers. A solution with equal concentrations of hydrogen and hydroxyl ions is called "neutral." It is neither acid (having more hydrogen ions) nor basic (having fewer hydrogen ions).

Rainwater is not pure; it comes in contact with and dissolves carbon dioxide, a natural component of the atmosphere. The solution of carbon dioxide

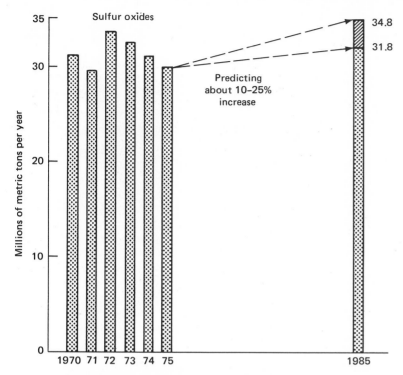

Figure 18-3   *Trends in air pollution emissions: sulfur oxides. (Source: U.S. Environmental Protection Agency, SEAS model, REPS model.)*

in water produces carbonic acid, a weak acid. The concentration of the hydrogen ions relative to the number of water molecules in unpolluted rain water containing dissolved carbon dioxide would be about 0.000001, or one hydrogen ion per million molecules of water. If you count the zeros before the decimal you will see that the hydrogen ion concentration has increased by a factor of 10. This level of hydrogen ions is presumed to be the natural condition for rainwater.

The rain falling in New England, on the other hand, has a ratio of about 0.0001 hydrogen ions per molecule of water (1 hydrogen ion per 10,000 water molecules). This is about a 100-fold increase above the expected concentration of hydrogen ions in rainwater. New England has acid rain because the area is downwind from the major industrial centers of the Northeast where sulfur-bearing fossil fuels are burned in enormous quantities (Fig. 18-4).

Not all the effects of acid rain are yet counted. Metal structures may corrode; rocks may be "weathered" more swiftly. Investigators in Sweden feel they have noted a reduction in forest growth in the past 15 years. The reduction is attributed to the acid rain carrying sulfur oxides from Germany, Holland, and other industrial countries. On the southern coast of Norway some streams have become acidic. As a result, the eggs of the salmon no longer survive in those waters and the salmon no longer breed there. Fish have disappeared from some of the lakes in the Adirondack mountains of New York State where high acid levels have built up.

Experiments have been performed in the laboratory in which plants were irrigated by water which had been made acid to the same extent as rainwater in the northeastern United States. The water was sprayed on the pine trees and tomato

Figure 18–4 *Regional impact of acid rainfall. Burning coal and oil releases sulfur oxides into the air. Carried over long distances, the sulfur oxides contribute to an acid condition in rainfall. The rainfall, in turn, may increase the acidity in bodies of water, making them unfit for fish. Timber stands may be stunted by the acid rain, and stone works corroded by it. (Source: Energy/Environment Fact Book Decision Series, U.S. Environmental Protection Agency, 1978.)*

plants to mimic the way rain arrives. Pine needles grew to only half their normal length in these experiments and fewer tomatoes were produced than normally.

(There may also be long-term changes in the productivity of the soil, because the acid rain may leach important nutrients out of the soil. This effect is speculated, however, and not observed)

Ways to Reduce Sulfur Dioxide Emissions

To scrub or not to scrub, that is the question. Whether 'tis more economical and efficient to

remove particulates and SO_2 from the end of the stack after the burning of coal

Or to clean coal before it is consigned to the flames. . . .

Environmental Science and Technology,
June 1974, p. 510.

Cleaning up the coal itself (In addition to the possibility of using expensive low sulfur coal, emissions of sulfur dioxide from power plants can be reduced by cleansing the coal of its sulfur, prior to use. There are two forms of sulfur in coal, inorganic sulfur and organic sulfur. Inorganic sulfur is the sulfur present as pyrites or metal sulfides such as iron sulfide (iron pyrite). Organic sulfur is sulfur that is chemically bonded to the carbon in the coal. Special washing steps are sufficient to remove the inorganic sulfur from the coal. The organic sulfur, in contrast, requires chemical treatment for removal. Sulfur as pyrites (metal sulfides) may range from 30%–70% of the total sulfur present in coal, but, on average, equal quantities of organic and inorganic sulfur are present.)

To cleanse the coal of the inorganic or mineral sulfur, the coal is first crushed to expose the mineral veins. The ground coal is then mixed with water in a large tank. Pyrite has a higher density than coal and hence sinks faster; the cleaned coal is skimmed from the top of the tank. Up to 500 to 1000 tons of coal can be processed per hour. Washing is most effective on coals with higher proportions of pyrites.

(Chemical cleaning of coal is a general name for a variety of processes all of which remove the sulfur that is bound to the carbon. Several dozens of processes are competing for research and development funds in this arena. Some of the chemical methods remove both the pyrite sulfur and a portion of the organic sulfur)

One promising cleaning method developed by the Batelle Memorial Institute mixes finely ground coal into water containing sodium and calcium hydroxide. Treated under pressure and high temperature, the coal is cleansed of most of the pyrites and half of the organic sulfur; these substances remain in the liquid phase. The coal is washed, then dried, and is ready for use. The method has the potential to make the high sulfur coal from the eastern United States an acceptable fuel in coal-fired electric plants. The coal could potentially be used without the need to "scrub" the sulfur dioxide from the stack gases, but a commercial scale application is needed to assess the cost.

(One process that appears to remove up to 60% of the organic sulfur is known as solvent refining of coal (SRC). Ground up coal is mixed with the organic solvent anthracene; nearly 95% of the carbon in the coal will dissolve in the anthracene. Exposed to high pressures and temperatures, the solution is then treated with hydrogen. The coal can be recovered either as a solid or liquid. The solid material, black and brittle, has less than 1% ash, though the original coal may have had anywhere from 8% to 20% ash. Sulfur is reduced to less than 1% as well, and the heating value has increased by 30% on a per ton basis. Solvent refining of coal, if it turns out to be commercially feasible, has bright prospects. With sulfur lowered sufficiently and ash nearly eliminated, it may be possible not only to avoid scrubbing the stack gases, but also to avoid electrostatic precipitation for particle removal.)

Scrubbers A number of methods are being and have been developed to "scrub" sulfur dioxide from the gas that exits the smokestack. The term scrub, unfortunately, is not scientifically descriptive; it simply means that the stack gases are cleansed of sulfur dioxide. The operations of the various "scrubbers" are based on chemical reactions with the sulfur dioxide in the stack gas of coal-fired power plants. The chemicals formed in these reactions may be waste products or marketable items.

Of the more than 50 control concepts, the one most thoroughly tested and reliable process is the lime-limestone wet scrubber. In this device, the flue gas is passed through a slurry mixture of limestone and lime in water. Sulfur dioxide is absorbed into the slurry and reacts to form calcium sulfite and calcium sulfate (gypsum). The flue gas is not only cleansed of about 80% of its sulfur dioxide but also 99% of the fly ash it carried. The flue gas has been cooled so much, however, that it must be reheated to make it buoyant enough to rise up the chimney for discharge to the atmosphere.

The resulting sludge must be disposed of in some way. One estimate is that a 1000-megawatt power plant can produce enough sludge in one week to cover one acre to a depth of three feet. Because it is a gloppy mass, it is not by itself a good landfill material. However, it appears that a hardener such as fly ash can be added to react with the slurry, producing a clay-like substance.

Because sludge disposal is seen as a serious problem, scrubbers that produce a marketable product instead appear very attractive. One such scrubber produces sulfur of high purity; another produces a dilute sulfuric acid. The sulfuric acid is not economical to transfer large distances; but the sulfur, which is used in such products as pharmaceuticals, industrial chemicals, and fertilizers, is quite compact. There is some concern that so many plants will be producing sulfur that the price will fall. The scrubber design known as the Wellman-Lord System is one that produces sulfur; it was tested in Gary, Indiana, and removed 91% of the sulfur dioxide while producing high purity sulfur. In early 1978, there were 29 scrubbers in operation, and another 28 were being built. A number of other installations had already been contracted for or were in the planning stages.

A recent concept combines both coal cleaning and scrubbing. Cleaning mineral sulfur from coal is relatively inexpensive but is limited to the fraction of the sulfur that is in pyrite (metal sulfide) form. The U.S. Environmental Protection Agency has suggested removing this fraction of the sulfur, then "scrubbing" the stack gas that is produced when the coal is burned. Since the stack gas will have a lower level of sulfur dioxide, it need not be as fully treated to meet emission standards.

During most of the decade of the 70s, portions of the utility industry kept up a steady criticism of the scrubbers. Financed principally by the American Electric Power Company, the largest electric power company in the country, a massive (3.6 million dollar) advertising campaign derided the scrubbers. They were called unreliable and expensive; the amounts of sludge were said to be enormous. However, as experience with scrubbers has grown, the reliability of the devices has proven to be very good. The mountains of sludge predicted have not materialized either.

The opponents of scrubbers advocated control of sulfur oxides by the use of very tall stacks that would disperse the gas more widely. When levels at the ground became hazardous, they suggested that the boiler burn a low sulfur fuel until the episode was past. Alternatively, it was suggested that power generation be cut back during such times. The opponents of scrubbers also suggested that the industry should wait to see how well coal cleaning will turn out before installing scrubbers.

Fluidized bed   Coal washing (physical cleaning) and scrubbers are existing technologies to eliminate sulfur dioxide fumes. In contrast, an experimental design to remove sulfur dioxide is the fluidized bed combustion system. The fluidized bed system burns coal as a conventional boiler does, but the coal is mixed with granular limestone and is layered on metal plates. Further, the air for combustion comes from below, passing through holes in the metal plates up past the coal. The air flow is so large that the particles of coal and limestone are lifted or floated (fluidized, as the name implies) above their bed on the plate.

The limestone reacts with the sulfur dioxide from the burning coal to form fine particles of calcium sulfate, which are carried off in the stack gases and will be removed with the fly ash, probably by the electrostatic precipitator. The temperature of the burning coal in the fluidized bed is lower than the temperature of burning coal in a boiler; and, as a consequence, fewer nitrogen oxides are formed. Thus both sulfur dioxide and oxides of nitrogen are lower for the fluidized bed combustion system. We should have results on the effectiveness of this experimental design by the mid-1980s.

PARTICULATE MATTER

Particulate matter is another serious air pollutant. Unlike the pollutants discussed earlier, particles are not of a single chemical type. Instead, numerous solid and liquid compounds are dispersed in the air from many sources (Fig. 18–5).

Particles Stem from Combustion and Industrial Activities

Although we mention them briefly here, two particle types are treated separately from the main discussion. Lead and asbestos have special properties and sources that require specific discussion. These substances are treated later.

The burning of coal produces not only ash particles (calcium silicates) and carbon particles, but also metal oxides such as calcium and ferric oxide. The metal oxide particles may react with the mist of sulfuric acid droplets. The reaction produces still other particles, the metal sulfates. The sulfuric acid droplets themselves are particles derived from the reaction of sulfur trioxide with water vapor. Both the acid droplets and the sulfate particles then, are derived in part from coal burning (see Figure 18–1).

The quantity of particles derived from coal burning is enormous. Fortunately, however, a very large proportion of the particles is removed from the stack gases. In 1975, about 3.2 million metric tons of particles were released to the atmosphere from coal-burning electric plants in the United States. Probably seven times that quantity was generated, but most of it was removed. The 3.2 million metric tons from electric plants, however, made up approximately one-fifth of the 16.4 million metric tons of particles reaching the air from *all* sources in 1975 (Fig. 18–6).

A small amount of ash does come from the incomplete burning of oil. Liquid hydrocarbons (compounds of carbon and hydrogen) and liquid derivatives of the hydrocarbons come from the incomplete combustion of gasoline and of diesel fuel. Still another kind of particle results from the photochemical reactions of nitrogen oxides and hydrocarbons in the air. These sunlight-stimulated reactions produce liquid organic substances which are scattered as tiny droplets in the air. The term "smog" has been coined to describe the resulting fog-like condition. Photochemical smog is particularly evident in cities such as Los Angeles where automobile use is excessive. In the early 1970s, up to 40% of the

airborne particles in Los Angeles stemmed from use of the automobile. Particles of lead are also emitted from automobile exhaust. These derive from the lead compounds put into gasoline to increase its octane.

Surface mining of coal and of other substances also produces large quantities of particles. The refining of ores and manufacture of metals are among the industrial processes releasing particles to the air. The grinding and spraying that accompany construction are also sources. Taken together, these industrial activities may account for more particle emission than occurs from coal combustion.

Asbestos escapes to the atmosphere from new building construction where it is being sprayed into place as an insulation; it also escapes in the demolition of older buildings. Finally, the incineration of solid wastes may, in some cities, be a significant source of particles when incinerators are located centrally.

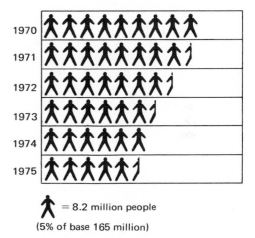

= 8.2 million people
(5% of base 165 million)

Figure 18–5   *Trends in national population exposure to particulate matter levels above the National Primary Ambient Air Quality Standard for 1970 to 1975. At least one pollutant, particulate matter, is finally coming under control. Nonetheless, in 1975, more than 40 million people, nearly 25% of the population, were breathing air whose particulate levels exceeded the health-based air quality standard. (Source:* Trends in the Quality of the Nation's Air, *U.S. Environmental Protection Agency, 1977.)*

CONTROVERSY

AIR POLLUTION

The good and sweeping intentions of many environmentalists are now an obstacle blocking those less-fortunate Americans who desire economic justice.

Bayard Rustin
The New York Times, 11 August 1975

... Using that resource (air) as an inducement to promote development is as short-sighted as would be the reckless use of any other resource.

Cubia L. Clayton
Science **193**, 953 (10 September 1976)

Faced with the high costs of pollution control, it is natural that one of an industrial leader's first impulses may be to relocate where the standards are less strict. This kind of reaction is, of course, viewed as a serious problem by labor and local political leaders who are understandably concerned with providing jobs and with lack of growth in their areas.

Bayard Rustin, a labor leader, writes:

> Many in the environmental lobby in a hysterical effort to reverse the effects of the industrial revolution in a few years, have supported sweeping and uninformed legislation which has had unnecessarily harmful economic consequences. The Clean Air Act is but one example. It has resulted in the closing down of hundreds of plants, from drop forges to specialty organic chemical plants. The environmental damage of the plants forced to close had been minimal and could have been easily corrected. The economic damage of the legislation has been severe.

> The poor black in the ghetto or the unemployed white worker would not find very much with which to identify in Udall's semi-revivalist call that we must no longer "over-indulge ourselves" or seek "to satisfy unlimited greed and desire for luxury." They would agree that "we must change our way of life," but they would take this to mean rather more luxury than they have been accustomed to in the past. The good and sweeping intentions of many environmentalists are now an obstacle blocking those less-fortunate Americans who desire economic justice. *

But this is not the whole story. An official of the New Mexico Environmental Improvement Agency has noted that:

> ...environmental quality is thus too often viewed as the tool with which to bargain.

> The difficulty, apart from environmental degradation, is that this approach fails to consider the reality of air quality as a natural resource which is as depletable as any other. Whether one agrees philosophically with national air quality standards, they do exist, and their existence means the end of the

age-old concept of an unlimited air resource. Hence using that resource as an inducement to promote development is as short-sighted as would be the reckless use of any other resource.

The imposition of a uniform, nationally designated Class II ceiling...actually ensures more development than would otherwise occur. This is because (i) those states that desire to use air quality as an inducement to development will not be allowed to develop at the expense of neighbors who are interested in maintaining as much of a quality environment as possible, and (ii) a tighter ceiling than that imposed by national standards will help impress on everyone that air is a depletable resource and that new industry must be required to utilize the best control technology in developing new energy supplies.

The question of available technology is the crux of the problem. Existing industry faced with the problems of the retrofit of control devices is finding the job difficult and expensive. In many cases, the result has been an unwillingness to accept the fact that the job of control is even possible. But a difficult job is not synonymous with an impossible one, particularly in the case of new industry where controls can be made an integral part of plant design.

The end result is that, rather than having an air shed used up by three or four inadequately controlled industries, more industry can be accommodated.[†]

Who is right? How would you balance air cleanup and jobs. Can you think of ways the poor and unemployed could benefit from air pollution control?

[*] Bayard Rustin, President, A. Philip Randolph Institute, The New York Times, 11 August 1975.

[†] Cubia L. Clayton, Science **193,** 953 (10 September, 1976).

If AQCR is

Above 98

61 to 98

0 to 61

Figure 18–6 *1975 Particulate emissions by Air Quality Control Region (AQCR), in thousands of tons per year. (Source: U.S. Department of Energy, December 1977.)*

Effects of Particulate Air Pollutants

Particles may settle on surfaces, leading to a dirty gray appearance. In addition to soiling, particles may cause corrosion, acting as centers from which corrosion spreads. The annual repair of these surfaces has a significant cost.

The sulfuric acid mist may damage plant tissue. Compounds from photochemical reactions produce a burn on the leaves of many vegetables. Beets, celery, lettuce, and pepper are among the many species susceptible to such damage. Inert, unreactive particles may simply soil plant surfaces.

The health of human beings is also affected by particles. The frequency of respiratory infections such as colds and bronchitis is seen to increase with particle levels. Furthermore, at high concentrations of particles, deaths in excess of the number expected for the time of year have been seen to increase. The health-based standard for particle concentrations has been set at 260 micrograms/ cubic meter averaged over a 24-hour period, and this level is not to be exceeded at any one point more than once per year. At a particle concentration of 375 micrograms/cubic meter, an air pollution "alert" is to be declared. Under an alert, industries might be requested to curtail or postpone their activities.

Particles have a subtle effect on the weather. They may act as nuclei upon which water vapor condenses. Lengthened periods of fog may result from high levels of particles in the air. Increased rainfall has also been seen where particle levels have risen. Furthermore, particles may be reflecting solar energy away from the earth. As such, they may be responsible for the small but noticeable cooling in the Northern Hemisphere of the last quarter century.

Ways to Control Particle Emissions

(Numerous devices are available to reduce particle emissions. These include the settling chamber, the after-burner to ignite and burn particles, and the electrostatic precipitator.) First applied to the collection of fly ash in 1923, the precipitator is now in use at nearly 1000 plants in the U.S. The operation of the precipitator consists first of attaching charges to fly ash particles. As the flue gas moves between the plates of the device, ions or electrons formed by a high voltage discharge will bombard the particles of fly ash and provide them with charges. The charged particles are attracted to and deposited on the grounded metal plates or collection electrodes. The plates must be cleaned periodically by vibrating them or rapping them with a mechanical device (Fig. 18–7). The devices can attain efficiencies of particle collection up to 99%, and can operate with little maintenance while drawing very little electric power.

(An alternative to the electrostatic precipitator is the "baghouse" composed of fabric bags to capture the fly ash.) The fabric filters have a long history of use in removing particles in industries such as grain mills, asbestos factories, and cement plants. Electric utilities have been purchasing these units on a trial basis in the last few years in order to evaluate them for use in power plants. (The principle of the baghouse is simple; it is a very large vacuum cleaner. Air is drawn up through the bags; the particles cannot pass out through the fine weave of the bag and are trapped inside it. Only clean air exits.)

Particulate control methods have been widely applied and are having an impact on air quality. Levels of particulates in the air have been steadily declining since 1970 (Fig. 18–8).

LEAD COMPOUNDS IN THE AIR

Where Lead Comes From

(Lead has been in gasoline for many years as either the compound lead tetraethyl or lead tetramethyl. These compounds are used to improve the

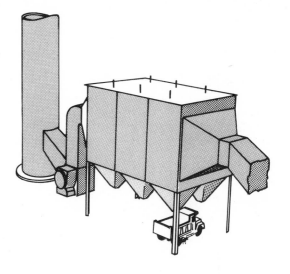

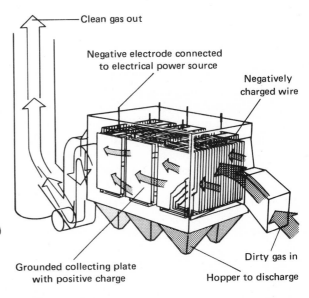

Figure 18–7 *Electrostatic precipitator. Dirty air flows between negatively charged wires and grounded metal collecting plates. The particles in the flowing air become charged and are then attracted to the plates which hold the accumulated dust until it is periodically knocked into hoppers. The clean air is then pumped out through the stack. (Adapted from* Controlling Air Pollution, *American Lung Association, 1974.)*

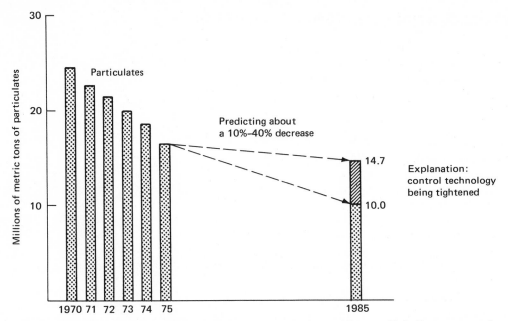

Figure 18–8   *Trends in air pollution emissions: particulates. (Source: U.S. Environmental Protection Agency, SEAS model and REPS model.)*

antiknock quality of gasoline; they eliminate the "ping" during engine operation. The substances are not absolutely necessary, however. A slightly higher priced refining process produces a fuel with the required antiknock properties. Lead need not be present in gasoline) In the past, lead levels have ranged from 2.5 to 4.23 grams of lead per gallon of gasoline. About 75% of this amount is released to the atmosphere in the exhaust stream of the car. In 1975, over 142,000 metric tons of lead were added to the atmosphere from gasoline combustion.

A Special Hazard to Children

(Lead is a cumulative poison, and its presence in the atmosphere adds to the burden of lead we carry in our body. Small children in poorer and older urban areas are exposed to another and more dangerous souce of lead: peeling paints which the unaware child may eat.) Since regulations were established in the early 1970s, most paints have become almost lead-free, although absolute control has not yet been achieved. However, paint manufactured before controls were set had from 5% to as much as 40% lead in the dried film.

The dangers in lead poisoning due to the eating of paint chips increase if there is lead in the atmosphere. Where lead concentrations in the air are highest, people carry a higher burden of lead in their blood and tissues. This is because a portion of the inhaled lead is absorbed and retained in the body. City residents then, are closer to the threshold levels of lead at which the symptoms of poisoning appear. Children have a threshold level about one-half that of adults so that they are even more susceptible to lead poisoning. Lead in the air is pushing lead tissue levels in children too close to the poisoning threshold. (This point is discussed in more detail on p. 394.)

(Although lead poisoning may lead to death, in milder cases it causes mental retardation. Even levels below the poison threshold appear to cause subtle inadequacies in learning.) The hazard to health is one of the reasons for the 1974 decision to eventually make gasoline lead-free. (The other reason is that lead poisons the catalytic converter which further burns the hydrocarbons and carbon monoxide remaining in exhaust gases.)

In 1977, the Environmental Protection Agency established a standard for lead in the atmosphere of 1.5 micrograms per cubic meter, averaged over one month. Lead emissions from automobiles will be declining each year as fewer and fewer of the cars on the road use leaded fuel (Fig. 18–9).

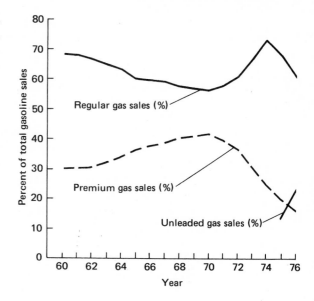

Figure 18–9   *Nationwide trends in regular, premium, and unleaded gasoline sales for 1960 to 1976. (Source: Air Quality Criteria for Lead, U.S. Environmental Protection Agency, December 1977.)*

The Health Effects of Sulfur Oxides and Particles

These two categories of pollutants, sulfur oxides and particles, are discussed together for two reasons. High levels of both of these pollutants have been measured during some of the worst air pollution episodes we have recorded. By worst, we mean those episodes with the largest number of deaths and illnesses. Sulfur oxides and particles in the air also potentiate one another. That is, the effects on health, when both substances are present at stated levels, is worse than if only one of the substances were present.

Why have sulfur oxides and particles appeared together in the past at high concentrations? The answer is that they have had a common source—the burning of coal. Coal is burned for electric power generation; before World War II it was used widely to heat homes in the United States.

Sulfur oxides and particles aggravate respiratory disease. When we first recognized that contaminants which fouled the air could cause death, sulfur oxides and particles were present at the scene. In the Meuse Valley of Belgium in 1930, in the tragedy at Donora, Pennsylvania, in 1948, and in the London Fog of 1952, a killer struck. Today, there is no longer any doubt that sulfur oxide pollution in the presence of particles was the killer.

For a time, we did not understand why in these pollution episodes it was always sulfur oxides <u>plus</u> particles. We have long known that particles act as a focus upon which water vapor condenses. We now understand that sulfur dioxide will quickly dissolve in the water droplet producing a highly acid, highly corrosive mist. It is this sulfurous acid mist that is thought to be responsible for so much damage to life and health.

In the London Fog of 1952, more than 4000 deaths were attributed to the excessive pollution levels. This is only the count of deaths. The number of individuals who became ill was obviously far greater. Sulfur dioxide levels may have reached 4000 micrograms[1] per cubic meter of air. The death toll was heaviest among the elderly, a population group already afflicted with lung and heart diseases. That appears to be the way sulfur oxides and particles act, by making worse existing respiratory and cardiac diseases. Although many deaths during and after the fog were due to such respiratory diseases as pneumonia and bronchitis, investigators were able to blame pollution levels because more deaths occured than had been expected for that time of year. The 4000 deaths attributed to the fog were those deaths in addition to the number expected for that time of year.

Since those early episodes, scientists have accumulated masses of data, some indicating health influences even at very low levels of air pollutants. In New York City, for instance, when sulfur dioxide levels (averaged over 24 hours) exceeded 500 micrograms per cubic meter of air, a small increase in deaths was noted. The daily standard set for sulfur dioxide is 365 micrograms per cubic meter, so the margin of safety between the standard and the level at which excess deaths have occured is not very large.

Individuals already suffering from bronchitis (excessive phlegm and difficulty in breathing) have been found to experience greater illness when smoke particle concentrations exceed 250 micrograms per cubic meter and sulfur dioxide concentrations exceed 500 micrograms per cubic meter, averaged over 24 hours. At annual average concentrations for each pollutant of 100 micrograms per cubic meter of air or greater, higher rates of bronchitis among adults and respiratory infections among children have been observed. The standard set for the annual average concentration of sulfur dioxide is 80 and for particles 75 micrograms per cubic meter, so that again, there is not a wide margin of safety. In the period 1960 to 1965, many major cities exceeded these standards. By 1975, only 5% of the monitoring stations were showing violations of the daily standard and 2.5% were showing violations of the annual standard (Fig. 18–10). However, many of the violations were in highly populated areas meaning that population risks were still high.

In London, since the devastating fog, particle levels have been cut from an annual average of 300 micrograms to 50 micrograms per cubic meter, and sulfur dioxide has been cut from 300 to 200 micrograms per cubic meter

---

1 A microgram is one one-millionth of a gram.

Figure 18-10   *Smoke from a power plant in Toronto, Ohio, drifts across the Ohio River to West Virginia. In order to reduce the sulfur dioxide content of the smoke to a level which will meet federal standards, the utility must burn low sulfur coal or oil, both of which are more expensive and less abundant than regular coal. (NYT pictures).*

averaged over a year. Investigators now find much less air pollution-related disease in London.

Particles in the air have been implicated as a cause of cancer. A carcinogenic substance is one that has been shown in the laboratory to produce cancer among experimental animals. Alternatively, the evidence that it produces cancer may come from observation of occupational groups inadvertently exposed to the substance. An example is the increased risk of lung cancer among asbestos workers.

A number of carcinogenic substances have been found in polluted city air, but there is no direct evidence that these compounds cause cancers in humans at the concentration observed. That is, we can say that carcinogens are in city air, but we cannot accuse them. In the language of the detective, the substances are "suspicious." They are suspicious because, in a number of countries, investigators have found more cases of lung cancer among urban dwellers than among people living in rural areas. Remember, however, that individuals who live in cities may have industrial jobs and different habits such as smoking and so on. Since it is now clear that smoking is a principal cause of lung cancer, scientists have had to eliminate or correct for smoking in their studies. Even after this is done, the conclusion remains that urban dwellers suffer more lung cancer than rural residents.

In addition to these surveys within countries, studies have been made on British migrants to the United States. Of the study participants, those born in Britain who came to reside in the United States had two-thirds the frequency of lung cancer as the permanent residents of Britain. However, the incidence of their lung cancer still exceeded that of the general population of the United

States. It bears mentioning that up to the present time, British life styles have differed from our own. Coal has been the main fuel for heating homes and buildings and for generating electric power as well as for the production of "town gas," a gas with properties resembling natural gas but manufactured from coal. Furthermore, the number of autos per 1000 population has been about half that in the United States. The airborne substances under greatest suspicion as potential carcinogens are polycyclic aromatic hydrocarbons, an example of which is benzpyrene. Known to cause cancer in experimental animals, these compounds have been found free or associated with particles of soot in urban air in the United States and in Britain.

Children and Lead Poisoning

Lead is found in water and in food in trace amounts. Lead is present in layers of old house paint. And lead is in the air because gasoline contains lead additives to make it burn more smoothly. These sources of lead, taken together, put urban children at great risk of developing lead poisoning. Children may begin to experience symptoms when the level of lead in their blood reaches 50 micrograms per 100 grams of whole blood, although the threshold for adults is about 60% higher. Surprisingly, many urban children have been found to have blood lead levels which are not far from the toxic limit. Evidence suggests that such levels just short of actual lead poisoning may be associated with learning difficulties.

Lead poisoning in the child is first seen as a variety of symptoms which include appetite loss, problems in discipline, and a lack of interest in play. The disease progresses to constipation, vomiting, seizures, and finally coma. A permanent loss of mental ability may result from a case of lead poisoning.

The risk of developing lead poisoning comes about in the following way. The food one eats has trace amounts of lead. It comes from the soil or from pesticides or from the settling of airborne lead particles on the leaves of vegetables. Lead is also in water. It comes from the weathering of rocks which contain lead substances and from lead particles which settled from the air. It is estimated that a one-year-old child ingests 130 micrograms of lead each day in food and drink.

City air is fouled with lead particles from the combustion of gasoline. The act of breathing moves these particles into the lungs. Lead has been measured in the air of cities at monthly averages up to five micrograms per cubic meter. At this concentration, a one-year-old will inhale and absorb almost as much lead from the air as he is absorbing from food and drink. The child's exposure to this toxic metal then can be nearly doubled because of the presence of lead in gasoline. The inhalation of street dust which has been found to have high levels of lead is still another route by which children are exposed to lead.

If lead in the air and in the diet were the only sources of lead, the threat to the urban child would be less severe. There is yet another source of lead, however, that exposes large numbers of children to the danger of lead poisoning. It is the lead in paint. Lead has long been used as a pigment in house paints, but since World War II titanium dioxide has been gradually replacing lead oxide in paint. Since the end of 1973, lead has been banned from most paints in the United States. Nevertheless, the houses from the pre-war era remain. Today many of these houses have fallen into disrepair, and the paint is crumbling from their walls. Mostly poor people inhabit these structures; their children may eat the poisonous paint chips.

Eating paint chips is tragically common. Small children of all social and economic classes are often observed eating nonfood items. It is the small child's way of sampling his environment. In this case, the sample is deadly indeed. In a one gram paint chip there may be up to 50,000 micrograms of lead. One estimate suggests that up to 2% of the children living in decaying pre-war dwellings may have levels of lead in the blood exceeding the toxic limit.

Lead poisoning is a public health problem of major importance. To guard against new cases, parents and children must be informed of the risk involved in eating paint chips. Detection programs are necessary to find those children already near the toxic limit. Where cases are found, the surface from which the paint peeled should be covered up or the paint should be completely removed.

Finally, lead should be removed from gasoline so that this extra source no longer contributes to the burden of lead that a child must carry. At present, lead is being removed from gasoline because it harms the catalytic converter. This is the device that has been added to new cars to burn hydrocarbons and carbon monoxide more fully. We may eventually see the day when nearly all gasoline is free of lead.

Questions

1. Where do sulfur oxides come from? Where do particles come from? What industrial activity using what fuel produces the most sulfur oxides?

2. Why does the gas sulfur dioxide cause mist to be acid?

3. What kinds of particles are produced by the auto?

4. Why are particles regarded as an "accomplice" of sulfur dioxide? (Enrichment)

5. What is the meaning of "excess deaths" and how does this concept fit into a study of the health effects of air pollution? (Enrichment)

6. Why do some individuals in the electric power industry say that tall smoke stacks are the correct response to air pollution from sulfur oxides? What does the Environmental Protection Agency suggest instead? What kinds of pollution will be prevented if the sulfur dioxide is removed from the stack gases?

7. We pointed out that salmon no longer run in certain streams in Norway where high acidity has prevented eggs from developing. That is, they no longer breed in and

hence run in such streams. The salmon does not, however, "make a decision." Instead, the life cycle of the salmon plays a role in this process of adjustment. From study of salmon life cycles, biologists have determined that the salmon, an ocean fish hatched in fresh water, somehow nearly always returns to spawn in the same freshwater stream in which it began its life. Explain how this remarkable homing instinct on the part of the salmon would eventually lead, in the presence of a high acid content, to a stream without salmon. Would you expect the salmon to be eliminated quickly or over several years?

8. List five ways that urban children are exposed to lead that may enter their systems. (Enrichment)

9. Explain how the lead from gasoline comes to contaminate air, water, and food. (Enrichment)

10. What is pica (you will need to look this up), and how is it related to the occurrence of lead poisoning?

Further Reading

Air Pollution Engineering Manual, 2nd ed. United States Environmental Protection Agency. U.S. Government Printing Office, Washington, D.C., 1978.

Although this is a highly detailed technical work; a number of portions are general descriptions which can be understood by the lay person: there are even discussions of the history of the technology. Whether all the details given here are of general interest, however, depends on the intent and training of the reader.

Air Quality Criteria for Particulate Matter, AP-49, U.S. Environmental Protection Agency, 1969.

Control Techniques for Particulate Matter, AP-51, U.S. Environmental Protection Agency, 1969.

Air Quality Criteria for Sulfur Oxides, AP-50, U.S. Environmental Protection Agency, 1969.

Control Techniques for Sulfur Oxide Air Pollutants, AP-52, U.S. Environmental Protection Agency, 1969.

While technical, these documents are very clearly written and very complete.

"Controlling Air Pollution," American Lung Association, 1974.

Kierig, P., "Scrubbing Coal," *EPA Journal,* January 1978

A general non-technical description of the Wellman-Lord scrubber system that produces sulfur as a product.

"Landfill Made by Scrubbers," *Business Week,* 16 January 1978.

An easy-to-read description of the problems associated with the sludge from scrubbers.

Likens, G. and F. Bormann, "Acid Rain: A Serious Regional Environmental Problem," *Science* **184**, 1117 (14 June 1974).

This article is a bit technical but is one of the earliest recognitions of the acid rain problem.

"Proceedings of the Conference on Health Effects of Air Pollutants," prepared for the Committee on Public Works of the U.S. Senate by the Assembly of Life Sciences, November 1973 (Serial number 93-15).

ReVelle, C. and P. ReVelle, "Lead," in *Sourcebook on the Environment*, Houghton-Mifflin, Boston, 1974, p. 134.

A view of all sources of lead exposure and their relative contributions to lead levels in humans.

Waldbott, G., *Health Effects of Environmental Pollutants*, 2nd ed., C.V. Mosby Company, St. Louis, 1978.

This well-illustrated (136 pages of illustrations) book has a wealth of information and documentation on all of the major air pollutants. It is a technical work, but clearly written. The book could serve as an excellent reference for the health aspects of air pollutants.

Chapter 19

Photochemical
Air Pollution

WHAT IS PHOTOCHEMICAL POLLUTION?

(A photochemical reaction requires light energy to take place. Certain pollutants in the atmosphere, nitrogen oxides and hydrocarbons, undergo photochemical reactions. These reactions produce new pollutants, including ozone, aldehydes, and exotic organic compounds. The new pollutants are referred to, in sum, as "photochemical air pollution" because they arise from photochemical reactions.)

The sources of photochemical air pollution then are the sources of nitrogen oxides and of hydrocarbons. By the same reasoning, the control of photochemical air pollution consists of controlling emissions of nitrogen oxides and of hydrocarbons.

Nitrogen Oxides and Hydrocarbons: By-Products of Combustion

(During the high-temperature combustion of fossil fuels, nitrogen and oxygen *from the air* react to produce nitrogen oxides.) This occurs whether the fuel is natural gas, coal, gasoline, or home heating oil. Approximately 90% of the annual emissions of nitrogen oxides stem from the combustion of fossil fuels. Probably about one-third of these emissions are from motor vehicle operation. Another one-quarter result from the burning of natural gas for heat and power. The use of coal for electric generation and in such industrial activities as the production of steel may account for an additional one-fifth of the nitrogen oxide emissions from combustion. The manufacture of explosives and nitric acid, two noncombustion sources, also result in emissions of nitrogen oxides (Fig. 19–1).

Hydrocarbons are released from many sources. Methane gas is present naturally in the atmosphere due to emissions from coal, gas, and petroleum fields; from fires; and from emanations from swamps. However, the methane is unreactive in the atmosphere. Thus, even though methane background levels may reach a milligram per cubic meter, its presence does not concern us, since it appears at these concentrations to be harmless. The hydrocarbons that do concern us are mainly by-products of the activities of civilization (Fig. 19–2). In 1975, one-third of the annual emissions of hydrocarbons stemmed from motor vehicle operation. In the cars of the 1960s, hydrocarbons evaporated from the carburetor and from the fuel tank. They also escaped around the piston during compression. Most important, however, incomplete combustion left unburned hydrocarbons as droplets and gases in the exhaust stream. Smaller sources of hydrocarbons were petroleum refining and transfer operations.

A Complex of Reactions

Levels of photochemical air pollution follow the pattern of use of the automobile. During the morning and evening rush hours, peak emissions of nitrogen oxides and hydrocarbons occur. These are

If AQCR is

Above 140

93 to 140

0 to 93

Figure 19-1 *1975 Nitrogen oxides emissions by Air Quality Control Region (AQCR), in thousands of tons per year. (Source: U.S. Department of Energy, December 1977.)*

the substances that react to produce the photochemical air pollutants.

Nitrogen and oxygen unite during the high temperature combustion of fuel in an automobile engine to produce the gas nitric oxide, which is then released to the atmosphere. In several hours' time, the level of nitric oxide in the air will decrease substantially. During the period of that decline, the level of nitrogen dioxide will be rising to a peak. Later, as the nitrogen dioxide level declines, the concentration of a third gas, ozone, will increase. Ozone levels, too, then decrease (Fig. 19-3).

The reactions that produce high concentrations of nitrogen dioxide and ozone are not perfectly understood. Nevertheless, certain key relationships have been identified. The presence of hydrocarbons and the photochemical properties of nitrogen dioxide make these reactions possible. These reactions

produce a typical pattern of pollutant levels over a period of time. Nitric oxide rises during the period of high emissions from motor vehicles (Fig. 19-4). Nitrogen dioxide rises and nitric oxide falls due to the reaction of hydrocarbons with nitric oxide. After nitric oxide levels have fallen, ozone levels rise due to the photodissociation of nitrogen dioxide. (These complex reactions are explained in more detail on p. 412.)

(Still other pollutants result from the consequences of these reactions. The hydrocarbons react with nitrogen dioxide to give peroxyacyl nitrate (PAN) compounds. Ozone reacts with hydrocarbons to produce aldehydes. Eventually, winds disperse all of the contaminants (Fig. 19-5))

(The term oxidant is applied to compounds capable of oxidizing substances that oxygen in the air cannot oxidize. Nitrogen dioxide, PAN com-

Figure 19-2 *1975 Hydrocarbons emissions by Air Quality Control Region (AQCR), in thousands of tons per year. (Source: U.S. Department of Energy, December 1977.)*

pounds, ozone, and aldehydes are oxidants. A concentration of oxidants is reported in terms of the weight of all such substances per cubic meter of air. The concentration does not indicate how much of any specific compound is present, although ozone is usually the largest component.)

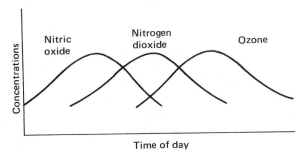

Figure 19-3 *Levels of nitrogen oxides and ozone during and after the morning "rush hour."*

Effects on Plants and Animals

(Hydrocarbons appear to have little direct effect on plants or animals although one hydrocarbon, ethylene, seems to stimulate plant growth. Elevated nitrogen dioxide levels, in one study, were seen to produce a higher incidence of human respiratory illness. (Health and biological effects are discussed in detail on p. 406). The gas also causes some dyes used on synthetic fabrics to fade.)

(The photochemical oxidants, on the other hand, appear to have a wide variety of effects on plants and man. Ozone and PAN compounds damage the leaves of common vegetables. Ozone also causes citrus trees to lose their fruit early. Ozone and aldehydes irritate the respiratory tract. The eyes may be irritated by aldehydes and PAN compounds. Individuals with asthma have found their disease aggravated by the oxidants.)

Figure 19-5   *The buildings in this city fade into a haze of pollution. In the city itself, respiratory problems are likely to be increasing. (Photo from the Environmental Protection Agency Center for Environmental Research Information, Cincinnati, Ohio.)*

Figure 19-4   *The automobile is the source of most of the carbon monoxide in the air of cities. In addition, the auto generously contributes nitrogen oxides and hydrocarbons to urban air. These are the substances out of which photochemical smog is formed. (EPA Documerica)*

The health-based standard for photochemical oxidants is 240 micrograms per cubic meter, averaged over one hour. It is not to be exceeded more than once per year. The standard was loosened to this level in 1979, from a level one-third lower, on the basis of more health effects data. Because hydrocarbons and nitrogen oxides levels are controlled in order to control the level of photochemical oxidants, we can expect a push from auto industry lobbyists to loosen the standards on hydrocarbons and nitrogen oxides.[1] In the interval 1973 to 1976, Los Angeles experienced one-hour oxidant levels exceeding 300 micrograms per cubic meter on over 15% of the days; that is, during about 60 days per year, on average, oxidant levels were 25% higher than the standard.

---

1 It happened once before in just this way.

CONTROL OF PHOTOCHEMICAL AIR POLLUTION

The control of oxidants is accomplished by controlling hydrocarbons and nitrogen oxides. Because the automobile is responsible to such a large degree for these pollutants, efforts are under way principally to decrease the amounts from this source.

The hydrocarbons that blow past the pistons are now being recycled to the combustion chamber. Those that evaporate from the carburetor and fuel tank are also recycled. Emissions of hydrocarbons from the exhaust system are decreased by using more air with the gasoline when it is burned. Beginning with the 1975 models, many new cars were equipped with a catalyst system. The catalyst system oxidizes the unburned hydrocarbons in the exhaust to carbon dioxide and to water vapor. Unleaded gasoline must be used by vehicles with the catalyst system to prevent "poisoning" the catalyst and making it ineffective.

Another catalyst system may be in prospect for controlling emissions of nitrogen oxides. This device would reduce the nitrogen oxides back to molecular nitrogen. At least one vehicle already appears to meet the standards for oxides of nitrogen (stated in grams per mile) without using a catalyst. The vehicle utilizes a "stratified charge" engine.)

Power plants that burn coal, gas, or oil are large sources of nitrogen oxides. A shift to nuclear electric power would reduce nitrogen oxides from this source but there are many problems with nuclear power (see Chapter 15). Decreasing the air intake in gas and oil boilers may also decrease emissions of oxides of nitrogen from these sources. (See Figures 19–6 and 19–7.)

A BRIEF SUMMARY: WHERE ARE WE ON THE ROAD TO CLEAN AIR?

The Environmental Protection Agency has published Primary Standards for air quality (Table 19–1). These standards are meant to protect people's health. (Other, secondary, standards are those meant to protect plants, buildings, etc.) How

well are we doing in terms of achieving the air quality standards set out by the EPA? How far have we come since 1970 when Congress passed the Clean Air Act, with high hopes of insuring clean, healthy air for all of us?

Now, as then, air quality varies from region to region. However, a few general observations can be made. Air quality is still, for the most part, better in rural areas than in cities. Taking the major air pollutants in turn, we can see some trends. Carbon monoxide emissions are waning. By 1975 only 13% of the air quality regions reported incidents in which carbon monoxide levels exceeded the primary standards (Table 19–2). The decrease in carbon monoxide emissions is expected to continue through 1985 (Fig. 17–6), because an effective control device, the catalytic converter, exists for the principal source of carbon monoxide, automobiles.

The picture is not as rosy for carbon dioxide emissions. Concern here is not with health effects but that increasing levels of carbon dioxide may adversely affect climate. Carbon dioxide levels in

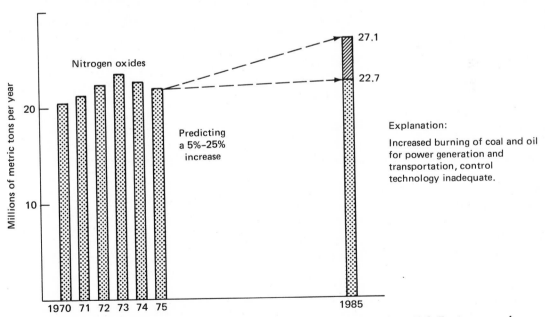

Figure 19–6 *Trends in air pollution emissions: nitrogen oxides. (Source: U.S. Environmental Protection Agency, SEAS model, REPS model.)*

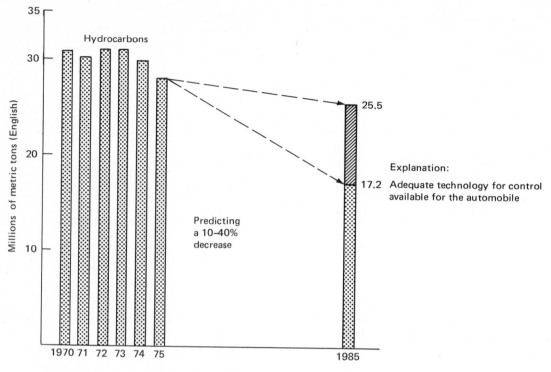

Figure 19-7   *Trends in air pollution emissions: hydrocarbons. (Source: U.S. Environmental Protection Agency, SEAS model, REPS model.)*

the atmosphere continue to rise (Fig. 17-4) as we continue to burn fossil fuels.

Sulfur oxide emissions are also expected to rise through 1985. Again the burning of a fossil fuel (coal) is the cause. In 1975, 16% of the air quality regions reported violations of the primary daily standard for sulfur oxides. Violations are expected to increase until either scrubbers are installed at electric plants or lower sulfur coal can be provided (Fig. 18-3).

The situation with respect to particulates is, on the other hand, improving rather than getting worse. Control technologies are available and are being used. Although approximately 44% of the air quality regions reported violations of the primary daily standards for particulates in 1975, by 1985, as much as a 40% decrease in particulate emissions is expected (Fig. 18-8).

A special note must be added about lead particles in the air, the primary source of which is leaded gasoline for automobiles. As more new cars come on the road and as more older cars are junked, the percentage of cars requiring unleaded gasoline is increasing, and at the same time, the amount of lead in the air is decreasing. In this area, at least, city air is becoming more healthy for children to breathe.

For photochemical pollutants, the situation in certain regions remains poor. In urban areas, such as Los Angeles, the growth in automobile use continues to offset any gains from the installation of pollution control devices. In this case, we are running very hard just to stay in one place. Overall, emissions of hydrocarbons, one of the classes of compounds responsible for photochemical smog, are expected to decrease through 1985 (Fig. 19-7)

TABLE 19–1 Primary (Public Health) Standards for Air Quality (As of January 1979)

Pollutant	Time Period of Standard[a]	Maximum Permissible Concentration
Suspended Particles	annual geometric mean[b]	75 micrograms per cubic meter
	maximum 24-hour concentration	260 micrograms per cubic meter
Sulfur Oxides	average annual concentration	80 micrograms per cubic meter
	maximum 24-hour concentration	365 micrograms per cubic meter
Carbon Monoxide	maximum 8-hour concentration	10 milligrams per cubic meter
	maximum 1-hour concentration	40 milligrams per cubic meter
Oxidants/Ozone	maximum 1-hour concentration	240 micrograms per cubic meter
Nitrogen Dioxide	average annual concentration	100 micrograms per cubic meter
Hydrocarbons	maximum 3-hour concentration	160 micrograms per cubic meter

[a] All standards based on concentrations over 24 hours or less are not to be exceeded more than once per year.

[b] The geometric mean is the antilog of x where x is the sum of the logarithms of the data divided by the number of data points.

because control technology, the catalytic converter, is good. Levels of nitrogen oxides, the other contributor to photochemical smog, will probably increase, however, due to increased fuel use and inadequate control technology (Fig. 19–6). In 1975, over 94% of air quality regions reported violations of the primary standard for photochemical oxidants.

In 1979, the standard was revised upwards, in a compromise by the government between what is possible and what costs too much. (The wisdom of this move is discussed on p. 411.)

Overall, we have made improvements in some areas and lost ground in others, although the air quality is undoubtedly better than if Congress had

TABLE 19–2 1975 Air Quality

Pollutant	Standards	Stations at which Standards were Exceeded No.	Stations at which Standards were Exceeded %	Air Quality Control Regions (AQCR) which Showed Violations No.	Air Quality Control Regions (AQCR) which Showed Violations %
Total Suspended Particulates (TSP)	Primary annual	437 of 2186	20	116 of 216	53.7
Total Suspended Particulates (TSP)	Primary 24-hour	311 of 4137	7.5	108 of 243	44.4
Sulfur Dioxide (SO₂)	Primary annual	35 of 1357	2.6	12 of 187	6.4
Sulfur Dioxide (SO₂)	Primary 24-hour	132 of 2631	5.0	37 of 229	16.1
Carbon Monoxide (CO)	Primary 1-hour	28 of 436	6.4	15 of 117	12.8
Carbon Monoxide (CO)	Primary 8-hour	232 of 436	53.2	77 of 117	65.8
Oxidants/Ozone (Ox/O₃)[a]	Primary 1-hour	356 of 416	85.6	96 of 102	94.1
Nitrogen Dioxide (NO₂)	Primary annual	19 of 824	2.3	5 of 128	3.9

[a] These are violations of the old oxidant standard of 160 micrograms per cubic meter. (Source: *Energy/Environment Fact Book*, U.S. Government Printing Office, 1977.)

not passed a Clean Air Act. One very special aspect of that 1970 legislation is that it aimed at a level of environmental quality Congress wanted, as opposed to a level that could be achieved, given the state of pollution control technology in 1970. Some of the controversy surrounding this kind of legislation is detailed in the Controversy.

Effects of Nitrogen Dioxide On Our Health

The combustion of fossil fuels in engines and furnaces produces such intense heat that pollutants are formed from two natural substances in the air, nitrogen and oxygen. These pollutants are the nitrogen oxides. Nitrogen and oxygen combine at the extremely high temperatures which exist in internal combustion engines such as that of the automobile. The combination also takes place in boilers such as those of the coal furnaces of electric power plants. About 90% of the nitrogen oxides are produced in the form of nitric oxide (one molecule of nitrogen plus one molecule of oxygen). The remaining 10% are in the form of nitrogen dioxide (one molecule of nitrogen and two molecules of oxygen).

Most of our information on health effects concerns nitrogen dioxide. The nitrogen dioxide initially composes only 10% of the emissions of oxides of nitrogen. However, a complex series of chemical reactions in the air converts much of the nitric oxide to the more hazardous nitrogen dioxide form. These reactions, which are not, strictly speaking, an oxidation by molecular oxygen, are stimulated by the presence of sunlight. Reactions induced by the presence of sunlight are termed "photochemical reactions." At this time, the details of the conversion of nitric oxide to nitrogen dioxide in sunlit air have not been fully understood. Whatever the process, we know that nitrogen dioxide is a product, and scientists have recently learned some of the hazards of this substance.

Nitrogen dioxide is a foul smelling gas. Even at levels as low as 120 grams in a __billion__ grams of air, about one-third of a group of volunteers were able to detect its presence. This is equivalent to about 0.23 milligrams[2] of nitrogen dioxide in one cubic meter of air (at sea level). The ability to detect the gas disappeared, however, after only 10 minutes of exposure, although people reported a dryness and roughness of the throat. Even these irritations vanished after prolonged exposure to levels 15 times the odor threshold discussed here.

Not only does nitrogen dioxide affect the sense of smell, night vision is altered by this gas. The ability to adapt one's eyes to see in the dark is decreased. This phenomenon has been observed at levels as low as 75 grams of

---

2 A milligram is 1/1000 of a gram.

nitrogen dioxide per billion grams of air (or 0.14 milligrams per cubic meter). The visual and olfactory responses to nitrogen dioxide may be called sensory effects. More important are the effects of nitrogen dioxide on disease and on the functioning of the human body.

Two effects of nitrogen dioxide on body functions have been noted. One is the greater effort required in breathing; medical investigators refer to this as increased airway resistance. The response has been observed in normal healthy individuals at gas concentrations as low as 30 grams per billion grams of air (0.056 milligrams per cubic meter). Individuals with chronic lung diseases experienced breathing difficulty at only 20 grams per billion grams of air (0.038 milligrams per cubic meter).

Another body function response to nitrogen dioxide has been suggested but remains to be verified. The measurements of a team of Czech scientists indicate that nitrogen dioxide can form methemoglobin in the blood. The investigators feel nitrogen dioxide gas can attach to hemoglobin, just as carbon monoxide can tie up hemoglobin, thus keeping it from carrying oxygen to the tissues. It is accepted that nitrites in food can attach to hemoglobin, forming methemoglobin, but this is the first suggestion that nitrogen dioxide gas can also.

A number of studies have noted increased respiratory disease in areas polluted by nitrogen dioxide, but the presence at the same time of other pollutants at relatively high levels make the results less useful. Firm conclusions on the causative agent become impossible. On the other hand, the study by Shy and others of over 4000 school children and parents in several portions of Chatanooga, Tennessee, does point toward nitrogen dioxide alone as causing increased respiratory disease. Although we say "causing," a more accurate statement would be that nitrogen dioxide made people more susceptible to the pathogens that cause respiratory disease. Nitrogen dioxide from a nearby TNT plant was thought to be the most concentrated by far of air pollutants in the study area. Researchers observed more colds, bronchitis, croup, and pneumonia among the population group exposed to these higher nitrogen dioxide levels than among a less exposed but otherwise similar population nearby. It was not clear, however, whether the increased respiratory illnesses were a response to sustained levels of nitrogen dioxide (as indicated, for example, by the average annual concentrations) or to short term concentrations (as indicated by average daily levels). Studies of the effect of nitrogen dioxide on the susceptibility of animals to respiratory disease have given similar results. These animal studies offer very useful support to an hypothesis about the response of humans to a pollutant.

Investigators have also sought to link nitrogen dioxide to increased mortality (death) in addition to increased morbidity (disease) rates. Statistical analysis indicates a positive relation between nitrogen dioxide concentrations and a higher mortality rate from heart disease and cancer. That is, areas with

AIR POLLUTION CONTROL: A "POLICY BEYOND CAPABILITY"?

We think this is a necessary and reasonable standard to impose upon the [auto] industry

Senator Muskie

Everybody wants to go to heaven, Miss Morgan, but nobody wants to die

Bobby Gentry, "Casket Vignette"

The 1970 Clean Air Act, called the "toughest air pollution law," was designed to promote the air quality we would like to have, not the quality of air we were sure we had the technology to achieve. Ever since then, scientists, industrial leaders, and politicians have been arguing over whether this was a good idea. Is it reasonable to base a law on what has been called "policy beyond capability," that is, what you hope to achieve rather than what you are fairly sure can be accomplished?

In the Senate floor debate on the bill, the following exchange occurred:

Mr. Griffin. Did the committee have any hearings in this session on this problem as to the state of the art—on the likelihood or possibility that this goal can be reached by 1975?

Mr. Muskie. Yes, we had testimony jointly before the Commerce Committee and before our committee from the automobile companies on the state of the art. With respect to this specific deadline, no.

Mr. Griffin. On this particular bill?

Mr. Muskie. No.

Mr. Griffin. No hearings?

Mr. Muskie. The deadline is based not, I repeat, on economic and technological feasibility, but on considerations of public health. We think, on the basis of the exposure we have had to this problem, that this is a necessary and reasonable standard to impose upon the industry. If the industry cannot meet it, they can come back....

Mr. Griffin. ...without adequate expertise, without the kind of scientific knowledge that is needed—without the hearings that are necessary and expected, this bill would write into legislation concrete requirements that can be impossible—and that will literally force an industry out of existence.... *

One scientist has commented:

It goes without saying that "policy beyond capability" has many risks....
Reaching accommodations with the regulated industries...carries with it the
risks of re-emergence of public concern, subsequent congressional
investigation, and, perhaps most important, only limited progress in reducing
air pollution. Enforcing the standards as written, on the other hand, will, in all
likelihood, greatly expand those to be regulated. As one example, meeting
the present federal air quality standards may well require controlling the
number of automobiles in our cities. Since automobiles are such an important
part of American life, regulating their use obviously can affect where we live,
how we get to work, how we spend money, how we entertain ourselves.
Whether public concern for pollution in 1969-1970 included such a broad
mandate is uncertain–and those who assume that it did are clearly taking a
risk in drawing those conclusions. Super-planning has never been super-
welcome in this country....In environmental matters, as with so many public
policy issues, we are well advised to heed the wisdom of Bobbie Gentry in her
song "Casket Vignette:." "Everybody wants to go to Heaven, Miss Morgan, but
nobody wants to die." [t]

What do you think? Does setting standards that we are not positive we can
meet act as a spur to the development of pollution control? Are we asking for
trouble with this type of policy? What about the air we have to breathe? How
much do you think Americans are willing to give up, if that is necessary, to
assure themselves of clean air?

* Congressional Record, 21 September 1970, pp. 5160-5195.

t Charles O. Jones, Annual meeting of the American Political Science Association, Washington Hilton Hotel, Washington, D.C., September 5-9, 1972.

higher nitrogen dioxide levels do have a greater number of deaths from these causes. In the two notable studies, though, the presence of other pollutants at the scene make it difficult for scientists to draw conclusions. In addition, a limited number of sampling stations in each study area leaves the pollutant profile in each area incompletely described.

Exposure and Susceptibility

Individuals with chronic (continuing) respiratory diseases, such as emphysema and asthma, and individuals with heart disease may be more sensitive to direct effects of nitrogen dioxide; we do not yet know whether this is true. What we do know, however, is that nitrogen dioxide is associated with increased cases of short term respiratory disease. Individuals with chronic heart and respiratory disease are more likely to develop complications from these short infections—dangerous complications, such as pneumonia. On the order of 10% to 15% of the population in the United States are thought to have some form of chronic respiratory disease.

This line of reasoning leads us to conclude that the standard for nitrogen dioxide should be set at a level that protects the population from increased respiratory infections. The standard set for nitrogen dioxide is an average annual concentration which should not be exceeded. Its value is 50 grams per billion grams of air or 100 micrograms per cubic meter of air.[3]

Long term exposure to nitrogen dioxide at levels of 500 grams per billion grams has been shown to decrease resistance to respiratory disease. Thus, the allowable concentration is set at one-tenth the level known to have effects on health. This is a fairly common margin of safety employed in setting standards. On the other hand, one of the statistical investigations—not fully accepted for reasons just cited—linked an increase in cancers to levels near the annual air quality standard.

At present, no short term standard (for instance, an average daily concentration) has been set, although a value was briefly proposed in 1971 and then abandoned. In the past, peak (instantaneous) levels of nitrogen dioxide in excess of 400 micrograms per cubic meter of air have been commonly observed in major cities.

Biological Effects of Photochemical Air Pollution (Smog)

The standard for oxidant concentration in the air, as pointed out earlier, is 240 micrograms per cubic meter, averaged over one hour. This concentration is not to be exceeded more than once each year. The new standard is for ozone alone, rather than all oxidants as the previous standard

---

3 One microgram is 1/1,000,000 of a gram.

was, even though other oxidants can account for as much as one-fourth of the total oxidants present in the air. The new standard is also 20% higher than the level at which "alerts" were called previously.

The standard was revised to this level in 1979 from 160 micrograms per cubic meter, a level set in the early 1970s. During 1974–75, the old standard was violated on 38% of the days at one or more of the stations in Los Angeles and in Philadelphia. Denver and Washington reported such violations on 27% of the days in 1974–75, and Cincinnati and Houston noted violations on 23% of the days in the same interval. Thus the relaxation of the standard from 160 to 240 micrograms per cubic meter will give the appearance of significantly better air quality.

The new standard appears to be a compromise between air quality and the expense of controlling pollution. It was at the level of the new standard that performance of a high-school cross-country team was observed to deteriorate. Asthmatics have been noted to have an increased frequency of attacks in the range of 300 to 500 micrograms per cubic meter, the lower level a mere 25% above the current standard. The fraction of the population afflicted with bronchial asthma has been estimated at 3% to 5%. Individuals with chronic lung disease also have been found to be affected by photochemical oxidants. Eye irritation is a significant effect of photochemical air pollution. At hourly averages as low as 200 micrograms per cubic meter, such eye irritation could be expected to begin. The standard, or rather the goal to be attained, would not exclude irritation. An unanswered question is "Is eye irritation a health effect?" Does it indicate that other effects are beginning?

A number of studies of laboratory animals were also used in setting the old standard. In the presence of oxidants in the air, mice became more susceptible to bacterial infections; red blood cells and the cells of heart muscles became misshapen. A three-month exposure to ozone levels at about two-thirds of the old standard produced a decrease in the weight of rats and disrupted the physiological chemistry of the animals. Hamsters were found to have damage to the chromosomes of white blood cells on short-term, low-level exposures.

Even before effects of oxidants on health were documented, investigators found damage to plants. Tiny dark spots called "stipples" were discovered on the leaves of plants in the 1940s. The condition was first observed in Los Angeles but has since been seen across most of the United States. The spots are the result of elevated ozone levels. Such levels result when excessive use of the automobile leads to high concentration of hydrocarbons and nitrogen oxides. Laboratory studies show that many common vegetables are damaged by ozone and also by PAN compounds. The gas also affects citrus trees by causing the fruit to ripen and fall at a smaller size. Many plants will show damage at oxidant concentrations between 80 and 160 micrograms per cubic meter, levels that are below the new standard.

Chemistry of Photochemical Air Pollution

Nitrogen oxides and hydrocarbons are the substances that react to produce the photochemical air pollutants. Nitrogen and oxygen unite during the high temperature combustion of fuel in an automobile engine. The product is principally nitric oxide; less than a hundredth of the product is in the form of nitrogen dioxide.

$$\underset{\text{nitrogen}}{N_{2(g)}} \quad + \quad \underset{\text{oxygen}}{O_{2(g)}} \quad \longrightarrow \quad \underset{\substack{\text{nitric} \\ \text{oxide}}}{2NO_{(g)}}$$

As the exhaust mixes with air, about one-tenth of the nitric oxide is oxidized to nitrogen dioxide.

$$\underset{\substack{\text{nitric} \\ \text{oxide}}}{2NO_{(g)}} \quad + \quad \underset{\text{oxygen}}{O_{2(g)}} \quad \longrightarrow \quad \underset{\substack{\text{nitrogen} \\ \text{dioxide}}}{2NO_{2(g)}}$$

However, once the nitric oxide is well diluted in the air, further conversion to nitrogen dioxide by this direct oxidation step will no longer take place. In most American cities, nitric oxide levels reach their peak during the morning rush hours. Even though further direct oxidation of nitric oxide to nitrogen dioxide does not take place, we note that, in several hours' time, the level of nitric oxide has declined considerably. Coincident with this decline, nitrogen dioxide levels are rising. As the concentration of nitrogen dioxide diminishes, a third gas, ozone, increases. Ozone levels, too, then wane. The reactions producing high levels first of nitrogen dioxide and then of ozone are not yet fully known or explained. What is certain, however, is that the presence of hydrocarbons and the photochemical properties of nitrogen dioxide make the reactions possible.

Hydrocarbons, which are simply compounds of carbon and hydrogen, are of two types. Saturated hydrocarbons have no room for additional hydrogen atoms. The carbon atoms are, in essence, completely reacted (or saturated). Unsaturated hydrocarbons, on the other hand, react readily with many substances. The reactions eventually exhaust the capacity of the carbon atoms to react. Methane, the background hydrocarbon mentioned earlier, is saturated and generally not capable of entering further chemical reactions in the atmosphere.

There is a series of reactions in which the primary pollutants, nitric oxide and hydrocarbons, are converted to the photochemical pollutants. These are nitrogen dioxide, ozone, aldehydes, and PAN compounds. The energy of light makes these reactions possible. Light falls on molecules of nitrogen dioxide produced as exhaust gases are emitted. In the presence of light energy, the nitrogen dioxide dissociates (comes apart) into a molecule of nitric oxide and a single free atom of oxygen.

$$NO_{2(g)} \longrightarrow NO_{(g)} + O_{(g)}$$

Free oxygen atoms are not normally present in the atmosphere because of their enormous reactivity. These free oxygen atoms react in two ways. One series of reactions is circular and leads back to the initial substances. The series, known as the "photolytic cycle," begins with the dissociation step. Atomic oxygen then reacts with oxygen molecules in the air to produce ozone. Ozone, in turn, reacts with nitric oxide to return nitrogen dioxide and oxygen. The cycle is shown as follows:

$$
\begin{array}{ccccc}
& \text{Dissociation} & & & \\
NO_2 & \longrightarrow & NO & + & O \\
+ & & & & + \\
O_2 & & & & O_2 \\
\text{Cycle} & \searrow & & & \downarrow \\
\text{completed} & NO & + & O_3 &
\end{array}
$$

While there is only a small decrease in nitrogen dioxide, the presence of ozone in the air is established by this continuously operating cycle.

The second series of reactions which follows the dissociation step is more important, however. The free atomic oxygen also reacts with unsaturated hydrocarbons to produce extremely reactive organic molecules called <u>free radicals.</u> These free radicals react with the nitric oxide initially present. Nitrogen dioxide is produced. This is a crucial reaction because more nitrogen dioxide is produced in this step than dissociated to begin the reactions. The result is a build-up of nitrogen dioxide levels. We cannot attempt to write balanced equations here because scientists still do not thoroughly understand the reaction. In words, the reactions are:

atomic oxygen   +   hydrocarbons $\longrightarrow$ free radical hydrocarbons

free radical hydrocarbons   + $xNO \longrightarrow xNO_2$ +   hydrocarbons

where $\underline{x}$ is the number of nitric oxide molecules that react with each free radical. The value of x is greater than 1.

Because of this sequence of reactions, nitric oxide levels fall and nitrogen dioxide levels rise. You may have noticed on smog-ridden days that the air takes on a brown tinge. This is due to the presence of brown nitrogen dioxide gas. Another mechanism for nitrogen dioxide build-up and nitric oxide depletion is through the reaction of nitric oxide with carbon monoxide and oxygen.

$$NO + CO + O_2 \longrightarrow NO_2 + CO_2$$

Remember that atomic oxygen is being produced all the while by the dissociation of nitrogen dioxide in sunlight. Further, ozone is being produced by the reaction of the atomic oxygen with the molecules of oxygen that are normally present. When nitric oxide concentrations become very low, the reaction of ozone with nitric oxides (see the photolytic cycle) cannot proceed. Hence, ozone levels build up.

In summary, a typical pattern of pollutant levels through the day is observed. Nitric oxide levels rise initially due to motor vehicle emissions. Then nitrogen dioxide concentrations rise and nitric oxide levels fall as the reaction of free radical hydrocarbons consume the nitric oxide and produce the dioxide. Ozone from the photodissociation of nitrogen dioxide then builds up because of the absence of nitric oxide.

Questions

1. What are the chemical categories of air pollutants produced by the automobile? List three categories. Which categories are known to affect our health?

2. Why did Congress require that hydrocarbon emissions from the automobile be decreased?

3. After 10 minutes of exposure to an atmosphere with 0.23 milligrams of nitrogen dioxide per cubic meter of air, volunteers lost the ability to notice the smell of the gas. People who visit Los Angeles often note the brown haze over the city and smell pollution. Residents notice the smell less. Why?

4. Nitrogen dioxide decreases the ability of the eyes to adapt to seeing in the dark. Can you think of a common activity where personal safety could be threatened by the decreased ability to adapt to darkness after brightness? Explain. Is the activity likely to be undertaken in air with elevated levels of nitrogen dioxide? Why?

5. Animal studies done in the laboratory indicate an increased susceptibility to respiratory disease in the presence of nitrogen dioxide. These studies support epidemiological evidence, that is, evidence obtained by observing the disease response of human population. Why are scientists careful about applying animal studies to setting pollution standards for humans? What animal might be advantageous to use from this standpoint? What is their principal disadvantage?

6. What are the names of the photochemical air pollutants? Why are they called "photochemical" air pollutants? How are they controlled?

7. The air quality standards for photochemical pollutants were set in such a way as to protect the health of a portion of the population that was particularly susceptible to the substances. Who are these people?

Further Reading

Air Quality Criteria for Hydrocarbons, U.S. Environmental Protection Agency AP-64, 1970.

Air Quality Criteria for Nitrogen Oxides, U.S. Environmental Protection Agency AP-84, 1971.

Air Quality Criteria for Photochemical Oxidants, U.S. Environmental Protection Agency AP-68, 1970.

Control Techniques for Carbon Monoxide, Nitrogen Oxide, and Hydrocarbon Emissions from Mobile Sources, U.S. Environmental Protection Agency, AP-66, 1970.

Although technical, these documents are very clearly written and very complete.

Waldbott, George, *Health Effects of Environmental Pollutants*, 2nd ed., C.V. Mosby Company, St. Louis, 1978.

This well-illustrated (136 pages of illustrations) book has a wealth of information and documentation on all of the major air pollutants. It is a technical work, but clearly written. The book could serve as an excellent reference for the health aspects of air pollutants.

Chapter 20

Oil Pollution

WHERE DOES THE OIL COME FROM?

Oil pollution. The words bring to mind pictures of wrecked tankers grinding over submerged rocks or of geysers of flaming oil shooting from well blowouts. Yet, historically, such dramatic happenings have accounted for only a minor portion of the total amount of oil spilled each year (Table 20–1). Some 5 million tons of oil are added each year to the world's waters. In 1975, the National Academy of Sciences assigned the following percentages to the various sources.

TABLE 20–1   Sources of Oil in the
World's Waters [a]

Used motor and industrial oil	50%
Normal shipping operations	37%
Normal oil refinery operations	4%
Tanker accidents	4%
Other ship accidents	2%
Offshore drilling	1%–2%

[a] Adapted from National Academy of Sciences, "Petroleum in the Marine Environment," Washington, D.C., 1975, p. 6.

Most of the oil spilled into natural waters has been the result, not of accidents, but of normal operations. This was true even in 1978 when the Amoco Cadiz was wrecked off the coast of Brittany. The tanker's entire cargo, some 69 million gallons of oil, escaped into the ocean. Yet this was still much less than the amount of used motor and industrial oil reaching the oceans each year.

Used Motor and Industrial Oil

Used motor and industrial oil enters natural waters through sewer outflows and from normal run-off. Although anyone found dumping oil into a waterway is libel to heavy fines, there are still few convenient depots for the collection of such wastes as used automobile oil. Thus, used oil most often finds its way into sewers or garbage dumps. From there it runs into nearby waterways.

Normal Shipping Operations

All ships collect both oil and water in their bilges. The simplest method of disposing of this oily water is to pump it into the ocean. This accounts for about 25% of the oil pollution from shipping operations. An international treaty, written in 1973 by the United Nations Intergovernmental Maritime Consultive Organization (IMCO), forbids the discharge of oily wastes near shores and limits the amount that can be dumped on the high seas. However, none of the 16 shipping powers who signed the treaty have yet ratified it. Some important shipping nations did not even sign the treaty. In areas where dumping is prohibited by the treaty, illegal dumping still goes on, especially at night when it is hard to detect.

A further portion of oil pollution from shipping operations is due specifically to oil tankers. Many tankers pump sea water into empty oil tanks as ballast and use sea water to wash out tanks between cargoes. When such a tanker nears an oil loading port it has tanks full of oily water to dispose of. For many years it was simply pumped back into the ocean. Newer ships are fitted with separate ballast and oil tanks or with slop tanks where oil can be recovered from ballast water. Further, some tankers now wash out tanks with oil rather than with water. However, all tankers and all ports do not have the special equipment needed for these procedures.

Refinery Operations

During the unloading and loading of oil at a refinery there may be small spills. In addition, the refinery uses water in many of the refining steps. If this water is then returned to natural waters, it will carry some of the oil with it. Refineries can be operated in an almost closed cycle, so that no oil escapes to the environment. Many, however, still pollute nearby waters.

Oil Well Blowouts and Tanker Accidents

Accidents to tankers and offshore oil well blowouts account for only a small percentage of the total amount of oil spilled, but they result in most of the visible damage. This is because a grounded tanker or an offshore blowout releases an enormous quantity of oil at one time. Because it is cheaper to ship oil in large quantities, the size of the tankers traveling the oceans is increasing rapidly. The average size supertanker now in use carries about 250,000 tons; it is as long as three football fields. Even larger ships have been built or are being considered, up to 1,250,000 tons. The possibilities for ecological damage if one of these tankers is wrecked are immense. Damage from the wreck of the 233,690-ton Amoco Cadiz included total loss of the oyster beds along the Brittany coast as well as effects on the fishing and resort industries. These ef-

fects may last up to 10 years in isolated coves and total costs could eventually exceed 30 million dollars.

According to a news article, during the week of December 29, 1976, crews were at work repairing the damages done by five oil tanker accidents in U.S. waters alone. Three tankers ran aground, one exploded and one leaked. Almost eight million gallons of oil were spilled in the incidents, which led to oiled beaches, oil-poisoned birds, and damage to fishing grounds.

Accidents in transportation are not the only things we have to worry about. As part of a program designed to meet the need for fossil fuels, the federal government has undertaken a program of drilling for oil under several areas of the continental shelf. It is estimated that some 82 billion barrels of oil are yet to be found in the United States, and about one third of this probably lies offshore under the continental shelf in the Baltimore Canyon, the Georges Bank area, the Southeast Georgia Embayment, off the Carolinas, Georgia, and Florida, and the Gulf of Alaska. According to a report by the Council on Environmental Quality, the Alaskan areas are unusually hazardous, with respect to offshore drilling. An earthquake of a magnitude of at least 7 on the Richter scale can be expected in the Gulf of Alaska every three to five years. (Gigantic waves, called tsunamis, caused by movements of the ocean floor, may reach heights of 35 feet in the area.) An earthquake in 1964 wiped out the original town of Valdez (the new town is across the harbor from the terminal end of the Alaska pipeline).

Most of what we know about the effects of oil in marine life is a result of offshore oil well blowouts or tanker accidents. A case in point is the blowout that occurred in 1968, while a Union Oil Company crew was drilling an oil well in the Santa Barbara Channel off the coast of California. On the morning of January 28, there was an explosion at Union Oil Company's Platform A. Oil and gas spurted up in a 100-foot column. Amidst a mist of oil droplets, the crew worked frantically to shut off the well. They were able to do so in a few minutes. But then there was a second explosion and poisonous gases began bubbling up from the northeast corner of the platform. Faced with this further danger, the crew abandoned the platform and made for shore in their life boats. Oil began to pour out of a rupture in the sea floor. The oil slick grew at 200,000 gallons per day. Seven days later, it finally came ashore on the beautiful Santa Barbara beaches during a series of storms. Workers, including many volunteers, began to spread straw on the beaches and rake it up again, to absorb the oil (Fig. 20–1). Others organized to try to save the sea birds, which staggered along the beaches or were washed into shore (Fig. 20–2). At least 3600 birds perished as a result of the Santa Barbara blowout.

BIOLOGICAL EFFECTS OF OIL IN THE ENVIRONMENT

Oil and Birds

(Birds suffer during an oil spill because the oil soaks into their feathers, ruining the waterproofing and insulating qualities. The birds can no longer keep warm or float. Estimates of the number of birds killed during an oil spill are often low because many birds simply sink out of sight. As birds try to preen away the oil they swallow it and are blinded and poisoned.)

(Oil contaminates or destroys sea birds' natural foods. Diving birds are especially affected, as they dive through the oil slick again and again in search of food.)

(Rescue attempts, in which oil is washed off the birds, also remove much of the natural waterproofing from their feathers (Fig. 20–3). The birds must then be kept in captivity until they molt and grow a new set of feathers. It is extremely difficult to keep and feed large numbers of wild birds for any length of time. Thus, bird salvage rates have averaged only a disappointing 1% to 15%.)

Figure 20–1 *(Top photo) A beach along the Santa Barbara Channel. This picture was taken a few years after the Union Oil Company well blow-out. (Standard Oil Company; R. Isear, photographer) (Bottom photo) Workers spread straw on the beach at Santa Barbara in an attempt to clean up oil coating the beach and floating on the water. (Massachusetts Audubon Society; Saunders, photographer)*

Toxicity of Various Kinds of Oil

Although the slick of crude oil at Santa Barbara washed in and out several times, which meant that the beach had to be cleaned and recleaned, marine life, aside from the birds, was not severely affected. Crude, or unrefined, oil is not as poisonous as certain refined oils. (The portions of oil that boil off at very low heat are the most toxic. Gasoline and home heating oil are rich in these

Figure 20-2 *The birds which wash ashore after an oil spill may be only a small percent of the number actually killed. Many lose their ability to float on water and die at sea. (Photo courtesy of Oil and Special Materials Control Division, U.S. Environmental Protection Agency.)*

phenols and aromatic hydrocarbons.) Since these substances also evaporate readily, the longer an oil slick can be kept away from a beach, the better. Part of these substances probably evaporated from the Santa Barbara slick before it reached the shore. In contrast, a spill of Number 2 home heating oil at West Falmouth, Massachusetts, in 1969, resulted in the death of numerous fish, bottom dwellers, and creatures in the intertidal zone (such as mussels, crabs, and starfish).

Detergents and Marine Life

When the Torrey Canyon went aground in 1967, spilling 30 million gallons of oil, it was the first major oil spill ever to occur. Not much was known about the eventual effects of oil on marine life. The main concern of almost everyone involved was that the oil would foul the vacation resort beaches along the coasts of England and France. Since the summer holiday season was due to begin, local hotel keepers and the government directed almost all their concern toward keeping the oil off the beaches. When this proved impossible, the next thought was to get rid of the oil as quickly as possible. Since oil and water do not mix well, detergents

were used to break up the oil into small droplets and enable it to mix into the water and wash away. Although this method was fairly successful in cleaning up the oily beaches, it proved disastrous to sea creatures.

(We now know that detergents make oil more toxic to marine life. The oil-detergent mixture sticks to wet surfaces, like fish gills, where oil alone would not stick. The detergent enables oil to penetrate deeply into the sand and kill even those creatures who might otherwise burrow to safety. Further, those early detergents used low-boiling petroleum hydrocarbons as solvents, exactly those fractions of crude oil most toxic to marine life! Because of the enormous amount of detergent used, two tons per ton of oil, a massive kill of marine life resulted. Except for a few anemones, almost everything within a quarter of a mile from shore and in waters up to 7 fathoms deep perished. In a few rocky coves that were too isolated or difficult to spray with detergent, the crude oil disappeared from view by itself in about two years.)

Figure 20-3 *Rescue workers attempt to clean oil from birds. Diving birds are especially affected by oil slicks because they dive into the slick again and again in search of food. There is some evidence that the birds are attracted to slicks, perhaps by sick fish near the oil. (Photo courtesy of Oil Spills and Spills of Hazardous Substances, U.S. Environmental Protection Agency.)*

New detergent formulas no longer contain toxic solvents. Nevertheless, they are still hazardous to marine creatures. The Environmental Protection Agency limits their use in the U.S. to oil fires or other cases where human life or the major portion of a bird population would be endangered by an oil spill.

Effects on Ecosystems

(In addition to its effects on individual aquatic organisms, oil affects whole ecosystems. There is a salt marsh in Southampton, England, where an oil refinery discharges, each day, 1500 gallons of water contaminated with very low levels of oil (10 to 20 parts oil per million parts water). This chronic pollution has killed off over 90 acres of marsh grasses in the area around the refinery. Once the grasses were gone, the sandy soil began to erode so that the affected area is now lower than its surroundings. Birds and other aquatic creatures that once found food in the area have been forced to move elsewhere. Thus, very small amounts of oil can, over a long period of time, have serious effects on an aquatic community.)

In areas where many oil spills occur, such as harbors or the Main Pass oil field in the Gulf of Mexico, some changes in the *kinds* of organisms that grow there are becoming apparent. Certain grasses and seaweeds cannot grow in oil contaminated water, while others "take over" in the absence of competition or the absence of the organisms that normally graze on them.

Oil in Food Chains

Many reports have been published about the tar lumps found floating far from shore in the oceans (Fig. 20-4). These lumps are what is left of oil spills after the low-boiling fractions have evaporated. The lumps range in size from tiny crumbs to pieces as big as a man's fist. There is a bacterial film on the lumps and even barnacles

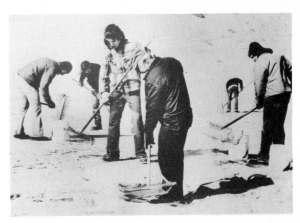

Figure 20-4   *On Nauset Beach Cape Cod, Massachusetts, workers use rakes to clean up oil pellets. Thirty miles of beach were littered with the tar balls, which ranged in size from raisins to footballs and washed ashore for a period of several weeks. The clean-up cost over $100,000. It is not known where the spilled oil, which led to the formation of the tar balls, came from. (NYT pictures)*

growing on them. Lumps have been found in the stomachs of saury, a surface feeding fish which is eaten by tuna. Thus the lumps are finding their way into food chains that lead ultimately to man.

(Both oil and oil tars contain some cancer causing substances. Several studies of shellfish grown in polluted waters have shown that they have an abnormally high number of growths that are similar to human cancer tumors. Oil, which is concentrated by shellfish such as oysters and clams, may be at least a partial cause of these tumors.)

Biological Recovery After an Oil Spill

(After oil is spilled, recovery time would certainly include the length of time it takes for all traces of the oil to disappear. However, it also means the time necessary for the polluted area to be repopulated with the kinds and sizes of organisms that lived there before the spill. If a spill does not kill all the resident organisms, those left will begin to repopulate the area, once poisonous parts of the oil have disappeared. Organisms from other areas will also begin to move in, either swimming and floating (e.g., larvae) or creeping in from nearby colonies (sea grasses). Competition among species and predation will begin to establish a balance between the different groups. But how long does oil persist in the environment after it has been spilled? Even experts disagree on this point.)

When the coastal barge Florida ran aground in Buzzards Bay at West Falmouth, Massachusetts, in 1969, there was one positive aspect. Scientists at nearby Woods Hole Oceanographic Institute were well equipped to investigate closely the effects of the spill. They discovered that even though oil may have disappeared from the surface of the water, its effects can still be far reaching and serious. In Buzzards Bay, heavy seas and onshore winds insured that the spill of home heating oil was well mixed with the water. In a few days no oil could be seen floating on the surface. Nine months later, however, oil could still be found in the sediments. Where the oil was found, there were fewer organisms than in neighboring regions. The oiled sediments had also eroded away to some degree. This seemed to allow the oil to spread further. The area in which oil was found was much larger than the original spill. It was now about one acre for each barrel of oil spilled. It may take more than a quick inspection, therefore, to judge the damage from an oil spill.

Other studies done in nearby marshlands seven years after the spill showed oil still present in the sediments in some areas. Certain of the sea creatures studied, such as the fiddler crab, were still absorbing oil into their bodies. This oil caused them to behave strangely and prevented the population of crabs from increasing to the levels that existed before the spill. (Scientists therefore warn that oil may have only short-term effects in some areas, but in others, such as marshes, the oil becomes mixed into the sediments and lasts for many years) (Fig. 20-5).

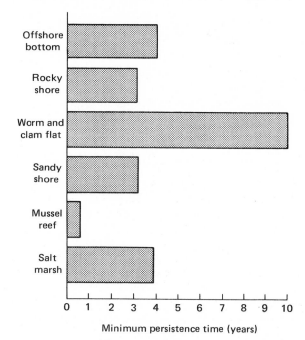

Figure 20-5   *Oil lingers for different times in different habitats. The times on the graph, however, should be taken as the minimum residence times of an oil spill, because most studies of oil spills are ended before the oil is actually gone from the environment. [Source: S.F. Moore, "Offshore Oil Spills and the Marine Environment,"* Technology Review, *p. 67 (February 1976).]*

ECONOMIC AND SOCIAL EFFECTS OF MIXING OIL AND WATER

Offshore Drilling

The highly populated states along the Atlantic coast are understandably uneasy about offshore drilling. The Northeast needs oil because it is located far from areas presently producing oil. But, while the area as a whole could benefit from the oil, the people living in the coastal states would bear all the environmental costs of producing the oil. These include possible oil spills from drilling and transporting the oil, as well as water and air pollution

OIL POLLUTION

Talk of an ecological catastrophe thus far is only talk. It has no factual basis.

Philip H. Abelson

...how about the victims? Hiroshima, too, has been rebuilt.

Marjorie Anchel

The Argo Merchant spilled five million gallons of crude oil off the coast of Nantucket in December 1976. In an editorial in <u>Science,</u> Philip Abelson criticized reporters for the way they had handled the story. He took the press to task for using terms like "ecological catastrophe" and proceeded to point out that there appeared to be no <u>lasting</u> damage to the environment.

> Obviously, when 28,000 tons of heavy oil are released to the environment damage must result. Already there has been loss of some sea birds. Later, part of the tar balls may reach land, where they will be a nuisance. Incidents of this kind should be avoided and the perpetrators should be forced to pay for any demonstrable damage. But talk of an ecological catastrophe thus far is only talk. It has no factual basis. *

Leaving aside his judgments about long-term effects, think about his assumption that the only <u>important</u> effects are those which are long term.

To the Editor:

> Three recent letters–by Gene Neary (Jan. 21), Norman Allstedter (Jan. 31) and W. A. Holman (Feb. 2)–constitute a disturbing series.

> Neary asks "...what were the irreparable consequences of the oil spilled on the seas during World War II?" Allstedter, public relations manager of Northeast Shell Oil, quotes the Interior Department: "Personal accounts indicate that beaches were cleaned and mostly free of tar by 1948." Holman (captain of an oil tanker) writes about the rescue of a frozen goose.

> None of the letters mentions the effects of spilled oil on thousands of geese and sea birds who drowned or slowly starved or froze to death. Ecological considerations are not necessarily the same as humane consideration. Even if there were no "irreparable" ecological consequences of oil spills, how about the victims? Hiroshima, too, has been rebuilt. †

Are the short term effects of oil spills important? If an area or a wildlife population appears to recover from an insult such as an oil spill (or a chemical spill or thermal pollution), can we term the spill a disaster or a catastrophe?

---

* Abelson, P. H., <u>Science</u> **195,** (14 January 1977).

† Anchel, M., <u>The New York Times,</u> 11 February 1977.

from onshore development of oil refining facilities. Localized over-crowding may occur as workers stream in to develop the oil fields. In some economically depressed areas, people may welcome the possibility of oil-related jobs. However, unless there is careful planning by local officials and agreements are drawn up between the oil companies and the towns, few local people may be hired. In such a case, the native population will suffer the inflation; housing shortages; overcrowding of schools, roads, and recreational facilities; and higher taxes, which accompany an oil boom, without sharing in any of the money it brings.

The Fairbanks newspaper, All-Alaska Weekly, admitted in an editorial on 10 May 1974, "In the past few years we have written so many editorials supporting the Trans-Alaska Pipeline project, that now that the project is beginning we feel a bit hypocritical in...complaining about its impact." The editorial pointed out that only people who were actually working on the pipeline were benefiting from its effects. "By and large the average citizen is being hit, and hit hard" by a "whirlwind of inflation," skyrocketing real estate prices, "unconscionable" rent increases or eviction, traffic jams, and the possibility of contaminated drinking water in wells outside the city, where building was proceeding rapidly.

Deep-Water Ports

The development of deep-water ports presents some of the same economic and social problems onshore as does drilling for oil on the continental shelf. A deep-water port is a complex of buoys and a platform. It is located in deep water, some distance from the coast. Such distance is necessary to provide deep enough water to float the supertankers bringing the oil (Fig. 20–6). Currently U.S. ports are served by tankers in the 50,000 ton range because the harbors are not deep enough to accommodate supertankers. From an environmental point of view, one benefit of the deep-water ports is that they can be located out of the mainstream of nor-

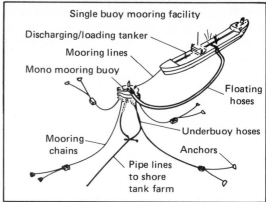

Figure 20–6 *A single point mooring (SPM) facility can load tankers of any size in any weather condition. Crude oil is fed from land through a 42-inch submarine line to this 2800-ton, 145-foot high structure standing in 100 feet of water two miles offshore. (An Exxon photo.)*

mal harbor traffic, thus reducing the possibility of tanker collisions.

One of the best sites for a superport is the Palau Islands in Micronesia (Fig. 20–7). The sociological effects of building a port there, however, are expected to be enormous.

LESSENING OIL POLLUTION AND ITS EFFECTS

Reusing Waste Oil

Used oil from automobiles, trucks, and industrial machinery probably accounts for one-half of all oil pollution. Thus schemes for reducing the pollution of water by oil must take this into account. Waste oil could be collected, refined, and then used again, although this is a complicated process. Many substances are added to oil to allow it to be used longer at higher temperatures and at increased bearing loads. Some examples of materials added to oil include compounds of barium, calcium, zinc, magnesium, phosphorus, and chlorine. Occasionally even vegetable and animal fats are added. Along with these additives, used oil picks up contaminants from the air and the machine itself.)

(However, new processes have been worked out that appear to remove all the contaminants and produce an oil that works as well as virgin oil. Because the cost of virgin oil is rising, it is becoming economically attractive to re-refine waste oil. One of the main problems involves collecting this waste oil so it can be reused. In some areas, citizens groups have organized waste oil collection campaigns. This involves setting up depots where car owners can drop off used oil after they change the oil in their cars; it also involves educating people about the need to dispose of the oil properly. The Federal Energy Administration has prepared a "Waste Oil Recycling Kit" to show citizens groups how to begin recycling centers for used oil in their communities.)

Moderating the Effects of Offshore Drilling

(One result of offshore drilling accidents such as the Santa Barbara disaster has been an improvement in the safety devices designed to prevent blowouts. Producing oil wells have safety valves designed to close off the well if storms or earthquakes destroy the platform. It is believed that these safety devices will work about 96% of the time.) Drilling for oil in Britain's North Sea oil fields has provided a good test for the safety of outer continental shelf oil production. The water is deep, and

Figure 20-7   *View of some of the islands that make up the Micronesian district of Palau, site of a proposed supertanker port. (Andrew H. Malcolm/NYT pictures)*

storms are frequent and rough. There have been no serious oil well blowouts since drilling started in 1970, although several wells containing natural gas have blown (Fig. 20–8).

(Still, because safety devices are not 100% effective and because there is always room for human error, there is a high probability that some spills and blowouts will accompany any offshore drilling.) For instance, the United States Geological Survey estimated that there is a 50/50 chance that at least one large oil spill will reach a beach or seaside park area in New England during drilling for oil off the coast. (The Outer Continental Shelf Bill, passed in 1978, includes strict safety procedures and also specifies that oil companies must set up a compensation fund for fishermen and landowners hurt by offshore spills.)

(There are also federal laws that can be used to control the effects offshore oil development has on the nearby shore areas. The Coastal Zone Management Act provides funds for the states to examine their coastal resources and decide which areas must be protected as wilderness, which are suitable for development, and so forth. The law also requires that once a state has developed a coastal zone management plan that is accepted by the federal government, the federal government cannot take actions that are not in line with the state plans.)

(The National Environmental Policy Act (NEPA) has also been used to prevent oil pollution

CONTROVERSY

OIL DEVELOPMENT AND NATIVE CULTURE

The superport would be an environmental disaster.... The people would trade a few years of money for their whole fragile environment and culture.

> Mr. Owen,
> Palau Conservation Officer

I can't accept all this so-called environmental concern.... These people want to keep Palau a human zoo so they can come and swim and take pretty pictures...

> Roman Tmetuchl,
> Palau Businessman

Two hundred islands in a chain almost 100 miles long make up the Palau Islands district of the U.S. Pacific Trust Territory. A few are inhabited; some are only a few yards across. But all are scattered among the coral reefs located in deep, clear, blue Pacific waters. The location of the islands and the depth of the surrounding waters would make the area an ideal port for supertankers carrying oil from the Middle-East to Japan. In Palau, the oil could be loaded off the supertankers and into small tankers, which are able to enter Japan's relatively shallow harbors. However, the dredging and blasting needed for construction of the superport would destroy parts of the coral reef and the very delicately balanced community of marine life it supports. Further, the construction workers, the oil port personnel and all of the accompanying hustle and bustle might overwhelm the predominantly farming and fishing culture of the native islanders. The following quotes are from an article by Andrew Malcolm in The New York Times, 7 February 1977:

"An outside project of this magnitude is too big for this little place. It would control our politics, our economy, our institutions, our lives. It is far beyond us. And after centuries under foreigners, we want to be free, man, free." (Moses Uludong, a leader of the Save Palau Organization.)

"Already our young people are getting the taste of money, and they like it. They will forget how to work with their hands. And I fear the arrival of greed." (Gloria Gibbons, Palau Princess.)

"Speaking as an individual, the superport would be an environmental disaster of the worst order. The people would trade a few years of money for their whole fragile environment and culture. The potential for environmental disaster here is much greater than even on the Alaskan pipeline." (Mr. Owen, Palau Conservation Officer.)

"I can't accept all this so-called environmental concern. The world has always had pollution and overcome it. These people want to keep Palau a human zoo so they can come and swim and take pretty pictures and then go home to their own lives while the people here starve without work. Save Palau, they say. For what? For whom?" (Roman Tmetuchl, Businessman.)

It is never easy to draw conclusions when environmental protection appears to be pitted against the human desire to upgrade living standards. What are your feelings about these questions: What environmental damage do you think the superport might do in Palau? What sort of disruptions might the construction crews cause? (Consider for instance, what services they will need, where they will live, what they will do for recreation.) How likely is it that native Palau islanders will either be hired by or provide services to the construction and operating crews of the port? Would you side with Mr. Tmetuchl or Mr. Owen? Supposing you were an islander, who would you side with?

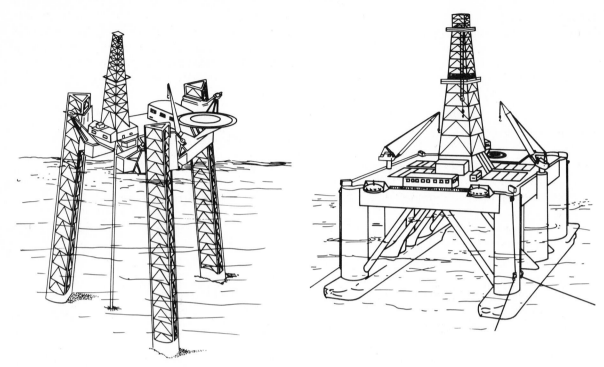

Figure 20–8 *In waters up to 300 meters deep, oil is produced from permanent platforms, fixed to sea bottom. Several safety values are attached to the system to prevent well blowouts. In deeper waters, or where storms are more frequent, floating platforms or even ones located almost entirely underwater may be used for drilling and production.*

damage in sensitive areas)such as the Santa Barbara Channel. This channel lies between the California coast and a series of islands. Along this coast lies one of the nation's longest stretches of recreational beaches. For this reason, there is a great deal of concern about the possible effects of oil spills and blowouts during drilling in the area. Proposed drilling in several areas was dropped after hearings required by NEPA. Although the hearings serve only an advisory function, it was brought out that one of the islands, San Miguel, is the main breeding ground left in southern California for sea lions, fur seals, and harbor seals.

Improving Tanker Safety

The U.S. Department of Commerce estimates that as many as 85% of shipping accidents are due to human error. Part of the problem may be due to the existence of "flags of convenience.")Certain countries, notably Panama, Liberia, Singapore, and Cyprus have much lower standards for the ships and crews to which they grant registration. Because it is cheaper to register in such countries (both because standards are lower and because taxes are lower than in countries like the U.S.), a large proportion of the world's shipping fleet is registered in them.

Although the U.S. Coast Guard has been reluctant to impose very strict standards on foreign ships and crews coming into U.S. waters because this might cause other countries to retaliate, in some way, Congress may pass new laws on tanker safety and crew training. Safety features such as double hulls and bottoms on tankers, special tanker lanes, and traffic controls should certainly be considered to prevent disastrous spills. Size limitation should also be explored for tankers serving the U.S.

Cleaning Up Oil Spills

If an oil spill does occur, what can be done about it? (The oil slick must first be kept from spreading with barriers such as floating booms (Fig. 20–9). The oil can then be soaked up with adsorbent materials or skimmed off the surface of the water. Various devices are available to skim the oil from the water (Fig. 20–10). However, booms and skimmers only perform well in calm waters, on small to moderate spills. Ocean currents can carry oil under the booms, and in rough seas, the oil will not form a slick at all, but droplets that mix into the ocean. In a storm, little can be done to prevent spilled oil from coming ashore on beaches. When oil does reach beaches, it is still best cleaned up by broadcasting straw and raking it up again.) According to a report by the Maine Department of Environmental Protection[1]

> Once the volume spilled exceeds the capabilities of present state of the art technology, man can then only realistically chip away at the oil with available recovery machines, walk behind the spill on the beaches physically recovering oil with rather primitive tools, attend to claims for various oil spill related losses, and measure the effects by counting nature's victims as they wash up on the beach.

1 Statistical Report—Oil Spills 1976, Maine Department of Environmental Protection, Bureau of Oil Conveyance Services. Quoted in *Maine Environ News* 3(13), 7 (11 February 1977).

Figure 20–9   *Corralling spilled oil. Workmen off a Perth Amboy, N.J., marina maneuvering part of a 13,800-foot boom in an effort to contain massive oil spill in the Arthur Kill and sweep it toward the shore and vacuum cleanup trucks. More than 250,000 gallons spilled from a punctured tanker. (William E. Sauro/NYT pictures)*

Figure 20–10   *Fishing boats tow an oil skimmer through an oil slick to demonstrate how the booms (V-shaped arms) funnel oil into the skimmer (at the point of the V) where it is skimmed off the surface of the water and collected. Note the darker area of clean water following in the wake of skimmer. Such systems do not work in choppy waters, where the oil slops over the top of the booms. (Photo courtesy of Clean Seas Incorporated, Santa Barbara, California.)*

(Thus one comes to the conclusion that oil pollution, like so many other kinds of pollution, is best dealt with by enforcing laws designed to prevent oil spills from occurring in the first place.)

Questions

1. What is the major source of oil pollution in the world's waters? Why are tanker accidents and oil well blowouts significant even though they contribute only a small percentage of all the oil spilled in water?
2. Suppose your class decided to start a waste oil recycling project in your community. How would you begin? How would you explain the need for such a project to citizens in your community?
3. What are the major effects of spilled oil on the water environment?
4. One of the disadvantages to building superports is that dredging must often be done to make deep shipping channels. The disposal of the dredged material can cause environmental problems. What are the advantages of deep-water ports and supertankers? What are other disadvantages?
5. How could a local government ensure that the rapid onshore development that comes with offshore oil production would benefit residents rather than hurt them?

Further Reading

Griner, L. A. and R. Herdman, Effects of Oil Pollution in Waterfowl: A Study of Salvage Methods, U.S. Environmental Protection Agency, Water Quality Office, 1970.

Kerr, R. A., "Oil in the Ocean: Circumstances Control Its Impact," *Science* **198**, 1134 (16 December 1977).

This article provides a good summary of several studies done to determine the long-term effects of spilled oil.

"The Language of Oil," Mobil Oil Corporation, 150 E. 42nd Street, New York, N.Y. 10017, (1974).

This small booklet, produced by Mobil Oil, clearly defines 100 commonly used terms relating to oil production and oil pollution. It is a handy reference to have while reading on the subject of oil pollution. Write to the above address for copies.

Oil and Gas in Coastal Lands and Waters, A Report by the Council on Environmental Quality, U.S. Government Printing Office #040-000-00386-0, 1977.

This is a well written explanation of oil pollution and its effects in waters near shore and on the shore itself. The economic and social costs of drilling for oil on the continental shelf are especially well covered.

Technology Review, **78**(4) (February 1976).

This issue contains several articles on offshore drilling. The biological effects of oil spills and the likelihood of offshore spills causing economic and biological damage are detailed.

Chapter 21

Thermal Pollution

WHAT IS THERMAL POLLUTION?

The generation of electricity produces waste heat which must be disposed of. When this waste heat is released into the environment, it can have harmful effects. For this reason it is classed as a pollutant—thermal pollution.

Review of The Power Generation Cycle

Whether a power plant operates by burning the fossil fuels, coal, oil and gas, or by nuclear fission, the plant first generates heat, which is used to boil highly purified water to steam. The steam, under very high pressure, is used to turn a turbine. The pressure of the steam when it leaves a turbine is markedly reduced. Such "spent" steam is no longer capable of turning a turbine. However, it is still very hot; that is, not all of the heat energy provided by combustion was transferred to the turbine. The energy that was not transferred is the origin of the "waste heat" which is discarded to the environment.

The water originally turned into steam is highly purified, to prevent solid deposits from building up. It is reused to avoid having to continually purify more water. Thus the low pressure steam is passed through a condenser after exiting from the turbine. In the condenser the heat in the spent steam is transferred to cool water drawn from a lake, a river, or the ocean. The steam condenses, and, at the same time, the cool lake water is warmed. The steam which was condensed is now liquid water and is cycled back to the boiler to be converted again to steam. The process is continuous in the sense that water is entering the boiler, being converted to steam turning a turbine and being condensed continuously (Fig.21–1).

Efficiency and the Second Law of Thermodynamics

A fossil fuel power plant, burning coal or oil, is said to be, at best, 40% efficient. That is, for every kilowatt of electricity produced (which is equivalent to 3400 British Thermal Units) the plant will produce 6000 BTU's of waste heat. Thus almost

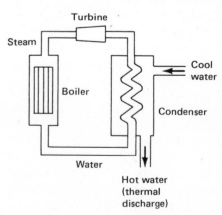

Figure 21–1 *How thermal pollution occurs from an electric power generating station is summarized in the above picture.*

two units of waste heat are produced for every usable unit of electricity. Nuclear power plants are even less efficient. Only about 32% of the energy they produce goes to generate electricity. A small portion of the waste heat from fossil fuel plants goes up the stacks. The rest of the waste heat from fossil fuel plants and all of the waste heat from a nuclear plant goes into the cooling water.

(Electric power generation provides an example of the second law of thermodynamics. That is to say, when energy is transformed from one type into another, some is always lost as heat. Here, the energy in fuels such as coal or uranium cannot all be used to produce electricity. Some is lost as heat.) (This point is explained more fully in Chapter 11.)

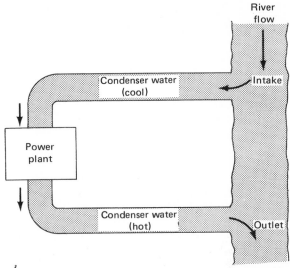

(**Figure 21-2** *Open-cycle (once through) cooling.*)

Cooling Cycles

(There are two basic cooling arrangements in electric generating plants. In the one we have been discussing, cool water is drawn from a nearby body of water, is used to condense steam, and then is discharged back to the body of water. This is called "once-through" or "open cycle" cooling. Thermal pollution of water is a result of open cycle cooling.)

(In the second basic arrangement, called "closed cycle" cooling, the heat which cooling water takes up to condense the steam is released to the atmosphere by a cooling tower. Since the cooling water circulates continuously through the plant, in closed cycle cooling, it would seem that no further water need be withdrawn or discharged after an initial quantity of cooling water is taken into the plant. This is unfortunately not the case; the cooling tower loses some water to the atmosphere and another small fraction of the water is drawn off and discharged (this is discussed further on p. 440). Water loss is made up by drawing new (make-up) water into the cooling system. The two cycles are shown in Figures 21-2 and 21-3.)

(In summary then, there are three, and only three, media into which the heat in condenser water may be directed. The heat can be disposed of, with

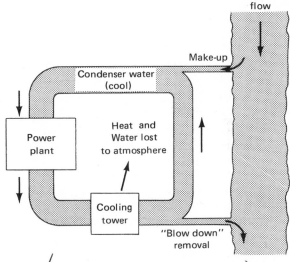

(**Figure 21-3** *Closed-cycle cooling.*)

no further steps, in the body of water from which the condenser water was drawn. This is the way heat was disposed of at most electric generating stations until the 1960s. The heated condenser water also can be applied without further steps to the land to furnish both moisture and warmth to crops. Such a use is uncommon, but has been experimented with on a small scale. Finally, the heat can be discharged to the atmosphere, through the use of cooling towers. This technique has been in favor since the first attempts to control thermal effluents.

In the next section we will discuss the effects of thermal pollution on aquatic life when once through cooling is used. Following that we will look at the advantages and disadvantages of other ways of disposing of waste heat.

BIOLOGICAL EFFECTS OF THERMAL POLLUTION

How Serious a Pollutant is Heat?

Off Turkey Point on Biscayne Bay, Florida, in the early 1970s, biologists measured an area of almost 75 acres that was barren of life. Surrounding this was another 100 acres that was only sparsely populated. Yet Biscayne Bay is known to yield over 600,000 pounds per year of marketable seafoods such as spiny lobsters and stone crabs. Sport fish abound in the Bay and a valuable bait shrimp fishery is located there. Many fish and shellfish use the bay as a breeding grounds, while a variety of wading birds feed in the shallows. What caused the apparent ecological disaster at Turkey Point in the middle of an area ordinarily teeming with aquatic life? The barren area was centered around an effluent canal from Florida Power and Light's Turkey Point power plant. At that time, the plant took cool water from the bay and returned heated water through the outflow.

Temperature is an important factor in the well-being of all organisms. For each species there is a range of temperatures which will support life. There are creatures adapted to living in the hot springs at Yellowstone National Park where temperatures may reach 70°C. There are even fish in arctic waters that have reportedly survived being frozen in ice. But for any particular species, the temperature range necessary for survival is relatively narrow; in some cases very narrow indeed. For

instance, some organisms that build coral reefs in the Caribbean can withstand no more than a few degrees change in temperature.

Warm blooded animals like humans have evolved a variety of mechanisms to keep their body temperatures in the proper range. The digestion of food produces heat. Sweating increases heat loss when the body is hot and needs to cool off. Creatures who can regulate their body temperatures are called homeotherms. Most aquatic creatures, however, are poikilotherms. They are not able to maintain a particular body temperature. These creatures must stay the same temperature as the water in which they live. Those that can swim, such as fish, move about to find suitable temperatures. This is called behavioral regulation of temperature. Those which cannot move, such as adult oysters or rooted plants are at the mercy of water temperatures. Outside certain limits, they simply cannot survive.

Effects of Temperature on Aquatic Organisms

Acclimation Both laboratory and field studies have shown that the temperatures that are lethal to an organism depend in part on the temperature at which the organism has been kept. Thus a fish held at 27°C for a few days or weeks may be able to survive longer when moved to 31°C than one which had been kept at 20°C. This improved ability to withstand higher temperatures, is called acclimation. Nonetheless, in a series of increasing temperatures, there is always a temperature at which no length of acclimation time at a lower temperature can help an organism survive. When raised to this "ultimate upper lethal temperature," the organism will die within a short period.

There are lower as well as upper temperature limits for survival. Organisms can acclimate in a downward as well as in an upward direction. Furthermore, if given a choice, creatures like fish, which can move about, spend most of their time at a particular temperature, one they seem to "prefer."

Thus in the winter fish may congregate in warm water outflows from power plants, not because they can't live in the naturally cold waters, but because they prefer their environment to be somewhat warmer. If the power plant must shut down during the winter for repairs or refueling, a fish kill can result. The fish will not be able to withstand the normal, cold water temperatures because they have acclimated to the warmer effluent temperature.

Survival and well-being of organisms Thermal fish kills are relatively rare events. Between 1962 and 1968, 18 fish kills resulting from thermal discharges were reported to the Federal Water Pollution Administration (now part of EPA). However, the number of kills may actually have been greater than this, since the reporting system was only voluntary. In the reported incidents, the number of fish killed ranged from 150 to over 300,000.

Although fish kills are dramatic evidence of the effects of thermal pollution, there are less obvious effects which can be even more serious. (Temperature can affect the ability of organisms to reproduce without actually killing directly.) For instance, trout require cool summer temperatures for the formation of eggs and sperm. Although the adult fish might be able to survive in warm summer water, they will not reproduce. For certain insects, hatching is triggered by increasing temperature. If waters are artificially warmed, the hatching temperature may be reached earlier in the year than is normal. At the Hunterston Generating Station in Ayreshire, Scotland, an inter-tidal sand dwelling copepod hatches earlier than usual, due to a heated effluent, and then dies off in greatear than usual numbers. There is not enough food available early in the year for the larvae.

(Fish that are not killed by high temperatures still may be unable to catch food. Even more subtle changes have been shown to occur that cause thermally shocked fish to be "picked out" by predators. Heat stressed fish may also be more susceptible to disease. In the long run, these sorts of effects can be just as devastating to the population as a direct thermal kill.)

The effect of excess heat on ecosystems (Temperature can affect the whole community structure of an aquatic environment) For instance, different species of freshwater algae compete for light, space and nutrients. (Temperature changes can alter the competitive position of different species, even though the changes are not severe enough to be lethal for any species.) At low temperatures (around 21 °C) yellow-green algae may predominate in a lake community. As the temperature is raised to 26 °C or 32 °C, green algae become more abundant, and finally the blue-green algae begin to dominate at very high temperatures (Fig. 21–4). In this manner, heat can seriously affect aquatic food chains, since the blue-green algae tend to be more resistant to grazing than other kinds of algae. In addition, blue-green algae characteristically have a greater mass than the species they replace.

As mentioned previously, heat can change the types of insects hatching at various seasons. Thus the available kinds of food can be changed by heat, leading to changes in the kinds of fish or other creatures a lake or river can support.

(Overall, the effect of heat is to simplify aquatic communities. That is, fewer species are found,

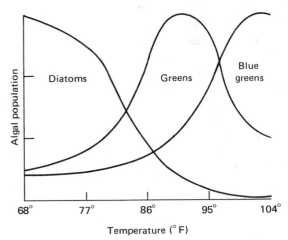

Figure 21–4 *Effect of temperature on types of phytoplankton. [Reprinted from Cairns, Jr., "Effects of Increased Temperature on Aquatic Organisms," Industrial Wastes 1(4), 150 (1956) with permission.]*

although there may be many individuals of each species.)One study found that fewer than half as many species were found at 31 °C than at 26 °C. Another 24 % disappeared at a temperature of 34 °C. Such a simple ecosystem is generally believed to be less stable than the original more complex one.

(Many natural waters are subject to seasonal variations in temperature. This variation allows different species to be dominant at different times, and therefore a greater number of species is able to compete for space and nourishment in a given area. A partial explanation for the effect of thermal pollution on populations may be that the thermal additions can even out this temperature fluctuation. Thus fewer species would be able to coexist in a given area.)

When temperature change causes the dominant species in an area to change or allows the area occupied by a particular species to enlarge, unfortunate results can occur. A case in point is the spread of shipworms into Oyster Creek. Although the creek waters would normally be too cold for shipworms, it is claimed that warm water from New Jersey Central Power and Light Company's Oyster Creek Plant has warmed up the creek, allowing the worms to spread. The shipworms are causing severe damage to all sorts of wooden structures in the area (Fig. 21–5).

(In freshwater, fish appear to be the most sensitive species. Protection of fish from excess heat would thus appear to protect freshwater ecosystems in general. In salt water, however, plant species may be more sensitive than fish are to heated discharges.)

(Many thermal effect studies have been carried out with respect to the location of power plants. In general, the studies seem to show that obvious harmful effects on ecosystems are more likely to occur near power plants located on naturally warm waters. This appears to be due to the fact that organisms living in warm regions are often already near their upper thermal limits. The additional heat from the power plant pushes the organisms over their thermal limits.)

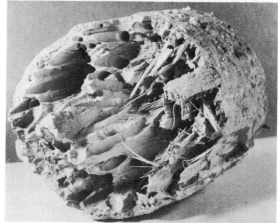

Figure 21–5 *Warm water effluents from a New Jersey Central Power and Light Company Nuclear plant on Oyster Creek appears to be allowing the spread of shipworms into Oyster Creek. The worms require warmer temperatures than are normally found in Oyster Creek. The larvae of shipworms eat their way into wooden structures like ship hulls or dock pilings, growing from pinhead size to more than two feet long and one-half inch diameter. (Carl Gossett/NYT pictures)*

Factors That Act in Combination with Temperature

(There are many factors that act in combination with temperature to affect the survival and well-being of organisms.)

Oxygen content of water (The amount of oxygen in water can affect how an organism reacts to heat. At lower oxygen concentrations, such as might be found in polluted water, lethal temperatures are lower and acclimation may not occur. Hot water may also stimulate the growth of bacteria in polluted water. The bacteria will then use up even more of the available oxygen. Thus when measured in terms of BOD,[1] heat can have the same harmful effect on water quality as a sewage outflow.)

Chemicals in water (Increasing the salinity, or salt content, of water changes the effects of temperature, making some organisms more resistant to heat and cold and others less so. Other chemicals can also make aquatic creatures less able to withstand high temperatures. This is true for phenol, hydrogen sulfide, a variety of pesticides, cyanide, copper, mercury, and cadmium. The presence of any of these chemicals as pollutants in water has the potential to make the effects of thermal pollution more severe.)

(Chlorine is another chemical which can act with heat to harm aquatic organisms. Chlorine is often used to prevent the fouling of cooling water pipes with organisms such as mussels or slime forming bacteria.)(This creates many problems because aquatic organisms are very sensitive to chlorine.) Studies have shown that oysters are sensitive to less than 0.05 parts per million chlorine. Levels as low as 0.12 parts per million have caused almost three quarters of the photosynthetic activity of algae to stop.(In some cases, power plant damage to nearby waters is probably due more to the chlorine used than to the heat given off.)This was demonstrated on the St. Croix River at the Allen S. King Power Plant, which chlorinates condenser water intermittently. Heat from the condenser outflow caused a 5% to 15% decrease in photosynthesis but, during periods when chlorine was added, the chlorine plus heat caused from 50% to almost 100% of photosynthetic activity to stop. Other techniques can be used instead of chlorine to prevent fouling. Examples are screens, proper water velocity, mechanical scouring, and short periods of high temperature in the condenser pipes (as little as 30 minutes per week of water at 38°C can be effective.) All have been used alone or in combination with chlorine. In any case it has been recommended that the amount of total residual chlorine[2] should not exceed 0.01 parts per million for more than two hours in any 24 hour period.

Entrainment (Microscopic floating plants, called phytoplankton, as well as fish, insects, and larval stages of various creatures, can be sucked into the condenser of a power plant along with cooling water. This is called entrainment. Although most of the harm to entrained organisms is probably due to water velocity and pressure, entrained organisms are exposed to higher temperatures than those in the receiving waters of the estuary, lake, or river. This can be a very serious problem in cases where the power plant takes a major portion of a river's flow during dry weather. Various designs of the intake and screens can reduce the number and size of organisms entrained in the condenser flow.)One study has estimated that the flounder population in Long Island Sound has decreased 6% as a result of 35 years of entrainment of the larval fish in local power plants. In another study, a power plant on Lake Erie was estimated to entrain and kill the equivalent of $30,000 worth of smelt (as larvae) per year.

---

1 Biological Oxygen Demand, see p. 180.

2 The measurement of residual chlorine is mentioned on p. 170.

THERMAL POLLUTION

One is somewhat appalled at the relative emphasis given to the thermal pollution abatement.

Frank L. Parker

A more appropriate determination of the costs of pollution control would balance those costs with the social and economic benefits of a healthier environment.

Douglas M. Costle

It is certainly true that we cannot have cleaner water without paying more than we do now. We will pay directly through taxes or indirectly through increased costs for goods and services. What level of benefits we want to gain and how much more we will have to pay is still being determined, however.

In a letter to the editor of <u>Science</u>, Frank L. Parker wrote: *

> ... industry will have to invest an additional $8 billion to meet 1977 requirements of "best practicable" water pollution control technology at existing plants. However, industry will have to spend an additional $9.5 billion to meet the 1977 standards for thermal discharges. The costs for thermal plant discharge elimination then exceed those from all other industrial wastes put together. When one compares the known incidence of disease and environmental destruction from the discharge of heavy metals, carcinogens, mutagens, pesticides, and so forth, to the known incidence of disease from thermal discharges (zero), one is somewhat appalled at the relative emphasis given to the thermal pollution abatement. A more realistic approach would be to evaluate each discharge in place and compare the benefits to be achieved by its elimination or reduction with the costs to achieve such benefits. From this a more realistic standard could be set.

Environmental Protection Agency administrator Douglas M. Costle gave a somewhat different argument in an article in the Environmental Protection Agency Journal: †

> ...a more appropriate determination of the costs of pollution control would balance those costs with the social and economic benefits of a healthier environment. For example, when natural systems become so contaminated that higher local and State taxes must be spent to clean them up, people have less to spend as consumers. Furthermore, corporations and consumers must pay higher health insurance premiums, greater production time is lost due to illness, more energy is needed to make water safe for drinking,

community income from recreational activities is lost–these are just a few of the considerations overlooked when the plea is made that environmental controls are unproductive corporate costs.

Frank Parker seems to suggest that the ability to cause disease is the main criterion on which to judge the seriousness of a pollutant. For this reason he feels that thermal pollution does not rank as a danger with pollutants such as mercury or pesticides. On the other hand, Costle mentions some problems that thermal pollution can cause: i.e., overgrowth of problem algae leading to increased costs for drinking water treatment or loss of recreational waters. What would you say are the most serious effects of thermal pollution? Are they as serious as effects from heavy metals or pesticides? Should we differentiate between health effects and other effects when it comes to paying for pollution cleanup?

Thermal pollution can have lethal effects on fish population or the population of other aquatic species. Should our concern for these organisms be equivalent to our concern for human health and well being?

* F. L. Parker, Science **185,** 568 (1974).

† D. M. Costle, EPA Journal **4(1),** 2 (January 1978).

BETTER WAYS TO DISPOSE OF WASTE HEAT

The thermal pollution problem can be approached in two ways. Ideally, the waste heat, which is after all a form of energy, could be made use of, rather than simply dumped into nearby bodies of water. Where this is not possible, and where the heat will cause harm to aquatic environments, technical solutions such as closed cycle cooling can be used.

Thermal Pollution Turned into a Useful Heat Source

Space heating One of the first proposals for using heated water from power plants was as a source of home heating and hot water. Because the power plant and the homes that are to use the water must be designed with this in mind, little use has so far been made of the concept. In West Germany, studies are being done to determine whether communities of over 40,000 people can be switched onto such a system. Four German cities and four towns will be involved in a pilot project because they already have suitable regional heating systems. In other cases, expensive pipes would have to be laid to carry the hot water.

Aquaculture Hot water can help food organisms to grow faster or farther north than normal. A number of projects using power plant effluents in aquaculture, or farming underwater, have been devised. For instance a hatchery at Vineyard Haven, Massachusetts, has used hot water to speed the growth of lobsters to marketable size in two years, one quarter of the time it usually takes. Farms are in operation that produce clams in the waters off the coast of England normally too cold for clams, and oysters in the waters of Long Island Sound, which is too far north for oyster farms. Both the clam and oyster farms use hot water from power plant effluents to maintain water at the necessary warm temperatures.

There are problems involved in using thermal effluents for aquaculture, however. Aquatic creatures often concentrate minute amounts of contaminants in water. Oysters and clams, which filter enormous quantities of water in order to sieve out their food, are especially likely to magnify toxic substances. Thus, if effluents from power plants are to be used in aquaculture, they must be free of such pollutants as copper from pipes, small amounts of radioactive materials, and pesticides.

In addition, the large number of organisms grouped together on a "fish farm" will produce wastes. It may be necessary to install some sort of sewage treatment facility to prevent polluting the fish farm water and also nearby areas. The fish farm itself is subject to disaster if the power plant must shut-down for emergency repairs. Placing farms in areas where two or more plants operate near each other could solve this problem.

Other uses The Environmental Protection Agency has attempted, in a series of studies, to stimulate uses of excess power plant heat. These studies found that the hot water can be piped underground to heat soil, and thus speed up the growth and size of many crops. The hot water can also be sprayed on fruit trees threatened by frost and can even cool vegetation in very hot weather (Fig. 21-6).

In short, hot water should be used in ways that enhance the quality of life rather than cause it to deteriorate. Table 21-1 lists some of the ways it has been suggested that power plant thermal discharges could be used.

Making Open Cycle Cooling Less Harmful

There are a number of engineering schemes that attempt to reduce the harmful effects associated with once through, or open cycle, cooling. Most work by spreading the heat evenly through the aquatic environment in order to avoid creating "hot spots," which could prove harmful to aquatic life. The most straightforward idea is to spread the

TABLE 21-1 Possible Beneficial Uses of Waste Heat from Power Plants[a]

Water desalinization
Agricultural uses: irrigation, frost protection
Waste disposal: desalinization or demineralization of sewage water, sterilization and
     drying of sewage
Sterilization of drinking water: using heat instead of chemicals
Refrigeration: using gas absorbtion refrigeration
Climate control: district heating and cooling, greenhouse heating, melting arctic ice
     and snow into irrigation water for arid regions
Heating intake waters at power plants: prevent fouling of pipes
Transportation: keep shipping lanes and harbors ice free
Rerefining of waste oil
Aquaculture: lure fish for catching
Power from new energy technologies: thermoelectric elements, etc.
Wildlife protection: warm water ponds for water fowl
Airport safety: defogging and de-icing runways
Mining: hot water and steam for hydraulic mining techniques

[a] Adapted from S. R. Fields, "Morphological Analysis of Beneficial Uses of Waste Heat from Power Plants,"
U.S. A.E.C. HEDL-TME 71-97 (1971), p. 8.

Figure 21-6 *Peach blossoms protected by ice formed when warm water was sprayed on them. Orchards protected this way yielded full crops in years when nearby unprotected crops were damaged by frost. (Research done under auspices of Corvallis Environmental Research Laboratory)*

waste heat in the spent steam through the largest possible volume of cooling water. This entails circulating the largest possible flow from the water source through the condenser. Another concept to avoid hot spots utilizes more than one point of outlet. A large diameter pipe is extended well into the receiving body. Holes for outlet of the heated cooling water are spaced several feet apart for the underwater length of the pipe. The pipe with its multiple hoses for release of the water is called a diffuser. Neither of these methods in any way diminishes the quantity of waste heat being put into the receiving water body.)

Another idea does transfer heat to the atmosphere to a certain extent. The warm condenser water can be floated on the surface of the receiving body because of its decreased density. This surface spreading maximizes contact between the colder air and the hot water, and evaporation, radiation, and convection lead to a rapid cooling of the floating condenser water. The surface spreading is likely to be confined to the discharge canal which leads from the plant to the ultimate receiving body, or else defects in the plan become apparent: fish might be reluctant to surface to consume insects, water birds might find the environment uncomfortable, etc.

Closed Cycle Cooling Schemes

Evaporation and cooling (The problems discussed so far have all been caused by the use of water to receive waste heat. Technologies have developed that can transfer much of the waste heat from cooling water to the atmosphere. These technologies provide two important benefits to the environment. If the heat can be transferred from the cooling water to the atmosphere, there is no need to discharge heated cooling water to the lake or river. In this way, damage to aquatic life is avoided. Not only is thermal pollution avoided, the withdrawal of cooling water from the lake or river can be cut to several percent of the amount normally withdrawn for once-through cooling. This is because the water that has been made cool can be reused to condense steam again and again. Thus, withdrawal of water from the main body can be cut dramatically, allowing the use of the water in other ways.)

(The technologies currently in use that cool condenser water are cooling towers and cooling ponds.)Both make use of the concept that when water evaporates a great deal of heat energy is absorbed by the evaporating water molecule to allow it to shift from the liquid to the vapor state.(In evaporative cooling, the evaporation of a small amount of cooling water withdraws a large amount of heat from the water which remains behind; this is the source of the cooling effect.)As an example, the rate of evaporation from a cooling tower is about 2.5% of the rate at which cooling water circulates through a plant; that is, 2.5 kilograms of water evaporate each minute out of every 100 kilograms per minute of circulating water. The temperature drop of water cooled by a cooling tower is on the order of 14 °C.

The use of the evaporative cooling principle is widespread. Bedouins keep their waterskins wet on the outside; the evaporation cools the water within the skin. In the American Southwest "evaporative coolers" are often used in place of air conditioners to chill homes.

Cooling ponds   In cooling towers and ponds the evaporative cooling principle is put to good use. Cooling ponds are technically simple; they are shallow ponds designed with very large surface areas to make evaporation take place more easily. The remaining water is cooled as a consequence of the evaporation. Radiation and conduction play lesser roles. Large areas of land are required if a plant cools its condenser water with such a pond. A design factor commonly quoted is one to two acres per megawatt of installed capacity. Thus, a modern 1000-megawatt plant could require two to three square miles of cooling pond area. Moreover, the efficiency of a pond is variable, depending on such weather factors as temperature and wind speed. In fact, the efficiency is least at precisely the time of the peak summer electric load. Mechanical sprayers may be used with ponds to increase the rate of evaporation and hence cooling.

In 1970 the U.S. Department of Justice brought suit against Florida Power and Light Company with respect to environmental damage from the Turkey Point Plant. The plant no longer discharges heated water into Biscayne Bay. Instead, cooling is accomplished by a variation on the cooling pond concept, consisting of a series of canals, described as resembling a giant automobile radiator. Although this system appears to be preventing environmental damage to Turkey Point, a large level area is required (6000 acres for 2320 megawatts of power). For many plants, such land areas are either not available or are too expensive (Fig. 21–7).

Wet cooling towers   Where land is too expensive for a cooling pond, towers are virtually the only cooling option available. Cooling towers are more complicated than the ponds, and there are several types. The natural draft evaporative cooling tower (often called a hyperbolic cooling tower because of its visual profile) is a massive concrete structure (Fig. 21–8). Warm condenser water is distributed in the tower by spraying it across the wood packing in

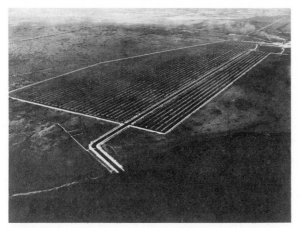

Figure 21-7   *Aerial view of the Turkey Point Cooling Canal System looking northwest. Power plant is in the upper right, and Homestead AFB is in top center. (Photograph courtesy of Florida Power and Light Company.)*

Figure 21-8   *A typical evaporative cooling tower. Note its enormous size relative to its surroundings. Such structures can be seen for miles. This is picture of the Trojan Nuclear power plant on the Columbia River near Prescott, Washington. (EPA Documerica)*

the lower portion of the tower. Air in the packing section of the tower is warmed and rises through the tower chimney because of its decreased density. Cool air is drawn in naturally at the base of the tower to replace the rising warm air. The steady movement of air past the fine, falling droplets produces conditions in which evaporation occurs easily. The rising warm air carries off the evaporated moisture as well as some fine water droplets (Fig. 21-9). Water that does not evaporate is cooled by its loss of heat to the water that does evaporate. No electric energy other than pumping the water to the height at which it is distributed is required. The natural draft tower is referred to as a "wet" cooling tower because a portion (up to 2.5%) of the recirculating water is lost (evaporated) to the atmosphere. The addition of the wet natural draft cooling tower to an electric plant may add about 5% to the cost of constructing the plant and about 1% to the residential electric rate.

Another type of wet cooling tower utilizes fans to move air from outside the tower into and through the structure. The mechanical draft tower requires electrical energy to operate, so that its operating cost is higher than the natural draft tower. In addition, the fans may be noisy and often need repair. Nonetheless, since the towers are much smaller (about 50 feet wide and high and 300 feet long), their construction costs are much lower. The evaporation process for both towers involves only water; salts in solution in the water remain behind. Hence, the continual evaporation of water results in an increased concentration of salts in the water. To slow the rate of salt build-up, a portion (about 1% to 3%) of the circulating stream is continuously removed and replaced by fresh water with its lower salt content. The removed water is called "blowdown" in the cooling tower trade. In addition, a mist of fine droplets may be carried in the air currents into the surrounding atmosphere. The sum of the losses due to evaporation, blow-

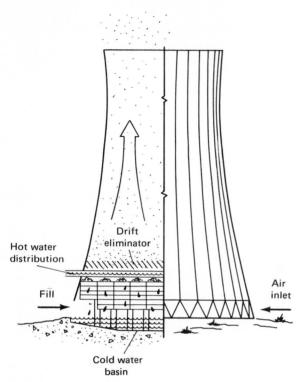

Figure 21-9 *Natural-draft, evaporative cooling tower. (Adapted from "Industrial Waste Guide on Thermal Pollution," Federal Water Pollution Control Administration, U.S. Department of the Interior, 1968.)*

down, and mist may account for 4% of the circulating water, and this quantity must be constantly replaced by a freshwater stream drawn from a lake, river, or other body of water. (Thus, even with closed cycle cooling, there is some demand for water by the plant. The water is required to replace water losses (1) to evaporation, (2) to mist, and (3) to blowdown removal.)

Dry cooling towers (Different from these "wet" towers is the "dry" cooling tower. The dry cooling tower does not use the process of evaporation to cool condenser water. Instead, it utilizes conduction and convection to transfer heat, although a cooling tower is still required.) Most of the dry towers are in use with plants smaller than 300 megawatts. Adaptation to the modern 1000 mega-

watt plant seems distant because of high construction costs, lower efficiencies of the towers, and large electrical requirements. In fact, the costs of a dry tower installation may approach the cost of the turbine-generator system itself.

The principle of dry cooling is simple; waste heat is transferred directly to the air. To accomplish this, air is passed upward through a vast heat exchanger; the fluid from which the air gains its heat passes downward through the tubes of the heat exchanger. The exchange takes place in a cooling tower; the tower is mechanical draft if the air is drawn upward by fans and natural draft (hyperbolic) if the air is drawn upward by density differences.

Are cooling towers an environmentally sound solution? (For several reasons, cooling towers are not perfect controls for thermal pollution. The towers are enormous in size and visual impact.) For a wet cooling tower, a height and base diameter of 400 feet would not be uncommon. Such a structure, located in rural areas next to a generating station, is the equivalent of a 40-story building and would dominate the landscape for many miles (Figs. 21–10 and 21–11).

The wet towers may also present other problems to the area in which they are located. To illustrate, a pair of wet natural draft towers on the Monongahela River at Fort Martin, West Virginia, are used to cool water from two 540 megawatt coal-fired power plants. The towers decrease the temperature of the condenser water from 45 °C to 32 °C. Of the 250,000 gallons/minute which are cooled by each tower, about 6000 gallons/minute evaporate. That quantity of evaporation is approximately equivalent to a room 10 feet on a side with a ceiling height of 10 feet; this quantity is evaporating every minute. The towers must be taller than the depth of the river valley which is 300 feet at this site. If they were not, mists could settle in the valley, enshrouding the area in a permanent fog. (Fog is a potential problem in cold weather when it could produce icy highways and low visibility. The taller natural draft tower has a lesser tendency to fog, because its moisture is released at

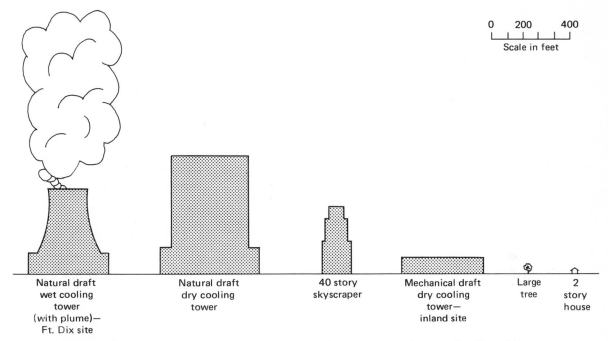

Figure 21-10 *A comparison of the sizes of cooling towers with some familiar objects. (Adapted from P. Meier and D. Morrell, "Issues in Clustered Nuclear Siting," Brookhaven National Laboratory, 1976.)*

such a high elevation. The dry cooling tower, which we discussed earlier, has no fog associated with it since water is not evaporated to provide the cooling.)

(Drift is like fog, an atmospheric problem that must be controlled. Drift, in contrast to fog which is evaporated moisture in the air, consists of particles of liquid water. These water droplets are captured by the upward flow of air through the tower and are dispersed throughout the area by winds.) The much shorter mechanical draft towers with their high velocity air flows tend to produce more drift than the natural draft towers where air flow is much slower.(Drift could cause corrosion of the power plant's electrical equipment, and would be unacceptable over densely populated areas.)Where the cooling water was drawn from a salt water source such as an estuary or the ocean, the drift from a tower would be salt water and could increase the salt content of agricultural land. Because

of its potential for damage, the designs of new cooling towers have been altered to decrease drift losses substantially.

(One further potential problem of cooling towers and waste heat disposal ought to be mentioned. Because of the growing scarcity of sites and for security reasons as well, active consideration has been given to "power parks." At such power parks, 10 to 15 nuclear plants each capable of producing 1000 plus megawatts would be constructed. Such parks would be capable of supplying the electrical demands of a state or several states. Currently, the total generating power of plants at any one site equals no more than several thousand megawatts.)

Recall that nuclear plants are 30%–32% efficient at converting heat energy to electrical energy. Thus, for every megawatt of electrical output, two megawatts or more of heat must be disposed of. A park producing power at the rate of 40,000 mega-

Figure 21-11   *Range of visibility of cooling towers and plumes. The towers and their plumes at a power park would be visible for many miles. In a study of the possibility of a power park (40 natural draft towers each 400 to 500 feet high) in New Jersey, the above map was constructed. It illustrates how far away the towers and plumes could be seen if the power park were centrally located. (Adapted from P. Meier and D. Morrell, "Issues in Clustered Nuclear Siting," Brookhaven National Laboratory, 1976.)*

watts would have to dissipate waste heat at a rate of more than 80,000 megawatts. (The waste heat could have an effect on weather patterns, since this rate of heat output rivals the rate at which the heat is dissipated by thunderstorms and volcanoes. The creation of new weather patterns by the disposal of this quantity of heat is not unthinkable. Rain and thunderstorm activity are possible. Since whirlwinds can result from the release of energy in a thunderstorm, tornadoes could occur as well. The mathematics for the prediction of weather cannot yet tell us whether these potential problems could become reality.)

Cooling Water and the Future

(Closed cycle cooling via towers or ponds has become the main way in which new steam electric plants dispose of waste heat.) The use of these cooling methods is required on new plants (with some exceptions) because of the almost total ban on heated discharges imposed in 1974 by the Environmental Protection Agency (EPA). Many of the plants built in the 1960s and early 1970s also utilize closed cycle cooling. An EPA ruling forced this action on many of the larger plants built between 1970 and 1974. The need to conserve water led still other plants in water short areas to adopt closed cycle cooling.

Although new steam electric stations built on inland waters are likely to employ closed cycle cooling methods, not all new plants will require it. Power plants on the cold ocean or on cold estuaries may escape the requirement. Power companies can win exemptions for their plants, if they can demonstrate that the heated discharges from them will not alter the ecological balance in the receiving waters. That is, they must show that the shellfish, fish, and wildlife normally native to the area will be able to survive and reproduce. Such a demonstration is most likely on cold and abundant natural waters such as the estuaries of the Atlantic North-East and the Pacific North-West as well as the oceans and bays of these areas.

Although we have discussed dry cooling as an alternative that avoids the problems of fog and of drift, it may become important as a means for condensing steam for power plants when virtually no water is available. Wet cooling requires water to replace losses due to evaporation, drift, and blow-down. Even this make-up quantity may, in some water short areas, become impossible to obtain. Under such circumstances, dry towers become the only alternative.

The need for dry cooling is expected to appear in the 1990s in a number of parts of the United States. In many of the states west of the Mississippi, particularly the Dakotas, Colorado, Wyoming, Utah, New Mexico, Arizona, and Texas, dry cooling is likely to become necessary in the 1990s. California is a special case. Although placing plants near the ocean would relieve the need for dry cool-ing, most ocean sites are too near the San Andreas fault for nuclear power plants. Water is largely im-ported to the lower portion of the state for munici-pal and agricultural use and may not be available for steam power plants. However, municipal waste water may be "reclaimed" by treatment and used for cooling. The concept is not new. Baltimore has long been selling its treated waste water to the Beth-lehem Steel Works at Sparrows Point for cooling. Even with reclaimed waste water being used for cooling, however, the rapidly growing power demands of California will probably make dry cooling necessary in the 1990s for a portion of its power generating capacity.

Questions

1. Most of the thermal pollution in this country is waste heat from power plants using once through cooling. Describe, from the point of view of the operation of power plants, why thermal pollution occurs.
2. Contrast winter and summer thermal fish kills.
3. Why is an actual "fish kill" not necessary in order to destroy a population of fish or other aquatic creatures?
4. Briefly outline the harmful effects of thermal pollution. Contrast the possible bene-ficial uses of waste heat.
5. What are the advantages of closed cycle cooling over open cycle cooling?
6. What are the disadvantages of closed cycle cooling?
7. Will thermal pollution become a problem if new energy technologies such as fusion, solar, wind, geothermal, and tidal power are used? Explain.
8. Table 21–2 lists a variety of suggested ways for using waste heat from power plants. Analyze the suggestions, noting where you feel other environmental prob-lems could result from these uses.

Further Reading

Biological Effects of Thermal Pollution

Barnett, P. O., "Effects of Warm Water Effluents from Power Stations on Marine Life," *Proceedings Royal Society of London B* **180**, 497 (1972).

A brief résumé of the effects of thermal pollution and one example of the harmful effects an operating power plant can cause.

Bush, R. M., *et al.*, "Potential Effects of Thermal Discharges on Aquatic Systems," *Environmental Science & Technology* **8** (6), 561 (1974).

The effects of heat on different species of organisms are compared in this paper which is intended to help in the choice of suitable sites for new power plants.

Gentele, G. H., *et al.*, Power Plants, Chlorine and Estuaries, U.S. Environmental Protection Agency, Ecological Research Series, EPA 600/3-76-055 (1976).

Contains a thorough, well written review of chlorination, its biological effects and its effects in receiving waters.

Jensen, L. D. *et al.*, The Effects of Elevated Temperature upon Aquatic Invertebrates, Edison Electric Institute, New York, 1969.

Engineering Aspects of Thermal Pollution

Furlong, D., "The Cooling Tower Business Today," *Environmental Science & Technology* **8** (8), 712 (August 1974).

This is a readable article, part advertisement for the author's corporation, but provides information on the wet-dry cooling concept.

"Industrial Waste Guide on Thermal Pollution," Federal Water Pollution Control Administration (now part of EPA), Pacific Northwest Water Laboratory, Corvallis, Oregon, 1968.

Relatively easy reading but technical publication; it includes a number of pictures and definitions.

Part 5

Natural Sources of Power and Energy Conservation

Let us ask the question: "Do methods of power generation have to be accompanied by degradation of the environment?" It is true that human activities will have some unavoidable effect upon the environment. After all, humans are part of a living world that is constantly changing. It would be practically impossible for us to have no effect on the environment in which we live. (This point was explained in terms of thermodynamic laws in Chapter 11.)

There are, however, natural balances, or righting mechanisms which tend to keep environments and the natural communities living in them in an equilibrium, where changes come only slowly. In many cases, human activities overbalance these mechanisms, causing rapid changes, with which neither humans nor wildlife tend to cope well. Conventional power generation, producing quantities of air and water pollutants, is such an activity.

In this fifth part we shall look at natural sources of power: wind, sun, tides, geothermal energy and hydropower. These methods of power generation seem to promise softer impacts on the environment than combustion of fossil fuels or fissioning of uranium. In addition, most of these energy sources are renewable, in the sense that nature makes them available virtually forever.

It is remarkable to think that the society that existed only two centuries ago had, besides human and animal energy, only three forms of energy at its disposal. All three of these forms could be traced to the sun. Hydropower was used to operate mills, which ground grain or wove cloth. Hydropower requires water running downward to the sea from the uplands where the water fell as rain. Of course, the sun evaporated the water in the first place. Wood was another source of energy; it was used for cooking, heating, and smelting iron. Again, the use of the sun's energy in the process of photosynthesis is responsible for the wood. And wind was used to pump water for irrigation and to fill the sails of great wooden ships. The giant windmills of Holland are vivid reminders of an early era when only such ingenuity and muscle were available to do our work.

In the past one hundred years, our industrial society has relied on the heat from fossil fuels, the heat from fission, and the energy from flowing water to power the enormous development that has taken place. To only a very minor extent has wind been utilized, or the sun's energy tapped or heat withdrawn from the earth. Yet these sources of energy, which stem from the earth's natural processes, are all around us.

In the last decade, we have increasingly been turning to these natural sources of energy because, in many respects, they provide energy without limit (Fig. A). Furthermore, the pollution that comes from them is often minor in comparison with that from fossil fuels and fission. We have been turning to these sources for another fundamental reason, though. As fuel supplies become less secure and

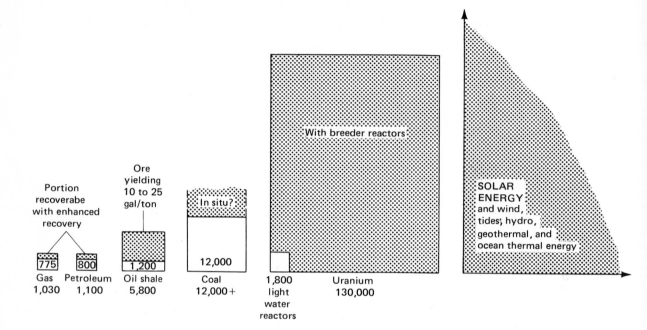

Figure A   *Available energy resources in the United States measured in quads. 1 quad = 1 quadrillion Btu = 1 million billion Btu. The total U.S. energy consumption in 1974 was 73 quads. [Adapted from "A National Plan for Energy Research, Development, and Demonstration: Creating Energy Choices for the Future," Energy Research and Development Administration (now part of the Department of Energy) 1975.]*

more expensive, these sources become more attractive and more economical. The rising prices of oil and gas are, in a large measure, responsible for our renewed interest in sun, wind, and water. It seems strange to be saying "Thank you" to the oil cartel, but it has made the point of dwindling energy resources more vividly than the arguments of any resource economist.

Thus, we are standing today on the edge of an exciting era; it is a frontier in much the same way that the western United States was a frontier (Fig. B). People will be laying claim to economic territory as a solar industry is built. Challenges of exploration and discovery in energy generation and energy conservation lie before us. Once again, we have the opportunity to become self-sufficient in

energy resources, this time by learning to use the natural energy sources which surround us. It is a period that will give new meaning to existence as we struggle to secure a stable, fruitful, and peaceful life in harmony with nature for generations to come.

In Chapter 22, we discuss how electrical energy is generated from falling water. We describe not only conventional hydroelectric generation but also the potential for small-scale hydropower installations and for the development of power from the tides. The environmental impacts of hydropower are treated as well.

In Chapter 23 we discuss the generation of electric power from the wind. We also show how heat stored in the interior of the earth can be used

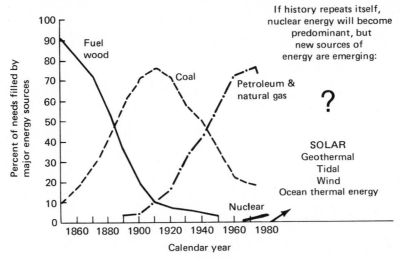

Figure B   *U.S. energy consumption pattern through time. [Adapted from "A National Plan for Energy Research, Development and Demonstration" Creating Energy Choices for the Future," Energy Research and Development Administration (now part of U.S. Department of Energy), 1975.]*

both for space heating and the generation of electric power; environmental impacts are indicated as well.

In Chapter 24, the many useful forms of solar energy are discussed. We include not only water and space heating but also solar electric generation of which there are several important forms. The conversion of solar energy to biomass fuels and the use of temperature differences in the layers of the ocean are also described.

Chapter 25 details methods of energy conservation. When energy was inexpensive, hot water from electricity was reasonable, and large heavy automobiles could still be operated at low cost.

Homeowners could afford not to have storm windows or adequate insulation. Electric heating of homes was even a possibility. Today we know that electric units for heating homes and water are expensive luxuries. People choose lighter, more efficient autos as gasoline grows steadily more expensive. Public transportation is being installed or upgraded in many parts of the country and car pooling is being encouraged. Insulation of homes and other energy saving improvements are rewarded by tax incentives, as we reach out to save energy in many ways. The savings are turning out to be large but to change old patterns of behavior takes time and requires a national commitment.

Chapter 22

Power From
Falling Water

CONVENTIONAL HYDROELECTRIC GENERATION

Water, an ancient source of power, remains today a good option for supplying electrical energy to our industrial civilization. The energy from falling water captured by the water wheel has been used directly to grind grain, to cut lumber, and to weave fabrics, but in 1895, the power of the mighty Niagra River was harnessed and used to produce electricity. This was the first large-scale effort at the production of hydropower in the U.S., although earlier hydroelectric plants had been built in Europe. No one will say that hydroelectric power is without problems (we shall discuss them in a moment). Nonetheless, when we consider the pollution and potential menace of other forms of power generation, hydropower begins to look much less menacing.

The principle of hydropower generation is simple. The kinetic energy of falling water is used to turn a turbine, which is linked to an electrical generator. Early hydroelectric plants were of the "run of river" type, in that the water flowing in the river was not dammed, but merely directed through a turbine. Large changes in river elevation were needed for these plants; the Niagara Falls development was of this type. Most modern hydropower installations, however, use dams to increase the volume of water which can be steadily discharged through the turbine (Fig. 22–1). Dams do more than provide a reservoir of water upon which to draw;

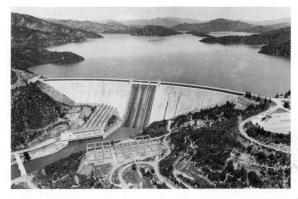

Figure 22–1 *Shasta Dam and Reservoir on the Sacramento River north of Redding, California. The Dam is a curved concrete gravity structure, 602 feet high (183 meters) and the reservoir stores over 4.5 million acre-feet (556,000 hectare-meter). Shasta Powerplant has five main generating units with total capacity of 442,310 kW. (Bureau of Reclamation)*

they increase the height of the surface of the water. The greater pressure provided by this higher water surface gives the falling water a higher velocity and hence more kinetic energy. The energy derived from this more powerful flow of water is, as a consequence, much larger.

In practice, water is drawn from the reservoir downward through a long smooth channel called a penstock and directed across the blades of a turbine, which rotate horizontally (Fig. 22–2). The turbine shaft is directed upward into the generator unit. Many turbine/generator units are needed at a typical installation. Efficiency factors on the order

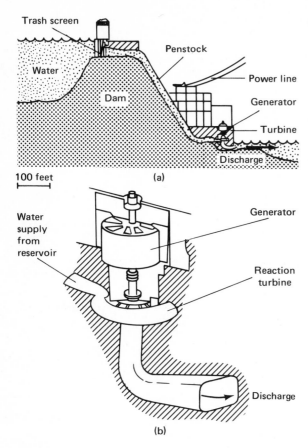

Figure 22-2 *(a) Components of a hydropower system. (b) Turbine-generator unit. (Adapted from D. Kash et al.,* Energy Alternatives, *Report to the President's Council on Environmental Quality, 1975.)*

of 60%–70% are common. That is, 60%–70% of the energy in the falling water is converted to electricity.)

(Hydropower units, though costly to install, require maintenance, but no fuel. The fuel is supplied by the sun, which evaporates water from low-lying oceans, lakes, and rivers. The water vapor condenses as rain, which falls in the highlands and flows downward to the sea. Hydropower units interrupt that flow and capture the energy in the water, which would otherwise be used up in carrying sediments to the sea.)

It is not actually necessary to dam a free flowing stream to create the benefits of an elevated water surface or a steady flow. Partial diversions of upstream water can be used to create a man-made lake off to the side of the river. Such a lake will have the benefits of both elevation and steady availability. As you read about pumped storage in the paragraphs ahead you will see that off-river storage lakes are actually quite common, but have not been thought of as replacements for reservoirs.

(The hydropower resources of North America relative to other continental areas are extensive (see Table 22-1), although most sites in the U.S. with a large potential have already been put to use. Nonetheless, much capacity at low-head sites remains to be exploited, as we shall discuss in a moment.)

PUMPED STORAGE

(Water can be used in another way in the electrical generating system. It can be used to provide supplemental power during times of peak electrical loads.)Ordinary reservoirs can be used in this way, but water can be stored off-river for this purpose. Pumped storage is the name given to the off-river reservoir which is drawn on at the time of peak loads. The pumped-storage reservoirs are specially created by pumping water to high elevations during times when power demands are low. At these times, when the full capacity of an electric station, perhaps a coal-fired plant, is not being used, energy can be generated to pump water uphill into the pumped storage reservoir. There the water remains until the time of peak power loads. Then, the water is allowed to fall through the penstocks and turn the turbine pump to generate electricity. The same pump which was used to force the water uphill is now turned in the opposite direction by the falling water to generate electric power. Essentially, the

TABLE 22-1   Regional Distribution of Hydropower Resources[a]

	Potential (1000 MW)	Percent of World Total	Developed (1000 MW)
North America	313	11	59
South Amercia	577	20	5
West Europe	158	6	47
Africa	780	27	2
Middle East	21	1	—
South East Asia	455	16	2
Far East	42	1	19
Australia	45	2	2
USSR, China, etc.	466	16	16
	2857	100	152

[a] Reproduced from M. King Hubbert, *Resources and Man,* 1969 with the permission of the National Academy of Sciences, Washington, D.C.

system uses spare generating capacity during periods of low demand and converts electrical energy into potential energy; this is the water stored at a high elevation. During peak demand times, the potential energy is converted back to electricity to supplement power from the base load plant, which is now taxed to its capacity (Fig. 22-3).

(There are two principal disadvantages of pumped storage. The first is that the high pool used for storage will have daily changes in elevation, going from nearly full to nearly empty (Fig. 22-4). The earth sides of the reservoir exposed by this daily fluctuation are quite unattractive, resembling dark desert mounds with an undulating shape. In addition, the pumped storage system is not fully efficient. Of the quantity of electric energy used to raise the water to the storage pool, only two-thirds is recovered. If coal is used to generate the electricity operating the pumps, the efficiency of conversion of the combustion heat to electricity is decreased by the two-thirds factor.)

Pumped storage can be costly to build. The unit in Ludington, Michigan, which pumps water from Lake Michigan 358 feet (110 meters) up to a 1.6 square mile (4.5 square kilometer) storage pool cost 340 million dollars to construct.

SMALL-SCALE AND LOW-HEAD HYDROPOWER

It is surprising to note, as the Corps of Engineers 1977 study did, that hundreds of hydropower sites, with dams already in existence, are no longer in use for electric generation. Many of these sites are small in capacity, five megawatts or less, but their potential is real. Abandoned during an era of cheap fuel for central electric power stations, the sites, if only put into service once again, could nearly double our capacity for hydroelectric power generation.

In the middle 1970s, about 57,000 megawatts of hydroelectric capacity were available in the United States. By restoring the abandoned dams and adding generating equipment to those dam sites never equipped for hydropower, 55,000 megawatts of capacity could be added. About half this quantity would be drawn from the small dams, five mega-

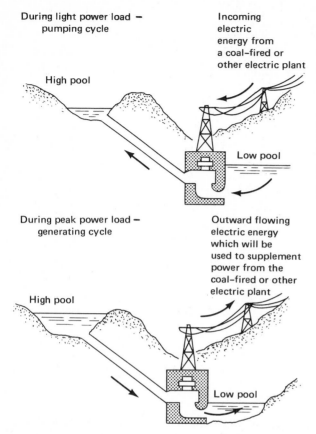

During light power load —
pumping cycle

Incoming
electric
energy from
a coal–fired or
other electric plant

High pool

Low pool

During peak power load —
generating cycle

Outward flowing
electric energy
which will be
used to supplement
power from the
coal–fired or other
electric plant

High pool

Low pool

Figure 22–3 *Pumped storage operation. (Adapted from D. Kash et al.,* Energy Alternatives, *Report to the President's Council on Environmental Quality, 1975.)*

Figure 22–4 *The Blenheim-Gilboa Pumped Storage Project. The high pool is atop Brown Mountain and falls 1000 feet (300 meters) when electricity is produced from the unit. The turbine-generators can produce power at a rate of 1 million kilowatts (1000 megawatts). (Photo courtesy of Power Authority of the State of New York).*

watts or less. Since a new thermal power plant is usually rated at about 1000 megawatts, adding these 55,000 megawatts is equivalent to building 55 major plants with very little more disruption than has already occurred because of previous development of the site (Fig. 22–5).

We mentioned earlier that the first hydroelectric developments were run-of-river plants; the sites chosen for these plants typically were waterfalls with sufficiently large changes in elevation to turn the turbines. When all such sites were exhausted, dams were built to create the needed changes in elevation. The types of turbine, the Francis, Kaplan, and Peyton wheels, all required large

"heads" of water, 100 to 2000 feet (30 to 610 meters) depending on the type.

In the early 1950s a French firm, Neyrpic, invented a new kind of turbine—the bulb turbine—which is so versatile it can generate power just from rapidly flowing water. Not even a dam is required. This development has dramatically changed the potential contribution of hydropower to electric power needs. This bulb turbine has made tidal power, with its low heads of 30 to 50 feet (9 to 15 meters), technically feasible as well. The tidal power plant on the Rance River Estuary in France uses the Neyrpic bulb turbine. (See Fig. 22–11 for diagram of bulb turbine.) The bulb turbine can be installed at sites previously too small to use and it can also be used to convert existing dams to hydropower installations. On existing dams, it could be installed just downstream of the outlet works in such a way that reservoir releases could be channeled through the turbine.

Another development in small-scale hydro has also taken place in France. The Leroy–Somer Company, a manufacturer of motors, has recently produced a miniature hydro plant they call Hydrolec. Marketed for $7800 in 1978, the smallest plant can

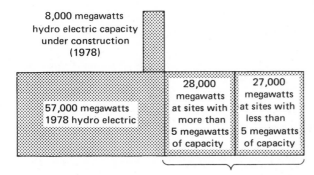

8,000 megawatts hydro electric capacity under construction (1978)

| 57,000 megawatts 1978 hydro electric | 28,000 megawatts at sites with more than 5 megawatts of capacity | 27,000 megawatts at sites with less than 5 megawatts of capacity |

55,000 megawatts dam sites with existing but unused hydropower capacity & sites with potential hydropower capacity

Figure 22-5 *Used and unused hydropower capacity in the United States (not including bulb turbine run-of-river possibilities).*

produce power at the rate of 5 kilowatts. A change in the water level of a stream of as little as 3 feet (0.9 meters) is sufficient to generate power as long as the flow rate exceeds 70 gallons (266 liters) per second. The smallest model weighs about 1000 pounds (454 kilograms) and stands 4 feet (1.22 meters) high. Operating full-time, the smallest Hydrolec could produce power worth over $2000 in a single year (at 5¢ per kilowatt-hour).

ENVIRONMENTAL IMPACT OF RESERVOIRS

(As flowing waters slow down, which indeed they must as they are brought to a halt by a reservoir, the sediment suspended in the turbulent waters will settle out of the water and drop to the bottom of the reservoir. Downstream from the reservoir, the clear water that was released into the river will erode earth from the river bank at a faster rate than the free flowing river would have, as if to make up for its loss of sediment load. Increased erosion downstream from a reservoir is a common occurrence.)

The bottom of the reservoir becomes coated with the sediments which have been transported from upstream areas (Fig. 22-6). The blanket of sediments will be exposed to view periodically as the level of water in the reservoir rises and falls in response to inflows and releases. The sediments gradually build up and, unless dug out occasionally, will begin to consume the storage volume of the reservoir. That is, a reservoir created to store water for water supply or to catch water for flood control can gradually lose its effectiveness unless the solids eroded from the upstream banks and deposited in its banks are excavated from the site.

These deposits of sediment may in part be prevented. Although erosion and the transport of sediment are natural and continuing events, farming, road development, subdivision construction, and forest cutting all accelerate erosion processes by exposing fresh and unanchored soil. Careful management of the soil (see Feeding the World's People) can help to reduce the burden of sediment carried by streams and thus can help prevent the rapid deposition of sediment in reservoirs.

The unsightly mounds of sediment which become visible during times of low reservoir

Figure 22-6 *A silted reservoir near Ithaca, New York. This reservoir has been drained because a new community water supply has been created on a nearby lake. The sediments, accumulated over some 50 years of operation, were consuming a large portion of the storage volume of the reservoir. As a consequence, a reasonable water level in the reservoir would look like a larger volume than was actually stored.*

storage are one reason that many individuals express distaste for dams. Another reason is more basic; there are valued lands lost to view and to use forever. There are valued animal and plant species that are destroyed. Not only are land species lost; fish that inhabit the stream may also be barred. Where once they swam upstream to spawn, the reservoir wall now blocks their path.

Reservoirs may be constructed so that fish are still allowed to pass upstream. Fish ladders consist of concrete steps down which a stream of water flows (Fig. 22–7). The ladder is so constructed that the fish can jump from one step up to the next and eventually to the reservoir and upstream.

The geologic history of a region may often be revealed in the layers of rock through which a river has cut its course. This history may be lost when a reservoir is thrown across a river. And the sheer majesty of the river's power which is demonstrated by the gorge may be lost, replaced by a quiet pool of water. Each potential reservoir site has its own characteristics. It may include a climax forest (trees that represent the last in a series of growths on the land before the succession of vegetation begins again) or it may possess a cluster of rare wildflowers.

The reservoir may flood out valuable farm lands, and may displace families and tribes from ancestral homes. Graveyards and tribal burial grounds may be forever obliterated. It is not clear whether a just compensation can be found for the people uprooted from such an area. Their loss is more than economic; how can one compensate an Indian nation for the loss of its heritage?

The reservoir may have still other impacts in addition to these. During certain times of the year,

Figure 22–7   *A fish ladder.*

the quality of water within the reservoir and the quality of the water released from it may be surprisingly poor. During the summer and early fall the lower layers of water in a reservoir can become very low in oxygen. The low oxygen condition is caused by a combination of two phenomena. The first is a lack of mixing of the waters in the reservoir during the summer and early fall. The second phenomenon is the consumption by bacteria of the dead algae in the bottom layers of the reservoir. The consumption causes a removal of oxygen from the water. (See below.) If this oxygen-poor water is released, fish and other aquatic creatures downstream from the reservoir can be harmed.

How Reservoir Waters Become Oxygen Poor

To understand how the decrease in the quality of reservoir waters comes about, we need to discuss the effects of temperature on the water in the reservoir. There is an annual cycle in the temperature profile of the reservoir. Deep lakes commonly undergo a similar cycle of events.

We describe the cycle, for convenience, as beginning in the late fall when the weather has begun to cool. The autumn and winter winds are transferring their energy to the reservoir in the form of waves, which mix into the body of the reservoir. The waters of the reservoir are well-mixed during this period. Thus a fairly uniform character in terms of quality and temperature of the water exists at all depths in the reservoir. That is, if one were to measure the temperature at all depths of the reservoir, one would find that, within limits, the temperature of the water is roughly the same everywhere. Such a picture continues through the winter months.

In the spring, however, sunlight warms the waters nearest the surface of the reservoir and warmer stream inflows enter the cold body of water. Above 4 °C, warm water is less dense than cooler water [1] and thus will float above it. Hence, the new warmer inflows will tend to form layers near the surface of the reservoir. Similarly, surface waters that have been warmed by sunlight tend to remain near the top. The wind's force at mixing is now less effective because the warmer, more buoyant water, though pushed into the interior, will rise again toward the surface.

Essentially, three layers of water at successively warmer temperatures are formed. The warmest layer is at the top; it is at the top because it has the lowest density. It is called the epilimnion. The coldest layer is at the bottom; it is at the bottom because it has the highest density; it is called the hypolimnion.

---

1 At temperatures at and not far from freezing, cooler water becomes less dense, more buoyant. You know this from your observation that ice accumulates first on the top of a pond rather than in its depths.

Between these layers is a layer in which there is a sharp temperature transition; the middle layer is called the thermocline. This formation of layers of differing density and temperature is known as stratification.

Stratification of the water into temperature layers is one of two phenomena that combine to cause a decrease in the quality of reservoir water. The other phenomenon is the growth of algae in the relatively clear waters of the reservoir. While some algae are growing, others are dying. The dead algae sink to the bottom layers of the reservoir; there, microorganisms in the water consume the dead algae as food. The microbes also remove oxygen as they grow and maintain themselves. Since the water in the reservoir has stratified and little mixing occurs between the upper and lower layers, once oxygen is removed in the bottom layer, it is not restored quickly. Thus, the bottom waters of a stratified reservoir may become very low in oxygen; these are the waters, we must point out, that are commonly released to downstream uses. In the section on Organic Water Pollution (see Chapter 8) we saw that waters that are low in oxygen do not support desirable aquatic life.

As colder weather arrives, the upper layers become cooler and cooler until they approach temperatures just less than the temperature in the bottom layers. At that point, the layering is made unstable, and the layers "flip." This event is called the "fall overturn" and the reservoir now becomes relatively well-mixed; that is, the water at all depths is of about the same temperature. The well-mixed character is sustained through the winter months.

POWER FROM THE TIDES

Franklin Roosevelt, President of the United States, 1932–1945, made his summer home on Campobello Island which lies along the western edge of the Bay of Fundy. From his home, he could see the rise and fall of the largest tides in the world and he recognized and spoke of the potential of this region for power generation. Though half a century has passed since Roosevelt's suggestion, his vision has not been fulfilled. Yet the promise in those towering tides remains.

Tides are the result of the gravitational pulls of the moon and, to a lesser extent, the sun on the great oceans. As the earth rotates, a portion of the waters of the ocean are lifted and held in position for a time by these gravitational pulls. When the swell of water in the grip of the moon reaches the land, as it must because of the rotation of the earth, it appears as a high tide. Further rotation of the earth releases the grip of the moon on that portion of the ocean and the tide falls away. Tides rise and fall twice each day, although the times shift with the season and the moon's position (Fig. 22–8).

The average height of the tidal swell is only a few feet (half a meter) *except* when the ocean tides move within relatively narrow bodies of water. There, an oscillation wave is set up which may be up to 10 to 20 times the normal height of the tidal swell. Bay of Fundy tides, the largest in the world, run up to 52 feet (16 meters). Between England and the European coast (France, Belgium, and Holland) such large tides are created as well. The highest tides of year occur when the moon and sun are most nearly in line and their separate gravitational

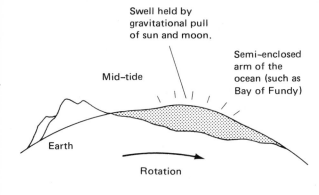

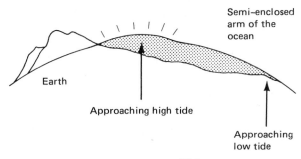

Figure 22-8 *Tides.*

pulls come together to increase the volume of water dragged across the sea.

The possibility of using the energy in the tides has been turned to reality in France where Électricité de France finished in 1968 a tidal power station on the Rance River which flows into the Atlantic Ocean. To understand how tidal power works, we can discuss the operation of the station on the Rance (Fig. 22-9).

The Rance River experiences tides nearly as high as those in the Bay of Fundy, up to 44 feet (13.5 meters). Although nearby coastal tides are lower, the Rance is a relatively narrow river and when the tide moves up and down within its banks, the water moves at a very high speed. A half-mile long dam has been built across the Rance and is

used to store the waters of the arriving high tide. When tidal waters are receding, the stored water is released to the ocean through turbines below the dam, and electricity is generated. To many people, this is the way tidal power is expected to operate. At La Rance Power Station there is more. Energy can be generated on both the falling tide and the rising tide.

The rising tidal waters are captured inland by opening a set of sluice gates, which allow the arriving tide to flow upstream in the direction of the river source. The gates are closed when the tide reaches its highest stage; then the water that was barricaded inside is allowed to flow seaward through turbines as the tide recedes. At low tide most of the water has been released. As the tide builds again, it does so against the closed gates, so

Figure 22-9 *La Rance Power Station lies across the estuary of the Rance River, which empties into the Atlantic Ocean between St. Malo and Dinard on the coast of Brittany. (Photo from the French Engineering Bureau, 1825 Jefferson Place, Washington, D.C. 20036)*

that water levels on the seaward side exceed those on the land side of the dam. When a sufficient head is built up, the water is allowed to flow upstream through the turbines, again generating electricity. Thus, electricity is generated both on the receding tide and on the arriving tide (Fig. 22-10).

La Rance Station near St. Malo, France, has 24 separate units capable of generating a total of 320 megawatts of electricity and capturing 25% of the energy possible to capture. Because the tide is moving up a river rather than along a broad coastline, the turbine and dam rest about three miles inland and are thus protected from the ravages of the open sea. Along the Bay of Fundy, the Passamaquoddy Bay on the easternmost coast of Maine could provide similar shelter for the dam and turbine. The potential for annual power production of the Passamaquoddy is more than five times that of the Rance project (Fig. 22-11).

In 1974, a tidal power station began operation in the USSR at Kislogubsk on the Barents Sea, about 30 miles (50 kilometers) from the port of Murmansk. The Soviet station, in contrast to the Rance Station, which was built in place, was built offsite on a nearby river sheltered from the sea and floated into position. Construction costs were cut considerably in this way. The Kislogubsk plant is an experimental unit; it has only 0.40 megawatts capacity, but its manner of construction, offsite assembly, and floating into position, is viewed by engineers as the model for construction of tidal power stations of the future.

Because of the enormous cost of these structures, governments are reluctant to invest in tidal power. The Rance project cost two and one half times the estimated cost of a river hydroelectric station of the same average power output, primarily because of the added cost of coffer-dams both ahead of and behind the project. Yet, once the initial investment is made, the production of power requires no fuel. Only maintenance of the structure is required. Thus, the costs of power will be held down. The Rance station is proof that the concept of tidal power can work on a large scale and it is

already influencing the public's views on tidal power. There are a number of sites around the world where the tides could be used to generate

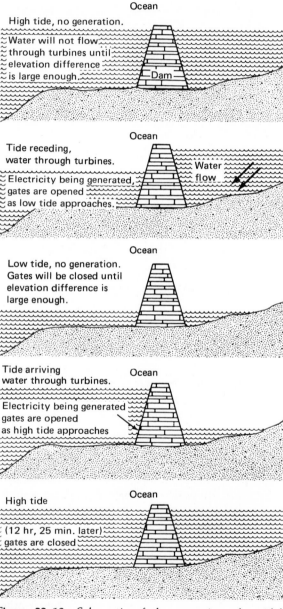

Figure 22-10 *Schematic of the operation of a tidal power station.*

electricity. Some of the most attractive are listed in Table 22–2.

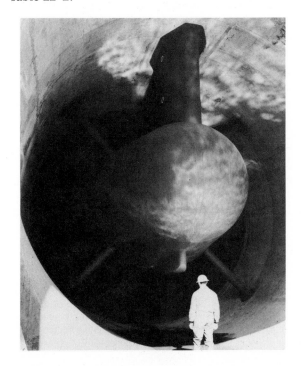

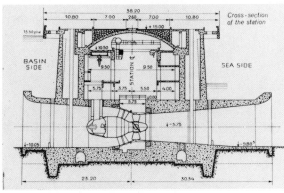

Figure 22–11 *(a) The bulb turbine at La Rance Power Station as viewed from the basin side. Note the size of the workman for scale. (b) A cutaway view of the power station showing the bulb turbine. There are 24 bulb sets at the La Rance Power Station. (Photographs courtesy of the French Engineering Bureau, 1825 Jefferson Place, Washington, D.C. 20036.)*

There are other drawbacks to tidal power besides cost. If the tidal project is a great distance from the nearest large load center, long and costly transmission lines will need to be built to provide the tide-generated electricity entry into the transmission network.

On the other hand, such long-distance transmission is becoming more common as the new and more efficient 765–kilovolt lines are installed. New York City may soon be getting power from Hydro-Quebec via a 765–kilovolt line which runs several hundred miles. Tidal power from the Passamaquoddy or Cobscook in Maine reaching down to Boston no longer seems so unthinkable. Furthermore, the distance of load centers may not be so important. Industries that are heavy users of electricity may find it desirable to locate near tidal power stations. Examples of such industries are aluminum companies, steel companies using the electric arc furnace, and chemical companies producing chlorine and caustic soda via electrolysis.

The distance from the load center aside, there is another drawback to tidal power; the production of electricity is not steady. A moment's reflection on the nature of the tides should explain this. In the ordinary production of tidal power, electricity is only generated when the tide is receding, that is, when the height of water stored in the reservoir exceeds that of the receding waters by a sufficient amount. As the tide drops and low tide approaches, the rate of power generation falls to zero since there is no height difference. If the tidal project has reversible turbines, as La Rance does, power can be generated on the incoming tide as well, but still, the power cannot be generated until the height of the rising tide exceeds the height behind the dam by a sufficient amount. As high tide is reached, the rate of power production again falls toward zero. Thus the curve of power production rises and falls twice each day for each of the two tidal cycles.

This cyclic production of electricity is unlikely to match the daily cycles of demand at load centers. Peak demands and peak production will occasionally occur together because the two tides shift in

TABLE 22–2   Selected Tidal Resources [a]

	Average difference between high and low tide		Average rate at which power could be produced
	Feet	Meters	Megawatts
Severn, England	32.7	9.8	1,680
Mont St. Michel, France	28.0	8.4	9,700
White Sea, USSR	19.0	5.7	14,400
Mozen Estuary, USSR	22.0	6.6	1,370
Passamaquoddy, U.S., Canada [b]	18.3	5.5	1,800
Cobscook, U.S. [b]	18.3	5.5	722
Annapolis, Canada, (N.S.) [b]	21.3	6.4	765
Minas-Cobequist, Canada (N.S.) [b]	35.7	10.7	19,900
Cumberland, Canada (N.S.) [b]	33.7	10.1	1,680
Petiteodiac, Canada (N.B.) [b]	35.7	10.7	794

[a] Source: J. McMullan, R. Morgan, and R. Murray, *Energy Resources,* Halstead Press, John Wiley and Sons, New York, 1977.

[b] All of the U.S. and Canada sites listed are along the Bay of Fundy.

time as the seasons progress, but more often the peaks of demand and production will occur at different times of the day. Somehow the tidal power must be fed into the transmission network at the proper rate. This means that other central power stations must usually be phased down in power output as the rate of tidal production reaches its maximum and phased up as the tidal production rate falls. A computer performs this task for Électricité de France at La Rance Station. The electricity from the tidal power station is, in effect, replacing on a fairly regular basis the electricity generated by other means. If the electricity replaced is from a coal-fired power plant, it is coal that is being saved.

A final word of caution is in order. As we look longingly at the tides and the awesome energy they carry, we must reflect as well on the environmental impacts of tidal reservoirs. The tidal reservoir is very likely to have an effect on the important biological area that extends along the shore of the ocean. The area, which is known as the intertidal zone, reaches from the point of the highest tide (or spray from the tides) to the lowest point exposed when the tides recede. (Both of these points are not fixed, but vary, within limits, with the seasons.)

Communities in this region consist first of those organisms that spend all or most of their time there. On sandy beaches there will be burrowing creatures such as crabs, shrimp, worms, and clams. On rocky shores there are organisms attached to the rocks such as mussels, oysters, barnacles, and the larger algae. Another set of organisms also occurs in the waters of the intertidal zone: the phytoplankton, the microscopic floating green plants such as diatoms, dinoflagellates, and the microflagellates, which are swept in with the tides.

Famous but dreaded members of the dinoflagellates are those responsible for "red-tides," which kill fish and sometimes make the flesh of shellfish poisonous to humans.

The phytoplankton along with the attached larger red, brown and green algae or "seaweeds" are the producers of the intertidal region. A variety of consumers also live in the intertidal zone. Some come in with the tide, among them zooplankton consumers, which spend their entire lives as plankton, and others which are the larval stages of crabs, jellyfish, sea urchins, snails, starfish, and other creatures. The intertidal zone is thus an important part of the sea's "nursery grounds." Still other consumers include the mature stages of crabs, clams, barnacles, snails, starfish, and other creatures which are present even at low tide.

(Tidal power schemes, especially those which involve maintaining a constant high pool (see below) change the natural periodicity of the tides. Such schemes will also affect water temperatures in the area. Furthermore, we are not at all certain of how the larval stages of marine species will survive passage through the turbine. Tidal power thus has the potential to change the relative balance among

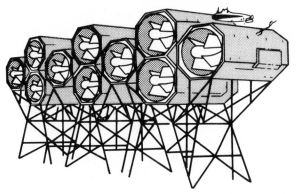

Figure 22-12 *Artist's conception of a turbine array placed in the Florida Current to convert the energy of the moving water into usable power. [Adapted from Louise A. Purrett, "Parachutes, Windmills, and Rivers in the Sea," NOAA Magazine 4(4), 17 (October 1974).]*

species that make up the communities of the intertidal zone. It is conceivable that nuisance species such as those responsible for red tide could be favored. Conversely, spawning of desirable species such as crabs or oysters could be harmed. In addition, we are not sure of whether erosion could be accelerated or the deposit of sediments hastened by such projects.)

Methods to Increase and Steady Tidal Energy

Engineers have devised some clever, even remarkable, ways to assist the tides in giving us energy. On the clever side, they first had to deal with the issue of steadying the production of electricity through the entire tidal cycle. To accomplish this objective, two side-by-side basins can be operated jointly. The high basin is allowed to fill from the ocean on the high tide. This high basin empties continually through a turbine to a second and lower basin. The second or lower basin empties on the low tide to the ocean. Operation always proceeds in the same direction with continuous power generated by the water flowing through a turbine from the high to the low basin. The idea remains a concept, to the best of our knowledge. Because of the increased cost of the development, no one has yet attempted a double-basin (Fig. 22-13).

A remarkable technique has already been put to use at the single basin at La Rance. During the final portion of the arriving tide and beginning portion of the receding tide, the difference in elevation between the water in the reservoir and that in the ocean may be only a few feet. During this time, electricity from some other source can be used to pump ocean water (using the turbines) up into the tidal basin. The water is pumped up against a difference of only a few feet so not much energy is required. When the tide has receded, that extra water falls through a distance of 20 to 30 feet, generating far more electric power than was used.

The same idea works at low tide except that the water is pumped out of the tidal basin into the ocean. The water level in the basin is thus reduced below sea level, and the incoming tide will fall a greater distance. This scheme of pumping at high and low tides and recapturing later in the tidal cycle more energy than was put in is sheer ingenuity.

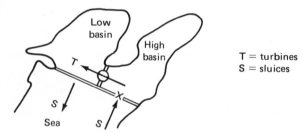

T = turbines
S = sluices

Figure 22-13   *Linked basins.*

Questions

1. What do we mean by "hydropower"? Where does the energy actually come from?
2. Is there a net gain of electric energy when a pumped storage system is used to generate the electricity? Explain. What is the advantage of a pumped storage generating system?
3. What are the environmental advantages and disadvantages of hydropower schemes? Briefly compare these to the environmental disadvantages of power from the burning of coal.
4. Explain how tides could be used to generate electric power. Where does the energy come from?
5. Discuss the possible economic and environmental advantages and disadvantages of tidal power. Compare to the disadvantages of generating power by burning oil.

Further Reading

Hydropower

Carter, L., "Con Edison: Endless Storm King Dispute Adds to Its Troubles," *Science* **184**, 1353 (28 June 1974).

This is the story through 1974 of the 14-year fight to block the pumped storage unit called Storm King.

Erskine, G., "A Future for Hydropower," *Environment* **20(2)**, 33.

An engineer converts his expert knowledge of hydroelectric generation and the bulb turbine into public property. A rare article.

Kash, D., *et al.*, *Energy Alternatives*, Report to President's Council on Environmental Quality, 1975, published by the University of Oklahoma Press as *Our Energy Future.*

A general, nontechnical description of the hydroelectric resource, oriented toward the large installations.

Kirschten, R., "Hydropower: Turning to Water To Turn the Wheels," *National Journal,* p. 672 (29 April 1978).

A political, newsy article on low-head hydro developments.

Tidal Power

Bernstein, Lev, "Russian Tidal Power Station," *Civil Engineering Magazine,* p. 46, (April 1974).

A well-written article on the Russian tidal station by the engineer in charge.

McMullan, J., R. Morgan and R. Murray, *Energy Resources*, Halstead Press, John Wiley and Sons, New York, 1977.

Tidal power is but one element of this paperback which is thorough if a bit technical/scientific.

Project Interdependence, Committee Print 95-33, Congressional Research Service, U.S. Senate, 1977.

There is a clearly written article on tidal energy without technical or engineering jargon. Recommended.

Chapter 23

Power from the Wind and Power from the Heat in the Earth

POWER FROM THE WIND

A Brief History

Wind has been in the service of humankind since primitive people first raised a sail above a fragile log canoe. The prevailing westerlies were the winds that powered the voyages of discovery to the new world, and wind carried the Spanish Armada to victory after victory. The Trade Winds caught the sails of the great clipper ships and opened India and China to commerce with the West.

Long before the wind was harnessed by the ingenious and durable Dutch, the ancient Persians had captured the wind to grind their grain. The Persian windmills turned on a vertical shaft, a design re-invented in the modern age and appropriately named for the Persian King Darius. The windmills in Holland were used not only to grind grains but also to pump water out of the low-lying lands (polders) so that they could be farmed; the Dutch windmills had the familiar horizontal axis (Fig. 23-1). The blades of these wooden windmills reached 80 feet (24 meters) in diameter, and the windmill could operate in winds of 25 miles per hour (42 kilometers per hour). Operators in the base of the mill oriented the blades to the force of the wind. Many of the Dutch windmills, now 500 or more years old are still in operating order, and courses are offered in Holland on windmill operation!

In the 1850s, the multivane windmill was invented in the United States. This windmill became common in rural America in the years that followed, finding its use first in raising water from wells. It was the multivaned mill that pumped water for the steam locomotives which began crossing the American continent in the 1870s. The multivaned windmill is still in production today, but steel blades are now used in place of the hand-made wooden blades used on the early models. With blades up to 30 feet in diameter, the multivaned fans could produce up to four horsepower (3 kilowatts) in a 15 mile an hour wind (25 kilometers per hour) (Fig. 23-2).

A new invention from Denmark gave windmills another chore on the American farm. In 1890, the Danes became the first people to generate electricity from a windmill. This windmill used a propellor with two or three thin blades instead of the multivaned design, and it could capture the wind's energy more efficiently and at higher speeds. Windmills brought electricity to the farms of America before the era of central station power and by the 1930s these one kilowatt generators were helping rural America "tune in" to an exciting medium, the radio.

The windmill generators were used to charge automobile-type batteries, which could then be used for lights or radio, even when the wind was calm. As central station electric power reached more and more of rural America through the public power programs like TVA, the use of such wind generators decreased. Now, except for experimental programs

Figure 23–1   *The two basic Dutch windmill designs. The design on the left had a "cap" which rotated so that the blades could be oriented to the wind. In the older design on the right, the entire building, except for the base, was rotated, to set the blades in proper position relative to the wind.*

usually sponsored by the government, only relatively remote settings rely on small-scale wind power. Such sites may be rarely visited yet require small amounts of power on a steady basis. It would be uneconomical to run power lines to such sites, so the small scale windmills are natural choices. Navigational beacons on islands at sea may often be powered by windmills.

Just for a moment though, about 40 years ago, it looked as though electricity from the wind could compete with electricity from fossil fuel generating stations. It was 1941, and a Smith-Putnam Generator rated at 1250 kilowatts in a 35 mile an hour wind (60 kilometers/hour) was installed at Grandpa's Knob, near Rutland, Vermont. Hooked into a power grid, the windmill delivered commercial power for 3½ years. The tower was 110 feet (33 meters) tall and its two blades weighed eight tons

apiece and were 175 feet (54 meters) in diameter. But in 1945 a sudden strong wind broke one of the blades, hurling it 750 feet (283 meters). A wartime shortage of materials prevented the windmill at Grandpa's Knob from ever being rebuilt.

Windmill Design

Windmills produce power when the wind pushes on the blades of the mill. It makes sense that the greater the "reach" of the blade, the more wind energy it can capture. It likewise makes sense that the greater the velocity of the wind, the greater the force on the blades and the greater the amount of energy captured.

The response to the diameter of the blade and the speed of the wind is not one-to-one. The power produced goes up with square of the diameter of the

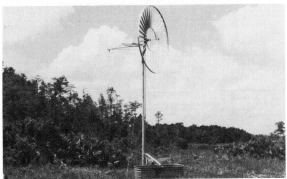

Figure 23-2   *The Multivane Windmill invented in the U.S. in the 1850s is still used on farms today to lift water from wells for livestock. Here, two modern adaptations of the multivane concept are shown. (a) The windmill atop the tower is the product of Dempster Industries Inc. (b) The windmill supported by the pole with guy wires is from Chalk Wind Systems.*

blade and with the cube of the velocity of the wind. Table 23-1 indicates how power output would change for a typical horizontal axis windmill at different wind speeds and blade diameters.)

Note that at a wind speed of 20 miles/hour (33 kilometers/hour), multiplying the blade diameter by 4 (from 50 feet to 200 feet) multiplies the power output by 16. Observe also that at a blade diameter of 100 feet (30 meters), a wind of 30 miles per hour (50 kilometers/hour) will generate 26 times more power than a wind speed of 10 miles/hour (17 kilometers/hour). This is why engineers lean toward big windmills and why they try to capture the higher winds.

Most of the large windmills now being built or in operation are designed to operate at wind speeds between 10 and 35 miles/hour (17 to 58 kilometers/hour). Winds less than 10 miles/hour produce little useful energy; winds higher than 35 miles/hour could wreck the windmill. To appreciate more fully the winds speeds at which the windmills operate, Table 23-2 lists events that occur at different wind speeds.

(In a sense, the accident of Grandpa's Knob was an early indicator of design problems in windmill technology, especially in durability and safety. The serious accident was an indicator that the huge propellor blades could "fatigue." "Fatigue" is a term from metallurgy, describing a metal that has undergone strong forces and whose structural strength is weakened as a result. Metal fatigue is what happened at Grandpa's Knob.)

Windmills should not be designed to capture gale winds. Even though such winds deliver far more power than low speed winds, they exert such a strong force on the blades that the machine may be destroyed. Furthermore, the proportional contribution of gale winds to total power output is very very small, making such risks not worthwhile. To combat the problem of gale winds, windmill blades are curved in such a way that they turn slightly to one side out of the direct force of the wind so that the full impact of large gusts will not damage the propellor. This old practice is known as "feathering". Newer materials which can withstand

TABLE 23–1[a]

Wind speed		Power output in kilowatts		
mi/hr	(km/hr)	Diameter 50 feet (15 meters)	Diameter 100 feet (30 meters)	Diameter 200 feet (60 meters)
10	(17)	3	14	54
15	(25)	11	46	182
20	(33)	27	108	432
25	(41)	53	211	844
30	(50)	91	365	1458

[a] Source: *Alternate Long Range Energy Strategies,* Joint Hearing before the Select Committee on Small Business and the Committee on Interior and Insular Affairs, U.S. Senate, 9 December 1976.

higher forces are also being used to prevent the breaking of blades.

There are other problems in windmill design and, by and large, they occur simply because of the nature of the system needed to capture the power of the wind. Windmills typically stand on tall towers which enable the blades to reach the stronger winds that occur at higher elevations. Near the ground, houses, trees, small hills, and the like interrupt and obstruct the wind. Thus tall masts are needed, but the heavy equipment, consisting of a propellor, a gearbox and generator, must sit atop the tall mast and this requires a very strong structure (Fig. 23–3). Another problem in using the power from windmills is the nature of wind itself. From little freshets to great gusts, the speed of the wind varies over a wide range. As a consequence, the cycles per second output from a windmill varies. To correct this, the alternating current power that is generated by the turning shaft is "rectified"; that is, it is converted into a steady, one-directional flow. For large windmills, this steady flow of current is fed to an "electronic inverter" which produces a stable alternating current, which can be fed into a power grid. Small windmills such as those used on isolated farmsteads or on islands at sea will feed the "rectified" output to a very large storage battery instead of to an "inverter." The batteries are essential to store

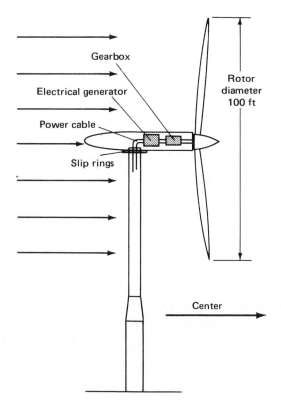

Figure 23–3   *Typical wind rotor system. (Adapted from D. Kash et al.,* Energy Alternatives, *Report to the President's Council on Environmental Quality, 1975.)*

TABLE 23–2[a]   Observations and Events at Increasing Wind Speeds

	Miles/hour	Kilometers/hour
Calm; smoke rises vertically.	1–3	2–5
Wind can now be felt on the face; leaves will rustle.	4–7	6–11
Leaves and small twigs move constantly.	8–12	12–20
Dust is raised and loose paper blows; small branches will move.	13–18	21–29
Small trees will sway; the waves on the water show crests.	19–24	30–39
Large branches are in motion. It is difficult to use one's umbrella.	25–31	40–50
The entire tree is set to swaying in the wind, and walking against the wind becomes difficult.	32–38	51–61
At these speeds, twigs are snapped off trees by the wind.	39–46	62–74
Structural damage starts to occur.	47–54	75–87
Trees may be uprooted; and severe structural damage is possible.	55–63	88–101
Such velocities are unusual inland; they cause widespread damage.	64–72	102–115
Hurricane conditions.	73–132	116–212

[a] Adapted from Industrial Instruments, Ltd., Stanley Road, Bromley, Kent, United Kingdom.

electrical energy for periods when the wind is too calm to produce any energy.)

The problem of cycles per second can be corrected, but the adjustment of power output is more difficult. Just as when we use tidal power, there will be times, with a windmill, when little or no power is being produced. At such times, conventional electric power will have to be increased elsewhere to meet power demands.

Wind Resources of the United States

Wind varies with the features of the natural landscape or urban terrain; it varies with the nearness to bodies of water, with the weather and the season, and with the height above ground. Winds over the land tend to be slower on average than those on the coast or over the ocean. Ocean winds are not only stronger, they are steadier as well. Thus some of our best wind resources are at sea. This led the engineer William Heronemus of the University of Massachusetts to propose several banks of steel windmills off the coast of New England. He envisioned that the electricity developed by the mills would be converted to a steady, direct current and could be used to electrolyze water to hydrogen. The pure water to be electrolyzed would be provided by the distillation of sea water also using the electrical energy developed.

The hydrogen would come ashore through a pipeline and be recombined with oxygen in a "fuel cell" to produce electricity and water. Heronemus's vision is, however, a long way off. For now, we need to look at the wind energy we can reach easily, the energy on the land.

Across the U.S., the average annual wind speed is about 10 miles/hour (17 kilometers/hour). It varies from a lowest average of 6.5 miles/hour up to 37 miles/hour, the average wind speed at the top of Mount Washington, New Hampshire. In fact, the winds on Mount Washington have been known to reach an incredible 150 miles/hour (250 kilometers/hour).

Interestingly, there is a steady wind across the Great Plains, though the Plains are relatively free of obstructions. For each square mile of the energy-rich Great Plains, Heronemus has proposed an 850–foot (283 meters) tower supporting 20 wind turbines. Each turbine would have blades with 50-foot (15 meter) diameters. About 1800 of these giant structures (one per square mile) would be needed to produce the same output as a 1000 megawatt electric plant, because of the way wind energy varies. How practical is such a scheme compared to a 1000 megawatt coal-fired plant, for example, with its comsumption of four million tons of coal per year and its air pollution and its need for cooling water? The question is not so heavily weighted against wind turbines as it may seem at first glance.

The President's Council on Environmental Quality has estimated that wind resources, if rapidly developed, could be used to displace between 4 and 8 quads of thermal energy by the year 2000. This is energy that would be consumed to generate electricity. Depending on how well we conserve energy, this could amount to between 5% and 7% of annual U.S. energy needs. The Council quotes[1] one study that estimates the wind potential of the U.S. at between 1 trillion and 2 trillion kilowatt

hours per year. U.S. consumption of electricity in 1976 was on the order of 2 trillion kilowatt hours, indicating excellent potential for wind power.

Environmental Problems and Costs of Wind Power

Does wind power cause air pollution? No. Does it cause thermal pollution? No. Does it consume fuels? No. It does cause noise and it does use land. It also has a visual impact, but the towers of long distance electric lines have a height near that of the tallest windmill presently being considered. And cooling towers are even taller.

There is one other impact of wind power. Large windmills rotate at about 30 cycles per second. This is also the synchronization speed of television in the United States. For a distance of up to 1 mile, these large windmills may interfere with television reception. With the use of fiberglass blades, which are proving to be lower in cost than metal ones, the distance falls to about 1/2 mile. Again this is for large windmills. It is not expected to be a problem for small windmills.

Birds may be hurt by the blades, but it is difficult to predict the extent to which this would occur. Certainly there are other environmental costs, in the mining of iron ore, in the manufacture of storage batteries, in the many more transmission wires and lines needed to collect electrical energy from multiple sources, but basically, when we count all the environmental costs, wind power comes out with a very high rating. What, however, is the economic cost of wind power?

Estimates of the cost of electrical energy from wind power vary almost as much as the wind itself. For purposes of comparison, a new 1000 megawatt conventional electric plant costs on the order of one billion dollars, or $1000 per kilowatt. The plant can operate about 90% of the time at its capacity, something a wind machine cannot do. Larger windmills are thought to produce power more economically than small ones. The federal government,

---

1 *Solar Energy: Progress and Promise,* Council on Environmental Quality, April 1978.

helped by the National Aeronautics and Space Administration (NASA) has been building large windmills since the middle 1970s. Their first effort, a 100 kilowatt machine, cost $5500 per kilowatt.

Later models cost about $2000 per kilowatt, but costs in the neighborhood of $750 to $1000 per kilowatt are the goal for these large windmills. Whether federal government experience on costs is meaningful is a debatable question. We note that a large windmill has been built by a private firm on Cuttyhunk Island off the South Coast of Massachusetts at a much lower cost than the government operation. Its entire cost may have been less than just the hub mechanism of the 100-kilowatt wind station built by the federal government in Sandusky, Ohio. Still the costs we have quoted are not yet comparable to central station power costs because of the varying nature of the wind. The average output from a windmill is about 15%–25% of its rated output. Costs per kilowatt of about $300–$400 would be needed to produce electricity at rates comparable to those for central station power. It is important to remember, however, that these are investment costs. Fossil-fired and nuclear power stations consume costly fuels which the windmill does not.

There is also a federal program to develop small windmills. It is thought that mass production of small windmills could reduce costs to $750 per kilowatt for an 8 kilowatt machine and $500 per kilowatt for a 50 kilowatt machine.

Program of Development for Wind Power

A number of large experimental wind generators have been planned or installed since the "energy awakening" began. A 100 kilowatt generator was first tested by the federal government at Sandusky, Ohio, in 1975. Although the station was beset by a number of difficulties, the Department of Energy has gone ahead with four additional sites at which to test wind turbines. These are at Clayton, New Mexico; Block Island, off the coast of Rhode Island; the Island of Culebra, a part of Puerto Rico; and Oahu, the largest and most populous of the Hawaiian islands. Each of these windmills will have a rated capacity of 200 kilowatts. The generator at Clayton has been delivering power to the town's municipally-owned utility since early 1978. The Culebra wind turbine began operation in July 1978 and is delivering electricity to about 150 island homes. The Block Island wind station began operation in 1979 (Fig. 23–4).

The structure for the wind turbine at Kaena Point, Oahu, began to rise in 1979, and, barring difficulties, should now be in operation. A still larger windmill began construction in 1978 near Boone, North Carolina. The mill stands atop a mountain, and a steel tower reaches 140 feet (42 meters) into the air. The 200-foot diameter twin blades can extract 2000 kilowatts of electric power from winds in the range of 25–35 miles/hour (40–56

Figure 23–4   *A new landmark now dominates the Block Island, R.I., skyline. Sitting atop a 100-foot tower, this experimental wind turbine with its graceful 125-foot blades is capable of meeting nearly half the island's power demands at peak operation during the windy winter months. Experience with this machine will be used by the Department of Energy and NASA to design wind turbines suitable for commercial use in the 1980s and 1990s. (Photo courtesy of U.S. Department of Energy; R. R. Peabody, photographer.)*

km/hr). At the time of its completion, the Boone wind station was the nation's largest windmill.

These experimental windmills were built with designs, technology, and materials that were not available in past decades. Thus, there is hope that these designs will "prove out" and deliver power economically, safely, and reliably. If they do, the stage will be set for an orderly program of wind power development. The program may someday reach those engineering dreams of banks of generators in the ocean or miles of generators stretched across the Great Plains, extracting electric power from nothing but the wind.

GEOTHERMAL RESOURCES

"Healthful and Refreshing Warm Baths," the advertisements read for the resorts of Lake County, California. It was the late 19th century, and residents of the bustling city of San Francisco, some 75 miles to the south on the California coast, would travel to these "geyser" baths to be "restored." The odor of hydrogen sulfide gas was so noticeable in the area of the hot springs that the local stream became known as Big Sulphur Creek. Restorative though the baths must have been, no one saw any other commerical potential in the hot springs until the 1920s when, perhaps inspired by success in Italy, drilling for steam began in the area (Fig. 23–5). But the drillers were unsuccessful in their attempts to sell the steam to electric companies.

By then, a geothermal electric plant had already been in operation in Larderello, Italy, for 20 years. It would still be another 30 years, until the 1950s, before two companies could interest Pacific Gas and Electric in using the steam that came bursting forth in northern California. By 1960, Pacific Gas and Electric was generating electric power at the rate of 11 megawatts. By 1972, Pacific Gas and Electric could produce power at the rate of 180 megawatts from its geothermal plant. In 1978, the utility announced plans to expand to·over 1000 megawatts by the early 1980s. This is the size of a modern coal-fired or nuclear power plant. Wells are

Figure 23–5 *The geysers–geothermal steam fields. Geysers such as these have been harnessed by the Pacific Gas and Electric Company to produce electric power in California. The steam, with any water droplets removed, is used to turn the turbine and generate electricity. An electric generating capacity of over 1000 megawatts has been built by the company. (Department of Energy)*

going ever deeper. Although the first steam was found at about 1000 feet, recent wells have been drilled to more than 9000 feet. In Italy, the Larderello plant had grown to 380 megawatts by the mid 1970s.

What is geothermal heat? How can it be used and where can we find it? Is it a clean source of power? Table 23–3 identifies existing and developing geothermal power plants around the world. Figure 23–6 indicates both plants and undeveloped resources.

Origin of Geothermal Heat

Simply put, geothermal heat is the energy from the earth's interior. Of course, the eruption of a volcano, such as Mt. St. Helens, is visible evidence of the enormous heat inside the earth. Scientists estimate the temperature in the core of the earth at thousands of degrees Centigrade. From the intensely hot interior of the earth (where molten

TABLE 23–3 Geothermal Power Installations in Operation or at an Advanced Stage of Development[a]

	Total Generating Capacity (Megawatts)	
	Installed	Planned addition
Chile		
El Tatio	—	15
El Salvador		
Ahuachapan	30	50
Iceland		
Namafjall	2.5	—
Italy		
Larderello region	406	—
Mount Amiata	25	—
Japan		
Matsukawa, N. Honshu	20	—
Otake, Kyushu	11	—
N. Hachimantai	10	—
Hatchobaru, Kyushu	—	50
Onikobe, Honshu	—	20
Mexico		
Cerro Prieto	75	75
New Zealand		
Wairakei	192	—
Kawerau	10	—
Broadlands	—	100
Philippines		
Tiwi, S. Luzon	—	10.5
Turkey		
Kizildere	—	10
U.S.A.		
The Geysers	600	400+
U.S.S.R.		
Pauzhetsk	5	7

[a] Source: F. Ellis, "Geothermal Systems and Power Development", *American Scientist* 63, 515 (September/October 1975).

metal and molten rock are thought to be the only forms possible) up to the surface of the earth, there is a steady decrease in temperature.

Only a few miles beneath the surface, one can occasionally find molten rock at 1000 °C or more. The more likely find, however, is hot solid rock at a temperature of perhaps 300 °C. At a number of points on the earth's surface, usually in areas of volcanic and earthquake activity, this immense heat comes bubbling to the surface in the form of water and steam at temperatures as high as 300 °C. This water and steam comes from ground water, which has wound its way from the surface down through porous rock and rock fissures into a region of very hot rock. Heated by the rock, even boiled, this water comes bursting to the surface under pressure as steam and hot water. We call this erupting column of water and steam a "geyser." Old Faithful in Yellowstone National Park is our best known geyser because of its towering spray and its insistent punctuality.

Two Important Uses for Geothermal Heat

Geothermal heat has the potential to be used in two basic ways: in the production of electricity and in the heating of homes, offices, and factories. Whether the heat will be used for electricity or for heating homes and factories depends on the form in which the resource makes its appearance. Sometimes the water comes billowing up from the earth as pure "dry" steam—that is, entirely vapor; no water droplets are mixed in the vapor. This dry steam can be used directly to turn a turbine and generate electricity. Condensed water may be reinjected into the earth or, if the quality is good enough, disposed of in a nearby body of water.)

The Geysers, a field in California which Pacific Gas and Electric uses, and the Larderello field in Italy are both examples of fields that produce dry steam. Japan has developed one small generating unit using geothermal steam in Matsukawa. Finding dry steam is still by chance, however; geysers led us to the California fields and others as well.

Figure 23–6   *The major high-temperature hydrothermal areas of the world. [Source: F. Ellis, "Geothermal Systems and Power Development," American Scientist* **63**, *515 (September/October 1975). Reprinted by permission of* American Scientist, *journal of Sigma Xi, The Scientific Research Society.]*

In some fields, the geysers spew forth a mixture of vapor and water droplets. Underground, these "wet steam" fields really have only water in them but the water is at unusually high temperatures (180°C to 370°C)[2] and is under very high pressure. The release of the pressure when the water comes to the surface causes some of the water to "flash" to steam. About 10%–20% of the flow is steam; the remainder is hot water.

This mixture cannot be used directly for the generation of electricity; the impact of the droplets will damage the turbine. In addition, geothermal water contains corrosive salts which make the use of wet steam even less advisable. The mixture must be separated into steam and water by a centrifugal separator, a device that whirls the heavier water droplets to the outer edge of the separator for col-

lection. The steam is then directed to the turbine for the generation of electricity, while the hot water is disposed of in various ways.

Although there are some novel ideas for using the rejected hot water, reinjection into the earth appears to be favored, because the water is often high in dissolved salts. The salts could damage a body of water into which the water was allowed to flow. The largest electric power plant in the world that uses wet steam is in Wairakei, New Zealand, where a plant with a power rating of 192 megawatts is in operation. The plant began operation in 1958. Whereas the Geysers plant and the plant at Larderello, Italy, are the models for power production with dry steam, the New Zealand plant is the model for power production with wet steam (Fig. 23–7).

A demonstration facility using water flashed to steam is under construction at Valles Caldera, New Mexico, about 60 miles north of Albuquerque. The Union Oil Company, teamed with the Public Service

---

2 Water boils under ordinary atmospheric pressure at 100°C.

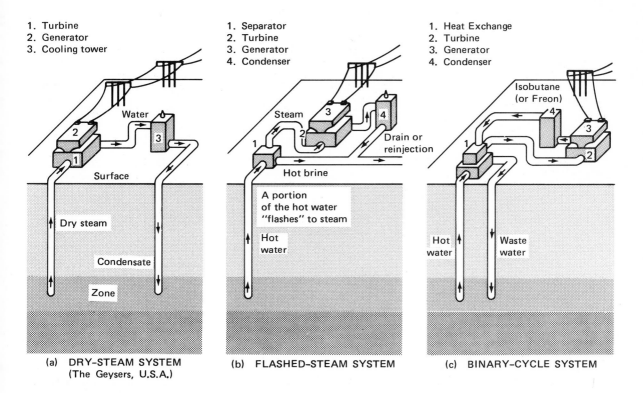

1. Turbine
2. Generator
3. Cooling tower

1. Separator
2. Turbine
3. Generator
4. Condenser

1. Heat Exchange
2. Turbine
3. Generator
4. Condenser

(a) DRY–STEAM SYSTEM
(The Geysers, U.S.A.)

(b) FLASHED–STEAM SYSTEM

(c) BINARY–CYCLE SYSTEM

Figure 23-7 *Electricity can be produced in three ways from geothermal resources. (a) Dry steam, when it is available, can be used to turn a turbine directly for electric generation. (b) When only hot water is available, a portion of the flow is allowed to "flash" to steam. The steam, after being separated from the water, is then used to turn the turbine. (c) Another alternative for electric generation is to use the hot water to boil a fluid like isobutane to a vapor. The isobutane "steam" is used to turn a turbine to generate electricity. (Adapted from "Western Energy Resources and the Environment: Geothermal Energy," Report to the Office of Energy Materials and Industry of the U.S. Environmental Protection Agency, April 1977.)*

Company of New Mexico, is building a 50-megawatt electric plant there which will tap geothermal water at 280 °C.

(There is a third type of water-containing field in addition to the dry steam and wet steam fields. It is a field that produces hot water only. Such fields are even more common than wet and dry steam fields. While at first glance, hot water alone, especially if it has a high salt content, does not seem terribly desirable, it does, in fact, find good uses.)

Reykjavik (Ra' ka vik), the capital of Iceland, with a population of 85,000 is heated almost entirely by hot water drawn from deep geothermal wells which lie under the city (Fig. 23-8). The use of geothermal water in Iceland began only in 1943. The water, after it has been used to heat the city, is sent to greenhouses which produce fresh vegetables for Icelanders. In the U.S., geothermal heat is used to heat homes in Boise, Idaho, and Klamath Falls, Oregon.

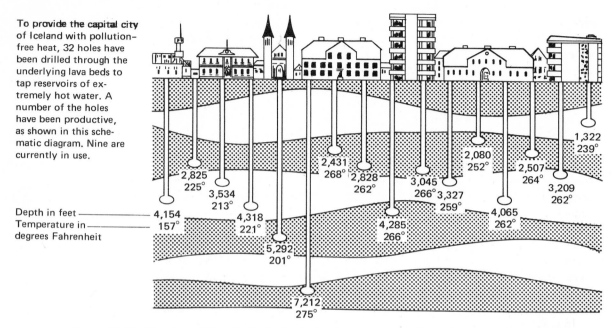

To provide the capital city of Iceland with pollution-free heat, 32 holes have been drilled through the underlying lava beds to tap reservoirs of extremely hot water. A number of the holes have been productive, as shown in this schematic diagram. Nine are currently in use.

Depth in feet ————
Temperature in ————
degrees Fahrenheit

1,322
239°

2,825
225°

3,534
213°

2,431
268°

2,828
262°

3,045
266°

3,327
259°

2,080
252°

2,507
264°

3,209
262°

4,154
157°

4,318
221°

4,285
266°

4,065
262°

5,292
201°

7,212
275°

Figure 23-8   *The city of Reykjavik, Iceland, has been using geothermal water to heat its homes, offices, shops and factories since 1943. (©1974 by The New York Times Company. Reprinted by permission.)*

Hot water may be used for more than heating homes and offices. The geothermal water can be used to produce a "steam" or vapor from a fluid that boils at a lower temperature than does water. Isobutane, one such fluid, might be boiled to an isobutane "steam" in a heat exchanger by using the hot water, and this "steam" used to turn a turbine. The spent isobutane vapor is condensed and returned to the heat exchanger to be boiled again. The U.S., Japan, and the USSR are experimenting with this approach, which is called "binary cycle."

There are still two additional ways to extract energy from the earth. One is not highly developed but shows promise. The other is only a concept at the moment. Geologists have noted that dry rock at temperatures as high as 300°C is about 10 times more abundant than water-bearing hot rock. Experiments indicate that this dry rock can be fractured by pumping down water under very high pressure. This technique, known as hydrofracturing, origin-

ated in the petroleum industry to get at gas deposits that were cut off from one another. Once the rock is fractured, water can be forced through the cracks and returned to the surface at a much higher temperature. The water produced in this way may be as hot as ordinary geothermal hot water, depending on the temperature of the rock from which it was produced.

There are also deposits not far below the surface where molten rock (known as magma) resides. There is at least one such deposit in the United States. The deposit of molten rock in Marysville, Montana, may be a little more than a mile below the surface. It apparently is the remainder of a magma flow that was pressing toward the surface to become a volcano, but never made it. The temperature in the molten mass may be on the order of 1000°C. Can the heat in the molten rock be tapped by drilling? We do not know. In many ways, the molten rock resembles the interior of an active volcano. Thus the question is even broader. Can we tap volcanic heat? Will it be possible to bring the heat to the surface in the form of steam? Sandia Labs, New Mexico, are charged by the Department of Energy with developing the answers to these questions.

Resources of Geothermal Heat

The U.S. Geological Survey estimates our reserves[3] of high temperature water and steam useful for generation of electricity at a quantity capable of producing 11,700 megawatts each year for the next 30 years. A plant of 1000 megawatts, the size of a modern nuclear or coal plant, is needed to supply an average city of one million people. These reserves then are capable of supplying the electricity needs of about 12 million people or about 5% to 6% of current demands.

In contrast to these *known* reserves, the as yet undiscovered resources of high temperature water and steam are estimated to be capable of producing 127,000 megawatts each year for the next 30 years (if they were immediately discovered and put to use). Some of these resources, however, might not yet be profitable to recover. The resource base we have described so far is only that portion useful for the generation of electricity. In addition, lower temperature water at 90°C to 150°C is abundant.

The intermediate temperature water systems are useful for heating, and the U.S. Geological Survey estimates the heating value of identified resources as equivalent to that of 14.3 billion barrels of oil. Given that the U.S. uses about seven billion barrels of oil each year *for all purposes* and has only about 30 billion barrels of reserves, this amount of heat is staggering. Furthermore, the value does not include as yet unidentified resources which the Survey suggests may be three times the value of the identified resource base. We must be cautious, however, in interpreting these numbers,

since population centers are not often on top of geothermal resources as Reykjavik, Iceland, is.

Two areas of the U.S. have unique geothermal resources. In the Gulf Coast region, there are huge deposits of extremely hot water under very high pressures due to sediment layers above them. We referred to these deposits as "geopressured systems" when we discussed them earlier in Chapter 14. The Geological Survey places these resources (in regions which they have assessed so far) at the equivalent of between 11.2 billion and 43.3 billion barrels of oil. Perhaps three times the energy resources are thought to exist in the areas as yet unassessed. In our discussion of natural gas we pointed out that methane is in solution in the geopressured water and that some geologists consider the quantity of recoverable methane to be enormous, far more by factors than our current natural gas reserves. Thus the geothermal deposits of the Gulf Coast are very attractive real estate for exploration.

In California, there are extensive deposits of very hot water in the Imperial Valley. An intriguing idea for this water-short region is to flash the hot water to steam, generate power, condense the steam and *use it for irrigation.* Though the water is initially high in salts, the condensed steam has, in fact, been distilled; that is, it has been made first into vapor and then condensed. Whereas the briny water would have laced the soil with salt if used as irrigation water, the condensate will not.

In summary, the geothermal potential of the continental U.S. appears to be very large and worth developing. And the estimates we provided here are only of underground water and steam systems. There is also the hot rock beneath the surface which may be useful for heating water which is pumped into it, and these resources are likely to be far more abundant than the water and steam systems.

Cautions and Environmental Impact

When geothermal energy is used, there is hot water to dispose of. This is true even if "dry steam"

---

3 These resources and reserves exclude those in national parks.

is used to generate electricity. Often this water has considerable amounts of salts dissolved in it.)

The brines from the Imperial Valley may be as much as 20% salt or about six times the salt concentration in seawater. At a geothermal deposit in El Salvador where salt content is about half that of seawater, the brine has been reinjected into a nearby well almost a kilometer deep. Up to 800 tons of water per hour are injected there, making it unnecessary to carry the water via pipeline to the sea, an earlier alternative. Nonetheless, we still are not certain that waste geothermal water can be reinjected for the long periods which may be necessary. The quantities of salt are astounding.

(Reinjection may prove to be more than a way to dispose of brine. It may also prevent the surface above geothermal wells from subsiding as the water is drained from the deposit. The water, reinjected into the earth via a well not far from the drawing well, will help to hold the earth intact. Not only may it prevent subsidence, it may also provide a "recharge" to the geothermal deposit. That is , if the rate of withdrawal exceeds the rate at which surface water descends through the earth to the deposit, the well could go dry. Reinjection may help to prevent such an occurrence. There is a possible drawback to reinjection; the water reinjected is typically more salty than that withdrawn, and this could lead to increasing levels of salt in the geothermal water.)

Geothermal electric plants require a source of cool or cold water as well as a source of heat for steam. The cool water is needed to condense the steam prior to disposal. In this regard then, a geothermal electric plant is no different from a coal-fired plant or nuclear plant. The water used to condense the steam may come from a nearby fresh water source such as a lake or river or it may come from the ocean. If it comes from a fresh water source, cooling towers will be needed so that the heated fresh water from the condenser can be cooled and be used again for condensing steam.

(The volume of cooling water needed by a geothermal power plant is larger than that for a nuclear or coal plant of the same electric capacity. This is because the geothermal electric plant is less efficient than its cousins, a defect often cited by critics of geothermal energy.)Even the Geysers plant in Lake County, California, which uses dry steam has this problem. The problem stems from the nature of the geothermal resource; the steam is under lower pressure and at a lower temperature than steam used in most electric plants. Because of this, the transfer of energy to the turbine blades is less efficient. At the Geysers, the thermal efficiency is about 22%. This compares with an efficiency of 30% for most new nuclear plants and nearly 40% for most new coal-fired plants. Accordingly, cooling water consumption is much larger for geothermal electric plants.

(Geothermal water is also likely to contain hydrogen sulfide. Hydrogen sulfide not only smells bad (it is the smell of "rotten eggs"), it may affect human health at high enough levels. Emissions from the Wairakei plant in New Zealand have caused silverware to blacken in a nearby village.)

(To summarize briefly; geothermal energy may have a real, even dramatic contribution to make to our energy resource base. It is worth exploring further, even with its drawbacks.)

Questions

1. How can the wind be used to generate power? Do you have any idea where the energy that is captured originates?
2. Why do wind and hydropower, together, make a good system?
3. What major environmental and economic disadvantages and advantages does wind power have? Compare these to the disadvantages of generating power from nuclear fuels.

4. Briefly describe how the energy in geothermal heat can be used as a substitute for other energy sources.

5. What environmental impacts are associated with the use of geothermal heat? Compare to the impacts of fossil fuel (coal, oil) power plants.

Further Reading

Wind Power

Merriam, M., "Wind Energy for Human Needs," *Technology Review* **79**(3), 28 (January 1977).

The author, an engineer, treats the subject of wind energy at the level of *Scientific American*. Although there are a few formulas and engineering graphs, most of the paper can be understood without these elements.

Metz, W. D., "Wind Energy: Large and Small Systems Competing," *Science* **197**, 971 (2 September 1977).

This clearly written and informative article is on the government's efforts to "demonstrate" wind energy.

Putnam, Palmer C., *Power from the Wind*, Van Nostrand, New York, 1948; Reprinted, Van Nostrand Reinhold, 1974.

This book is listed for historical as well as practical interest. Note the author's name, and recall the generator at Grandpa's Knob.

Wade, N., "Windmills: The Resurrection of an Ancient Energy Technology," *Science* **184**, 1055 (7 June 1974).

Recommended for its clarity on the topic of how windmills work.

Geothermal Energy

There are a number of good general references on this subject. The ones listed are all useful and clearly written.

Barnea, J., "Geothermal Power," *Scientific American* **226**(1), 70 (January 1972).

Hammond, A., "Geothermal Energy: An Emerging Major Resource," *Science* **177**, 978 (15 September 1972).

Kiefer, Irene, "Earth Boils Below While We Scratch the Surface for Fuel," *Smithsonian* **5**(8), 82–88 (November 1974).

Wheeler, Romney, "The Geothermal Option," *Panhandle Magazine* No. 4, pp. 2–10, (1978).

Chapter 24

Solar Energy

INTRODUCTION AND HISTORY

Legend tells us that Archimedes saved his home city of Syracuse in Greece with solar energy. Ordering a thousand soldiers to turn their shields to the sun in the shape of a parabola, Archimedes focused the sun's rays on the sails of the ships of an invading navy and burned them. No further practical applications of solar energy are recorded until the 19th century, when inventors began to experiment with the sun's ability to heat and even boil water.

When the 20th century began, it was still an open question as to what fuels would power society. Although fossil fuels jumped ahead because they were then inexpensive, the pattern is now beginning to reverse. There were, you see, early and exciting efforts to harness solar energy.

In France, a solar steam-engine was invented by Mouchot; his collector and engine were displayed in the 1878 World's Fair in Paris (Fig. 24–1). By the early 1900s, farmers in California and Arizona had constructed solar irrigation pumps. By focusing the sun's rays on a boiler, they were able to produce steam which was used to provide the turning power of a pump. Such a steam engine was even introduced in Meadi, Egypt, in 1912. Built by Shuman and Boys of Philadelphia, this engine used a long parabolic collector that could be turned to track the sun. The collector provided heat sufficient to boil steam for a 100-horsepower piston engine.

Such pumps and engines were only one aspect of a budding solar industry. Another dealt with solar hot water.[1] Around the turn-of-the-century, residents of Southern California found themselves paying dearly for the coal needed to make hot water. To reduce these costs, they painted tanks black and placed them so that the tanks would absorb heat from the sun. On most days, the water would not be warm enough for showering until afternoon and then would cool quickly in the even-

Figure 24–1 *Solar collector driving a steam engine. This was exhibited by Mouchot at the 1878 World's Fair in Paris. [From M. Wolf, "Solar Energy Utilization by Physical Methods,"* Science **184,** 383 (19 April 1974) *©1974 by the American Association for the Advancement of Science.]*

---

1 The discussion of the history of solar hot water is drawn from two articles: K. Butti and J. Perlin, "Solar Water Heaters in California, 1891–1930," *Co-Evolution Quarterly,* Fall 1978, p. 4; K. Butti and J. Perlin, "Solar Water Heaters in Florida, 1923–1978," *Co-Evolution Quarterly,* Spring 1978, p. 74.

ings. The Climax solar water heater, invented and patented in 1891 by Clarence Kemp of Baltimore, gave better results. With tanks packed in an insulated box with a glass window to let the sun in, the Climax Water Heater was often roof-mounted or attached to a wall. The water in the black tanks would warm more quickly and retain its heat longer because of the decreased heat losses from the box. The Climax Water heater sold for $25 in Pasadena around 1900.

An even more efficient design was invented by Frank Walker of Los Angeles in 1898. Although double the cost of the Climax, Walker's heater was nonetheless popular because it was hooked into a conventional water heating system so that hot water could be drawn at all times, not simply late in the afternoon. Walker's design was mounted in the building so that the glass window cover was flush with the roof; water was drawn from the top of the tank where the hottest water naturally collected. Even though hot water could be drawn at all times of the day, the need to use coal or gas for early morning hot water made the system expensive to operate. The sun's energy had somehow to be stored if one wanted hot water early in the day without using coal or gas.

To accomplish this, William Bailey created the basic elements of the modern solar hot water system—in 1909! Bailey's design, called the Day-and-Night water heater, separated the component used for water heating from that used for heat storage (Fig. 24–2). While a collector with copper tubes much like the modern collector was mounted on the roof, an insulated tank within the attic of the building was used to store the heated water. Water rose from the collector to the tank by the thermosyphon principle, which we shall discuss later. Bailey's system could hold heated water through the night and into the next day. Although the Day and Night sold for $100, it could cut gas heating bills by 75% or about $25 per year. In four years, the water heater paid for itself.

In 1913, however, a record deep freeze hit some parts of Southern California. The water froze

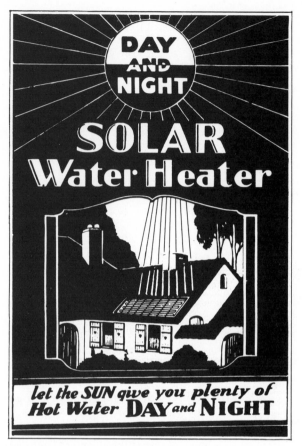

Figure 24–2 *Day & Night brochure of 1923. (From* Co-Evolution Quarterly, *Fall 1977.)*

and cracked the metal tubes, destroying the plumbing of the system; water leaked into houses. Bailey's fertile mind was called upon to invent a system that would not freeze. The design simply mixed alcohol with water, creating an anti-freeze, which was used as the collecting liquid, and the collector loop was separated from the hot water loop. The collector liquid, now resistant to freezing, passed through a coil in the storage tank and heated the water in this way. By the end of World War I, more than 4000 units of the Day and Night had been sold.

When natural gas discoveries in California made solar hot water less attractive in the 1920s, Day and Night built and sold gas water heaters. In

1923, Bailey sold the rights to manufacture the solar heater to a Florida businessman, H. M. Carruthers. In 1932, an employee of Carruthers, Charles Ewald, added many new design features to the basic Day and Night model. One important change was to use soft copper tubing in the collector. The soft copper tubing, which proved resistant to cracking even in Florida freezes, allowed Ewald to avoid the alcohol-filled loop that Bailey invented and to heat the water directly for use.

Since the basic patents on the Day and Night had expired by the 1930s, other companies entered the business as well. In 1941, when America entered World War II, sales of solar water heaters in Miami were twice that of electric and gas water heaters. More than 15,000 solar hot water units had been placed in service in Miami by that time. The war brought the business to a quick halt, though; civilian uses of copper were forbidden.

The industry did not recover after the war because electric rates were falling while costs in both labor and materials were shooting up. Although the industry has yet to recover, thousands of solar hot water units are still in service in Miami. Now as electric rates rise, solar hot water is once again looking attractive in Miami. (See "Prospects for Solar Energy" and Table 24–1.) From coast to coast in south Florida, new homes are once again being offered with solar hot water. In a very real sense, it is something of a "homecoming."

APPLICATIONS OF SOLAR ENERGY

The Sun as a Source of Heat for Water and Buildings

To collect the sun's rays efficiently for the production of heat, special solar collectors much like the one that Bailey invented have been designed. The designs are meant to increase the capture of light energy and to decrease losses of heat energy. Typically, the collector is flat and stationary. This "flat-plate collector" is set on a roof at the angle

that most nearly allows the rays of the sun to fall perpendicular to the surface of the collector.

Many different designs are possible for the flat plate collector. The principles behind the designs are, however, much the same. We will discuss a design in which parallel metal tubes cross the collector; the metal tubes are welded to a metal plate. The heat from the sun's rays is captured by the tubes and plate and stored in the liquid which flows in the tubes. Figure 24–3 illustrates the design of the collector.

When the rays of the sun fall on the metal sheet, the sheet is warmed. The metal pipes, which

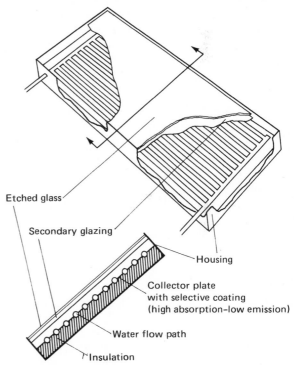

Figure 24–3 *Flat plate solar collector. Cold water is pumped in at the bottom pipe of a row of parallel metal pipes. Warm water exits from the top pipe. The pipes are bonded to a blackened metal sheet. Below the sheet and around the outside of the box in which the pipes are housed may be glass wool insulation. Two glass plates cover the top of the box. (Adapted from "Solar Energy Handbook," Honeywell, Form 74-5436, p. 9.)*

are welded to the sheet, are warmed in turn. The liquid pumped through the pipes receives this heat and carries it off. The pipes and backing sheet are of metals such as aluminum, copper, and iron, chosen because they are good conductors of heat. Both pipes and sheet may be blackened by a paint containing carbon black in order to decrease reflection of the sun's rays and to increase their capacity to absorb heat.

Above the pipes are two panes of glass. While the glass will absorb some of the light as heat and reflect some small portion (perhaps 10%) of the sun's rays, it nevertheless prevents back radiation to the sky. Just as the glass panes of a greenhouse prevent the warmth inside from being radiated out of the structure, the glass cover of a solar collector traps heat within the device. The panes of glass do more besides. Because glass is a poor conductor of heat, it provides a barrier between the warm interior and colder air outside the collector. Thus, losses of heat by conduction to the outside air are decreased. Convection losses, as winds sweep by the collector, are also decreased because the air, which is next to the panel and which is carried off by the wind, will have a smaller content of heat. Plastics may also be used over the heat collecting surface in place of glass.

The box in which the metal collector is housed should be made of a low conducting material such as wood or plastic. The low thermal conductivity of these materials of construction will help to prevent heat losses to the surrounding air and structures. Insulation should be placed in the bottom portion of the box. The box itself is usually mounted on the roof in a stationary position.

The flat plate collector we have just described can be incorporated into the heating systems of new buildings including private homes, by replacing portions of the roofing that would otherwise be required (Fig. 24–4). While the cost of these collectors has been high, assembly-line mass production may bring their price down.

The collector itself is the exterior symbol of a home or building which uses solar energy. Inside

Figure 24–4   *Solar heating. The top photo is the Town Elementary School in Atlanta, Georgia (Westinghouse photo). The bottom photo is of a house in District Heights, Washington, D.C. (U.S. Department of Energy /Photo by Jack Schneider). On both structures are solar panels that collect the sun's energy for use within the buildings. Provision is made in both buildings to store heat energy for cold, cloudy periods when sunlight is not available.*

the building, attached to the collector, is a system of heat exchangers and storage tanks, which store and transfer the sun's heat. Diagrams of the system look slightly complex because the liquid in the collecting loop is not water but a mixture of water and antifreeze. We will describe a solar heating system that uses a loop containing a water and antifreeze mixture and another loop containing water alone for the heat transfer and storage functions. (Other systems use air and hot rocks, but we will mention these only briefly.) Ethylene glycol is the antifreeze now used in cars, but ethanol (ethyl alcohol or grain alcohol) has been used in the past as well.

A mixture of 80% water and 20% antifreeze (ethylene glycol) remains liquid to −8°C (18°F); a mixture of 70% water and 30% antifreeze remains liquid to −15°C (5°F). This liquid, however, obviously cannot serve as a water supply as the circulating liquid of the earlier systems did. Further, we would not wish to store great quantities of this heated liquid because antifreeze is very expensive. The liquid we store in great quantity must be cheap, and water is the logical choice. To heat the water, the conventional method is to send the heated mixture of water and antifreeze through a heat exchanger in a large water storage tank. The water heated in the tank is stored rather than the water/antifreeze mixture.

The water in the storage tank heated by the water/antifreeze mixture can now be used in several ways. It can be used to provide hot water for such uses as dish washing, showers, etc. Or it can be used to provide hot water for heating the home.

Taking the case of domestic hot water first, we can see in Figure 24–5 that city water enters the dwelling and is passed through a set of heat exchange coils in the storage tank. The flowing city water extracts heat from the hot water storage tank. This heated water is then fed to the conventional hot water tank where it can get a "boost," if necessary, to reach the temperature desired by the household—say 65°C (150°F). The "boosting" can be by burning gas, oil, or by using electric resistance heating. The heat exchanger through which the city water passes may be a finned tube or simply a small tank inside the larger tank.

Note that the sun's heat has been transferred from the collecting liquid to the water in the storage tank and then from the water in the storage tank to the city water which makes up the domestic water supply. The water in the storage tank has both accepted and given off heat. There is a reason to use the stored water between the two loops of moving liquid. If the collecting loop should develop a leak

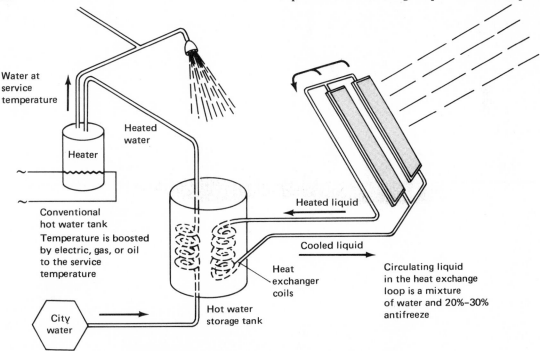

Figure 24–5 *Schematic of a solar hot water heater.*

in its heat exchanger, the water/antifreeze mixture will not leak into the home's hot water supply, but into the stored hot water. Unless both heat exchangers were to leak, a relatively unlikely event, the antifreeze will not reach the hot water supply. Even if both were to have minor leaks, the antifreeze level in the hot stored water would be very small and that reaching the hot water supply much smaller yet.

The other use for the hot water in the storage tank is for home heating (Fig. 24–6). In contrast to the domestic hot water supply system, the home heating system uses a closed loop of water which continually passes through the storage tank. This closed loop takes heated water from the storage

tank and delivers it to a heat exchanger attached to a hot air furnace. Except for the water-to-air heat exchanger, this furnace is the same as a conventional hot air furnace. The heat exchanger on the furnace uses the hot water to heat the air that the blower fan forces through the heating system. An automatic thermostat in the house senses the temperature in the dwelling. If the air heated by the hot water is not warm enough, the thermostatic controls turn the burner on so that an "assist" is given to the solar heating system.

As you can see, the solar hot water supply system and the solar home heating system both rely on the storage tank. Both systems draw their heat from the storage tank. It makes sense, then, that if a

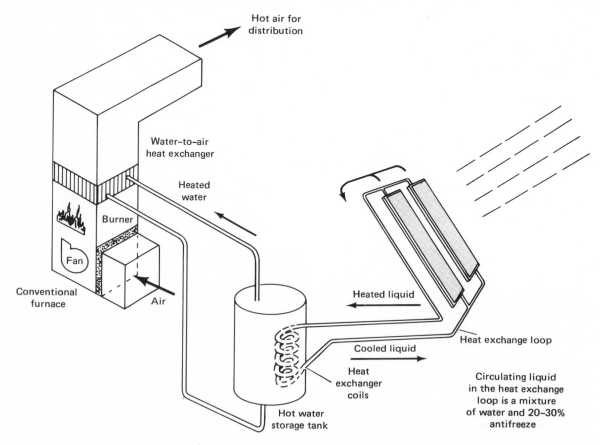

Figure 24–6 *Schematic of solar heat for the home.*

family is installing a solar home heating system, they should install a solar hot water system as well, since only a heat exchanger and some additional collector area are needed in addition to the equipment and piping already in place.

If one is going to heat a home or produce hot water with solar energy, the storage tank is an essential feature. The heat stored in the water can carry over the sun's energy from today until tomorrow or the day after or the day after that, depending on the volume of water that is stored. The storage tank frees the home from the variability in sunshine. It is much like a reservoir that stores water from periods of high flow until periods of need when the flow is low. The storage tank puts away solar energy from sunny days to days of need when there is little or no sunshine. When the stored heat runs out and there is still no sunshine, the conventional furnace goes on.

A more distinctly different system for solar heating relies on air as the collector medium. Air passes through the collector and absorbs the sun's energy. A flat metal plate with finned projections may replace the plate and tube arrangement. Of course, this air cannot be stored the way water can be; the volume of hot air that would have to be stored would be enormous! An easy substitute for hot air storage is a pebble or rock bed. The hot air from the collector is passed through a tank or a concrete or even wooden structure filled with rocks and pebbles. The rocks and pebbles are heated by this flow of warm air and act to store the heat until the dwelling requires it. When heat is required by the house, a blower will force house air through the warm pebble bed. The air which exists from the bed is heated and is distributed through the house. If the temperature in the house is not kept high enough by the solar heat, a conventional furnace will then turn on and will assist in heating the air.

(The systems described are what might be termed "active" systems in which pumps, heat exchangers, valves, and the like must operate to ensure that the sun's heat is captured and distributed to the home. In contrast to these "active" solar heating systems are designs that utilize solar energy with no operating equipment. That is, the designs are able to capture and distribute the heat without any motors or mechanical devices. Such systems are termed "passive.")

(There are several good ideas for the passive collection of solar energy. One uses the ordinary collector panels with tubes filled with the water/antifreeze mixture. Ordinarily, a pump is needed to circulate the liquid through the tubes to the heat exchanger in the storage tank. In the passive system, however, the collector is placed below the storage tank. When the liquid in the collector tubes is heated, its density decreases. Being lighter, the liquid will rise naturally up into the storage tank where its heat is withdrawn. The cooled liquid falls through the return piping to the collector panels and there is heated again. The system is known as a "thermosiphon." While this does eliminate one pump, the remainder of the system will require active assistance.)

Another design avoids collecting heat from the sun during the summer and captures the sun's energy in the winter when heat is needed. Large-paned windows are installed on the south wall of a building. These windows face the sun and the rays of the sun would be expected to enter the window both summer and winter. To block the rays of the summer sun, a roof overhang extends out over the windows (see Fig. 24–8). The overhang shades the window in summer when the sun would otherwise contribute unwanted heat, but does not block the rays of the winter sun. The winter sun hangs low in the southern sky throughout the day and can furnish welcome heat to the dwelling.

The overhang concept can be extended to multiple floor office buildings (see Fig. 24–9). Here an overhang extends over the outer rim of every floor, blocking the sun in summer but allowing its entrance in winter. Interestingly, for the individual dwelling, a deciduous tree can serve much the same purpose as an overhang. Again, assume we have large windows on a south facing wall. The leaves of a deciduous tree growing in front of the window will block

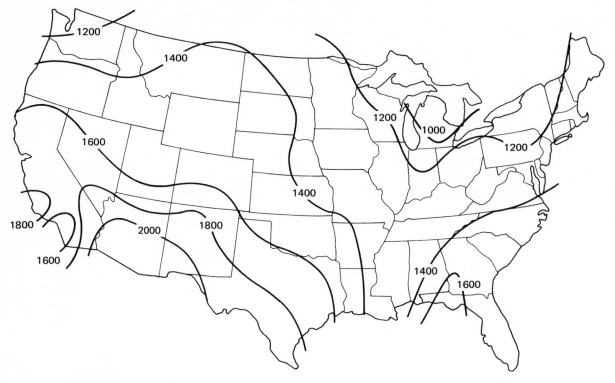

Figure 24–7   *Distribution of solar energy over the United States. The figures give solar heat in Btu/ft² per day. Although some sections of the country have more sunshine than others, the attractiveness of solar energy depends also on the cost of alternate energy souces, such as electricity, oil, etc. (Source: D. Kash et al.,* Energy Alternatives, *Report to the President's Council on Environmental Quality, 1975.)*

out much of the searing summer sun, but in winter the sun's rays can penetrate through the bare branches to enter the window. Pine trees elsewhere around the house help to cut the winter wind and thus decrease the loss of heat from the house.

A clever design of excellent potential using no moving parts is the Trombe Wall, a solar collector which is a part of the house itself. In a sense it, too, is a large panel of windows facing south to catch the sun, but it is also more. The large windows allow ready entrance of the sun's rays, but a few inches behind the windows is a concrete wall. Behind the concrete wall is the living space of the home. The side of the wall that faces south is painted black so

that the maximum amount of sunlight is absorbed as heat. The wall heats up and the air space between the wall and the window heats up as well. Openings at the bottom of the wall allow cool air from the floor level of the living area to enter the space. Openings near the top of the wall allow the warm, less dense air in the air space to rise up and out to the living area. That is, natural convection carries warm air into the room and cool air out. The dense concrete structure acts as a heat storage device as well. Clever as the Trombe Wall is, you the reader can probably improve the concept by devising a way to prevent the wall from radiating heat to the outside in the cold night-time hours.

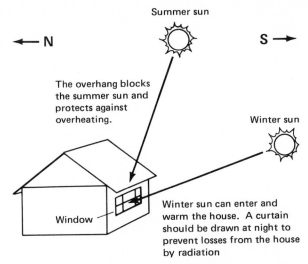

Figure 24-8 *Proper orientation and design of a house can decrease heating and air conditioning requirements. The south-facing house has an overhang which blocks the sun's rays in summer but does not block them in winter. (The house illustrated is in the Northern Hemisphere.)*

Electric Generation and Sunlight

Direct conversion of sunlight to electricity (The process of converting sunlight directly to electricity is known as photovoltaic conversion. The devices called semiconductors made from silicon, which have so transformed the computer industry, are the basis of solar cells. Light falling on a solar cell can "boil" electrons out of the cell, leaving behind a "hole" from which the electron came. Solar cells are so constructed that portions of the cell contain either an excess of holes or an excess of electrons. Light falling on the cell releases electrons, which are attracted to the portion of the cell with holes, creating an electric current. The current is a directed (or direct) current (dc) and so requires conversion to alternating current (ac) before its use by a consumer.)

The solar cells have been used in dramatic applications thus far, where cost has not been a mat-

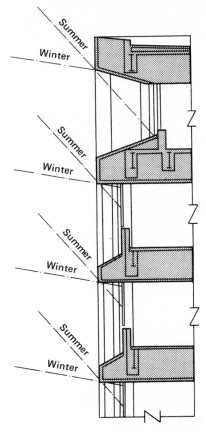

Figure 24-9 *Cutaway of a south-facing wall in a building which excludes the summer sun but allows the winter sun entry. (Adapted from "Designing an Energy Efficient Building," September 1975, General Services Administration.)*

ter of concern. For instance, solar cells furnished the power for the Apollo moon rocket and the Viking orbits. Their cost, however, remains very high because of their low efficiency of conversion, only 12%–14% of arriving solar energy to electricity. A recent estimate was $15,000 for the solar cell array needed to deliver a kilowatt of electric generating capacity. In contrast, a kilowatt of central station electric power, which can be provided at all times, costs about $1000.

Thus, the photovoltaic system has a long way to go before it becomes useful for generating central station power or for turning on our lights at home. This is not to say that photovoltaic conversion is not promising or attractive. In a world in which technology advances at a rapid pace, there is hope that the semiconductors can be markedly reduced in cost.

If cost barriers can be overcome, however, photovoltaic electricity becomes very attractive indeed. No steam is generated; nothing is burned or heated or fissioned; no fuel is mined and transported. No parts move and wear out, and the cells are expected to last a long time when enclosed in glass or plastic.

Even though the solar cell is not economic yet, the Department of Energy is going ahead with a number of projects to demonstrate photovoltaic conversion. The cost goal of the Department of Energy program is $500 per kilowatt at peak operating time, with this to be achieved by 1986.

Interior lighting from sunlight   Although no electricity is generated in this next application of sunlight, we discuss interior lighting here, because interior lighting is provided in most situations only by electricity. We always knew that the sun gave us light, and homes are often built with numerous windows to take advantage of this gift from nature. In industrial buildings and in high rise office buildings, however, interior rooms are often built without access to sunlight, because windows are simply impossible to provide. Instead, incandescent and fluorescent bulbs furnish the light required for work. Artificial lighting in residential and commercial buildings has been estimated to account for 15% of U.S. electrical consumption, and this fact motivates us to explore how we can turn the sun's rays to use in lighting our way.

The idea of "piping" sunlight through a building was given currency in 1974 when the Hyatt Regency Hotel in Chicago brought in sunlight to decorate the glass ceiling of their lobby. In 1976, two physicists at the Sandia Laboratory in New Mexico "piped" in sunlight to light their windowless office. To their pleasure, they found the lighting scheme not only economical but pleasing as well in the quality of the light they obtained. The faint image of the sun was reflected on the walls, and clouds could be seen passing over the image.

The system to light interior rooms with sunlight has only one moving part, a flat mirror, which is controlled by a small computer to track the sun through the daytime hours.

Relative economics for solar versus electric lighting appear favorable because incandescent lights are only 10% efficient in converting electrical energy to light. Fluorescent bulbs, while more efficient, still convert only 20% of electrical energy into light. Sunlight directed into a building provides heat as well as light. Since lights furnish a portion of a building's heating load in winter, the use of sunlight for interior lighting can potentially save heating fuel as well.

Central station electricity from solar energy   The heat from the sun's rays can be focused by lenses to produce exceedingly high temperatures. It can also be concentrated in a single point by properly adjusted mirrors. This is what Archimedes is reputed to have done when he directed the solar assault on the warships approaching Syracuse. The large mirrors used to collect and aim the sun's rays are known as heliostats. These mirrors can rotate to follow the sun in the sky and can reflect and focus the rays of the sun in unison on a single point. With this enormous concentration of energy it becomes possible to boil water to steam, and the steam can be used to produce electricity. It works in the following way.

A tall tower, perhaps 600 feet or more (200 meters) in height, is erected facing a field of heliostats. The heliostats may cover several square kilometers or more, depending on the size of the plant. The heliostats concentrate the rays of the sun on a boiler mounted near the top of the tower (Fig. 24–10). There, at the top of the "power tower," as the device is known, water boils to steam and the

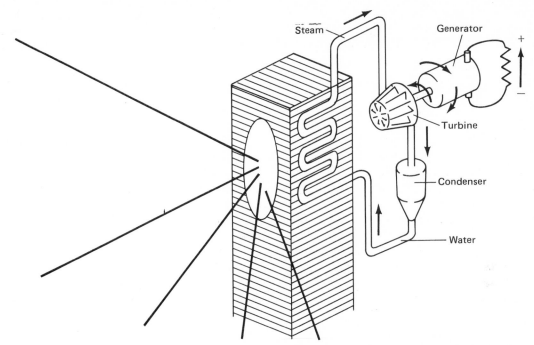

Figure 24-10   *The power-tower concept. Water is boiled to steam by the concentrated rays of the sun. The steam is used to turn a turbine. (Adapted from D. Kash et al.,* Energy Alternatives, *Report to the President's Council on Environmental Quality, 1975.)*

high temperature steam is used to turn a turbine and generate electricity. Electric plants with a capacity up to 100 megawatts are envisioned.)The U.S. Department of Energy and the Electric Power Research Institute have sponsored the construction of a demonstration solar electric plant which was finished in 1978 (Fig. 24–11). Located in Albuquerque, New Mexico, the plant should be able to deliver about 2 megawatts of electricity. The power tower is 200 feet (60 meters) tall; 222 heliostats developed by the Martin Marietta Company concentrate the sun on the power tower. Each heliostat is built out of 25 individual mirrors which are four feet (1.2 meters) square (Fig. 24–12).

A larger, solar steam electric plant capable of producing 10 megawatts of electricity began construction in Barstow, California, in 1978. By the mid-1980s, the Department of Energy hopes to have operating a plant with the capacity to deliver 100 megawatts of electric power.

Figure 24-11   *The solar power tower test facility. This demonstration facility at Sandia Laboratories, Albuquerque, New Mexico, was completed in the late 1970s. Sunlight is reflected and focused by the 222 heliostats onto the face of the 200 foot tall power tower. The intense heat boils water to steam. The steam turns a turbine to generate electricity. About 2 megawatts of electricity can be delivered from the installation which is a joint undertaking of the U.S. Department of Energy and the Electric Power Research Institute. (Department of Energy)*

Figure 24-12 *The heliostats (mirrors) used to reflect and focus the sun's rays on the power tower. Developed by Martin Marietta, each heliostat consists of 25 mirrors, each of which is 4 feet (1.2 meters) square. [Adapted from* EPRI Journal, *p. 17 (March 1978).]*

Electric power from space   The most distant concept proposed for solar generated electricity is the orbiting satellite proposal. The Satellite Solar Power Station would use solar cells to convert sunlight to electricity. The satellite's orbit could be adjusted so that it received light 24 hours a day. The electricity would be converted to microwaves and an antenna would direct a microwave beam back to a receiving antenna on earth. The microwave beam would be converted to direct current and then fed to a direct current network (Fig. 24–13).

Costs of building the satellite station, including the solar cells and other equipment, plus the cost of launching the payload into orbit suggest that we are unlikely to see this form of power generation in the near future. Smaller, more efficient, mass-produced solar cells at lower costs could change the picture.

Biomass: Biological Conversion of the Sun's Energy (alias Photosynthesis)

Of all the conventional and widely used energy sources we discussed, only nuclear power was not linked to the sun. Oil and gas are thought to be the buried remains of ocean plankton, species which derived their energy directly or indirectly from the sun. Coal is the converted substance of what was once woody plants buried beneath the sediments in geological history. Hydropower can be traced to the sun's evaporation of water in the hydrological cycle. Thus, our most common energy resources are linked closely to the sun.

Now, in an era of awakening to the finiteness of our energy resource base, we are turning to the sun to convert its rays directly to heat, hot water, electricity and mechanical energy. That is, we are attempting to use our mechanical and engineering skills to gather the sun's rays directly to our purposes without any intermediate processing by nature. It looks as though we will be reasonably successful at this conversion, but the fact that oil, gas, and coal all owe their energy to the sun reminds us strongly that nature already does an excellent job at converting the sun's rays to energy. We all know the process as photosynthesis, the conversion by use of the sun's energy of carbon dioxide and water to energy-rich organic molecules. Why not take advantage of the energy being captured by green plants *here and now?* It is an idea that has already been used in earlier times.

Most American families cooked and heated their homes with wood-burning stoves until the 20th century. Some still heat their homes with wood furnaces. They are rural people typically, largely concentrated in Maine, Vermont, New Hampshire, and parts of the rural South. Wood was always available to them cheaply and it made sense to be economical. Now we are emulating them; over half a million new wood burning stoves have been sold since 1974. Wood is also the fuel most used by traditional societies throughout the developing world.

The saying goes that wood heats twice, once when you cut it and again when you burn it. Wood stoves, as opposed to drafty fireplaces which can cost you additional heat, are also a valuable backup to conventional heat. If an ice storm were to snap transmission lines, as one did in New England in the winter of 1973, the loss of electricity would

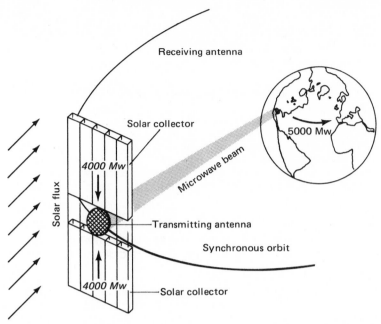

Receiving antenna

Solar collector

4000 Mw

Solar flux

Microwave beam

5000 Mw

Transmitting antenna

Synchronous orbit

4000 Mw

Solar collector

Figure 24-13 *Satellite solar power station. The concept of orbiting a space station whose solar cells would generate electricity for use on earth has been the subject of much discussion. The station would transmit its power to earth via a microwave beam. While a bold concept, many obstacles, including enormous cost, stand in the way of its development. (Adapted from "Space Shuttle Payloads," Hearings of the Committee on Aeronautical and Space Sciences of the U.S. Senate, 31 October 1973.)*

mean little or no heat in forced-air heating systems because no fan could blow heat through the house. A wood stove with fuel would be a welcome feature at such a time.

Wood is a fairly clean fuel; apparently sulfur does not become incorporated in wood until it has reached the stage of advanced peat formation. Its heat content is also high; a cord of wood (a stack 4 ft × 4 ft × 8 ft) has about 20,000,000 Btu on average; a ton of coal has 24,000,000 Btu. The price of a ton of coal and a cord of wood vary by region and season but are not far apart.

The fuel for wood stoves, if derived from scrap or from downed trees which would otherwise be wasted or would rot, does not diminish our resource base. Indeed, it takes pressure off the oil and natural gas markets. If the wood is drawn from woodlots that are not being replenished by the planting of new trees, however, there is an environ-

mental effect. Natural regeneration of woodlots, depending on the locale, may take as little as 25 years or as long as 100 years. Where the wood comes from does make a difference. One good place to get wood is from State Forests and National Forests where removal of downed timber has a positive effect. Fire hazard is reduced by this removal and sunlight can reach seedlings more easily, thereby stimulating new growth.

Thousands upon thousands of new acres of trees would have to be planted to make wood biomass a major fuel, because our present tree resources are largely devoted already to timbering and recreation. One proposal is to grow sycamore trees for only five years to immature size, harvest them mechanically and replant. Poplar is another fast-growing variety that may produce high yields. Although the burning of wood would add carbon dioxide to the atmosphere just as the burning of oil,

SHOULD ENERGY SOURCES BE CENTRALIZED OR DECENTRALIZED?

Amory Lovins, a noted independent energy policy analyst, has suggested that there are

> ...two energy paths that the U.S. (and, by analogy, other countries) might follow over the next 50 years. The first, or "hard" path is...high technology, centralized, increasingly electrified, and reliant chiefly on depletable resources (coal and uranium). The second, or "soft" path is...relatively low technology and decentralized, electrified only where essential, based on renewable resources, and fission free.... *

Alvin Weinberg, director of the Institute for Policy Analysis of the Oak Ridge Associated Universities, takes issues with Lovins' suggestion that other forms of energy can be employed as effectively as electrical energy. He feels that

> ...Lovins ignores electricity's advantages....The question of whether the advantages of electricity–its convenience, its cleanliness (compared with decentralized fossil fuel systems) and its costs are worth sacrificing...(because of) its admittedly poorer thermodynamic match with its end use and its possible social consequences if it is generated by fission. †

In the same article, Weinberg cites a speech by Franklin Roosevelt in 1936 to support his views on central station electric power:

> Sheer inertia has caused us to neglect formulating a public policy that would promote opportunity to take advantage of the flexibility of electricity; that would send it out wherever and whenever wanted at the lowest possible cost. We are continuing the forms of over centralization of industry caused by characteristics of the steam engine, long after we have had technically available a form of energy which should promote decentralization of industry....

But David Lilienthal, former Chairman of the Tennessee Valley Authority, one of the largest suppliers of electricity in the nation, disputes what was once accepted wisdom

> We were persuaded to accept the fashionable idea that great new generating stations and huge, regionalized transmission systems would deliver electric energy more efficiently at lower cost to the public than small, local decentralized ones. Nobody foresaw what would happen to the cost of oil and gas and coal, to the cost of transporting these fuels, or to the cost of constructing huge generating plants and mighty transmission lines. No one foresaw the rapid diminution of the world's reserves of oil. We placed major reliance on nuclear energy without being fully aware of its costs or its hazards. ‡

Alvin Weinberg favors central station electric power; Amory Lovins advocates dispersed siting of power stations and individual energy generation.

Their debate in many ways is "the" debate on centralized <u>vs</u> dispersed power sources; their exchanges are "the" exchanges of two communities. Both brilliant, both articulate, they have squared off, combatting each other's views. To call one a conservative and the other a liberal, or one a traditionalist and the other modernist would probably be incorrect. Weinberg, however, has referred to Lovins and others as energy radicals. It is a most inappropriate tag because of its political implications.

Both points of view have a long history. Individual energy generation is perhaps the oldest tradition, but the last half century has been an era of growth in centralized power generation. Thus, the debate is not one of tradition with new ideas, but one between two points of view that were both valid at different times in history. Each point of view says that current problems can be solved its way. Whose way do you think is most efficient and why? Or is it possibly a new time or a transition time when both points of view are valid? The issue of whose way is best should be approached by attempting to construct a list of advantages and disadvantages in each concept.

To answer a question such as this probably requires focusing on a single use of energy and answering specific questions. Since nearly half of America's homes heat hot water with electric energy, and since solar hot water is feasible and comparable in cost to electric hot water, it would be useful to compare these two energy uses. Recall, in making your comparison, that solar hot water may require electric energy to boost its temperature when the sun's rays are hidden for too long. Of course, burning oil or gas could be used to boost the temperature of the water as well.

The questions you might use to make your comparison include:

Which is more reliable: hot water from electricity from a huge central power station, or hot water from solar energy with an occasional electric assist? Which is safer? Which is least subject to disruption? Which is least costly? Which is more convenient? Which is cleaner? Or, more accurately, where is the mess? If you add up the environmental problems, which method has the least impact? Which uses more resources?

Are there intangibles in your comparison between central and decentralized power? That is, are there elements in the one that you favor that you cannot qualify but that you simply prefer? What are those elements, if any? Can you identify a tradition that has the same point of view on what is preferred?

---

* Alternate Long Range Energy Strategies, Joint Hearings Before the Select Committee on Small Business and the Committee on Interior and Insular Affairs, U.S. Senate, Interior Committee Serial No. (94-47) (92-137), 9 December 1976, for sale by the Superintendent of Documents, U.S. Government Printing Office.

† Review of <u>Soft Energy Paths: Toward a Durable Peace,</u> by Amory Lovins (Ballinger, 1977) appearing in <u>Energy Policy,</u> p. 85 (March 1978).

‡ Quoted in <u>National Journal,</u> p. 674 (29 April 1978).

gas, or coal does, the growing trees would be withdrawing carbon dioxide from the air for photosynthesis. A rough balance of carbon dioxide addition and removal seems possible.

Sugar cane is another crop that gathers up the sun's energy efficiently. In fact, sugar cane may be the best biological converter we have. Its energy can be made available for human use by fermenting it to ethyl alcohol. Sugar cane does not grow as well in this country as do other crops, particularly corn. Brazil, on the other hand, is well-suited for sugar cane and is seriously investigating the commercial production of fuel alcohol from sugar cane.

One unusual species of plant, the Hevea rubber plant, first found wild in Brazil, produces hydrocarbons which can be used to make rubber. Grown almost exclusively in Indonesia and Malaya, the plant yields almost a ton per acre. Though this is half the yield of sugar cane, the growers think Hevea yields can be tripled. Although the Hevea plant has been our source of natural rubber for many years, enthusiasts see the hydrocarbons it produces as substitutes for oil. Hevea has competition, however, from a wild desert shrub known as guayule (pronounced gwy-oo'-lee). Native to the deserts of Mexico and the American Southwest, guayule has been cultivated in the U.S. before. In the early 1900s, guayule was one of our principal sources of rubber, but rubber from the Hevea plant replaced it in the marketplace. During World War II when Southeast Asia was dominated by the Japanese, we again established a rubber industry based on guayule, but the industry collapsed when the rubber producing nations were liberated. Now, a number of scientists want to give guayule another try after selecting the best strains for cultivation.

Corn and corn wastes have been suggested as a source of biomass fuel. The burning of husks and stalks may be used in grain-drying operations, a use which presently calls on propane and liquified petroleum gas for heat. Corn can also be used to produce one component of the blended fuel popularly known as gasohol. Gasohol is a mixture of 90% unleaded gasoline and 10% grain alcohol. Corn carbohydrates can be fermented to produce the grain alcohol, which is then distilled to a high purity alcohol. The technology of fermentation and distillation is well known and is the basis of the manufacture of whiskey from corn. The process is so simple that it has been done at home in illegal stills for many years.

The potential for biomass fuels is very large, but the impact of this development is uncertain. In many ways, where use is made of otherwise discarded materials, producing biomass fuels is only good conservation practice, reflecting the ethic "waste not, want not." In other ways, where new crops or new processes or new acreages are proposed, we tread on less certain ground. We do not know if we can afford the acreage for energy crops at the expense of food crops. We do not know if we can devote enough land to wood for burning because of the large demands for timber and paper. Also, the processes do not have the advantage of some of the other solar technologies because they require central processing. Solar house heating, in contrast, is a decentralized activity, carried out in each dwelling.

Of perhaps greatest concern, however, is the soundness of biomass proposals from a biological point of view. Planting only a single crop has a tendency to deplete soil of particular nutrients, to attract insect pests, and to reduce the survival of a diversity of animals. Caution is needed lest tropical forests be destroyed to produce a source of energy. We definitely need field crop research on these ideas in order to evaluate biomass further. In the same way that the pioneers could not evaluate the soil without growing crops, we cannot evaluate the biomass ideas until we try them.

Ocean Thermal Energy Conversion (OTEC)

Another idea to exploit the sun's energy draws on the temperature differences that exist between the surface and the depths in the tropical oceans.

The temperature differences are very large in some places, especially in the areas in which the warm Gulf Stream flows; between the surface and waters at a depth of 2000 feet (600 meters), the temperature difference may be as great as 22°C (40°F).

(The principle of OTEC is to use these waters alternately to boil and condense a "working fluid." In between, the vapor is used to turn a turbine. The fluids most studied and discussed are ammonia and propane. OTEC works this way: warm surface waters are used to boil the liquid working fluid to a vapor. The vapor is used to turn a turbine and generate electricity; cold waters are brought up from the ocean's depths to condense the vapor in another heat exchanger (Fig. 24–14).)

Of course, all of this is to be carried out at sea, so the electricity would require transport to the land. Underwater cables would be used for this purpose. On the other hand, the electricity could be used as it is generated for such electric intensive industries as aluminum or electric arc steel.

Many problems must be solved; corrosion by sea water is one problem area. Ocean organisms and slime may clog the heat exchanger, requiring the use of large quantities of antifouling chemicals. Still another potential difficulty is the design of the huge heat exchangers themselves. Environmental questions such as impact on climate by diverting Gulf Stream waters still need to be answered. Nonetheless, vast quantities of electricity may be generated in this way; one estimate by the Department of Energy suggests that power equivalent to that from 200 electrical plants of a 1000 megawatt rating could be produced. Of course, the testing is proceeding in stages; a one megawatt plant was under construction in 1979, and a 100 megawatt plant has been tentatively scheduled for completion by the mid-1980s (Fig. 24–15).

PROSPECTS FOR SOLAR ENERGY

(Solar energy can benefit people in a number of basic ways. First, by replacing fossil fuels, air and water pollution are decreased. Second, the replacement of fossil fuels means a decrease in fuel imports, especially of oil, and this will help to secure

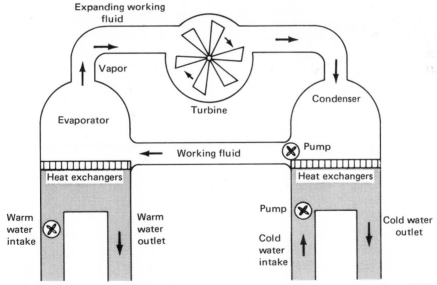

Figure 24–14 *Schematic diagram of a closed cycle ocean thermal power plant. (Adapted from John R. Justus, "Renewable Sources of Energy from the Ocean," in Project Interdependence, Committee Print 95-3, U.S. Congress, November 1977.)*

Figure 24-15 *Artist's conception of an OTEC factory ship. Ocean thermal energy conversion (OTEC) would exploit the temperature differences between the layers of the tropical oceans. Here, an artist provides a vision of what an OTEC factory ship moored at sea might look like. (The Johns Hopkins University Applied Physics Laboratory)*

the value of the dollar. Third, by replacing nuclear fuels here and abroad, the threat of the spread of nuclear weapons is decreased. Finally, solar sources can provide us some protection by making our fuel supply less subject to interruption.

What portion of our energy needs can solar energy contribute? There are certainly optimistic views of its potential. The number of houses using solar heating systems leaped from about 30 in 1973 to thousands by 1978. Both the President's Council on Environmental Quality[2] and the Project Independence Report[3] suggested that under conditions of "accelerated development," solar power

could contribute nearly 25% of the nation's energy requirements by the year 2000. Both of the reports had a wider definition of solar energy than we use here; solar included energy from wind and from hydropower. If we remove the potential of these two energy sources from the solar category, their projection is about 13% for the contribution in the year 2000 of solar energy in the form of heating, cooling, electric generation at central station, photovoltaic electric energy and biomass. Thirteen percent is a very large number for an economy that will be using between 80 and 120 quadrillion Btu of energy in the year 2000.

It is good that such a major contribution is in sight, but what are the conditions of "accelerated development?" "Accelerated development" has no precise definition—it means many things. Basically, however, it means some form of government intervention in the market for solar technology. Without such involvement, the paths that the solar market can take will differ dramatically.

Government intervention in markets is nothing new. Before the OPEC era, oil from overseas was brought in on a quota system for many years to prevent the domestic oil market from being flooded with cheap imported oil. Strategic metals have been stock piled for many years both to have a store of the metals and to shore up these markets. Farmers have received price supports from the Department of Agriculture on basic grains for many years and have been paid to keep acreage out of production. The federal government, through the Atomic Energy Commission, has done most of research and development on nuclear power and still provides the insurance for a nuclear catastrophe. The first nuclear fuel reprocessing plant as well as one under construction were heavily supported by government money. In fact, the entire nuclear industry has been backed by government involvement from the start. Thus, the suggestion that the government stimulate the solar market is not at all a radical proposal.

Incentives are necessary because active solar heating systems cost a lot. A house with an area of

2 President's Council on Environmental Quality, Solar Energy: Progress and Promise, April 1978, Government Printing Office, Washington, D.C.
3 Federal Energy Administration, Project Independence Blueprint, Final Task Force Report: Solar Energy, Government Printing Office, Washington, D.C., 1974.

1500 square feet (135 square meters) and two to three days of heat storage may need 750 square feet (68 square meters) of collector area. At a cost of $10 per square foot ($110 per square meter), the cost of the collectors alone is $7500. The volume of the tank in gallons would be about twice the collector area. The 1500 square foot house will then need a tank of 1500 gallons storage.[4] Installation and other equipment may push the cost to $10,000 and more for the entire solar heating system.

(Basically two kinds of cost need to be brought down to help people buy solar technology. The cost of collectors can be brought down by mass production, that is, by expansion of the market to many thousands, even millions, of units. The cost of installation, pipes, plumbing, and the instruments for solar heating can be brought down by having a number of experienced companies willing to compete with one another for jobs. Again a mass market is the key. These two costs, collectors and installation, make up the bulk of the cost of a solar heating system.)

The National Energy Act of 1978 took a number of steps to increase people's access to solar technology. One was the authorization for the federal government to purchase nearly 100 million dollars worth of photovoltaic arrays to generate power at remote federal sites such as those used for research and atmospheric monitoring where fuel can only be taken with difficulty. A purchase of this size is designed to help bring about mass production of the cells which had been individually assembled.

(Another mechanism the government is using to influence people to buy solar is by tax credits.) The National Energy Act of 1978 established a tax credit procedure which works in the following way. A credit is given against taxes owed of 30% of the first $2000 of investment in solar energy or energy conservation. If the investment is more than $2000 but less than $10,000, the excess over $2000 is credited against taxes at the rate of 20%. Thus, a family that invests $6000 in solar heating is able to deduct from taxes owed 30% times $2000 plus 20% times $4000 for a total of $600 + $800 or $1400. This amount is subtracted from taxes owed, effectively reducing the investment in the solar equipment by $1400 from $6000 to $4600.

There are many more variations on these basic devices to stimulate use of solar energy, but this is basically what the CEQ means when they say "accelerated development"; they mean a market for solar equipment which is given government support through low-interest loans, tax credits, deduction, or similar concepts. Since the precise effect of any of these programs is unknown, the future is difficult to predict. Nevertheless, such a set of programs is sure to provide a welcome extra growth to our already expanding solar market.

One of the greatest potential areas for solar energy, of course, is in the heating of water and buildings. Figure 24–16 shows three possible futures for solar heating, a conservative future, a moderate future, and a strong future. The conservative future (Scenario 1) views solar energy as little more than an occasional choice by purchasers of new homes. Solar systems will be less expensive on a new home or building than on an existing dwelling because the collector panels can replace a portion of the roof and the roof slope can be designed at the correct angle for efficient collection. Also, no separate mortgage is necessary, as the solar portion of the dwelling is included in the single mortgage for the building. The moderate future (Scenario 2) assumes that in addition to use in new homes some current homes and buildings will be converted to solar energy as well. The strong future sees solar systems as an option chosen often by builders, buyers, and businesses, and as a choice favored for a substantial number of existing dwellings and other buildings. In this strong future, 5.5 quads of energy (5.5 million billion Btu) will be delivered by solar energy by 1990. This is roughly the equivalent of a billion barrels of oil, or 20% of our total oil use in the middle 1970s.

---

4 A cylindrical tank with a 3-ft diameter and 7-ft height would provide roughly this volume.

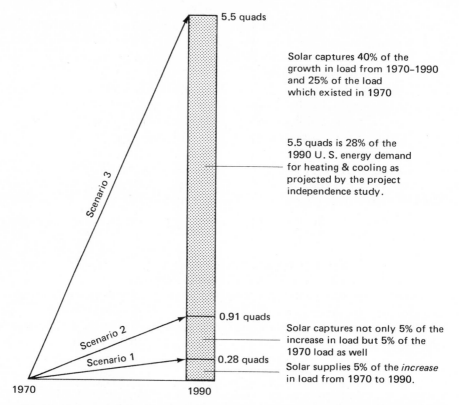

5.5 quads

Solar captures 40% of the
growth in load from 1970-1990
and 25% of the load
which existed in 1970

5.5 quads is 28% of the
1990 U. S. energy demand
for heating & cooling as
projected by the project
independence study.

Scenario 3

0.91 quads

Scenario 2

Scenario 1

0.28 quads

Solar captures not only 5% of the
increase in load but 5% of the
1970 load as well

Solar supplies 5% of the *increase*
in load from 1970 to 1990.

1970   1990

Figure 24-16   *Solar energy for heating and cooling by 1990: Three alternative futures. Note that
1 Quad=1 million billion Btu. (Source: Arthur McGarity, "Solar Heating and Cooling: An
Economic Assessment," National Science Foundation, 1976.) Note that slightly different percent replacement and capture of air conditioning requirements are actually given, but the
results do not shift by more than a few percent.*

Table 24-1 lists a representative sample of the
cities where solar heating competes with electrical
energy. In a study of solar heating,[5] one analyst
chose 20 representative cities to investigate for the
economic potential of solar heating. Note that at
1975 prices, New York City homeowners would be
saving money if they heated with solar plus electric
rather than electric alone. Future developments in
terms of cost reduction for solar and cost increases
for electricity only make solar systems more attractive in New York.

If we simply assume that mass production
lowers collector costs by about 20% and installation by 30%, six more cities in the list find solar
heating with electric backup better than or competitive with electric heat alone at 1975 electric
rates. These cities are Alburquerque, New Mexico;
Boston, Massachusetts; Grand Junction, Colorado;
Los Angeles, California; Santa Maria, California;
and Washington, D.C. Of course, these are not all

5 Arthur McGarity, "Solar Heating and Cooling: An
Economic Assessment," National Science Foundation,
Washington, D.C., 1976.

TABLE 24-1   Where Solar Heating is Competitive

	(1) Solar + Electric Backup Better Than Electric Heating Alone. (1975 Local Electric Rates & 1975 Costs of Solar Equipment and Installation [a])	(2) Solar + Electric Backup Better Than Electric Heating Alone. (1975 Local Electric Rates & Mass Production of Solar Equipment [b])	(3) Solar + Electric Backup Better Than Electric Heating Alone. 1975 Electric Rates Grow at a Real Rate (non-inflationary) of 5% Per Year & Mass Production of Solar
Albuquerque, NM		Albuquerque (0.85)	Albuquerque (0.96)
Atlanta, GA			
Boise, ID			Boise (0.80)
Boston, MA		Boston (0.50)	Boston (0.50)
Charleston, SC			Charleston (0.99)
Cleveland, OH			
Grand Junction, CO		Grand Junction (0.65)	Grand Junction (0.80)
Indianapolis, IN			Indianapolis (0.70)
Los Angeles, CA		Los Angeles (0.80)	Los Angeles (0.80)
Madison, WI			Madison (0.74)
Miami, FL			Miami (1.00)
New York City, NY	New York City	New York City (0.45)	New York City (0.80)
Oklahoma City, OK			Oklahoma City (0.90)
Phoenix, AZ			Phoenix (0.96)
Rapid City, SD			Rapid City (0.85)
San Antonio, TX			
Santa Maria, CA		Santa Maria (0.77)	Santa Maria (0.90)
Seattle, WA			
Washington, D.C.		Competitive in Washington, D.C.	Washington, D.C. (0.80)

[a] 1975 Solar Costs are $12 per square foot for collector plus about $5000 installation.

[b] Improved Solar Costs are $9.66/sq. ft plus $3500 for installation.

Numbers in parentheses are solar fractions, the fractions of annual heating loads met by solar energy. This chart is prepared for a 1500 sq. ft. house with heating loads specific for the city.

the cities where solar would be advantageous, but simply those out of the selected group. If mass production lowers collector and installation costs as above and electric rates rise at a "real" as opposed to inflationary rate of 5% per year, homeowners in 15 out of the 20 study cities discover that solar plus electric backup saves them money over electric heat alone. If tax credits are offered and low-interest loans made, solar energy becomes more attractive yet.

Questions

1. Briefly, sketch how the sun's energy can be collected and used to heat water for an average house. Why might you need a conventional "backup" heating system?
2. Add a simplified solar home-heating system to the sketch you made for Question 1.
3. Contrast active and passive solar energy systems.
4. Describe how the sun's energy might be used to generate electricity using solar cells. What environmental advantages does such a system have over generating electricity by burning coal or oil?
5. Suppose instead that electricity was generated by focusing the sun's rays to produce steam. Now compare the environmental advantages of solar versus conventional electric generation.
6. What environmental problems can you see arising if large tracts of land are devoted to growing trees for wood to be used as a fuel source? (Part of the answer is in Chapter 30, but think about it yourself, first.) What about problems associated with burning the wood?
7. Discuss the economics of solar energy from the point of view of a homeowner. What are the major costs? What are the savings? How can the government help homeowners?

Further Reading

General References on Solar Energy

Hammond, Allen L., "Solar Energy: The Largest Resource," Research News, *Science* **177**, 1088–1090 (22 September 1972).

Morrow, Jr., Walter E., "Solar Energy: Its Time is Near," *Technology Review*, pp. 31–44 (December 1973).

"Solar Energy: Progress and Promise," Council on Environmental Quality, U.S. Government Printing Office, Washington, D.C. 20402, Stock No. 041–011–00036–0, April 1978, 52 pp.

Electric Power from Space

Von Braun, W., "Energy from Space," *Popular Science*, p. 65 (September 1975).

Glaser, P., "Solar Power from Satellites," *Physics Today*, p. 373 (February 1977).

O'Neill, Gerard K., "Space Colonies and Energy Supply to the Earth," *Science* **190**, 943–947 (5 December 1975).

Ocean Thermal Energy Conversion

Metz, William D., "Ocean Temperature Gradients: Solar Power from the Sea," Research News, *Science* **180**, 1266–1267 (22 June 1973).

Metz, William D., "Ocean Thermal Energy: The Biggest Gamble in Solar Power," Research News, *Science* **198**, 178–180 (14 October 1977).

Othmer, Donald F. and Oswald A. Roels, "Power, Fresh Water, and Food from Cold, Deep Sea Water," *Science* **182**, 121–125 (12 October 1973).

Photosynthetic Conversion of Solar Energy

Broad, William J., "Boon or Boondoggle: Bygone U.S. Rubber Shrub Is Bouncing Back," *Science* **202**, 410–411 (27 October 1978).

Burwell, C. C., "Solar Biomass Energy: An Overview of U.S. Potential," *Science* **199**, 1041–1048 (10 March 1978).

Griffin, D., "Natural Rubber has a Future After All," *Fortune,* p. 78 (24 April 1978).

Hammond, Allen L., "Alcohol: A Brazilian Answer to the Energy Crisis," Research News, *Science* **195**, 564–566 (11 February 1977).

Hammond, Allen L., "Photosynthetic Solar Energy: Rediscovering Biomass Fuels," Research News, *Science* **197**, 745–746 (19 August 1977).

Lipinsky, E. S., "Fuels from Biomass: Integration with Food and Materials Systems," *Science* **199**, 644–650 (10 February 1978).

Maugh II, Thomas H., "Guayule and Jojoba: Agriculture in Semiarid Regions," *Science* **196**, 1189–1190 (10 June 1977).

Photovoltaic Conversion and Interior Lighting

Hammond, Allen L., "Photovoltaic Cells: Direct Conversion of Solar Energy," Research News, *Science* **178**, 732–733 (17 November 1972).

Hammond, Allen L., and William D. Metz, "Solar Energy Research: Making Solar After the Nuclear Model?" Research News, *Science* **197**, 241–244 (15 July 1977).

Kelly, Henry, "Photovoltaic Power Systems: A Tour Through the Alternatives," *Science* **199**, 634–643 (10 February 1978).

Metz, William D., "An Illuminating New Use for Solar Energy," *Science* **194**, 1404 (24 December 1976).

Solar Heating

Hammond, Allen L., "Individual Self-Sufficiency in Energy," *Science* **184**, 278–282 (19 April 1974).

Hammond, Allen L. and William D. Metz, "Capturing Sunlight: A Revolution in Collector Design," Research News, *Science* **201**, 36–39 (7 July 1978).

McGarity, Arthur, "Solar Heating and Cooling: An Economic Assessment," National Science Foundation, Directorate for Scientific, Technological and International Affairs, Division of Policy Research and Analysis, Washington, D.C. 20550, 1977.

Solar Thermal Electric Power

Metz, W., "Solar Thermal Electricity: Power Tower Dominates Research," Research News, *Science* **197**, 353 (22 July 1977).

"Spinning a Turbine with Sunlight," *EPRI Journal,* p. 14 (March 1978).

Chapter 25

Energy Conservation

"We must extend our resource base to replace the fuels now becoming scarce." That is the theme of the energy companies, the businesses that profit from selling fuel and electricity. There is another way to extend our energy resource base and to prevent costly and risky fuel imports. It does not make profits for the energy companies, however. In fact, they earn less and consumers save money by finding ways to save energy.

Energy conservation ranges from simple and inexpensive to costly and complex. It can be as simple as keeping outside doors closed in winter, turning down the thermostat, turning out unnecessary lights, wearing a sweater, showering instead of bathing, using a fan instead of an air conditioner, or riding the bus instead of driving. Conservation can be accomplished by such relatively simple purchases as insulation for an attic, storm windows, weatherstripping for a door, or a flow restricting device for a shower. Investments can go still larger: high efficiency appliances or fuel-efficient automobiles may be purchased.

Energy conservation saves money; the dollars we do not spend for energy are in our pockets. Energy conservation can decrease the need to import oil and gas from overseas. As we discussed earlier, this helps the value of the dollar in the international money market and helps to prevent inflation. Energy conservation postpones the days of scarcity and keeps fuel prices from rising too fast.

Finally, energy conservation means less of the air pollutants that are by-products of combustion. It may also mean fewer of the poisonous products of fission to dispose of or to guard for centuries.

Energy conservation also becomes a focus for attention when we observe the average quantity of energy consumed per capita in the industrial world (Figs. 25-1 and 25-2). In 1950, the United States had a population of 152 million which in that year consumed on average about 210 million Btu *per person.* (Btu stands for British thermal unit and is the quantity of heat required to raise the temperature of 1 lb of water by 1°F.) By 1975, the U.S.A. had reached a population of 214 million people, roughly a 40% increase. Energy consumed in that year had soared to 329 million Btu *per person,* roughly a 55% increase in per person consumption alone. Similar per person increases were noted in Canada. The combination of a population rise and an increase in the per person consumption caused the total annual energy consumed to jump from 31 quadrillion Btu in 1950 to 69 quadrillion Btu in 1975. One quadrillion is a million billion or 1,000,000,000,000,000.

These two aspects of energy use, per person consumption and population, reveal the two basic ways to control the growth in energy use in the United States. Conservation can decrease the energy consumption per person, and a stable population can limit the total energy consumption.

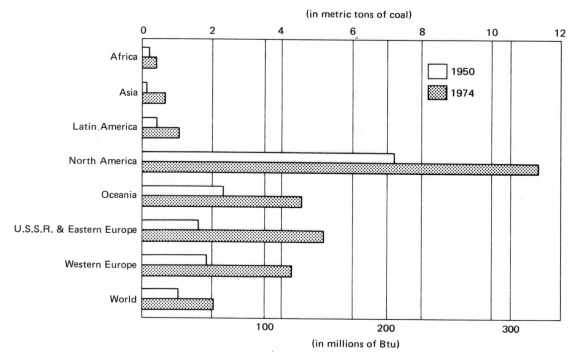

Figure 25–1   *Energy consumption per capita by world regions for 1950 and 1974. The people of North America are, far and away, the world's largest energy users on a per person basis. The 329 million Btu consumed by the average person in 1974 in North America is equivalent to the energy in 11.5 metric tons of coal (per person per year). (Sources: Department of Economic and Social Affairs, United Nations: "World Energy Supplies, 1950–1974," 1976; and "Selected World Demographic Indicators by Countries, 1950–2000," May 1975.)*

What specifically can energy conservation do? The Council on Environmental Quality[1] has described two alternate futures. Although both futures include a maximum commitment to solar energy, the first future assumes that conservation will be given sustained attention and emphasis. Future I, "The Conservation Route" is contrasted to Future II, "Business as Usual" in Figure 25–3. ("The Conservation Route" will make it possible to decrease our reliance on oil and gas due to the combination of conservation and the development of solar energy.) Our reliance on coal and nuclear power need in-crease to only a small degree. In stark contrast, the "Business as Usual" future, which projects energy demands increasing at 1.9% per year, would require a sixfold expansion of nuclear power and more than a doubling of our reliance on coal, even with an expansion of solar energy to its maximum possible contribution.

The relative impacts of these two energy futures are shown in Figure 25–4. The annual requirements for coal mined in the "Business as Usual" future are about double those in the "The Conservation Route" as are strip-mining requirements. And in the "Business as Usual" future, the number of coal-fired plants is double that in "The Conservation Route." The role of nuclear power in "The Conservation Route" measured in terms of the number of nuclear electric plants is less than half that in the "Busi-ness

_____

1 *The Good News about Energy,* 1979, available from the Superintendent of Documents, U.S. Government Printing Office, Washington, D.C.

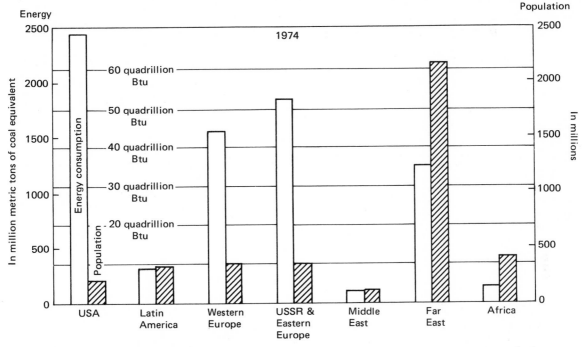

Figure 25-2 *Energy consumption by regions of the world for 1974. The United States had about 5.5% of the world's population in 1974, but consumed about 30% of all of the energy consumed in that year. We used about 30% of the oil consumed in that year, 20% of the coal, and 49% of the natural gas. (Source: Population Reference Bureau, 1974.)*

as Usual" future. Spent fuel generated is likewise about halved.

What are the means that are envisioned in "The Conservation Route"? They include more insulation in homes, apartments, public buildings, and factories; more efficient appliances and machines; better insulated refrigerators; gas stoves without pilot lights; fluorescent lamps instead of incandescent bulbs. Total energy systems in industry may both generate electricity and process heat, as well as space heating. Automobiles can be made (and are becoming) far more fuel efficient. In short, energy conservation, if we can achieve it, holds out as great an opportunity as discovering new sources of energy.

We have selected two areas on which we shall focus attention in our discussion of energy conservation. These are (1) the way we heat buildings and make hot water and (2) the way we transport people and freight. Building heating and water heating together account for about 20% of our annual use of energy. Transportation consumes about 25% of the energy we use each year.

There are choices in these two areas that can be made now by the consumer or business person which can influence energy consumption. New technology is not necessary in these areas; methods presently available are sufficient to make dramatic differences both in energy consumption and in pollution. For instance, if the cars on the road in the U.S. in the mid-1970s had on average the efficiency of the cars on the road in Europe, our annual travel could have been accomplished with 42% less gasoline. Because choices are now available in these areas which can make a big difference in our energy use, we discuss these two areas in detail.

Each of these two areas is what is called a "point-of-use." Other improvements in points-of-

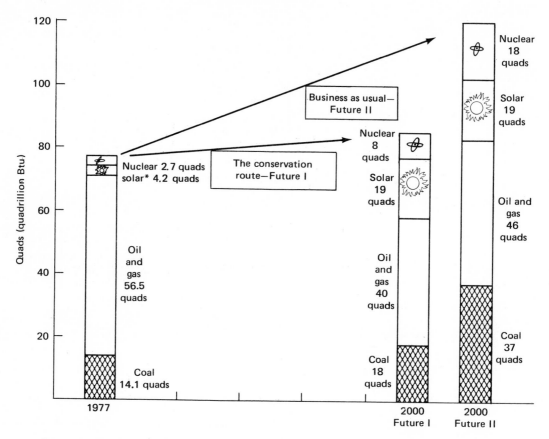

Figure 25-3 *Two possible energy futures: "The Conservation Route" and "Business as Usual." Note that both futures include a maximum commitment to solar energy development. (Data Source: "The Good News About Energy," Council on Environmental Quality, 1979.) *Includes 1.8 quads of biomass; remainder is hydropower.*

use are possible. Some we have already mentioned, such as natural and fluorescent lighting in place of incandescent lighting. More efficient appliances are possible including spark-ignition gas stoves replacing pilot light gas stoves and better insulated refrigerators replacing current models. Labeling appliances with their energy consumption per unit time will be a valuable aid to consumers, not only helping them save energy but helping them save money as well.

In addition to improved point-of-use efficiencies in energy consumption, the penalty of using the "electric middleman" can be softened as well. Cogeneration using a jet engine and a conventional boiler one after another or using magneto-hydrodynamic generation in conjunction with a conventional boiler can, as discussed in an earlier chapter, increase the efficiency of electric generation. "District heating" in which waste heat from electric generation is put to use in heating buildings makes the same fuel do two jobs. All three of these energy conservation methods have been treated fully earlier, although we did not then point to these methods as being ways to save energy. Thus, effi-

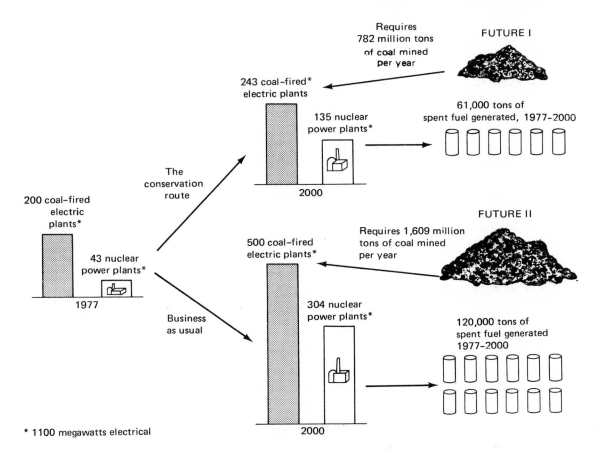

Figure 25-4 *Some of the impacts of two possible energy futures. (Data Source: "The Good News About Energy," Council on Environmental Quality, 1979.) *Electric plants are assumed to be 1100 megawatts electrical.*

ciency of conversion in addition to efficiency at the point of use offers opportunities for energy conservation.

Nonetheless, we return to the two areas mentioned earlier because of their importance in energy consumption and because of their potential for change.

THE WAY WE HEAT HOMES AND MAKE HOT WATER

This is a simple choice to make in most cases, at least from the standpoint of what to reject if pos-

sible. In the recent past, there have only been three options, but now solar energy is entering the picture. These three options are natural gas, oil, and electricity. Let us look at these choices carefully. Suppose you are in the market to purchase a home and you have seen three houses, all of them affordable, and similar in other respects, except in the fuel used for heating. One heats with natural gas, another with distillate oil (home heating oil), and the third with electricity. In each house, hot water is produced in the same way that the home is heated; that is, the gas-heated home heats its water with natural gas, etc. Which home is preferable from the point of view of energy conservation? From the point of view of monthly costs of energy? Assuming that natural gas is allowed to rise in price, it will become, in not too many years about as costly to heat a home with natural gas as it will be with oil. Gas has been much cheaper for some years but the pattern is changing as the government allows the price of natural gas to rise to correct shortages. The energy efficiencies of a hot air heating system that is oil-fired and one that is gas-fired are about the same. Either new system will be 80% efficient. That is, about 80% of the energy derived from burning the fuel goes into raising the temperature of the house.

Electric heating, on the other hand, even when the house receives extra insulation as the electric company suggests, still costs more than heating by gas or oil. How much more depends on the local electric rates. It also depends on how much electricity is used in all activities, since the cost per kilowatt hour decreases as more is used. In most cases, the electric home is considerably more expensive to heat than the home heated by gas or oil. The cost of creating hot water using electricity is large as well. A home heated with oil or gas but using an electric hot water heater will find its electric bill about double that of a home that heats water by gas or oil. Solar energy is rapidly becoming an available alternative for heating homes and hot water in many sections of the country. Its economics relative to oil, gas, and electricity are discussed under solar energy, where we show that the sun is quickly becoming a serious rival for the other methods of heating water and homes.

Not only is the electric home more costly to heat, in several instances it is extremely wasteful of resources. We mentioned in an earlier chapter that the second law of thermodynamics gives rise to an upper limit on the fraction of heat energy that can be converted to work or to some other form of energy. For our purposes, the other form of energy is electrical energy, and the impact of the second law is that only about 40% of the heat liberated by

burning coal, oil, or natural gas can be converted to electrical energy. The unconverted remainder of the heat from burning the fuel is wasted by the electrical plant into the environment, either to the air or to the water, the latter being the phenomenon we know as thermal pollution.

Let us suppose that the electric generating station is oil- or gas-fired. If we assume that the plant is 33% efficient, a reasonable efficiency figure, then two-thirds of the heat energy liberated by combustion is lost into the environment. The remaining one-third in the form of electrical energy can be used to heat a home. If there is a 5% loss in transmission and heating at the home, we have about 31% or 32% of the original energy in the fuel left to heat the home.

In contrast, if oil or gas is burned in the home to provide heat, the efficiency is much greater. New oil and gas furnaces attain 80% efficiency; but after some use, the heat exchanger efficiency falls and the conversion of combustion heat to useful heat decreases to 60% to 70%. It is reasonably clear in such circumstances that electric heat produced by a plant that burns oil or gas is wasteful of that fuel. The same conclusion follows immediately for heating water. Similar comparisons are not useful if the electric plant is coal-fired or nuclear, since these fuels are not used to heat homes (coal use for home heating is very limited).

The "heat pump" could change both the economics and energy cost of electric heating (Fig. 25-5). The electrical device, a sort of refrigerator or air conditioner in reverse, is capable of heating a home during many winter days; it exhausts cold air to the outside and warm air to the inside of a house. Measurements of the performance of these units have indicated that two units of heat energy can be produced for every one unit of energy (as electricity) delivered to the house. This cuts the cost of electric heating in half. Better performances of heat pumps may be "around the corner" as more people try them as alternatives to conventional heating. One reason for their growing popularity is that the same unit that heats the home in winter can be used

Chapter 25 Energy Conservation 517

WINTER

1. Cold indoor air return
2. Cooled refrigerant out
3. Refrigerant heated
4. Cold outdoor air entering
5. Colder outdoor air discharge
6. Heated refrigerant in
7. Heated indoor air to house

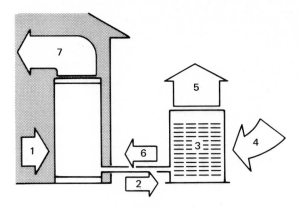

SUMMER

1. Hot indoor air returned
2. Heated refrigerant out
3. Refrigerant cooled
4. Warm outdoor air entering
5. Hot outdoor air discharge
6. Cooled refrigerant in
7. Cooled indoor air to house

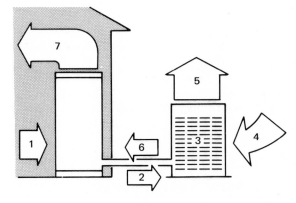

Figure 25-5 *How a heat pump works. (Adapted from GE Weathertron® Heat Pump Brochure.)*

to cool the home in summer. Thus, one investment, not much more costly than central air conditioning alone, can provide a unit which can both heat and cool.

Why are electric heat and electric hot water heaters seen so frequently in new homes if their economics are so unfavorable? It is a matter of what are known as "first costs." An electric water heater may cost about $150 installed; an oil-fired heater, $750 installed; a solar water heater with collectors runs several thousand dollars. If gas is available, gas water heaters will probably be installed since their purchase price for the builder is about that of electric water heaters. If gas is not available, a builder will almost always give customers electric water heaters because they are cheaper to buy; a smaller investment is needed by the builder, and he can sell the house at a lower price.

MANNER OF TRANSPORT OF PEOPLE AND GOODS

Passenger Transport

No nation has more automobiles per person than the U.S. nor drives as far. We have a 55 mile an hour speed limit, however, and there are countries where people drive faster, much faster. Still, no society consumes as much energy per person in auto travel as our own. With so many cars, we are able to cover many more miles. Americans are often accused of wastefulness in this regard compared to Europeans. The accusation is both true and false.

If we compare Europe to the United States, we see a very different society. Population centers are much closer together in European countries; families are, in general, less widely separated. The huge distances between cities and towns in the American countryside are not characteristic of Europe. Nor is car ownership anywhere near as extensive. In the United States there is slightly more than one car for every two people; in the Western European countries (the wealthier countries) there is about one car

DAYLIGHT SAVINGS TIME VERSUS 64 MILLION POUNDS OF CANDLES EVERY YEAR

Benjamin Franklin was astonished. "An accidental sudden noise waked me about six in the morning, when I was surprised to find my room filled with light. I imagined at first that a number of lamps had been brought into the room; but rubbing my eyes I perceived the light came in at the windows," he said in a letter to the Journal of Paris. "I looked at my watch, which goes very well, and found that it was but six o'clock; and still thinking it something extraordinary that the sun should rise so early, I looked into the almanac where I found it to be the hour given for the sun's rising on that day. Those who with me have never seen any signs of sunshine before noon, and seldom regard the astronomical part of the almanac, will be as much astonished as I was, when they hear of its rising so early; and especially when I assure them that it gives light as soon as it rises. I am convinced of this. I am certain of my fact. One cannot be more certain of any fact. I saw it with my own eyes. And, having repeated this observation the three following mornings, I found always precisely the same result." * †

This discovery gave rise to "several serious and important reflections." Dr. Franklin considered that had he slept to his usual rising hour, he would have slept six hours by the light of the sun, and in exchange have been up six hours the following evening by candlelight. Since the latter was a much more expensive light, he was induced by his "love of economy" to estimate how much could be saved by using sunshine instead of candles. For Paris, where he was residing, he calculated a saving of over 64 million pounds of candles every year. "It is impossible that so sensible a people, under such circumstances, should have lived so long by the smokey, unwholesome, and enormously expensive light of candles, if they had really known that they might have had as much pure light of the sun for nothing." * †

Daylight savings time was first proposed in a serious way as a method by which to save energy in England in 1907. By 1918, the U.S. had adopted the procedure of setting the clocks ahead one hour during the spring, summer, and fall. The procedure was dropped, however, after the end of World War I. It was not until 1942, the year after the U.S. entered World War II, that the nation returned to daylight savings time, except that this time it was called "wartime" and it was in effect year-round. Turning the clock ahead brought a decrease in the peak demand for electric power that occurred in early evening. The war over, the nation again abandoned its manipulation of the clock, but now daylight savings time had become popular enough for a number of states to adopt the plan on their own.

By 1966, to correct the lack of uniformity in times across the country, Congress enacted a bill that put all states (with several exceptions) on daylight savings time from the last Sunday in April to the last Sunday in October. In 1974, year-round daylight savings times went into effect again, this time for about a year and a half as an energy saving response to the oil embargo. Public opinion on the potential hazard to school children in the dark morning hours was apparently responsible for the abandonment of this plan.

---

* Quoted from B. Franklin, "An Economical Project," Letter to "The Journal of Paris, 1784.

† David Prerau, "Changing Times: National Time Management Policy," Technology Review **79(5),** 54 (March/April 1977).

---

for every four people. In Europe, auto purchases are heavily taxed; road taxes levied on car owners increase the burden still further. Furthermore, the price of gasoline in European countries is two and one-half to three times higher than in the United States.

Public transportation developed in Europe before the era of the auto. Because of this, the networks of the bus, train and streetcar systems are excellent; service, moreover, is frequent. Thus, Europeans generally have available to them a public transport system which, by appropriate transfers, can link almost any place in Western Europe with any other, nearly door-to-door. Such a journey would begin by walking to the end of the block to catch a subway, bus, or streetcar. From there, one proceeds to the central station where long distance buses and intercity trains are available. The need to use the automobile, therefore, is less frequent.

Even though such excellent systems are available, the importance of the auto to European travel is growing at a phenomenal rate. It is growing as fast as the growth of personal wealth allows it. Figure 25–6 illustrates the trends in the Netherlands, a typical European country with regard to auto use. There all the growth in passenger kilometers is going into auto travel. Similar patterns are seen for other Western European nations.

Having defended Americans, we now must also admit the truth. We have come to "need" our automobiles. We need them not simply because we have built a dispersed society in which the auto is a necessary means of communication. We need them for personal reasons as well. Kenneth Boulding, an economist, explained our relation with the auto in the following way,[2]

> The automobile, especially, is remarkably addictive. I have described it as a suit of armor with 200 horses inside, big enough to make love in. It is not surprising that is it popular. It turns its driver into a knight with the mobility of the aristocrat and perhaps some of his other vices. The pedestrian and the person who rides public transportation are, by comparison, peasants looking up with almost inevitable envy at the knights riding by in their mechanical steeds. Once having tasted the delights of a society in which almost everyone can be a knight, it is hard to go back to being peasants. I suspect, therefore, that there will be very strong technological pressures to preserve the automobile in some form, even if we have to go to nuclear fusion for the ultimate source of power and to liquid hydrogen for the gasoline substitute. The alternative would seem to be a

---

2 Kenneth Boulding, "The Social System and the Energy Crisis," *Science* **184,** 255 (19 April 1974).

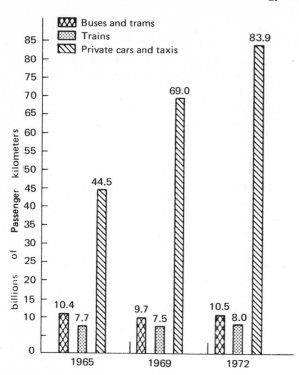

Figure 25-6 *Surface, motorized passenger kilometers by mode in the Netherlands. In the 7-year interval from 1965 to 1972, passenger kilometers on transit systems such as trains, buses, and trams (electric streetcars) barely changed. Total passenger kilometers jumped from 72.6 billion in 1965 to 113 billion in 1972—an increase of 40.4 billion. Of this increase, 97% was accounted for by the increase in passenger kilometers in private cars and taxis. (Source: C. ReVelle, "Public Transport and the Netherlands, Implications for Transport Policy in the United States," Center for Metropolitan Planning and Design, The Johns Hopkins University, 1977.)*

society of contented peasants, each cultivating his own little garden and riding to work on the bus, or even on an electric streetcar. Somehow this outcome seems less plausible than a desperate attempt to find new sources of energy to sustain our knightly mobility.

Boulding may be suggesting that our "habit" is incurable, that the best we can hope for is a substitute fuel, such as a diesel "fix" or an electric "fix." He may be correct that the habit is beyond break-

ing. The growing use of the auto in Europe under such costly conditions gives us an idea of how enticing is the siren call of the automobile (Fig. 25-7).

If the habit cannot be broken what can we do? Try to imagine a society with 110 million private automobiles, traveling as many miles as we do and using 42% less gasoline. This is where the Europeans have learned a trick we are only starting to learn. European cars are, on average, much more fuel efficient than American cars. An accepted figure for the efficiency of European autos in the mid-1970s, averaged *across all cars on the road*, is 10 kilometers per liter or 22.5 miles per gallon. These high efficiencies are the response of motorists to the high prices of gasoline, since more efficient cars are less expensive to operate. The comparable figure, the average mileage for all cars in the U.S. in the mid-1970s was 13 miles per gallon. If our fleet of cars had the average efficiency of European autos, each mile our fleet logged would require 42% less fuel.

Put another way, a car that travels 13 miles on a gallon of gas will require 769 gallons to go 10,000 miles (an average yearly figure). The car that travels 22.5 miles per gallon will need 444 gallons to

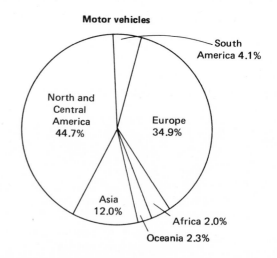

Figure 25-7 *World distribution of motor vehicles in 1977. (Source: Motor Vehicles Manufacturers' Association, Facts and Figures, 1979.)*

go the same distance. At the current price of gasoline, the 325 gallons saved over a year amounts to a healthy sum. It seems that more fuel-efficient cars have a significant impact on our budgets as well as on energy conservation.

How do we arrive at an era of fuel-efficient automobiles? It is reasonably clear that high gasoline prices have encouraged the production of efficient cars in Europe. These high prices include, in general, taxes that amount to half or more of the total price of a liter of gas. Gas prices in the United States have risen rather sharply since the oil embargo and the rise of the cartel. Gasoline prices in 1980 were nearly four times what they were before the embargo. Since taxes on gasoline have not risen appreciably, most of the price increase is going into paying for more expensive crude oil. This rise in prices, which has channeled so many dollars to OPEC members, has had a desired effect in terms of new automobile manufacture. More energy saving cars are coming down the production line and are entering the fleet of cars on the road as consumers see the value of saving gasoline and dollars.

The American public's recognition of the importance of fuel economy is mirrored in legislation passed by Congress in 1975. Congress set standards for the average fuel economy of autos manufactured after 1977. From a 1974 average value of 12.9 miles per gallon, auto manufacturers were able to achieve an average mileage for all 1977 cars of 17.8 miles per gallon in anticipation of meeting the first actual mileage standard, 18 miles per gallon on average, in 1978 (Fig. 25–8).

Auto manufacturers complained of the costs of retooling to meet the later standards and warned consumers of car price increases to come. They also argued that actual redesign of drive trains and new advanced engines would be necessary. In the first few years, they were able to increase the average mileage of their cars by selling smaller and hence lower weight cars. They also achieved lower weights by substituting plastic and aluminum for steel in larger cars. General Motors introduced diesel engines on a number of its models. It is of interest that the new cars manufactured in 1981 only

had to achieve a mileage standard that approaches the *average of all cars* on the road in Europe. Obviously the technology already existed for that achievement.

Even though more efficient new cars will be sold as time goes on, it will take time for energy efficient cars to replace those that are inefficient. The state of the nation's economy in terms of unemployment and inflation will influence how fast new cars are purchased and how fast older cars are retired. Even though energy efficient cars will save energy day by day, they cost more to buy initially, and because they are efficient, they will depreciate in value more slowly. That means that lower-income individuals will hold their older cars longer before they can afford an efficient used car. Thus, both the economy and the higher cost of the fuel efficient cars could delay a transition to a fleet of cars with higher mileage.

What can we do in the meantime? Individual action and responsibility will help a great deal. First, and most simply, there is a speed at which automobiles and trucks are most efficient in terms of fuel consumption. For automobiles that speed is 35 to 40 miles per hour; for trucks it is a bit higher. During World War II, a nationwide speed limit was imposed of 35 miles per hour. That speed limit was

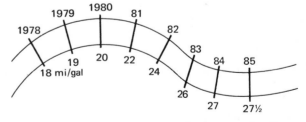

Figure 25–8 *Fuel economy standards as set by Congress and the U.S. Department of Transportation. These standards are to be achieved "on average" by all manufacturers. That is, the average fuel economy of all vehicles sold by a particular manufacturer in a given year must meet the standard for that year. The standards for 1978, 1979, 1980, and 1985 were set by Congress as part of the Energy Policy and Conservation Act of 1975, while those for 1981, 1982, 1983, and 1984 were established in June 1977 under the authority of the Act.*

HOW DO WE GET FUEL-EFFICIENT CARS ON THE ROAD?

To stimulate oil conservation, we need a generation of high-mileage cars, but how do we get them?

We should adopt a surtax on all cars weighing more than 3,000 pounds escalated upward to as high as $1,000 tax on the big luxury cars weighing over 5,500 pounds.

John Quarles, in a Speech before the Coal and the Environment Conference, Louisville, Kentucky, October 1974.

I would like to describe a proposal for a stiff gasoline tax, specifically, an increase in the tax to $2 a gallon (with rebates on a per adult basis). . . . A $2 tax is not much higher than present taxes in countries like France and Italy.

Robert Williams, appeared in The Dependence Dilemma, edited by Daniel Yergin, published by Harvard University, 1980.

We must be cautious to see that a national policy of energy conservation does not unfairly burden the poor and those living in rural areas.

C. ReVelle, "Public Transport and the Netherlands: Implications for Transport Policy in the U.S.," Center for Metropolitan Planning and Design, The Johns Hopkins University, 1977.

We can tax new car purchases by their weight or by their horsepower; we can tax gasoline; we can give income tax rebates to individuals who purchase efficient new cars; we can ration gasoline, allowing one or two gallons per car per day. The many possible ways to direct people's attention to fuel-efficient cars are not all equal in their market effect; nor are they equal in their impact on the upper income, middle income, and lower income families.

There are four basic options listed above; they are (1) tax purchases of in-efficient new cars, (2) tax gasoline, (3) give rewards for purchasing efficient new cars and (4) ration gasoline. We can assemble many plans from these basic options, but for each plan there is the same set of questions to be answered:

1. How will it work? That is, what is the effect on the rate of sales of different kinds of autos? On the sale of gasoline today and in the future?
2. (a) If the government pays, where should the money come from? From general tax revenues? From a special fund? (b) If a tax raises money that goes to the government, what should be done with the money? Should it go into general tax revenues with no special earmarks? Or should it be set aside in a special purpose fund. To give you an idea of what amounts of money are involved, each cent of federal gasoline tax raises one billion dollars each year.
3. Who is affected? Are the poor hurt more than the well-to-do? That is, is the policy "progressive," taking dollars from those better able to afford the expense? Can you think of ways that the impact on the poor can be softened?
4. What are the political chances?
5. Is there an impact on the national economy? On local economies?

To focus the way you answer these questions, here are some hypothetical individuals all of whom are very much concerned about such policies. Explain how the various conservation policies could affect the following people:

Juan Lopez and his family live in Chicago; he maintains a set of apartments in the inner city in return for his own apartment. He just bought a 1971 Zonker Grand Marquis.

Arthur Williams and Sally Adam-Williams have moved back to their home town of Wayside, Nebraska, but the nearest job is in Omaha, some 70 miles away. The do not want to leave their rural environment for a city, but they need to make a living. They are about to buy their first car.

Calvin and Nancy Beale own and run a diner in Ellsworth, Maine. Their diner, the Ellsworth Eatery, is well-patronized and locally famous for homemade pies and the "Maineburger." The bulk of their business comes in the summer when tourists flock to nearby Acadia National Park. Tourists with tents and motor homes come from all up and down the Eastern Seaboard, even from as far south as Virginia. No trains serve Ellsworth.

Can you think of other scenarios showing how people will be hurt by the various gas conservation policies? Can you think of people who will hardly be affected at all by the listed methods?

designed to conserve fuel for the war effort. The 55 miles per hour limit currently mandated by the federal government represents a compromise between the speed that conserves the most energy and the speed people apparently prefer to go. The 55 mile per hour limit, while initially planned to save energy, turned out as well to be a life-saving device. Motor vehicles fatality and accident rates have fallen significantly since the speed limit was reduced (Fig. 25–9). This additional reason to maintain the nationwide limit helps to support energy conservation.

There are other features in the way we drive that influence mileage. Underinflated tires can decrease mileage by a mile per gallon; radial tires can increase mileage by a similar amount. Idling wastes gasoline. If a car idles for a half hour, about a quart of gasoline is needlessly burned. Racing starts, uneven acceleration, and braking in city traffic can reduce the mileage a car will obtain by up to 30%. Driving in lower gears takes more gasoline, as does an untuned engine. Short trips in cool and cold weather use more gallons per mile while the engine is warming up. Longer trips take advantage of the time spent warming the engine.

So far, we have argued principally for changes in the automobile itself, not in the way our society transports itself. Public transportation (buses, train systems, and the like), if properly implemented, can have an important role in reducing energy consumption. Figure 25–10 illustrates the relative efficiency of various means of transportation. As can be seen, between-city travel is typically most efficient because vehicles are more fully loaded and because fewer starts and stops save energy.

Not shown in the figure is the energy cost of airplanes. If an airplane is 50% full, its efficiency is estimated at 1.2 passenger miles per 10,000 Btu. Because airfares for trips planned well in advance have come down by one-third or more in recent years, planes are now flying more fully loaded. At 75% capacity, which many flights are currently achieving, the passenger mile per Btu is increased considerably.

Another vehicle not listed is the "auto in city traffic with a load equal to 75% of capacity." Such a vehicle is known as a car pool. People in car pools most often live near one another or near the line of travel. They also work near one another. The four-person car pool with an average load equal to 75% of capacity has an energy efficiency of about 3.33 passenger miles per 10,000 Btu, an efficiency level about that of a train traveling between cities.

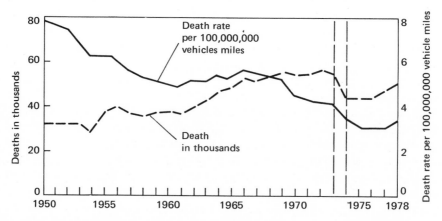

Figure 25-9 *Traffic deaths and death rates. Between 1973 and 1974, the year that the 55 mi/hr limit was first enforced, the annual death toll dropped by 9000 and it has remained lower than its previous levels every year since. (Source: National Safety Council,* Accident Facts, *1979.)*

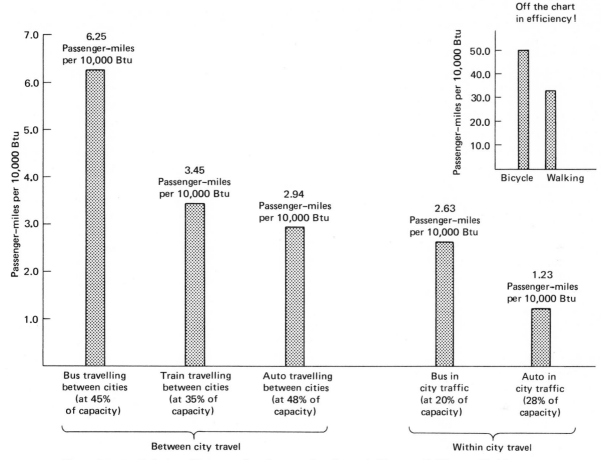

Figure 25-10 *Relative efficiency of various modes of travel. [Source: E. Hirst and J. Moyers, "Efficiency of Energy Use in the United States," Science **179**, 1299 (30 March 1973) ©1973 by the American Association for the Advancement of Science.]*

To obtain a large proportion of people who will ride the bus or subway, a number of factors must be positive. Routes must cover a large portion of the geographic area, so that access to public transport is not difficult. Stops must be frequent enough and spaced closely enough on the route that the time to walk to a stop plus the time to wait for a bus is not too great. Equipment must be clean; conditions must be safe and comfortable; costs must be reasonable when compared to use of the automobile. Routes cannot be too winding or the time from origin to destination will put off potential riders. In this regard, the use of express bus lanes on super highways has a positive effect in reducing travel time.

Even when all these conditions are favorable, ridership on public transport may still be so low that the costs of providing the service in equipment, personnel, gasoline, and maintenance cannot be paid by the revenues generated from fares. Even in Western Europe where public transport runs quite full, governments make up the difference between revenues and the costs of actually maintaining and operating the system.

If the costs of public transport must be supported in part by the government, we need to justify this expenditure. It can be justified on the basis of energy savings, but, of equal importance, air pollution will be reduced by diminishing the number of cars on the road. Those who continue to drive have some benefits in terms of less crowded streets. Where should the government find the money to make up the costs of public transport that are in excess of the system's revenues? Should income taxes be raised to pay this bill? Or should some other form of taxation be used?

Freight Transport

About 30% of the energy we spend each year on transportation goes into the movement of goods; the remainder is expended on moving people. Of the various modes of freight transport, railroads, pipelines, and waterway transport all have comparable levels of efficiency. Trucks consume four to five times more energy per ton mile than these other modes (Table 25-1).

The transport of goods, especially between cities, has evolved in much the same way as the transport of people. Where initially trains carried the bulk of goods moving between cities, trucks have been increasing their share of the load quite steadily. Thus, the energy-efficient mode of transport—rail—is losing ground to an inefficient mode, the truck. Certain bulk goods such as coal,

however, are unlikely to ever give way to truck transport. Instead, pipeline transport appears a possibility to capture a portion of the coal transportation market. (See page 278 for a discussion of coal-slurry pipelines.)

The reasons for the growth in trucking are rooted in the door-to-door nature of truck service. Once goods are loaded, they often need not be reloaded before arriving at the final destination. If the particular goods are not moving in sufficient quantity to justify a rail track from door-to-door, truck transport, though costly, becomes a very attractive alternative. A routing that involves first a truck then a train and then a truck may be low in transport cost because of the rail savings, but the two transfers consume time and money.

TABLE 25-1   Energy Cost of Freight Transport Modes[a]

Mode	Btu/ton-mile
Pipeline	450
Railroad	670
Waterway	680
Truck	2,800
Airplane	42,000

[a] Source: E. Hirst and J. Moyers, "Efficiency of Energy Use in the United States," *Science* 179, 1299 (30 March 1973).

Peak Load Pricing–Decreasing the Need for New Power Plants

According to the Edison Electric Institute, in the 10-year period from 1978 to 1988, utilities are planning to add nationwide about 300,000 megawatts of new electrical generating capacity.[3] This is the capacity projected as necessary to meet the growing demands for electrical energy. The consumer will pay

3 National Electric Reliability Control, 8th Annual Review, August 1978.

dearly for this new construction. The cost of a single new 1000 megawatt electrical plant is now approaching one billion dollars. Yet not all of these plants are truly necessary; most electric utilities are well aware of an economic method to decrease their need for new power plants. The method, known as peak load pricing, has already been adopted in much of Europe, but most utilities in the United States have ignored its existence.

The number and capacity of the power plants scheduled for addition is a response to the future demands foreseen by the electric utilities. The additions, however, do not respond to the growth in the total use of electric power. Instead, capacity additions are planned to meet a high proportion of the peak or highest power demands.

At certain times of the year, depending on the geographic area, electric demands, averaged over the day, reach their highest levels. Further, within a season at certain times of the day, the rate of power drain from the central station may peak momentarily. As an example, during the hot summer, Los Angeles and surrounding areas may require enormous quantities of electrical energy for air conditioning. On any particular hot day, loads will reach their highest levels in the late afternoon when air conditioning is used to counteract the heat that has accumulated in homes, offices, and factories and when home appliances are in active use.

The electrical generating station must be able to supply even the highest peak demand levels, even though such momentary levels may occur for only several hours out of the year. To meet these demands, special equipment may be utilized to generate supplemental power. The special equipment may be used because the boiler/turbine system of the electric plant is not large enough to supply power at the desired rate. Turbine engines that resemble the engines in jet airplanes may be used to generate the additional electric power needed at such times, as may pumped storage hydropower (see Chapter 22). Natural gas when it is available or home heating oil may be used to power the turbine engines.

If the power demand exceeds the capacity of the boiler/turbine system, the daily peaks must be met by this turbine engine generation or from pumped storage. At such times, the boiler/turbine system will be operating at full capacity. On the other hand, in slack seasons where daily power demands are much less, the boiler/turbine system will not be operating anywhere near full capacity. Since the investment in an electric plant is considerable, it makes sense that the plant should not have idle capacity at some times and be taxed to its limits at others. It makes sense also that the plant should not be built with more capacity than is absolutely necessary, because of the enormous cost.

What we are saying is that maximum cost-effectiveness occurs when the load is as level as possible. Telephone companies decrease long-distance calling rates at night and in the evenings to increase the use of equipment that would otherwise be idle at those times. They are trying to level the load on their facilities and make use of slack capacity. Utilities can level their loads as well with "peak load pricing."

You already know how peak load pricing works; the telephone company has even taught you to use it. Long distance calls in the daytime are usually of a business nature and the volume of such business calls falls in the evening and at night. On a daily basis, the phone company decreases the rates for long-distance, direct-dialed calls at night to encourage use of the phone at hours when it otherwise might not occur. The lower rates also shift some calls that would have been made in daytime into the evening and nighttime hours. This decrease of the peak calling rate during the day means that fewer long distance trunk lines will need to be installed to carry the flow of calls cross country. The telephone company also lowers long distance rates on weekends and holidays when the decreased volume of business calls leaves much of the company's network idle. The lower night rate system used by the phone company is a benefit to consumer and company alike.

Electric power companies should be encouraged to price electric energy in the same way; that is, to lower the price of electricity at times when the rate of use is low, and, as necessary, raise the price during times of peak demand.

At night and in the late evening when electric power plants are usually operating at only partial capacity, the rate per kilowatt-hour of electricity would be lowered (Fig. 25–11). People using dishwashers, electric driers, washing machines, etc. would probably postpone these activities from peak hours of use into those relatively slack times with the lower rates. Electric hot water heaters would be put on timers so that water heating would occur during those slack hours. Industries might be encouraged to schedule some of their operations for the evenings. During the day, electric rates would be somewhat higher to recover revenues from uses that have shifted to the evening hours. All in all, capacity would be used more efficiently, peak rates of use would be clipped off, and the need for investment in new generating capacity to meet the peaks will be diminished. Peak load pricing may also be applied on a seasonal basis. That is, higher prices may be used in the peak air conditioning season so that loads will be "clipped" in the summer and so that investment in new capacity can be decreased.

With such advantages, why are power companies reluctant to go to peak load pricing? The fault probably lies in the nature of the beast.

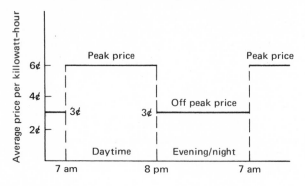

Figure 25–11   *Peak-load pricing: A simplified example.*

Utilities are regulated by governments because they are the sole suppliers of electric power for the regions they serve. If they were not regulated in the prices they charge, they could charge very high prices without losing customers because they are the only suppliers in the region. Thus, Public Service Commissions limit the prices the utilities can charge. The commissions allow the utilities to charge prices high enough to earn a "fair rate of return" on the investments made in generating and transmission facilities.

By the rate of return we mean the ratio of profit to invested capital. If you put $100 in a savings bank, the $5 or $6 earned each year would be the return on the $100 investment. Your rate of return is 5% or 6%. Since utilities are generally allowed to earn about 12% on their invested capital, they have an incentive to spend on their facilities as much money as is reasonably possible. This is one reason that nuclear plants have probably been so popular with utilities; the greater cost to build a nuclear power plant means that the fair rate of return applies to a bigger investment base. Since peak load pricing would decrease the number of new power plants that are needed, the utility will make a smaller investment. This means, in turn, fewer dollars of income for the utility, since the rate of return will apply to a smaller investment base. Thus, utilities have been reluctant to use peak load pricing.

Peak load pricing, however, does more than benefit the consumer's pocketbook; by decreasing the number of power plants required, the area of land as well as the water flows that must be committed to these enterprises are decreased. Hence the earth is spared unnecessary disruption and despoiling. Fewer transmission lines will crisscross the landscape, and the number of new pollution sources will decrease. Furthermore, the massive energy commitments required to build power plants are delayed. Since peak loads are clipped off, less oil and natural gas are used to generate the inefficient turbine engine power used at such times. Peak loak pricing is a simple but powerful concept that can benefit the environment considerably.

Questions

1. Compare heating water for the home by gas, electricity, and solar energy in terms of: initial cost of the equipment, cost to run the system, efficiency of energy use. Which system would you choose if you expect to live in your home three more years? Ten more years? The rest of your life?

2. What is (are) the purpose(s) of "daylight savings time"?

3. How is it that people in some European countries manage to travel almost as many miles by private auto as we do in the U.S., but they use less than 60% as much gasoline?

4. How can individuals, all by themselves, reduce the amount of gasoline used?

5. What energy savings exist in mass transit?

6. If you do not ride a bus to work or school, what would it take to get you to do so? If you do ride one, why do you choose this method of transportation?

7. Recycling is another method of energy conservation. Using references from your library (see references listed at the end of this chapter, for a start) answer one of the following:

   a. Calculate the savings when aluminum beer and soft drink cans are recycled instead of made from virgin ore.

   b. What uses are there for recycled paper? What conditions have to be met before paper can be recycled? Is the recycling of paper economically justifiable in present day society?

   c. What uses are there for recycled glass? Have glass-recycling projects been successful?

   d. Does your local sanitation system involve any trash separation and recycling? What possibilities exist for public or commercial recycling of solid wastes?

   e. What savings exist in the area of automobile salvage and recycling?

Further Reading

The subject of Energy Conservation has received enormous attention since the oil embargo, as the following list of articles from *Science* and *Technology Review* demonstrate. Articles listed as appearing under Research News in *Science* (marked with an *) are unlikely to be very technical (in the sense of having equations and analysis). Generally, feature articles in *Science* will be readable even if they do have a mathematical component.

Berg, C. A., "Conservation in Industry," *Science* **184**, 264–270 (19 April 1974).

*Boffey, P. M., "How the Swedes Live Well While Consuming Less Energy," *Science* **196**, 856 (20 May 1977).

Hafele, W., "Energy Choices that Europe Faces: A European View of Energy," *Science* **184**, 360–367 (19 April 1974).

*Hammond, A. L., "Conservation of Energy: The Potential for More Efficient Use," *Science* **178**, 1079–1081 (8 December 1972).

*Hammond, A. L., "Energy: Ford Foundation Study Urges Action on Conservation," *Science* **186**, 426–428 (1 November 1974).

*Hammond, A. L., "Energy Needs: Projected Demands and How to Reduce Them," *Science* **178**, 1186–1188 (15 December 1972).

Hirst, E. and J. C. Moyers, "Efficiency of Energy Use in the United States," *Science* **179**, 1299–1304 (30 March 1973).

Hirst, E., "Transportation Energy Conservation Policies," *Science* **192**, 15–20 (2 April 1976).

Landsberg, H. H., "Low-Cost, Abundant Energy: Paradise Lost?," *Science* **184**, 247–253 (19 April 1974).

Mazur, A. and E. Rosa, "Energy and Life-Style," *Science* **186**, 607–610 (15 November 1974).

Pullen, J. J., "Energy from Architecture," *Country Journal*, pp. 76–87 (April 1977).

Ross, M. H. and R. H. Williams, "The Potential for Fuel Conservation," *Technology Review*, 49–57 (February 1977).

Schipper, L. and A. J. Lichtenberg, "Efficient Energy Use and Well Being: The Swedish Example," *Science* **194**, 1001–1013 (3 December 1976).

Small, J., P. Achenbach, and S. Petersen, "Energy Conservation in New Housing Design," *Science* **192**, 1305 (25 June 1976).

Starr, C., "Is Sweden More Energy-Efficient?" *Science* **196**, 121–124 (8 April 1977).

For an understandable and directly useful discussion of methods you can apply to save energy, we recommend:

"Tips for Energy Savers," U.S. Department of Energy, Asst. Secr. for Conservation and Solar Applications, DOE/CS-0020, March 1978.

Automobile Facts and Figures

Motor Vehicle Manufacturer's Association publication: "Motor Vehicles, Facts and Figures, 1979." This book is reissued each year and is available from MVMA.

(U.S. Department of Energy)

Part 6

Toxic Materials, Cancer, and the Environment

A startling and disturbing fact, appearing from the mass of scientific data on cancers, is that 60% to 90% of human cancers are probably caused by environmental factors. "Environmental" is used, in this case, in the broadest sense. The term covers such things as eating and smoking habits as well as exposure to possible carcinogens (something that causes cancer) in air, water, and food.

About 400,000 Americans die each year from cancer and, at the same time 700,000 new cases are found (Fig. A). A large portion of these cancers must be attributed to smoking, and another portion to working in certain industries. There is also evidence that the general public may be endangered by

toxic substances which are widely distributed in the environment (Fig. B). In Parts 2, 3, and 4, we discussed a number of water and air pollutants, emphasizing where the various air and water pollutants come from and why they are found in our environment.

Now we take a closer look at certain toxic substances and factors in the environment. Of particular concern here are those factors or substances suspected of contributing to, or causing, cancer deaths.

In the first chapter, we discuss some of the data that has focused attention on the search for carcinogens in the environment. One special fea-

Figure A *U.S. deaths from various causes. (Each figure represents 10,000 deaths.) [Source: "The Implications of Cancer-Causing Substances in Mississippi River Water," The Environmental Defense Fund, 1525 18th Street, N.W., Washington, D.C. 20036, (1974), p. 15.]*

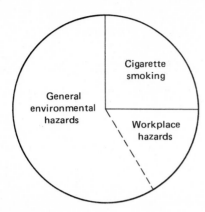

Figure B *Environmentally caused cancer deaths. Approximately 25% of cancer deaths are due to smoking. Another large (but we don't know how large) portion is due to exposure to carcinogenic hazards where people work, e.g. radiation caused cancers among nuclear workers, chemically induced cancers in workers in the petroleum and chemical industries, and asbestos caused cancers in building trades workers. The third category includes hazards to which the general public is subject, e.g. organic chemicals in drinking water, carcinogenic drugs, cosmetics and food additives, air pollutants; or hazards to which members of the public subject themselves, e.g. sunbathing and high fat diets.*

ture of much of this information is that it comes from the study of diseases (such as cancer) in groups of people, or populations. This is the science of epidemiology.

In the following chapter, air and water pollutants suspected of being major contributors to the burden of carcinogens are discussed. Individuals have little control over their exposure to some of these pollutants, for instance arsenic, asbestos, chlorinated hydrocarbons, or radioactive particles in air. In other instances, for example, smoking and x-ray exposure, the individual has complete control. Occupational exposure to carcinogens must fall into a third category, since people usually have some choice about what jobs they take, but not much control over the work environment once that choice is made.

In the next chapter there is a general description of the kinds of toxic materials found in foods, with special emphasis on those suspected of causing cancer. Again, there are two types of problems. There are those over which people have some individual control, for instance the consumption of drugs and alcohol or dietary habits in general. And there are those problems over which the individual can exercise little control, as the contamination of foods with possibly carcinogenic pesticides.

The last chapter addresses the problem of protecting people from toxic substances in the environment. There has been relatively little attempt at government control of some of the voluntary factors that may increase the occurrence of cancer, such as choice of diet or smoking habits. However, many laws have been passed in an attempt to control toxic substances in air, water, and foods. These laws are the focus of the concluding chapter.

Chapter 26

Tracking Down Carcinogens in the Environment

THE EPIDEMIOLOGIST AND CANCER

In 1977, the U.S. Public Health Service published an atlas of cancer mortality for the 20-year period 1950 through 1969. The atlas is made up of a series of maps of the United States, on which the rates of death from various forms of cancer are shown, county by county. Some striking variations can be found in the occurrence of the various forms of cancer. For instance in the Northeast (New Jersey, southern New York, Connecticut, Rhode Island, and Massachusetts) there is a very high rate of mouth, throat, esophageal, laryngeal, and bladder cancers. The rates are high for males but not for females in this area, which suggests that the cancers may be related to the jobs at which the men work. And, in fact, the area is highly industrialized, with a large concentration of chemical industries. Several manufactured chemicals (2-napthylamine is one) are known to cause bladder cancer. In Salem County, New Jersey, one-fourth of the work force is employed in the chemical industry or related industries. The rate for bladder cancer in Salem County is the highest for any U.S. county with a reasonably large (10,000 or over) population. On the other hand, colon and rectal cancer is high for both males and females in this area. Although working in the chemical industries may eventually be shown to increase a person's chance of developing bladder cancer, some other explanation must be sought for the high incidence of colon cancer.

Lung cancer cases are concentrated along the coast of the Gulf of Mexico and along the Atlantic coast from northern Florida to South Carolina. This pattern is interesting because it does not match the pattern of highly industrialized urban areas where air pollution and cigarette smoking are most common. That there are concentrations of various cancer types is clear, however. Thus there is a great deal of support for the idea that these cancers are caused by some condition or some combination of conditions in the environment of the different areas. This method, of matching diseases in populations with factors that might have caused them is part of the science of epidemiology. Such figures as those presented in the atlas cannot prove what causes a cancer, but they can provide clues as to where to look further. For instance, on the Baltimore census, there is an area where the cancer mortality rate is 4½ times that of the city as a whole. The tract surrounds a former Allied Chemical plant which manufactured arsenic compounds for 100 years. Arsenic is a known carcinogen and former plant employees have a lung cancer rate 14 times higher than residents in other areas of the city. When investigators tried to see whether emissions from the plant could have been responsible for the high neighborhood rate of lung cancer as well, they were at first puzzled by the pattern of where cancer victims lived with respect to the plant. Most victims seemed to live east and north of the plant, while prevailing winds blow north and northwest. If

536

plant emissions contributed to the occurrence of lung cancer, one might have expected the pattern of cancers to follow the wind direction, the way the pollutants would have blown. Eventually, former plant employees told researchers that there was once a railroad spur that ran to the plant from the north. Arsenic compounds were shipped to the plant along this line. If the cars were not sealed after unloading at the plant, arsenic could have blown out and settled along the rail tracks. In fact, a high concentration of arsenic was found in the soil along the old rail bed, decreasing away from the plant.

Actually some 150 to 200 chemicals were manufactured at the plant at one time or another. Thus, it may never be possible to know for sure what specific substances caused the high cancer rate at the plant and in the neighborhood. This case points out another problem involved in determining the causes of cancers: a long time usually goes by between exposure to a carcinogen and the diagnosis of cancer. (See p. 538 for more details.) For instance, it is now known that lung cancer caused by breathing asbestos dust does not appear for 20 years or more after exposure to the dust. In one study, an average of 39 years elapsed between the time workers breathed the asbestos and the time they developed lung cancer (Fig. 26–1). Because there is such a long time period between exposure to a carcinogen and the proof that harm has been done, many scientists and legislators feel that we must have some control over substances released into the environment. Otherwise we may find out in 20 or 40 years, when it is too late, that some commonly used chemical is a carcinogen.

HOW POTENT ARE VARIOUS CARCINOGENS?

We know that, in general, some substances are able to cause cancer at lower concentrations than others. However, we do not yet have the information that would make it possible to state the relative potency of most carcinogens for humans. In part, this is due to the fact that most of the information comes from animal experiments and we do not know if it can be directly applied to humans. The effects of carcinogenic materials on humans have been studied well enough in only six cases[1] to rank their relative strengths. In these six cases, the carcinogenic effect in humans is roughly comparable to that obtained in rat experiments. It is on this basis that results in rats are often applied to humans (Fig. 26–2). Animal data, however, may not be a valid basis to use in assigning human potency to carcinogens, because carcinogenic effects differ among animal species. To give two examples: aflatoxin is not carcinogenic in adult mice, although it is in rats; 2-napthylamine is not a rat carcinogen, although it is carcinogenic to humans.

In 1973, an argument arose over chemicals in hair dyes, which some researchers felt were shown to be carcinogenic in a test called the Ames Assay, which uses bacteria. Hair dye manufacturers disagreed vigorously. The various tests for carcinogens are described on page 539. It appears that, at the moment, carcinogens in terms of their effects on humans can only be classed as very potent or very weak or something in between.

Another problem is that some materials which are not carcinogenic themselves seem to cause cancer in combination with other substances. In addition there is evidence that small amounts of carcinogens add together to increase the total carcinogenic effect.

Of the almost 2 million chemicals known, only about 6000 have ever been tested to determine whether they were carcinogenic. Of this 6000, 1000 have given some evidence that they do cause cancer. Of the 1000, a few hundred are considered proven carcinogens.

---

1 Benzidine, chlornaphazin, diethylstibesterol (DES), aflatoxin B_1, vinyl chloride, and cigarette smoke.

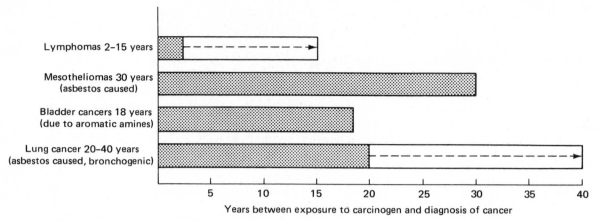

Figure 26-1 *Latency period of some cancers. There is a relatively long time period between the time a person is exposed to a carcinogenic substance and the time when the cancer it caused can be diagnosed.*

Carcinogenesis

A great deal is not yet understood about how a carcinogen affects cells. However, there are generally felt to be three stages involved. The first is called initiation. The carcinogen causes some permanent change in a cell. The most likely site for this change is the cell's DNA. Second is a stage called preneoplasia. This stage is long, often lasting 20 years or more in humans. Little is known about what happens during this long, apparently quiet period, but there is evidence that some cells are repaired, or revert to normal during preneoplasia. Vitamin A may play a role in repairing damage to cells, as do a variety of enzymes in cells. Cells that are not repaired go on to stage three, or transformation. Now they show the characteristics of tumor cells and begin to proliferate. Much research is being directed toward finding substances that can help cells repair themselves during preneoplasia in order to undo the effects of exposure to carcinogens.

A further complication is the existence of chemicals called promoters. These are substances that do not cause cancer by themselves, but they can cause cancer when combined with very low doses of known carcinogens—doses that are otherwise too small to cause cancers. The artificial sweeteners saccharin and sodium cyclamate are both examples of compounds that are either very weak carcinogens or not carcinogens at all, but promoters. In animal experiments, sulfur dioxide in combination with benzo(a)pyrene causes cancer while neither of the two chemicals alone does so.

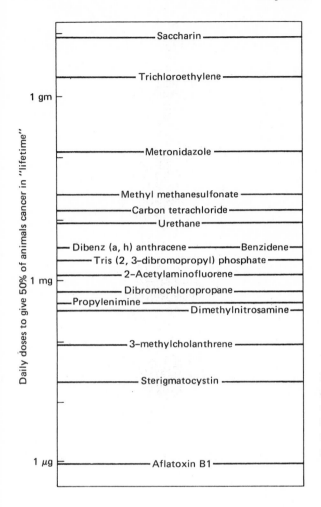

Figure 26-2 *Relative strength of carcinogens in rats and mice. The doses shown are those which cause cancer in half of a group of test rats or mice when they receive it daily over their lifetime. As you can see, aflatoxin is over a million times more potent than saccharin. [Source: T.H. Maugh II, Science* **202**, *38 (6 October 1978) ©1978 by the American Association for the Advancement of Science.]*

Testing for Carcinogens

Perhaps the most difficult problem to solve concerning the regulation of carcinogens is how to identify carcinogenic substances before they enter the environment.

Although we are attempting to protect humans, we cannot experiment on humans to obtain the answers directly (except perhaps when such information surfaces as a result of using chemicals to treat illnesses. Certain drugs have been found to be human carcinogens.) Because human cancers often take 20 or

30 years to develop, even methods which involve comparing groups of people who were unintentionally exposed to chemicals (pesticides workers, industrial employees, construction workers) and groups of people who were not exposed, are of limited value. Information can be gained this way, but it comes too late.

Experiments on animals such as monkeys or dogs, while they provide useful results, are extremely expensive and also take years to produce useful results. Such experiments may be justified in cases where many people are being exposed to suspected carcinogens. Some of the information we have on smoking and lung cancer or chemicals and bladder cancer comes from this type of research.

However, there are 9000 synthetic chemicals manufactured on a large scale today. Some 500 to 1000 new ones are introduced each year. What is needed is a quick, cheap way of screening chemicals to see which ones are or might be carcinogenic before they are released into the environment. The accepted method at the moment is to use two species of rodents—rats and mice or hamsters. The animals are given the test chemical in the highest doses possible without causing them to become sick or die (from some cause other than cancer). After a few years, the animals are examined to see whether more than the normal number of cancers can be found in the treated group of animals. Very large numbers of animals may be needed to detect a cancer-causing substance. For instance, suppose a chemical caused cancer in 1 out of 10,000 humans exposed to it. If the entire U.S. population was exposed, 20,000 people would develop cancer. To detect this hazard in an animal experiment, 10,000 rats would be needed to find one cancer. At least 30,000 would be needed to call the results significant. Further, humans may be more sensitive to some chemicals than rodents. For instance, humans are 60 times more sensitive to the effects of thalidomide than mice and 100 times more sensitive than rats. Nonetheless, this type of test remains the most accurate and quantitative available today.

Several new methods have been suggested to provide a faster, cheaper way to screen chemicals for carcinogenicity than using conventional animal test systems. One group of methods uses animal cells grown in test tubes. Changes in cells exposed to chemicals can be correlated to the development of cancers. Results are promising, but much work remains to be done on this group of methods. Ideally the animal cells used should be human cells but this is not yet possible.

Another method called the Ames Assay, which is already in use, involves adding the chemical to cultures of the bacterium Salmonella typhimurium (Fig. 26–3). The bacteria are then examined to see if the chemical has caused a mutation. Mutations are heritable changes in the composition or arrangement of genes, which are composed of DNA. In this particular system, researchers look to see if bacteria that require the nutrient histidine to grow have changed, or mutated, so they can grow without it. (This is called a back-mutation.) Although the system detects not carcinogens, but mutagens, it has been found

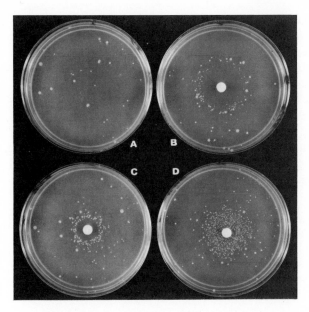

Figure 26-3 *The "spot test" for mutagen-induced revertants. Each petri plate contains, in a thin overlay of top agar, the tester strain TA98 and, in the cases of plates C and D, a liver microsomal activation system (S-9 Mix). (Plate B did not require the liver system.) Mutagens were applied to 6-mm filter-paper discs, which were then placed in the center of each plate: (A) control plate: spontaneous revertants; (B) plate showing revertant colonies produced by the Japanese food additive furylfuramide (AF-2) (1 μg); (C) by the mold carcinogen aflatoxin B₁ (1 μg); (D) by 2-aminofluorene (10 μg). Mutagen-induced revertants appear as a circle of revertant colonies around each disc. [From Bruce N. Ames, Joyce McCann, and Edith Yamasaki, "Methods for Detecting Carcinogens and Mutagens with the* Salmonella/ Mammalian-Microsome Mutagenicity Test," *Mutation Research* **31**: *347-364 (1975). Reprinted by permission. Copyright ©1975 by the Elsevier Scientific Publishing Company, Amsterdam. Permission: Elsevier/North Holland Biomedical Press.]*

that there is a good correlation between the two. That is to say, 90% of the carcinogens tested have been found to be mutagens. Furthermore, all of the known human carcinogens give a positive result in the Ames Assay. Of the noncarcinogenic substances tested, 87% were negative in the Assay while 13% gave a false positive result (Fig. 26-4). It is possible that these percentages may improve even more as the method is refined. For instance, it has been shown that certain substances undergo chemical changes in the body, which turn them into carcinogens. Thus benzo(a)pyrene is a carcinogen after conversion to an active form by microsomes. If benzo(a)pyrene is used in the Ames Assay, it is not mutagenic. If, however, a preparation of liver microsomal enzymes is also added to the test system, mutations occur. Further refinements such as this

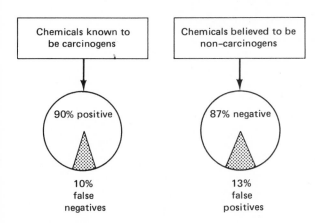

Figure 26-4 *Results of the Ames Assay. When chemicals which have already been tested for their ability to cause cancer in animals were tested in the Ames assay for their ability to cause mutations in bacteria, these results were found. Ninety percent of the known carcinogens caused mutations in bacteria. It is possible that 100% of carcinogens are actually mutagens. The other 10% may first be transformed in the body to some other substance which is the ultimate carcinogen. Of those materials believed to be noncarcinogenic, 87% did not cause mutations. The 13% which did cause mutations are undergoing further animal tests to see if some of them are really carcinogens which were missed in previous tests.*

will undoubtedly increase the accuracy of the Ames Assay and of the other "quick tests," such as those which use cell cultures. The Ames Assay is quick (about three days to run a test) and inexpensive (about $200 per chemical compared to $100,000 per chemical in conventional rat and mouse assays). It is the possibility of false negatives, however, that limits the usefulness of the test. That is, up to 10% of the compounds that are carcinogenic may not be mutagenic in the Ames Assay. Thus, no chemical could be given a clean bill of health on the basis of the Ames Assay alone.

Probably some combination of testing methods will be worked out to ensure that we catch all of the potential carcinogens before they can be distributed in the environment. One such system is shown in Figure 26–5. The

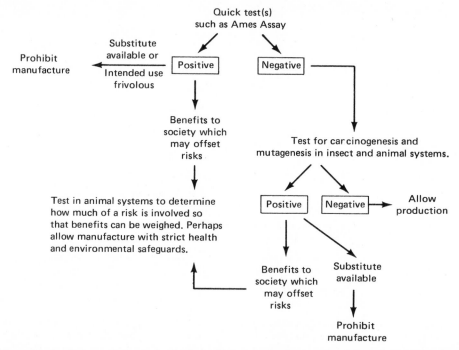

Figure 26–5 *One possible method of combining animal assays and "quick tests" is shown here (adapted from B.A. Bridges,* Environmental Health Perspectives, No. 6, 1973). *All chemicals would be subjected first to the cheap and rapid "quick tests." Chemicals that proved to be mutagenic would be presumed to be hazardous and either rejected out of hand, or if they were felt to be important enough to be worth some risk, they would be tested further in animal and insect assays to determine just how great the risk would be. This would allow regulatory agencies to determine whether the benefits outweighed the risks and what production and distribution safeguards were necessary. Compounds that were not mutagenic in quick tests would be tested further in animal and insect systems. If they passed these tests, manufacture could be approved. If they proved to be carcinogenic or mutagenic in these assays, further detailed tests could be done, again, to allow a risk versus benefit decision.*

Ames test and those like it are certain to be important tools, however, if we are to meet the goal of the Toxic Substances Control Act of 1976 to determine the safety of all new chemicals <u>before</u> they go on the market.

Questions

1. Explain why epidemiological studies, such as the study of the occurrence of cancer according to locality, cannot prove that certain substances or occupations lead to cancer.
2. Why is it a problem that a long time goes by between exposure to a carcinogen and the clinical diagnoses of cancer?
3. Why do we have to use animal studies to determine whether substances are carcinogenic? What problems does this cause?
4. What is the Ames Assay?

Further Reading

Ames, B. A., "Identifying Environmental Chemicals Causing Mutations and Cancer," *Science* **204**, 587 (11 May 1979).

Carter, L. J., "Chemical Carcinogens: Industry Adopts Controversial 'Quick' Tests," *Science* **192**, 1215 (18 June 1976).

Smith, A. M., "Searching for Carcinogens: Does the Ames Test Work?," *New Engineer*, p. 25 (April 1977).

These articles detail some of the controversy about tests designed to determine whether substances are carcinogenic or not.

Carter, L. J., "Cancer and the Environment (I)," *Science* **186**, 239 (18 October 1974).

Gillette, R., "Cancer and the Environment (II)," *Science* **186**, 242 (18 October 1974).

Winklehaus, C., "EPA's New Toxics Control Strategy," *Journal of the Water Pollution Control Federation*, p. 2352 (December 1977).

These three articles provide an interesting historical perspective on controlling carcinogens in the environment. The first two papers detail cases in which stronger laws are shown to be necessary to control toxic substances. The third paper explains the results of a lawsuit by environmental groups to force a government agency to make stricter policies in regard to toxic substances.

Chapter 27

Carcinogens in Water and in Air

TOXIC ORGANICS AND THE NEW ORLEANS WATER SUPPLY

Water Treatment Causes a Problem

In 1967, the Environmental Protection Agency began to study why drinking water drawn by the city of New Orleans from the lower Mississippi River smelled and tasted "fishy" or "oily." Although the results of the study were not encouraging (many organic chemicals were found in trace amounts in the water), the source of the problem seemed fairly clear. A number of petroleum-chemical and coal-tar products industries discharge wastes into the Mississippi. Thus, it was no surprise that the 1972 report on the problem, issued by the EPA, stated that chemicals in the wastes from these industries were responsible for the tastes and odors in New Orleans drinking water.

The report also noted, however, that chlorination of drinking water might be adding to the problem. This is because chlorine, in the form used to disinfect drinking water, can react with natural or pollution-related organic materials to form chlorinated hydrocarbons, some of which are carcinogenic.

Since that report was issued, this last aspect of the problem has caused the greatest controversy. Chlorinated hydrocarbons and other organic compounds have now been found in many water sup-plies. In one survey of 129 municipal drinking water plants that chlorinate water, chlorinated organic compounds at levels of a few parts per billion were found in the finished water of almost every plant. Chemicals found included chloroform and carbon tetrachloride, both known carcinogens. (In the experiments that showed these chemicals to be carcinogenic, however, the doses of chloroform and carbon tetrachloride were much higher than humans could get by drinking New Orleans water. The problems involved in deciding whether the low levels in water are hazardous are explained on p. 549.)

In 1974, the Environmental Defense Fund published a report noting that the number of cancer deaths in New Orleans was higher than expected. The report suggested this was due to carcinogens, both chlorinated and nonchlorinated organics, in the drinking water.

A Possible Solution

No one is recommending that we stop chlorinating drinking water. The immediate risk of incurring epidemics of water-borne bacterial diseases such as typhoid fever and viral diseases such as polio is too great. Instead, the focus has been on removing from drinking water organic compounds that come from industrial wastes, pesticides from farmland, urban wastes, and even natural sources. Many of these compounds are known to be toxic or carcinogenic. In addition, water treatment methods have been changed to reduce the chlorinated organics formed.

The EPA is requiring water suppliers to reduce the level of the group of chemicals called trihalomethanes, which include chloroform. This reduction can be accomplished in several ways. (1) The source of raw water can be changed to a less polluted one. This would decrease the amount of organic material able to be chlorinated to trihalomethanes. (2) Adjustments can be made in the water treatment scheme, such as changing the point at which chlorine is added. (3) Other methods of disinfection, such as use of ozone (p. 171), can be tried.

In addition, because there are too many different possible organic contaminants to set standards for each one, it was decided to require the use of a technique (granulated active carbon) that would remove most of them (Fig. 27-1).

Cities that can show, by sampling and monitoring, that they draw water from unpolluted sources (unpolluted deep wells, for instance) do not have to treat water for toxic organics. Initially the regulations were written to apply only to large water supplies (those servicing over 75,000 people). Smaller water supplies are being phased in more slowly as administrative, technical, and economic problems are worked out. The new regulations have met a good deal of opposition from the people who are supposed to put them into effect.

The other way of attacking the problem is, of course, to reduce the amount of toxic organic material added to natural waters in the first place. This is part of the focus of toxic substances control efforts, discussed on p. 613.

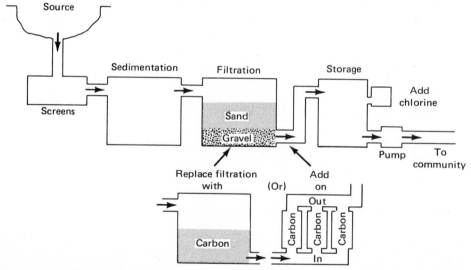

Figure 27-1 *The granulated activated carbon filter. This diagram of a typical water treatment scheme shows where a GAC filter is added on in order to remove organic pollutants.*

The Epidemiologist As A Detective

Although no one argues with the fact that carcinogenic materials were found in New Orleans drinking water, it is not an easy matter to decide how

much of a hazard this represents. According to the 1974 EPA study, various organic chemicals were present in the parts per billion range. In fact, except for a few chemicals,[1] most were present at less than 1 part per billion. We do not know the effects of these very low concentrations of toxic chemicals, day after day, over a person's whole lifetime.[2]

Another approach is to look at statistics on cancer deaths in New Orleans and compare them with statistics on people drinking less polluted water. This was done by scientists from the Environmental Defense Fund along with researchers at Case Western Reserve University and Resources for the Future. Ten parishes in Louisiana and one-third of another draw their drinking water primarily from the Mississippi, the rest use less polluted ground water or other surface water supplies. When these two populations are compared, it can be shown that there are more cancer deaths among those drinking Mississippi River water. Unfortunately, this is not proof that the water is causing excess cancer deaths. Some other factor might be the true cause. The association with drinking water might be chance or associated in some way with the true cause (for instance, some food in the diet of people living in the affected area could be the cause). Researchers thus checked a variety of other possibilities.

Cancer deaths were divided into groups by race and sex. The deaths were still found to be associated with the source of drinking water for white and non-white males and for non-white females. Dividing people into groups by income made no difference, nor did dividing them by occupation. Cancer deaths were not correlated with living in southern versus northern parishes. The southern parishes are different socioeconomically from the northern parishes and thus might be expected to have differences in diet, etc. Even elevation was checked, since air pollution could be expected to be worse at low elevations than at high ones. No correlation was found, however.

Cancer rates in the city itself might be falsely high if people have moved there for better medical care after the disease is diagnosed. However, subtracting the New Orleans data from that of all other Louisiana people drinking Mississippi River water did not change the positive relationship with cancer deaths.

The epidemiologist must be even more of a detective, however. Why are cancer deaths not correlated with the source of drinking water for white females? What about the role of smoking or alcohol consumption? Can you think of other factors to check?

Studies such as this one cannot prove a link between drinking water and cancer. However, they provide us with fascinating and tantalizing clues as to the causes of environmental cancers.

---

1 Chloroform was found at 113 ppb, 4 others were found at 1–8 ppb.

2 By the way of comparison, in one experiment mice were given 0.0001 milliliters of carbon tetrachloride for each gram of body weight every fourth day for a total of 30 times. Twelve out of 15 mice developed cancers. The others died from other causes.

HOW MUCH DO WE WANT TO SPEND FOR CANCER PROTECTION?

... This translates into 62.8 excess deaths from cancer per year.

National Academy of Sciences

The individual would die of drowning before he got a tinge of neoplasia [tumor transformation].

Herbert Stokinger

I am not convinced it [special water treatment] is worth it at this moment.

Oscar Adams,
Director of Engineering for
the Virginia Health Department *

Who has informed public utilities that average householders are not willing to pay an additional $7 per year for clean water?

D. J. Baumgartner,
EPA Corvallis Environmental Research Laboratory †

Perhaps nothing illustrates the difficulty of setting standards for trace pollutants in drinking water as well as the controversy over chloroform. This chemical is in some cases an industrial pollutant, but by far the greatest amount of chloroform comes from the chlorination of drinking water. Scientists, however, are not in agreement over whether the amount of chloroform in drinking water poses a threat to health.

---

* O. Adams "U.S. Cities Face Tapwater Dispute," Baltimore Evening Sun, p. A-1 (27 December 1977).

† D. J., Baumgartner, "Drinking Water: Sources and Treatment" Science **197,** 324 (22 July 1977).

Recent studies show that 95% to 100% of chlorinated water supplies contain chloroform at levels of 20 micrograms per liter to 311 micrograms per liter. Several studies have shown that chloroform (at doses much higher than are found in water supplies) causes cancers in animals such as rats and mice. The National Academy of Sciences, using these experimental results, calculated that:

> at a concentration of 10 ppb (10 micrograms per liter) during a lifetime of exposure this compound would be expected to produce one excess case of cancer for every 50,000 persons exposed. If the population of the United States is taken to be 220 million people, this translates into 4,400 excess lifetime deaths from cancer, or 62.8 per year.

> In view of (this potential in humans), and taking the risk estimates into account, it is suggested that very strict criteria be applied when limits for chloroform in drinking water are established. [‡]

Not everyone agrees that you can use this type of animal data to calculate human risks, however. Herbert Stokinger, Chief of Experimental Toxicology at the National Institute for Occupational Safety and Health says:

> Why is EPA concerned about the presence of such minute amounts of a substance shown to be carcinogenic only in animals and at levels that would correspond to a daily intake of $CHCL_3$ of from 17,000 to 34,000 mg/day?...

> Extrapolating animal data to man for the purpose of estimating the incidence of cancer in human populations can prove a fallacy. Biologic response to a toxic agent is not the same at low and high dosages—an assumption made by data extrapolators. The body has built-in antagonists that can counteract the toxicity of foreign substances, including potential carcinogens. Natural and dietary sources of anticarcinogens also exist. [§]

The Environmental Protection Agency, directed by Congress to set standards that will protect the American public has decided to err on the side of caution. It is requiring the removal of chloroform from drinking water by activated carbon filters.

(Controversy continued on next page.)

[‡] Drinking Water and Health, National Academy of Sciences, Washington, D.C., 1977.

[§] H. Stokinger, "Toxicology and Drinking Water Contaminants," Journal of The American Water Works Association, p. 399, July 1977.

For large supplies installing Granulated-Active-Carbon (GAC), we estimate the average annual cost for a family of three to be in the $6 to $10 range.... This is about the cost of one night at the movies for a family that size. ||

Although the EPA held hearings at which interested people could state their views, the public was not asked directly, for instance by a ballot referendum, whether they were in fact willing to pay extra for the added protection. In contrast, when fluoridation of drinking water (to prevent tooth decay) was an issue, in the 1960s, and early 1970s, the Safe Drinking Water Act was not yet in effect. At that time, the EPA did not set drinking water standards. Each state set its own standards. The fluoridation issue was decided, in general, by a separate vote in each community. For this reason, we now have a patchwork of fluoridation practices. Some communities add fluoride and others are still vigorously opposed to the idea.

Oscar Adams, Director of Engineering for the Virginia State Health Department was quoted in the Baltimore Sun, 27 December 1977:

> I think most cities are going to be put into a real hard position in that to meet the standards is going to require considerable capital costs....
>
> If I felt this were a true public health problem, I would be for it. But I have trouble equating this danger against all the other dangers in our environment that might cause cancer.... I am not convinced it is worth it at this moment.
>
> The real kicker is that it would mean a raise in water rates of 27 percent a year...or $3.91 per person....

Is there justification for setting standards, and, as in this case, costing the consumer money, when there is no clear cut demonstration of risk? Do you feel that the protection you gain from toxic organics is worth a raise in your water bill? Although the cost is not high, the jump in the average water bill, for a system that installs GAC, will be noticeable. (Compare it to your yearly water bill.) Does the cost make a difference in your attitude? Suppose it was a few cents a year, a few hundred dollars?

Do you feel you should be more directly consulted about matters relating to your pocketbook and the protection of your health? Would you rather have the EPA decide them?

What do you suppose the cost of 62.8 excess cancer deaths per year might be to society as a whole? Compare this to the cost to the U.S. population as a whole each year if everyone was paying for a GAC System. Is this comparison valid?

|| Douglas M. Castle, EPA Agency Administrator, quoted in Environmental News, 25 January 1978, USEPA, Washington, D.C., 20460.

ASBESTOS

Diseases Caused by Breathing Asbestos

A great deal of time and effort were spent before people discovered that the microscopic creatures we call "germs" cause diseases like pneumonia or scarlet fever. Imagine how hard it would be to find the reason for a sickness if the symptoms did not show up until 20 or 40 years after a person were exposed to the causative agent. This is the case, however, with diseases brought on by breathing asbestos.

Asbestosis is a disease in which breathing is made difficult by the presence of asbestos fibers in the lungs. The tissue around the fibers becomes tough; oxygen cannot be transferred to the blood by such tissues. Asbestosis usually shows up 20 years or more after a person starts working with asbestos.

Breathing asbestos dust can also cause cancer both of the lungs and of the membranes covering the lungs. It seems that even a short exposure to asbestos can cause cancer 20 to 40 years later. In one case, a woman went to school for a short while near an asbestos field in South Africa. She and her classmates used to slide down piles of asbestos wastes on the way home from school. After her family moved away, she was never exposed to asbestos again. Fifty years later she died from a rare type of lung cancer (mesothelioma) caused by the dust she breathed when she was five years old. Men who have worked in shipyards are at an increased risk of developing mesothelioma. This is true even if they only worked there for periods as short as a few weeks and even if they did not directly handle the asbestos insulation used in ships. We do not yet know how little asbestos it takes to cause lung cancer.

Uses of Asbestos

The mineral nature of asbestos ensures that it does not burn. For this reason, asbestos is useful in fire-proofing materials. It is used in firemen's clothing, in heatproof gloves and mats, for oven linings, and for furnace ducts. Asbestos is also included in ceiling and floor tiles and in roofing tar and cement. In some products, the asbestos is firmly bound so that it is unlikely to contaminate the environment. Floor tiles containing asbestos, for instance, fall into this category. In other cases, asbestos is used in ways that permit its release into air and water (Figs. 27–2 and 27–3). In 1979, some brands of hairdryers were found to be lined with asbestos. The asbestos fibers could blow out with the hot air and contaminate users' homes.

In some cases asbestos is found where it is not meant to be. For instance, powdered talc, which is used in a variety of products such as talcum powder, can contain as much as 80% asbestos.

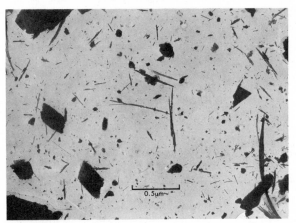

Figure 27–2   *Asbestos fibers are found in several products used by the home handyman. The above photograph, taken with an electron microscope, shows asbestos fibers in a consumer spackling product. Asbestos can contaminate the air when the spackling is mixed or when the dried spackling is sanded. [Photograph from A. N. Rohl et al.,* Science **189**, *551 (15 August 1975) ©1975 by the American Association for the Advancement of Science.]*

ASBESTOS

When the pollution-oriented health administrators and the public alike begin to focus clearly on the enormity of the bill that would be required to reduce pollution to meet unnecessarily severe standards....then will come the day of reckoning....

H. E. Stokinger, Chief
Laboratory of Toxicology and Pathology,
National Institute for Occupational Safety and Health

But we paid dear and we will pay dear.

Billie Walker,
widow of asbestos worker

The price of environmental quality is a recurring question. How much are we willing to pay to improve our environment?
H.E. Stokinger presents one view:

When the pollution-oriented health administrators and the public alike begin to focus clearly on the enormity of the bill that would be required to reduce pollution to meet unnecessarily severe standards...precipitously prepared from undigested, dubiously related facts...on which the public has been ill-advised or misled...then will come the day of reckoning and rude awakening to the folly of past antipollution actions. Already industry has felt the bite; shortly, the public will. Hardest hit are the mineral and chemical industries. On top of multimillion-dollar outlays for air pollution control, and sums of similar magnitude for water, are multibillion-dollar legal suits that stagger the imagination, cripple large industry, and eliminate small industries. Two consequences of profound economic importance are the increased price of basic chemicals and the loss of employment. Already a number of small manufacturing plants have been forced to close, unable to bear the burden of meeting pollution standards. Heavy industry, unable to survive on repeated annual financial losses or to continue on less than a 4 to 6 percent profit margin, will ultimately pass the needless charge on to the consumer.

It thus should be evident that such actions, with their unbearable consequences, should only be taken when it is clear beyond a shadow of scientific doubt that human health is in imminent danger. *

Until a few years ago, there was an asbestos plant in Tyler, Texas. Workers there were exposed to asbestos dust from 1954 to 1972. Although public health officials told plant managers that the plant violated health standards for working with asbestos, the management did not install the necessary safety equipment. Neither the public health officials nor the plant management told the workers of the dangers involved. Some 25 to 40 of the 900 workers at the plant have died from breathing the asbestos dust and the death toll may reach 200. The remaining workers sued the plant owners and the government. A 20 million dollar settlement was agreed upon, to be provided by the owners, the government and the asbestos suppliers.

William Morris, a 50 year-old former asbestos worker, is afraid that he is dying. And he is angry, because for years neither his employer nor visiting Federal health inspectors warned him that clouds of dust he was sucking into his lungs at work caused cancer.

"I may live for another six months," he said, panting heavily as he sat in his darkened living room, a shoe box full of pain killers and other drugs at his feet.

"It was pretty damned dirty of them not to let us know"....

Reports of the Tyler settlement have reached many of the workers and their families. "It sounds like a whole lot of money," said Billie Walker, a 51-year-old widow whose husband worked at the plant for almost 11 years before dying in November of 1973. "But we paid dear and we will pay dear," she added as she sat at her kitchen table in Tyler with her two teen-age sons. [†]

We might take the viewpoint that industry (and society) will have to pay in one way or another: either for pollution control or for the consequences of not controlling pollution. Compare ways in which 20 million dollars involved in this case was spent and could have been spent.

[*] H. E. Stokinger, Science **174,** 662 (12 November 1971).

[†] The New York Times, 20 December 1977.

WHEN DO WE HAVE ENOUGH INFORMATION TO ACT?

If that agency (EPA) has forbidden the use of quarried material on the basis of the evidence described...it is a disgrace.

John T. Hack

The agency is to be applauded and its scientists commended for their judgment...

A. N. Rohl

In 1976, a high school science teacher began to wonder whether stone quarried near his home in Maryland might be a source of asbestos fibers in the air. Don Maxey, a rock collector and a person concerned about the environment, knew the stone was serpentine rock, a type known to contain asbestos. The crushed stone was being used to surface roads, playgrounds and parking lots all around the Rockville, Maryland, area. When samples of dust and rock that he took appeared to confirm his suspicions, he called in reseachers in the field of asbestos pollution.

There was indeed a very high concentration of asbestos in the air around Rockville: almost 1000 times higher than the average of levels in 49 American cities. The use of the asbestos-containing crushed stone from the Rockville quarry appears to be the most likely explanation for the unusually high concentration of airborne asbestos. (Early measurements by the Environmental Protection Agency showed levels to be as high along some roads as they were at Reserve Mining Company's taconite ore processing plant at Silver Bay, Wisconsin.) When the problem was made public, there was a range of responses. Some parents kept their children home from a school near where a road was being resurfaced with crushed stone. A few real estate agents suggested hushing up the matter lest property values decline. There was scientific controversy, too, about the effects of exposure to asbestos dust at the levels found in Rockville. John T. Hack wrote in a letter to <u>Science</u>:

> The report "Environmental Asbestos Pollution Related to Use of Quarried Serpentine Rock" by Rohl, Langer, and Selikoff (17 June, p. 1319) appears to be an example of sensational writing...
>
> ...(L)ittle information is available about any actual danger that results from exposure to airborne dust containing such minerals in the kind of environment described. Even the authors of the report say only that there is a possibility of danger.
>
> The medical aspect of the problem is more to the point than the mineralogy of the quarry. If some medical research group would come up

with some hard data that related illnesses or mortalities to degree of exposure to the mineral dust then it might be possible to judge the hazard....

I do not know the reasons for the recent Environmental Protection Agency decision relating to this matter, but if that agency has forbidden the use of quarried material on the basis of the evidence described in the <u>Science</u> report, it is a disgrace. *

Hack thus appears to be calling for more evidence before action is taken on the problem.

Rohl, Langer, and Selikoff replied:

Hack says there appears to be little information available about "any actual danger that results from exposure to air-borne dust containing such minerals in the kind of environment described." Unhappily, there is much information available about asbestos-related deaths–from mesothelioma, lung cancer, cancer of the esophagus, stomach, colon, rectum, and other cancers. But most of these have been elsewhere, in other environments–mines and mills, factories, construction sites, shipyards, households of asbestos workers (contaminated by dust brought home on clothes and shoes), and neighborhoods around asbestos mines and factories. Do chrysotile ores have deadly potential when excavated and crushed in Quebec and not in Rockville? Should the state of Maryland, Montgomery County, and the Environmental Protection Agency (EPA) allow children in schoolyards to play on surfaces spread with gravel containing asbestos? Should "hard data"–sought after a 20- to 30-year latency period–from Bethesda, Silver Spring, Chevy Chase, Baltimore, Alexandria, College Park, and Washington be added to the experiences of Long Beach, Paterson, London, Dresden, Newark, Rochdale, Johannesburg, Thetford Mine, Barking, and Sverdlovsk before environmental asbestos air pollution is controlled? †

In fact, the EPA did take action by formally advising the county not to use the stone any longer. The agency also recommended that the county repave areas in which the rock had been used.

The problem, however, may not be limited to the Rockville area. Similar rock is found in many places in the United States. Rock quarried in all these areas may be high in asbestos and thus unsafe to use for paving (Fig. 27-4). Decisions must be made in this case, as in many others dealing with environmental pollutants, on information that suggests a hazard to human health but is not sufficient to prove one.

Do you feel the EPA should wait for more information before issuing regulations outlawing the use of this particular kind of rock for paving? (Such regulations would, of course, be a great economic hardship to quarry owners.) Or, do you think the possible effects of the continued use of asbestos containing rock are so serious that its use should be banned immediately? What should we do when the evidence is not conclusive?

---

* John T. Hack, <u>Science</u> **197,** 716 (19 August 1977).

† A. N. Rohl, A. M. Langer, and I. J. Selikoff, <u>Science</u> **197,** 716 (19 August 1977).

Figure 27-3 *Asbestos-containing coatings have been used on ceilings in a number of schools and public buildings. The coating has, in some cases, begun to flake off, scattering asbestos fibers into the air. In the above photograph, parents at Ramtown School in Hopewell, New Jersey, examine the flaking ceiling that has scattered asbestos onto the hall runner (lower right). A similar situation in a building at Yale University was improved when the entire buiding interior was scraped and repainted at great cost. In New Jersey, where the problem has been studied most thoroughly, 10% of the schools were found to be painted with the asbestos coating. (Frank Dougherty/NYT pictures)*

Several prescription drugs as well as some brands of beer and gin have been found to contain asbestos fibers. Most likely the asbestos fibers are washed into these products when the liquid passes through filters made of asbestos. Asbestos fibers contaminate the drinking water in Duluth, Minnesota, and Superior, Wisconsin. Both cities draw their water from Lake Superior. For 22 years, Reserve Mining Company dumped 67,000 tons of mining wastes into Lake Superior every day. The wastes are taconite tailings, the material left after iron has been extracted from taconite rock. The major component in the tailings is asbestos.

Estimating the Risk

Although asbestos is a known carcinogen when breathed in, the effects of drinking asbestos are not clear. Twenty-two years is not long enough to tell whether an excess of cancers will develop among residents of the two cities that have contaminated water.

Low levels of asbestos fibers are definitely present in the air in cities. The main cause appears to be construction or demolition of buildings built with asbestos fireproofing and asbestos wallboard. Fibers are also found in the air where asbestos is mined or milled or made into various products. Unfortunately, there is not a good method for determining low levels of airborne asbestos fibers.

We do know that almost all of us—not just people who work with asbestos—have asbestos bodies in our lungs. (Asbestos bodies are fibers of asbestos that the body coats with a special substance, a protein, after they are breathed into the lungs.) Unfortunately, we do not know whether the presence of asbestos bodies means that cancers or other asbestos-related diseases will some day develop. Remember, cancers caused by asbestos may not become visible for 20 to 40 years. Because we know so little at the moment about the effects of small amounts of asbestos, it seems wise to prevent as much asbestos as possible from getting into air, water, and food.

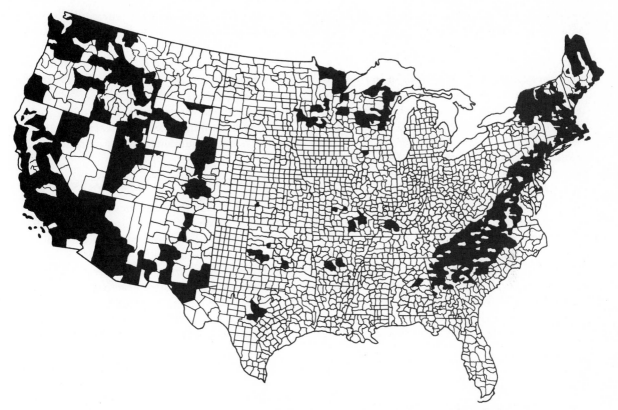

Figure 27-4   *The mineralogy of the counties and regions blacked in above is such that quarries found in them may produce asbestos-bearing rock. The map was prepared by the Environmental Defense Fund from information derived by the Mining Enforcement and Safety Administration from reports by Batelle-Columbus and the U.S. Geological Survey. [Source: L. J. Carter, "Asbestos: Trouble in the Air from Maryland Rock Quarry," Science **197**, 237 (15 July 1977) ©1977 by the American Association for the Advancement of Science.]*

ARSENIC

Arsenic is more often thought of as a poison favored by authors of detective fiction than as a pollutant in the environment. Arsenic is found naturally in all soils, in water, and in air. There are, however, several ways in which harmful amounts of arsenic can come to contaminate the environment.

Effects of Arsenic

Arsenic in high concentrations (0.2 to 10 milligrams per liter) in drinking water has poisoned whole communities and, in some of the cases, caused epidemics of skin cancer. It is also believed to be the substance that causes the higher than normal incidence of lung cancers among workers in industries such as smelting. As is the case with most cancers, long periods of time pass before these diseases develop. Between 15 and 20 years may go by before skin cancers appear and 30 to 40 years may pass before lung cancer is detected.

Summary of the Problem

At the present time it seems that there are two sources of arsenic pollution that pose a threat to the environment. One is the emission of arsenic from coal-fired power plants.

In the U.S., all the power plants and coal burning industries together emit between 1000 and 3000 tons of arsenic per year as a result of burning coal. The actual figure depends on how efficient smoke-stack control devices are. The most commonly used device, electrostatic precipitators, are felt to be about 70% to 90% efficient at removing the dust containing arsenic. There have been reports of hearing damage to children living near a coal-fired power plant in Czechoslovakia, due to arsenic emissions. This problem can be controlled by anti-pollution devices and responsible monitoring by public agencies.

The second problem involves the increasing use of arsenic-based compounds in agriculture.

The major portion (80%) of the arsenic used in this country is used in agriculture. Arsenic is used on cotton to wilt the leaves so the cotton can be picked mechanically. This has led to air pollution downwind from cotton gins. When pesticides such as DDT and other chlorinated hydrocarbons are outlawed, arsenic may be used instead.

The effect of unnatural, but still low level concentrations of arsenic on aquatic life is not clear. Arsenic does seem to accumulate in living organisms, but its concentration does not increase as it passes up food chains.

OZONE

An Essential Gas

High above the earth, in that part of the atmosphere called the stratosphere, there is a relatively little-known gas essential to life on earth. That gas is ozone. Each molecule of ozone is made up of three atoms of oxygen. Ozone in the stratosphere absorbs over 99% of the rays of ultraviolet light that comes from the sun. These ultraviolet rays are often welcomed because they cause skin to tan. They can also cause it to burn and, over long periods of time, ultraviolet rays cause skin cancers.

High concentrations of ultraviolet rays are harmful to many forms of plant and animal life. They are sometimes used to sterilize objects because, in high concentrations, they kill bacteria. It is generally believed that life on land did not develop until the earth's protective ozone layer formed. Because ozone in the stratosphere is essential to life, reports that there are a number of human activities that can destroy ozone are understandably alarming.

Threats to the Ozone Layer

The first reports were about supersonic transport (SST) planes, that fly directly in the stratosphere. These planes release two contaminants, water and oxides of nitrogen, that can destroy ozone. Fears about the possible effects of SST's on the ozone layer were a strong factor in the U.S. decision to stop development of a fleet of SST's. However, the actual effect that a large fleet of SST's would have on the ozone layer is still hotly debated.

In 1974, a more serious threat was uncovered. Ozone is known to be destroyed by chlorine in the atmosphere, and a source of chlorine had been discovered. Compounds called chlorofluorocarbons[3], widely used as propellants in spray cans and in refrigerators and air conditioners, were found to be slowly diffusing into the stratosphere. These compounds do not react easily with other materials. For this reason, they are ideal for use in spray cans. However, it has been found that they are broken down by ultraviolet light, with the release of chlorine. This chlorine can then break down ozone. How much ozone are the chlorofluorocarbons destroying? Unfortunately, there are many things we still don't know about ozone and the stratosphere. The best guesses are that chlorofluorocarbons may cause 1% to 20% decrease in the ozone layer.

To this must be added the effects of other substances that destroy ozone. Nuclear explosions in the stratosphere destroy ozone by the release of

---

3 This is sometimes shortened to "fluorocarbons."

oxides of nitrogen. Oxides of nitrogen from a stratospheric nuclear blast are rapidly carried all over the globe. Thus, in the event of a nuclear war, the resulting increase in ultraviolet rays could be as serious a problem as nuclear fall-out. Some scientists feel that nuclear weapons testing during the late 1950s caused a decrease in the ozone layer. Fertilizers in soil release nitrogen oxide, which can react with ozone. The use of fertilizers is increasing yearly as we try to feed an increasing world population. At least one scientist feels that fertilizers will come to pose as serious a threat to ozone as the chlorofluorocarbons. Bromine is known to destroy ozone in the laboratory. As methyl bromide, it is widely used as a fumigant in agriculture. How much bromine escapes to the stratosphere is not yet known. Similarly, the effect of man-made quantities of chemicals such as carbon tetrachloride and methyl chloroform is not clear although it is felt that they could release significant amounts of ozone-destroying chlorine.

Effects of Decreasing the Ozone Layer

One of the uncertainties involved in predicting effects on the ozone layer is knowing how much ozone is actually present in the stratosphere. The average concentration is probably about three parts per million. However, ozone concentrations vary geographically. In addition, the ozone concentration can vary, normally, by as much as 30% from day to day. Over a period of years, the variation averages out to 10%. These yearly changes seem to be connected to the 11-year cycle of sunspots, and there appears to be a natural balance between the formation and destruction of ozone. There are two major concerns about the additional decreases caused by human activities.

A decrease in the ozone layer would allow more ultraviolet rays to reach the earth, which would almost certainly increase the occurrence of skin cancer. (Skin cancers are different from most other cancers. For one thing they rarely spread to other parts of the body. More about skin cancer is

on p. 560). It seems quite certain that a proportion of skin cancers are caused by exposure to sunlight over long periods of time. Fair-skinned people are much more apt to develop skin cancers since the pigments in dark skin help to screen out ultraviolet rays. People living in tropical climates are more subject to skin cancer, apparently because they are exposed to more ultraviolet light.[4] Estimates are that each 1% decrease in the ozone layer will cause a 2% increase in skin cancer. Skin cancer is almost always curable, although the treatment is not pleasant. Plants and animals also show sensitivity to ultraviolet rays. An increase in ultraviolet light could be expected to cause the extinction of some microscopic life forms and to damage other species or decrease the living space available to them.

The second concern is that changes in the stratosphere might cause changes in climate. Increases in ultraviolet light may possibly increase the melting of polar ice, with resultant flooding. Chlorofluorocarbons, in addition to effects they have on ozone, absorb infrared radiation. This could trap heat in the atmosphere and cause a warming of the earth. At the present time, there is a great deal of uncertainty about the final effect of chlorofluorocarbons and changes in the ozone layer on climate. Predictions actually range from a new ice age to temperature increases of up to 1°C. What should be remembered is that life on earth has evolved under the present set of conditions. Lasting changes in one direction or another will have profound effects on life as we know it.

How the Future Looks

Because chlorofluorocarbons are believed to be a serious threat to the ozone layer, U.S. government agencies have decided to prohibit unnecessary[5]

---

4 Look again at the photograph of the sunbathers on page 532. Many doctors now recommend that a lotion which screens out ultraviolet rays should be used at the beach or pool in order to reduce the risk of skin cancer.

5 A few drug products, certain pesticides, some aircraft, and some electrical maintenance products are exempt.

uses. The use of chlorofluorocarbons as spray can propellants was outlawed after 1979. This accounted for about three-fourths of the use of chlorofluorocarbons in 1973. The gases are still used in refrigerators and air conditioners, although a search for substitutes is under way. Even with these strict controls, the quantity of chlorofluorocarbons in the stratosphere will continue to increase for some time. This is because those compounds already released are still around and are diffusing very slowly upward. Remember, they are not broken down except by ultraviolet light in the stratosphere.

The possible destruction of ozone by nitrous oxides from fertilizers appears to be a much more difficult problem to solve. Dr. H. S. Johnston, who pointed out the problem, notes: "Obviously if the choice is between eating and long-term cancer risk, people are going to choose to eat."[6] Thus the ozone

---

6 *Science* **195**, 658 (18 February 1977).

problem becomes tied up with problems of rapid population growth, the "green revolution," and the crisis in providing enough food for everyone.

Other countries seem, in general, unconvinced that a serious threat to the ozone layer exists. The U.S. accounted in 1973 for one-half of the world use of chlorofluorocarbons. However, only Sweden has followed the U.S. lead in banning the use of chlorofluorocarbons in spray cans.

SMOKING: A PERSONAL FORM OF AIR POLLUTION

The Problem

A scientist who begins a study on the health effects of air pollution often decides that he must separate the people in his study into two groups, those who smoke and those who do not. This is because a person who smokes is adding a variety of

Skin Cancer

In addition to uncertainties about the amount of ozone that will be destroyed by human activities, there are uncertainties as to how much skin cancer ozone destruction will cause. Skin cancers are of three types. The two most common types rarely spread to other parts of the body. The rate of cure is 95% or greater, by chemotherapy (using drugs such as fluororacil), surgery, x-rays, or cauterization. The third type of skin cancer, melanoma does spread rapidly to other parts of the body. It is a much less common type, however, and is not as strongly linked to exposure to sunlight. For instance, the rate of occurrence of melanoma is 75% greater in southern states than in states on the Canadian border, while the rate of the other two types of cancer is 250% higher.

The rate of skin cancer in the U.S. is variously reported to be 200,000 to 600,000 cases per year. Each 1% decrease in ozone is believed to cause a 2% increase in ultraviolet radiation and as much as a 2% increase in skin cancer. Proportionately then, each 1% decrease might cause 4000 to 12,000 extra cases of skin cancer, each year, in the United States alone.

pollutants to the air he breathes. Some of these pollutants are already present in city air; some are new. It has been said that smoking is a personal form of air pollution. While the effects of breathing in some of the substances in tobacco smoke are known, the effects of others are not. Still, there appears to be a growing acceptance in this country of the fact that smoking is hazardous to health. Over the past 10 years, surveys by the National Clearinghouse for Smoking and Health have noted that a smaller and smaller percentage of adult Americans smoke. When smokers are divided into groups by age, a decrease since 1964 can be seen in the percentage of Americans who smoke in most age groups. (Among women 21–24, women 55 and over, and men 65 and over there has been a slight increase.) (See Fig. 27–5).

The 1975 survey also noted that 61% of those who smoke have tried seriously, at least once, to stop smoking. In fact, 9 out of 10 people say they

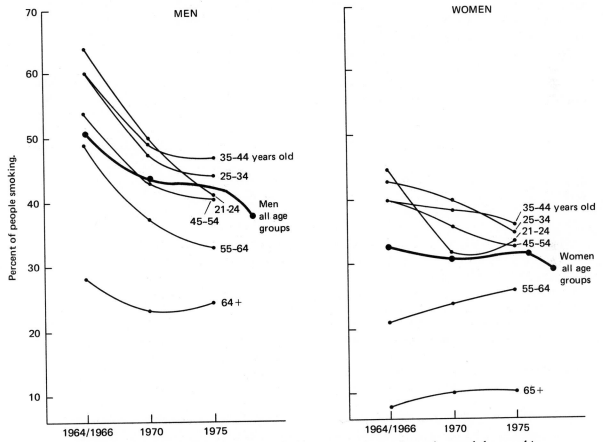

Figure 27-5   *Percentage of Americans smoking by age groups. A trend towards less smoking is apparent in all age groups except men 65 and over, women 21–24, and women 55 and over. Recent data shows that among teenagers there is an increase in the percent of smokers. Similar trends are seen all over the world. (Source: National Clearing House for Smoking and Health and from the 1978 Health Interview Survey.)*

would stop if it were easier to do. However, slightly more than half of those asked expected to be smoking 5 years from now.

Scientific studies on the effects of smoking fall into two general groups. First, there are those in which people who smoke are compared with people who do not smoke. People who smoke are found to suffer not only more lung diseases, such as lung cancer and emphysema, but also more heart disease.

Second, there are studies in which tobacco smoke is examined to see what chemicals it contains. These chemicals are then often given to animals to determine whether they are harmful. An examination of the makeup of tobacco smoke shows that it contains a number of chemicals known or suspected to cause cancer. Other harmful chemicals such as lead and carbon monoxide are also found in cigarette smoke.

What's in a Puff?

People who smoke expose themselves to rather high levels of carbon monoxide. Carbon monoxide, which is also a pollutant in city air,[7] binds to the hemoglobin in blood. Hemoglobin so bound is incapable of carrying oxygen to tissues in the body. Hemoglobin binds 200 times more tightly to a molecule of carbon monoxide than to a molecule of oxygen. Thus, the body tissues in a smoker will receive a lower-than-normal supply of oxygen. The air quality standard for carbon monoxide is determined by averaging carbon monoxide concentrations over an eight-hour period. This standard, 10 milligrams of carbon monoxide per cubic meter of air, is the level that should not be exceeded more than once a year. When we live in such a concentration year round, about 2% of our hemoglobin is tied up as carboxyhemoglobin (the combination of carbon monoxide with hemoglobin). Compare this percentage to that of the heavy smoker (about a pack a day). This person may have further tied up

an average of about 6% of hemoglobin. Thus, we might say smokers have greatly decreased the quality of the air they breathe. How much of a problem is this? It probably depends on where they live and what they eat! Most cities already have a high concentration of carbon monoxide in the air. Concentrations are highest in traffic jams, underground garages, and tunnels. Levels in these areas can reach 70 milligrams per cubic meter or more. Someone breathing air at the 70 milligram level for eight hours would have a blood concentration of about 10% carboxyhemoglobin. At this level, tests have shown a decreased driving ability. People cannot respond as quickly to such stimuli as brake lights and changes in the speeds of cars. Since smokers have inactivated 6% of their hemoglobin due to the carbon monoxide in cigarette smoke, we can see that in certain situations they may well have decreased further their ability to act quickly when they need to be alert and quick.

Increased blood levels of carbon monoxide may also increase the risk of fatal heart attacks. This is explained later.

What does what we eat have to do with the problem? Nitrites, which are used to cure meats such as hot dogs and corned beef, also tie up hemoglobin.[8] Nitrites react with hemoglobin to form methemoglobin. Methemoglobin, like hemoglobin when it is tied to carbon monoxide, is incapable of carrying oxygen. A large corned beef sandwich (about one-quarter pound of meat) could inactivate 1.5% to 5.7% of an adult's hemoglobin by converting it to methemoglobin. Thus, the effects of air pollution, a food additive, and smoking may all combine to inactivate a significant portion of a person's hemoglobin. The combined, or synergistic, effect of smoking and various air, food, and water pollutants are just beginning to be uncovererd.

Cigarettes also contribute particles of nickel, arsenic, cadmium, and lead to the lungs of a smoker. Individually and in high quantities, these

---

7 Chapter 17

8 Nitrites in food are covered on p. 594.

metals have known and harmful effects on humans. Together and in lower quantities, their effects are not clear (Table 27-1).

In an earlier era, the pesticide lead arsenate was used on tobacco. Because neither lead nor arsenic is broken down in the soil, they are still found wherever lead arsenate was once used. Tobacco plants grown in that soil today absorb lead and arsenic. The lead level in tobacco leaves is variable, depending on where the tobacco is grown. One study, however, estimates the lead in a cigarette at an average of 13 micrograms. Of this quantity, about 1.5 micrograms of lead appear in the smoke. This is the quantity inhaled by an individual as he smokes one cigarette. Twenty cigarettes in a day means 30 micrograms of lead inhaled. About one-third of this quantity is absorbed into the blood.

Lead has many adverse effects on health.[9] Smoking alone will not lead to levels that cause apathy, sluggishness, and brain damage. However, lead is also found in food and in water. It is also in air, due to lead additives in gasoline. Smoking a pack a day adds about 50% more lead to the daily respiratory intake for an individual who lives in a polluted city.

Arsenic is a cumulative poison; that is, many small doses can accumulate over a period of time until a poisonous level is reached. Tobacco smoke does not contribute enough arsenic to kill anyone in the classic manner. However, it is suspected that at lower levels arsenic is a carcinogen.

A pack of cigarettes contains about 30–40 micrograms of cadmium and 85–150 micrograms of nickel. In large enough quantities, cadmium has several effects in the body. It interferes with the body's use of calcium. It may also contribute to high blood pressure and heart disease. A smoker inhales about 2 micrograms of the cadmium in a pack of cigarettes, and about ½ to 1½ micrograms of nickel. The effect of these metals on human lungs is

not clear. Studies in animals, however, have shown that both nickel and cadmium can, under certain circumstances, be carcinogenic. Actually, a larger portion of both the cadmium and nickel are found in the "side-stream smoke," that is, the smoke not inhaled by the smoker. Thus, about 9–13 micrograms of cadmium and 12–21 micrograms of nickel per pack of cigarettes are added to the environment around the smoker. Nearby non-smokers are therefore also exposed to possibly harmful levels of these metals. In this respect, smoking may not be such a "personal" form of air pollution after all.

The Health of Smokers

When groups of smokers are compared to groups of non-smokers, one finds a number of ways in which smokers appear less healthy than their counterparts who do not smoke.

Heart and circulatory problems According to a 1979 mortality study by State Mutual Life Assurance Company of America, people who smoke suffer death rates double those of nonsmokers at any age. Smokers are at a greater risk with respect to both heart disease and lung disease. For instance, people who smoke are approximately twice as likely to suffer an immediately fatal heart attack as those who do not smoke. When a person stops smoking, the risk decreases. In 10 years, the risk to former smokers is about the same as for non-smokers. It is suspected that there is a relationship between the carbon monoxide in cigarette smoke and an increased risk of sudden death. It is possible that carbon monoxide, because it ties up hemoglobin in blood, reduces the amount of oxygen available to the heart muscle itself. This puts a strain on the heart and makes it pump harder. This strain on the heart may explain the higher likelihood of sudden death from heart attacks in groups of smokers.

Smoking also seems to increase the possibility of cerebral hemorrhages. It further increases the occurrence of peptic ulcers and the chance of dying from them.

---

9 These effects are discussed in Chapter 18.

Effects on unborn babies   Women who smoke tend to have smaller babies. There are more spontaneous abortions, more still births, and up to one-third more new-born deaths among babies born to mothers who smoke during pregnancy. In general, it can be said that there are more frequent complications of pregnancy and labor among women who smoke. These effects seem to be due to a shortage of oxygen in the blood during pregnancy. Again, carbon monoxide in cigarette smoke, which ties up the oxygen carrier, hemoglobin, is the culprit.

Effects on lungs   It is not surprising that many of the adverse effects of smoking involve the lungs. This is characterized by a great deal of mucous and coughing. Smokers suffer more ordinary respiratory illnesses and have been shown to have more complications, such as pneumonia, after surgery than non-smokers. Smoking is a major cause of emphysema. This is a disease in which the alveoli, tiny sacs at the end of each bronchiole, do not expel air as they normally should in exhaling (Fig. 27–6). People with emphysema have trouble getting the used air out of their lungs so that they can take a fresh breath. Those who stop smoking usually improve.

The disease most commonly associated with cigarette smoking is lung cancer, although smokers also have more of a chance than non-smokers of contracting cancer of the larynx, esophagus, mouth, bladder, kidney, and pancreas.

Cigarette smoking is the major cause of lung cancer in males in the United States. Smoking is also a cause of lung cancer in women. In fact, the rate of lung cancer is increasing so rapidly among women that within the next 10 years it is expected to surpass the rate of breast cancer, now the major cancer in women.

When people stop smoking, their risk of developing lung cancer drops to that of non-smokers over a period of 10 to 15 years. The choice of a filter-tip cigarette or a low-tar brand also seems to reduce the risk of lung cancer, although not to the same rate as for non-smokers. Tobacco smoke can

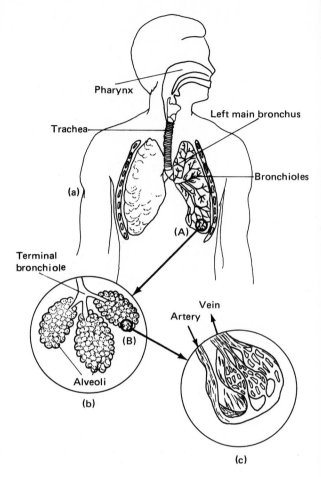

Figure 27–6   *The respiratory tree. (a) Air is breathed in through the nose and mouth, moves down the trachea into the bronchi, bronchioles, and finally into the alveoli (A). (b) The alveoli are tiny sacs at the end of the bronchioles. Oxygen enters the bloodstream at this point (B). (c) Capillaries carry blood around the alveoli, allowing oxygen to diffuse from the alveoli into the blood.*

increase a persons risk of developing cancer from other causes. For instance, a person who both smokes and works with asbestos has a nine times greater chance of developing asbestos-caused lung cancer than an asbestos worker who does not smoke.

TABLE 27–1 Possibly Carcinogenic Chemicals Found in Cigarette Smoke[a]

Components	Estimated concentrations in 100 cigarettes (85 mm nonfilter)
Polynuclear aromatic hydrocarbons	10–30 μg
1. Benzo(a)pyrene	3.9
2. Dibenz(a,h)anthracene	0.4
3. Benzo(b)fluoranthene	0.3
4. Benzo(j)fluoranthene	0.6
5. Dibenzo(a,i)pyrene	Trace
6. Benz(a)anthracene	0.3
7. Chrysene	2.0
8. Indeno(1,2,3-cd)pyrene	0.5
9. Benzo(c)phenanthrene	Trace
10. Methylbenzo(a)pyrenes	0.1
11. Methylchrysenes	2.0
N-heterocyclic hydrocarbons	1–2
1. Dibenz(a,h)acridine	0.01
2. Dibenz(a,j)acridine	1.0
3. 7H-dibenzo(c,g)carbazole	0.07
N-nitrosamines	1–10
1. Dimethylnitrosamine	0.4
2. Diethylnitrosamine	Trace
3. Methyl-n-butylnitrosamine	Trace
4. Nitrosopyrrolidine	0.4
5. Nitrosopiperidine	Trace
N-alkyl-heterocyclics:	
1. I-methylindole	Present
Pesticides and fungicides:	
1. TDE	10–100
2. o,p-DDD	10–100
3. DDT	10–100
4. Maleic hydrazide	10–100
Beta-naphthylamine	2–3
Polonium 210	1–50 picocuries
Nickel compounds	Present
Volatile phenols	20–30 mg
1. Phenol	
2. Cresol	

TABLE 27–1   (Cont.)

Components	Estimated concentrations in 100 cigarettes (85 mm nonfilter)
Nonvolatile fatty acids	20–100 mg
1. Stearic acid	
2. Oleic acid	
N-alkyl heterocyclics:	
1. 9-methylcarbazole	Present

[a] This table lists chemicals found in cigarette smoke that are known to cause cancer or are strongly suspected. Some of them are not present in a high enough quantity to be responsible for cancer by themselves. It should be remembered, however, that they are all in there together. Little is known about the combined effects of carcinogenic substances. Source: The Health Consequences of Smoking, A Report to the Surgeon General; 1971, United States Department of Health, Education, and Welfare, Public Health Service, Health Services and Mental Health Administration, p. 265.

The Effects of Smoking on Non-Smokers

Two-thirds of the smoke from a cigarette is not inhaled by smokers but rather goes into the environment around them. In addition to nickel and cadmium, this sidestream smoke contains twice the tar and nicotine, three times the 3, 4 benzopyrene and five times as much carbon monoxide as the smoke the smoker inhales. Studies on "passive smoking" (simply breathing in a room where people are smoking) show that non-smokers can inhale as much nicotine and carbon monoxide in an hour, in a very smoky room, as if they smoked one cigarette themselves.

Children exposed to tobacco smoke at home may suffer more respiratory illness, according to several studies. Furthermore, some doctors recommend that people with heart disease avoid smoky places because they could inhale enough smoke to injure their health.

Smoking Laws: Should There be More?

Separate sections for smokers and non-smokers are required by federal regulation in public vehicles transporting people between states. Some 70% of the people queried, in a 1975 study by the National Clearing House on Smoking and Health, felt that smoking should be prohibited in more places, although only 51% of smokers felt this way. A total of 78% felt that smoking should be prohibited in places of business if the management wishes. And three out of four people believe that doctors, teachers, nurses, and others in the health professions should set us all an example by stopping smoking themselves. Should government take a more active role in trying to stop people from smoking? See the Controversy, p. 568.

RADIATION AND CANCER

What is Radiation?

Origins of radiation   Since the time of Pierre and Marie Curie, we have known that certain substances emit particles and rays during the reactions that occur within the nuclei of their atoms. And, as time has gone by and understanding has increased, our concern about the biological effects of these radiations has grown. From a scientific curiosity, uncontrolled radiation has become recognized as a distinct hazard; the allowable exposure for workers and the public to radiation has been steadily decreased. Each passing decade seems to bring to light new evidence of effects upon humans at levels earlier thought to be safe.

In this discussion, we hope to answer a number of basic questions. What is the nature of radiation and what are its common sources? What are the effects of radiation on people who are alive today and on future generations? How are the technological sources of radiation exposure being controlled?

Nature of radiation   Radiation stems from several sources. The nuclear decay of unstable elements gives rise to particles and rays. The unstable elements, which we refer to as radioactive, emit alpha particles, beta particles, and gamma rays. The alpha particle is relatively heavy. It barely penetrates several sheets of paper, and hence, even if one stood next to a flask containing an element that emitted alpha particles, there would be no danger. If the element were inhaled, however, and the alpha particle emitted in the human body, the danger is thought to be great. Alpha particle emissions from plutonium are discussed in the section on nuclear power, under the title "The Hot Particle Problem" (Chapter 15).

A beta particle may penetrate several millimeters of aluminum or more; therefore its handling requires greater caution and shielding. The gamma ray has the greatest penetrating ability of all the radiation from radioactive decay. It may pass through several centimeters or more of lead without a significant weakening of its energy. Those who work near substances that emit gamma radiation must exercise the greatest caution to limit their exposure.

In addition to radiations from unstable elements, scientists have learned how to produce high energy photons. Photons are the infinitely small particles that compose a beam of light. High energy photons are called x-rays and are commonly used to examine the internal structure of the human body in order to diagnose and treat disease or injury. Physicians and dentists regard x-rays as one of their most important tools for diagnosis. Like gamma rays, x-rays are highly penetrating; several centimeters or more of lead are necessary to block these beams.

People and Radiation

Sources of human exposure   Radioactive elements exist naturally in the earth's crust. In some places they are more concentrated than in others. Nonetheless, since the era of the atom bomb and the beginning of nuclear electric generation, the sources of exposure to radioactive elements have multiplied many times.

Whereas exposure to radiation from radioactive elements and x-rays is largely the result of human activities, there is still another form of radiation that is both natural and almost inescapable. Cosmic rays, which consist of both particles and electromagnetic radiation, are constantly bombarding us from outer space. Although a portion of cosmic rays can be blocked out by several thicknesses of lead, another portion penetrates even into the deepest mines. The intensity of cosmic rays increases at higher altitudes, so much so that jet crews might one day conceivably be classified as radiation workers. Cosmic ray intensity also increases as one moves toward the polar latitudes.

Human exposure to radiation comes via one or more of these routes: decay of radioactive elements, x-rays, cosmic rays. We measure exposure to radiation most commonly in "rems" and "millirems" (one-one thousandth of a rem), units reflecting both the intensity of the radiation and its effect on human tissues. Radiation standards are also set in rems and millirems. If we exclude radiation from medical x-rays and that which results from nuclear electric power activities, we have what is referred to as natural background radiation. This is the radiation we would receive if there was no exposure to man-made radiation.

SMOKING

If we have to make a mistake we should make a mistake in the direction of protecting public health.

Gus Speth,
Council on Environmental Quality

...He should first remove the products about which there is no mistake.

Rodney Adair

The New York Times quoted the Chairman of the Toxic Substances Strategy Committee as follows:

The Carter Administration is now preparing to move aggressively against cancer-causing chemicals in the home, the workplace and the general environment. Today cancer is our second leading killer, and a concerted, Government-wide attack on the substances that cause much of it is essential.

This strategy of prevention is bound to fail, however, if those whom it is designed to protect do not understand the basis for chemical regulation. The recent controversy on the proposed Food and Drug Administration ban on saccharin treated us to a dangerous amount of hilarity about the high dosage levels used in animal tests and demonstrated the prevalence of misunderstanding in this area.

So far, we have no scientific basis for setting a safe threshold dose for a carcinogen. Until such a scientific basis is demonstrated, if we have to make a mistake we should make a mistake in the direction of protecting public health. And that means we assume, for the present, that there is no safe level for a carcinogen.

In a matter so grave as cancer, makers of social policy cannot wait for our scientists to give us the precision we would prefer in regulating hazardous chemicals. Until we have precise methods for determining chemical safety, prudence must prevail: We must regard suspect chemicals, indicated on the imperfect evidence of animal tests, guilty until proved innocent. *

Rodney Adair replied:

Mr. Speth should surely realize that most of the hilarity is cynical, and is extremely well justified, because of the attitude of Government toward the carcinogenic effect of tobacco smoke. We have the Surgeon General's report, the statistical evidence of 100,000 new lung cancer cases a year, of which 80 percent are caused by smoking, and the approximately 90 percent fatality rate associated with lung cancer.

How can Mr. Speth and the F.D.A. expect anything more than cynical laughter toward a Government that subsidizes the tobacco industry and at the same time threatens the removal of other products which are often only marginally suspect and on which there is no statistical evidence of any undue incidence of cancer among its human users?

If he truly believes his own statement in reference to removal of carcinogenic materials from the market, i.e., "if we have to make a mistake, we should make a mistake in the direction of public health," he should first remove the products about which there is no mistake. This would probably occupy so much of his department's time that the making of mistakes would become less and less of a problem as science produces the answers for further action on a proven basis. [†]

What is the responsibility of the government to its citizens when there is a proven health hazard such as smoking? Do you agree with Mr. Adair that the government should remove tobacco products from the market?

---

[*] Gus Speth, member Council on Environmental Quality and Chairman on the Carter Administration's toxic substances strategy committee. The New York Times, p. A-21, 9 March 1978.

[†] Rodney Adair, The New York Times,, p. A-26, 29 March 1978.

Background radiation exposure in the United States is most commonly in the range of 100–150 millirems per year. Leadville, Colorado, at two-miles (3.3 kilometers) above sea level, has one of the highest annual radiation exposures in the U.S. at 160 millirems, due to the high intensity of cosmic rays reaching the city. Radioactive elements such as potassium-40 (a form of potassium) and radium in the earth's crust are also contributors to background radiation.

X-rays ordered by physicians and dentists are a common source of exposure to radiation. Although such exposure has been estimated at 90 millirems per year, this is an average figure. Many individuals receive far higher doses and some none at all (Fig. 27–7). The annual chest x-ray for tuberculosis is now thought to be not such a good idea. Much of the older x-ray equipment is still in use and delivers radiation doses far higher than necessary. Another type of x-ray exposure occurred in the early 1950s when shoe stores were using machines to x-ray the foot and shoe to assure a good fit. Children would play repeatedly with the machines without supervision.

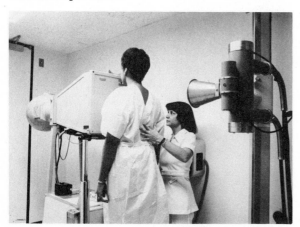

Figure 27-7 *Dental and medical x-rays account for about 90 millirems of radiation per person per year, on the average. This is approximately equal to radiation from natural background sources. (American Cancer Society)*

During the era when atomic and hydrogen bombs were tested in the atmosphere, radioactive elements were scattered around the globe in clouds of particles. Rain washed these particles from the atmosphere in "fall-out." The fall-out in areas very near atomic explosions, such as in the Pacific Ocean islands, was large enough to leave measurable levels of radiation in the soil, but the minor amount of continuing fall-out is negligible when compared to background radiation exposures.

Exposure to radiation from radioactive elements is a concern that has grown with the arrival of the nuclear electric power industry. Routine discharges of low levels of radiation from electric power plants were allowed in the first 15 years of the nuclear industry. The older nuclear plants were to be operated so that no individual in the general population would receive more than 500 millirems of radiation per year, and the average exposure of people away from the site would be kept below 170 millirems per year. Such operation reflected the then current standards for exposure of people in the general population. In the middle 1970's, the Atomic Energy Commission was under attack for these standards and the response was a tightening of the allowed emissions of radioactive elements from nuclear plants. From 500 millirems, the largest annual allowable dose was reduced to 5 millirems and the average dose to less than 1% of natural background, or 1 millirem.

Under these new exposure levels, the normal operations of nuclear power plants are quite clearly not a concern; yet radiation remains a problem. One remaining source of radiation exposure involves fuel reprocessing plants (see Nuclear Power, Chapter 15). Another source is the worked and discarded uranium ore (tailings). Of greatest concern, however, is the possibility that terrorists or warring nations could hijack spent reactor fuel or fissionable material and build nuclear weapons: an atom bomb from a fissionable material such as plutonium, or a dispersal device that spreads the deadly contents of a spent fuel rod. The detonation of these weapons could bring exposure to radiation on a massive scale.

Finally, mention should be made of the fact that a number of radioactive elements can be concentrated in food chains (biologically magnified). An example is phosphorus-32, which was found to be concentrated as much as 5000 times above levels in the water in whitefish downstream from the Hanford Atomic Power Plant on the Columbia River. Blue gills and crappies in the same river were found to contain 20,000 to 30,000 times the concentration in water of phosphorus-32, while filamentous algae were found with up to 100,000 times the water level of phosphorus-32. Zinc-65, iron-59, and iodine-131 are other examples of radioactive elements concentrated by living organisms. These are sources of radiation exposure to the general public when certain foods are eaten. Fish may contain phosphorus-32, oysters and clams zinc-65, and milk iodine-131. Unfortunately, not enough is known about the paths such elements take through food chains to estimate the hazards involved.

Biological effects of radiation The effects of radiation on the health of people alive today can be divided into two categories. The acute symptoms that result from intense short-term exposures occur within days or weeks. Such exposures are very unlikely except as a result of a nuclear war or a severe, accidental occupational exposure. The effects of long-term low-level radiation exposures are likely not to be seen for a number of years. These delayed symptoms cannot be distinguished from the common diseases of aging, especially cancer. We know that radiation can cause bone cancer, lung cancer, thyroid cancer, and leukemia. These diseases have been observed in individuals exposed to radiation levels of 100 rems or more in an accident or catastrophe.

Although radiation can induce cancer, a person who has cancer cannot ordinarily point to radiation as the cause. There are so many other causes of cancer that its presence in the population is not proof of exposure to damaging levels of radiation. The only way one can show that low levels of radiation cause cancer in humans is to study a fairly large population (1000 or more in some industry) that has been exposed to low levels of radiation. One then needs to compare the incidence of cancers of various types in this population with the incidence of these same types of cancers in the general population. If the number of cancers is significantly larger in the population under study, one can infer that the low levels of radiation—or other job-related factors—contributed to the increase. Even if low-level radiation can be linked to cancer, a particular individual in the study will generally be unable to prove that the origin of his cancer was radiation. Shipyard workers were facing just such a problem in 1979 and the early 1980s. It should be noted that children, infants, and unborn infants are especially vulnerable to damage from radiation. Doses that would not harm an adult could lead to early cancer in these individuals.

Radiation may do more than shorten the life span of the immediate victim; children born to people who have been exposed to damaging levels of radiation may suffer from genetic defects. Many of these children will not survive because such defects are usually too damaging. Those that do survive carry damaged genetic material.

Should radiation standards be lowered? On the basis of studies of populations of atomic energy workers and others exposed to low-level radiation, some experts conclude that the standard for occupational exposure of 5 rems per year should be cut to 0.5 rems or 500 millirems per year. A recommendation for such a standard was made in 1978 by a Committee of the National Academy of Sciences. The recommended level is only about twice the level of a chest x-ray exam.

Nearly 5000 of the 70,000 atomic energy workers were exposed to levels greater than 2 rems in 1977. In addition to the employees of nuclear power plants, the nation has about 170,000 x-ray technicians. Although they should be limited to less than the occupational standard of 5 rems per year, their actual exposures are largely unknown.

Uranium Mine Tailings: An Unnecessary Hazard

Uranium miners are exposed to radiation on their jobs and apparently as a result have experienced a high rate of lung cancer. The radiation to which the miners are exposed comes from the natural radioactive decay of uranium. After a number of steps, beginning with the initial uranium decay, a radioactive gas known as radon is produced. Atoms of the element radon are dispersed thoroughly in the air of the mines because radon is a gas.

The decay of radon results in new radioactive elements in particle form. Since the particles are only single atoms, rather than being of any size, they remain suspended in the air for a long while. This contaminated air is inhaled by the miners, and the radioactive "daughters" of radon are deposited in the miners' lungs. Masks or filters are ineffective in capturing particles consisting of a single atom. Hence, the only way to control exposure of the miners is to replace the air in the mine with fresh air on a frequent basis.

Miners are not the only people exposed at this step in the atomic fuel cycle. Wastes, called tailings, remain after the uranium oxide has been extracted from uranium ore. This rubble still contains some uranium which could not be removed. Since radon is a product of uranium decay reactions, the air in the vicinity of the rubble also contains radon gas at levels up to 500 times the natural background. The air near the tailings is likewise contaminated with radon "daughters."

Through oversight and a lack of awareness of the hazards, the tailings have been used to make concrete for homes and buildings. Rubble from the Old Climax Mill in Colorado is about 0.03% uranium; it was used as building material and as landfills for homes in the town of Grand Junction, Colorado. The tailings were not only used to level land surfaces; they were also used in mortar, concrete, and backfill around basement walls.

Levels of radon decay products in the air of these homes and buildings are up to five times the concentrations permitted for uranium miners. Hence, the exposure of people who live in those homes to a potential cancer-causing agent has been substantial. A number of homes have had foundation materials, even walls and chimneys replaced and backfill removed and replaced; the federal government and the state of Colorado assisted with the costs of these repairs.

Most of the deposits that are rich enough in uranium to be mined are found in the western portion of the United States. The states of New Mexico, Wyoming, Colorado, and Utah are the principle suppliers of ore, and here mounds of tailings may still be found. When milling operations moved on to more radioactive pastures, the piles of tailings were simply abandoned. In this way piles of tailings fell into private hands. Often, the new owners were ignorant of the dangers of the material. In eight states, such piles still exist, some very near large communities. The smallest such pile is only two acres in area; the largest is 107 acres in extent. One pile, next to an abandoned mill, is within 30 blocks of downtown Salt Lake City, Utah. In all, in the Western states there are known to be 23 uranium tailings piles over which no control is exercised. The mills have closed; the piles remain (Fig. 27–8).

Controlling the problem is difficult. Several feet of soil can be deposited over the piles and vegetation established. This will prevent the wind from lifting the dust-like particles into the air. However, it is thought that 10 to 20 feet of earth will be necessary to prevent the escape of radon gas. This suggests the immediate need to bury the tailings that are currently being produced from continuing mining and milling operations.

Figure 27–8   *Radiation physicist standing on a 250-foot pile of uranium tailings. The city of Durango, Colorado, is seen clearly on the other side of the pile. (Photo courtesy of* Oak Ridge National Laboratory Review)

TOXIC SUBSTANCES IN THE WORKPLACE

Types of Toxic-Substance Hazards on the Job

To most people, job safety brings to mind safeguards against immediate hazards to life and limb —machinery accidents or chemical burns. And of course these are hazards workers rightfully expect their employers to help them guard against. There are other, less visible, but equally crippling job hazards, however, and the extent of these problems is only now becoming known. A study published in 1977 by the National Institute of Occupational Safety and Health (NIOSH) stated that one out of every four U.S. workers may be exposed during their working lifetimes to a toxic substance that can cause disease or death. This means that almost 22 million workers have been or are being exposed to toxic substances —solvents, mercury, lead, or pesticides, etc. Furthermore, some 880,000 of these workers, or 1% of the entire work force, are currently being exposed to substances known to be carcinogenic, such as asbestos, chromate compounds, arsenic, or chloroform (Tables 27–2 and

TABLE 27–2   Cancer-Causing Chemicals[a]

Most Dangerous Chemicals[b]	Most Dangerous Chemicals in Workplaces[b]
N-nitrosodiethyl amine	asbestos
thallium	formaldehyde
chromium	benzene
asbestos	lead
nickel	kerosene
coal tar pitch volatiles	nickel
methyl methane sulfonate	chromium
acetamide	coal tar pitch volatiles
yellow OB	carbon tetrachloride
ethyleneimine	sulfuric acid

[a] Eighty-six commonly used industrial chemicals were ranked in terms of their ability to cause cancer in a 1977 NIOSH study. The 10 most dangerous chemicals are listed in the left column. However, if chemicals are ranked not only in terms of their ability to cause cancer but also in terms of how many workers are exposed to them and for how long and in what ways, a new list of most hazardous industrial chemicals is found. These are shown on the right.

[b] Source: *Science* **197**, 1268 (23 September 1977).

27-3). Often the workers or even employers were not aware of the hazards because the toxic substances were in products known by trade name only.

It is not only blue collar workers who are in danger (Fig. 27-9). White collar workers are not exempt. Secretaries and management employees may be exposed to toxic air pollutants at plants (as may the families of asbestos, lead or pesticide workers exposed to these substances from the workers' clothing). Anesthetists and other operating room personnel are more likely than the general population to suffer kidney and liver diseases or cancers or to have babies with birth defects (Fig. 27-10).

Medical Detective Work

Matching diseases with toxic substances   How could a problem of this size have been ignored for so long? One reason, of course, is the use of trade

Figure 27-9   *Old-time miners such as those shown were concerned mainly about hazards to life and limb. However, breathing dust is also a serious health hazard. Among coal miners, it leads to black lung disease (Chapter 13). Workers in the cotton industry may contract brown lung disease from breathing cotton dust, while uranium miners are at an increased risk of lung cancer from breathing uranium dust. (U.S. Department of Energy.)*

TABLE 27-3   Industries Using Cancer Causing Agents[a, b]

Industrial and scientific instruments	solder, asbestos, thallium
Fabricated metal products	lead, nickel, solvents, chromic acid, asbestos
Electrical equipment and supplies	lead, mercury, solvents, chlorohydrocarbons, solders
Machinery	cutting oils, quench oils, lube oils
Transportation equipment	formaldehyde, phenol, isocyanates, amines
Petroleum and products	benzene, napthalene, polycyclic aromatics
Leather products	chrome salts, tanning organics
Pipeline transportation	petroleum derivatives, welding metals

[a] When industries are ranked according to how likely workers are to be exposed to a carcinogenic material, some surprises are found. Industries commonly thought to be "clean" show up in the top 10 on the list. For instance, manufacturing of industrial and scientific instruments requires very little in the way of carcinogenic materials. However, those hazardous substances that are used are involved in hand assembly of machines; thus worker exposure is great. In contrast, the chemical industry, which may manufacture tons of carcinogenic materials, may employ only a few people to carry out the operation. Thus total worker exposure is relatively small.

[b] Source: *Science* **197**, 1268 (23 September 1977).

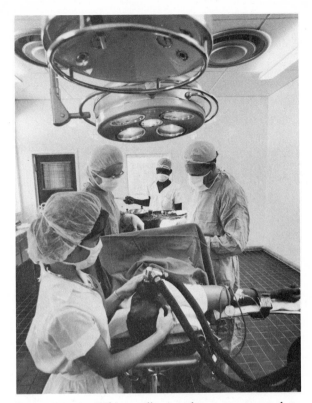

Figure 27–10 *White collar workers are exposed to hazards in industry, research laboratories, and hospitals. People who work as anesthetists in hospitals have an increased risk of developing cancer and other serious diseases, apparently due to exposure to chemicals in their work environment. (©Marc & Evelyne Bernheim/Woodfin Camp & Associates)*

names and the absence of labels that list ingredients by common or chemical names. The 1977 NIOSH study took two years to complete, partly because 70% of the substances found at workplaces were identified only be trade name. NIOSH had to contact 10,000 manfacturers to find out the ingredients. Still, this problem can be remedied.

A more difficult problem is to determine which of the known materials in the workplace are toxic. Direct testing of toxic materials is (intentionally) only done on animals. Effects on humans are detected by studying groups of people who have been unintentionally exposed to toxic materials. This type of epidemiological work is necessarily a science of indirect investigations, since the information must be taken as it is found. Neat experiments cannot be planned to test interesting theories.

In the past, workers have often been the accidental guinea pigs for toxic materials. Asbestos, arsenic, and vinyl chloride were all recognized as carcinogens because of studies showing that workers handling these materials developed more cancers than comparable groups in the population. However, some diseases, such as cancer, do not appear for many years after exposure to the toxic material. Even if a company keeps careful records on employees, it is easy to lose track of workers who are no longer employed and who may become sick after they leave. In some cases, companies have been reluctant to release data on employee health, for a variety of reasons. Fear of being sued by employees or confined by additional government regulations are probably among these reasons.

Laws Protecting Workers: Do They Hurt More Than They Help?

The Occupational Safety and Health Administration (OSHA) is by law responsible for protecting the health of workers. It is OSHA that sets limits for exposure to various toxic materials or determines what safety equipment is needed by workers. Furthermore, OSHA inspects workplaces to see that employers are following the regulations. The problems involved in proving that certain materials are hazardous to workers have in the past made it difficult to set standards. A great many regulations were challenged by manufacturers in hearings or in court. This resulted in long and costly fights among OSHA, manufacturers, and workers rights groups. A new plan, agreed upon by OSHA, the Environmental Protection Agency, the Food and Drug Administration, and the Consumer Product Safety Commission, will possibly change this. Under the

CONTROVERSY

WOMEN'S RIGHTS AND THE WORKPLACE

If American Cyanamid...can get away with removing women of childbearing age from these jobs, we will have established the principle of altering the worker to the configuration of the workplace instead of altering the configuration of the workplace to protect the worker.

> Anthony Mazzocchi, Vice President,
> Oil, Chemical, and Atomic Workers Union. *

It is not that the female is unique, but that we are attempting to protect the fetus. We are talking about exposures that won't hurt the mother but to which the fetus is sensitive.

> Dr. Bruce W. Karrh,
> DuPont Corporate Medical Director *

A number of companies have started to exclude or remove women workers of child-bearing age from jobs where they might be exposed to teratogenic substances (substances that cause birth defects). In one case, four women workers at American Cyanamid claimed they had themselves sterilized because talks with company officials led them to believe they might lose their jobs otherwise. Although some companies are transferring women to less hazardous positions with salaries equal to their former jobs, in other cases women allege they have been discharged or offered only lower paying positions.

Anthony Mazzocchi, vice president of the Oil, Chemical, and Atomic Workers Union, notes that the reproductive capacity and genetic makeup of men as well as women can be affected by many hazardous substances (e.g., the pesticide DBCP, which can cause sterility in males). Furthermore, lead workers, asbestos workers, and men who worked at a Kepone plant have brought home enough dust in their work clothes to affect other members of their families.

Dr. Karrh agrees that cleaning up workplaces to make them safe for everyone would be best:

> But...first of all it is not technically and economically feasible to clean it up to a safe level. And, second, we don't have the data to know what is a safe level. But I will say that this is a problem that occupational health specialists are starting to look at very seriously.

Sue Nelson, director of the Office of Policy Analysis for the Occupational Safety and Health Administration adds:

> What this is really about is that the employers are trying to save themselves from expensive lawsuits....It is easier for a woman worker than a man to bring a lawsuit against a company on behalf of a fetus.

Do you think women able to bear children should be excluded from jobs involving exposure to hazardous chemicals? What about men able to father children? Should they be excluded? Supposing there is some doubt about whether a chemical is teratogenic, who should make decisions (the company, the government, the worker) about whether women and/or men who could be affected should be excluded from jobs involving contact with the chemical? Or should everyone be protected in any case? Suppose we aren't sure how to do that? Suppose it would cost so much in protective measures that the company might decide not to produce the product?

---

*   The New York Times, p. A-1, 15 January 1979. The following extracts are from the same article.

CONTROVERSY

WORKPLACE HAZARDS

...There is now the urgent problem of surveillance and management and treatment of people who [are] at increased risk of developing cancer.

Dr. Irving Selikoff

...It could cost as much as $54 billion to provide warnings and health surveillance services...

The New York Times, 3 October 1977

An area that needs attention is the notification of workers who were exposed to toxic materials in the past but may not know it. Dr. Irving J. Selikoff said in an interview in the November-December 1977 EPA Journal, p. 8:

> ...Both in the workplace, and in the environment in general, there is now the urgent problem of surveillance and management and treatment of high risk groups–people who inadvertently were exposed in the past to agents which we now know places them at increased risk of developing cancer in the future. At present, there is little surveillance or care for them. I consider this a social lapse and I strongly urge that attention be devoted to this as rapidly as is possible.

However, this could cost a great deal of money. According to an article in The New York Times, p. 1, 3 October 1977:

> In discussing the costs involved, according to one Government analysis circulating within the Carter Administration and Congress, it could cost as much as $54 billion to provide warnings and health surveillance services to the 21 million Americans now believed to be exposed to harmful conditions while working.

Yet this cost must be offset against the cost to society of treating occupational diseases. Each year three to five billion dollars is spent on treatment for cancer victims and about 12 billion is lost in wages. Early diagnosis can cure or prevent the spread of many cancers.

Do you feel that the government should attempt to ferret out and warn all workers who have been exposed to toxic materials in the past and who may develop a disease from this in the future? (Remember how much this will cost you in increased taxes. Remember, too, that there are savings to be gained from decreased health and welfare costs.) Do we owe these people treatment as well as warnings?

Can you think of benefits to yourself from keeping track of the health of these people?

plan, all toxic materials will be assigned a category, and the hazard to workers and the public will be estimated. Normal procedures will still be followed in deciding the lowest possible exposure limit that can be met by industry. Still to be decided is whether there will be any comparisons between the risk of a particular substance and the benefits to be gained from its use.

Economics of Controls

Understandably, manufacturers feel that worker protection devices will increase their production costs. We can expect these costs to be passed on to consumers in one way or another. However, society already pays costs comparble to these in medical care, lost productivity, and welfare payments to miners and millers crippled with lung diseases, to cancer victims, and to those poisoned by lead, mercury or pesticides.

Unions also worry about new restrictions. They fear that increased manufacturing costs will cause plants to close, especially small ones, with the result that many jobs will be lost. To this, Douglas Costle, Environmental Protection Agency Admin-istrator, wrote in the December 1977 issue of the *EPA Journal*, p. 4.

"It must also be recognized that environmental regulations create jobs. The facts show that more people have been employed now than would have been without the major pollution control programs. Approximately 19,000 job losses have been attributed to pollution control compared to perhaps half a million jobs that were generated because of cleanup efforts. Such jobs are generated in three ways. First, construction of equipment and plants required by environmental programs create the largest number of jobs....

The second way in which jobs are created is in the pollution control equipment manufacturing industry.

Finally, many more indirect jobs are stimulated by these expenditures.

I want to stress that when we talk about the issue of jobs *versus* the environment we are caught in the old mindset of looking at pollution contols as unproductive, profit-decreasing expenditures. Rather, we need to explore calculating productivity in a larger and more meaningful perspective, one that includes protection of workers' health."

Questions

1. How can chlorination of drinking water cause a possible cancer problem?
2. Should we stop chlorinating drinking water? What are the possible solutions for this problem?
3. Is it reasonable to require only large water supplies to remove chlorinated organics? Aren't the people in small communities important, too?
4. What are the sources of arsenic pollution which cause the most concern?
5. What problems can high arsenic levels cause?
6. Briefly describe the diseases caused by breathing asbestos.
7. What is the effect of smoking while working with asbestos?
8. List the ways you, personally, might be exposed to asbestos fibers. If you can only think of one or perhaps none, make up a hypothetical person and list as many sources as possible from which that person might be exposed to asbestos dust.
9. How would you prevent the person in question 4 from breathing so much asbestos dust? (You can think of ways to avoid the dust or to avoid breathing it.)

10. Why is the ozone layer important?

11. What are the major threats to the ozone layer?

12. Will the human race save itself by accident? It has been predicted that several pollutants will affect the earth's climate, in opposite directions. Perhaps the effects will all cancel. Briefly contrast the possible effects of a decrease in the ozone layer, an increase in the carbon dioxide level and an increase in the level of particulates in the air. How much of a gambler are you?

13. Suppose a friend asked you whether you thought smoking was really dangerous. What would you tell him? Why?

14. How would you answer some of these questions which were asked in the National Clearing House on Smoking and Health Surveys.

    (a) Should cigarette advertising in magazines, newspapers and on T.V. be banned? Why?

    (b) Should management be able to prohibit smoking in places of business?

    (c) Should smoking be prohibited in other public places, i.e., schools, eating establishments, courtrooms, jails, bars, airplanes and trains? (The American Lung Association can supply a booklet "Second-Hand Smoke" which has interesting information for discussion.)

15. What are the three most common sources of radiation?

16. What is meant by natural background radiation?

17. How do average radiation releases from Nuclear Power Plants compare with background radiation levels in the U.S.?

18. Suppose there was a continuing low-level of radiation exposure in a particular industry. How would you suggest health authorities determine if the exposure level is safe?

19. Suppose the frequency of leukemia in the male workers in a particular industry with radiation exposures was three times that for males of the same age in the general population. How would a worker who has cancer prove that the cancer was the result of his exposure on the job?

20. Explain why it is/is not important to try to detect all possible carcinogens before they reach the marketplace. Are we trying to develop a risk free environment? Is this good? Is it possible?

21. What sort of evidence exists that some jobs carry an increased risk of developing cancer or reproductive defects?

22. Should the benefits of certain products be balanced against risks to workers, if exposure to carcinogens is involved? Give an example. Where does the concept of personal freedom fit in with your answer?

Further Reading

Arsenic

Arsenic, Committee on Medical and Biological Effects of Environmental Pollutants, National Research Council, National Academy of Sciences, Washington, D.C., 1977.

This book is a clear summary of the arsenic problem. Included are chapters in the distribution of arsenic in the environment, the chemistry of arsenic, and the biological effects and the metabolism of arsenic.

Toxic Organics

Donaldson, W. T., "Trace Organics in Water," *Environmental Science and Technology* **11** (4), 349 (April 1977).

In part, what chemicals you find in water depends on what methods you use to look for them. This serious problem is explained by Donaldson.

Draft Analytical Report New Orleans Water Supply, U.S. Environmental Protection Agency, Dallas, Texas, November 1974.

The Implications of Cancer-Causing Substances in Mississippi River Water, The Environmental Defense Fund, Washington, D.C., November 6, 1974.

Page, T., R. H. Harris, and S. Epstein, "Drinking Water and Cancer Mortality in Louisiana," *Science* **193**, 55 (2 July 1976).

These three references cover the New Orleans controversy about drinking water and excess cancer deaths. The EDF report, especially, contains a great deal of background on the possible environmental causes of cancer.

Asbestos

Environmental Health Perspectives, Volume 9, December 1974.

A major portion of this volume is devoted to the problems of asbestos in the environment and its effect on health. Many review articles are included. There is also an annotated literature review on asbestos.

Rohl, A. N., *et al.*, "Asbestos Exposure during Brake Lining Maintenance and Repair," *Environmental Research* **12**, 110 (1976).

Sawyer, R.N., "Asbestos Exposure in a Yale Building," *Environmental Research* **13**, 146 (1977).

These papers detail two of the ways in which the general public becomes exposed to asbestos as an air pollutant.

Smoking

Adult Use of Tobacco 1975, Center for Disease Control, National Clearing House for Smoking and Health, Bureau of Health Education, Building 14, Atlanta, Georgia, 30333.

The results of an extensive study of smoking habits of U.S. citizens is published in this booklet.

Introduction to Lung Diseases, American Lung Association (1975).

The American Lung Association has a number of booklets, pamphlets and posters on smoking. Call or write to the association in your state for a listing of what is available.

Report to the Surgeon General on the Health Consequences of Smoking, U.S. Department of Health, Education and Welfare (1971).

Report of the Advisory Committee on Smoking and Health to the Surgeon General, U.S. Department of Health Education and Welfare (1979) available from Office on Smoking and Health, Park Bldg. Rm 158, Rockville, MD 20857.

The volumes summarize the huge amount of data that has now been collected on smoking and health.

Occupational Health

Hoover, R. and J. F. Fraumeni, Jr., "Cancer Mortality in U.S. Counties with Chemical Industries," *Environmental Research* **9**, 196 (1975).

This is an interesting case in point, illustrating the problems and strengths of epidemiology as a tool for determining the environmental causes of disease.

"OSHA on the Move," *Environmental Science and Technology* **11** (13), (December 1977).

A good history of the problems of regulation of toxic substances in the workplace up to this date.

Radiation Hazards

Carter, L. J., "Uranium Mill Tailings: Congress Addresses a Long-Neglected Problem," *Science* **202**, 191 (13 October 1978).

The problem of uranium tailings piles near populated areas is detailed along with possible legislative solutions.

Oak Ridge National Laboratory Review, Summer 1976.

This extremely hazardous, but not well known environmental problem is detailed in this article.

Chapter 28

Toxic Substances In Foods, Drugs, and Cosmetics

DIET AND DISEASE

Many nutritionists now agree that the foods and drinks Americans choose may be contributing to the development of disease. As much as 40% of cancer deaths may be related to eating or cooking American-style. Broiling, frying, and charcoal grilling meats may cause carcinogenic substances to form. High fat diets may cause the production of excess bile and lead to intestinal cancers. Excess fats can also stimulate the production of hormones that could promote breast cancer.

Alcohol is suspected of causing birth defects and also of increasing the risk of respiratory and digestive tract cancers. Excessive amounts of salt can contribute to high blood pressure while excess sugar leads to tooth decay. Furthermore, we are all concerned about various other additives or contaminants that are found in small amounts in foods, drugs and cosmetics and which may cause diseases such as cancer.

INTENTIONAL FOOD ADDITIVES

Why Do Food Manufacturers Use Additives?

A trip to the supermarket should convince anyone that many additives are used to color, preserve, or otherwise "improve" food, drugs, and cosmetics. Over 1800 different chemicals are used in foods alone. The term "additive" covers a wide variety of materials. In foods there are three main groups. The first includes natural substances such as sugar, salt, or vitamin C. A second group consists of laboratory copies of natural substances, an example of which is vanillin, the chemical that is the main flavoring in natural vanilla bean extract. There are also the substances that are entirely synthetic, or invented in the laboratory; BHA, EDTA, and saccharin among others (Fig. 28–1).

Additives are used for many reasons, all of them understandable. However, some are more justifiable than others. Many are used to increase a product's appeal to consumers. Drugs are flavored to help hide bitterness or other unpleasant tastes. Foods may be colored to help identify flavors (yellow for lemon candies, pink for strawberry ice cream). But colors and flavors are also used as substitutes for expensive ingredients that are left out of cosmetics or foods. For instance, real fruit is often missing from artificially colored and flavored soft drinks.

Modern methods of selling food have made certain additives necessary. Chemicals that kill mold and keep foods moist mean that bakery products and candies can be shipped across the country and still remain fresh tasting for long periods of time. Antioxidants, which prevent fats from becoming rancid, make convenience foods such as boxed cake mixes possible. In fact, the whole group of convenience foods and special diet foods probably could not exist without the food ad-

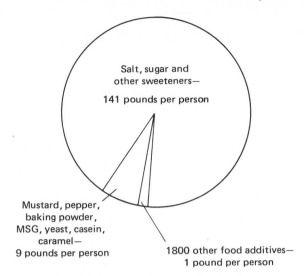

Salt, sugar and
other sweeteners—

141 pounds per person

Mustard, pepper,
baking powder,
MSG, yeast, casein,
caramel—
9 pounds per person

1800 other food additives—
1 pound per person

Figure 28–1   *Americans consume about 150 pounds of food additives per person each year. Most of this is salt, sugar and other sweeteners.*

ditives that flavor, color, and stabilize them. In some cases, additives make a wider variety of foods possible. Certain foods could not be canned, frozen, or packaged for shipment or out-of-season selling without additives.

There is not complete agreement, of course, that all or even most additives are necessary. For instance, it has been argued that the consumer does not gain from the use of preservatives:

> Even assuming that these chemicals are harmless, the advantage in selling bread that does not go stale for a week or more, to take an example, seems to lie more with the baker and retailer than with the consumer.[1]

It should be noted, however, that a shorter shelf life would mean higher costs to manufacturers. In a competitive market, this would be passed on as higher food prices for the consumer.

In another instance, at Senate hearings in 1972, Dr. Michael Jacobson, a co-leader of a consumer

action group, demonstrated several dozen pairs of similar food items, one of which had no chemical additives while the other had preservatives, emulsifiers, dyes, or anti-foaming agents. He stated, "This shows that such chemicals are rarely, if ever, needed in the production of quality foods if manufacturers use good manufacturing practices"[2] (Fig. 28–2).

However good the reasons for using food, drug or cosmetic additives may seem, we would like to be sure that the additives are safe.

Protecting the Public

Food, drug, and cosmetic laws   The safety of food, drug, and cosmetic additives is governed by the Federal Food Drug and Cosmetic Act of 1938 and its amendments. Adulteration of food is prohibited. This means harmful substances cannot be added and also that unexpected ingredients (such as low-cost fillers) cannot be substituted for materials the consumer would normally expect to find in a product.

In 1958, a food additives amendment was passed. Under this law, the manufacturer is required to prove the safety of an additive before it can be marketed. It is estimated that it costs a manufacturer between $200,000 and $3,000,000 to test a new food additive and prove it safe enough to market.

Also in the 1958 amendment is the Delaney clause, which forbids the use of an additive, in any amount, if it has been shown to cause cancer in humans or animals. This clause has caused a great deal of argument because it is felt by some people to prevent reasonable judgments about risks compared to benefits. Some of the arguments for and against the clause are detailed in the section on artificial sweeteners.

Still unsure that food additive laws fully protected the public, Congress passed color additive amendments in 1960. These amendments

1 Dr. W. Lijinsky, testifying at Senate hearings on food additives. *The New York Times,* 22 September 1972, p. 1.

2 *The New York Times,* 22 September 1972, p. 1.

FOOD ADDITIVES: WHAT IS A SAFE ADDITIVE?

There are no safe or unsafe food additives, merely safe and unsafe levels of use....

Manufacturing Chemists Association

Scientists Hear Reports Vinyl Chloride May Be More Dangerous Than Realized.

Headline in Wall Street Journal
13 May 1974.

One way proponents of food additives often justify the use of these substances is to point out that "everything is made up of chemicals, even people" and that food additives are, after all, just other chemicals. The Manufacturing Chemists Association states in a booklet entitled Food Additives:

> Regardless of whether an additive comes from foods or from other substances, it is chemical in nature. Failure to appreciate this fact causes a great deal of misunderstanding regarding the food supply. Too often, people without scientific background believe that the word "chemical" means something dangerous, or, at best, "unnatural." They fail to realize that everything in the world is essentially chemical–from the concrete and steel in the Empire State Building to the vitamins in our food. The average homemaker would look incredulous if she were told she had just fed her family triglyceride esters of palmitic, oleic, linoleic and stearic acids. But when she learns that words are the chemists' terms for the chief components of cooking fats and shortenings, she begins to understand that foods are made of many chemicals.
>
> It is a new concept to many people when they realize that not only food but the elements that go into clothing and shelter, and even the earth itself and all its inhabitants, can be described in terms of chemicals. *

Perhaps this is a bit simplified. Certainly some chemicals are safer for humans to eat than other chemicals. Another booklet from the MCA covers this point:

> Remember that there are no safe or unsafe food additives, merely safe and unsafe levels of use or ways of using them. The study of such levels, and the effects of processing, cooking and ingesting chemical compounds put in foods by man or nature is the work of toxicologists, analytical chemists and other scientists. This is a never-ending task. Advances in one area stimulate study and

greater learning in others. Some substances–such as selenium salts–earlier declared toxic at all levels have since been found not only safe but, below specific levels, are essential for health. After years of use, cyclamates were banned as food additives; more reliable tests may prove these sweeteners entirely safe. The reverse is also true.[†]

Whether we agree or disagree that "there are no safe or unsafe food additives" perhaps comes down to whether we believe that we are presently able to decide on safe or unsafe levels of chemicals that may or may not be toxic. Faith in our capability to make these decisions is continually shaken by articles such as the one below on the use of a vinyl chloride compound to wax fruit. You may remember that vinyl chloride is a chemical used in producing polyvinyl chloride plastics. Workers in this industry were recently shown to suffer a rare type of liver cancer, apparently caused by breathing vinyl chloride. Vinyl chloride was also used as a propellant in spray cans of pesticides, hairsprays, and a variety of other consumer products. These uses were all ended when it was decided by government scientists that the safety of vinyl chloride had not been proven at any level.

> Polishing apples has become as obsolete as rug-beating. These days, apples...and certain other fruits and vegetables, are often waxed....
> ...the coatings most frequently used on fruits and vegetables are carnauba, which is derived from palm fronds, paraffin, and mineral oils.
> Food and Drug Administration regulations, however, allow the use of other chemicals for coating produce. Among these is a substance with the chemical name vinyl chloride.
> According to Gerad McCowan, the assistant to the director of the F.D.A.'s division of food-color additives, the vinyl chloride compound has never been used for coating citrus fruit. He said that the agency had originally approved the use of the substance upon the application of the manufacturer of the substance.[‡]

* "Food Additives, What They Are/How They Are Used," Manufacturing Chemists Association, 1974, p. 16.
† "Food Additives Who Needs Them?," Manufacturing Chemists Association, 1974, p. 8
‡ Frances Cerra, The New York Times, 17 December 1975, p. 62.

Figure 28-2 *In some schools, parents who are concerned about the amount of artificial flavors, colors, and preservatives that children eat and drink have banned the sale of foods that contain additives in the school lunch room and vending machines. Soft drinks are replaced by milk and fruit juices. Other snack foods are replaced by fruit or additive-free cookies and potato chips. (Edward Hausner/NYT pictures)*

direct the FDA to set tolerances, or safe limits, for the amount of colors used in foods, drugs, and cosmetics. Furthermore, the FDA was told to continue to *certify* colors. In this procedure, FDA laboratories safety test a sample from each batch of food color. The batch is then given a numbered certificate. The manufacturer must account for what happens to each pound of food color in the certified batch. A few colors are declared safe without certification by the FDA. These are usually colors from natural sources such as grape skin extracts or red beet juice.

These are Federal laws, which apply to foods that are sold across state boundaries. Foods sold only within a state are subject to the laws of that state.

The GRAS list   When Congress passed the 1958 Food Additive Amendment, which required manufacturers to prove the safety of food additive chemicals they wished to manufacture, a list of some 600 chemicals was prepared and designated the GRAS List (Generally Regarded As Safe). These chemicals were exempted from the amendment. That is, they were allowed to be added to food, even though they had not undergone a great deal of testing to prove their safety, since they had been in use for a number of years. Chemicals were put on the list if it was the opinion of experts in nutrition and toxicology that no unhealthful effects had ever been traced to these additives. Since then, some chemicals have proved to be less safe than was believed. In 1969, the Food and Drug Administration began an extensive testing program to determine whether the items on the GRAS list were indeed safe.

About 300 items on the list have been tested so far. Of these, 80% have been declared safe. Another 14% are probably safe but are undergoing further study. About 4% were declared uncertain and are also being studied. Two percent were found to be unsafe and have been removed from the market or restricted in use. This last group includes carbon black, red dye #2, monosodium glutamate, and saccharin.[3] The sections that follow cover in detail some of the additives, intentional and unintentional, found in foods, drugs, and cosmetics.

---

3 In 1978, the Center for Science in the Public Interest published a table in which common food additives are grouped according to estimates of their safety. The table can be ordered as a poster from C.S.P.I., 1755 S Street, N.W., Washington, D.C. 20009.

ARTIFICIAL SWEETENERS AND THE DELANEY CLAUSE

Substitutes for sugar have come to seem not only desirable but necessary to a nation beset by weight problems. In the United States, almost five million pounds of the artificial sweetener saccharin were sold in 1976 (Fig. 28–3). Thus, it came as unpleasant news in March 1977 that the FDA was considering a total ban on the use of saccharin in foods. New studies had shown that saccharin caused bladder cancer in rats.

The Domino Effect

Saccharin was not the first artificial sweetener to be declared unsafe. In 1969, the FDA banned cyclamates. Sales of this group of artificial sweeteners had reached around a billion dollars per year when it was found to cause bladder cancer in experimental rats. At the time of the cyclamate ban, manufacturers of diet foods were upset but not desperate. After all, they had saccharin to fall back on. Diet soda recipes were changed to leave out cyclamates and rely on saccharin. In 1975, aspartame, a sugar substitute that was approved for use

Figure 28–3 *Saccharin is used in products where sugar would be too bulky, e.g., in pills to make them easier to swallow. Saccharin is also used to flavor some toothpastes, because sugar is felt to contribute to the formation of dental caries (cavities). (NYT pictures)*

in 1974 was withdrawn from the market when it was shown to cause brain seizures in monkeys.

When the saccharin ban was proposed, it began to seem like a game of dominos. Doubts about the safety of one artificial sweetener caused scientists and officials to look harder at the safety of others, which were in turn found to be possible hazards.[4]

Mail flooded congressional offices when the saccharin ban was announced. One Representative said he had received more mail about saccharin than about any other subject during the three years he was in Congress. Most of the mail opposed the saccharin ban. In addition, soft drink manufacturers began a vigorous lobbying campaign. Three quarters of the saccharin used in 1976 went into diet soft drinks. The controversy that followed highlights many of the problems involved in regulating food additives and drugs.

Experimental Evidence

In the first place, experiments showing that saccharin causes cancer were challenged. The experiments were done on rats, a common experimental animal. The rat, however, concentrates its urine more than other species before excreting it. Rat bladders were thus exposed to higher concentrations of saccharin than might be the case in other animals.

Saccharin does not cause mutations in bacterial cells (the Ames Assay) but it has been shown to cause mutations in test-tube cultures of hamster cells and human cells. Although this does not prove that it is carcinogenic, it is cause for suspicion. (These tests are described on p. 539.)

In the rat experiments, very high doses of saccharin were used, much higher than humans would be expected to consume. The rats were fed saccha-

---

4 Eight months after the proposed saccharin ban, xylitol, an artificial sweetener used in gum, was also found to cause cancer in rats.

rin as 5% of their diet for two generations. This prompted a number of people to calculate that a human would have to chew 6700 pieces of artificially sweetened bubble gum daily or drink 800 diet drinks each day to receive the same dose the rats did. Are these reasonable comparisons? Not really. Rats live for only 2½ years compared to humans' 70–80-year lifetime. Thus, researchers using high doses are trying to compensate for the longer exposure time humans have (and thus recreate the higher accumulated dose). Furthermore, we are interested in finding out if a chemical is carcinogenic even if it only causes one case in 100,000 people per year. (That would be 2000 cases of cancer in the United States each year caused by saccharin.) In order to detect a carcinogen of this strength at normal human levels of exposure, scientists would have to run experiments on hundreds of thousands of rats. The expense and work involved are enormous. Thus higher doses are used in the belief that there is a relationship between dose and rate of cancer causation: the higher the dose, the more likely a carcinogen will cause cancer. There is evidence that this is reasonable to assume.

Social Effects of a Saccharin Ban

Leaving aside questions about the experiments themselves, many people felt that a saccharin ban smacked of the prohibition era. They argued that, if there was a risk, they would rather be told about it and left to make their own decision.

The Delaney Clause

The FDA, however, was on firm legal grounds in proposing a saccharin ban. A 1958 amendment to the Food, Drug, and Cosmetic Act states "no additive shall be deemed to be safe if it is found, after tests which are appropriate for the evaluation of the safety of food additives, to induce cancer in man or animal" [Section 409(c)(3)(A)]. This is known as the Delaney Clause after the Representative who proposed it. There has been much criticism of the

Delaney Clause. In the main, this is because it allows administrators no flexibility. That is, they are not given the opportunity to decide whether a very small amount of a carcinogenic substance might either be harmless, or worth the risk. Second, foods such as meats and dairy products usually contain detectable amounts of pesticides such as DDT. This is a carcinogenic chemical that has become widespread in the environment. It is incorporated into foods without being intentionally used in their production. The Delaney Clause could be used to justify stopping the interstate shipment of such foods. In practice, the clause is not used this way, but it is felt by some to be a law that could not be enforced and, therefore, one that is invalid. Those in favor of retaining the Delaney Clause point out that we have no evidence to show that small amounts of a carcinogen can be harmless. The safest course is to assume that any amount of such a chemical can cause some cancers, the number of cancers just becoming smaller as the dose is decreased. Since even weak carcinogens can act together to cause greater hazards, the Delaney Clause provides a safeguard against the additive effects of small amounts of carcinogens in the environment.

Another important feature of the Delaney Clause is that it allows decisions to be made on fairly black and white issues. Either a substance causes cancer in experimental animals or it does not. This prevents the kind of delays in removing carcinogens from the marketplace that would occur if regulatory agencies had to fight long drawn-out court battles over risks versus benefits. Nonetheless, the saccharin issue has provided the strongest challenge to the Delaney Clause since it was passed in 1958.

Artificial Sweeteners and Diabetics

At congressional hearings on the proposed saccharin ban, several people argued that artificial sweeteners are necessary not only for weight con-

CONTROVERSY

DOES EVERYTHING CAUSE CANCER?

We must not ban the universe—or even paper it over with warning labels.

Congressman James G. Martin

It is argued that any substance in large enough amounts is carcinogenic. For instance, Representative James Martin,* speaking before the Manufacturing Chemists Association said:

> The rapidly growing abundance of over 2,000 suspect chemicals seems limited only by the resources of rat breeders and feeders. It seems probable that for every chemical there is a dose above which the metabolic defenses and DNA repair and immune systems of some cancer-prone rat can be overwhelmed, especially if concentrated <u>in utero</u> by a transplacental effect. Ironically, the only chemicals unlikely to be overdosed to a carcinogenic level are those that are lethal poisons. If cyanide were a sweetener, it would pass.

This type of argument has been used to protest bans on artificial sweeteners, hair dyes, and food colors, but experiments do not really bear out the claims made. When the National Cancer Institute tested 140 chemicals suspected to be carcinogens (by feeding them to animals at the maximum amount which did not kill or poison them directly) less than 10% of the chemicals caused cancers. If this small a proportion of actually suspected chemicals causes cancer, it seems reasonable to assume that the ability of a chemical to cause cancer is a relatively rare phenomenon.

Suppose you were debating Martin on the floor of the House as he attacked regulations banning a popular food additive suspected of being a weak carcinogen. Could you explain the arguments that everything does not cause cancer, and that dosing animals with high levels of suspected carcinogens is a reasonable way to prove whether or not these chemicals are carcinogenic?

---

* Mr. Martin is also a Ph. D. chemist. The extract above is taken from his speech in October 1978.

trol but also for diabetics who must limit the amount of sugar they use. It was stated that it would be a special hardship to these people not to have a sugar substitute. On the other hand, it was also pointed out that there are no medical studies that show diabetics who use saccharin do a better job of sticking to their prescribed diet than those who do not.

Is a Saccharin Ban the Right Solution?

In the end, Congress bowed to strong public pressure and directed that the proposed ban be put off while the evidence was sifted. Meanwhile the FDA required stores selling saccharin containing products to display a warning sign stating that saccharin had been shown to cause cancer in rats. This warning was also required on the packages of foods containing saccharin. The original House version of the bill did not include a warning label, which prompted Representative Andrew Maguire to jokingly propose that manufacturers should include an "assurance" label stating:

> Assurance: Saccharin does not cause cancer, in the opinion of your Congressman, despite all scientific evidence that it does.[5]

A study committee of the National Academy of Sciences and the Institute of Medicine[6] concluded that saccharin was a "moderate" carcinogen.[7] Nonetheless, the committee did not recommend a total ban on the sweetener. Rather, the committee suggested consumers be warned of the possible hazards of saccharin use. Further, since studies in 1973–1978 showed that one-third of children under

---

5 *The New York Times*, 18 October 1977, p. 16.

6 Food Safety Policy: Scientific and Policy Considerations, National Institute of Medicine and National Academy of Sciences, Parts I and II (1979). (Available from National Technical Information Service.)

7 See Figure 26–2 for an assessment of the relative carcinogenic potential of saccharin.

10 years of age consumed food products containing saccharin, mainly diet soft drinks, the committee recommended that steps be taken to reduce children's exposure to saccharin.

In contrast, the Canadian government did ban saccharin in foods in 1977. Doctors there say that this has had no visible effect on the health of diabetics in Canada. Further, the diet soda industry has developed two new types of diet sodas for the Canadian market. One is a group of low-sugar sodas that have about 60 calories per can, about half that in the regular product. The second type are no-sugar sodas that can be sweetened with saccharin tablets, which are still available to consumers. The president of the Canadian Diabetes Association noted that most people use only one or two 15-milligram tablets per can. This should be compared to the 120 milligrams per can used in commercially prepared diet soda.

ARTIFICIAL COLORS IN FOODS

A fact known to every food manufacturer is that grape soda tastes better when it has a rich purple color. Ice cream seems creamier if its slight yellow color hints at thick cream and egg yolks. And no one wants to buy the oranges that are pale orange or slightly greenish. These and other problems and effects call for the use of food colors; substances added to food to cover-up or improve natural colors. Any material added to food to color it is called artificial color, although the material used might be called "natural." For instance, carotene, the substance that makes carrots orange, is often used to color other foods orange. The food must then be labeled artificially colored.

Food colors have a somewhat unpleasant history. Before strict regulations were enforced, it was not uncommon for colors to be added to cover up spoiled or adulterated food (e.g., food to which some unlabeled, cheaper ingredient had been added). Furthermore, the food colors used were sometimes poisonous. Mercuric sulfide and red lead

CONTROVERSY

ARTIFICIAL SWEETENERS

All day long we've been taking calls from people, some of them in tears, demanding that we leave saccharin alone.

FDA Official *

Somebody's trying to pull off something funny, and we're the ones who are going to get hurt.

Alvin Brown, packer at
Sweet 'n Low Company *

Establishing "low level" level residue tolerances for any one carcinogen in food would, by precedent permit unlimited numbers of other carcinogens also to be added.

Dr. Samuel S. Epstein, †
testimony before Senate
Select Committee on
Nutrition and Human Needs,
20 September 1972.

Dr. Morris Crammer, head of the National Center for Toxicological Research has written, about saccharin:

> I think most of us are unprepared, in the absence of persuasive evidence, to replace all those items of human convenience, even though an ultraconservative mathematical hypothesis may imply a relatively high level of risk. ‡

Do you feel that the FDA should ban saccharin on the basis of the Delaney Clause (no substance which has been shown to cause cancer may be added to food) or should people simply be warned about the hazard and left to make their own decision? This is the way alcoholic beverages (known to cause birth defects) and cigarettes (known to cause lung cancer) are treated. Are there basic differences between alcohol, tobacco, and foods which would mean they should be regulated differently?

---

* Quoted in The New York Times, 11 March 1977, p. 28.
† Testimony before Senate Select Committee on Nutrition and Human Needs, 20 September 1972.
‡ Quoted in The New York Times, 20 July 1978, p. A17.

were on occasion used to color cheese. Candies were often colored with lead chromate, red lead, or white lead. Forty-six percent of the candy sampled in Boston in 1880 contained one or more of these poisonous coloring agents.

Despite this rather criminal past, it is considered today that food colors can serve some reasonable uses as food additives.

There are two general groups of food colors. The first group, called uncertified colors, consists of colors that are for the most part natural products. That is, they are usually extracts of various plants.

The second group of colors is called Food Drug and Cosmetic Act Certified (FD & C) colors. These are synthesized or extracted from nonplant materials, such as coal tar.

The industry using the largest amount of certified color is the beverage industry. About 30% of all soft drinks sold contain certified color. Sixty percent are colored with caramel coloring (the cola drinks), while 10% are uncolored (club soda and lemon-lime drinks).

In 1963 Congress directed the FDA to check the safety of the certified and uncertified dyes in use. Nine certified food colors were taken off the approved list after this was done because it was decided they were unsafe or their safety was not proven. Since then, three more dyes, red #2, red #4 and carbon black, have been removed from the approved list when expriments showed they might be carcinogenic.

PRESERVATIVES

Keeping Foods Fresh

One of the major problems in the food industry involves keeping foods from becoming stale or spoiling between the time they are made in the factory and the time they are bought and served. The reaction of fats with oxygen leads to the forma-

tion of a variety of bad tasting, smelly compounds. Oxidized fats are said to be rancid. Antioxidants are added to foods to prevent such reactions. BHA and BHT[8] are examples of antioxidants used in foods. Other common antioxidants are propyl gallate and ethoxyquin.

The Special Case of Nitrates and Nitrites

Nitrites and methemoglogin   Nitrates (NO_3^-) and nitrites (NO_2^-) are a special case among the food preservatives. They were originally added to meats and fish, before refrigeration was available, to prevent them from spoiling. Over the years, however, people have come to like the cured flavor these additives give to meat. Thus, we now eat cured meats not only because they are a safe way to keep meat but because we like the flavor.

The combination of sodium nitrate and sodium nitrite has three effects on meats. It prevents the growth of bacteria which cause various kinds of food poisoning such as botulism. It creates the typical pink color, of cured meats, such as ham. And nitrates and nitrites give meat that special "cured" flavor. Meat was originally cured by adding only potassium nitrate. It was later found that bacteria were converting some of the nitrate to nitrite. The nitrite is actually the chemical that prevents the growth of bacteria and gives the meat a pink color.

Nitrites (NO_2^-) are not harmless compounds. Hemoglobin in the blood, which carries oxygen to all the body cells, reacts with nitrites to form methemoglobin. Methemoglobin cannot carry oxygen. When 70% of the hemoglobin in the blood is converted to methemoglobin, death results from suffocation. At lower levels of methemoglobin, there may be problems such as dizziness or difficul-

---

8 BHA, butylated hydroxyanisole; BHT, butylated hydroxytoluene.

ty in breathing. For this reason, there are legal limits to the amount of nitrite which can be added to meat or fish to cure it. Cases have occurred in which children have been poisoned by bologna and frankfurters containing more nitrite than the legal limit.

Although most of the nitrite (NO_2^-) we eat is from cured meats, the nitrate (NO_3^-) we eat comes in about equal amounts from vegetables and cured meats. Spinach, beets, eggplant, radishes, celery, lettuce, and greens all can have high levels of nitrates. Since nitrates (NO_3^-) are not as poisonous as nitrites (NO_2^-), because they do not oxidize hemoglobin, problems do not usually arise unless bacteria change the nitrates to nitrites. This has happened in some cases, for instance, when opened jars of baby food spinach have been left unrefrigerated. Infants have been poisoned in this manner.

Nitrites and cancer   Besides the fact that nitrites (NO_2^-) lead to the formation of methemoglobin, there is another problem with respect to adding nitrites to foods. Nitrites can react with certain amines, chemicals that are also present in foods, to form nitrosoamines. In one study, 75% of a variety of nitrosoamines tested were carcinogenic, that is, they caused cancer in test animals. Cancers were produced in all species of animals tested (dogs, monkeys, rats, parakeets, hamsters, guinea pigs, mice, and trout). Nitrosoamines also seem to act with weak carcinogens to make them stronger. A 1974 EPA report concluded "as a family of carcinogens, the nitrosoamines have no equals."[9]

Nitrosoamines are also found as air pollutants, especially in cities. It is believed they are formed from nitrogen oxides pollutants in urban air. It has been suggested that the presence of nitrosoamines could explain the higher occurrence of cancers in urban areas. Further, nitrosoamines are found in

cutting oils, used in industry where metals are cut or ground, and in some pesticides. The total effect of nitrosoamines in food and in the environment is not yet known. Certain scientists, however, feel that the effect, in terms of causing cancer, will eventually be found to be very large.

Should nitrites be banned as food additives? Before 1978, it was illegal for cured meats such as sausage, ham or bacon to contain more than 200 parts per million of nitrites. This limit was determined by considering the effects of nitrites on an average person. At this level, 100 grams of meat (1/4 pound of corned beef on rye) would convert from 1.5% to 5.7% of the consumer's hemoglobin to methemoglobin. If the person was also a heavy smoker, living in a city, a further amount of hemoglobin would be inactivated by the carbon monoxide in the cigarette smoke and urban air. The combined effects of the air pollution, smoking, and corned beef sandwich would be just short of causing the first signs of oxygen starvation, headaches.

In 1975 it was shown that when bacon containing legal levels of nitrites was cooked until it was crisp, nitrosoamines formed at the parts per billion level. There is some evidence that vitamin C can prevent this reaction. The FDA considered a ban on the use of nitrites in cured meats such as bacon. However, the question of safety from food poisoning was also involved. That is, would people be in more danger of dying of food poisoning if nitrites were not allowed in cured meats than they would be in danger of suffering a nitrosoamine caused cancer if nitrites were allowed to be added?

Since the main problem appeared to involve cooked bacon, the FDA lowered the amount of nitrite allowed in bacon. Another preservative, potassium sorbate, is added to prevent botulism food poisoning. You can see from the ingredient list on the package that some manufacturers of sausage meat and bacon no longer add nitrites to their product. However, in at least one recent study, nitrites all by themselves, appeared carcinogenic. If these

9 See Figure 26–2 for a comparison of one nitrosoamine, dimethylnitrosoamine, with various other carcinogens.

results are confirmed, the FDA will move to phase all nitrites out of foods.

ADDED VITAMINS AND MINERALS

A number of manufactured food products contain added vitamins and minerals. In some cases, this is required by law to replace nutrients lost during processing. Thus white bread is enriched with thiamine and riboflavin, vitamins lost when wheat is milled to white flour. In other cases, vitamins are added to commonly used foods to ensure good nutrition. For instance, ready to eat breakfast cereals, which are of little nutritional value but which are eaten for breakfast by millions of children, are enriched with vitamins. Artificial orange breakfast drinks are enriched with vitamin C because they are widely used as a substitute for orange juice, a natural source of the vitamin.

Such supplements are not without hazard, however. Certain vitamins, notably A and D, can accumulate to toxic levels in the body. Children given large doses of vitamin A have developed a deformity in which one leg is shorter than the other. Some pregnant women who received calcium and vitamin D supplements, drank vitamin D enriched milk and sunbathed (a natural source of the vitamin, which is formed on the skin by the action of sunlight) have borne babies with excess calcium deposits in the skull. Some scientists have voiced concern that excess iron may cause people to become susceptible to disease. Such an effect has been seen in animal experiments.

The FDA is responsible for ensuring that people do not receive excess doses of vitamins and minerals from foods to which they have been added. In the face of the uncertainty that exists over what is and is not an excess dose, however, the consumer would probably be wise to avoid foods that must be enriched to be nutritious. Instead, it would be safer to rely on less processed foods, traditional safe sources of the vitamins and minerals necessary for good health (see Controversy p. 598).

UNINTENTIONAL FOOD ADDITIVES

There are three groups of toxic substances in foods that are not added to them on purpose. The first is naturally occurring toxins. The second group includes traces of pesticides (residues) left from the spraying of crops. Third, there may be drug residues in meat, left from chemicals added to animal feeds.

Actually there is a fourth category that could be included under unintentional food additives: additions that occur because of some accident during the manufacture or shipment of food. The PBB disaster, discussed on p. 604 provides an example of such an accidental contaminant.

NATURAL TOXINS

Some toxic materials are formed in foods by natural processes during storage. Mycotoxins are produced by molds, especially under damp storage conditions. Aflatoxins, produced by the mold Aspergillus fluvus, are potent carcinogens that have been found in rice, peanuts, corn wheat, and food products from Asia, Africa, and North and South America. (See Figure 26-2 for an estimate of the carcinogenicity of aflatoxins.) In Georgia in 1977, a drought was followed by an aflatoxin plague that destroyed the value of the remaining corn crop. Corn cannot be sold in interstate commerce if it contains more than 20 parts per billion of aflatoxin.

Charcoal broiling meats may even be hazardous. Two types of materials appear to be produced. One is related to the tars in cigarette smoke, the other is a breakdown product of proteins in the meat. There is some evidence that these materials might be carcinogenic.

PESTICIDE RESIDUES IN FOOD

Regulation of Pesticides in Food: A Problem for EPA

In 1978, the House Commerce Committee began hearings on cancer-causing chemicals in foods. Many problems in regulating carcinogenic substances such as pesticides,[10] in foods, were brought out in testimony before the committee.

How do pesticides, toxic chemicals intended to kill pests, get into food? In many cases, small amounts of pesticides used to treat crops or animals remain on produce or in meat when they are marketed. These small amounts are called *residues.* There are government regulations stating exactly how high residues can legally be and listing procedures that must be followed to keep residues below the legal limit.

In other cases, however, foods become contaminated with pesticides that were not even used to grow food crops. This is a process of secondary contamination in which pesticides in dust or in rain land on food crops. For instance, most food products from animals (milk, meat, cheese) contain measurable amounts of the pesticide DDT. This comes about because of widespread contamination of the environment with DDT, which is picked up by wind on dust particles and blown around the world. Rain washes DDT out of the air and onto pastureland. Grazing cattle eat DDT contaminated grass and incorporate the DDT into their meat and milk.

In the United States, the Environmental Protection Agency is responsible for protecting society from the risks involved in using pesticides. The most recent major pesticides law, passed in 1972, is the Federal Environmental Pesticide Control Act. Under this law, manufacturers must supply the EPA with data showing that a pesticide they wish to sell is effective against pests and will not harm humans or the environment. If the EPA is satisfied with the data it will *register* the pesticide. In some cases a *tolerance* is allowed. That is, a certain amount of residue of the pesticide may safely remain on food or feed when it is marketed.

EPA also has the power to *cancel* the registration of a pesticide it decides is harmful in some way. The manufacturer may protest cancellation and continue to ship and sell the product until the issue is finally decided by EPA. However, if EPA feels there is an immediate danger to human health and welfare, the registration of a pesticide can be *suspended,* in which case all shipments must stop at once.

Regulation of pesticides has proven to be a major problem for the EPA. By 1976, $2.4 billion in pesticides was sold every year. The industry is expected to grow to $3.3 billion by 1984. Industries on this scale have political muscle. EPA, itself, estimates that one-third of the 1500 active ingredients in pesticides are toxic and up to one-fourth may be carcinogenic. Critics point out that although a number of pesticides are restricted to use by certified operators (Fig. 28-4), only a few pesticides have so far been banned except for emergency use.[11] Part of the reason for this record may be that industry as well as other special interest groups are almost certain to challenge regulations through law suits and by stimulating congressional pressure to reverse decisions. For instance, when aldrin/dieldrin was suspended, both Shell Oil, the manufacturer, and the Environmental Defense Fund filed suit, because neither group was happy with the decision (although for exactly opposite reasons). In the same vein, when Mirex was banned for control of fire ants, congressmen from the southern states where the ants are a problem

---

10 Pesticides and the problem of growing enough food for the world's people are discussed in Chapter 34. Pesticides as water pollutants are noted in Chapter 6.

11 DDT, for instance, and heptachlor/chlordane, Mirex, DBCP, and aldrin/dieldrin.

LIFESTYLE AND DISEASE

Don't let any food faddist or organic gardener tell you there is any difference between the vitamin C in an orange and that made in a chemical factory....

Dr. Frederick J. Stare
Chairman of Department
of Nutrition,
Harvard School of
Public Health

The guru of food faddism is not Adelle Davis, but Betty Crocker.

Nutrition Action

The American diet has become a controversial subject. One group of people, sometimes labeled "health food faddists," argue that we should eat natural foods, grown without fertilizer or pesticides and packaged without additives or fortifiers of any kind. Others reply that there is no difference between "natural foods" and those grown with fertilizers and pesticides, or between vitamin C from rosehips and that produced in a laboratory.

The composition of our bodies can be stated in terms of water, protein, fat, carbohydrates, vitamins, and minerals. And so can the composition of an orange or a loaf of bread. And don't let any food faddist or organic gardener tell you there is any difference between the vitamin C in an orange and that made in a chemical factory and added to grape or apple juice–(or) between the B-vitamin thiamine in pork and that purchased from the chemical factory and added to flour as a part of the enrichment procedures. *

And according to Dr. Bernard Oser, former member of the National Academy of Sciences Food Protection Committee, not only are some additives harmless, they are essential.

Were it not for food additives, baked goods would go stale or mold overnight, salad oils and dressings would separate and turn rancid, table salt would turn hard and lumpy, canned fruits and vegetables would become discolored or mushy, vitamin potencies would deteriorate, beverages and frozen desserts would lack flavor and wrappings would stick to the contents.[1]

The other side of this argument can be seen in the following discussion:

Food faddism is indeed a serious problem. But we have to recognize that the guru of food faddism is not Adelle Davis, but Betty Crocker. The true food faddists are not those who eat raw broccoli, wheat germ, and yogurt, but those who start the day on Breakfast Squares, gulp down bottle after bottle of soda pop, and snack on candy and Twinkies.

Food faddism is promoted from birth. Sugar is a major ingredient in baby food desserts. Then come the artificially flavored and colored breakfast cereals loaded with sugar, followed by soda pop and hotdogs. Meat marbled with fat and alcoholic beverages dominate the diets of many middle-aged people. And, of course, white bread is standard fare throughout life.

This diet–high in fat, sugar, cholesterol, and refined grains–is the prescription for illness; it can contribute to obesity, tooth decay, heart disease, intestinal cancer, and diabetes. And these diseases are, in fact, America's major health problems. So if any diet should be considered faddist, it is the standard one. Our far-out diet–almost 20 percent refined sugar and 45 percent fat–is new to human experience and foreign to all other animal life....

It is incredible that people who eat a junk food diet constitute the norm while individuals whose diets resemble those of our great-grandparents are labeled deviants....[‡]

How do you feel about "health foods?" What does make up a good diet?

* Frederick Stare, "Food Additives, What They Are/How They Are Used," Manufacturing Chemists Association, Washington, D.C., 1971, p. 16.

† B. Oser, Food Additives–Mountains or Molehills, a speech given before the National Newspaper Food Editors Conference, San Francsco, California, on 25 September 1970.

‡ Editorial, Nutrition Action, Center for Science in the Public Interest, April 1975.

Figure 28–4 *Farmer studying a manual on pesticide use during a pesticide applicator training course. A number of pesticides have been classified for restricted use only. This means that only persons who have passed a state test and received a permit may use them. Almost all commercial operators and about one-half of the nation's farmers are certified to use restricted pesticides. Training sessions for applicators stress the hazards to people and to the environment of certain pesticides. Safe use, storage, and disposal of these pesticides are taught. Examples of pesticides in the restricted use class are: endrin, paraquat, sodium cyanide, and strychnine. (USDA photo)*

pressured EPA into allowing the use of ferramicide, a mixture of Mirex and two other chemicals. Ferramicide was alleged to breakdown more quickly in soil than Mirex, but no hard data existed on whether it actually did so. EDF filed a successful suit to delay the use of ferramicide until further tests were made. In a climate such as this, knowing a long legal battle will result, officials may be reluctant to propose new restrictions.

This sort of pressure has, to some extent, caused a situation in which EPA only seems to make regulations when it is forced into complying with toxic substances laws by lawsuits brought by public interest groups.

Problems in Setting Tolerances

The House Commerce committee has been critical of EPA's tolerance setting program. In many cases, pesticide tolerances have been set without all the needed information. On pesticides registered more than 10 years ago, we need information on carcinogenicity and data on birth defects and mutations. Furthermore, EPA relies on safety data supplied by the pesticide manufacturer. The committee, noting that the manufacturer is hardly a disinterested party, suggests that a better program to check the accuracy of such data is needed.

In addition, problems arise when EPA tries to calculate how much of a pesticide people are exposed to. When setting tolerances for pesticide residues on foods, EPA usually divides the total amount of a particular food sold in the U.S. by the total U.S. population. This is the yearly per capita consumption. An obvious problem with this method is that it averages consumers with nonconsumers, making no allowance for the wide variation in dietary habits in the U.S. The procedure leads to underestimates of the amounts of certain foods consumed. For example, a 1965–66 study showed that only 1 in 20 families eats fresh pears. However, the tolerance for pesticides used on pears is set using the assumption that everyone in the U.S. eats pears. What this means is that, since those who do eat pears are eating more of them than the tolerance calculation assumes, they are exposed to higher residue levels than calculated (Table 28–1). A better system for estimating consumption of various foodstuffs is needed.

Problems in Monitoring Pesticides

The U.S. Department of Agriculture is responsible for checking that meat and poultry do not contain more than the legally allowed residue of hazardous chemicals such as pesticides, drugs (e.g., antibiotics or hormones)[12] and other substances

---

12 More about why these are used and resulting problems on p. 603

TABLE 28–1 EPA Estimates of Consumption for Certain Foods: Less than 7.5 ounces of the Following Foods per Year per Person

Artichokes	Eggplant	Musk mellons	Swiss chard
Avocadoes	Figs	Nectarines	Tangelos
Barley	Honeydew melon	Okra	Tangerines
Black-eyed peas	Hops	Plums	Turnips
Blueberries	Horseradish	Radishes	Walnuts
Brussel sprouts	Kale	Raspberries	Winter squash
Coconut	Mangoes	Rye	
Cranberries	Molasses	Safflower	
Dates	Mushrooms	Summer squash	

It is easy to see that, in many of these cases, if someone eats the food at all, they will eat more than 7.5 ounces per year.

(e.g., lead or mercury).[13] The Food and Drug Administration checks all foods except meat and poultry. The FDA is also the agency that is supposed to actually take contaminated foods off the market if the USDA or FDA monitoring programs find excessive residues. In addition, FDA is the agency that can prosecute growers for sending contaminated foods to market.

However, neither FDA nor USDA monitors for all the toxic substances known to occur as residues in food. This is partly a matter of economics. Both agencies rely mainly on tests that detect whole groups of chemicals at one time. The tests for some chemicals are difficult and take a great deal of time to carry out. Furthermore, in some cases there are no good tests for a particular chemical. Thus a number of hazardous chemicals are not tested for at all.

Even if USDA finds high levels of a chemical in a meat sample the meat may not be taken off the market. This is because meat leaves the slaughter house where samples are taken within 24 hours, while test results are not received for up to a week. Since meat is not tagged after leaving the slaughter house, there is no way to retrieve a contaminated

carcass by the time the test results come back. In 1976, the USDA sampling program found that 5% to 15% of meat and poultry samples contained excess chemical residues. Since almost none of the products from which the samples came were removed from the market, it is estimated that 1.9 million tons of beef and 1.1 million tons of pork were sold in 1976 with illegal chemical residue levels. The FDA has similar problems in locating and removing contaminated shipments of other foods.

Do We Need More Regulation?

How serious a health problem is this apparent pattern of tolerance violations and contaminated food reaching the marketplace? It is probably impossible to say at the present time. We are dealing with quantities that are almost always too small to make people sick immediately. Rather, we are concerned that some of the chemicals may be carcinogenic, that they may have additive effects with other carcinogens in the environment, and that we probably will not know the final effects until after the long time it takes for cancers to develop.

The Commerce committee recommended more personnel and better monitoring procedures for sampling and detecting illegal chemical residues. In

---

13 The human health hazard of mercury is discussed on p. 134, that of lead on p. 390.

GOVERNMENT PROTECTION DURING AN ILLEGAL ACTIVITY?

The United States Government has a responsibility to insure that its actions do not foreseeably endanger the health and safety of its citizens.

Senator Charles Percy

They [marijuana smokers] don't have to smoke.

Lee Dogoloff, Office
of Drug Abuse Policy

In 1977, many marijuana smokers were jolted by the news that the pot they were smoking could be contaminated with a poisonous herbicide, paraquat. This came about because the Mexican government, supported and encouraged by the U.S., began a new program to stamp out marijuana farming in Mexico. Scouts located fields of marijuana by plane and then destroyed them by spraying with the herbicide paraquat, a program for which the U.S. government helped to pay. Paraquat is absorbed by the marijuana plants and kills them by preventing photosynthesis. However, paraquat does not always act immediately. It can take up to two to three days to kill plants. Thus, some enterprising farmers managed to harvest their crops and get them on the market even after they were sprayed. Paraquat levels in some samples of marijuana were found to be as high as 2264 parts per million. At this level, one to three cigarettes a day for a few months could cause irreversible lung scarring.

Public reaction to the problem varied. Marijuana is, after all, an illegal substance. An official in the White House Office of Drug Abuse Policy stated:

the government does feel some responsibility to smokers, but individuals do have some responsibility and choice in the matter–they don't have to smoke. *

What do you think? Does the government have a responsibility to protect the health of its citizens even if they are doing something illegal? Does the government have a responsibility to at least not harm its citizens' health directly? (The fact that the U.S. government helped pay for the spray program can be viewed as causing harm.) Does the fact that the government was attempting to stamp out trade in an illegal drug have any bearing on this case?

---

* Science **200,** 418 (28 April 1978).

addition to this, a method of retrieving contaminated meat and produce from the market is needed. The committee also recommended that *no* residues should be allowed for chemicals even *suspected* of being carcinogens, of causing birth defects or mutations. This would be a very broad extension of the Delaney Clause, which only forbids tolerances for *known* carcinogens.

DRUGS IN ANIMAL FEEDS

Toxic Residues Left in Foods

A number of different chemicals are added to animal feeds to make the animals gain weight more quickly. This is a matter of concern for two reasons. The first is that small amounts (or residues) might be left in the meat and thus be eaten by humans. The hormone diethylstilbesterol or DES has been used as a growth promoter in cattle. DES is known to have caused cancer in children born to women who took the drug while they were pregnant. There is also evidence that it has increased the risk of cancer to the women themselves. Whether DES should be allowed as a feed additive depends in a large part on whether any remains in the meat when it is marketed. Arsenic, another known carcinogen, is approved as a feed additive on the basis that it is all excreted by the cattle before they are slaughtered.

Farmers are encouraged to look for growth promoting additives because of the high cost of cattle feed. Anything that causes the cattle to put on weight faster or shortens the time to market means higher profits. This has stimulated some unusual additives (Fig. 28–5). An example is the cement kiln dust one group of farmers dumped into their cattle feeding troughs, on impulse, while they were using it to lime their fields. The cattle, to everyone's surprise, both grew faster and ate less feed. One wonders what harmful materials may be in the cement dust, however. For instance, both mercury

and arsenic are sometimes found in the gases and particles given off during cement manufacture.

According to testimony in 1978 before the House Committee on Interstate and Foreign Commerce, the Department of Agriculture is having a "major problem with antibiotic residues in dairy cattle and calves...We find 10 to 15 percent of the dairy cattle and four percent of the calves we test with violative levels of residues...The causes of this are not completely clear, but it appears to be primarily a situation where farmers are treating sick animals with antibiotics. When the animals fail to respond, some farmers may send them to slaughter without first observing FDA drug withdrawal requirements."[14]

Drug Resistance Risk

A second reason for concern also involves the use of antibiotics in cattle feed. These drugs are used regularly in cattle feed to improve the animals' health and to enable farmers to raise large herds of cattle or flocks of chickens in close quarters without

Figure 28–5   *Feeding dairy cows in Maryland: Farmers found antibiotics sharply cut feed bills. (USDA photo)*

---

14 C. T. Foreman, Asst. Secretary of Food and Consumer Services, USDA, "Cancer-Causing Chemicals in Food," Report to the Committee on Interstate and Foreign Commerce, 95th Congress, December 1978.

the occurrence of epidemics of animal diseases. It is known, however, that this increases the possibility of developing strains of bacteria resistant to the antibiotics used. Whether this resistance could be passed on to bacteria causing human diseases is unknown. There was a case in England in which the injection of large doses of antibiotics into veal calves appeared to start a human epidemic of antibiotic resistant salmonellosis. Since then, antibiotics for animals in England have been available only by prescription. It is not known whether the smaller quantities of antibiotics in feeds pose the same risk.

PBBs: STORY OF AN ACCIDENTAL DISASTER

During the summer of 1973, 10–20 bags of the flame retardant, Firemaster, were mixed in with a shipment of the feed additive Nutrimaster. Firemaster is the Michigan Chemical Company's trade name of PBB (polychlorinated biphenyl). The contaminated feed was eaten by animals on dozens of Michigan farms in what is now viewed as a classic example of how a toxic chemical can cause an accidental environmental disaster. Fredrick Halbert, a dairy farmer whose farm is near Battle Creek, Michigan, received a shipment of the PBB contaminated feed in late summer, 1973. His cows ate the feed for about 16 days, consuming up to one-half pound of PBBs, each, in the feed. During this time they began to show signs of serious illness: loss of weight, runny eyes and noses, overgrowths of the hoofs, and sharply decreased milk production (Fig. 28–6).

Halbert and the veterinarians he called in were unable to determine what was wrong. They sent all sorts of samples out for analysis, but no one could find anything unusual in the cattle or in their feed. Halbert by this time had changed his cattle feed because he was still suspicious that there might be something toxic in it. All told, Mr. Halbert estimates that he spent $5000 tracking down experts and sending out samples. Finally, he located

Figure 28–6 *Effects of PBB poisoning in this cow in Michigan include wrinkled skin, humped back, overgrown hoofs, and a damaged udder. (Jane Brody/NYT pictures)*

George Fries, a scientist at USDA's Agricultural Research Center in Maryland, who, by lucky chance, was familiar with PBBs. Fries was able to determine that the feed was contaminated with large amounts of PBBs. The men also realized that Michigan Chemical Corporation manufactured both a fire retardant containing PBBs and Nutrimaster, a feed additive with a similar appearance. In the investigation that followed, it was found that Firemaster was usually packaged in bags with red lettering. However, due to a shortage of pre-printed bags, for a time in 1973 both chemicals were packaged in plain brown bags with the names stenciled on in black. The final link was the finding of partially used bags of Firemaster at the Farm Bureau Services mill where Halpert's feed had been mixed.

Although the authorities were now aware that there was a serious problem, some 11 months had gone by since Halbert had received his contaminated shipment of grain. Meanwhile, other farmers had received some of the contaminated feed and had fed it to their animals. State agricultural investigators visited farms suspected of receiving contaminated feed and quarantined those animals which had high PBB levels.

By this time several thousand Michigan farm families and their neighbors had eaten contaminated meat, eggs, and milk. An undetermined but presumably lesser amount of PBBs had entered the food supply of the country as a whole.

Although the effects on the most heavily contaminated cattle were obvious, it is still not clear what effects the PBBs have had on people who ate contaminated food. Researchers are having difficulty sorting out symptoms and levels of exposure. In part, this is because the investigations are being carried out two to four years after people were exposed. Perhaps it is also true in part because individual susceptibility to PBB poisoning may vary. Since related chemicals, PCBs, are suspected carcinogens, groups of individuals exposed to PBBs are being followed for the next 20 years to determine the long term effects of the exposure.

The state of Michigan set the limit for PBB in meat at 0.3 parts per million. Thirty thousand cattle, 6000 hogs, 1500 sheep and 1.5 million chickens died or were condemned. Further 18,000 pounds of cheese, 2500 pounds of butter, 34,000 pounds of dry milk products and 5 million eggs had to be destroyed.[15] Many farmers whose animals tested out below the legal limits reported sickness among their animals and their families. They were not able to destroy sick animals and receive compensation from the state because their animals were below the legal PBB limit. Some state officials have blamed these particular problems on poor animal husbandry. Some officials also point out that no one has been able to find a correlation between blood levels of PBBs and symptoms in humans. The farmers claim the state did not want to lower the PBB limit because the government didn't want to pay the additional compensation to farmers whose herds would be condemned under the new limit.

Gerald Woltjer, 39, who had a $1 million investment in his dairy farm, lost everything at auction and still owes $500,000. Unable to work because of a wide range of health problems, including painful joints, dizzy spells, extreme fatigue and blurred vision, Mr. Woltjer, his wife and five children are forced to live on welfare.

A man with a big smile who was once, according to his wife, jolly and even-tempered, Mr. Woltjer says he is now very moody, anxious, irritable and is also extremely impatient with the children.

Mr. Woltjer and a number of others examined by Dr. Sidney P. Diamond, the neurologist on the Mount Sinai team, were suffering from a loss of confidence and ambition and diminished sexual activity.

"How much of the problems are organic and how much is emotional overlay may be a valid scientific question, but in terms of the lives of these people, it's really not a relevant issue," Dr. Diamond remarked.

"These people's lives are destroyed, and that is as important as pain in their joints," he continued. "You can't take fluid out of the soul and show PBB levels."[16]

Almost all of the PBBs have probably disappeared from the country's food supply by now, although many farm families still live with the disastrous effects on their lives and health. But Dr. Irving Selikoff has this to add:

...The disaster we have seen was caused by at the most only 2,000 pounds of PBB...But Michigan Chemical made and sold 12 million pounds of Firemaster, and they aren't saying where it went.

Most of the chemical is believed bound up with plastics to reduce flammability. But, Dr. Selikoff asks: "How much PBB escapes into the general evironment? Will we discover 40 years from now that we have another problem like PCBs on our hands?"[17]

---

15 In October 1974, levels of PBBs were found as high as 595 parts per million in milk; in poultry, as high as 4600 ppm; in eggs, up to 59.7 ppm; and in meat, up to 2700 ppm.

16 *The New York Times,* 8 November 1976, p. 1.
17 *The New York Times,* 12 August 1976, p. 20.

And Dr. Corbett, of the University of Michigan, pointed out:

> "If we had a strong toxic substances control act, this never would have happened...Michigan Chemical would not have been allowed to make a food supplement on the same grounds as toxic substances, nor would the company have been permitted to use such similar brand names or to package the chemical in plain brown bags distinguished only by the brand name label."[17]

TOXIC SUBSTANCES IN DRUGS AND COSMETICS

Hair Dyes

In the past, cosmetics were viewed simply as materials that stay on the outside of the body. Cosmetics were thus felt to be much less able to cause harm than foods and drugs. We now know that this view is too simple. Many substances are in fact, absorbed through the skin.

A case in point involves hair dyes. When the 1938 Food, Drug, and Cosmetic law was passed, hair dyes were specifically not included in the regulations. This was because they were known to cause allergic reactions in some people and it was felt that this might, under the wording of the law, be enough to ban them from the market. So they were exempted, with the provision that warning labels should be put on them to tell consumers about possible hazards. This action set the stage for a later problem. In 1975, certain hair dyes were found to cause mutations (changes in the genetic material) in bacteria (the Ames Assay). This indicated that the dyes might be carcinogenic. Studies by the National Cancer Institute showed that a least one dye did, indeed, cause cancer when fed to laboratory animals. The FDA was already aware that the dyes were absorbed through the scalp. In fact there are many complaints on file that hair dyes have caused a temporary color to appear in the urine of consumers. The industry, however, argued that the feeding studies were irrelevant since hair dyes are not eaten but are applied to the skin. They further pointed out that studies of hairdressers showed no excess occurrence of cancers. Nonetheless, the FDA probably would have had enough evidence to ban certain dyes if they had not been specifically exempted from the law. The agency had to settle for warning labels on products containing 2,4 DAA (2,4 diaminoanisole), 4–MMPD (4-methoxy-m-phenylenediamine), or 4-amino-2-nitrophenol.

The Case for Testing Before Marketing

In past years, the FDA has been able to ban several cosmetic ingredients, including hexachlorophene, an antibacterial agent used in soaps, deodorants, and skin creams. Hexachlorophene was found to cause brain damage in infant monkeys. Another material banned was vinyl chloride, now known to cause birth defects and otherwise rare liver cancer. Vinyl chloride was once widely used as a propellant in aerosol cans of hair spray and pesticides.

In general, cosmetics are not required to be proven safe before they are marketed, as are food additives and drugs. Rather, after the substances are on the market, the government must prove that they are unsafe and should be withdrawn. Evidence is piling up that this is the wrong way around.

However, even premarket testing by manufacturers may not assure the safety of a product. Such testing depends in part on good faith by the manufacturer. Cases have been uncovered in which a manufacturer has submitted false data or incorrect conclusions. A case in 1976 involving the Searle Company led to the tightening of FDA regulations on animal testing procedures as well as a more careful review of data submitted to the FDA.

Alcoholic Beverages

Another toxic material, sometimes classified as a drug, that is exempted from certain provisions of the Food, Drug, and Cosmetic Act, is alcohol. The authority to label alcoholic beverages is given by law not to the FDA, as are the labeling of all other foods, drugs, and cosmetics, but to the Bureau of Alcohol, Tobacco, and Firearms in the Treasury Department.

This has caused two recent problems. In the first case, the FDA wanted to require the listing of ingredients on beer, wine, and liquor labels. This would make it easier to recall products containing an ingredient found to be harmful (a food coloring, for instance). In addition, some people are allergic to a few of the ingredients used in alcoholic beverages. Labels listing ingredients would make it possible for these people to avoid the products that might cause an allergic reaction. Some ingredients that might cause such reactions include, according to FDA Commissioner Donald Kennedy, yeast, fruit, malt, molasses, spices, preservatives, egg whites and fish glue (the last two are used as clarifying agents in wine).

Winery owners and distillers opposed the labeling requirements. In part, this is because although they know what goes into making any particular batch of beverage, without expensive tests, they can not be sure what is left in the final product or what new compounds have formed during fermentation or aging. The FDA lost a federal court suit in 1976 in which it was confirmed that only the BATF was in charge of the labeling of alcoholic beverages. BATF had earlier refused to require ingredient listing.

The problem again came up when new data was found on the effects of alcohol on the unborn child. As little as three ounces of alcohol a day, or one drinking binge, during pregnancy can cause fetal alcohol syndrome. Babies with the syndrome may show reduced growth, mental retardation, smaller head size and defects in other organs. In addition, it is not known whether lesser amounts of alcohol might cause smaller but still unwanted effects.

Again the FDA was unable to require labeling, this time warning of the dangers of drinking alcohol during pregnancy. Eventually FDA and BATF agreed to a compromise in which FDA dropped its request for a warning label about fetal alcohol syndrome in return for BATF's agreement to work out requirements for ingredient listings on alcoholic beverages.

Weighing Risks and Benefits

As a final caution it should probably be noted that even if all the regulations and procedures work properly, the consumer can never be completely sure that approved drugs and cosmetics are safe. There is uncertainty in the testing procedures themselves, and certain effects, notably the production of cancers, take many years to become apparent.

New data comes to light periodically on the hazards of drugs once thought to be quite without serious side effects. For instance, 15 years after the use of "the pill" became common, benign liver tumors (which grow but do not spread to other organs) were found to be a risk to women taking certain of the oral contraceptives. (Those containing mestranol are suspect. This includes Enovid, Ovulen, Ortho-Novum, Norinyl, and Norquen.) As other examples, birth defects are now believed to result from the use of the common tranquilizers Valium, Miltown, and Librium if they are taken in early pregnancy. Phenacetin, an ingredient in several common analgesics, is suspected of causing renal tumors in humans. The use of diethylstilbesterol (DES) was found to cause vaginal cancer 15 to 20 years later in a percentage of daughters born to mothers who had taken it to prevent miscarriage during pregnancy.

It seems only reasonable that consumers should know as much as possible about the benefits and possible harmful effects of the drugs or cosmetics they are using. In this way they can take some part in the decision about long-term risks compared to present benefits.

CONTROVERSY

TRIS, PAJAMAS, AND CHILDREN'S SAFETY

The use of an untested chemical (tris-BP) as an additive to pajamas is unacceptable.

Arlene Blum and Bruce Ames *

Sleepwear treated with tris-BP was being sold in late 1977 (after the ban).

Marian Gold, et al. †

There are two problems a regulatory agency must face when it considers taking a potentially dangerous product off the market. In the first place, something just as dangerous may be substituted. Second, the product may be sold anyway but not labeled. Both problems occurred when tris-BP was banned as a flame retardant in childrens pajamas.

In 1975, the Consumer Products Safety Commission established standards for the flammability of childrens pajamas. Aruguing that as many as 3000 injuries and 100 deaths each year resulted from flammable childrens sleepwear, the commission required all fabrics used in childrens sleepwear to pass a standard test proving the fabric would not burn.

Since making a piece of clothing fabric flame-retardant adds 10%-30% to its cost, it was felt that many people, especially the poor would not, by choice, buy flameproof garments. In a large part to protect children in families with low incomes, then, the commission did not simply require a label stating whether a fabric was flammable or not.

Most manufacturers met the new standards by treating sleepwear with tris-BP. This was the most effective and least expensive way to meet the standards. All treated garments were to be labeled fire retardant and the material used was to be specified. Up to 5% of a fabric's weight consisted of tris to make it fire retardant. However, in 1977 it was discovered that tris-BP was mutagenic in bacterial assays (The Ames Assay). This indicated strongly, although it did not prove, that tris-BP might be carcinogenic. Further, tris painted on rabbit's skin at doses of 2.27 grams per kilogram caused male rabbits' testicles to atrophy. As a result, tris-BP was banned for use in childrens' sleepwear and all treated garments were recalled.

This is not the end of the story, however. Researchers checking in late 1977 found tris-BP treated garments still being sold–unlabeled. There is evidence that a replacement flame retardant, Fyrol FR2, may also be mutagenic. ‡

Thus, we have a dangerous situation which may be partly caused by a regulation. That is to say, sleepwear is required by law to be flameproof, but the safety of the chemicals being used to meet the regulation are in doubt.

Adding flame-retardant chemicals to almost all children's pajamas, as a consequence of the Consumer Product Safety Commission's standards, most probably is reducing the number of burns and deaths due to children's nightwear catching fire, although statistics are unavailable. As we have indicated, there are also other ways of reducing fire injuries.

The risk of the exposure of tens of millions of children to a large amount of a chemical must be balanced against the risk of fire. A calculation...suggests that the risk from cancer might be very much higher than the risk from being burned. Flame retardants (and most other large volume industrial chemicals) either have not been tested or have not been adequately tested for carcinogenicity. The use of an untested chemical as an additive to pajamas is unacceptable in view of the enormous possible risks. *

Can you think of other ways to help solve the problem of fire-related in-juries? Do you think children's pajamas should be required by law to meet flammability standards? (You may want to read the two articles quoted above for more information.)

---

* A. Blum and B. Ames, Science **195,** 21 (7 January 1977).
† M. Gold et. al., <u>Science</u> **200,** 785 (19 May 1978).
‡ The manufacturer has now removed this substance from the market.

Questions

1. How would you feel about the government's banning all but natural plant extracts as flavoring agents and food colors? Flavorings and colors would be available but their cost might be so high that many foods (soft drinks, candies, cake icings, fruit drinks, etc.) could not be economically produced. Think about the artificially colored and/or flavored products you normally use, which might not be available. Would this bother you?

2. In some countries, bread is sold almost exclusively without preservatives. Most people buy it fresh each day. Would that work in the U.S.?

3. What is the Delaney Clause? Are you in favor of letting it stand as is or would you rather see agencies allowed to balance the risks from carcinogens against the possible benefits from their use?

4. Do you believe the FDA should forbid the use of nitrites in foods (remember, you could then have only gray hot dogs or cold cuts and there might be an increased risk of food poisoning) on the basis that they are at least suspected if not proven carcinogens? Or do you feel the consumer should be able to choose whether or not to buy nitrate containing products?

5. Are there people who would not be able to make a choice about nitrites intelligently?

6. Can you think of ways in which the food industry could obey a nitrite ban and still provide safe products? Consider economic effects of any alternatives you suggest.

7. What do you see as the advantages, if any, to "health foods" as sold in a typical health food store? What are possible problems, if any?

8. What sort of toxic substances in foods could be called unintentional additives?

9. How do pesticide residues get into foods?

10. Do you believe the Department of Agriculture and the FDA should try to monitor all of the possible toxic residues in foods? What about the cost of this?

11. Should the EPA set food tolerances for substances suspected of being carcinogens?

12. Why might EPA's method of calculating the average daily consumption of foods be especially hazardous for milk?

13. What rules would you make about toxic chemicals manufacturing to protect the public against a disaster similar to that which occurred with PBBs? What kinds of information and employees would you want on hand so that you could determine quickly if a dangerous accident had occurred?

14. Who should determine the levels at which contaminated livestock must be condemned? Do the farmers deserve compensation and if so, from whom?

15. Do you think alcoholic beverages should carry a label warning about the dangers of fetal alcohol syndrome? Can you think of any problems this might cause?

16. Would you find it useful to have the ingredients listed on cosmetics and alcoholic beverages? Why?

Further Reading

Toxic Substances in Foods, Drugs, and Cosmetics

Reimode, G. O., "Food Additives," *Scientific American* **226(3)**, 15 (March 1972).

Food Colors, Committee on Food Protection, Food and Nutrition Board, National Research Council, National Academy of Sciences, Washington, D.C., 1971.

These are easily readable summaries of the history, uses, and laws concerning food additives. They do not give any feeling, however, for the controversy concerning the safety of our food supply.

Brody, J. E., "Food Additives, Do They Hurt?", *The New York Times*, 12 July 1978, p. C-10.

Cancer-Causing Chemicals in Food, Report by the Subcommittee on Oversight and Investigations of the Committee on Interstate and Foreign Commerce, 95th Congress, December 1978.

Lyons, R. S., "U.S. Additive Laws Face New Attack," *The New York Times*, 22 September 1972.

The above articles give something of the other side.

Food Additives, Manufacturing Chemists Association, Inc., 1825 Connecticut Avenue, Washington, D.C.

Food Additives, Who Needs Them, Manufacturing Chemists Association, Inc., 1825 Connecticut Avenue, Washington, D.C.

Industry's side of the argument is presented in these two well-written booklets.

Chapter 29

Controlling Toxic Substances

LAWS CONTROLLING TOXIC SUBSTANCES

TOSCA

AN EXPENSIVE SOLUTION?

LAWS CONTROLLING
TOXIC SUBSTANCES

Some of the environmental factors that contribute to diseases such as cancer are controlled only by the individual. Each of us chooses whether to smoke or drink alcoholic beverages or follow a particular kind of diet. However, some of the other environmental factors that can affect human health seem likely candidates for regulation by laws. Congress has passed the Clean Air Act, the Federal Water Pollution Control Act, the Safe Drinking Water Act, and the Pesticides Control Act in an attempt to reduce air and water pollutants in the environment. The Occupational Safety and Health Act is designed in part to protect workers from exposure to toxic substances where they work while the Food, Drug, and Cosmetic Act is meant to assure consumers that harmful quantities of toxic substances do not appear in foods, drugs, and cosmetics.

TOSCA

Despite the fact that all these laws were already on the books, in 1976 Congress passed another toxic substances law, the Toxic Substances Control Act (TOSCA). TOSCA was designed to fill gaps where other laws did not apply in the regulation of toxic substances. Basically, TOSCA's main provisions fall into two categories, the first dealing with gathering data on hazardous substances so that informed regulatory decisions can be made. The second concerns premarket review of chemicals. These provisions attempt to ensure that hazardous substances are identified before they reach the environment, rather than afterwards. This is a major change from previous laws, which were generally designed to control toxic substances after they were discovered in the environment. Pesticides, radioactive substances, and food additives are not covered by TOSCA because they are regulated by other laws.

Under TOSCA, the Environmental Protection Agency is required to keep a national registry of hazardous chemicals. This registry lists where the substances are produced and how much of each is produced. The agency will then be better able to assess the dangers these chemicals pose to the workers who manufacture them and the people who live near the factory.

Companies are required to keep records of any harmful effects, either to the health of their workers or to the environment, caused by the chemicals they handle. Manufacturers must test compounds they expect to market for their effects on health and environment (this is already required of pesticides manufacturers, under the Pesticides Control Act of 1974). The EPA must also be notified 90 days before any new chemical is marketed, to give the agency time to determine its safety. Furthermore, the EPA is empowered to seek court injunctions to prevent the manufacture or distribution of any chemical that is, in the agency's view, an imminent danger to health or the environment.

AN EXPENSIVE SOLUTION?

Estimates abut the cost of the premarket screening for new chemicals vary from the Manufacturing Chemists Association's high estimate of between $360 and $1300 million per chemical, to the Environmental Protection Agency's $80 to $140 million. This should be compared to the chemical industry's total sales of $72 billion in 1974 post-tax profits of $5.5 billion, and $2 billion spent on research and development.

Another law that has come under fire because of its cost to industry is the requirement, under the Water Pollution Control Act, that industries now pretreat hazardous waste materials before discharging them. At first only industries that discharged wastes directly into natural waterways had to do this. Starting in 1979, EPA has also required pretreatment by industries discharging wastes into municipal sewer systems. Such pretreatment should, in large part, help prevent toxic substances from entering the environment. However, economically reasonable methods are not available for complete removal of all hazardous substances from industrial wastes. Table 29–1 gives EPA's estimates of the predicted increase in toxic chemicals in water from 1975 to 1990.

The controversy (p. 616) surrounding this particular regulation is of general interest because it highlights the way in which air and water have been treated, in the past, as free resources. One result of this treatment is the current level of toxic substances found in our environment.

TABLE 29–1   Future Increases in the Pollution of Water with Toxic Chemicals 1975 to 1990[a]

Toxic Material (Dissolved solids)	Industrial Chemicals	Other Chemical Products	Nonferrous Metals	Steel	Fabrication	Electric Utilities
Cadmium			8%	0%[b]		
Chromium	170%	0%		29%	0%	
Lead			0%	5%		47%
Mercury	7%		2%			
Zinc		149%	2%	4%	0%	
Aluminum			0%	0%		
Copper	45%		0%		0%	18%
Cyanide				9%	0%	
Ferrous metals				0%		
Fluoride			5%	33%		
Selenium			17%			
Titanium oxide[c]	24%	0%	0%			
Titanium oxide[d]	1%					

This table shows how the discharge of toxic chemicals to water is expected to increase by 1990. For instance, by 1990 the manufacture of industrial chemicals is expected to increase mercury pollution by 7% over 1975 levels. Improvements in control methods could greatly decrease these values. (Source: U.S. Environmental Protection Agency, 1978.)

[a] Data are 1990 values as percents of 1975 values.

[b] Value in 1990 is reduced to negligible amount from 1975 level.

[c] Suspended solids.

[d] Dissolved solids.

Questions

1. Why did Congress pass the Toxic Substances Control Act when there were already a number of laws dealing with toxic substances?
2. What do we mean when we say that air and water have been treated in the past as free resources? How do regulations requiring that industries pretreat waste change this way of looking at air and water?
3. Instead of requiring industry to pretreat wastes to a certain level we could tax them for the amount of pollutant they emit. Can you see any advantages to this method? Any disadvantages?

Further Reading

Toxic Substances Control

Walsh, J., "EPA and Toxic Substances Law: Dealing with Uncertainty," *Science* **202**, 598 (10 November 1978).

Culliton, B. J., "Toxic Substances Legislation: How Well Are Laws Being Implemented?" *Science* **201**, (29 September 1978).

The political and procedural problems of regulating toxic substances are noted in these two articles.

EPA Journal **4**, (8), (29 September 1978).

This entire issue is devoted to toxic substances control. Articles are clearly written but on a more elementary level than the two above.

CONTROVERSY

Should we not ask the EPA to show us that the benefits of its regulations are at least as great as the costs?

R. W. Crandall

Conventional estimates of gross national product are . . . meaningless so long as they fail to account for the cost of . . . uncorrected levels of pollution.

R. D. DuBoff

The Environmental Protection Agency is now requiring industries that have been discharging toxic wastes into sewer systems to set up systems to pretreat the wastes. * In this way, materials such as toxic organics will be removed before they can reach municipal sewage treatment plants. The measure should help reduce toxic organics in drinking water by stopping the discharge of these materials into natural waters that might be used as sources of drinking water. In addition, toxic wastes often "knock out" sewage treatment plants by killing microbes that are a vital part of treating sewage. An instance of this occurred in Louisville, Kentucky, when chemicals dumped into a sewer knocked out the city's sewage treatment plant for 45 days. This cost the city five million dollars. Of course, pretreatment is not without cost to industry. Robert W. Crandall, a Senior Fellow at the Brookings Institution, took issue with the EPA's regulations in a letter to <u>The New York Times,</u> 3 July 1978.

Unfortunately, I had the time to read the remainder of the article, in which Jorling is quoted as saying that he did not know how much these industrial standards would cost but that it would possibly be in the "low billions" of dollars. This rather cavalier assessment of costs is not uncommon among environmentalists who tend to view G.N.P. as a stock which does not change with increasing regulation. Perhaps someone should point out to Mr. Jorling that there are not many "low billions" of dollars in the entire industrial output of the economy. His assurance that the costs of those latest regulations are in the "low billions" of dollars may be likened to the executioner's assurances to the condemned that he will only have to fall "a few feet" with the noose around his neck.

Recently a colleague, Edward Denison, measured the effect of environmental expenditures by business upon economic growth and found that by 1975 they had reduced economic growth by more than 10 percent of its recent average. The total cost of these regulations in 1975 was in the "low billions" of dollars–$9.6 billion, to be exact. If a single set of new regulations will cost a few billion more, what can we expect when a large number of new air-quality standards are promulgated in the next few months? Before we proceed, should we not ask the E.P.A. to show us that the benefits of its regulations are at least as great as the costs?

This was answered by Richard B. DuBoff, an economist at Bryn Mawr College, on 10 July 1978 in The New York Times:

Perhaps, to borrow Mr. Crandall's own language, someone should point out to him, and to Mr. Denison, that their thoroughly conventional estimates of gross national product are overstated, or meaningless, so long as they fail to account for the costs of present, uncorrected levels of pollution–of air, water and industrial inputs (the cotton "brown lung" disease but the latest case). Even a nonradical economist like Paul Samuelson warns students who use his textbook that "clearly we must adjust for any such 'bads' that escape the G.N.P. statistician whenever society is both failing to prevent pollution and failing to make power (or water or air or cotton) users pay the full costs of the damage they do." Once we make these adjustments, we see that net economic welfare grows more slowly than (conventionally measured) G.N.P.

Mr. Crandall's reproof of "environmentalists who tend to view G.N.P. as a stock which does not change with increasing regulation" is another fine example of the tendency of orthodox economists to leap to policy "tradeoff" conclusions without any probing of underlying economic relations (let alone their broader political context).

Should industry be required to pretreat wastes? Who will pay for this, eventually? In what ways do we pay if industries do not pretreat wastes?

---

* The phrase used to describe pretreatment procedures is "best available technology economically achievable." What is economically achievable is decided separately for each standard.

(Shelly Grossman ©1978/Woodfin Camp & Associates)

Part 7

Land Resource Issues

I conceived that the land belongs to a vast family. Of this family, many are dead, few are living, and countless members are still unborn...
A Nigerian Chief*

An often heard phrase is "the quality of life." It refers to the society's and an individual's well-being. One kind of well-being is economic; another is social. Elements of economic well-being include jobs, pay, and the cost of food, housing, clothing, transportation, medical care, and other basics. An assessment of social well-being asks how civilized and stable a society is; that is, whether there is wide-spread crime, alienation, or prejudices. Well-being also takes in our concept of freedom to move, to change jobs, and to travel. Beyond the economic and social components of quality of life is still another concept, less distinct and less measurable, but nonetheless real and important. It has to do with the quality of our environment.

The quality of the environment does not lend itself to neat characterization. Instead, a set of questions might be used to grope toward the meaning of quality of the environment. One might ask, for instance, if there is unspoiled forest, nearby, that is open to the public to hike or backpack and observe wildlife? Are there parks where families can picnic? Are there coastal beaches within reach that have not been locked up in private ownership? Are historic buildings preserved so that people can see the origins of their settlement? Are farms preserved nearby both for the food they supply and their contribution to our need for green? Or does surburbia with shopping centers and strip development sprawl to infinity? Is there a place where children can buy their pumpkins from the farmer who grew them? Or better yet, where the children can grow them themselves? Each individual is likely to have a different set of questions to reveal the quality of the environment. What we ask obviously reveals our own personal measures of quality of environment.

One element stands out as a key in all these questions; one element touches every one of these concerns: how we use the land. If we use the land badly, the questions will be answered in a negative way. If we use the land wisely, the questions can be answered in a positive way.

The subject of land use is an unusual blend of issues. To improve the use of the land, we need to know how we can influence the numerous private decisions that go into creating a pattern of land use. We also need to know where private decisions are likely to fail to bring desired patterns of use; that is, where is it necessary for the public to acquire and preserve the land and its qualities? And when the public comes to own such lands, how can their qualities best be protected?

Part 7 begins with a study in Chapter 30 of the natural habitats and communities found on land. The following chapters discuss how humans use land areas and how we are attempting to preserve some of these areas and the natural communities living there. Chapter 31 treats private land use decisions and the methods evolved to influence them. In Chapter 32, we take up the issue of public lands and their purposes.

---

* Quoted in *Land Use and the Environment,* Environmental Protection Agency, Office of Research and Monitoring, Environmental Studies Division, Virginia Curtis, ed., 1973.

Chapter 30

Lessons from Ecology: Land Habitats and Communities

SUCCESSION AND CLIMAX

A Tall Grass Prairie Park

When the first pioneers moved westward, they found vast expanses of grassland, the American prairies. Their covered wagons, moving through seemingly endless seas of waving grasses, were called prairie schooners. These prairies once covered more than a million square miles, from southern Canada all the way to the Gulf of Mexico and from the Eastern forests to the Rockies (Fig. 30–1). The prairie grasses range in size from short grasses a few inches high to tall grasses such as big blue stem, which grows taller than a man. At one time there were 400,000 square miles of tall grass prairie, but now barely 1% of that remains. The rest, along with 50% of the short grass prairies, has been farmed or paved over to make roads, housing developments, and shopping centers.

Tall-grass prairie forms a stable ecosystem composed of the characteristic grasses: big blue stem, Indian, little blue stem, and switch grass, which grows to heights of 3 to 8 feet. Many wildflowers are mixed in with the grass and 80 species of mammals including deer, bobcats, and badgers find a home there (Fig. 30–2). In addition, there are 300 bird species and well over 1000 kinds of insects. Once the grass is plowed up it may be difficult or impossible to reestablish the system in any reasonable length of time.

Several conservation groups are working to establish a Tall Grass Prairie National Park before there is nothing at all left to save.

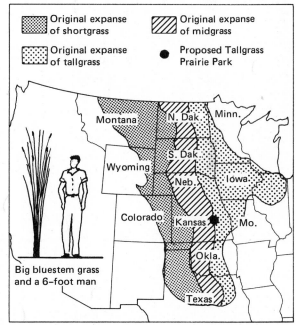

Figure 30–1 *The American prairie grasslands once occupied a million square miles across the entire midsection of the U.S. Tall grass prairie, where grass grew more than 6 feet tall, covered the easterly portion and blended into a mid-region of mid-height and mixed-height grasses. Farthest west was the short grass prairie. (Source: Save the Tallgrass Prairie, Inc.)*

Figure 30-2 *A coyote roams through tall-grass prairie in spring before the grass has reached its full height. (Leonard Lee Rue/Monkmeyer) "There was only the enormous, empty prairie, with grasses blowing in waves of light and shadow across it, and the great blue sky above it, and birds flying up from it and singing with joy because the sun was rising. And on the whole enormous prairie there was no sign that any other human being had ever been there." (Quote from L. I., Wilder,* Little House on the Prairie, *Harper and Row, 1935, p. 40.)*

Succession

You may wonder why the grassland community established itself in the first place. What factors determine the kind of vegetation and wildlife that will be found in various parts of the world? The prairie grasslands are part of an ecosystem that is both characteristic of that given area and also stable in time. That is, the kinds and proportions of the various plants and animals do not change much in time. This sort of stability is not found in all communities. Consider the fate of an abandoned farmer's field in a temperate climate such as the northern U.S. The first year after the farmer gives up the field, it is filled with annual weeds, the kind of weeds that grow up each year, produce seed, and die. Perennial weeds, which are dormant during the winter months, start to appear the second year. Eventually small shrubs and trees grow among the weeds and then take over the field. As the shrubs and trees grow larger and begin to shade the ground, the weeds which once grew in the field no longer find conditions suitable for growth. The kinds of animal species found in the field change as the vegetation alters. This rather orderly process is called *succession*. During succession, each community of plants and animals actually changes the environment in which it lives so that conditions favor a new community, one usually involving larger species. The changes may involve such factors as the temperature and acidity of the soil, the amount of water and sunlight available. For example, pines may be the first trees to grow in a field. Pine branches eventually shade the ground, however, and pine seedlings do not do well in shade. Deciduous tree seedlings, such as oak and hickories which can grow in shade, may then take over the forest floor and supplant the pines themselves.

Succession occurs not only in abandoned fields but in all kinds of habitats. For instance, after a volcano erupts, tiny plants called algae begin to colonize the cooled lava. Over the years, through a series of replacements, the communities inhabiting the area before the volcano erupted will appear. Succession also occurs in ponds and lakes.

Climax Communities

In many areas, a community finally appears that is not supplanted by another one. This last community does not change conditions to make them unsuitable for itself. Such a community then survives, perhaps for centuries, unless other factors such as disease or climate change or human activities are introduced. This relatively stable community is called the *climax community*. Major areas of the world have characteristic or regional climax communities. Regional climax communities, also called *biomes*, appear to be determined by temperature and rainfall within a given area. Other factors such as soil type and soil age also influence the kind of vegetation that grows up (Fig. 30-3).

WORLD BIOMES

Tundra

(In the far north, the top few inches of soil thaw only in the summer. Below this thin layer, the soil remains permanently frozen year round. The frozen layer is called *permafrost* and the whole area is known as the *tundra biome*. Roots cannot penetrate the icy layer and so vegetation is limited to mosses, lichens (such as reindeer moss), grasses, and small woody plants (Fig. 30–4). Overall, growth on the tundra is very slow. For this reason, when the tundra is disturbed, as by tire tracks, the scars do not heal. In the summer, the tracks fill with water and mud and in the winter they freeze to ice. This particular problem was of much concern during the planning of the Alaska pipeline, which crosses so many miles of tundra.)

Forests and Grasslands

(Below the tundra, there is an area called the *taiga*, or *northern coniferous forest biome*. Although the area is cold in winter, when summer comes the soil thaws completely; there is no permafrost (Fig. 30–5).)

(South of the taiga there are several different biomes, depending on rainfall and soil type. In areas with 30 to 60 inches of rainfall, moderate temperatures, and definite winter and summer seasons, the *deciduous forest biome* occurs. This includes most of the eastern United States, central Europe, and parts of Asia (Fig. 30–6).)

(Where rainfall is too sparse to support trees, grasslands develop. Grasslands need between 10 and 30 inches of rain a year compared to forests which require between 30 and 60. The taller grasses generally need more rainfall and deeper, richer soils than the short or midheight grasses (Fig. 30–7). Grasslands also occur in tropic areas where rainfall may total 40 to 60 inches. However, in these areas

Figure 30–3 *Lake succession. With time, lakes and ponds may begin to fill up with sediments and plant debris. The waters become more and more shallow and less and less clear. (This is the process of eutrophication.) Eventually the lake or pond may fill so completely that it becomes a bog or marsh. Following this, a succession of land plants appear until the normal climax for the region establishes itself. Succession starting from an aquatic environment, such as a lake or pond, or from a terrestrial environment, such as an abandoned field, tends to lead to the same climax vegetation in a given area, although the path the succession takes may be very different. This supports the idea that the climate of a given area determines what the climax vegetation will be. The pond in the photo is a good distance along the road to becoming a bog. The margins are filled with spongy accumulations of plant material. Only a small area of open water remains. (Grant Heilman Photography)*

the rain comes all in one season, followed by a dry season.)

(Areas of the tropics where there is little seasonal variation in temperature but very abundant rainfall (60 to 90 inches a year) will usually produce tropical rain forests. Many species of both plants and animals live in these forests. In fact, tropical rain forest biomes (and coral reefs) contain the greatest diversity of life found in any biome. That is to say, the largest number of different species per square meter are found there (Fig. 30–8).)

Deserts

(In areas where rainfall is only 10 inches per year or less, *desert biomes* occur. Only those organisms adapted to long periods of drought and widely varying day–night temperatures can survive these conditions. Because there is little or no vegetation to moderate the sun's energy, days on the desert tend to be very hot while, as soon as the sun goes down, night temperatures plummet (Fig. 30–9). The biomes described here are only the major biomes found in the world. Many sub-types exist as does a variety of biomes covering smaller areas.)

SOME CHARACTERISTICS OF COMMUNITIES

Diversity

(Generally, during succession, the number of species increases in each successive community.) The first communities that colonize a lava flow or sand dune or field are simple; they contain few species. As more plants appear, there are niches for more species of insects and animals. These creatures in turn, by feeding on plants, cause pressures for new species of resistant plants to develop, and so on, through the various stages of succession(In this way, diversity increases as succession goes forward. However, diversity of species is not greatest in the climax community. Rather, diversity seems to peak sometime before the climax community appears.)

Energy Use

(Another community measure that increases during succession is energy use. As different groups of plants succeed each other, taller species grow up. The amount of biomass, or the weight of living material, both plant and animal, increases. Thus more and more biomass is produced for the same energy input (e.g., sunlight) as succession progresses. However, energy use, too, peaks before the climax community occurs.)

Climax Community Characteristics

In climax communities, other factors seem to be more important than those leading to species diversity and increasing biomass. An example is the increasing size of individuals. This gives species the ability to store nutrients or water against seasonal shortages. This and other factors lead to increased competition between species and a loss of species in the climax community.

(The important point about a climax community, however, remains that it replaces itself, rather than changing the environment to be more suitable for a new community of plants. Thus, the climax community tends to be stable over a long period of time.)

Productivity and Yield

Field ecologists spend part of their time determining the weight and number of organisms found in climax communities and in the succession stages leading up to the climax. Water environments can be described in terms of their productivity, that is, the number of fish or other seafood caught per year. This is sometimes called the yield. Another term used is the standing crop. This means the actual number of fish in the body of water.

(These terms are also applied to land environments. Here *standing crop* is either the number of

Figure 30-4   *Arctic tundra. (Pro Pix/Monkmeyer Press Photo Service) "Across the northern reaches of this continent there lies a mighty wedge of treeless plain, scarred by the primordial ice, inundated beneath a myriad of lakes, cross-checked by innumerable rivers and riven by the rock bones of an older earth. They are cold bones into which an eternal frost strikes downward five hundred feet beneath the thin skin of tundra bog and lichens which alone feel warmth under the long summer suns; and for eight months of the year this skin itself is wrinkled by the frosts and becomes part of the cold stone below....*

Yet of all things that it may be it is not barren. During the brief arctic summer it is a place where curlews circle in a white sky above the calling waterfowl on icily transparent lakes...It is a place where minute flowers blaze in microcosmic revelry, and where the thrumming of insect wings assails the greater beasts, and sets them fleeing to the bald ridge tops in search of a wind to drive the unseen enemy away. And, not long since, it was a place where the caribou in their unnumbered hordes could inundate the land in one hot flow of life that rose below one far horizon, and reached unbroken past the opposite one." (From Farley Mowat, The Desperate People, *Boston, Little, Brown and Company, 1959, p. 21.)*

individuals or the weight (biomass) of a population at a given time while *productivity* refers to the amount of living tissue produced per unit time by a population. The amount harvested is called the *yield.* Thus in a cornfield, the standing crop is the total weight of all the corn plants in the field at the time of measurement. The productivity is the amount of corn plant tissue produced each day or per season, while the yield is the weight of the harvest.

INFLUENCES ON SUCCESSION

The discussion of succession ignored two factors that may have a strong influence on the type of vegetation in a specific area. One factor is, of course, human activities. Wherever we live, we humans turn the land to our own uses, cutting down or burning those species for which we have little use and encouraging those species we need. The way humans manage forests for the producing and suitable for pulp. Such stands are much more subject to severe insect damage than the normal, mixed climax community. For this reason, quantities of pesticides must be used to protect the trees. Manufacturing and applying pesticides, of course, requires energy. Pesticides can also have many adverse effects upon the environment, some of which are detailed in Chapter 34.

Grasslands are used by humans to graze cattle or to grow corn and wheat. Up to a point, these are

Figure 30-5 *Taiga. The northern coniferous forest biome is made up of conifers such as spruce, fir, and tamarack along with some deciduous trees such as birch. Bears, wolves, moose, squirrels, lynx, and many other mammals are common as are many species of birds in the summer (U.S. Department of the Interior; National Park Service Photo)*

tion of wood provides an example. (We noted earlier that climax communities do not produce the greatest biomass per unit of energy input. Productivity is not as great in climax ecosystems as it is in some earlier stages of succession.) In order to increase production of food or wood, humans will try to keep an area at a younger successional stage. These younger stages, however, are less balanced, less able to maintain themselves than an older climax stage. Large inputs of energy and ingenuity are needed to maintain them. For instance, in some areas, paper companies plant vast tracts of forest land with soft-wood tree species that are fast grow-

Figure 30-6 *Deciduous forest biome. In different areas, the mixture of deciduous trees making up the forest will vary. For instance, in the North Central region, the forests are predominantly beech–maple while in the West and South, the forests are composed mainly of oak and hickory. Deer, bear, squirrels, and foxes roam the woodlands while many bird species such as woodpeckers, thrushes, and titmice find a home there. In the fall the leaves turn color and then fall to the ground. (Grant Heilman Photography)*

Figure 30-7   *Grassland biome. Native big bluestem and wildflowers in the Flint Hills of eastern Kansas. This 1000-acre tract of virgin tall-grass prairie was donated to Kansas State University by the Nature Conservancy and is near the site of a proposed national park. Many grazing animals live on the grasslands. In the U.S., buffalo and pronghorn antelope were once numerous. Burrowing rodents such as gophers and prairie dogs are found, along with their predators—coyotes, kit foxes, and badgers. (Nature Conservancy).*

reasonable uses. It is not strange that the American farmer has done well farming ploughed grassland, since the two plant communities, grains and grasses, involve similar ecosystems. Grasslands, however, can be overgrazed, especially tropical grasslands. In these areas, and in tropical rain forests, most of the plant nutrients are cycling in the plants themselves and do not spend much time in the soil. Thus, removal of too much vegetation by overgrazing removes almost all the nutrients, leaving soils that no longer support plant life.

Similar to the problem of overgrazing is that of overuse of slash-and-burn farming in areas of tropical rain forest. In this type of agriculture, the farmer cuts and burns the natural vegetation before planting his crops. Such plots can only be used for a few consecutive years and then must lie fallow for 10 or even 30 years to regain their fertility. As populations increase in these areas, farmers are tempted to reuse a plot too soon.

A special influence humans have on vegetation is the result of strip mining. If the stripper is not careful to save the topsoil and spread it out on top again after he fills in his mining trenches, a wasteland can result. This is explained by looking at a profile or cross sectional cut of soil layers. Figure 30-10 shows the soil profile for a deciduous forest and a grassland. The top layers of soil, which con-

Figure 30-8   *Tropical rain forest biome. Typically, there are three layers of vegetation. A sprinkling of tall trees, which lose their leaves during dry seasons, pokes up through a canopy of broadleaved evergreens and plants. Below this is a third layer of plants that flourish wherever there is a hole in the canopy. Lower levels of the forest are very humid and have a constant temperature. Climbing vines and plants growing on other plants are common. So many species are found in tropical rain forests that it may be hard to find two trees of the same species for a distance of several acres. Animals and insects are likewise numerous and diverse. (©1972 Carl Frank; Photo Researchers, Inc.)*

Figure 30-9   *Desert biomes. Desert plants are adapted to survive long periods without water. Some have few, small, leathery leaves, which can be dropped during the dry season to reduce water loss. Examples are sagebrush and mesquite. Others can store quantities of water in the plant tissues. These are the cacti and the euphorbias. A third group includes the annuals which spring up, flower, and set seed in the short period after a rain. Desert animals are similarly adapted to the conditions of their environment. Many are active only in the cooler night periods, burrowing into the ground during the day to remain cool. (Verna R. Johnston; Photo Researchers, Inc.)*

tain the nutrients cycling from dead plants and animals to living ones, are the most important from the point of view of supporting growth. When the strip miner turns the layers over, the deeper nutrient-free layers are turned uppermost. This, in addition to the acidity produced by coal wastes, gives rise to a soil in which almost nothing can survive. Only imported topsoil or the slow, centuries-long processes of soil formation and washing away of acid will restore the original vegetation to land treated in this manner. Erosion, as described in Chapter 34 can produce similar results because the top soil is washed away by rains. These are examples of how growing human populations, with their increasing needs for food, fiber, and mineral resources, can unbalance their environment.

On the positive side, people have turned deserts into gardens by adding the missing factor—water. Here again there are problems, however. The rapid loss of water by evaporation leaves behind minerals. Unless this is accounted for in irrigation schemes, minerals can accumulate until nothing will grow.

The other factor to which little attention has been given is fire. Many ecosystems are adapted to the natural occurrence of periodic fires. African grasslands, for instance, are composed of fire tolerant species because fires are common during the long dry season. The topic of fire and forest management is covered in Chapter 32.

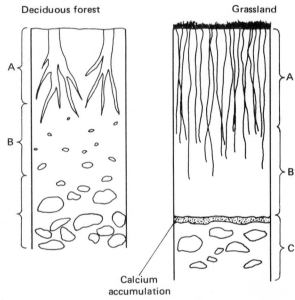

Deciduous forest   Grassland

Calcium accumulation

Figure 30-10   *Soil profiles. This is how soil would appear on the sides of a trench cut into the ground. The upper layer or A horizon is topsoil and contains plant and animal debris, which is being turned into humus by soil microorganisms. In the B horizon, the organic material has been changed into its inorganic components. The third of C horizon is parent material; that is, soil characteristic of the rock from which it was formed. Climate and vegetation act on parent rock material over the ages to form soil characteristic of a given area.*

Questions

1. Describe what is meant by *succession*.
2. What is the main characteristic of a climax community? What are some other characteristics it has compared to earlier stages of succession?
3. What are the major factors influencing which climax community grows up in an area?
4. UNESCO has a program to identify and preserve examples of all the world biomes. What is a biome? Of what possible use could it be to preserve an example of each type?
5. Explain how clearing forest land for farming is an example of increasing food production at the cost of unbalancing the environment.

Chapter 31

Private Land Use Decisions

A CHANGING TRADITION

It has been tradition that decisions about land use made by private individuals and companies have been "let alone" to a very large degree. This tradition is changing, however, as we discover that our land resource is limited and that the impact of private decisions have a way of adding up. In the past, a new factory meant jobs and purchasing power. Now we also recognize that certain kinds of industry may bring air and water pollution and high volumes of traffic. We are faced with difficult choices about what is desirable or needed and what is undesirable or can be done without.

Stores, shopping centers, and the like are also beginning to face controls on their freedom of action as citizens express their concern about the noise and traffic caused by new commercial development.

Finally, we now realize that the development of the suburbs is consuming prime farmland at an agonizing rate. The consumption of prime agricultural land for homes is a trend that must somehow be altered.

ENFORCING URBAN/SUBURBAN LAND USE PLANS

Zoning by Use Classification

The most common form of land use control is that practiced at the local level by community and county governments. Referred to as "zoning," it is another way of saying that land has been classified as to its appropriate uses. By and large, such zoning is not concerned with air and water pollution but with the uses of neighboring land. Is the proposed activity a reasonable "neighbor" to the activities that already exist in the area?

The idea of zoning is young. The first comprehensive municipal zoning law in the United States was established in 1916 by New York City. Its appeal was enormous and immediate. In only 10 years, more than 400 American cities had adopted zoning laws. It was in that year, 1926, that the Supreme Court of the United States upheld zoning for the Village of Euclid, Ohio, as a legitimate means to control nuisances. The legal justification of zoning is that it protects the health, safety, and welfare of the citizens, but in fact, zoning most often is used to protect the values of privately held land.

Zoning laws Zoning begins when the state government passes legislation that "enables" local and county governments to create land use plans if they wish to do so. The local governments then draw a map which labels each area or zone of the locality as to its allowable uses. Typically, the uses fall into one of four categories in addition to agriculture. These categories may be further divided as well. Figure 31–1 shows the four categories in an arrangement that has been termed "cumulative zoning." Within this planning framework, a given category

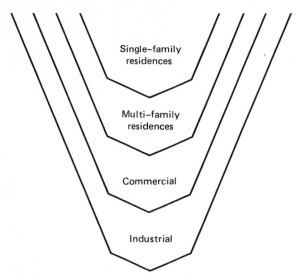

Figure 31–1 *The cumulative zoning concept. Single-family and multi-family residences and commercial activities are allowed within an area zoned industrial. Single-family and multi-family residences are allowed within an area zoned for commerical activities. Single-family residences are allowed in an area zoned for multi-family units.*

of use can accept all uses above it in the chart. Land zoned in the "commercial" category, for instance, can be used for stores and offices, its designated use, but single and multifamily dwellings are also allowed. On the other hand, land earmarked as being in the "multi-family" zone can not be used for either commercial or industrial purposes. The "single family residences" category is often referred to as the "highest," or most restrictive, grouping because no other uses are allowed.

Zoning used to protect a land use pattern   In addition to stating the allowed uses of the land, zoning may be utilized to enforce a uniform pattern of use. As an example, in the single family residence category, the minimum size of the lot may be specified, or the distance of the dwelling from a sidewalk (setback) may be given, or the minimum space between dwellings may be spelled out.

Zoning is typically not cast in concrete. Although changes in the ordinances can be made by

the body that enacted them, more often those who are putting land to "non-conforming" uses may apply for "variances" or "special exceptions." Such exceptions are considered by a planning board or zoning commission.

While having a procedure to consider non-conforming uses is quite reasonable, the results of the procedure are often not. Commercial interests, apartment developers, and the like often have obtained exceptions in the past for their activities simply by pleading hardship or by pointing to existing non-conforming uses. When a sufficient number of exceptions have been made, the original labeling of the area becomes meaningless and further exceptions become easy to obtain. Zoning, then, as a barrier to certain kinds of development, resembles a windbreak. It can help to shelter an area from the winds of change but it cannot always protect it.

Enforcement of zoning laws   Although zoning is interpreted and administered by a planning board or zoning commission, it is enforced by the courts. Those who would violate the zoning statutes can be enjoined from doing so; that is, they may be given a court order to prevent their violation. The court's muscle is its ability to fine those who disobey its orders.

We noted, though, that zoning sometimes fails. Failures may occur as individual instances or they may accumulate to produce wholesale changes in a neighborhood. Such failures may be due in large part to the way variances have traditionally been granted. Requests for rezoning a given parcel of land are heard, as we noted, by a zoning board or commission; their meetings are generally open to the public. The commissioners or board members take note at the meeting of members of the community who come to protest the "exception" which has been requested. A lack of protest by citizens is often interpreted as meaning that the residents of an area do not object to the proposed exception in the use of the land. In fact, citizens may not know the meeting is being held, many have prior commitments, or may be unable to attend. If no one appears at the meeting to protest the change in land use, the board may grant the exception to the person or firm requesting it.

The publicly proclaimed purpose of zoning is to maintain stable patterns of settlement, but the goal of zoning in practice is quite different. People use zoning primarily to protect the value of the residental property they own. Nearby factories and traffic are seen as threats to property value, and zoning is used to exclude uses that could diminish property value. Zoning, then, has not generally been an instrument to protect the general environment—only the environment of some people.

Shortcomings of zoning   It has been argued that some forms of zoning amount to economic discrimination. Consider, for instance, the common requirement that lot sizes be an acre or more. Such a requirement will cause a builder to erect expensive homes in order to earn enough profit on his investment in the land. Expensive homes are bought by people with upper levels of income; lower income people will be unable to afford these dwellings.

It has also been argued that zoning can lead to a monotonous conformity in the landscape. The diversity of having shops, playgrounds, and public buildings near residential sites can create an atmosphere of interest and appeal. Those who have visited Europe will have seen neighborhoods with a wider variety of activities than in the suburbs of the United States. In Europe, each neighborhood, in addition to its homes and apartments, is likely to have a school, a bakery, a vegetable store, a meat market, a drug store, a candy store, a cigar store, and so on.

Residential areas in the United States that exclude commercial shops also tend to waste energy. A purchase at the grocery store by a resident of the suburbs often requires a trip by car. Most Europeans can find a sufficient selection of food within a few blocks of their homes or apartments. They not only save on gasoline by walking to market but get useful exercise as well. Thus, the exclusion by zon-

ing of any kind of commerce in a residential area results in an added use of energy.

Zoning as a Tool for Environmental Improvement

Zoning, however, for all its shortcomings, is firmly in place in the United States. And despite its flaws, zoning can be used in clearly positive and beneficial ways. New zoning concepts are at the root of the improved possibilities for zoning.

Natural area zoning   One new concept is Natural Area Zoning. First adopted in New York City in 1974, natural area zoning could be used to protect wetlands such as swamps, bogs, ponds, and creeks, as well as forested areas and rugged landscapes. The green-belt area of Staten Island, a borough of New York City, was the first area to be classified by the New York City Planning Commission as a protected natural area, and hence not open to development.

Cluster zoning   We noted that zoning has been criticized for the economic barrier it often sets up. A concept that would not set up such economic barriers is Cluster Zoning. This concept allows homes to be built much closer together than conventional zoning allows so long as the total ratio of dwellings to land area remains at the desired level. The homes may be built with shared exterior walls, common walkways, and small yards. Such common wall construction requires fewer materials, shorter lengths of water and sewer lines, less sidewalk area, and less pavement. The dwellings can be sold at a relatively moderate price as a result.

As an example, a community that insisted on one-half acre per home (or two homes per acre) might be willing to accept 40 homes clustered on 10 acres if the developer was willing to leave undeveloped 10 additional acres in the area surrounding the clustered dwellings. For the entire area of 20 acres, the ratio of homes to acres would be the required two to one. The added open space can enhance a community and the clustered homes may be less expensive than individual dwellings.

Zoning to Protect the Public

Zoning has also been used to protect the public from hazard. Although hazard zoning is much less common than general land-use zoning, its most common example is flood-plain zoning. The flood plain of a particular river is the area outside the river's normal channel over which its flood waters may flow. An area may be classified by zoning as flood plain, and certain land uses may thus be banned. These prohibitions make a great deal of sense because local, state, and federal governments bear much of the burden of flood damages. Such burdens include rescue, cleaning up and restoration of homes, shops, and factories devastated by flood. Governments may even go to the extent of building flood control structures such as levees and costly dams in order to protect dwellings that ought never to have been located on the flood plain in the first place.

Some years ago, Professor Gilbert White recognized that the problem of flood control was being approached much like that of the proverbial horse who escaped from the barn. Only after the horse's escape was the barn door being closed. Far better, reasoned White, to avoid the misery, hardships, and enormous expense of restoration and of flood control structures. These could be avoided by simply barring new development in the path of a potential flood. The hardships and expense could also be avoided by removing flood threatened dwellings from their current sites.

White's ideas were translated into federal law in 1968. Flood insurance was provided under the National Flood Insurance Program enacted in that year, and this insurance was used to entice local communities to adopt flood-plain zoning. This desirable, low-cost insurance which the Act made available is paid for, in part, by the federal government. However, it can be made available only to residents of communities that take active measures to prevent further development in the flood plain. By the mid 1970s, more than 13,000 communities had indicated a desire to participate in the insurance programs and hence prevent further development in the flood plain.

Growth Management Plans

There are still other ways to influence private decisions on the development of real estate. Several of these fall under the heading of "growth management," and their history is quite recent. While growth management plans appear legitimate for the moment, they are nevertheless controversial. Although general land use zoning has been approved in principle through decisions of the U.S. Supreme Court, challenges to growth management plans have to this time never been heard by the Court. Their present status could, therefore, be changed.

First in Ramapo, New York, and then in Petulama, California, communities attempted to control the rate at which they grew. Ramapo, under pressure for housing from the New York Metropolitan area, developed an 18-year plan that included an orderly expansion of its road network and water and sewer utilities. The costs of rapidly expanding this network to go wherever developers chose to build would have been prohibitive to the community. Hence their plan was not only aimed at preserving their rural environment but also at preventing the explosion of local taxes. Developers who wished to build in areas as yet unimproved had to bear the costs of roads and of water and sewer lines themselves. Challenged in court by the developers, the town's plan was upheld in 1972.

Petulama, California, under pressure for housing from the San Francisco Metropolitan area, adopted in that same year a plan that limited the construction of new houses to 500 per year. This figure was far lower than their growth rate of nearly 2000 homes per year in the recent past. Petulama's plan was at first challenged in Federal District Court, where the plan was ruled unconstitutional, as infringing peoples' "right to travel" where they wished. When the case was brought to the Court of Appeals, however, the ruling was reversed, and Petulama's authority to limit its growth was upheld.

Two features have been common to all methods to influence private land use decisions we have discussed so far. First, they have been aimed at urban and suburban land-use decisions, as opposed to decisions on rural land. Second, the methods have all involved using the police power of government to enforce a set of rules that private individuals are bound by law to follow, unless exceptions are granted in response to specific requests.

PRESERVING RURAL LAND

In contrast to the methods that depend upon government enforcement power are techniques that use economic means to convince or require private owners to make better decisions on land use. These economic methods fall short of actual purchase of the land, but in terms of accomplishing their more limited objectives, they are very powerful. Moreover, they are generally more suitable for use on rural land, especially farm land, than the enforcement power of government.

Why are we interested in rural land preservation? We will answer the question from a purely humanistic viewpoint, but it is also true that the development of rural and wild land will destroy the habitats of many valued species.

Farmlands Are Becoming Suburbs

The rural land surrounding most of our cities is in genuine danger of extinction as the suburban fringe expands foundation by foundation, out into the countryside. We value and admire the farmland that surrounds us because it reminds us of an earlier era in which certain aspects of the quality of life were better—when traffic, noise, and congestion

were less and when the proportion of our population that worked the land was far larger than today. The suggestion of "a ride in the country" is a very common reference to our feelings for rural land. The tranquil, less hurried atmosphere and the visual pleasure it may impart touch a responsive chord in many of us.

There is more, however, beyond our feelings for rural land. The rural lands on the fringe of the suburban front contain among them the richest of the agricultural lands near our cities. Some refer to such land as "prime." It must be surprising to be told that the rural farmland being gobbled up by suburban sprawl is among the best and most productive in the country. It seems almost perverse that the rural land most in need of protection for our own economic well being is also the land most threatened. The explanation is simple.

When our cities and towns were first established, dirt roads connected them one to another, and dirt roads reached out into the surrounding countryside where vegetables, fruit, and animal stock were grown. Naturally these roads went to the richest farmlands, for these were the roads farmers would use to transport their animals and their produce to the town and city markets. This fact of historical geography has brought us to the present in which the same roads are now widened and paved. These roads which once carried produce-laden wagons into the cities, now carry commuters in steel-clad vehicles back and forth between homes and factories. The result is that the rich farmland along these roads is taken for new housing developments. Thus, more and more farm products come from further away at greater expense. This increased expense is due to the increased distance of shipment and to the fact that the produce may possibly be grown on less productive land. It is as though we were eating not merely the bread but the bread basket as well.

How severe is this problem? How fast is prime farmland being gobbled up? While the statistics are patchy, the fact of farmland development can be observed by simply a car ride to the city fringes. An estimate by the President's Council on Environmental Quality in the late 70s placed the agricultural land consumed by suburban "sprawl" at 1,000,000 acres (400,000 hectares) per year. This is the equivalent of 1520 square miles (4000 square kilometers). This figure represents only the loss of agricultural land. In fact, some 16 million acres of all kinds were lost to urbanization between 1967 and 1975, a rate of about two million acres per year. Furthermore, the loss rate appears to be increasing. Only about 1.2 million acres of all types were lost to urbanization each year in the decade from 1958 to 1967. California, at about the same time, estimated its annual rate of agricultural land conversion at 21,000 acres (8400 hectares) or about 3% of the annual loss in the entire country. These figures, however, must be viewed in the perspective of how much land we have and how it is currently being used.

The land area of the United States, excluding Alaska and Hawaii, is 3,522,000 square miles (9,100,000 square kilometers). This is equivalent to 2.27 billion acres (0.91 billion hectares). Of this quantity, about 2.7% of this land area is built over with homes, commerce, and industry. About 17% (384 million acres) of our total land area is considered by the·U.S. Soil Conservation Service to be "prime farmland."

Not all of the 384 million acres of prime farmland are in production; only about 250 million acres are so used. Of the remainder, about 86 million acres are unlikely to be farmed because they are already committed to other uses or are too distant from the needed transportation network. The last 48 million acres could be used for farming but for one reason or another presently are not.

We noted earlier that about 1,000,000 acres of agricultural land were being converted to urban use each year. An annual loss of 1,000,000 acres out of the 250 million acres of prime producing farmland is about 0.4% of our prime producing farmland lost each year. This loss is the more significant because of its location. The cities are the obvious focal point for urban expansion. Suburban development spreads

in much the manner as the ripples caused by a stone falling into a pool of water. Cities are also consumers of agricultural products. Continuing loss of farmland in the urbanized East means higher expenditures for produce because of the increased transportation costs and the higher consumption of energy. It also means greater dependence on other portions of the country for food (Fig. 31–2).

Taxation is a Root Cause of Farmland Loss

We indicated that the roads of agricultural commerce are the paths along which people from the city move to suburbia; this is but one reason for the threat to farmlands. Another and more powerful reason is the way we tax land. While the states and the federal government derive most of their income from taxes on industrial profits, on wages, and on certain luxuries, such as alcohol, the counties and local governments draw much of their income from taxing the land.

Figure 31–2 *Farmlands are being converted to commercial building, dwellings and pavement. Prime producing farmlands are being consumed as the suburbs of the cities sprawl across the countryside. In Massachusetts, for instance, if present trends continue more than half of its prime farmland existing in 1977 will be consumed by the year 2000. (Photo by Lester Fox)*

To tax the land, the local government must first attach a value to the land. The attachment of the value is called assessment of the property; the value is known as the assessed value. Practices differ from locality to locality, but the basis of the assessed value is very frequently the price the land would bring if sold on the market. Some local governments may assess a property at value equal to the value in the marketplace; others may assess a property at 50% of the market value, and so on. The tax on the property is determined by multiplying the assessed value by the tax rate. The tax rate is in dollars per $1000 of assessed value and is a uniform value throughout the taxing locality.

As housing sprawls, land once distant from the suburbs comes closer and closer to spreading development, and land values soar. Land once of value only for farming may become many times more valuable for homesites. The increase in the value of rural land is a natural outcome of suburban expansion. Because the market value has increased, the assessed value will increase and the taxes on the land will rise. Farming will become less profitable because of the farmer's obligation for higher taxes. At the same time that farming is becoming less profitable, the opportunity for profits in the real estate market begin to tempt the farmer who lives on the suburban fringe. Land in the family for generations or purchased decades earlier can become a source of wealth for a farm family. Land is often now a farmer's money in the bank for retirement. A farmer would have to love the land a great deal not to be tempted. At first he might sell only a lot or two on the main road and then some less used acreage. Each sale makes his land more valuable for homesites, makes his taxes higher, and the temptation to retire from farming greater. Some people jestingly call farmland "the farmer's last cash crop."

Ways of Preserving Rural Land

Assessing land on the basis of use   The vicious circle of increasing temptation for the farmer has been observed for some time, and efforts have been made to interrupt the economic forces which bear down on farmland. Many states have decided not to value rural land on the basis of its most valuable use; instead they are assessing it on its actual use. That is, even if a parcel of farmland could be sold for a high price, its assessed value reflects its use as farmland. Property worth $4000 an acre for homesites but only $1000 an acre as a farm will be taxed as a farm so long as that use continues. The tax threat to the farmer's pocketbook from an unexpected rise in value of the land is thereby eliminated.

Although the tax threat is pushed aside by such a system of taxation, the temptation from the increased value of the land remains. If the profits from the sale of the land are banked, they draw interest. When the annual interest on the proceeds of the sale begins to approach the annual profit the farm family earns through their labor, farming may be abandoned, whether there is a lower tax or not. Furthermore, the method of putting lower taxes on land used for farming can have unintended results. The farmer could sell his land to a speculator who might simply hold it or rent it out for farming while the price of the land continues to rise. The speculator may be rewarded with low taxes while he waits to make a killing on the real estate market. The speculator may, in fact, be attracted to the property because it will cost him so little to hold it. Special property tax treatment for the farmer, it is now generally agreed, cannot by itself halt the expansion of suburbia.

One idea that has brought some success involves a combination of lower taxes for farmland with a penalty if the land is sold. If the land is sold for development purposes, the penalty the owner must pay is the difference between the taxes that would have been owed if the land had been valued at its market price and the taxes actually paid while the property was valued as farmland. This amount added over all the years during which the special tax was applied must be paid if such a sale is made.

Incorporating family farms   There is another reason that farms are declining in number on the suburban fringe and it also is rooted in taxes, but the taxes are not on land; they are on inheritances. When a farmer dies, estate taxes must be paid on the value of the estate. Since the farmlands in the path of advancing suburbs become increasingly valuable, the value of the estate soars. Federal taxes on the estates of the deceased farmers can be considerable, amounting to as much as 25% of the value of the estate. The sons and daughters who inherit the farm have to pay such taxes to receive the inheritance. A farm in Cutchoque, New York, valued at only $800 per acre for agriculture was valued by the Internal Revenue Service (IRS) at $3500 per acre. The tax came to about $1200 an acre. As a neighbor described the situation, "... they [the Federal Government] forced its sale, very simply."[1] (See Fig. 31–3).

To preserve their family farms, the farmers have increasingly incorporated. As the owners of their corporations and holding the stock in them, they are able to make stock gifts each year of up to $3000 to individuals, in this case their sons and daughters. These gifts can be made in the form of stocks, and no taxes need be paid as long as the value of the gift is less than $3000 annually. If they were simply to give deeds for the land to their children, the decreased value of their property would affect their credit at the bank. The banks would see the farm decreasing in value and would be less likely to make loans the farmers need for seed, fertilizer, and equipment. Giving stock does not dilute the corporation, which remains a single entity despite having many owners. Incorporation obviously does not answer all the farmer's estate tax problems. Clearly some change in federal taxation of farm estates is needed.

Purchase and transfer of development rights by the public The problem of the loss of farmlands is occurring nationwide. Around nearly every major city, the suburbs lick greedily at the farms on the

rural fringe. At a few places in the nation, however, local governments are attempting to preserve their rural character by an interesting and powerful method. The method involves separating the right to develop the land from the land itself.

The idea of outright public purchase of development rights was first put into practice on Long Island, which feeds commuters by the hundreds of thousands into the nation's largest city, New York. Suffolk County, one of the several counties on the Long-Island, was once noted for fine agriculture, but through the decades of the 60s and 70s, it came under increasing pressure from real estate developers. Farms were gobbled up by the apparently unstoppable movement of New York City's suburbs.

In 1972, John Klein, county executive, proposed that it would be in the citizens' interest if the county government could somehow halt the conversion of farms to homes. He argued persuasively on economic grounds that the loss of farmlands would bring higher tax rates to the community. Because of the increased school-age population that would result, he argued, new schools would be required, and taxes would have to be raised to pay for them. It took the county legislature until 1977 to pass Klein's program.

In September of that year, John Klein sat down at a card table in a potato field with Nathaniel Talmage and signed an agreement guaranteeing that Mr. Talmage's 133-acre farm would remain undeveloped forever (Fig. 31–4). Klein's program is aimed at preserving 15,000 acres in Suffolk County and is the model that counties, states, and even nations are watching.

When Klein and Talmage sat down at the card table, Mr. Talmage transferred to Suffolk County the ownership not of his land but of the "development rights" to his land. He was paid a substantial amount for giving up those rights to the county, but nowhere near the amount he would have received if he had sold the land itself to real estate developers. With the signing of the agreement, the ownership of the Talmage property was now split

---

1 "Estate Taxes Drive Farmers Off the Land," *The New York Times*, 14 May 1972.

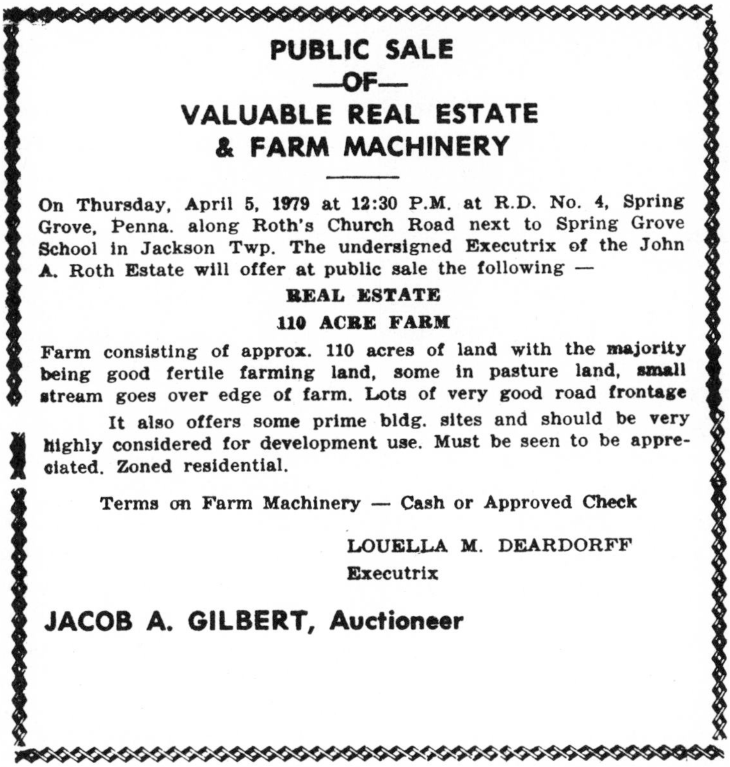

```
PUBLIC SALE
—OF—
VALUABLE REAL ESTATE
& FARM MACHINERY
```

On Thursday, April 5, 1979 at 12:30 P.M. at R.D. No. 4, Spring
Grove, Penna. along Roth's Church Road next to Spring Grove
School in Jackson Twp. The undersigned Executrix of the John
A. Roth Estate will offer at public sale the following —

REAL ESTATE

110 ACRE FARM

Farm consisting of approx. 110 acres of land with the majority
being good fertile farming land, some in pasture land, small
stream goes over edge of farm. Lots of very good road frontage

It also offers some prime bldg. sites and should be very
highly considered for development use. Must be seen to be appre-
ciated. Zoned residential.

Terms on Farm Machinery — Cash or Approved Check

LOUELLA M. DEARDORFF
Executrix

JACOB A. GILBERT, Auctioneer

Figure 31-3 *A farm falls to suburbia. Note the advice to developers that the land is likely to be useful for a subdivision. Such advertisements as this are common in many newspapers.*

into two pieces. Nathanial Talmage still possessed the land and all dwellings and improvements on the land. He retained the right to farm it or even to let it go wild. He may not subdivide it to build homes on the lots, but he may, if he wishes, sell off parcels of land. The buyer of those parcels, however, will not be able to build new dwellings because the right to develop the land has passed to the hands of Suffolk County. As a consequence of the transfer of the development rights, the market value of the land has decreased because of the restriction on how the land may be used.

There are distinct benefits for Mr. Talmage, in addition to the money he was paid, benefits that go far to compensate him for the decreased market value of his land. First, as he clearly planned to continue farming, he has lost no privilege that he

needs. His choice of crops, whether to sow or not, his choice of equipment, these items remain his decisions. Second, the loss in the market value of the land means lower real estate taxes to pay, since taxes are calculated by multiplying a tax rate times some portion of the market value.

Third, Mr. Talmage's children are more likely to be able to continue farming this land after his death. This is because the federal inheritance taxes on the Talmage estate will be much lower than on land with the potential for development. The estate or inheritance taxes are essentially determined by a tax rate multiplied by the market value of the estate, and the reduced market value of the land, now shorn of its development rights, will lead to lower estate taxes. A tax burden that could have brought the family to dispose of the farm is now

Figure 31-4 *Saving a farm. John V.N. Klein, right, Suffolk County Executive, and Nathaniel A. Talmage signing land agreement in a potato field on Mr. Talmage's farm. (Louis Manna/NYT pictures)*

lightened, and it is more likely that the family will be able to continue farming.

It cost Suffolk County $357,000 to obtain the development rights to the Talmage land; the county had to raise the money by selling bonds, but they still have a bargain. Their payment to Mr. Talmage was 80% of the market value of his land. That is, the possibility for development of the land is by far the most important factor in determining its market value. With development rights for the new land in the hands of the county, the value of the land is expected to fall to a bare 20% of the earlier market price. Put another way, the value of the land when sold for homesites is five times its value as farmland only. And as time goes by, that purchase will become more of a bargain. Pressures for new homes in the county will only increase with time as the suburbs continue to expand. Such pressures will drive the market value of land which can still be developed for homes ever higher; the value of the same land used only for agriculture will rise more slowly.

Maryland has a related program in which farmers may donate development rights to the state. In return for their donation, they are rewarded by being taxed on the value of the land for agricultural use rather than on its value for development. New Jersey has also developed a program to preserve farmlands; the state has been willing to purchase development rights of farms on the urban fringe. Not much land may be preserved, however, because the price offered per acre is low. Other states including Wisconsin, Indiana, Connecticut, Massachusetts, and California have forest and cropland protection laws.

Purchase and transfer of development rights by private corporations The purchase of development rights is not restricted to governments, although many states are watching the Suffolk County experiment. Such rights may also be purchased by private individuals, by corporations, and by non-profit institutions. In fact, any legal entity that can hold property can hold development rights as well.

Of what use could these development rights be to corporations? To understand how a home-building corporation could make use of purchased development rights, we need to reemphasize the nature of the transaction on Long Island. There Suffolk County purchased the development rights on certain farms; those development rights now in the hands of the county are likely to *never* be exercised. It is possible, however, that the development rights purchased by a corporation on a parcel of farmland in one place can be used by that company *elsewhere* than on that parcel. Once used elsewhere, the land from which they were withdrawn can no longer be developed and is thereafter taxed only as agricultural or undeveloped land.

The first use of this concept to preserve rural lands took place in the small town of Saint George, Vermont. This town, which then had no commercial development whatsoever (no store, gas station, post office, etc.), was in the path of the spreading suburbs of Burlington, and its population had gone from about 100 in 1960 to 500 in 1970. To guide its future growth, the town purchased 48 acres of land to be used as the nucleus for future homes and commercial enterprises. These 48 acres were made available to developers for building but only if the developers agreed to keep land open elsewhere in the town.

A description of how the town will use this 48-acre project area to guide the expected development reads:

> To achieve the objective of concentrating settlement and preserving the rural character of most of the rest of Saint George, the town may oblige a developer to transfer to the town development rights purchased from owners outside the project area in exchange for the opportunity to develop in the core village area. For example, a developer wishing to construct twenty units of housing in the village area would have to purchase twenty acres of land zoned at one family to the acre elsewhere in Saint George and transfer his acquired right of twenty units of housing to the project area. The twenty acres from which the rights were transferred will remain open land in perpetuity or until the town releases it to meet future needs. The land will be taxed only at its value as undeveloped land.[2]

It should be noted that those 20 acres purchased elsewhere need not have been land suitable for building. It could be swamp, ravine, mountaintop, or hillside. The point is that a one-for-one trade of

_____

2 L. Wilson, "Precedent Setting Swap in Vermont", *Journal of the American Institute of Architects,* **61(3)**, 51 (March 1974).

preservation area for housing development is taking place.

Conservation easements State and local governments have a number of options available to them, short of outright purchase, to prevent the development of wild land. One such option is to secure a conservation easement of the land, a statement attached to the deed specifying the uses to which a parcel of land may (or may not) be put. These may include whether buildings may be erected, the maximum height of buildings, the cutting of trees, the use of billboards, and the like.

How can one go about attaching such statements to deeds? The simplest way is for the community to buy the property outright when it comes up for sale. The local government attaches the statement on allowable uses and then puts the land back on the market. The land will probably sell for less than the community paid for it because its use is now restricted. The loss in value, however, may be viewed as the amount the community had to pay to put the restriction in the deed.

Although the method of purchase/resale is the easiest way to attach a deed restriction to the land, desired parcels are not likely to come on the market just at the time when a community is ready to act. More than likely, the local government will have to convince the owners that adding the restriction will bring them benefits. The benefits may be direct payment for the attachment by the government, or the benefits may be tax advantages in terms of federal estate taxes, federal income tax deductions, or reduction in assessed valuation and hence in property tax.

Questions

1. What is the purpose of zoning? Contrast the stated purpose and the actual purpose.
2. What social and environmental shortcomings are often associated with zoning? How can zoning be used to protect the environment and people?
3. How do growth management plans differ from traditional zoning?
4. Explain how taxation and the growth of suburbs are "eating up" prime farmlands in the U.S.

5. Describe some ways of keeping privately held land rural. What advantages do such methods have? That is, why not just have the government buy the land outright?

Further Reading

Bosselman, F. and D. Callies, *The Quiet Revolution in Land Use Control.*
Prepared for the Council on Environmental Quality, Washington, D.C., 1971, available from the Superintendent of Documents, U.S. Government Printing Office, Washington, D.C. ($2.75).
This and most of the following entries are good general references.

Kidder, T., "The Battle for Long Island," *Atlantic Monthly,* November 1976, p. 47.
An interesting recounting of one region's "fight against urban sprawl" with a discussion of the farm preservation program of Suffolk County.

Leopold, A., *Sand County Almanac,* Oxford University Press, 1949. Reprinted, Sierra Club/Ballantine, 1970.
Leopold observes nature and humans through the 12 months on his Wisconsin farm. This highly readable, classic work has a fine essay "The Land Ethic" in which the author describes the relationship that ought to exist between people and the land.

Platt, R., *Land Use Control: The Interface of Law and Geography.* Resource Paper 75-1. For sale by Association of American Geographers, 1710 Sixteenth Street, N.W. Washington, D.C., 20009.

Stover, E., ed., *Protecting Nature's Estate,* Bureau of Outdoor Recreation, U.S. Department of the Interior, 1976. For sale by the Superintendent of Documents, U.S. Government Printing Office, Washington, D.C.

Untaxing Open Space, the President's Council on Environmental Quality by the Regional Science Research Institute, 1976. Available from the U.S. Government Printing Office.

Whyte, W. H., *The Last Landscape,* Doubleday and Company, Garden City, NY, 1968.

Chapter 32

Preserving Public Natural Areas

THE PUBLIC PRESERVATION MOVEMENT

Why Should Lands Be in Public Hands?

State and National Parks, State and National Forests, National Wildlife Refuges, National Monuments, National Seashores, National Lake Shores, Wild and Scenic Rivers—the list of ways that the public sets aside land for recreation and preservation is longer still.

Wetlands may be rescued from development because they serve migrating birds on their annual journeys up and down the continent. Wild fowl may winter at such places or simply stop at them enroute. A forest or desert may be set aside for its unique vegetation. The Redwood National Park and numerous California state parks are devoted to preserving the two species of Redwood tree, the largest living species. The Joshua Tree National Monument and the Saguaro (cactus) National Monument are used to preserve remarkable desert plants. Some parks are used to preserve stunning natural landscapes. Death Valley National Monument, Yellowstone National Park, and Grand Canyon National Park are examples that come immediately to mind. Other parks such as the Everglades are chosen to preserve whole ecosystems of plant and animal species. And for every preserved area, it would not be an exaggeration to say that there is another area yet to save.

Save from what? From timbering in the case of the Redwoods; from vacation homes in the case of the wetlands; from mining in the case of Death Valley; from commercial exploitation in the case of Grand Canyon and Yellowstone; from being drained or made into a jetport in the case of the Everglades. The threats are abundant. Although sometimes innocent, often they are from those ignorant of the values that other people place on such natural resources, and the threats continue to occur.

The road to preservation of our landscape and of our species is marked by several disasters, not all in the distant past. Our first spectacle was Niagara Falls which fell quickly to commercial development in the 1820s to 1840s. Mills were built at the foot of the Falls; the surrounding forests were cut; and tourist operators set up at the rim of the Falls. Visits to the Falls were highly commercialized. One pair of English visitors who observed the commercial activities at and around the Falls during this period called for protection of the Falls as "the property of civilized mankind." In California in the 1850s, one of the largest of the giant sequoia trees in the mountains of Southern California was cut down and transported to New York and to Britain for exhibition. Such acts led to pressure to put the Yosemite region in public hands.

As late as the 1960s, over 100 years later, the awesome and beautiful rock walls of Glen Canyon were sealed finally from view by construction of a massive dam. And even into the 1970s, we find mining of existing claims officially approved in

Death Valley by act of Congress. Preservation seems to require not only the initial act of setting aside but continuous surveillance as well.

A Brief History of the Federal Preservation Effort

Creation of the federal parks   In 1864, the tradition of federal creation of park lands began. The Yosemite Act gave 44 square miles (113 square kilometers), which had been in the possession of the U.S. simply as public land, to the State of California to preserve for the public. The Act was introduced by a senator from California who justified his unusual proposal by noting that the lands were "for all public purposes worthless," by which he meant that no mineral resources or usable water power could then be derived from the land. The preserve included the Yosemite Valley, the towering mountains that formed its sides, and one other tract. The giant sequoia was at last protected. In 1890, the park would be taken into federal possession by Act of Congress.

Eight years later, Congress created a park in the Yellowstone region of Wyoming. This time the park was put directly under federal direction, presumably because Wyoming was still only a territory. Only a year before the creation of Yellowstone Park in 1872, the fabulous geysers had been in danger of being claimed as private property, apparently because of their value as a tourist attraction. Nevertheless, the oratory in Congress that supported the Act stressed the "worthlessness" of this park land. Again, the lack of value referred to a lack of mineral resources or of potential for agricultural use. An 1895 act prohibited hunting in Yellowstone National Park, but the thrust of wildlife protection by setting aside preserves devoted only to animals was still in the future. Indeed, the protection provided by National Wildlife Refuges evolved in a tradition entirely separate from that of the National Park System.

The low commercial value of the lands committed to parks was one theme heard again and again. Another theme was that the areas set aside represented one aspect of America's claim to greatness. The beauty of the parks was compared frequently to the architecture and art of Europe. Though we did not have the wealth of history of the European continent, we had nonetheless our own natural history treasure that was worth preserving. In its day, this viewpoint served us well, for the notion of preserving an ecosystem (then still undefined) could never have occurred to people in that era.[1]

No one was concerned then with preserving a forest for we had at the time what seemed an abundance of land, species, timber and mineral resources. Wilderness was not far from anyone's doorstep. Indeed, in those early years the nation saw its business as taming and subduing the land rather than preserving it intact. The realization that land and forest was a limited resource dawned only gradually; and when it did occur to people, it was in the perspective of profit making. Mark Twain poked fun at the madcap commercial atmosphere of land grabbing that occurred after the Civil War. "Buy Land; they ain't making it any more!" one of his characters suggested.

Forest preserves and the U.S. Forest Service   In the period 1880–1910, however, a genuine appreciation of the role of natural resources and the public land was beginning to surface. Individuals like Franklin Hough, Bernhard Fernow, and Gifford Pinchot helped to shape a policy of *forest protection* as opposed to scenery or wildlife protection. Thus began a tradition different from the National Parks Movement and different from the movement to protect wildlife. Their efforts, which came out of the Divi-

---

1 The discussion of the history of the National Parks Movement is drawn from "The National Park Idea" by Alfred Runte, *Journal of Forest History* **21**(2), (April 1977). The article is insightful, exciting, and beautifully illustrated. It reaches back and recaptures an early era in the environmental movement.

sion of Forestry within the Department of Agriculture, were directed at forest protection and at soil and water conservation.

With the passage of the Forest Reserve Act of 1891, the President was empowered to set aside areas of land in the public domain as public reservations to ensure that adequate timber would be available to the nation in the future. Although President Harrison created, within two years, 15 forest preserves with over 13 million total acres (5.2 million hectares), no means was actually provided to protect these reserves until 1897. In that year, Congress passed an amendment to an unrelated Appropriations Bill that did offer protection for the reserves.

The Pettigrew Amendment, as it is known, directed the Secretary of the Interior to make rules for protection of the reserves and authorized the sale only of mature or dead timber within the reserves. The timber was to be marked for removal before cutting could begin. The amendment of 1897 shaped the nation's forest management policies for 63 years until the Multiple Use Act of 1960 expanded its provisions.

Effective protection, however, did not exist until 1905 when the administration of the Forest Preserves was transferred from the Department of the Interior to the Bureau of Forestry within the Department of Agriculture. The man who was given the responsibility for the 63 million acres (25 million hectares) now in the Preserves was Gifford Pinchot. He renamed his agency the U.S. Forest Service, the name still in use today. The preserves were renamed as well; they became the National Forests. By 1907, there were over 150 million acres (61 million hectares) of preserves in his national forests.

Wildlife protection and the refuge system Still another tradition of land preservation was growing at about the same time. Whereas the purpose of the national parks was to preserve magnificent scenery and the purpose of the national forests was to preserve timberland, the purpose of wildlife refuges was the preservation of animals by protecting their habitat.

The federal government did not become active in the preservation of wildlife until the turn of the century. In 1900, Congress passed a law that instructed the Secretary of Agriculture to take steps to protect and restore game and wild birds. This was the era in which the nation was sadly witnessing the disappearance of the passenger pigeon and this loss undoubtedly influenced the passage of the law.

In 1903, the United States created its first preserve for animals, a refuge on Pelican Island off the Florida coast. This was soon followed by the establishment of wildlife ranges in the Wichita National Forest (1905) and the Grand Canyon National Forest (1906). And in 1906, Congress prohibited hunting on all of the preserves that were now to be set aside for the protection of wildlife. In 1908, Congress set aside a National Bison Range in Montana for the threatened buffalo.

The fate of the passenger pigeon helped to move Congress to pass the Migratory Bird Treaty Act in 1918, which placed migratory game in the protection of the federal government. Two years later, the Supreme Court declared that the federal government's power to regulate migratory wildfowl came before that of the states. Justice Holmes,[2] writing for the majority, noted that without federal regulation, "...there soon might be no birds for any powers to deal with... It is not sufficient to rely upon the states. The reliance is vain."

The 1918 Act, however, had not provided for the acquisition of bird habitat. This defect was corrected by the 1929 Migratory Bird Conservation Act, which authorized the Secretary of the Interior to create refuges for migratory wildfowl and to operate them as "inviolate sanctuaries" (Fig. 32–1). Funds to purchase wildfowl sanctuaries from states and private owners have come, as we mentioned in

---

2 Quoted from Evolution of National Wildlife Law, prepared for the Council on Environmental Quality by the Environmental Law Institute, 1977.

our discussion of endangered species, from the sale of hunting stamps. Revenue for other types of refuges other than wildfowl refuges has not been so readily available.

ADMINISTRATION OF FEDERAL NATURAL AREAS

A Fragmented System

These three traditions, the National Parks, the National Forests, and the National Wildlife Refuges were all to evolve over the years. Uses and purposes have changed and blended as new needs and new pressures have emerged. Today these three systems are still managed by different agencies, in keeping with their differing purposes at the outset. The National Park Service administers the National Park System within the Department of the Interior; the U.S. Forest Service manages the National Forests in the Department of Agriculture, and the U.S. Fish and Wildlife Service is responsible for the system of National Wildlife Refuges in the Department of the Interior.

Table 32–1 shows land areas these agencies administer as well as the supply of state parks and forests.

Not all the land the federal government holds for the public is in the National Parks, the National Forests, or the National Wildlife Refuges. A considerable portion of the government's holdings are in the National Resource Lands, some 118 million acres (48 million hectares) of public land in the lower 48 states and another 330 million acres (134 million hectares) in Alaska. These lands, which are supervised by the Bureau of Land Management (BLM), are the remainder of the "public domain" (Fig. 32–2).

The bulk of the lands we refer to as the public domain are those lands that remain from the great land purchases of the past, such as the Louisiana Purchase. Much of these lands were granted to the railroads, and settlers acquired a portion of them by homesteading. The Forest Preserves and the National Parks and Wildlife Refuges withdrew other lands from the public domain. But the land that has never been sold and which remains in the hands of the BLM is vast and often trackless.

Some of the land held by the BLM is timberland. By a quirk of history, some 2.4 million acres (about one million hectares) of the most productive timberland in the U.S. fell into the hands of the Bureau of Land Management. Two large tracts of

What a Few More Seasons Will Do To The Ducks.

Figure 32–1   *The fate of wild birds as seen from the 1930s (courtesy of the Des Moines Register). This series of cartoons was the work of J. N. Darling, a nationally syndicated cartoonist with the Des Moines Register. In the middle 1930s, because of his interest in wildlife protection, Darling was appointed Chief of the Bureau of Biological Survey in the Federal government. His efforts were instrumental in building and strengthening the National Wildlife Refuge System.*

land granted to railroads in Oregon reverted back to Bureau of Land Management when a violation of the grant occurred. These are the 2.4 million acres of highly productive timberland that the BLM manages today. They apparently are administered by the BLM from much the same viewpoint as the Forest Service.

The National Resource Lands, as BLM lands are called, are available for mining, grazing, and timbering. Within the last decade, the recognition of valuable wilderness in the National Resource Lands has prompted an evaluation of these lands for their wilderness and recreation potential. In the pages to come we will survey the management of natural areas by the federal agencies involved in their protection.

National Wildlife Refuges

Although originally set aside as "inviolate sanctuaries," the use of the National Wildlife Refuges has evolved over the years to include other purposes. The use of the Migratory Bird Hunting Stamp Act of 1934 as a means to finance the purchase of refuges for wildfowl may have influenced decisions by Congress in 1948 and 1959 to allow the Secretary of the Interior to authorize hunting on the refuges. Although hunting was originally limited to no more than 25% of the refuge area by the 1948 legislation, the limit was raised to 40% in 1959. The Secretary was allowed to use his/her judgment to decide if hunting on a particular refuge was com-

patible with the purposes of the refuge. No challenge to this use has yet been successful in a court of law. Three years later the Refuge Recreation Act of 1962 authorized the Fish and Wildlife Service to open the wildlife refuges to public recreation. The authorization gave to the Secretary of the Interior the responsibililty to determine whether recreation at a particular site would interfere with the basic purpose of the refuge.

In 1966, the wildlife system which consisted of game ranges, wildlife management areas, wildlife refuges, and other units was consolidated into the "National Wildlife Refuge System" under the administration of the Fish and Wildlife Service (Fig. 32–3). The allowable uses of these areas were expanded as well. Now, in addition to bird hunting and public recreation, the uses, when found to be compatible with the central purpose of the refuge, could include hunting, fishing, and public accomodations. Just as the National Forests were expanded to become multiple-use lands in 1960, the National Wildlife Refuges were now becoming multiple-use, with the added proviso that their original purposes as species habitat should not be compromised. This notion of accompanying but secondary uses of the refuges has led to the refuges being referred to as "dominant-use" lands as opposed to "multiple-use."

National Forests

The multiple-use concept   The term "multiple-use" has come to be applied most frequently to the

TABLE 32–1   **Land Areas Administered by State and Federal Agencies**

Agency / Office	Name of Land Holding	Millions of acres[a]	Millions of hectares[b]
Bureau of Land Management	National Resource Lands	448	181
U.S. Forest Service	National Forests	183	74
National Park Service	National Park System	31	12.5
U.S. Fish and Wildlife Service	National Wildlife Refuges	32	13
State and Local Parks and Forests	—	about 25	about 10

[a] One acre is a square area about 210 feet on a side.

[b] One hectare is a square area 100 meters (328 feet) on a side.

management of the National Forests. Indeed, an act of Congress known as the Multiple-Use-Sustained Yield Act was passed in 1960 to define in a formal way the meaning of multiple-use in the National Forests.

The Act stated that the National Forests were to "be administered for outdoor recreation, range, timber, watershed, and wildlife and fish purposes." It further stated that "The establishment and maintenance of areas of wilderness are consistent with the purposes and provisions of this act." The

National Forests were also to be managed for a "sustained yield" of forest products. "Sustained yield" was defined as "the maintenance in perpetuity of a high-level annual or regular periodic output . . . without impairment of the productivity of the land."

Surprisingly, the Act, which is now the principle policy directive of the Forest Service, was opposed both by conservation organizations *and* the timber industry. In fact, the Forest Service saw the law as a means of making legitimate the manage-

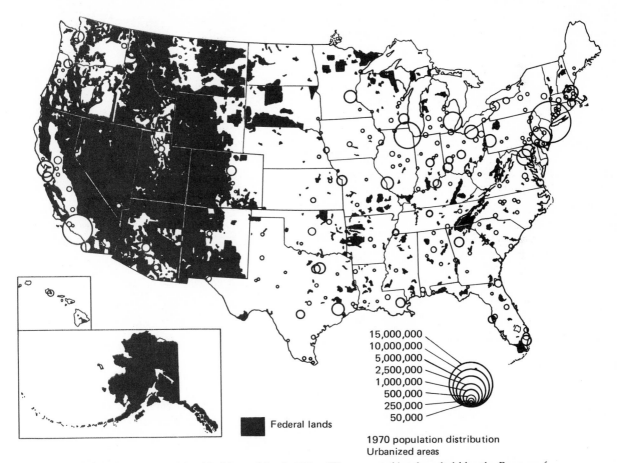

15,000,000
10,000,000
5,000,000
2,500,000
1,000,000
500,000
250,000
50,000

Federal lands

1970 population distribution
Urbanized areas

Figure 32–2 *Federal land holdings. Nearly 700 million acres of land are held by the Bureau of Land Management, the U.S. Forest Service, the National Park Service and the U.S. Fish and Wildlife Service. Much of the land, however, is in the western U.S. and Alaska, distant from major concentrations of population. [Source: Bureau of Outdoor Recreation (now the Heritage Conservation and Recreation Service).]*

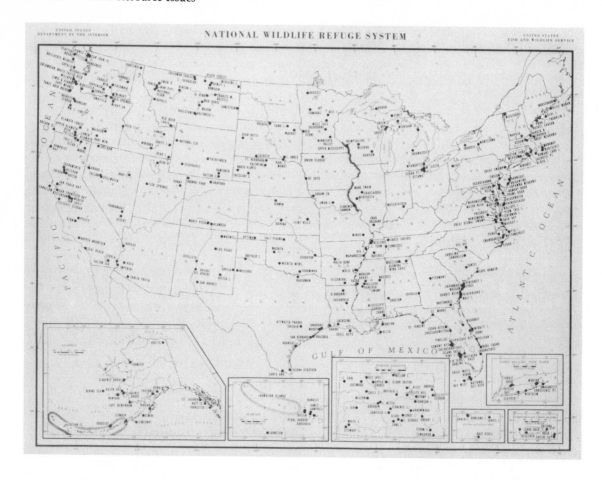

Figure 32–3 *Map of the National Wildlife Refuge System, 30 September 1978. Most but not all of the Wildlife Refuges managed by the Fish & Wildlife Service are included in this map. The Fish & Wildlife Service publishes a directory of the nearly 400 refuges. The directory lists refuges by state and county, noting its size and date established and the primary species protected in the refuge. (Map courtesy of the Fish & Wildlife Service, U.S. Department of the Interior.)*

ment practices it had followed since the earliest days of the Agency. The National Forests had always been administered for all these purposes and mining in addition. The opposition by both conservationists and by timber people stemmed from their suspicion of the motives of the Forest Service.

Conservationists were concerned that the Act would be the tool by which the Forest Service could authorize extensive new cutting in the National Forests. Because of pressure from the Sierra Club, the statement on wilderness was added to the Act, but the Sierra Club never supported the bill. The opinion of the Sierra Club of the motives of the Forest Service in supporting the act is explained by Harold K. Steen[3]:

3 H. K. Steen, *The U.S. Forest Service: A History*, University of Washington Press, Seattle, 1976, p. 302.

Several events coincided during the 1950s which cost the Forest Service the support of many conservationists. Since Pinchot's time, the national forests had been held in reserve to supply lumbermen when privately owned timberlands had been cutover and were in the regrowth cycle. Year after year and chief after chief reaffirmed the policy to log a substantial portion of the national forest system—at the appropriate time. The appropriate time followed World War II, when increases in population and affluence concurrently multiplied recreational pressures on all public lands. Couched in terms of timber famine and devastation, decades of propaganda by the Forest Service to justify federal regulation of logging had created the popular image of rangers protecting forests against rampaging, greedy lumbermen. Now the postwar public came to the forests—camping, fishing, hiking—and "discovered" logging in the national forests. To many of the public, national forests and national parks were the same, and logging either was bad. Implementation of long-term timber management plans collided with the results of an extremely effective public relations program against "destructive" logging. As a result, the public felt deceived.

Industry, in contrast, saw the bill as giving weight to public uses which they felt could only work against the interests of the timber companies.

Four years later conservation interests were finally successful in passing the Wilderness Act (1964), making it possible for areas within the National Forests to be specifically set aside as wilderness areas only. The Act was essentially a statement of distrust in the motivations of the Forest Service. The Multiple Use Act of 1960, while it affirmed wilderness as a legitimate "use," did not order the establishment of such areas. The Wilderness Act gave Congress the right to designate areas of the National Forest as wilderness and hence protect them from logging. By the mid 1970s, 12 million acres (4.9 million hectares) had been included in the National Wilderness Preservation System;

over 90% of the 12 million acres were in the National Forests.

Timber management in the national forest In addition to the order from Congress to the Forest Service to set aside wilderness areas, court cases arose over the Forest Service's application of the Multiple Use Act. Specifically, these cases claimed that the Forest Service was not giving due consideration to wilderness in its timber sales. Controversy on the way in which the Forest Service has administered sales of timber has focused not only on where timber sales have been allowed, but also on the manner in which trees have been cut. One specific method has been attacked again and again; it is the practice known as *clear cutting.* Clear cutting means simply the cutting of all the trees in an area regardless of size, quality, or age. The practice of clear cutting on National Forests has grown through the years. Lumber companies prefer it because it is obviously more economical than selective cutting of mature and dead trees. Since there is a cost of moving lumber-jacks and equipment to a site, the companies prefer to take all the timber they can get before moving on to a new area.

Although there are added costs to the company when it must do selective cutting, there are also costs to the public when clear cutting takes place. In addition to the visual impact of a forest cut to the ground, clear cutting can affect rainfall retention in the area and erosion. More rapid run-off over wide areas can lead to widely fluctuating stream flows, including damaging flows. And high run-off rates can speed the erosion process as well. Further, species habitat is completely wiped out by clear cutting. Wilderness values are, of course, destroyed in the region of the clear cut (Fig. 32-4).

The timber companies argue for clear cutting for other reasons than economics. One commercial timber species, the Douglas fir, is thought to grow best in direct sunlight. Company foresters point to clear cutting as the way to assure optimum growth of a new forest of Douglas fir. In 1974, however, a scientific study by Franklin and De Bell disputed

Figure 32–4 *A poorly designed clearcut can ruin the view of a forested mountain area. This photo was provided by the Bureau of Land Management, but it does not necessarily depict activities on BLM lands. Clearcutting has been a recent practice on Forest Service Lands.*

the claims that direct sunlight was necessary for the regrowth of the Douglas fir.

One aspect of clear cutting deserves mention because of its bearing on the ecology of the harvested site. Recall that the clear cut virtually wipes out the entire habitat. The species that return to an area regrown from a clear cut are expected to be somewhat different from the species that existed prior to the clear cut. This is because the regrowing trees will all be of about the same age. A natural forest, one which has had time to mature through several growing cycles will have trees of many different ages. The forest regrown from a clear cut is referred to as an even-aged stand; the forest with trees of different ages is called an uneven-aged stand. The uneven-aged stand is thought to be ecologically superior in terms of the balance of species present. While clear cutting is displeasing visually, certain areas are definitely harmed by the practice.

A clear cut on a steep slope may lead to erosion of the best soil off the slope; a very long time is required for the forest to return. An area that in-

cludes a stream channel can also be damaged if the trees are cut too close to the banks. Erosion of the stream banks damages both stream and forest. The opening of the stream to sunlight will increase the temperature of the flowing water, as well (Fig. 32–5).

Other methods of cutting are used by the timber industry. These include *selection cutting, seed-tree cutting,* and *shelterwood cutting. Selection cutting* removes mature timber, usually the oldest and largest trees. The cutting cycle involves a return to the area for timber cutting every 5 to 20 years to remove the newly generated mature timber. This method of cutting produces an uneven-aged stand, and has been widely practiced.

Seed-tree cutting is a poor name for this logging method, which actually *leaves* seed trees (Fig. 32–6). The seed trees left to regenerate the stand may be left singly or in groups. Although seed-tree cutting has been used for commercial harvesting in Montana, it is still relatively experimental.

Figure 32–5 *Construction of a logging road near a creek. This exposure of loose, unanchored soil will cause rapid erosion from the land. The particles will enter the creek, producing soil-choked waters. Sediment in the creek can decrease the ability of the stream to carry high flows, causing flooding downstream. (This photo was provided by the Bureau of Land Management, but it does not necessarily depict activities on BLM lands.)*

Shelterwood cutting proceeds in three stages. In the first or "preparatory" stage, dying trees, defective trees, diseased trees, and trees of unwanted species are removed, leaving space for new trees to grow. Ten to 15 years later, a stage called "seed cutting" opens the stand further so that seedlings can receive adequate sunlight and heat. In the final phase, called shelterwood removal cutting, when seedlings have been established, the remainder of the mature trees are cut.

It is well to note that without supervision most logging systems can produce unwanted results. As an example, selection cutting, unsupervised, can be used to select the best or most marketable trees as opposed to the mature trees for which it is intended.

In 1973, clear cutting was taken to court, and it was found that the 1897 law which formed the basis for management of the forests still has an impact in the present. In Izaak Walton League *vs.* Butz,[4] a district court examined the timber harvesting that the U.S. Forest Service was allowing in the Monongahela National Forest in West Virginia. The court declared that clear cutting in the National Forest, the removal of all trees in an area regardless of age, was in violation of the 1897 law. The 4th Circuit Court of Appeals upheld that decision in 1975, citing the rules of the 1897 Act, which limited cutting to the trees individually marked and trees of mature growth or dead. In 1976, the Forest Management Act finally replaced the original law, putting in the hands of the Forest Service the responsiblity to manage the National Forests as they judge best. The result has so far been a return to the practice of clear cutting so clearly prohibited in the 1897 law. (See The Tragedy of One-Shot Forestry.)

How much timber comes from public lands? In the 1900s, virtually all the lumber produced in the U.S. was cut from private forest lands. After World War II, housing pressures forced the government to open the National Forests to companies in a serious way, although private logging had been going on in the National Forests to a minor extent since the establishment of the preserves. The share of the annual production of lumber from public lands had reached 15% by 1950. By the early 1970s, the lumber from public lands accounted for 40% of annual timber production.

The Forest Service has obviously "upped the cut" and the explanation lies in part in the initial mission of the service. The preserves, now the National Forests, were set aside for the day when timber supplies on private lands had fallen so low that the preserves would be needed to meet demands. Rising demand for lumber in the decades after World War II was the signal to open the forests. Now with the prices of lumber at near record levels, there is pressure from many directions to keep the national forests producing. Lumber companies, homebuilders, construction workers, and consumers want lumber at relatively low prices. Management of the forests is a very large problem in the national economy.

If we listen to the debate between conservationists and the timber companies, two terms are heard again and again; these are "allowable cut"

Figure 32–6 *Seed-tree cutting prior to log removal. Some trees are left to seed and regrow the cutover area. (Bureau of Land Management.)*

---

4 Butz was then Secretary of Agriculture and thus in charge of both the U.S. Forest Service and the National Forests.

and "sustained yield." A "sustained yield" cut is a quantity of timber less than or equal to the natural increase (in cubic feet of lumber) occurring in the area in the past time period. The "allowable cut" is the quantity of timber that a company is given the right to harvest from an area over a set period. A forest area that has an allowable cut equal to the new growth will, when the cut is complete, still have the quantity of standing timber it had before the new growth took place. That quantity of new growth can be expected again, and the yield from the forest can be sustained. Should the rate of cutting in the National Forests be allowed to approach the new growth? Remember that the National Forests are our final reserve of timber, to be cut when the yield from private forests is insufficient to meet demands. Although we are now cutting extensively in the National Forests, we must ask if it is because of high demand or because private forests have

been poorly managed. Private forestry efforts must be encouraged so that the use of the National Forests for meeting timber demands can be decreased. There is, in addition, the issue of whether we should allow timber that is cut from the National Forests to be sold in export to foreign nations as it now is.

If the cut in the National Forests is equal to sustained yield, the stock of standing timber will not decrease, but it will be standing still in the face of growing demands. Unless private forests are substantially regrown, the pressure for cutting in the National Forest can only increase. Allowing the cut to approach sustained yield in the National Forests makes sense only if the private forests are being regrown. It may be that decreasing the allowable cut in the National Forests could help to encourage adequate private forestry efforts. The problem is that people want houses now and forests take a minimum of 25 years to grow.

THE TRAGEDY OF ONE-SHOT FORESTRY

The developing pattern of forest management in the 50 eastern national forests is like a personal tragedy to me. I am puzzled and dismayed by the five new management plans I have seen for national forests from Vermont to South Carolina. They call for clearcutting and even-aged management for most of the mixed hardwood types of forest area close to heavily populated areas. The plans show a relative insensitivity to nontimber values and propose an almost exploitative, primitive silviculture, with the objectives of short-term financial returns to the timber operator and ease of administration and future management by the Forest Service....Our bureaucrats will designate a boundary of timber, and the operator will simply harvest all the trees, large and small, regardless of quality or species, and the next clearcut will be 80-120 years hence.

How can this be? The Forest Management Act of 1976 has noble and true words about the importance of environmental values, about the relative unimportance of dollar values alone, and says in so many words that clearcutting shall not be used unless it is the "optimum" method. But there is a catch. The discretion and the judgment are left to the professionals in the Forest Service....The "discretion" has been exercised all in one direction–cheap timber production on the bulk of the forests. How can any professional agency prescribe <u>one</u> silvicultural system for the vast diversity of types, topography and climates that constitute eastern United States forests?...now the Forest Service seems to believe it has a legal basis for almost universal clearcutting, and it is acting accordingly....

What are the consequences of large-scale block clearcutting of eastern hardwoods?...The consequences...are greater waste of timber, lower environmental values and greater danger of damage to soil, site and water. Let me explain. Most of the individual trees in eastern hardwood forests are now below mature saw-timber size. But clearcutting harvests these smaller trees, just when they are growing fastest. This is a waste . . . .

This crisis in forestry has been caused by failure to reconcile and balance the values of the timber as a commodity, the forest as an environment, and the integrity of the forest-site-soil-water ecosystem. Such a balance cannot be achieved by the one-shot system of clearcutting and even-aged management now being proposed for the eastern national forests....If the Forest Service will not or cannot reconcile and balance forest values, then Congress or the courts will have to act again.

Leon S. Minckler

Appeared in Sierra Club Bulletin, July/August 1978

Leon S. Minckler is now retired from the U.S. Forest Service;
he has taught environmental science and forestry at
Virginia Polytechnic Institute and the State University of New York.
Mr. Minckler is considered one of the shapers of forestry opinion.

Fire, enemy of the forest, or is it? A careless match drops glowing on the forest floor amid fallen and dried needles. The tiny embers slowly ignite the nearby needles; at first there are only embers, but dried needles are everywhere thick beneath the match; the embers grow hotter and tiny flames lick at the jumble of dead twigs. The twigs ignite and a flame leaps higher to catch fallen and dead branches. For a moment it looks as though the fire will go nowhere; then the bark on the dry branch crackles and flame encircles it. The branch itself begins to burn.

Now the fire spreads rapidly on many fronts; the small dry materials on the ground spread it quickly to more branches. At first the living trees and fallen logs do not catch fire, but as the heat grows, the leaves, twigs, and stems of the live trees start to ignite. As the intensity of the fire increases further, fallen logs catch fire and even the trunks of live trees begin to burn. Though the live trees could not catch fire by themselves, in the heat of the forest fire, they will burn.

The fire front moves quickly through the dry underbrush, but fallen logs continue to burn long after the fire front has passed. Thus, though the fire may be stopped at a fire break, the fire within the forest continues. The embers from these burning logs may later feed a new phase of the fire.

Fire is no stranger to the forests. Lightning may set a forest on fire as well as a match. When rain from high clouds evaporates before reaching the earth, a lightning bolt that strikes dry ground may ignite a blaze. Forest ecologists have been able to study the frequency of fires by observing fire scars embedded in the annual growth rings of the tree. By careful counting of both rings and scars, they have been able to show a remarkable phenomenon. A study in a California forest found that forest fires reoccurred in about eight year cycles as far back as 1685.

A study in the Boundary Waters Canoe Area of Minnesota, now part of the Wilderness System, found through a study of lake sediments that fires have reoccurred again and again in these forests for thousands of years. In the jack-pine forests of this region, fire is thought to be the agent that regenerates the stand, since the cones on the jack-pine are opened by intense heat. The seeds fall on a ground where competing plants have been burned away and where sunlight will find better entry. The fire has made it possible for new jack-pine seedlings to grow.

Fire may be both an agent to regenerate a forest and a means to protect the forest from a holocaust. How can a fire protect a forest? It can do so by consuming the underbrush and preventing a major buildup of deadwood that could lead to a truly deadly fire. If a light surface fire burns quickly through a small buildup of underbrush, the trees in such a forest may be only scarred, not destroyed. These are the scars ecologists counted in the tree rings in order to establish the frequency of fires. Because the buildup of underbrush is reduced by light surface fires, any individual fire may not reach the searing intensity that consumes the living trees themselves. Fire can be an agent of forest survival and of regeneration; after years of warning from Smokey the Bear, it may seem hard to accept.

Smokey's warnings are sincere nonetheless; in order for a fire to be of value to the forest, the forest must be in a condition that will not lead to complete destruction. If large amounts of dead fuel have accumulated over decades of careful protection, the hazard to the forest is very great, for the fire is likely to reach an intensity that will consume the live trees as well as the underbrush. No one but the professional forester or forest ecologist can know the hazard or value of fire to a forest. Smokey is still telling the truth.

How is the tool of fire to be utilized? Prescribed or controlled burns have been used in some Western forests to prevent buildup of dead underbrush, but in other forests, the careful protection of the Forest Service since the turn of the century makes prescribed burning impossible to use. Any attempt at burning in such areas would lead not to surface fires but to major fires, which are difficult to control. Such major fires would occur because of the large quantities of dead plant materials accumulated at ground level. In an area where a major fire has already occurred recently, however, controlled burning might be used in following years because of low levels of deadwood buildup on the forest floor.

National Park System

The National Park System, now grown to more than 31 million acres (12.5 million hectares), is administered by the National Park Service. The Park Service was created in 1916 to administer a system of 15 national parks and 22 national monuments already in existence. The Park Service was charged with conserving

> the scenery and the natural historic objects and the wildlife therein and to provide for the enjoyment of the same in such manner and by such means as will leave them unimpaired for the enjoyment of future generations.[5]

By the end of World War I, the Park Service was able to take over protection of the parks from the U.S. Army whose troops had been used to build roads, fight fires, and generally protect the areas.

---

5 *Preserving Our Natural Heritage,* prepared for the U.S. Department of Interior by the Nature Conservancy, 1977, p. 32.

Although the Park System lists 287 units, the Park Service administers other areas as well. The more familiar responsibilities of the Park Service include the 38 National Parks, 61 National Monuments, 52 National Historic sites, 16 National Historical parks, 16 National Recreation Areas, and 14 National Seashore and Lakeshores. In 1974, the Park Service recorded 217 million visits to its units, about an 80% increase from a decade earlier (Fig. 32–7). The less familiar responsibilities of the Park Service include areas which the Service does not even consider part of the Park System: the National Natural Landmarks, National Environmental Education Landmarks, and National Historic Landmarks.

Over the years, the Park Service has accepted other tasks in addition to marshalling people carefully through their holdings. One of their important functions is preservation of historical landmarks, both natural and man-made. They not only preserve but also restore early settlements and interpret early events for visitors. For instance, at DeSoto National Memorial, a relatively out-of-the-way site in Florida, a camp much like that of the early Spanish explorers has been set up. Weapons, such as the crossbow and musket, are displayed and demonstrated by the staff and a movie recreates the DeSoto expedition. Such educational efforts are common in other units of the National Park System.

The Park Service also conducts surveys of historic buildings and engineering works and maintains a register of sites significant for historical and educational reasons. In addition, the Service provides grants for surveys of sites possibly worthy of preservation. Under agreements with other agencies, the Park Service supervises recreational use of land not in their system.

New lands enter the National Park System by a number of methods, the most basic of which are (a) purchase, (b) condemnation, (c) gift, and (d) exchange or transfer. Lands have had to be purchased frequently in recent years, as the system of National Seashores has grown. Often, where no purchase can be negotiated, the Park Service has had to resort to condemnation. In legal terms, condemnation is made possible by the power of "eminent domain," the privilege of the government to take lands for its use. Though lands may be condemned, they must still be purchased. Just compensation, to use the legal term, or "fair market value," must be paid for lands so taken.

Over the years, the National Park System has benefitted significantly from gifts from states and from private individuals. Two gifts from the Rockefeller family have been especially important. One formed the nucleus of Acadia National Park in Maine, the first National Park in the East. The other gift provided land on the Island of St. John, in the Virgin Islands (a U.S. territory) for the Virgin Islands National Park, a treasure of beach, ocean, and coral reef.

Lands may enter the system by what might be called "barter," in which land in the park system is traded for land outside. More important, however, is the transfer of land from the jurisdiction of another federal agency to the park system. Voya-

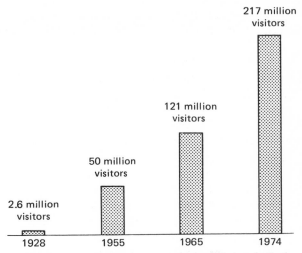

Figure 32–7 *Visits to units of the National Park System. Though the National Park System has been growing in size, the growth in visitors is principally at the larger and more famous members of the system. (Source: Statistical Unit of the National Park Service.)*

THE BAXTER FIRE—WOULD YOU LET IT BURN?

Just think if you can, but don't strain yourself, do-gooder, of all the comfortable homes that could have been built with that wasted lumber.

E. D. Chasse *

If you could look at the park as something besides a giant woodlot you might not be so concerned about the burn. This beautiful park has burned many times in the past, and it will burn in the future.

Melvin Ames, †

As we have seen, fire can play two important functions in the forest ecosystem. First, light surface fires clear away dead underbrush which, if it were to accumulate, could cause massive and intense forest fires. In this century, our very success in preventing forest fires has led to conditions in many forests that could lead to terribly severe fires, fires far more severe than might otherwise have occurred. "Prescribed" burning has been used by professional foresters in some forests to prevent such buildups. Prescribed burning is undertaken with caution; days chosen are moist and windless; all forest uses are curtailed.

The second function of fire is replacement or renewal of forest stock. Fire may act as nature's instrument of forest renewal and has acted in this fashion for hundreds and thousands of years without the aid of the forester. Only recently have we discovered the significance of fire to the forest ecosystem. Such knowledge of the role of fire will increasingly be put to use in "managing" forests, but it is not yet clear precisely how the knowledge will enter into future decisions. In the summer of 1977, a massive fire occurred in Baxter State Park in Maine. The fire was started by lightning; its intensity was fed by downed timber. Let us, for a moment, enact the scene. Prime recreation land is being destroyed but the wilderness is renewing itself. Will you let the fire burn?

You will want to know all the circumstances of the fire, how it spread and how it is being fought, what resources may be destroyed. The reasons people are giving to fight it or let it burn will interest you as well.

The story begins many decades ago when the will of Percival Baxter, former Governor of Maine, was opened by his lawyers. To the people of Maine, Baxter bequeathed a 200,000 acre (81,000 hectares) forested tract with 46 mountain peaks and ridges. The jewel of the tract is towering Mount Katahdin. The mountain at 5240 feet (1600 meters) is the northern-most point of the Appalachian Trail, the wilderness foot path that runs from Georgia to Maine. Governor Baxter's will instructs that the land is

> forever to be held by the State of Maine in trust for public park, public forest, public recreational purposes, and scientific forestry, the same also forever shall be held in its natural wild state and except for a small area forever shall be held as a sanctuary for wild beasts and birds.

Over the years, the park had been developed, so that by 1977 it included an automobile road and seven campgrounds. The camps included space for trailers and tents as well as bunkhouses and shelters. Because of the popularity of the park, reservations were suggested for those who needed to be sure of space. When the fire struck Baxter Park in 1977, it was a foregone conclusion that the park authority would fight it. The battle, however, was ex-ceedingly difficult because of the rough terrain and the presence of numerous downed trees.

These downed trees were from a fierce winter storm that had struck the southwest slope of Mt. Katahdin in 1974. Wind and ice had combined to snap tall trees in two, leaving a pileup of dead logs on the slopes. The Great North-ern Paper Company had been allowed into the park to harvest the blowdown for use as pulpwood in the hopes of lessening the risk of fire, but the harvest was never completed.

Legal action was taken by the Baxter Park Defense Fund, and a judge's order brought the harvest to a halt in August 1976. While the judge did not for-bid the removal of the downed trees, he directed that the heavy equipment being used had to be removed from the park. The heavy equipment was cut-ting into the forest floor, leaving gouge marks that might be used as roads or that could reroute streams into the depressions. Increased erosion from scarred areas would be likely as well. The equipment was removed by the paper company and the harvest of the blowdown came to an abrupt end. The following summer, when lightning ignited the blaze, much of the tangle of deadwood still lay on the ground.

The primary means chosen to fight the fire was to scrape out a 16-mile (27-kilometer) fire line around the perimeter of the fire. Bulldozers and skid-ders were borrowed from the Great Northern Paper Company to cut the

(Controversy continued on next page)

12-foot (3.6 meters) wide fire line; the company, which owns forest adjacent to the park, sent firefighters as well. Beaver sea planes were also used in an effort to control the fire. The planes loaded water from nearby ponds and dropped the water in 150-gallon (564-liter) loads on fire spots. On the ground, firefighters carried hoses and pumps into the forest. The pumps were set up on streams and ponds; the hoses, connected to the pumps, were dragged to the fire front.

In the heat of the blaze, the Baxter Defense Fund, which had prevented removal of the blow-down, let it be known that it planned once again to go to court. Their aim, as before, was to force out the heavy pieces of equipment that were scarring the slopes of the park and crisscrossing the mountain streams in an effort to create the fire line. Although public reaction to the threatened suit was intense, the Defense Fund would have gone ahead with their legal action had they not seen that the fire was in fact being brought under control faster than court action could be taken. The legal steps to cause the removal of the machinery were not undertaken, and hence no injunction against the firefighters was obtained.

The fire line was completed by the eighth day of the blaze, and the fire was brought under control within those lines. Nonetheless, weeks of watching were required to be certain that the fire was truly out, since fire can smolder in the dry litter that makes up the forest floor. The Baxter Park Authority, in an effort to slow erosion, built water bars in the deep gouges left by the bulldozers. These miniature log dams slowed the flow of water in the depressions and allowed earth particles to settle rather than be swept away in the flowing water.

Although the fire was out, the controversy raged on. Letters and editorials in the newspapers of Maine insulted the environmentalists. Letters also appeared expressing sympathy for the concerns of the Baxter Defense Fund. The quotations that began this discussion were from such letters as these. There were after-shocks, including court action by the Defense Fund to insure a natural reclamation of the scars left from the firefighting; but eventually the controversy, too, appeared to go out. Like the remains of a forest fire itself, however, it is only smoldering beneath the surface. If the fire rages again, would you let it burn?

---

* Letter to Bangor Daily News, 27 July 1977.

† Letter to Bangor Daily News, 19 August 1977.

geur's National Park in Minnesota is being acquired by gifts, transfer, purchase, and condemnation. Minnesota is donating about 50% of the 220,000 acres (89,000 hectares) of land and water in the park; the Forest Service, which controls 10% of the proposed park, is transferring another component to the Service and the Park Service will have to purchase the remaining 40% of the area. Those purchases may be willing sales or may require condemnation.

The National Park System still bears the imprint of the early laws with which it began. Justified as the preservation of spectacles and grandeur, the parks were to serve people in a way that would enable them to see the most memorable scenes within the park.

Since the parks were initially remote, this required development at the park itself. Such development included accomodations, food, information, and the like. Initially, such development was not intrusive, since relatively few people had the means to visit the parks. Now, however, the Park Service is increasingly called on to usher vast crowds past its most important possessions. Balancing its need to serve and its need to preserve is becoming increasingly difficult for the Park Service.

Wilderness

The Wilderness Act of 1964 cut across agency boundaries and called for the preservation of wilderness in all the public land holdings of the government. Units that entered the wilderness system could be areas in the National Parks, National Forests, National Wildlife Refuges, and the National Resource Lands (Fig. 32–8). Private holdings could be acquired as well. The language of the Wilderness Act explained its purposes in this way,

In order to assure that an increasing population accompanied by expanding settlement and growing mechanization does not occupy and

modify all areas within the United States and its possessions, leaving no lands designated for preservation and protection in their natural condition, it is hereby declared to be the policy of Congress to secure for the American people of present and future generations the benefits of an enduring resource of wilderness. For this purpose there is hereby established a National Wilderness Preservation System to be composed of Federally owned areas administered for the use and enjoyment of the American people in such manner as will leave them unimpaired for future use and enjoyment as wilderness, and so as to provide for the protection of these areas, the preservation of their wilderness character, and

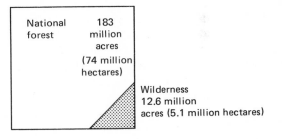

National park system
31 million acres (12.5 million hectares)

Wilderness
1.12 million acres (.45 million hectares)

National wildlife refuge system
32 million acres (13 million hectares)

Wilderness
0.72 million acres
(.29 million hectares

Figure 32–8 *National Wilderness Preservation System as a proportion of other Federal recreation lands, 1977. (Sources: Preserving Our National Heritage, 1975 and Outdoor Recreation Action, No. 44, Summer 1977.)*

for the gathering and dissemination of information regarding their use and enjoyment as wilderness. . . .[6]

Congress, to guide recommendations on what to include in the Wilderness Preservation System, attempted to define wilderness. They suggested that

A wilderness, in contrast with those areas where man and his own works dominate the landscape, is hereby recognized as an area where the earth and its community of life are untrammeled by man, where man himself is a visitor who does not remain. An area of wilderness is further defined to mean in this chapter an area of undeveloped Federal land retaining its primitive character and influence, without permanent improvements or human habitation, which is protected and managed so as to preserve its natural conditions and which (1) generally appears to have been affected primarily by the forces of nature, with the imprint of man's work substantially unnoticeable; (2) has outstanding opportunities for solitude or a primitive and unconfined type of recreation; (3) has at least five thousand acres of land or is of sufficient size as to make practicable its preservation and use in an unimpaired condition; and (4) may also contain ecological, geological, or other features of scientific, educational, scenic, or historical value.[6]

The 1964 Act first instructed both the Department of Interior and the Department of Agriculture to study their holdings for their suitability as units in the National Wilderness Preservation System. The Act then quickly designated 54 areas totalling over nine million acres (3.6 million hectares) as the first components of the Wilderness System. In 1975, the Eastern Wilderness Areas Act supplemented the first Act by focusing attention on Eastern wild areas that had recovered from earlier impacts of human activity. A number of Eastern areas scarred by fire or by logging and now largely regrown or reforested were placed in the Wilder-

ness System. Other Eastern areas were designated for study for possible inclusion.

The Wilderness Act, in historical perspective, ranks with the setting aside of Yosemite and Yellowstone, the Forest Reserve Act of 1891, and the establishment of wildlife refuges. Each of these earlier acts responded to an emerging awareness. The first of the national parks set aside grandeur; the Forest Preserves were aimed at preserving our timber resources. The wildlife refuges were created to protect species from extinction. The Wilderness Act went further; it demonstrated both how far our intelligence of our environment had come and the concern we had to preserve not one item but nature in its entirety. Grandeur was not enough; timber resources were not enough; species alone were not enough. The preservation of nature's intricate pattern as seen through wilderness was required.

The wilderness system has grown since 1964. By 1977, more than 100 additions with a total area of over five million acres (2 million hectares) had joined the original 54 designated areas. In that year, another 70 areas totalling nearly 25 million acres (10 million hectares) were still under consideration by Congress (Table 32–2). And still more areas were to be studied for the Wilderness System by the Forest Service, Park Service, the Fish and Wildlife Service, and the Bureau of Land Management.

National Wild Rivers and National Trails

Four years after the Wilderness Act of 1964, the preservation of wilderness was again before the Congress, this time in two related acts. The wilderness under consideration in 1968 was not simply land and mountains. The remainder of the nation's free flowing rivers whose banks were still free of development were the subject of the Wild and Scenic Rivers Act. The historic wilderness footpaths, the Appalachian Trail, the Pacific Crest Trail, and others were the subject of the National Trails Systems Act. The passage of these two Acts, while

---

6 Wilderness Act of 1964.

TABLE 32–2   National Wilderness Preservation System as of March 1, 1977[a]

Components of the System		Units	Acres
National Forest Wildernesses		94	12,605,405[b]
National Park Wildernesses		17	1,120,213
National Wildlife Refuge Wildernesses		52	718,087
	Total	163	14,443,705[b]

Wilderness Proposals Before Congress		Units	Acres
National Forests		17	3,163,887
National Parks		22	14,018,285
National Wildlife Refuges		34	7,520,142
	Total	73	24,702,314

Data source: *Outdoor Recreation Action*, No. 44, Summer 1977. Published by the U.S. Department of the Interior, Bureau of Outdoor Recreation (now Heritage Conservation and Recreation Service).

[a] For a listing of the areas by name, location, size, and managing agency, see the issue of *Outdoor Recreation Action* indicated as the data source.

[b] Some acreages estimated pending map compilation; also, totals in all charts do not add because of rounding acreage in some instances. Totals from U.S. Forest Service, National Park Service, and U.S. Fish and Wildlife Service.

they had defects, strengthened the wall that preservationists were building around the wilderness.

National wild and scenic rivers   The 1968 Act began the process of setting aside the yet unspoiled rivers of the nation. Eight rivers were tagged for immediate entry into the system: these were the

1. Clearwater River, Middle Fork, Idaho
2. Eleven Point River, Missouri
3. Feather River, California
4. Rio Grande River, New Mexico
5. Rogue River, Oregon
6. Saint Croix River, Minnesota and Wisconsin
7. Salmon River, Middle Fork, Idaho
8. Wolf River, Wisconsin

By 1978, the system had been expanded by congressional, state, and federal agency action to include 28 rivers.[7] Taken together, these 28 rivers cover some 2300 miles (3800 kilometers). The Park Service has sole administration of five of the 28 and shares two others with the Forest Service. The Forest Service administers 10 of these as the single guardian, and shares other rivers with the Bureau of Land Management. Five of the rivers are administered solely by the states and one by the Bureau of Land Management. A river may be recommended for inclusion by individual state legislatures. Such a river, if approved by the Secretary of the Interior, becomes part of the system without further Congressional action. The Little Beaver River in Ohio

7 A listing of the rivers, giving the managing agency and length of protection, is available from the Heritage Conservation and Recreation Service, U.S. Department of Interior, Washington, D.C. 20240.

CONTROVERSY

All this land is being set aside for non-use at a time when public interest in the maintenance of single-use wilderness is evaporating.

R. L. Brown

The U.S. Forest Service should think in terms of generations if not of centuries.

Editorial, Business Week
29 January 1979

Additions to the National Wilderness Preservation System are surrounded by controversy. A study by the U.S. Forest Service made public in 1979, considered the potential uses of 62 million acres of roadless land in the National Forests. Known as RARE II, for the second Roadless Area Review and Evaluation, the study concluded that the areas in question should be largely given over to other uses than wilderness. Of the 62 million acres, only 15 million acres were proposed for wilderness, while 36 million acres were to be "open" for multiple-use activities. These activities include "mineral entry" (prospecting and mining), grazing of stock, and timber cutting. Eleven of the 62 million acres were not designated either "open" or for wilderness but were recommended for further study. The next step after the Forest Service proposal was for Congress to consider the study and then allocate areas to wilderness as they saw fit. Some people thought the area of land which the Forest Service had recommended for wilderness to be quite ample and perhaps even too generous.

In a letter to the New York Times, R. L. Brown asserted:

All this land is being set aside for non-use at a time when public interest in the maintenance of a single-use wilderness, which appears to have been a fad of some three or four years' duration, is evaporating. Entry into wilderness areas last year totaled only seven million man-days, a fraction of the entry by Americans into national parks and onto other types of Federal land. Sporting-equipment manufacturers report that sales of tents, sleeping bags and other equipment used by hikers for treks into the wilderness are decreasing....

It is not true that all use of land results in permanent loss of wilderness characteristics. Most exploration for mineral deposits, for an example, fails. All signs of preliminary exploration are usually obliterated by new growth in a very short time. Even clearcut log areas reforestrate, and wilderness groups now suggest that many reforested areas are in fact wilderness and should be so designated. *

(To put some of Brown's figures in perspective, the National Parks include about 31 million acres. Visitor days to the parks in the late 1970s reached a total of about 215 million.)

Other people felt the Forest Service had recommended too little wilderness.

> The U.S. Forest Service should think in terms of generations if not of centuries. But continuing pressure from lumbermen, mining companies, and enterprising recreational developers makes it hard for Forest Service officials to think beyond the day's schedule of appointments. The result is a built-in bias in favor of early utilization, which shows in the recommendations the Forest Service has just drawn up for classifying some 62 million acres of undeveloped land in the national forests. [†]

> They (the recommendations) would designate only 15 million acres as wilderness and 11 million for future study, while opening 36 million for various kinds of development. There are strong arguments for at least reversing these proportions, designating sufficient wilderness areas besides barren ice and rock, and being sure possibilities are not overlooked for both preserving wilderness and attaining necessary development. [‡]

Is Brown right that wilderness use is a passing fad? Brown implies that wilderness is only set aside for human uses such as hiking and camping. Is that a correct interpretation? What criteria should be used in deciding how much public land should be set aside as wilderness compared to the amount of land on which multiple uses are allowed? Are visitor-use days a good measure of the value of parks or of wilderness?

* R. L. Brown, Letter to the Editor, The New York Times, 9 May 1979, p. A24.

† Editorial, Business Week, 29 January 1979, p. 134.

‡ Editorial, Christian Science Monitor, 8 January 1979, p. 28. Reprinted by permission from The Christian Science Monitor ©1979. The Christian Science Publishing Society. All rights reserved.

entered the system in this way and is administered by the state of Ohio. In all, five state-administered wild rivers have joined the system, the Allagash in Maine, the Little Miami and Little Beaver Rivers in Ohio, the New River in North Carolina, and the Lower St. Croix in Minnesota and Wisconsin.

The rivers or portions of rivers included in the system are to be kept free of dams or other structures; that is, their free-flowing character is to be preserved. Further, a corridor on both sides of the river is to be kept free of all development, thereby maintaining the wilderness setting of the river. The rivers placed in the system are classified into one of three categories according to the legislation which created the system:

(1) Wild river areas—Those rivers or sections of rivers that are free of impoundments and generally inaccessible except by trail, with watersheds or shorelines essentially primitive and waters unpolluted. These represent vestiges of primitive America.

(2) Scenic river areas—Those rivers or sections of rivers that are free of impoundments, with shorelines or watersheds still largely primitive and shorelines largely undeveloped, but accessible in places by roads.

(3) Recreational river areas—Those rivers or sections of rivers accessible by road or railroad, that may have some development along their shorelines, and that may have undergone some impoundment or diversion in the past.[8]

The protection provided a river in the national system is of several kinds. First, the Federal Power Commission is prohibited from licensing new power dams on the protected portion of the river. Rivers under study for inclusion are also protected in this way while they are being studied. Power developments above and below the protected portions of the rivers are not prohibited, but if they conflict with the flow in the protected stretch they may still be prevented.

---

8 Title 16, Section 1273 of the U.S. code.

In addition to banning dams and diversions, the U.S. government can no longer sell or exchange public lands within the boundaries specified for the rivers. Although mineral claims in these areas that were established before the Act were not voided, new mineral-related activities are under the regulation of the Secretary of the Interior to prevent either pollution or other degradation of the area.

National scenic and National historic trails Although the Appalachian Trail is today under the primary administration of the Secretary of the Interior and the Park Service and the Pacific Crest Trail under the primary administration of the Secretary of Agriculture and the Forest Service, it would be a mistake to assume that the federal presence is the reason these trails exist today. In particular, the bulk of the Appalachian Trail is not a result of federal effort. The federal presence does now provide some added protection to the trails and helps to ensure their preservation, but at the time the trails were assembled, ordinary citizens and private organizations furnished the creative force.

The original proposal for the Appalachian Trail came in 1921 in an article by Benton MacKaye, a forester and regional planner, in the *Journal of the American Institute of Architects*. MacKaye was suggesting the linkage of a number of trails as well as building new trails. Hiking and outdoor clubs expressed great interest in MacKaye's idea, which initially was for a trail without end, and they undertook to build MacKaye's dream. Within a year of the proposal, the first section of the trail was opened in the Palisades Interstate Park in New York and New Jersey. Of the more than 2000 miles ultimately assembled, only 350 miles of trails existed at the outset. All of the trails that already existed were in New England and New York. The formation of the Appalachian Trail Conference, a federation of clubs, government agencies, and individuals, in 1925 gave hope that the trail could become a swift reality. The complexity of assembling the trail, however, slowed its progress, especially south of Pennsylvania where hiking clubs were still to be organized.

Completed in 1937, the Trail runs along the mountain backbone of the East, the Appalachian mountains, from Mount Katahdin, Maine, to Springer Mountain, Georgia (Fig. 32–9). Public land in the national forests and in the national parks from Virginia south helped to make the final assembly possible (Table 32–3). To protect the portion of the trail through public land, the conference signed an agreement with the Park Service and Forest Service as well as most of the state governments involved in 1938. Under its terms, the federal government promised not to construct incompatible projects within a one-mile corridor on either side of the trail; the states agreed to a quarter-mile strip on each side.

Nevertheless, establishing the connections between such sections proved to be a large task, as the trail often wound its way across private lands. Permissions had to be requested of owners; at the time, simply an oral agreement was usually needed. As the pressure from use grew, however, individual owners became more reluctant to extend their hospitality. The trail was threatened by its very popularity.

At about the same time the Appalachian Trail was beginning to be assembled, Clinton Clark of

Figure 32–9   *The Appalachian Trail near Mt. Katahdin (© Russ Kinne 1974/Photo Researchers, Inc.*

TABLE 32–3   Approximate Division of National Scenic Trail Jurisdictions

Agencies with Jurisdiction	Appalachian Trail		Pacific Crest Trail	
	miles	kilometers	miles	kilometers
U.S. Forest Service	719	1158	1856	2989
National Park Service	215	346	249	401
Bureau of Land Management	—	—	204	329
States	289	465	43	69
Private	805[a]	1296	106	170

[a] No cooperative agreements for use exist on 75% of the 805 miles (1296 kilometers) in private ownership.

California proposed (1932) a similar footpath extending from Canada to Mexico on the West Coast (Fig. 32–10). The Pacific Crest Trail was also planned as a mountain range path. Its 2600 miles (4300 kilometers) includes the Cascade Crest Trail and the Oregon Skyline Trail in Washington and Oregon. The Pacific Crest Trail follows the Lava Crest Trail, the Tahoe-Yosemite Trail, and the John Muir Trail, among others, in California. Fully 85% of the Pacific Crest Trail is on federal lands, such as national park or national forest, so that the trail's existence is less threatened by development than the Appalachian Trail.

After a number of years of effort, Congress passed and President Johnson signed the National Trail Systems Act in 1968. The Act created a system of scenic and recreation trails across the country and made the Appalachian Trail and its sister trail, the Pacific Crest Trail, the first members of the system. Federal purchase of the needed right-of-ways for the trails was authorized by the 1968 Act but funds were not appropriated for another 10 years. If negotiations with private land owners proved unsuccessful, the government was given the right to acquire the land by "condemnation."

The Land and Water Conservation Fund administered by the Heritage Conservation and Recreation Service also will assist states in purchasing portions of the corridor needed for the trail, but this federal effort was not a part of the National Trail Systems Act. Individual trail clubs and the Appalachian Trail conference still maintain funds to buy trail right-of-way, and they continue to make such purchases to the present time. The struggle to create the trail and to preserve it still falls largely to individual citizens and the trail clubs.

The National Trail Systems Act did more than promise to stabilize the historic mountain backbone trails of the east and west coasts. While the Appalachian Trail and Pacific Crest Trails were designated the first National Scenic Trails, the Act directed that other trails be studied to see if they would qualify for inclusion in the Scenic Trail System. The Continental Divide Trail has since been designated as a National Scenic Trail (1978). From Glacier National Park in Montana, the trail follows the continental divide down to New Mexico and the Mexican border (Fig. 32–11).

Recognizing the historical significance of a number of the proposed trails, in 1978 Congress created a new category of trail, the National Historic Trail. Four trails have been taken into the system as National Historic Trails. The Oregon Trail (2000 miles from Missouri to Washington), the Lewis and Clark Trail (3700 miles from Illinois to Oregon), the Mormon Pioneer Trail (1300 miles from Illinois to Utah) and the Iditarod (an Alaskan Gold Rush trail, 2000 miles from Seward to Nome) thus entered the National Trail System. Other routes are still under study.

The trails will not necessarily follow the original routes; development has in some cases

Figure 32-10   *The Pacific Crest Trail. (U.S. Forest Service)*

wiped out entire sections of the original trail. Thus, some rerouting will be necessary to achieve unbroken trails. In addition, federal protection is provided only to those portions of the trails which are on federal lands.[9]

For a number of years, the promise of the National Trail Systems Act to preserve the trails of the nation was a hollow commitment. The hope that the Appalachian Trail could be finally secured by federal purchase of threatened areas has not yet been fulfilled. Congress finally appropriated money under the act for such purchases in 1978, 10 years after the initial commitment to acquire the needed land. During these 10 years, development marched closer to the Trail in many places, and the cost of acquiring right-of-way rose considerably. Now, as before, the Trail's continued existence as an unbroken footpath is the result of the pooled private efforts of individuals and hiking organizations. There is a lesson in the history of the Appalachian Trail.

---

9 For a listing of information on these trails, write to the Heritage Conservation and Recreation Service, U.S. Department of the Interior, Washington, D.C.

PROBLEMS OF ACCESSIBILITY

A stone and mortar walkway with wooden handrails guides the visitors away from the picnic area down along a steeply sided stream. The visitors witness a quickly changing scene of rock cliffs and evergreen-covered earth banks. Then the valley opens with a thunderous roar; a magnificent waterfall confronts the visitors. It is a spectacle worth remembering; it is a vivid demonstration of the remarkable power of flowing water, cutting through the rock over countless centuries. The visitors, an elderly couple, would have been unable to see it had not the walkway and handrails been present to guide and steady their steps. The walkway was the legacy of a work crew from the Civilian Conservation Corps who, almost 50 years ago, helped to make parks more accessible to less athletic people. It was an effort of noble proportions and one of benefit to millions of people.

Yet such development at park after park, year after year, leaves its mark on the wilderness setting. Construction of roads, walkways, picnic grounds, swimming areas, restaurants, campgrounds, shops, service stations, and even hotels begins to take on a pattern. Nature is being enclosed in an amphithe-

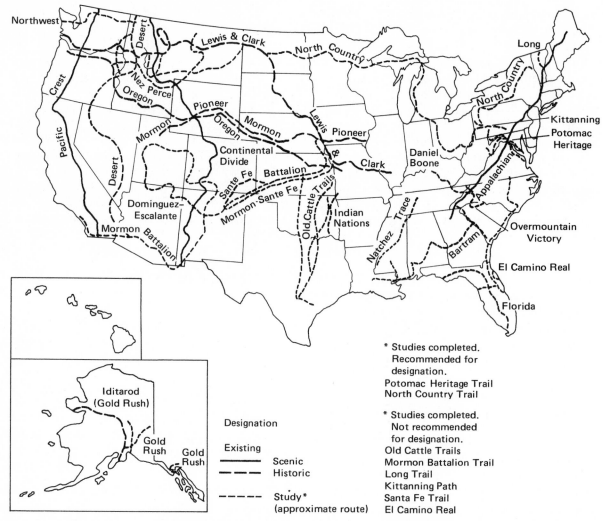

Figure 32–11   *National Trail Systems: National Scenic and National Historic Trails. (Source: U.S. Department of the Interior, 1978)*

ater. A supporting staff of ticket takers, ushers, waiters and waitresses are needed for an orderly display. Each "improvement" makes the park more accessible and accommodates more and more people. And the crush of people justifies the improvements; the park authorities are simply attempting to meet "demand" for public recreation.

Many argue that such developments, which are designed to meet the need for public recreation,

degrade wilderness values to an unacceptable extent. Indeed, the word recreation when broken into its parts means "to be created again." Only by usage has its meaning evolved to include amusement, exercise, or educational activities.

The measurement of success in wilderness preservation is very difficult. It is far easier to count visitors and use the annual visitor count as the measure of success of a park operation than somehow assess how well the land has been preserved. At a famous Midwestern site managed by the National Park Service, one of the goals of management is to keep the average time of a visit to less than 50 minutes. If the average time a party takes there is much more than 50 minutes, backups on the access road to the site will occur and motor homes will boil over. The Park Service needs to move people in and move them out.

Although an increased visitor count is a possible measure of success in a park, in a wilderness area an increased visitor count may be a danger signal. Congested trails in the John Muir Wilderness of California led the U.S. Forest Service to go to a permit system for trail camping there. The number of hikers on the trail to Mount Whitney was limited to no more than 75 per day. Since two days are usually needed to hike the 8-mile trail, the number of overnight campers is generally kept to less than 150 on any given evening. In the early 1970s, more than 1200 hikers had been on the trail during one day of the labor day weekend before the permit system was begun.

Often the development of tourist facilities at the site of the spectacle itself diminishes the experience of visiting. Old Faithful in Yellowstone with its parking lot and visitor center is often mentioned as a natural area that has been over-developed. The development in the Yosemite Valley is also seen as overdone. Such developments began in a era when visitors to the parks were few and the modest facilities needed for the visitors did not intrude on the visitors' appreciation of the park's values even if the facilities were built at the site itself. But as the number of visitors grew, the development required to serve them had to expand as well. It was easiest apparently simply to "add on."

The alternative to development at the site, worthy of careful consideration, is a visitor center at or near the park gate. Such a visitor center, remote from the scenes of the park itself, was wisely built in the early 1970s at Acadia National Park in Maine. At other parks, however, existing facilities are unlikely to be torn down to be rebuilt at less scenic locations; the expense is seen as too great. Nonetheless, the concept is attractive because not only development but also litter that accompanies tourism can, to a degree, be kept from the park interior.

Wilderness and recreation come in conflict in more ways than number of people and development. Certain human activities allowed in a natural area can also diminish wilderness values. The prime examples are motor boating, off-road vehicles operation, and snowmobile use. The activities are supported, however, not only by a large public but by business interests who organize lobbying efforts so that their products can continue to be used.

One area where wilderness has come into conflict with motors is the Boundary Water Canoe Area of northern Minnesota, one of the areas recently brought into the wilderness system. The Boundary Waters have other problems in addition, which we shall mention briefly. The landscape of the Boundary Waters Canoe Area consists of a rich mixture of lakes and forested islands. Although the Wilderness Act of 1964 is thought to prohibit motors in wilderness areas, a special paragraph was inserted in the Act applying only to the Boundary Waters. The paragraph allows motor boats in areas where they were already being used. The same paragraph allows timber cutting to continue as well. Of the one million acres of wilderness in the Boundary Waters, about half is still untouched by the axe and the chainsaw. Mining may well be allowed in this land which has been designated as wilderness. The International Nickel Company has been longing to explore for copper and nickel deposits believed to exist there.

Motor boats and snowmobiles in the Boundary Waters are the issues we wish to emphasize here though. Snowmobiles had been allowed into the area until 1976 when the Forest Service decided that their operation could cause damage to the land-scape and prohibited their use. Several bills have been introduced in Congress to resolve these conflicts over allowed uses. One would make about

THE ORV FIGHT

Just put your gang on Suzuki's DS trail bikes. And head for the boonies. Doesn't matter where you go. Peaks or valleys, it's all the same to those rugged off-road machines.

Suzuki advertizing copy
Cycle World **17(3)**, 110-
111 (1978)

St. Francis of Assissi himself while driving an off-road vehicle on wild land could not avoid diminishing the recreational experience of many non-ORVers in the same area.

CEQ report, 1979

One of the most bitter controversies over the use of public lands involves off-road vehicles (ORVs): trail bikes, snowmobiles, dune buggies and four wheel drive vehicles. Some 25% of the American public enjoys the use of these vehicles. ORVs allow their users the physical thrill of conquering rugged terrain and also allow them to penetrate far into wilderness areas for hunting and fishing and camping. But ORVs can destroy that very wilderness as well as ruin the wilderness experience for hikers, backpackers and other non-motorized users.

ORVs destroy the vegetation they ride over, leaving soil exposed to erosion from wind and rain. They compact soil, killing plant roots and decreasing the soil's ability to absorb water, which further increases erosion. In the arctic, ORV use leads to the melting of permafrost and the formation of water-filled tracks. The noise from ORVs can drive away wildlife or even damage vital hearing ability in certain species.

Non-ORV users are not only worried about these environmental effects but also extremely angry at the intrusive noise ORVs produce.

...(O)ne ORV operator can effectively restrict a large public area to his own use through the emission of loud engine noise, obnoxious smoke, gas and oil odors and dangerously high speeds. Whereas previously many persons of all ages and wealth could observe the beauty of unspoiled land, now a single ORV can reign supreme. *

Consider the snowmobile...and the often-heard argument that this machine makes it possible to "get way back in there, away from it all." There are, of course, several other ways to "get back in there," including snowshoes and skiis. Maybe if you need an engine to get there, you don't belong there in the first place! To my mind, "getting away from it all" means, foremost, getting away from our society's overdependence on the combustion engine. [†]

(We need to) provide some place on God's green earth for man to spend some time without hearing a damned motor. [‡]

I hope there is some way we could outlaw all off-road vehicles, including snowmobiles, motorcycles, etc., which are doing more damage to our forests and deserts than anything man has ever created. I don't think the Forest Service should encourage the use of these vehicles by even suggesting areas they can travel in....I have often felt that these vehicles have been Japan's way of getting even with us (for World War II). [||]

In some areas an attempt has been made to restrict ORVs to particular areas or trails. TVA for instance has tried this on an experimental basis. TVA ranger Scott Seber says,

I used to hate ORVs. Now I feel they can be worked with. We have demonstrated that they don't have to be running amuck everywhere. [§]

This has not satisfied everyone, however.

(We) object to the continual enhancement of non-ORV recreation at the expense of the off-road vehicle enthusiast. We do not feel that all compromises should be made at the expense of off-road motorcyclists. [**]

How do you feel about ORV use? Do you think all public lands should be open to ORV users? Or, do you feel their use should be forbidden on public lands; perhaps restricted to privately run ORV parks?

---

[*] Gary A. Rosenberg, Environmental Affairs, 1976.

[†] Jerry Buerer, professor of sociology, Marquette University and organizer of a group to protect the rights of non-ORVers, 1975.

[‡] Ben Huffman, Vermont Department of Forests and Parks, 1974.

[§] Off Road Vehicles on Public Land, Council on Environmental Quality, 1979, p. 14.

[||] Senator Barry Goldwater, letter to William D. Hurst, Regional Forester, Region 3, U.S. Forest Service, Albuquerque, New Mexico, 23 March 1973.

[**] Robert Rasor, AMA, 1976, Environmental Impact Statement, supra note 19, at 260.

half the park a "recreation area," in which logging and motorized sport are legitimate activities. The bill would essentially be a vote of confidence for the concepts applied to the area at the present time by the Forest Service. Another bill would make the full one million acres of the Boundary Waters Canoe Area into wilderness with no intrusions allowed— no logging, no mining, no motors. The only exception would be previously settled areas such as the town of Ely.

The town of Ely itself has been divided into two camps on the issue. The supporters of wilderness there are up against a coalition of loggers, resort owners, sawmill owners and their employees, and merchants. All of the latter group earn their living from activities that would cease or be curtailed if the area became a full wilderness.

It is true that fewer people are prepared to use the wilderness than would come to resorts or campgrounds. The Boundary Waters, however, are only a part of Superior National Forest, a two million acre expanse out of which the Boundary Waters were selected for wilderness designation. All of the multiple uses are allowed in those additional one million acres of Superior National Forest that have not been and would not be designated as wilderness.

The issue in the Boundary Waters is not unique. In many areas, local interests oppose the setting aside of land for park or wilderness. Somehow we must balance the interests of many future generations against jobs and income in the present.

Expanding the System of Natural Areas

The Heritage Conservation and Recreation Service

The Land and Water Conservation Fund is a new and powerful means to expand public natural areas from the local level to the national level. Created by Congress in 1965, the Land and Water Conservation Fund furnishes federal money to match local money for the purchase of natural areas. The Fund is administered by a federal office distinct from the Park Service, the Forest Service, or the Fish and Wildlife Service. The office, the Heritage Conservation and Recreation Service (HCRS), housed in the Department of the Interior, manages no land, but it plans and assists in recreation-related efforts at all government levels.

The office, formerly known as the Bureau of Outdoor Recreation, was created in 1962 by the Secretary of the Interior at the request of President Kennedy. The Land and Water Conservation Fund Act of 1965 gave clout to the mission of the office. The Act established a fund of money to be allocated to state and federal agencies for the purchase of land. The fund is fed by revenues from the sale of surplus properties, from the leasing of the outer continental shelf for oil and gas, and from other sources including direct appropriations by Congress. Of the fund's revenues, 60% are for the use of the states. Three functions are supported by the fund: the planning related to a specific project, the acquisition of the area, and development of recreation facilities and resources at the site. The fund will supply up to one half of the total cost of a specific project. For a state to be eligible to apply for money, it must have already developed a comprehensive outdoor recreation plan that includes an assessment of outdoor recreation needs and recreation resources in the state. Rules for these grants are contained in the HCRS's <u>Outdoor Recreation Grants-in-Aid Manual.</u> The remaining 40% of the fund is for the use of federal agencies involved in outdoor recreation.

Contributions of the fund to purchases are often made in cooperation with the efforts of state governments and conservancies. Federal laws giving tax benefits for donations of land are often involved in the transactions. After describing the functions of conservancies, we will illustrate the acquisition process with several examples. These examples will show how conservancies, state governments, and the Heritage Conservation and Recreation Service have been combining their talents and using federal tax law to enrich the public with new natural areas.

Donations and Tax Benefits

We mentioned that the National Park System has received a number of gifts, but we did not describe in any detail how gifts are made. To someone interested in giving a valuable property to the public, the monetary advantages that can be obtained from the gift are important.

The simplest type of gift, called an outright donation, transfers the land directly from owner to recipient at the time of agreement. Depending on whether the recipient is a conservancy (a non-profit land trust) or a government agency, different but nonetheless attractive tax benefits to the donor are available. Land may also be given in a will and thus transferred after the death of the owner. In still another type of gift, the transfer can be made in such a way that the donor retains for his lifetime the privilege of living on the land even though the land has already changed hands. Still another popular form of gift is the bargain sale in which the owner sells the land to a conservancy or government agency for less than its fair market value. The difference between what would have been obtained by sale on the open market and what actually was obtained in the bargain sale represents a gift or donation, and the value of the difference will determine the tax benefits to the donor.

It is this bargain sale procedure that we will mention further in describing how the HCRS, conservancies, and state governments have been functioning together to expand the system of natural areas. We need first, however, to dwell for a moment on the tax benefits of donation because these benefits are a part of the argument that convinces would-be donors to become the generous individuals they wish to be. The value of the land is first established through an appraisal by a professional appraiser. The appraiser normally considers what similar tracts of land in the geographical vicinity have been sold for on a per-acre basis and applies that to the land in question with adjustments for the specific factors of its location, such as road frontage, distance from water and sewer, etc. The appraised value reflects the price the property would bring if placed on the open market.

The specific federal tax benefits differ depending on whether the property is given to a conservancy or to a local government or agency of the federal government. If the donation is made to a conservancy or land trust, the donor may deduct from his taxable income either the value of the gift or 30% of his adjusted gross income for that year if the gift is larger than this amount. If the value of the gift exceeds 30% of his adjusted gross income, the donor may carry over the remainder and deduct that remainder in the following year subject to the same percent limitation. Any excess may be carried over to the following year for a total of six years.

Here is an example.

Mr. Sandune donates a beach property worth $75,000 to the Nature Conservancy. His adjusted gross income is $130,000 per year. He can deduct from his taxable income up to $0.3 \times (130,000) = \$39,000$ of the $75,000 gift. The income on which he must pay taxes is now reduced to $130,000 - $39,000 or $91,000. His tax savings are considerable. The following year, if he has the same level of income, he can deduct up to $75,000 - 39,000 = 36,000$, the remainder of his contribution. Again, he obtains large tax savings.

Tax benefits are even greater if the gift is to a local government or government agency. The full value of the gift may be deducted from the adjusted gross income without any limitation calculated as above. In the simplified example above, Mr. Sandune can deduct $75,000 from his adjusted gross income; that is, he will not need to pay any tax on $75,000 of his income because of the gift. Corporations may also donate land, but special rules apply for the tax treatment of their donations.

Conservancies or Land Trusts

A number of private non-profit organizations exist which assist in the process of obtaining land for the public and which themselves obtain lands for future transfer. Most prominent among these is the Nature Conservancy, a private, non-profit organization that buys ecologically and geologically valuable land. The Conservancy has purchased nearly 1800 parcels of land totalling 1.1 million acres since it was founded in 1950. About two-thirds of these purchases have been transferred to government agencies, mostly to state governments, for their permanent protection as natural areas. The remainder are managed by the Conservancy itself.

The Conservancy, which has many state branches, does more than buy and transfer land. It is active in identifying and inventorying natural areas across the country. Because of the Conservancy's stature in preservation activities and because of the tax advantages that come to donors of land, the

Nature Conservancy is often successful at obtaining land by simply describing to potential donors their purposes and the tax advantages from gifts.

A 138-acre parcel at Stevens Creek, South Carolina, was obtained by the Conservancy in this way from the Continental Group. The land, which harbors several endangered species, was then transferred to the protection of the state of South Carolina. In a similar fashion, the Conservancy was able to obtain a portion of Jupiter Island off the East Coast of Florida. The manatee, a remarkable sea mammal resembling nothing so much as a large cigar in shape, lives at Jupiter Island. This distant relative of the elephant, known more commonly as the "sea cow," is one of America's endangered species. The 500-acre tract on Jupiter Island will aid in protecting the manatee.

Another example of the Conservancy's work is the preservation of the 200-acre farm of Dr. and Mrs. Wright near Albany, New York. Given to the Conservancy outright by Dr. and Mrs. Wright, the area is now maintained as a nature preserve. One of the most significant of the Conservancy's purchases is the Dewey Ranch in Kansas. The 7200-acre tract preserves a portion of the tall grass prairie. The prairie that drew so many pioneers west for settlement is now virtually gone. Only small pockets of this unique ecological landscape still exist, so the Conservancy's purchase is an especially important one. The area is to be leased at present to Kansas State University for ecological research. (See Chapter 30 for a discussion of the tall grass prairie.)

Coordinated Federal and Private Actions

Federal tax laws, as we pointed out, often make it economically attractive for people to donate land for natural areas. The Nature Conservancy helps to explain these advantages and may assist in the transactions themselves.

Take the case of a 3100 acre parcel in Rutland County, Vermont, which was prime land for a vacation home development. This parcel, which adjoins a state forest, was put up for sale in 1971. Its appraised value was $610,000. Vermont's Agency of Environmental Conservation had no money to acquire the property and not even enough to go halves with the Land and Water Conservation Fund in purchasing the land. The Agency proposed to the owners that they sell the property to the state for one half of its market value, or $305,000, money they felt they could obtain from the Land and Water Conservation Fund whose guidelines allow them to contribute up to half of the market value for a tract of land. The difference between the full market value $610,000 and the bargain sale price $305,000 they were suggesting could be treated as a donation for federal tax purposes. Although interested, time was important to the owners. They wished to move faster than the grant from the Land and Water Conservation Fund could be received. The Nature Conservancy stepped in to protect the potential sale.

They offered the owners $305,000 in cash and pointed out the tax advantages of the donation of the remainder of the property's market value.

The owners could deduct up to 30% of their individual shares of the donation from their taxable incomes. Furthermore, they would save realtor fees and save taxes on a portion of their investment profits as well. All of these advantages were theirs, said the Conservancy, in addition to public recognition of their gift.

The owners sold the property to the Conservancy and the Conservancy held the land in its name until the State Agency obtained its grant from the Land and Water Conservation Fund. With the grant from the Heritage Conservation and Recreation Service in hand, the Agency purchased the land from the Conservancy. The price was the original $305,000 plus miscellaneous interest and legal fees. Here the Conservancy both acted as an influential spokesman for land preservation and actually took possession of the land on a temporary basis to insure that it would not be sold to developers.

The linkage and the coordination of the Nature Conservancy, the states, the Land and Water Conservation Fund, and federal tax laws is a very powerful tool in saving land.

Questions

1. Why are so many different agencies involved in administering public lands? Briefly outline the agencies and their responsibilities.
2. Is fire always a bad thing in a forest? Explain.
3. Are the National Forests a subsidy to timber companies even though the timber companies must bid against one another to log a particular parcel of forest? Should the National Forests be logged at all? Should we allow timber cut from National Forests to be sold in export? How could we prevent this if logging continues in the National Forests?
4. Are trail shelters compatible with a wilderness designation? That is, should trail shelters be built or allowed to continue to exist on land which has been given wilderness status? In Olympic National Park, Washington, such a controversy arose when the National Park Service began to remove trail shelters. Users of the wilderness complained and were conducting a campaign to have the shelters restored.
5. Which do you think is to be preferred: federal or state ownership of a public natural area?
6. Question on Enrichment: Suppose you wanted to see a particular parcel of land, now privately held, become part of the public lands system. Describe how you might go about it if (1) you are the owner. (What tax advantages are available to you?) (2) you do not own the land.

Further Reading

Dodge, M., "Forest Fuel Accumulation: A Growing Problem," *Science* **177**, 139 (14 July 1972).

This well-written article can lead you into the literature of this controversial area.

Fitzsimmons, A., "National Parks: The Dilemma of Development," *Science* **191**, 440 (6 February 1976).

A non-technical article with details of park development problems and methods to deal with intrusions on the setting.

Krieger, M., "What's Wrong with Plastic Trees?", *Science* **179**, 446 (2 February 1973). See also Letters and Reply in *Science* **179**, 813 (25 May 1973).

We don't all think alike and this article proves it. But can you argue with the point of view? The letters show you how to begin.

Marsh, G. P., *Man and Nature*, Scribner's, New York, 1864. (Reprinted by Harvard University Press, 1965.)

From an historical standpoint, this is the first warning of how badly we were using the land. Marsh points to overgrazing of land and overcutting of forests as having the potential to "destroy the balance which nature has established."

Moore, W., "Fire!", *National Wildlife Magazine*, p. 4 (August 1976).

A readable account of how fire is used in forest management.

Preserving Our Natural Heritage, Volume I, Federal Activities, prepared for U.S. Department of the Interior by The Nature Conservancy, 1975, for sale by the Superintendent of Documents, U.S. Government Printing Office, Washington, D.C. 20402.

No other book we have found brings together so much information on the role, holdings, and management practices of the federal agencies concerned with land preservation. Because it is so well organized and devoid of pictures and maps (these are referred to but never reproduced), there may be a problem with maintaining an adequate attention span. If you can remember your question on federal activities in conservation long enough, the answer is quite likely to be in this book. Either that or a reference on the subject may be cited.

Steen, H. K., *The U.S. Forest Service: A History*, University of Washington Press, Seattle, 1976.

Wonderfully detailed historical account of the beginnings and evolution of forest management from the post Civil War Era to the early 1970s. Recommended for those keenly interested in the past of the forestry movement and its influence on the shape of the present.

Journals in which you may wish to browse on land preservation are the *Journals of Forest History, American Forests, Journal of Forest Ecology,* and *National Parks and Conservation Magazine.*

For information on trails and trail activities, write for the "Directory of Sources of Trail Information" which is available from

Heritage Conservation and Recreation Service
U.S. Department of the Interior
Washington, D.C. 20240

This free document can lead you to the information you need on specific trails and the management/maintenance of them. As implied by the title, the Directory only says "where to go for information", but that much is still valuable in any area where little in the way of centralized information exists.

(WFP photo by T. Page)

Part 8

Feeding the World's People

It seems fitting to begin an environmental studies book as we did, with a chapter on population problems. The pressures generated by rapidly expanding populations cause or at least aggravate many environmental problems. A growing population strains the capacity of normal water supplies and generates quantities of waste that natural disposal systems cannot handle. Air and water quality deteriorate as fossil fuels are burned to supply larger numbers of people with power and transportation. Even the technological approaches to correcting problems bring about new problems; nuclear power leads to nuclear waste disposal and chemical pesticides cause chemical contamination of the environment.

Population growth must be viewed as a central cause of all of these and other environmental issues, because with fewer people the natural regenerative capacity of resources such as air, water, and land would work much, much better. A few humans in a wilderness can drink from clear streams, burn wood for warmth, and bury their wastes without making any significant impact on their environment. Thousands of tourists visiting Yellowstone Park, obviously, cannot do the same. In a similar way, growing human populations affect their environments.

Growing the food needed to feed ever increasing populations causes or aggravates many of the environmental problems discussed in other sections of this book. Energy resources are consumed in growing food, because natural gas is used to make fertilizer and because the machines that plant and harvest the crops use petroleum. Water must be found to irrigate vast tracts of potentially fertile but waterless land. Water is also easily polluted with chemical pesticides or fertilizers. Human needs for farmland conflict with the need of wildlife for secure habitats. Meanwhile, overuse or poor management of soils can cause this precious resource to wash away in torrential rain storms. Even the earth's climate itself may be affected when forests are cut down to make way for the plow.

Ensuring adequate food for all the world's people involves many issues, some of them as technical as the ones just mentioned. But other issues are primarily economic, social, or political, as is for instance the child health problem caused by the decline of breast feeding in developing countries. Because so many environmental issues underlie the food problem it seems appropriate to use this as a summary study.

In the following chapters, we first look at the scope of the food problem. Is there a food supply crisis? How can farmers, especially those in developing countries where population growth is most rapid, grow more food? How can we handle soil problems such as erosion? What are the benefits and hazards of chemical pesticides? But these questions deal with only part of the story. Social and political realities control, to a very large extent, how much food is actually produced and where it goes (Fig. A). The last chapter in this Part tries to give a feeling for these additional factors that affect our ability to provide enough food for all the world's people. Then an attempt is made to summarize the optimistic and the pessimistic views of future world food supplies. This last chapter also deals with the issue of world grain reserves, because here may be the solution to both unnecessarily high food prices and catastrophic famine.

Figure A   *Feeding the world's people involves many problems, some technical, but many economic, social and political as well. Here school children are fed in a special government program to improve nutrition and also attract children to school. (WFP photo.)*

Chapter 33

Food Resources

IS THERE REALLY A FOOD CRISIS?

A friend recently said that he had gone to a dinner to publicize the food crisis. When someone asked if a dinner was really an appropriate way to highlight this particular problem, he replied that only rice and tea were served—and every third person got nothing to eat.

Was this an accurate reflection of the world food situation? Is there really a food crisis? People can be found to argue both yes and no.

Are People Dying from Starvation?

It is undoubtedly true that there are too many people in the world who do not get enough to eat, by any standard. There are areas of chronic starvation, where people, day after day, eat fewer Calories or less protein than they need. Over the long term, this leads to both mental and physical crippling. There are also times of actual famine when people die in large numbers. The aftermath of the war for independence, combined with serious flooding, led to mass starvation in Bangladesh in 1974. In the years prior to 1978, the African Sahel region was struck by drought that led to famine for the wandering tribes inhabiting the area.

But problems arise in estimating how many people are starving or malnourished and in defining what a reasonable standard of nutrition should be. In the first place, it is uncommon for someone to die of starvation. In most cases, a person who has too little food will be more likely to develop diseases than someone who eats well. Furthermore, the undernourished victim is more likely to die of diseases from which a well-nourished person could recover. Typhus, cholera, smallpox, plague, influenza, tuberculosis, and relapsing fever all commonly strike persons weakened from lack of food. What this means is that the actual cause of death will be listed as something other than starvation, even though it is clear that starvation is an underlying or major cause of death.

Children and Poor Nutrition

Children suffer from two diseases specifically related to malnutrition. Kwashiorkor occurs when a child eats a diet that may have enough Calories but has too little protein. A major symptom of the disease is edema or swelling of the abdomen (Fig. 33-1). Kwashiorkor usually occurs after children are weaned. Marasmus, which involves a shortage of both protein and Calories, occurs when infants less than one year old are fed overdiluted formula from unsterile nursing bottles. Severe diarrhea almost always occurs in both diseases. Infant marasmus can be prevented if mothers nurse their babies. Human milk provides needed protein for the child. In addition, because human milk is a completely sanitary food source, it is less likely that the child will be infected by diarrhea. Even kwashiorkor is uncommon if a child is nursed through the second year or longer. Furthermore, nursing has an incomplete but significant contraceptive effect.

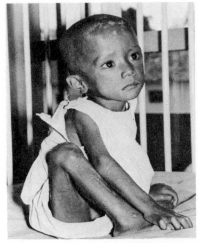

Figure 33-1 *(a) Suffering from acute malnutrition, kwashiorkar, a small boy sits on his hospital bed in Djkarta (UNICEF photo by Jack Ling.) (b) Infantile maramus (UNICEF photo by Lynn Millar.)*

Women who nurse are *less likely* to conceive again while nursing. This can be a positive effect in terms of population control where other birth control methods are not available or are unacceptable.)

Unfortunately, in some developing countries, formula companies have engaged in campaigns to sell parents on formula feeding of infants. Because of the high cost of infant formulas compared to workers incomes [1] and wherever sanitary conditions are poor, infant and child health will suffer. (Children with marasmus and kwashiorkor have higher death rates and may also suffer permanent brain damage.[2])

Estimating Calorie Requirements

Estimates of the number of people who receive less food than they need are complicated by a lack of agreement on how many Calories are needed in tropical versus temperate climates (people living in warmer climates may need fewer Calories). Also a problem is that nutritional estimates, based on what people eat in the developed countries, may be too high, since people there are generally regarded as overfed. Furthermore, economists who make the estimates tend to take into account only those crops moving through the marketplace, for instance, the cereal grains, or those animals that can be easily counted such as beef cattle. This method probably underestimates how much people, at least in rural areas, are getting to eat since family gardens and other local sources of food do exist.

On the other hand, the food supply in an area has sometimes simply been divided by the number of people there. This method takes no account of the fact that income is unevenly divided, thus purchasing power is uneven. Starvation can exist in a land of plenty if there are people who cannot afford to buy food. This situation exists in all countries, undeveloped or developed, including the U.S.

Nevertheless, the Food and Agriculture Organization of the United Nations estimated in the 1970s that 25% of the people in undeveloped countries are malnourished (over four million people).

---

1 In developing countries, commercial formula if fed at the proper strength would cost one-fourth to one-third of a worker's income. For this reason the formula is often over-diluted.

2 The social and health aspects of breast feeding are discussed further in the article by B. Winikoff, "Nutrition, Population and Health: Some Implications for Policy," *Science* **200**, 895 (26 May 1978).

These people lack one or more nutrients (such as protein) needed in a healthy diet. A smaller percentage is undernourished; that is, they do not get enough Calories to maintain their body weight with normal activity. Most undernourished people of course are also malnourished.

The Food Crisis in Historical Perspective

There has always been periodic famine. The concept of "food crisis" seems to imply something more. Are we reaching the point in terms of population growth where we simply cannot increase food production to meet the demand created by new mouths to feed? Might large segments of the world population starve to death in a nightmarish solution to overpopulation?

Since World War II, just that sort of prediction has been made and then withdrawn several times. Perhaps something is to be learned from past history of the world's food supply. After the war, when death rates were reduced by public health measures in many undeveloped countries and population growth rates began increasing, many writers predicted that worldwide famine was on the way. This feeling was bolstered by Food and Agriculture Organization reports, which gave the impression that one-half to two-thirds of the world's population was malnourished. This was probably an overstatement. Statistics on food supplies were scanty and unreliable and high values were used for the number of calories required in a healthy diet.

Developing countries were nonetheless making slow but steady progress toward feeding their populations until 1965–1966. In those years, India suffered two droughts that markedly decreased her food output. Because India is home to such a big chunk of the world's population (one out of every six people in the world is Indian), these crop failures loomed large in the totals of world food production. Many people felt that the world famine had arrived.

Immediately after this, however, favorable weather and the introduction of high yielding grain varieties in India and other parts of Asia reversed the trend. World food production rose and a mood of optimism prevailed through 1971. The Green Revolution had begun. In the U.S., the government paid farmers not to produce grain to prevent prices from falling too low. However, in 1972 the Russian grain crop failed. The Soviet Union and eastern European countries purchased 28 million tons of grain, reducing grain stocks in the U.S. and Canada to a 20-year low and causing high grain prices all over the world. Demands from other countries also rose, in part due to the fact that rising incomes in countries such as Japan and in Western Europe were allowing more people to afford meat. Increased meat production means grain consumption by increasing numbers of cattle.

About this time, the Peruvian anchovy fishery had collapsed as a result of changes in the normal water temperature and overfishing. Harvests that had totalled 12.3 million metric tons in 1970 fell to 3 million metric tons in 1973. Visions of feeding the world from the oceans began to fade in the hard light of reality (Fig. 33–2).

Then the weather demons struck again. In 1974, drought and early frosts caused the worst growing season in the U.S. in 25 years. Grain harvests were 20% below the expected levels and

Figure 33–2 *Sacks of fish meal awaiting shipment abroad line a warf at a Peruvian port. The meal is used to enrich feeds for poultry and other live stock. However, overfishing and environmental factors have decimated this once bountiful harvest. It is not known if the fishery will recover. (Embassy of Peru)*

grain reserves hit new lows. World famine was again predicted.

In addition, food prices hit new highs. (One of the problems associated with intermittant scarcity of grains is that the resultant rising food prices set off an inflationary spiral in which wages, and then other prices, must rise as people demand higher salaries to purchase the higher priced food. When the following harvests are good, however, there is not a corresponding decrease in prices. Even if the farm price of food falls, wages are still set at the new high levels as are prices of goods that have risen because of wage increases. This is called the "rachet effect" of increasing food prices (Fig. 33–3).)

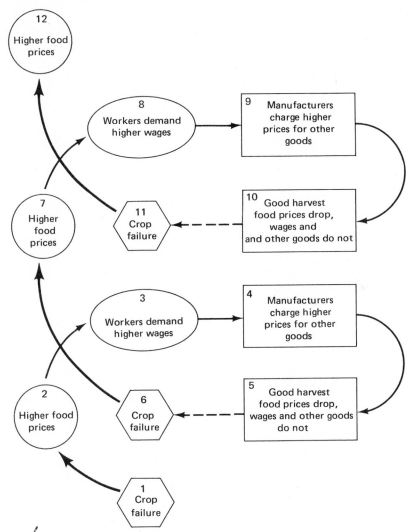

Figure 33–3 (*The "rachet effect" of rising food prices. When food prices rise, they cause an increase in wage demands and also in the prices of other goods. When food prices drop, however, wages and other prices do not follow suit. Thus, if food prices lurch from high to low and back again, they have the effect of "jacking up" the cost of living.*)

There were certainly areas of famine in the 1970s, but the predicted world food crisis did not occur. Indeed, since 1975, favorable weather in most areas of the world has led to a new build-up of grain stocks. By the end of the 1978–1979 growing season, world reserves totalled about one-fifth of a year's consumption.

The Distribution Problem

(Even when there is a reasonable surplus of food, however, the world is troubled by uneven distribution of food riches.) One-half of the grain reserves in 1979 were in one country, the U.S. Many experts now seem to agree that a world food crisis does not have to materialize in the near future. Problems are seen as ones of transport, getting food surpluses to where the needy people are, and of changing income distributions so that the countries and the people who are now too poor to purchase enough food can afford a good diet. Although on the face of it, this may sound simple, some of the social, political, and economic problems involved are brought out in the last chapter.

HOW CAN WE GROW MORE FOOD?

Traditional Farming Versus the Modern American Farm

The Igorot In the central highlands of the Phillipine island of Luzon are the villages of a fierce mountain people. The Igorots were once noted as head hunters, but their fame now rests on their outstanding ability to farm the steep mountain slopes of the island's central region. In the incredibly short period of a few hundred years these people have built a series of terraces on the mountainside, by hand labor with only a few primitive tools (Fig. 33–4). On these terraces, using no artificial fertilizers, they grow rice in amazingly high yields. A single hectare of land (2.47 acres) yields a whole year's supply of rice, their main food, to a family of five.

Figure 33–4 *Igorot rice terraces are marvels of engineering and skill practiced by a people with no modern tools or equipment. (Photo by Charles Drucker)*

The high yields are due partly to the enormous amount of labor the Igorot put in. On the terraces, rice is grown by water culture. The terraces must be weeded continually to keep the water retaining walls from crumbling. Another important factor appears to be nitrogen-fixing algae that grow in the water along with the rice. These algae provide nitrogen, which fertilizes the rice crop while the rice provides carbon dioxide and some necessary shade for the algae. The Igorots have developed an environmentally sound and self-renewing agricultural system that supports them well. Along with the agricultural system, a social system has developed.

Villages are organized into work groups that construct irrigation canals, terrace new slopes, plow the fields, and harvest crops, far more efficiently than solitary workers could (Fig. 33–5). From these work groups the whole social structure of the villages has grown.

They sponsor rituals, and they help to resolve disputes; they form political parties and in times of trouble they become military units. They enter into virtually every phase of the communities' activities. Perhaps most importantly, they provide the Igorots with a stability of association, continuity in their personal relationships that in our own affluent, mobile society is becoming ever more rare and valued. The interdependence of the working groups has made the Igorots so culturally conservative that almost a century of contact with Western Society has resulted only in superficial changes in their way of life."[3]

The wet-culture of rice is a special type of farming. All over the world there are farmers growing crops such as this, developed over long periods of time by methods that use much manpower but little or nothing from modern agricultural technology. Such crops are usually well adapted to the climate, soil type, and the availability of water where they are grown. But in many cases, these farming operations are at or near the subsistence level. The farmer grows enough food for his own family and some to trade for the other goods they need to live (Fig. 33–6). What hope does technology offer

---

3 C. B. Drucker, "The Price of Progress in the Phillipines," *Sierra*, p. 22, (October–November 1978).

Figure 33–5 *Among the Igorots, work groups are much more than temporary labor gangs: they form the basis for society itself. Here work groups transplant rice seedlings. (Photo by Charles Drucker)*

Figure 33-6 *A farmer in India irrigating his field. Agriculture in undeveloped countries uses much human-power and few machines. If farms are mechanized to increase agricultural production, where will these people go to find jobs? (William Borders/NYT pictures)*

for bringing these farmers to a point where they can grow enough food for the rapidly growing urban populations in their countries?

Modern American farms   Before answering this question, it will be helpful to describe the modern American farm. There appear to be two kinds of U.S. farms (Fig. 33-7). Some 20% of the farms produce 75% of the food and fiber. The rest of the farms yield less than $20,000 per year in sales. Presumably these families farm only part-time. Profitable farms are big. This may mean 600 to 800 acres in the cornbelt, where one family can work this much land alone with the proper machinery and a small amount of help at harvest time. On a dairy farm, there must be 40 cows per worker to see a profit because, again, a great deal of machinery is involved.

Still, except in certain fields, such as raising broiler chickens, large corporations are not farm owners. Most farming does not lend itself to corporate management, since managers would have to put in long hours and have knowledge of optimum farming techniques for each separate locality. Nor is farming as profitable, in terms of return on investment, as many other business ventures.

The next few sections detail some of the agricultural developments, such as synthetic fertilizers, irrigation schemes, and high-yield crops, that have increased U.S. food production to a point where North American farms produce most of the food exported in the world today (Fig. 33-8). The promise of similar techniques in developing countries is examined.

Fertilizer, Energy, and Food Production

With few exceptions, the American farmer uses large amounts of artificial fertilizers. In fact, 30% to 40% of increased U.S. productivity in recent years is attributed to increased use of fertilizer. Nitrogen fertilizers, which account for half of all the fertilizer in the world, are produced mainly from fossil fuels such as natural gas and coal. Energy is also needed for the manufacturing process itself. Shortages of fossil fuels have raised the cost of fertilizers over threefold in recent years, which has in turn contributed to the higher cost of food. An especially unfortunate result of higher fertilizer prices is that poorer countries can now afford less fertilizer, yet these areas are precisely where it would do the most good. This is because as more and more fertilizer is added to a field, the increased crop yield per pound of fertilizer begins to decrease. The field is, so to speak, saturated with fertilizers. Most large farms in developed countries, such as the U.S. would fall into this category. In contrast, farms in undeveloped countries, where little or no fertilizers have been used in the past show a much larger increase in yield for a smaller amount of fertilizer. In terms of the total world food supply, then, the distribution of fertilizer is uneven.

Energy is used in modern agriculture not only to produce fertilizer but also to run the machines that plant and harvest crops and in the after-harvest processes: handling, storage, processing and distribution (Fig. 33-9). One further point should be made in relation to energy and agriculture. Most people prefer to eat meat if they can afford it. Yet, as the second law of thermodynamics predicts, meat is an inefficient way to obtain food energy. As a general rule, an animal eats 3 to 10 pounds of grain to produce one pound of meat. A much more efficient use of food resources is for people to consume the grain directly, as is now done in most undeveloped nations (Fig. 33–10).

Resources of Land and Water

Irrigation In many areas, notably the American Southwest, fertile lands that receive little rainfall are sucessfully farmed when irrigation systems are developed. Some 15% of the world's farmland is irrigated and 30% of the world's food is produced on these acres. Again, energy is a necessary component of the system. In most cases, fossil fuel is needed directly for pumping or draining systems.

In many areas, however, irrigation brings hazards along with the benefits. In tropical countries, irrigation canals serve as a breeding place for the snails that carry the disease, schistosomiasis. This is a non-fatal disease, but one that takes a great toll of strength from people who live in areas where the disease did not exist before irrigation (Fig. 33–11).

Furthermore, irrigation water evaporates to some degree, leaving behind salts that were in the water. In some areas, including parts of the American West, this has left soils so salty that crops will no longer grow there. Soils can also become water logged if drainage systems are not included in irrigation plans.

Some scientists have expressed concern that within the next 100 years, irrigation may become so widespread that the flow of major rivers into the sea may be completely stopped. The Nile and other rivers entering the Mediterranean are most likely to suffer that fate. The environmental effects of such an occurence are unknown, but might include such things as decreased fisheries, as the flow of

Figure 33–7 *Settlers in 1887 in front of their sod home. Although transformed by modern agricultural technologies, the family farm remains the characteristic unit in the Corn Belt and Wheat Belt. However, the look of the farm has changed quite a bit since that time. In 1850 there were 1.5 million farms in the U.S. with an average size of 196 acres. By 1920 there were 6.5 million farms, on average 149 acres each. By 1959, however, there were only 3.7 million farms and the average size was 303 acres. This trend has continued to the present. There are fewer and fewer farms and their average size is still increasing. (USDA photo)*

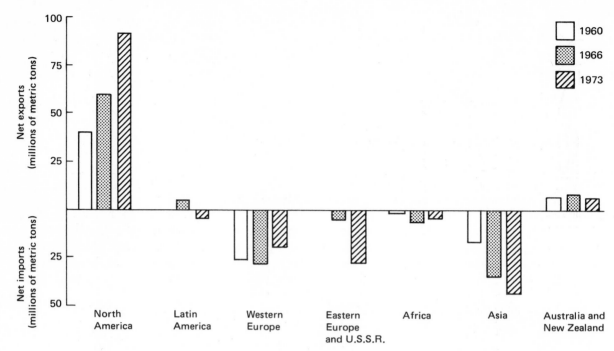

Figure 33-8   *Eugene B. Skolnikoff, Director of M.I.T.'s Center for International Studies calls North America "the Saudi Arabia of food" and this chart reveals the growing dependence of almost all the rest of the world on grain exported from the U.S. and Canada. [Source: N. Scrimshaw, Technology Review, p. 19 (December 1974).]*

nutrients from land into sea is stopped. Water as a scarce resource is discussed in Chapter 4.

Irrigation schemes may even have adverse social effects, such as causing overpopulation in newly irrigated areas.[4]

The supply of arable land   There is no shortage of land physically suited for growing crops. There may be as much as twice again the amount of arable land in the world as is now farmed. Some important qualifications have to be made, however. In the first place, most of the land that is easily cropped is already being farmed. The remaining land needs a great deal of energy and labor for such

tasks as the clearing of forests or building irrigation systems. Furthermore, the land that is expected to be most fertile is also already in cultivation. Some of the land not now being farmed may be only marginally fertile. Thus, if new land is farmed, expected yields may not be as great as on land currently being farmed. However, to offset this last point, land can be improved by farming if the proper techniques and energy are available. Indian farmers cultivate almost the same number of acres as U.S. farmers, but produce only 40% as much crops, partly because of a lack of fertilizers and machinery. In many other countries, farmers keep their fields fertile by shifting cultivation every few years. When the land is allowed to lie fallow for a few years or a few decades, natural processes such as nitrogen fixation renew essential plant nutrients. In addition,

---

4 For more on this see the reference by E. Barton Worthington.

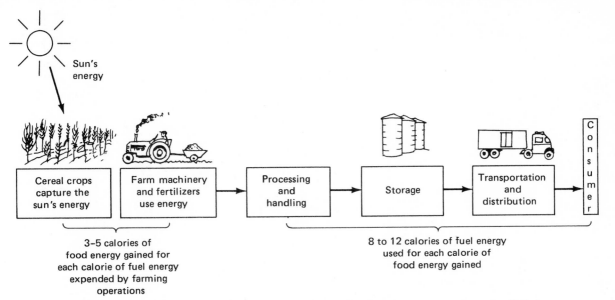

Sun's energy

Cereal crops capture the sun's energy | Farm machinery and fertilizers use energy | Processing and handling | Storage | Transportation and distribution | Consumer

3–5 calories of food energy gained for each calorie of fuel energy expended by farming operations

8 to 12 calories of fuel energy used for each calorie of food energy gained

Figure 33–9 *Energy costs for producing food in a developed country such as the U.S. In farming itself there is a net gain of energy because more of the sun's energy is captured by plants than is used by farming procedures. After this however, processing, handling, storage, and distribution of food use up this energy gain and more, so that, by the time food reaches the consumer, 5 to 7 calories of fuel energy have been used to produce each calorie of food energy.*

populations of insect pests and weeds, which take a heavy toll of the crop, die down during the fallow period. These same ends can be achieved to a large extent if artificial fertilizers and pesticides are available.

Plans for clearing new lands for farming, however, put experts interested in increasing food production on a direct collision course with those concerned about preserving endangered species of plants and animals. In South America, a large part of the land that could be farmed is now under forest cover and providing habitats for large numbers of animals and even larger numbers of plant species. The concept of preserving a wide diversity of species in the world comes into conflict with the vision of providing enough food for everyone. As human populations increase until more land is necessary for food production, other desirable uses, such as species protection, are edged out of the picture.

High-Yield Grains: The Green Revolution

Perhaps the biggest hope technology has held out to farmers in developing nations is the promise of high-yield grains. New varieties of wheat and rice have been developed that far out-produce traditional varieties. In theory, this is a marvelous advance. Using the new varieties, a farmer can grow more grain even without increasing the amount of land under cultivation. This has become known as the "Green Revolution."

In practice, however, there are difficulties. The new grains are very responsive to fertilizers. In fact,

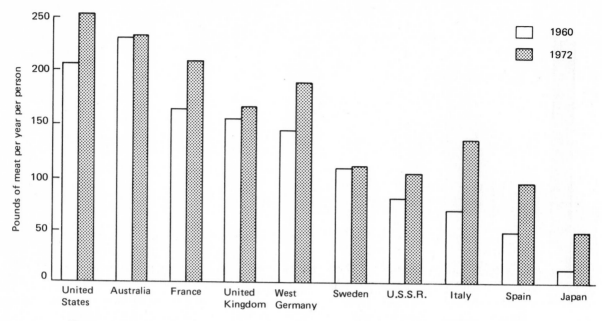

Figure 33-10 *World meat consumption. As people earn higher incomes, one of the things they spend more money on is meat. The combination of rising food prices and more meat consumption in the wealthier countries acts to make food grains less available to people in poorer countries. [Source: N. Scrimshaw,* Technology Review, *p. 18 (December 1974)]*

high yields depend on the use of large amounts of fertilizer. In addition, to reach their full potential, the new varieties need more water than older varieties and in some cases they are more susceptible to disease and insects. What all of this means is that a farmer must have both the technical knowledge of how to use the new varieties and the money to spend on necessary supplements such as fertilizers, pesticides, and irrigation mechanisms. Without these supplements, the new grain varieties are no better and sometimes worse than the older ones.

This is not to say they are without value. In a period of six years, from 1965 to 1971, farmers in the northwest area of India achieved spectacular increases in food grain production, in some cases averaging almost 10% per year. A combination of high-yield grain, more fertilizer and good rainfall led India to announce in 1971 that she soon expected to be self-sufficient in food. However, the failure of the monsoon rains in 1973 and 1974 destroyed these hopes.

In summary, agricultural technology has a great deal to offer the world in terms of increasing food production. But the benefits it promises are very dependent on a series of other factors: energy for fertilizer and to mechanize food production; rainfall or extensive irrigation, which will probably also require energy; pesticides and the technical knowledge of how to use them; and finally, money or credit, because the farmer must invest in all of these technological items before he ever harvests a crop.

It is also apparent that the Green Revolution brings with it problems that, if ignored, could cause the whole system to collapse. One example is the build-up of salts in irrigated soils. Another is the erosion of soils from cultivated lands. Pesticides also present a two edged sword. In Chapter 34, we shall look at the problems associated with erosion and pesticides.

Figure 33-11 *Irrigation project in Khuzestan, Iran. Canals such as this are good breeding spots for schistosomiasis. This disease is spread by a snail that lives in slow-moving water. Children playing in the water are quickly infected. (Paul Hebert, 1975.)*

Questions

1. How can we say people rarely die of starvation? Don't people die from too little food every day in the poorer countries?
2. What special effects does poor nutrition have on children?
3. Suppose you were in charge of a program to encourage breast-feeding in developing countries. On what arguments would you base your campaign?
4. Why is it sometimes said that the food crisis is a problem of distribution rather than of supply?
5. Imagine you are a farmer in a developing country. Describe the values and drawbacks you see in synthetic fertilizers, a local irrigation scheme, high-yield grains, a government plan to clear the forest near you for the planting of crops (you would share in the increased farmland).

Further Reading

Idyel, C. P., "The Anchovy Crisis," *Scientific American* **228** (6), 22 (1973).

Many people still believe that a limitless harvest waits in the oceans. The collapse of the anchovy fishery illustrates how actions stemming from such beliefs may cause a complete loss of what ocean resources exist.

Intercom: the international population news magazine. Population Reference Bureau, Inc., 1337 Connecticut Avenue, N.W., Washington, D.C. 20036.

This publication carries a capsule prepared by FAO on the world food situation each month. Good for up to the minute information.

Jensen, N. F., "Limits to Growth in World Food Production," *Science* **201**, 317 (28 July 1978).

> The possibilities for further technological increases in food production, as well as the social and economic limits that might exist, are explored in this article.

Science **188**, (9 May 1975).

> This issue of *Science* is devoted entirely to food problems. It provides a great deal of background material as well as the history of world food problems up to 1975.

Winikoff, B., "Nutrition, Population and Health: Some Implications for Policy," *Science* **200**, 895 (26 May 1978).

> In this article, the author discusses the interrelationship between the food and population problems, and also advantages and social aspects of breast feeding in developed and undeveloped countries.

Worthington, E. B., "The Greening of the Desert: What Cost to Farmers," *Civil Engineering—ASCE*, p. 60 (August 1978).

> In this article, the social effects of irrigation projects, often overlooked by planners and engineers, are pointed out.

Chapter 34

Erosion and Pesticides

699

EROSION

What Is It?

Storm clouds gather and a brisk wind stirs the poplars in the hedgerow. A tiny drop of rain falls on the freshly plowed earth. Then another drop, and another, and the storm begins in earnest. Each drop hits the earth like a miniature meteor; soil particles spray in a circle around the drop and a small crater is formed. As more droplets reach the earth, their arrival rate begins to exceed the rate at which the water can be absorbed into the soil. Puddles form, then rivulets. The looser and finer soil, the best soil, is picked up by the rivulets and carried downslope. The velocity of the rivulets increases; more and more soil is carried in suspension. Rivulets join to form temporary streams, which flow in gullies, and the rapidly flowing water, enroute to a true stream, cuts sharply into the earth.

The event is not isolated or uncommon; it is any rainstorm on a plowed field. Multipled by thousands of farms or millions of acres, as much as four billion tons of soil may be carried off each year by water erosion in the U.S., about three quarters of it from farmland. Wind may scour away another billion tons of soil. A recent report by the Council for Agricultural Science and Technology has estimated that one-third of all crop land in the U.S. was being eroded so fast that its productivity would suffer. Such losses as these cannot be without con-

sequences. What are the effects of erosion on the land and water? How can erosion be slowed?

Kinds of Erosion

Soil scientists divide erosion into three components: sheet, rill, and gully erosion. In sheet erosion, surface water moves downslope in a wide flow; it may roll soil particles with it, bump them along, or completely suspend them in the water, depending on the size of the particles and the velocity of the water. Since sheet erosion occurs over wide areas in a field, it may not be noticeable at first, and awareness may dawn only after much damage has been done to the soil.

In rill erosion, in contrast to sheet erosion, the water joins into rivulets (Fig. 34–1). Flowing at higher velocities through narrow miniature valleys, the water will have greater erosive power. However, because rivulets are small and can be plowed up with the rest of the soil and smoothed over, they are also easy to miss in farming. You may be able to see large rills forming in a hill of fresh earth at a construction site.

Gully erosion is not as easy to miss as sheet or rill erosion; the rivulets join into such large flows that tractors cannot smooth them over (Fig. 34–2). Gulleys resemble ordinary streams with soil banks, except that water flows in them only after rainstorms; at other times they are dry. The banks of the gully are usually only loose soil, which the water can erode away. Earth overhangs may be ex-

Figure 34-1 *Severe rill erosion in the state of Washington. A fine tilling of the summer's cover crop combined with poor growth in the fall of the wheat crop left this steep slope in a highly eroded condition. The county crew which came to clear the road was shooed off by the farmer who told them "I'll haul the soil back up the hill." (USDA—Soil Conservation Service; Photographer: Fred Wetter.)*

posed by the water that undercuts the soil banks; since they are structurally weak, it is common for the overhangs to collapse into the gully, accelerating the erosion process.

Wind also erodes the soil, lifting and transporting the finer particles and organic matter. Such erosion leaves behind the less desirable coarser soils which, because they retain water less well, are themselves more likely to erode.

All these forms of erosion carry away soil, nutrients, and plant residues, essential components of a productive earth. Pesticides are also adsorbed onto the soil particles and can later dissolve in waters into which the soil is carried.

Environmental Effects

It is well to remember that erosion is a natural process. It has been observed for centuries; it even has beneficial effects. The rich delta lands, lands built up at the mouths of the Mississippi, the Rhine, and the Nile, all exist because soil was carried in

enormous quantities off the land. The richness of the deltas is a two-edged sword, however. If the delta soils are rich, the soil that remains on the land upstream must have been robbed of some of its value. That, in fact, is one of the principal concerns we have about soil erosion, the decreasing productive capacity of the soil.

Although erosion is a natural process, the massive extent to which it is presently occurring is the result of human activities. The farming of more land, only in this century made productive by irrigation, the grazing of cattle, the construction of buildings and highways, and the activities associated with mining–all disturb the soil and accelerate erosion. In addition, the pavement which now overlays so much of America contributes to faster erosion. Because water cannot be absorbed by paving materials, it collects and may run swiftly off the pavements. Its increased volume and velocity will erode and suspend soil, carrying it into stream beds.

Loss of topsoil The three billion tons of soil eroded from U.S. farmland each year come from individual farms, where erosion could be more or

Figure 34-2 *Gully erosion in Montana. This gully cuts deeply through a hay meadow. The farmer will have to do significantly more plowing to prepare his land—now in two fields instead of one. (USDA—Soil Conservation Service; Photographer: D. J. Anderson.)*

less than some average figure. From sloped land that is unprotected and has been freshly plowed, under an intense rain, annual erosion losses could reach 60 to 100 tons per acre. The higher figure is equivalent to the loss of about 0.7 inches of top soil per acre. This is not an annual loss but the loss due to a single rainstorm. On a national basis, losses average about 9 to 12 tons per acre per year, but even this is an enormous quantity since new soil formation occurs at only about 1.5 ton per acre per year.

Such losses create numerous problems. The productive capacity of the soil may be decreased depending on the depth to which the top soil extends. One figure cited is that a loss of two inches of top soil will cause the yield per acre (measured in bushels, let us say) to decrease by 15%. The loss in yield may be due to changes in the soil structure. When best soil has been carried off, the remaining soil may not absorb water as well nor be as porous to the air. Hence, more run-off is likely to occur, making less water available to the crop. Nutrients may be dissolved or carried off with the suspended soil, leaving a less productive subsoil. Since the subsoil is generally less permeable, less water can be stored between rains, making the likelihood of damage from droughts greater.

The less desirable subsoil remaining is also more difficult to till; the rills make plowing more difficult yet; and the gullies may block plowing entirely, resulting in the complete loss of portions of the land. Not only is the land lost, but the fields are cut into smaller blocks by the gullies. Hence, the time and effort to cultivate an area increase. It is akin to having to mow the grass on an acre (about 200 feet by 200 feet) of land in blocks that are 10 feet on a side instead of making cuts that are 200 feet long. (A final effect of water erosion on farmland is the damage to seedlings. These immature plants may be dislodged and carried off. Or they may be buried by sediment deposited on them.)

Effects on streams, reservoirs, and aquatic life (The soil particles carried in suspension and bumping

along on the bottom of the stream bed will eventually settle when the velocity of the stream decreases sufficiently. As sediments accumulate on the bottom of the stream channel, the carrying capacity of the stream will be decreased. The carrying capacity is the maximum flow rate the stream can carry without overflowing its banks. Run-off will be increased from eroded areas because of a loss in the absorptive capacity of the soil. Thus, more water will be coursing down a stream channel which is unable to handle it. And, compounding the problem is the fact that the volume of run-off is multiplied again by the volume of the soil particles carried in suspension.[1] Flooding by streams flowing through badly eroding areas is more likely. Not only are stream channels filled up by sediments, reservoirs are filled up as well. These structures, which are used to store water for irrigation, for hydropower, and for water supply, bring the flowing waters to a standstill. When this occurs, most of the sediments, which were in suspension, drop down. The build-up of sediments in reservoirs will decrease the quantity of water that can be stored and hence the quantity of water available on a reliable basis for water supply. Large reservoirs are actually designed with extra capacity to hold the sediments. Of course, this added capacity drives up the cost of the reservoir. A 1968 survey of nearly 1000 reservoirs in the U.S. showed about 40% of them filling at over 3% per year. These were most often smaller reservoirs, less than 100 acre-feet.

(Besides using up the capacity of rivers and reservoirs, sediments carried in suspension affect aquatic life. Sediments may bury the habitat of bottom feeders. The turbid water allows the entry of less sunlight, hindering the growth of aquatic plant species.)

---

1 One extreme case of erosion has been documented on the Yellow River in China. At flood stage, the flow in that river has been as much as 50% sediment by weight. The yellow color of the river is due to the color of the soil particles eroded from the uplands.

(Finally, we must list among the effects of water erosion the poor water quality caused by the substances carried into our streams in run-off.) Fertilizers and animal wastes are one category. Both fertilizers and animal wastes use up vital oxygen in the water and destroy the habitat of aquatic life. Pesticides enter the water through erosion as well. In fact, the Commission on Pesticides, of the Department of Health, Education, and Welfare, estimated that erosion is the most common route of entry of pesticides into water.

Effects of wind erosion (Wind erosion has its own set of effects. Of course, its effect on the productivity of the soil is similar to the impact of water erosion. The lighter more granular soils with high organic content will be scoured away by the wind, leaving a coarser, less absorbent subsoil exposed. The impact of wind erosion on human activities, however, is far different from that of water erosion.)

Dust storms, when the winds lift and carry soil particles on a massive scale, are fearful indeed. One episode is recorded in which nearly 1300 tons of particles were entrained per cubic mile of air. Annual soil losses due to wind erosion have been measured at levels up to seven tons per acre. People and livestock have difficulty breathing in dust storms and develop respiratory and eye ailments. Crops, especially seedlings, cannot survive such conditions. In the 1930s, the Midwest went through such a severe series of dust storms and blowing soil that the Southern Great Plains were nicknamed the "Dust Bowl" (Fig. 34–3). The storms seem to come in bunches. For instance, there were 120 storms recorded near Dodge City, Kansas, in 1936–1937. Another cluster of storms struck the area about 20 years later. Apparently there is some combination of wind, weather, and soil conditions that triggers these storms.

(Although the storms are distinct and noteworthy events, blowing soil is common. Its effects are the same as those of dust storms though slower and less dramatic. Scour of plant leaves by soil particles may reduce crops. If plant roots are uncovered, the crop will not survive. Soil is stripped of its richer portions.)

Methods of Controlling Erosion

(Perhaps the most important single activity that brings about soil erosion is farming.) When soil is turned over, exposing a surface with no cover, with no protection whatsoever, the possibilities for erosion are greatly increased. One set of alternatives to control erosion, then, concentrates on the way in which the soil is tilled (Fig. 34–4). Since cover by plants hinders both water erosion and wind erosion, a technique known as minimum tillage farming is sometimes used. For instance, rye may be planted in a field in the fall and then killed by a herbicide in the spring before corn is planted in the field. The remains of the rye crop are left on the field to reduce soil erosion by decreasing the impact of rain drops and slowing the flow of water across the field. Further, when crops are rotated, by planting cash crops such as wheat alternately with grasses or legumes, the soil gains protection from erosion and may be nourished as well. Unfortunately, the high price of wheat in the late 1970s drew many farmers away from wheat-fallow rotations and wheat-sorghum-fallow rotations [2] to continuous planting of wheat; no intermediate plantings were used to firm the soil and restore its nutrients. The problem of making farmers act in their long-term interests as opposed to short-term interests is a basic one that we shall address in a moment.

(Another important aspect of erosion is the slope of the soil. The greater the slope, the more severe erosion is likely to be, other factors such as rainfall and soil types being the same. This has led to the development of soil management practices, such as contour plowing or terracing. Such methods are designed to interrupt, divert, and ab-

---

2 A fallow crop is one that nourishes the soil. Sorghum, although grown as a grain, is a grassy plant.

Figure 34–3   *In the 1930s the Midwest suffered a series of severe dust storms caused by winds which blew soil off fields during dry weather. So many dust storms hit the Southern Great Plains that the area became known as the "Dust Bowl." (Reprinted by permission from* Erosion and Sediment Pollution Control, *by R. P. Beasley. Copyright 1972 by Iowa State University Press, Ames, Iowa 50010.)*

sorb the flow of water in its path down a field (Fig. 34–5).

Two other major influences on the extent of erosion, the frequency and amount of rainfall, are not generally controllable by human activity.

We conclude by noting that methods involving the use of vegetation can protect the soil not only from water erosion but from wind erosion as well. Protection against wind erosion, however, may also involve planting trees such as willow or poplar, two fast growing species. Arranged in long lines, called shelter belts, the trees decrease the wind speed downwind from the belt. Earth banks, wooden fences, and rock walls may also be useful in checking wind erosion.

Encouraging Soil Conservation

One would think that soil conservation was in the farmer's best interests. It may be, but only in the long run. In the short run, the farmer must survive to plant another year. We indicated a moment ago that crop rotations in raising wheat may be abandoned in the quest for immediately higher profits. Soil conservation measures that involve restructuring the land cost money as well. And land taken out of production is land that will not produce a crop and an income. Furthermore, the loss in fertility of the soil due to erosion can temporarily be overcome by liberal doses of fertilizer.

The U.S. Department of Agriculture (USDA) has sponsored soil conservation programs since the middle 1930s, when the Dust Bowl era dramatized how very fragile our soil resource is. Recent studies have shown, however, that the programs are not working. A 1977 study by the General Accounting Office of the U.S. government investigated the conditions and practices on 283 farms, chosen on a random basis. The farms, which were in the Midwest, Great Plains, and Pacific Northwest, consisted of

Figure 34-4 *This aerial view of a field of peach trees in New Bridgeton, New Jersey, resembles nothing so much as a tufted bedspread. Yet the pattern actually reflects good soil conservation practice. Between the rows of peach trees are strips of clover sod which simultaneously enrich the soil and help to anchor it in place. (USDA—SCS photo by Clarence Deland)*

119 farms that had had conservation plans prepared by the Soil Conservation Service of the USDA and 164 farms that had not. Of all the 283 farms, 84% were losing soil by erosion at an annual rate of five tons per acre or more; it is thought that a loss of five tons per acre each year cannot be sustained without a loss of productivity, even by deep soils. Of the 119 farms with conservation plans, fewer than half were actually making use of them.

Although the USDA does have a cost sharing plan (with about $200 million dollars to share each year) to stimulate conservation measures by farmers, many of the uses to which the money has been put turn out to be measures to improve the yield of crops rather than stabilize the soil. A more radical proposal, one that definitely limits the farmer's freedom of action, would link farm benefits to the soil conservation measures that have been put into effect. That is, the farm owner would only be able to participate in the price support, crop insurance, and farm loan programs if a defined program of soil conservation had been or was being carried out on the farm. There is precedent for such a linking of programs. The Federal Housing Administration will only insure loans for houses that meet a minimum standard of quality. The U.S. Department of Transportation allocates federal money only to states that enforce a safety measure, the 55 miles per hour speed limit. Thus, the linking of conservation programs with farm benefit programs is not unthinkable.[3] The more appropriate question is whether such a concept could ever be legislated, given the many farm states in the nation and their representation in Congress. It might take another Dust Bowl.

Figure 34-5 *Conservation farming in Pennsylvania. One of the first Soil and Water Conservation Districts in the U.S. was founded in Lancaster County, Pennsylvania in 1937. From a wasteful pattern of straight line furrows which climbed up hillsides, the pattern of cultivation evolved to its present state. This aerial view shows plowing which follows the contours of the hills and it shows strip cropping, two soil conservation practices which check erosion and hence preserve the soil. (USDA—SCS photo by Don Schuhart).*

---

3 More on this in T. Barlow, "Solving the Soil Erosion Problem," *Journal of Soil and Water Conservation* **32**(4), 147 (July–August 1977).

PESTICIDES AND FOOD PRODUCTION

The Need for Pest Control

Each year an estimated half of the world's critically short food supply is consumed or destroyed by insects, molds, rodents, birds and other pests that attack foodstuffs in fields, during shipment and in storage.[4]

Experts believe that if the insects and diseases that attack crops such as cereal grains in the field could be controlled, some 200 million extra tons of grain would be available each year. This amount of grain would feed one billion people. Moreover, if we could stop the pests, such as rats and insects, that eat or destroy food once it is harvested, there would be 25% more grain available to the world's hungry people, without any increase in the amount of food actually grown (Fig. 34–6). In parts of India, estimates are that as much as 70% of stored foods are lost to pests each year.[5]

The challenge, and it's not a simple one, is to solve these pest problems without harming people or the environment. In India, stored grain is often illegally treated with the pesticide DDT to kill insects. As a result, people in India have more of this poisonous and carcinogenic chemical in their body fat than people in any other country in the world.

Problems in Using Pesticides

(Any material used to kill pests is called a pesticide, and usually any organism that competes with humans for food, fiber, or living space is called a pest.) The use of chemicals to kill pests is not a new idea. For centuries, farmers have used minerals such as arsenic, lead, and mercury or natural plant substances such as pyrethrum, obtained from a daisy-like plant, to kill insects and other pests. However, in 1945 a new era in pest control began— DDT came into widespread use to control the lice and fleas that plagued the armies of World War II. DDT was the first widely used synthetic, or laboratory produced, pesticide. In the years that followed, many synthetic pesticides were introduced. In a number of cases, spectacular increases in yields resulted where the new pesticides were used(By 1967, half of all of the pesticides used were, like DDT, in the chemical family of chlorinated hydrocarbons.)

Slowly, however, evidence began to accumulate that the synthetic pesticides were not an unmixed blessing.

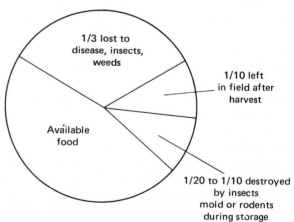

Figure 34–6 *In the U.S., more than one-third of the crops grown are lost to insects, disease, and rodents in the field and during storage. One-third is lost in the field, another 10% is left in the field after harvest and 5% to 10% is lost during storage. In other parts of the world, these figures are even higher. On average, as much as one-half of the world's food crop is lost to disease, insects, and rodents during growing, shipping, and storage.*

---

4 J. E. Brody, *The New York Times,* 28 October 1974, p. A-1.

5 The use of pesticides creates many environmental problems. This section focuses on the use of pesticides in growing foods. The problem of pesticides in water is covered in Chapter 6, while the effects of a specific pesticide, DDT, on wildlife are detailed in Chapter 3. In addition, the possible effects of pesticides in food on human health are considered in Chapter 28.

Biological magnification (Problems arose in three main areas. In the first place certain pesticides[6] tend to accumulate in living organisms. In some cases, pesticides not only accumulate in organisms to levels greater than are found in the environment, but concentrations keep increasing as they move up food chains. This is the effect known as biological magnification (Fig. 34-7; see also p. 22).)

Persistence (The second area of concern involves the length of time pesticides remain in the soil or on crops after they are applied.) The chlorinated hydrocarbons, such as DDT, and pesticides that contain arsenic, lead, or mercury are all known as (persistent pesticides. This means that they are not broken down within one growing season by sunlight or bacteria.) The half-life of DDT, for instance, may be as long as 20 years. It takes 20 years for one-half of an amount of DDT used to be broken down to simpler compounds. Mercury and arsenic are never really broken down. They are moved around or buried in muds.

Persistent pesticides can build up in the soil if a farmer applies them year after year. In some orchards, where lead arsenate has been used to control insects, arsenic in the soil has built up to a level that kills the fruit trees. (The long life of persistent pesticides is a major factor in the process of secondary contamination, whereby foods that were never treated with pesticides still become contaminated.) An example can clarify this process. DDT sticks to soil particles after it is applied. In many areas, dust storms have picked up this contaminated dust and blown it, literally, around the the world. Rain washes this dust out of the air in places where pesticides and even farming are practically unknown. Thus seals, penguins and fish in the Antarctic all show traces of DDT in their fat. One author points out that during the heyday of DDT use, in the 1960s, 40 tons of DDT were

Figure 34-7 *Bald eagle. In 1972, the use of DDT was banned in the U.S., in large part due to the proven effect the pesticide had on reproduction in predatory bird species such as eagles, ospreys, pelicans, and falcons. DDT accumulates in these birds, causing them to lay eggs with thin shells, which provide no protection for the growing chick. In many parts of the country, populations of these birds had declined or become extinct. Since 1972, DDT levels have steadily declined in the U.S. and at the same time some of the disappearing bird species have made a comeback. Ospreys on the East Coast are raising increased numbers of young; larger numbers of bald eagles have been sighted from Colorado to Maryland; and brown pelicans appear to be doing better along parts of the California coast. (Bureau of Sport Fisheries & Wildlife; Luther C. Goldman, photographer)*

dumped on England, every year, in rainfall. In this way, pastures, and crops of all kinds receive an unwanted and unintended treatment with DDT.

---

6 Mainly those, such as DDT, in the chlorinated hydrocarbon family and those containing mercury.

SHOULD THE U.S. SELL DANGEROUS PESTICIDES TO OTHER COUNTRIES?

States have the responsibility to insure that activities within their jurisdiction or control do not cause damage to the environment of other states.

United Nations Conference on
the Human Environment,
Stockholm, 1972.

You can bet we're not going to tell Brazil not to build a dam on the Amazon—particularly if the Germans or somebody else will build it if we don't.

Official of the Import-Export Bank. *

One of the ways to ensure that the U.S. minimizes, or is at least aware of, possible environmental damage that its actions could have abroad, is to require environmental impact statements for projects abroad similar to statements required for government projects here in the U.S. The United States now sells, overseas, some pesticides that are banned in the U.S. (except for public health emergencies). Examples are DDT, Aldrin, Dieldrin, Heptachlor, and Chlordane. In 1976, several environmental groups successfully sued to stop these sales until environmental impact statements were prepared. Even after the statements are prepared, however, agencies such as the Agency for International Development can go ahead and finance sales. The impact statements serve only to make the agency aware of the possible results of their actions.

What do you think? Should we allow sales, to other countries, of pesticides we restrict severely in our own country? Do we have the right to make this decision for other peoples? Do we have the responsibility? What about the argument, similar to the one the export-import official made about a Brazilian dam, that if we do not sell pesticides to other countries, someone else will?

---

* <u>Business Week</u>, (22 January 1979) on the subject of environmental impact statements for international projects.

Resistance (But biological magnification and persistence are not the only problems associated with the use of pesticides. A third serious problem is that pests can become resistant to pesticides—the pesticide will no longer kill them. This can come about because of mutations that occur in some of the enormous numbers of insects hatching each season. Mutations are changes in an organism's inherited genetic material, the material that determines the organism's characteristics.)

A pesticide may kill off most of the insects in a certain area. However, a few who have mutated so that they have slightly different characteristics than the others, may survive. These few can repopulate the sprayed area, passing on their resistance to the pesticide to their offspring. In some cases, pesticides themselves may cause the development of resistance by stimulating the production of enzymes that act to break down pesticides. Those insects that have been stimulated to produce the largest amounts of the enzymes will be the most likely to survive and reproduce. In several generations, insects with sufficient enzyme to break down any reasonable amount of pesticide will appear. This sort of effect has led to the development of houseflies resistant to all of the major types of pesticides used.[7]

(The commonly used "broad-spectrum" pesticides, which kill a wide variety of insects, make the development of resistance simpler. They ensure that the newly resistant insects will have little competition for food. The pesticides also kill off many predators that would eat the insects and parasites that would sap their strength. Thus the resistant insect is given a good chance to establish itself, free from normal controls. There are a number of cases in which pesticides have created new pests in this manner.) For instance, the ladybug normally controls a type of citrus scale insect. If an orchard is sprayed with DDT, however, the ladybugs are killed while the scale insect is not. The scale insect then can multiply unchecked and cause severe damage to the citrus crop.

In a number of cases, extensive use of pesticides has left farmers worse off than before. For instance, cotton has long been grown in Peru in the Cañete Valley. Before World War II, insects were controlled with arsenic and nicotine sulfate. Yields of cotton were about 470 bushels per acre. In 1949, crops were severely damaged by cotton bollworms and aphids. This caused growers to try the new chlorinated hydrocarbon insecticides such as DDT and toxaphene. At first it seemed to be a successful move. Yields nearly doubled. However, by 1955, the picture had changed. Not only had insect pests become resistant to synthetic insecticides, but new pests were appearing due to the fact that the insecticides killed off the beneficial insects that had preyed on pests. Chemicals were unable to control insect populations and, by 1956, the cotton growing industry had collapsed.

A similar sequence of events occurred in the Rio Grande Valley in Texas. Such large amounts of costly pesticides became necessary in Texas that it was no longer possible to grow cotton and sell it at a profit.

Even the successful public health uses of pesticides are threatened by insect resistance. DDT and dieldrin are used in many countries to spray walls inside houses. For months after the spraying, mosquitoes that land on the walls are killed. In this way, diseases like malaria and yellow fever, which are spread by mosquitoes, can be controlled. However, in recent years, a number of species of mosquitoes, including those which carry malaria have become resistant to various pesticides. Worldwide, malaria is the largest single cause of death and debilitation. The disease is characterized by chills and fever. In the 1950s, campaigns to eradicate malaria were begun. Based mainly on spraying with chemical pesticides such as DDT, the campaigns were at first successful. The mosquitoes

---

7 Which pesticides are safest to use? See p. 715 for one view.

however became resistant to DDT and then to pro-poxur, the spray that replaced DDT. Now cases of malaria are once more on the rise. In 1955 when eradication programs began 250 million cases were reported and 2.5 million people died. By 1965, only 107 million cases were reported but in 1976 the number rose to 150 million. Furthermore, the malarial parasite is becoming resistant to drugs used to treat the disease itself.

Thus, synthetic chemical pesticides have not proven to be as easy to use nor as beneficial as was originally hoped. Insects and other pests, such as rats, have shown they are able to become resistant to chemical pesticides, while pesticides that are persistent can accumulate from year to year in the soil and in the muds at the bottom of lakes and rivers. Fat-soluble pesticides are absorbed from water or grass or crops and become concentrated in the milk and body fat of many animals, including man.

The Organic Farming Solution

If pesticides are now seen as a somewhat tarnished miracle, what might be a better strategy in growing food and fiber? One possibility is to use no synthetic pesticides at all and to use only manure or special crops for fertilizer. This is, of course, organic farming. Once the only possible type of agriculture, organic farming became associated with health food faddists and back-to-nature cultists after the development of synthetic pesticides and fertilizers. Organic farming is usually now more charitably viewed as a fine ideal for the home gardener but not a reasonable goal for large scale farmers. In 1971, Earl Butz, then Secretary of Agriculture, said in an interview, "Before we go back to an organic agriculture in this country somebody must decide which 50 million Americans we are going to let starve or go hungry." Nonetheless, evidence is accumulating that in some situations, and when properly carried out, organic farming may be a match for what is now called conventional farming, i.e., using synthetic pesticides and fertilizers (see Controversy). The USDA estimates that if no pesticides at all were used in the U.S., farmers would lose some 70% of their crops to pests, but other experts do not agree with this estimate. Professor David Pimentel of Cornell University counters that the use of crop varieties resistant to pests and other nonchemical techniques, such as changing the kind of crops grown in different areas, could cut total losses to 16%. Certain crops, such as cabbage, potatoes, and some fruits, might suffer heavier losses.

The Integrated Pest Management Solution

Cultural techniques and resistant crops (At the present time, most experts agree that a combination of organic farming techniques and the careful use of some synthetic pesticides and fertilizers offers hope for maximum agricultural production with a minimum of environmental damage. This combination has come to be called integrated pest management. Under this system, a farmer first uses a variety of methods that decrease the possibility that his crops will be attacked by insects or disease (Fig. 34-8).

Before synthetic pesticides were invented, farmers used many cultural practices to reduce insect damage. Some of these old techniques and some newly developed ones are very useful. A book published in 1860 lists this control for apple worms. "The insect that produces this worm lays its egg in the blossom-end of the young apple. That egg makes a worm that passes down about the core and ruins the fruit. Apples so affected will fall prematurely and should be picked up and fed to swine. This done every day during their falling, which does not last a great while, will remedy the evil in two seasons. The worm that crawls from the fallen apple gets into crevices in rough bark, and spins his cocoon, in which he remains till the

Figure 34-8   *The boll weevil can be controlled by integrated pest management. Early maturing cotton is planted and harvested before boll weevil populations became large and spray programs are carefully designed not to harm helpful predators that control other cotton pests. (USDA photo; Lewis Riley, photographer)*

following spring."[8] Plowing alone can destroy up to 98% of the corn earworm pupae that winter in the soil. Rotating the kind of crops grown in a field each season or mixing crops in a field rather than growing large fields of the same kind of crop year after year can prevent the buildup of pests. Destroying crop residues after harvest destroys the winter home of pests such as the boll weevil. By growing certain kinds of hedgerows around fields, a farmer can increase the number of predatory insects, which eat other insects, in a field. This can lead to better control of the plant-eating insects that damage crops.

Plant varieties are now being bred which are resistant to major insect pests. In some cases, this is accomplished by breeding plants that look or taste unattractive to pests. Wheat resistant to Hessian fly has been available for 30 years. As with pesticides, however, new insect mutants may develop and

begin to attack the resistant species, making it necessary to again breed new crop varieties.

Monitoring and treatment ⟨Next, in integrated pest management, the farmer monitors his acres to watch for the beginnings of insect attacks. Traps baited with a variety of substances lure insects and give the farmer early warning of a pest invasion (Fig. 34-9). Only when something must be done to prevent serious crop damage does the farmer or his advisor choose a pest control treatment. This need not be a chemical spray in all cases since a number of methods of biological pest control are known.⟩

Biological controls ⟨An examination of the way insects reproduce and are related to each other has identified several points at which humans can interfere in order to reduce insect damage to crops. Insect pheromones are chemicals synthesized by insects. Some are sex attractants, which help them to find mates. Others mark paths to food sources. If large amounts of a pheromone can be obtained, the

Figure 34-9   *A scout collects insects from a trap in a field. This trap serves as an early warning of insect pests for farmers in the area. (USDA photograph.)*

8 J. H. Walden, *Soil Culture*, Saxton, Barker and Company, 1860, p. 22.

CONTROVERSY

ORGANIC GARDENING

You would never have believed it, we outyielded our neighbors by 100% or better on everything.

K. C. Livermore, Nebraska
organic farmer

...The result of abandoning the use of pesticides?...Silent Autumn.

<u>ChemEcology,</u> January 1978

Although most agriculturists remain unbelievers, there is a small but growing number of people who believe that even large farms can be run, profitably, without synthetic pesticides and fertilizers. The high prices of fertilizer and pesticides, due partly to the energy crunch, has led many farmers to experiment with using fewer synthetic chemicals.

K. C. Livermore is a Nebraska farmer who grows alfalfa, oats, soybeans, and corn on 260 acres. He says:

> We've done much better without chemicals. We hurt some at first when we switched over because we had to get the soil back in balance, get the poisons worked out of it. But in our fourth year there was a big turnaround and now we're outyielding our "chemical neighbors" by far.

Mr. Livermore says about insect and weed problems,

> ...We don't have an insect problem like our chemical neighbors do. We don't have an altered plant. Our plants are natural and healthy. They pick up antibiotics from the soil, which turns insects away as nature intended. And we have insects, like ladybugs, which fight off the enemy insects. Ladybugs thrive on our farm.
>
> Also, as soon as you get a natural, healthy soil, there isn't any weed problem. Nature put in weeds to protect the soil. Weeds grow down in the soil and pick up trace minerals, and as they die they deposit these minerals on the soil's surface.
>
> And when you have your soil in balance, weeds just don't grow as fast and you don't grow as many of them. Another thing is that when we used chemicals we had a clotty soil. Now it will run through you hands just like flour at times. Earthworms and other life in the soil are alive and can loosen it. It's easy to push the weeds right over when we cultivate. *

Other farmers, such as Mike Shannon who farms 30,000 acres in San Joaquin, California, use a minimum of pesticides. His S-K ranch actually owns a pesticide supply company and a crop dusting service but has cut its own pesticide usage by two-thirds. "It costs money," he says. "We have the planes, but I'd rather not touch them." The S-K ranch is advised by Richard Clebenger, a man trained in integrated pest management † who admits, "there still are a lot of farmers...who can't sleep right unless they've given their fields a good spray." Still, his business is now worth $400,000.

Dow chemical, on the other hand, has published a book, Silent Autumn, which describes the plagues of Biblical times as well as a variety of famines caused by uncontrolled pests in recent years. The book emphasizes the idea that careful use of pesticides is necessary to feed people in an ever more crowded world.

For the experimenting organic farmers, though, K. C. Livermore sums it up:

> We'd like to see this thing get turned around...We'd like to see the wildlife and the birds back here like it was in the 1940's and 50's. Is that a profitable way to farm? You bet it is. We use one-fourth less input and get as much or more back than anybody else. That should be real easy to calculate in your mind...

* All quotes from EPA Journal, pp. 24-25 (March 1978).
† Which attempts to use as little pesticide as possible.

pheromone can be used to bait traps.) This has been done with the Gypsy moth, whose sex-attractant pheromone has been artificially synthesized. Traps baited with pheromone can be used either to catch insects and kill them with a poison or to give early warning of a pest's appearance in the field.

Insects and weeds are sometimes imported from other countries without their natural parasites, predators, or diseases. Such pests may then spread unchecked and cause great damage. Successful attempts have been made to control many of these undesirable immigrants as well as some native pests by introducing parasites, predators, and diseases. A tiny wasp, *trichogramma*, controls cotton bollworms by parasitizing bollworm eggs. Two kinds of leaf eating beatles have been able to keep Klamath weed, once a serious problem in California, under control. (In such cases, of course, it is necessary to determine in advance that the new species will not begin to eat crops or kill desirable insects if it runs out of pests.)

Bacteria that attack only pest species can be used as a control measure. Bacteria causing milky-spore disease will kill Japanese beetles and can be applied much like a synthetic insecticide.

(Certain kinds of insects can be grown in large numbers and then sterilized by radiation or chemicals. Sterile males of these species, when released, mate in the wild with normal females. Females who mate with the sterile males lay eggs that do not hatch.) Screw worms have been eliminated from a Caribbean island, Curacao, and controlled substantially in the southwestern United States by the release of sterile males. Not all insects can be grown artificially in large enough numbers or sterilized easily enough to use this method, however.

Economics of IPM (For a number of crops, integrated pest management is cheaper than conventional techniques, which involve spraying with chemicals at regular intervals.) Even though farmers must pay to have their crops monitored, cotton, apples, and citrus fruits all can be grown more eco-

nomically using IPM methods. (The savings are mainly due to the use of less chemical spray, although sometimes yields are increased also (Fig. 34–10).) In one example, 600 acres of tomatoes in California were being sprayed with chemicals four or five times a season at a cost of $20 to $30 per acre. In spite of this, fruit worms continued to damage the crop. A company specializing in IPM was able to control the fruit worms, as well as reduce costs to $8 to $10 an acre and, during the second year of the program, no chemical sprays were necessary at all. As the cost of fertilizer and pesticides goes up due to fossil fuel shortages, the savings in such IPM programs will become more and more attractive.

Figure 34–10 *A Michigan apple grower inspects his apple trees. By waiting until pests are actually found and timing spraying for the most vulnerable period in a pest's life, farmers can use much less chemical spray than if they treat crops on a regular schedule. (Andrew Sacks/NYT pictures)*

Pest Control Experts

Understandably, farmers are slow to change their farm practices. Their income depends on the crops they raise and so they want to be sure of the effectiveness of a new method before they risk using it. Nonchemical pest control methods tend to be more complicated to use than chemical sprays and they usually do not have the immediate and obvious "knockdown" effect that sprays have. Furthermore, farmers get a good deal of advice from pesticide manufacturing companies, which are hardly disinterested parties. These factors, as well as the fact that there are so many pest control techniques from which to choose, have led to the need for people specifically trained in pest control and also for "scouts," people who monitor fields for the presence of insect pests. A number of small firms have already been started to provide pest management services. The Department of Agriculture is attempting to work out certification programs to ensure that persons calling themselves pest control experts actually have the necessary training. Several university programs are being developed in pest control. They will eventually produce people who will know the advantages and disadvantages of chemical sprays, cultural practices and biological controls. These trained specialists can then help to choose, for each situation, the combination of methods which will best solve a pest problem.

Which are the Safest Pesticides?

In order to say whether or not a pesticide is safe, a number of questions have to be asked. How poisonous is the pesticide to humans and to wildlife? How long does the pesticide persist in the environment and does it accumulate in living organisms? Jerome Weber, an agricultural scientist at North Carolina State University, has developed a simplified rating scheme to determine pesticide safety on the basis of the answers to questions such as these (Table 34–1). According to this table, the all-around most dangerous pesticides are the chlorinated hydrocarbons and mercury compounds, such as ethyl-mercurychloride. It is easy to see why the EPA has forbidden or restricted the use of these chemicals in the U.S.

The organophosphorus pesticides are very toxic to rats and humans and are therefore dangerous for the people who use them. But, since they do not accumulate in the environment or in living organisms, their total hazard is somewhat less than that of chlorinated hydrocarbons. Herbicides, such as the triazines and organic acids listed in the table, are widely used in agriculture to kill weeds and to clear roadsides. These compounds score the lowest, generally 7 or lower.

Among the factors this rating scheme does not take into account are the potentials these chemicals have as mutagens (causing changes in heredity), carcinogens (causing cancer), or teratogens (causing birth defects). Such properties are more difficult and expensive to determine than the four factors considered, but are equally important in the long run.

TABLE 34–1   Relative Toxicities of Selected Pesticides

Common name	Toxicity to rat [a]	Toxicity to fish [b]	Longevity [c]	Bioaccumulation [d]	Total
Chlorinated hydrocarbons					
Aldrin	3.2	3.9	4.0	3.1	14.2
DDT	2.7	3.7	4.0	4.0	14.2
Dieldrin	3.1	3.9	4.0	3.0	14.0
Endrin	3.5	4.0	4.0	2.8	14.3
Lindane	2.7	3.4	4.0	1.5	11.6
Mean value	3.0	3.8	4.0	2.9	13.7
Organophosphorus chemicals					
Azinphosethyl	3.5	3.4	1.2	1.0	9.1
Dichlorvos	3.0	2.7	1.3	1.0	8.0
Disulfoton	3.9	3.3	1.1	1.0	9.3
Malathion	1.8	3.2	1.1	1.0	7.1
Parathion	3.6	3.3	1.3	1.0	9.2
Phorate	4.0	3.7	1.1	1.0	9.8
Mean value	3.3	3.3	1.2	1.0	8.8
Carbamates					
Carbaryl	2.1	2.4	1.1	1.0	6.6
Carbofuran	3.6	2.9	1.4	1.0	8.9
Mean value	2.8	2.6	1.2	1.0	7.6
Triazines					
Ametryn	1.8	2.4	1.2	1.0	6.4
Atrazine	1.7	2.0	3.6	1.0	8.3
Cyanazine	2.5	2.5	1.2	1.0	7.2
Prometone	1.7	2.5	4.0	1.0	9.2
Mean value	1.9	2.3	2.5	1.0	7.7
Organic acids					
Dalapon	1.5	1.5	1.1	1.0	5.1
Dicamba	1.8	1.0	1.3	1.0	5.1
Endothall	2.9	2.5	1.1	1.0	7.5
Picloram	1.1	2.4	3.6	1.0	8.1
2,4-D	2.4	1.4	1.1	1.0	5.9
2,4,5-T	2.5	2.8	1.4	1.0	7.7
Mean value	2.0	2.1	1.6	1.0	6.7
Miscellaneous					
Alachlor	1.7	2.0	1.3	1.0	6.0
Bromacil	1.4	1.6	2.5	1.0	6.5
Captan	1.0	3.0	1.0	1.0	6.0
Copper sulfate	1.7	3.0	4.0	1.0	9.7

TABLE 34-1   Relative Toxicities of Selected Pesticides

Common name	Toxicity to rat [a]	Toxicity to fish [b]	Longevity [c]	Bioaccumulation [d]	Total
Diuron	1.6	2.3	2.7	1.0	7.6
Ethylmercury chloride	3.3	2.4	4.0	3.0	12.7
Pyrazon	1.6	1.8	1.2	1.0	5.6
Maximum value	4.0	4.0	4.0	4.0	16.0
Minimum value	1.0	1.0	1.0	1.0	4.0

[a] How poisonous a pesticide is to rats (based on the LD_{50}, the amount that will kill half of a group of test animals) is a good indication of how toxic it is to mammals in general. On a scale of 1 to 4, 4 is the most toxic.

[b] How poisonous a pesticide is to fish (based on the LC_{50}, the concentration that will kill half of a group of test fish) is a good indication of how toxic it is to other aquatic life. On a scale of 1 to 4, 4 is the most toxic.

[c] How long a pesticide persists in the environment is indiciated by how long it remains in the soil after it is applied.

readily broken down	15 weeks or less
moderately broken down	15 to 45 weeks
slowly broken down	45 to 75 weeks
persistent	75 weeks or longer

On a scale of 1 to 4, 4 is the most persistent.

[d] Bioaccumulation can be measured by exposing a test species such as oysters to the pesticide for a length of time and then measuring how much accumulates in the oysters. On a scale of 1 to 4, 4 represents the highest accumulation.

Questions

1. Explain what erosion is and what its major causes are.
2. Summarize the effects erosion has on the environment.
3. What basic principles are involved in controlling erosion?
4. Why do you think farmers do not seem particularly interested in erosion control?
5. What are the major environmental problems associated with the use of pesticides?
6. If pesticides cause so many problems, why do farmers use them at all?
7. If you were a farmer, would you be an organic farmer, would you use integrated pest management, or would you use chemical pesticides whenever you felt you had an insect problem? Why?

Further Reading

Barlow, T., "Solving the Soil Erosion Problem," *Journal of Soil and Water Conservation* **32**(4), 147 (July–August 1977).

Beasley, R. P., *Erosion and Sediment Pollution Control*, The Iowa State University Press, Ames, Iowa, 1972, 320 pages.

Written by an expert with 30 years of experience, this clearly done book offers both a general and technical treatment of the topics. Early portions of the book are more descriptive and general than the later sections where formulae and charts are used for calculations.

Carson, R., *Silent Spring*, Houghton Mifflin, Boston, 1962.

Graham, F. Jr., *Since Silent Spring*, Faucett Publication Inc., New York, N.Y. 1970.

The classic book by Rachel Carson began a public controversy over the effects of pesticides which continues today. Carson was eventually proven correct in many of her claims although, as detailed in F. Graham's book, they were attacked violently when her book was first published.

DDT: A Review of Scientific and Economic Aspects of the Decision to Ban its Use as a Pesticide, EPA report to Congress, 1975. Available from EPA News Services Division, Room 329, West Tower, 401 M Street, S.W., Washington, D.C. 20460

DDT represents something of a success story in pesticides regulation. The chemical was banned because of its known harmful effects on reproduction in birds and the suspicion that it might have long term effects on humans. Since the ban, monitoring studies confirm that DDT levels in the environment are decreasing and some affected bird species appear to be recovering their ability to reproduce. This report gives some background and results of the DDT ban.

EPA Journal **4**, (March 1978).

Most of this issue is devoted to a series of articles on the methods and promise of integrated pest management.

"Methods and Practices for Controlling Water Pollution from Agricultural Non-Point Sources," U.S. Environmental Protection Agency, October 1973, EPA-430/9-73-015.

A very good description of erosion control measures with numerous pictures.

Pimentel, D., *et al.*, "Land Degradation: Effects on Food and Energy Resources," *Science*, (8 October 1976).

A well-researched but somewhat technical viewpoint on erosion.

Pimentel, D., ed., World Food Pest Losses and the Environment, AAAS Selected Symposium 13, 1978, American Association for the Advancement of Science, Washington, D.C.

This volume contains a series of papers on food production and pest control. Both the hazards of synthetic pesticide usage and the promise of IPM are examined. More data is included than in the following reference and the writing is more technical.

Report of the Secretary's Commission on Pesticides and Their Relationship to Environmental Health, U.S. Department of Health, Education and Welfare, December 1969.

This massive study covers just about everything known about pesticides up to 1969 and helped form the basis for the 1972 Pesticides Control Act. Much background information is summarized here.

"To Protect Tomorrow's Food Supply, Soil Conservation Needs Priority Attention," U.S. General Accounting Office 1977, CED-77-30.

This is the GAO's discouraging report on how well soil conservation measures have been working.

Chapter 35

Food and the Future

ECONOMIC AND POLITICAL PROBLEMS IN FOOD PRODUCTION

Technology offers farmers in developing countries a great deal of hope for increasing food production. Whether the farmers can take advantage of this help depends in a large part on political and economic conditions. For instance, before small farmers in a developing country can take advantage of improved technology, credit must be available so they can purchase seeds, pesticides, and other equipment. Large banks, however, are usually wary of lending money to small farmers. In part, these farmers may turn to local moneylenders, friends, or relatives. The energy crunch is seen as making the credit problem worse, since prices for fertilizers, fuel, and related goods are rising. In the long run, price increases will be a curb on the success of the Green Revolution.

In the long run, too, land reform may be necessary in countries where a few wealthy landowners hold most of the political power. This is because such landowners are able to obtain new technological information for themselves fairly easily and they see no need to make such information available to the small farmer. Yet the small farmer must have help, especially education in the technology of using high-yield grains, if the developing countries are to significantly increase agricultural production.

Another political restraint on agricultural production in the past has been that politicians have often tried to hold down food prices. This is done to keep living costs down, in some cases, so that the large numbers of city dwellers will continue to vote for the politicians in power. The effect of this policy is to remove any incentive for farmers to grow more food than their family needs. Market prices are simply too low to justify the work and investment involved in more crops.

For instance, in the Sahelian states it has been a practice to assure a job to any high school graduate. Such a policy has led to vast increases in the number of civil servants, whose salaries can be kept very low because food prices are kept at low levels. This has acted to slow food production because the farmers see no reason to grow more crops at current prices.

The management of irrigation schemes, too, can have important effects on crop production. Managers are often political appointees, living in the cities, too far away from the scene to know how to manage the water resource effectively. Public water managers have tended to focus on avoiding disagreements about the proper distribution of water rather than on making hard decisions on what is really the most efficient way to use the available supplies. Furthermore, the water itself is usually undervalued. This is shown by the success of private water distribution schemes in countries such as India, even when they compete directly with public water supplies. That is, farmers are willing to pay the higher prices for additional water from private sources even when some public water is available.

OPTIMIST VERSUS THE PESSIMIST

Population Growth May Outstrip the Food Supply

As the history of the world's food supply shows, predictions about the ability of agriculture to feed the world are usually colored by the current stock of food on hand. When harvests are bad and grain surpluses fail, the pessimistic view prevails. Calculations are made showing that we simply cannot boost agricultural production to a rate significantly higher than the population growth rate. When harvests are good and surpluses accumulate, the optimistic point of view catches hold and predictions are that a good diet for everyone is just around the corner.

Historically, the pessimists can point to the fact that food production has just barely kept ahead of population growth. With so many of the world's people already undernourished, keeping up is not good enough (Fig. 35–1). Population growth in developing countries continues, although there are signs that growth may be leveling off in some. In terms of increasing the actual supply of food, the Green Revolution is being held back by the high prices for energy, which prevents developing nations from obtaining the fertilizer and fuel necessary to grow high-yield grain varieties successfully.

Furthermore, the pessimists warn us that the high-yield grains themselves are susceptible to disease and insects. As the use of these varieties becomes more common, the stage may be set for a crop failure similar to the potato blight that devastated 17th century Ireland. A great variety of seed types in some measure protects us from widespread crop failure due to disease or insect pests.

In socio-economic terms, as incomes rise in some countries and people begin to be able to afford meat, food purchasing power is shifted even further away from poorer countries where incomes remain low. One result of the energy crisis has been to separate the undeveloped countries into two groups: the wealthy undeveloped OPEC countries,

Figure 35–1 *In many countries food production barely keeps up with rapid population growth. Unexpected bad weather can cause crop failures that lead to famine, when the margin between the amount of food produced and the amount of food consumed is small. The United Nation Food and Agriculture Organization operates WPF (World Food Programme) to provide emergency famine relief and to help increase food production in developing nations. Almost half of WFP's resources go to projects to foster land improvement, help new settlers, raise better livestock. (WFP photos. Top photo by J. Bradford; bottom photo by B. Bhansali)*

who are the oil exporters, and all the others who
have no oil. The oil exporting countries will be able
to buy the food needed to upgrade the diet of their
people. For this reason, they stand in a somewhat
different position with respect to possible food
shortages than the other undeveloped countries.)

The Optimistic View

(Optimists support their point of view by
noting that in most undeveloped countries the
Green Revolution has barely started.) New grain
varieties and attention to those crops such as the
root vegetables, on which relatively little research
has been done, promise great increases in yields.
The United States, itself, could at least double
agricultural production if the returns were great
enough. (And here lies a major *if*. If population
growth can be slowed, if incomes can rise, then the
historic problem of poor nourishment can be
solved. In the simplest terms, people are undernourished because they are poor.) With jobs, and
reasonable incomes they could buy food. And, according to most experts, the world's capacity for
producing food could probably meet this demand.
The paradox is that developing countries must find
a way to provide more jobs in rural areas so that
people there will not migrate to the cities (where
there are no jobs, either) while modernization of
agriculture may cause reduction in the number of
people required to run farms (Fig. 35-2). At the
same time, agricultural and other development
schemes are threatening to destroy the social and
economic fabric on which present incomes are
dependent.

The Philippine government is proposing to
build four dams in the central mountains of Luzon.
The dams are intended to provide hydroelectric
power for a variety of Philippine development
schemes. They would also flood as many as a
dozen Igorot villages, submerging thousands of
hectares of laboriously built rice terraces and forc-

Figure 35-2 *In developing nations new, "appropriate
technology" solutions to farming problems hold out the
hope of increasing yields without displacing workers.
Farm machines do not have to be labor saving to be
useful. For instance in many developing countries,
farmers must wait for the rainy season before plowing.
Machines to help farmers plow the hard caked earth
before the rains fall, would allow them to have their
crops planted and ready to take advantage of the monsoon rains. Above are pictures of a human-powered
tractor used in Philippine rice paddies and an easily built
and repaired rice thresher designed to thresh Philippine
rice. (Courtesy of J. K. Campbell, Cornell University,
Department of Agricultural Engineering.)*

ing thousands of these mountain people to move: to cities where jobs are scarce; to a government resettlement area where they would have to begin building all over again; or to other Igorot villages where overcrowding could destory the carefully worked out social fabric.

(There are really, then, two categories of food problems. First, can we actually produce enough food to provide an adequate diet for everyone? This question is one of technology, with certain added social and political features. The answer, for the near future is probably yes. Second, can we ensure that everyone actually receives enough food? And the problems in this category may be the more difficult to solve. Based on social inequalities and economic variables, these problems and their solutions penetrate to the very core of whether humans, across the world, will care and take responsibility for each other)(see Controversy).

(One way that is suggested for the short term relief of famine and long term relief from rising food prices is the establishment of grain reserves.)

ARE GRAIN RESERVES A SOLUTION?

...Let Pharoah proceed to appoint overseers over the land, and take the fifth part of the produce of the land of Egypt during the seven plenteous years....That food shall be a reserve for the land against the seven years of famine which are to befall the land of Egypt, so that the land may not perish through the famine.[1]

History of the Idea

So spoke Joseph to the Pharoah of Egypt some 3600 years ago. In ancient China, too, we are told of the Confucians who created a "constantly normal granary." In the modern era, Henry Wallace, Secretary of Agriculture for Roosevelt and Presi-

---

1 Genesis 41:34, 36.

dential aspirant in the 1940s, was the leading advocate of grain reserves. Though Wallace had been urging an "ever-normal granary" since 1912, it was not until the 1930s, when farms were being destroyed by dust storms and drought, that he was able to create a government agency that would store grain. Even then, the publicly announced objective behind the government purchasing a grain stock was to stabilize farm incomes.(The stabilizing of farm incomes was a goal for two reasons. First, the political power of farmers made a program of insured farm incomes important. And second, preventing farmers from going out of business because of losses in bad years did help to make the supply of food more secure.)Wallace's program used crop insurance, paid for by a contribution of a portion of the crop itself, limited direct payments, acreage allotments, and price support loans which if defaulted on did not require repayment. Although farm incomes were kept up, output grew and the stocks of grain grew, bringing criticism of the program.

The criticism ended, however, when World War II broke out; the stocks of wheat, cotton, and corn became a vital military resource, as the United States undertook to aid its allies with badly needed contributions of food. By the close of the period of reconstruction after the War, all current U.S. farm production was being consumed and the stocks were gone. Stocks began to build again as farm productivity grew to new levels and government purchases at price supported levels continued.

(In 1954, Congress passed Public Law 83–480, a temporary measure now become permanent, which was aimed at reducing the growing stocks in a way that benefitted both poorer nations and American foreign policy. The government was allowed under this Law to sell some of its surplus stocks to nations in need of grain. Emergency relief could also call forth shipments from the overflowing U.S. cupboard. Famines and natural disasters were among the situations to which the U.S. could respond with food aid. Voluntary relief organizations and

THE LIFEBOAT ETHIC

This obscene doctrine...

Roger Revelle, Harvard
University

It is just as obscene to let people die in the future as it is to let them die now.

J. D. Martin, Sociologist, Lakehead
University, Ontario, Canada

There is a certain argument, related to population growth and the supply of food in the world, which is known as the "lifeboat ethic." The argument is based on an analogy. In the analogy, so many people crowd onto a lifeboat that it sinks and all are lost. If fewer people had been in the boat, the argument goes, those few might have reached shore safely.

Roger ReVelle states in an editorial in <u>Science</u>: *

The specter, unseen by some and ignored by others, looming over the World Food Conference this week in Rome is the continuing rapid population growth of the world's poor countries. Some scientists and publicists have seriously advocated a "lifeboat ethic," saying that nations which do not <u>compel</u> human fertility control (by what means is never stated) are endangering the survival of our species–hence they should be starved out of the human race by denying them food aid. This obscene doctrine assumes that men and women will not voluntarily limit their own fertility when they have good reasons and the knowledge and means to do so.

The sharp decline in birthrates during the past decade in a dozen developing countries belies the assumption. But one thing is clear from this experience: environmental changes can bring down birthrates only if they affect the people who have the children–the great mass of the poor who now have little hope for a better life.

Many people sent letters to the editor in reply to this editorial. J. D. Martin, a member of the sociology department at a Canadian university wrote:

The idea of letting people die, Revelle says, is "obscene." Well, if so, it is just as obscene to let people die in the future as it is to let them do so now.

...I am also more than 90 percent convinced that non-Western populations will keep growing until we can't feed them, even at great cost to the quality of our soils.

We Westerners brought it on ourselves, by saving lives through medical skill and humanitarian generosity. Nobody seemed to foresee the demographic consequences of drastically reducing death rates; or those who did either hoped for some vague miracle or couldn't (wouldn't?) be heard. The millions of lives saved by our medical help became the hundreds of millions of lives that are due to be lost in famines.

...Shall we impoverish the West in order to make the problem even worse, and in the process weaken both our land and theirs?

I think not; this is the essence of the "lifeboat ethic" which Revelle criticizes. Let too many people into a lifeboat and all will sink. The same may be true of our spaceship called Earth.[†]

F. A. Cotton, a member of the chemistry department of Texas A & M University supports Martin's position:

Nobody can look without horror on the prospect, let alone the actual spectacle, of fellow human beings starving to death. It is a monstrous thing, but we live in an age of monstrosities–some still latent but imminent, unless actively forestalled–and it is literally necessary to consider not only relative degrees of monstrousness, but the fact that some monstrosities are qualitatively more ghastly than others. Overpopulation and starvation are interdependent monstrosities, but of a qualitatively different nature. I believe that the former is far more dire than the latter.

If a quarter of the people in the world starved to death next year, the human condition, in the larger sense, would not be basically or permanently changed. After a few generations, this calamity would leave no basic imprint on our collective consciousness, any more than did the deaths of one-fourth of the people of Europe in the great plague of the 14th century.

However, if the population of the world goes on increasing at the present rate for very much longer, the human condition will be basically and catastrophically altered, in an irreversible way. [‡]

What do you think of Martin's argument that "we Westerners brought it (a food crisis) on ourselves" by providing medical aid and knowledge to developing countries. Should we have denied them this aid in the first place? Do you agree with Cotton that worldwide starvaton can be equated with deaths from plague in the 14th century? Do you agree with ReVelle that the lifeboat ethic is an obscene one, or do you feel that the horrors of over-population call for such a drastic response?

After you have thought about these questions, you might want to read The New York Times editorial by Barbara Ward, reprinted in the Controversy on p. 728. Another way of looking at this issue is to ask: Is the nutrition problem really a population problem, as Revelle's critics contend, or is the population problem, at least in part, a nutrition problem? Could better nutrition help people in undeveloped countries solve population problems themselves? Beverly Winikoff (in Science **200,** 895 (26 May 1978)) considers some of the ways these questions can be answered.

---

* Science **186,** 589 (15 November 1974).
† Science **187,** 1029 (21 March 1975).
‡ Science **187,** 1030 (21 March 1975).

governments were the recipients who then distributed the food. Under 83–480, the U.S. participates in the World Food Program. We have been a consistent and generous supporter of this Program, a UN-FAO initiative that supplies food aid for both emergency situations and to nations carrying out development projects.)

To restrain grain production further and to dampen the build-up of stocks, Congress established the Soil Bank in 1956. Under this short-lived program, farmers were paid for the acres they kept *out of production.* Public criticism of possible abuses led to an early end of the program in 1960 before it had been fully tried. By 1961, stocks of feed grain and wheat had reached nearly 1.5 billion bushels and the federal government was paying out a million dollars a *day* in storage costs. Congress felt the need to direct the Department of Agriculture to dispose of some of its stocks.

Surpluses and contributions of food aid continued through the 1960's, but the early 1970s brought a confused situation with alternating years of surpluses and shortages. The combination of a lower-than-trend production of grain combined with heavy export demand for wheat and the inflated price of wheat led the government to dispose of nearly all its grain stocks by mid-1973. The stocks were not to be rebuilt. The high prices brought about by the high level of demands as measured against current production brought welcome profits to farmers. If government sales had been made from stock piles, lower market prices and lower farm profits would have resulted.

Farmers and Grain Reserves

(Farm interests argue against a national grain reserve firstly on the basis that the U.S. has for the past three decades consistently been able to export grains and still meet domestic demands. The argument that our national grain demand can surely be met, is only partly true. In the presence of heavy demand for export, grain prices will rise and U.S.

consumers will have to pay the higher price. Further, the same high grain prices that annoy the affluent citizens of this country are a deadly blow to individuals in developing nations.)

In nations where perhaps 60% of the average family's budget is spent on food, a price rise for wheat or rice can make the difference between poor nutrition and malnutrition. (The need for a grain reserve becomes more important as a means to dampen price increases for people in the developing nations, not simply as a source of emergency grain.)

(It is at times of shortage and high prices that a grain reserve is needed and that farmers will argue most strongly against it. The reason is that a grain reserve would dampen prices and profits in times when production levels are down in relation to demand. Farm interests are largely unwilling at present to forego profits at such times even in return for protection of profits at times when production relative to demand is high. The protection is discounted and dismissed because farmers see the near future as a time when all their products will continue to be absorbed on the world market.)

(It is when production exceeds demand, causing prices to fall, that a farmer will want the government to buy up his surplus so that prices fall no further. Thus, the ideal time from a political point of view for discussion of building a grain reserve is in the period when farm prices are low due to unusually high levels of production.)

In the United States, government purchases of grain for a federally held reserve will serve to protect farmers' income in times of production excesses. What would happen if developing nations were to build grain reserves from their own harvests? Such reserves could make a very big difference in nutritional levels of the population and could also aid the national economy of a developing country by eliminating the need for food imports, thus preventing a deficit in the balance of trade. The World Bank through the use of loans and the FAO are encouraging individual nations to build grain reserves, but the costs of purchase and storage are considerable.

OFEC TO FIGHT OPEC?

We don't need America, it is they who need us. They want our oil.

Ayatollah Khomeini
The New York Times, 21 May 1979

Let's meet economic warfare with economic warfare.

John D. Goode

The United States and Canada, together, supply about 90% of the world's exported grain. Whenever a few countries control a resource that many other countries need, the possibility for economic blackmail exists, as we have seen in the case of oil. In fact, a few U.S. citizens have suggested that we counter threats of OPEC oil boycotts or price increases by forming OFEC–The Organization of Food Exporting Countries.

The only way to avoid a continuous sting by OPEC is by the establishment of OFEC–the Organization of Food Exporting Countries.

Let's ration gasoline, start mass production of ethanol and methanol to add to gas, recover old oil, and exploit oil shale.

But, let's also form an "Organization of Petroleum Importing Countries" (OPIC). Let's agree on very, very high prices for food export items such as wheat, corn, soybeans, rice, meats, etc., to the OPEC nations. Let's meet economic warfare with economic warfare. *

If present trends continue, Middle East oil producing countries are expected to increase grain imports. As incomes rise in these countries, demand will rise not only for grain but also for meat from grain-fed animals.

Explain why you feel the two organizations (OPEC and OFEC) would be or would not be equivalent. Explore both the economic and moral consequences. What sort of environmental consequences resulted from the formation of OPEC? What might you expect, environmentally speaking, from OFEC?

---

* J. D. Goode, Letter to the Editor, Business Week, 30 April 1979.

NOT TRIAGE, BUT INVESTMENT IN PEOPLE, FOOD, AND WATER

Barbara Ward, an economist and writer, was quoted in The New York Times. *

Now that the House of Representatives has bravely passed its resolution on "the right to food"–the basic human right without which, indeed, all other rights are meaningless–it is perhaps a good moment to try to clear up one or two points of confusion that appear to have been troubling the American mind on the question of food supplies, hunger, and America's moral obligation, particularly to those who are not America's own citizens.

The United States, with Canada and marginal help from Australia, are the only producers of surplus grain. It follows that if any part of the world comes up short or approaches starvation, there is at present only one remedy and it is in American's hands. Either they do the emergency feeding or people starve.

It is a heavy moral responsibility. Is it one that has to be accepted?

This is where the moral confusions begin. A strong school of thought argues that it is the flood tide of babies, irresponsibly produced in Asia, Africa and Latin America, that is creating the certainty of malnutrition and risk of famine. If these countries insist on having babies, they must feed them themselves. If hard times set in, food aid from North America–if any–must go strictly to those who can prove they are reducing the baby flood. Otherwise, the responsible suffer. The poor go on increasing.

This is a distinctly Victorian replay of Malthus. He first suggested that population would go on rising to absorb all available supplies and that the poor must be left to starve if they would be incontinent. The British Poor Law was based on this principle. It has now been given a new descriptive analogy in America. The planet is compared to a battlefield. There are not enough medical skills and supplies to go round. So what must the doctors do? Obviously, concentrate on those who can hope to recover. The rest must die. This is the meaning of "triage."

Abandon the unsavable and by so doing concentrate the supplies–in the battlefield, medical skills; in the world at large, surplus food–on those who still have a chance to survive.

It is a very simple argument. It has been persuasively supported by noted business leaders, trade-unionists, academics and presumed Presidential advisers. But "triage" is, in fact, so shot through with half truths as to be almost a lie, and so irrelevant to real world issues as to be not much more than an aberration.

Take the half truths first. In the last ten years, at least one-third of the increased world demand for food has come from North Americans, Europeans and Russians eating steadily more high-protein food. Grain is fed to animals and poultry, and eaten as steak and eggs.

In real energy terms, this is about five times more wasteful than eating grain itself. The result is an average American diet of nearly 2,000 pounds of grain a year–and epidemics of cardiac trouble–and 400 pounds for the average Indian.

It follows that for those worrying about available supplies on the "battlefield," one American equals five Indians in the claims on basic food. And this figure masks the fact that much of the North American eating–and drinking–is pure waste. For instance, the American Medical Association would like to see meat-eating cut by a third to produce a healthier nation.

The second distortion is to suggest that direct food aid is what the world is chiefly seeking from the United States. True, if there were a failed monsoon and the normal Soviet agricultural muddle next year, the need for an actual transfer of grain would have to be faced.

That is why the world food plan, worked out at Secretary of State Henry A. Kissinger's earlier prompting, asks for a modest reserve of grain to be set aside–on the old biblical plan of Joseph's "fat years" being used to prepare for the "lean."

But no conceivable American surplus could deal with the third world's food needs of the 1980's and 1990's. They can be met only by a sustained advance in food production where productivity is still so low that quadrupling and quintupling of crops is possible, provided investments begin now.

A recent Japanese study has shown that rice responds with copybook reliability to higher irrigation and improved seed. This is why the same world food plan is stressing a steady capital input of $30 billion a year in third-world farms, with perhaps $5 billion contributed by the old rich and the "oil" rich.

(What irony that this figure is barely a third of what West Germany has to spend each year to offset the health effects of overeating and overdrinking.)

To exclaim and complain about the impossibility of giving away enough American surplus grain (which could not be rice anyway), when the real issue is a sustained effort by all the nations in long-term agricultural investment, simply takes the citizens' minds off the real issue–where they can be of certain assistance–and impresses on them a nonissue that confuses them and helps nobody else.

Happily, the House's food resolution puts long-term international investment in food production firmly back into the center of the picture.

And this investment in the long run is the true answer to the stabilizing of family size....the whole experience of the last century is that if parents are given work, responsibility, enough food and safe water, they have the sense to see they do not need endless children as insurance against calamity....

Go to the root of the matter–investment in people, in food, in water–and the Malthus myth will fade in the third world as it has done already in many parts of it and entirely in the so-called first and second worlds.

It may be that this positive strategy of stabilizing population by sustained, skilled and well directed investment in food production and in clean water suggests less drama than the hair-raising images of inexorably rising tides of children eating like locusts the core out of the whole world's food supplies.

But perhaps we should be wise to prefer relevance to drama. In "triage," there is, after all, a suggestion of the battlefield. If this is how we see the world, are we absolutely certain who deserves to win–the minority guzzlers who eat 2,000 pounds of grain or the majority of despairing men of hunger who eat 400 pounds?...

Is this the battlefield we want? And who will "triage" whom?

• <u>The New York Times,</u> 15 November 1976.

An International Grain Reserve

(There is another possibility that is less costly and more efficient but more difficult to achieve politically. An international grain reserve, one not belonging to a particular country, could be created.) Proposals for such a reserve were in the air in the mid-1970s. Resistance was great, but the benefits large. (Consider how an international agency operating a grain reserve might function: two prices might be used to trigger the operation of the reserve. A price for grain that fell below a preset low price would be a signal for the agency to buy grain. The low price signal to buy serves two functions. First, low prices are the time to set grain aside; holding or investment costs for the agency are kept low in this way. The low prices also mean an abundant harvest relative to demand, indicating that now is the time to buy and store grain without driving the price up significantly. Second, the purchase of the excess harvest will act as a price support for grain. Farmers will find that a good portion of their excess harvest will thus be absorbed without the price falling still lower. Thus, the presence of a grain reserve will help to ensure the income and prosperity of farmers in times when too much good weather works against them.)

(In a similar fashion, when the price for grain rises above some predetermined high limit the agency will offer its grain on the international market. The grain will be available for purchase by those poor nations for whom higher prices would be a burden. The price at which the grain is sold essentially puts a lid on the international price of grain. Who will pay more than the high signal price if grain can be purchased at that price from the international agency? It is true that farm profits are dampened when the agency decides to sell its grain, but the profits that are dampened will not be the normal profits that keep a farm in business, but the profits that are built of hardship and malnutrition for the poorest. The farmer selling grain at the grain reserve price will still be making a sturdy profit.) Joseph seems to have had quite an idea.

Questions

1. Explain how economic and political problems can be as important in terms of food production in developing countries as introducing new technologies.
2. How would grain reserves help even out fluctuations in food prices?
3. What effect could this evening out have on the inflationary spiral?
4. Suppose you are an American farmer. Are you in favor of a grain reserve? Why?
5. Are you optimistic or pessimistic about future world food supplies? Why? Explain why you do or do not feel the answer to this question is very important.

Conclusion

TWO VOICES

In Part 7, we noted the words of a Nigerian Chieftain who, sensing the briefness of our own span of years compared to the infinity of years belonging to our ancestors and descendants, said

I conceived that the land belongs to a vast family. Of this family, many are dead, few are living and countless members are still to be born.[1]

And in Chapter 3, we quoted Henry Beston, a biologist, who provided us a unique view of nature and its relationship to humankind.

For the animal shall not be measured by man. In a world older and more complete than ours they move finished and complete, gifted with extensions of the senses we have lost or never attained, living by voices we shall never hear. They are not brethren, they are not underlings; they are other nations, caught with ourselves in the net of life and time, fellow prisoners of the splendour and travail of the earth.[2]

We quoted these two people not only because of their eloquence but also for the wisdom in their messages.

1 *Land Use and the Environment,* V. Curtis, editor, EPA, Office of Research and Monitoring Environmental Studies Division, 1973.
2 Henry Beston, *The Outermost House,* Viking Press, New York, 1962, p. 25.

Our Narrow Niche in Time

The concept of the Nigerian Chieftain places our years on earth in some perspective. Being aware of the finiteness of our time span makes us sensitive to the fact that the depletable resources such as the fossil fuels that we use and consume are lost to future generations. The land we consume in dwellings, by paving, by surface mining, by allowing unchecked erosion, can only be restored over centuries. The species destroyed by human activities are lost forever. It is true that technology is constantly finding new resources, but this is not a process upon which we can depend. Nor should we depend on it, since each new application of technology seems to carry new environmental burdens. Shale oil may one day be produced in large quantities to power the engines of society, but the upheaval of the land and pollution of the water from producing shale oil make it a potentially damaging technology. The commercial actions that maximize our well-being today may lead to degraded conditions for future generations. The sacrifices that parents make for their children are an example of how the human species instinctively looks to the protection of future generations.

Our Narrow Niche in Space

The second concept is that we share our space. We are not alone, as Beston points out, nor can we survive alone. We are part of a wider ecological system. We cannot chart all the interrelationships in this system and so we cannot see its limits, just as we cannot see to the limits of the universe itself, no matter the power of our telescopes. Sadly, because we cannot see the limits of our ecosystem and find all the species of plants and animals upon which we depend, and upon which these species in turn depend, some of use are tempted to believe that what we can see is the whole of the system. Europeans of the 15th century felt that way about the extent of the earth, until Columbus sailed beyond the limits of contemporary knowledge. In the same way, new Columbuses show us, every day, that our knowledge of the environmental relationships on which we depend is incomplete. We are not alone nor are we free to be independent of the environment that surrounds us.

A Thank-You Across Generations

These ideas on sharing time and space lead us to consider actions that have long-term benefits. That is, the benefits of the actions that we may take today are distant in time and are to be enjoyed principally by future generations, rather than by us. These ideas also lead us to consider the prevention of actions that will have long-run negative effects, even though the effects may fall outside of our own time span. We sense the necessity of protecting the welfare of future generations, but we have no way to measure the resulting satisfaction and comfort of those future generations. Few of us will be present 100 years from now to hear a "thank you," though it be a thunder that echoes from every succeeding generation. Yet, if you have ever seen the Redwoods in California, or hiked the Pacific Crest Trail or the Appalachian Trail, or viewed the waterfall and geysers of Yellowstone, you may yourself have said a thank you to generations past. The wisdom that inspired the setting aside of these lands is both an inspiration and an instruction.

Long-run benefits stem from other actions than setting land aside. If population growth is slowed and finally halted, our future impact on the earth can be slowed and controlled. When we turn away from the fossil fuels and nuclear power to limitless sources of energy such as the sun and wind, we preserve petroleum and coal for future uses. These steps also limit the degradation of the land from surface mining and help to check oil pollution of the oceans. Turning away from nuclear power can lessen for future generations the burden of pollution by radioactive elements, and it can slow the spread of nuclear weapons, perhaps the most awesome threat to the future of life on earth.

Concern With the "Here and Now"

Concern about the narrowness of our niche in time and in space is complemented by a third concern, which also motivates us to improve the environment. The third concern is with the "here and now." In some cases, human health and lives are threatened by pollution and toxic substances in the environment. In other cases, it is the welfare of plants and animals that is threatened by pollution and human activity. To remove these threats requires action in the present. Such actions as controlling automobile emissions or scrubbing sulfur oxides from the stack gases of power plants have present benefits in decreasing the frequency of respiratory illnesses such as bronchitis. Removing such hazardous substances as asbestos and vinyl chloride from the workplace will lower the number of cancer deaths among the work force. Preventing oil spills in the oceans will protect waterfowl and aquatic species from being destroyed in local environments. Using closed-cycle cooling on power plants will decrease the quantity of water required by power plants and prevent species from being sucked through a condenser at enormous speeds and from experiencing thermal shock. Treating waste water on a particular river to remove organic wastes will protect the oxygen content of the river and thereby ensure the survival of fish and other aquatic species.

This concern for the present is one point of focus for improving the environment. Indeed, a large portion of this book was devoted to the subject of cleaning up pollution in water, in the air, and in the workplace. Fortunately, at the same time that we are making present-day improvement in the environment, we can also make long-term changes for the better. Reducing the pollution burden today is plainly a commitment to maintain the quality of the air and water and workplace for future generations. Furthermore, since many of the toxic chemicals can also cause birth defects and changes in the genetic information, the removal of toxic substances from the workplace has long-term as well as short-term benefits.

Because of this implied promise to maintain the quality of the environment once it has been improved, our short-term interest in improving our surroundings merges with our long-term concern for the quality of the environment of future generations.

RESTORING THE ENVIRONMENT—PERSONAL CHOICES

A Choice of Methods

We began the book by observing some of the thoughts people have when faced with the challenge of restoring and preserving the environment. Now we examine some of the ways that people act on behalf of the environment. It would be far too simplistic for us to offer you a formula for environmental progress. What works in one time and place may be terribly inappropriate in another. And it would be foolish and arrogant for us to advise you in what manner to act. People, as you can see in the statements that follow, respond to the challenge in ways that differ not only in their impact, but also in their intent.

"My contribution to the Sierra Club is helping to influence govenment action on environmental/conservation problems."[3]
Sierra Club Member

"Fix it up
Wear it out
Make it do
Do without
Old Yankee Saying

"Save the Whales, Don't Buy Japanese Products,"
Bumper sticker

"We want to build a self-sufficient community, complete with organic gardens, solar heat, wind power, and—especially—the chance to provide an environment God intended for raising healthy families...."
From the advertising section of Mother Earth News, No. 36, November 1975.

"We should applaud and look up to those who adopt life styles that are modest in terms of the amount of space they monopolize or the amount of materials and energy they consume..."[4]
Maurice Strong

"In 1976, we tried to save (the harp seal) pups by spraying their coats with a harmless green dye...In 1977, we were back again, placing our bodies over the pups to save their lives."[5]
Greenpeace

Political action groups work within the system and, if they are successful, become a part of the system. Their implied purpose is to influence the government's position on environment and conservation issues and possibly therefore influence the lives of many other people. The choice of an alter-

---

3 Sierra Club Environmental Survey, *Sierra Club Bulletin* (March/April 1979).

4 Maurice Strong, first executive Director of the United Nations Environment Program, now Chairman of Petro-Canada. From *Mazingira*, N. 3/4, 1977.

5 Greenpeace, letter requesting support for their activities, January 1980.

nate life-style, on the other hand, is an act that usually influences only one or a few lives. Although it does not touch the system, it has its special satisfactions. Civil disobedience, in contrast, threatens or blocks the system itself. It draws attention, but requires large sacrifices. Still another way to influence the environment is to choose a profession that is concerned with environmental improvement. These are the ways people have chosen to influence the environment:

> political action groups
> consumer boycotts
> alternate life-styles
> civil disobedience
> a conserving ethic.
> an environmental profession.

Environmental Decisions

How you decide to influence environmental progress is up to you. What is important to remember, however, is that decisions about the environment are being made right now. Decisions are being made by politicians, by administrators, by company executives, and by individual citizens. It is true that these decisions often require inputs from experts on matters such as the biology and technology of a situation, the effects of pollutants at various levels, the cost of cleaning up, and the people who gain or lose benefits when a substance or technology is changed or set aside.

Nevertheless, we must all decide the level of costs we are willing to bear, and who should bear them; what benefits we can give up, and what risks we are willing to endure. Experts can only lay out the choices; you must help make the decisions. No one is more qualified to decide social, moral, and economic issues—which come up again and again—than you, the individual citizen. This privilege and this burden are yours in a democratic society. We have exposed you to the clash of opinions over values, over risk, and over science to show you where and how your own opinions and views are needed and are valuable in the environmental debate. And we have tried to give you as much information as possible about present-day environmental problems. In the future, however, you will have to face new problems, problems now only dimly seen or, perhaps, problems not yet dreamed of even by the most far-seeing environmental expert. We can only leave you with a plea to enter the decision process because of the importance of your views. We ask that you look carefully at actions and solutions affecting the environment.

Glossary

acclimation the biochemical changes which enable an organism to withstand changed temperatures (either higher or lower).

acid mine drainage water which has dissolved iron pyrites (ferrous sulfide) which were left behind from coal mining operations. The water becomes acidic and will deposit ferric hydroxide (yellow boy) on stream bottoms. The acid makes the water undrinkable and is harmful to aquatic life. The acid waters will also be low in dissolved oxygen, a substance needed by aquatic life for survival.

acid rain the rainfall downwind of major fuel burning areas has been observed to be acidic. Sulfur oxides from the burning of fossil fuels are the culprit. Acid rain may stunt the growth of plants and turn lakes and streams acidic, driving out the normal aquatic species.

activated sludge plant (a secondary treatment process) a device for removing dissolved organics from waste water. The plant is a well aerated tank in which microbes using oxygen convert the dissolved organics to simpler substances.

algae simple, often microscopic, plants which live in water or very moist land environments.

amino acids chemical compounds from which proteins are made.

anaerobic in the context of water pollution, a condition of water in which all the dissolved oxygen has been used up or removed. Only a few specially adapted species can survive in anaerobic (oxygen depleted) waters.

arable able to be farmed.

area mining of coal in this surface mining method, the overburden is removed and laid up in successive parallel rows; the technique is used on relatively flat lands.

ballast ships need weight in certain areas of their cargo space to make them stable and easy to steer. Oil tankers use as ballast either the oil they carry or seawater.

barrel of oil one barrel of oil equals about 43 gallons. A ton of oil varies in volume according to what kind of oil it is (e.g. gasolines are lighter than diesel oils) but as a general statement one ton of oil is about seven barrels or 300 gallons.

biochemical oxygen demand (BOD) the amount of oxygen that would be consumed if all the organics in one liter of polluted water were oxidized by bacteria and protozoa; it is reported in milligrams per liter. The number is useful in predicting how low the levels of oxygen in a stream or river may be forced to go when organic wastes are oxidized by species in the stream.

biodegradable able to be broken down by living organisms.

biological controls control methods which use natural predators, parasites or diseases to control pests or which rely on the use of naturally produced chemicals such as insect pheromones.

biological magnification the process by which certain, often toxic, materials become more concentrated as they move up food chains. That is, organisms at the top of the food chain contain more of the substance than do organisms on the bottom of the food chain, or than does the environment itself.

biomass the weight of the living creatures in a given area.

737

biomass energy   energy to be derived from crops (trees, sugar cane, corn, etc.) by either direct burning or by conversion to an intermediate fuel such as alcohol.

biomes   climax communities characteristic of given regions of the world.

birthrate   the number of babies born each year per thousand people in a population is the birthrate for that population.

bituminous coal   the most plentiful form of coal in the U.S. Its high-heating value and abundance also make it the most widely used coal in the U.S. Its principal use (about 2/3) is in steam electric power plants.

black lung disease   the inhalation of coal dust over a relatively long period and the implanting of the particles in the lung leads to a condition in which the elasticity of the lung is destroyed. Breathing becomes severely labored for the coal miner who falls victim to black lung disease, and many coal miners are permanently disabled by the disease.

BLM   *see* Bureau of Land Management.

bloom   a rapid overgrowth of the water plants called algae.

blowdown   the water removed from that circulating in a cooling tower to prevent solids from building up in the tower water. This water is replaced by fresh water called "make-up."

blowout   explosive release of gas and/or oil from an oil well.

BOD   *see* biochemical oxygen demand.

boiling water reactor (BWR)   the water passing through the core of this nuclear reactor is heated by the fission process and is converted directly to steam which turns a turbine. The BWR is one of the two major types of nuclear reactors used for electric production. See Pressurized Water Reactor for contrast.

Btu   the British thermal unit, the quantity of heat needed to raise the temperature of one pound of water one degree Fahrenheit. One Btu is equivalent to 1.054 kilojoules (a common metric energy unit).

bulb turbine   a French invention, this is a turbine whose blade face is oriented perpendicular to the direction of water flow. In contrast to the Peyton, Francis and Kaplan wheels which require water elevations of 100 feet or more, the bulb turbine can generate electricity from only rapidly moving water. It thereby makes possible both tidal power and hydro development at many sites not previously feasible.

Bureau of Land Management   the managing agency of the National Resource Lands. The Bureau is within the Department of Interior.

BWR   *see* Boiling Water Reactor.

Calorie   a measure of the heat or energy content in food. Human food requirements are usually measured in kilocalories (the amount of heat needed to raise the temperature of 1000 grams of water by one degree centigrade) and the term is written with a capital c, "Calories."

carbon cycle   carbon dioxide is produced when plants and animals respire and it is consumed by green plants during photosynthesis. Carbon dioxide also dissolves in the oceans and precipitates as limestone or dolomite.

carbon monoxide   the principal source of this major air pollutant from the incomplete combustion of carbon is the automobile. The carbon monoxide molecule is one atom of carbon bonded to one of oxygen. Its action on human health is a result of its "tying up" of hemoglobin, the protein which carries oxygen to the cells.

carcinogen   something that causes cancer.

carnivores   meat-eaters.

catalytic converter   the device used on automobile exhaust gases to combust carbon monoxide (to carbon dioxide) and hydrocarbons (to carbon dioxide and water). An additional converter may be used to convert oxides of nitrogen back to oxygen and nitrogen.

chlorinated hydrocarbons   chemicals composed mainly of carbon and hydrogen plus one or more atoms of chlorine. Examples are the pesticides DDT, aldrin, dieldrin, chlordane and heptachlor.

chlorinated organics   chemical compounds composed of carbon, hydrogen and oxygen and one or more atoms of chlorine. These chemicals can be formed in drinking water by the action of the disinfectant chlorine on organic chemicals found in some water supplies.

chlorination   the controlled addition of chlorine to water destined for drinking and to waste waters being discharged into receiving bodies. Chlorine kills bacteria and viruses and so renders the water safe for human consumption and use. The process is also known as disinfection.

cholera   an intestinal disease caused by specific bacteria. The disease can be spread by water polluted by human wastes.

clear cutting the practice of cutting all trees in an area, regardless of size, quality or age. The practice hastens erosion, is visually displeasing, and leads to a loss of species habitat. Better practices are *selection cutting*, *seed-tree cutting* or *shelterwood cutting*.

climate a complex of factors affecting the environment. Climate includes: temperature, humidity, amount of precipitation and rate of evaporation, amount of sunlight and winds.

climax community the characteristic and relatively stable community for a particular area.

coagulation a process for the removal of suspended material from drinking water. A "floc" of insoluble material is created by the addition of alum or ferrous sulfate. When the floc settles in a detention basin, suspended material is captured in the floc and is settled out as well.

coal cleaning the removal of sulfur from coal by washing and chemical steps. Cleaned coal is less likely to require "scrubbers" to remove sulfur oxides from the stack gases.

coal gasification the conversion of the carbon in coal to a gas which can burn. Coal is burned in the presence of oxygen and steam to produce carbon monoxide and hydrogen gases. This low-Btu gas can be burned directly or can be converted to a high quality methane by the addition of hydrogen in further reactions. The latter product can be substituted for natural gas.

coal liquefaction the conversion of coal to a hydrocarbon liquid. Pyrolysis is one methodology for this conversion. A modification of coal gasification will also produce a hydrocarbon liquid. Solvent refining of coal is still a third possible process.

coal slurry pipeline a technology with wide potential in which coal is transported via pipeline as a mixture (about 50/50) of coal particles and water.

coliform bacteria these are the "indicator organisms;" their presence in a sample of water is taken to mean that the water is contaminated by human wastes and may contain organisms which cause disease. The coliforms are a group of bacteria found normally in the human intestine, but the bacteria are rarely themselves the cause of diseases.

combined cycle power plant in this concept, oil or gas is first burned in a turbine (jet) engine, generating turning power in the engine for electric generation. The hot gases are then used to boil water to steam to turn a conventional turbine for power generation. The efficiency of conversion of heat to electricity is increased by this two-stage process.

combined sewers most American cities have one sewer system that carries both domestic waste water and the water from rain storms. During storms, the total of the flows is too large to be treated and hence enters the water body virtually without treatment.

community all of the living creatures, plant and animal, interacting in a particular environment.

condense to change from a gas to a liquid, as when steam is condensed to water in a power plant condenser.

consumers organisms who eat other organisms. Primary consumers eat producers, secondary consumers eat primary consumers and so on.

contour mining of coal in this surface mining method, L-shaped cuts are made into the hillside in long curving arcs which follow the contour of the hill.

contraceptive a device, chemical or action which prevents pregnancy.

cooling tower a structure designed to cool the water that was used to condense steam at the power plant.

core (of a nuclear reactor) the concrete and metal shielded structure which houses the nuclear fuel in a reactor; the place where the fission process and heat production is occurring.

crude oil oil as it comes from the ground, in its natural state. Crude oil is a mixture of many chemicals compounds.

DDT a member of the chlorinated hydrocarbon family of pesticides, DDT was banned in the U.S. in 1972 because it was interfering with reproduction in certain bird species.

death rate the number of people who die each year per thousand people in a population is the death rate for that population.

deciduous forest biome most of the eastern U.S. is part of this biome, characterized by trees that lose their leaves each fall.

demographic transition pattern of change in which birth rates fall as, or after, death rates fall. After the transition a country's birth rate is closer to its death rate and the population does not grow rapidly.

detritus   dead organic matter, composed of plant and animal remains.

developing nation   a term applied to countries which have little or no technological development.

development rights   the rights, most often accompanying ownership of the land, that enable the owner to construct buildings, build roads and sewers, and otherwise alter the land. Such rights can be sold or transferred to other parties, separating the rights from the land itself.

digestor   a water pollution control device which is used to further degrade and stabilize the organic solids that arise from primary and secondary waste water treatment.

dissolved oxygen (DO)   the amount of oxygen dissolved in water, reported in milligrams per liter. Levels of 5 milligrams per liter or above indicate a relatively healthy stream. The maximum level of dissolved oxygen is ordinarily 8 to 9 milligrams per liter, depending on the water temperature.

district heating   the spent steam from a steam-electric power plant can be carried in underground pipes to homes, offices, factories, etc. for heating during the winter months.

diversity   a measure of the number of species in a given area. The more species there are per square meter, the higher the diversity.

DO   *see* dissolved oxygen.

drift   particles of liquid water which escape from cooling towers and cause corrosion or other environmental problems.

ecosystem   all of the living organisms in a particular environment plus the nonliving factors in that environment. The nonliving factors include such things as soil type, rainfall and the amount of sunlight.

electric power plant (thermal)   the heat (hence the word "thermal") from a burning fossil fuel or from the fissioning of uranium is used to boil water to steam and the steam is used to turn a turbine, generating electricity.

electrostatic precipitator   a device which removes particulate matter from stack gases; the particles are given an electric charge and then attracted to a collecting electrode. The precipitator is very efficient and is widely used on electric power plants.

emergency core cooling system   a spray system designed to inject water rapidly into a core of a nuclear reactor which has experienced a loss of coolant accident.

endangered species   a species which has so few living members that it will soon become extinct unless measures are begun to slow its loss.

enrichment   the process of converting uranium from 0.7% uranium-235 to 3% uranium-235, the higher concentration being needed for the fuel rods of the reactor of a nuclear electric power plant.

entrainment   organisms which are caught in the condenser water pipes and drawn into a power plant are said to be entrained.

EPA   Environmental Protection Agency.

epidemiologist   a scientist who studies how diseases affect whole groups of people.

erosion   the loss of soil due to wind or as a result of washing away by water.

estuary   a coastal body of water, partly surrounded by land, but having a free connection with the ocean.

eutrophic   water which has a high concentration of plant nutrients.

FDA   Food and Drug Administration.

filtration   in the context of water supply, this is the practice of forcing the water through beds of sand to trap the bacteria which cause disease and, thereby, preventing their presence in drinking water.

first law of thermodynamics   energy can be changed from one form to another, but it cannot be created and it cannot be destroyed.

fission   when the nucleus of certain heavy atoms is struck by neutrons, it breaks apart into two or more fragments (fission products) with the production of heat and more neutrons; fission means "breaking apart."

food chain   a picture of the relationship between the predators in an area and their prey, i.e., who is eating whom. When the relationships are simple and few creatures are involved it is a food chain.

food web   in most environments food chains are connected into complicated food webs which are made up of many organisms, with many interrelationships.

fossil fuels   fuels like coal, oil and natural gas derived from the remains of organic matter deposited long ago.

fuel cell   a device in which oxygen combines with hydrogen or carbon monoxide producing direct current electricity. Fuel cells are expected to be relatively clean producers of electricity unless coal is burned to provide the carbon monoxide.

fuel reprocessing   spent nuclear fuel rods are broken apart. Then the highly radioactive fission products are separated and uranium and plutonium are recovered to be re-used. No reprocessing plant operated in the United States through most of the 1970s, although reprocessing is taking place in Europe.

fusion   the combination of the deuterium atom (a form of hydrogen) with either another deuterium or a tritium atom (still another hydrogen form). The combination releases enormous heat which scientists hope some day to capture to produce electrical energy. Fusion is the basis also for the hydrogen bomb.

geopressured methane   natural gas known to be dissolved in salt water in deep caverns at extremely high pressures and temperatures. It is not known if geopressured methane can be recovered profitably.

geothermal heat   heat from the earth's interior carried to the surface as hot water or steam. The steam and hot water can be used to heat homes, offices, and factories or can be used to generate electricity.

geothermal power plant   an electric generating station which uses hot water or steam from the earth's interior as the energy source. Releases of steam from the earth can be used directly to turn a turbine. Alternatively, hot water releases can be "flashed" to steam for such uses. Or hot water can be used to vaporize a fluid such as isobutane which will then be used to turn a turbine.

greenhouse effect   the effect of carbon dioxide in the atmosphere is compared to that of the panes of glass in a greenhouse. Sunlight passes the glass into the greenhouse, without effect, but heat radiated outward from inside the greenhouse is blocked. Sunlight reaches the earth without effect, but outward heat radiation is blocked by carbon dioxide molecules which absorb the energy. The greenhouse effect suggests the earth will undergo a warming trend as carbon dioxide from fossil fuel combustion accumulates in the atmosphere.

green revolution   the term given to the new developments in farming, including the use of high-yielding grains, which promise to enable farmers to grow much more food on the same number of acres than with conventional techniques and older crop varieties.

growth promoter   a substance which makes an animal grow better or more quickly.

habitat   the physical surroundings in which an organism lives.

half-life   the time required for one-half of a given quantity of a chemical to disappear from the environment (or to be excreted by the body, if it is a chemical absorbed by a living organism).

heat pump   the reverse of a refrigerator or air conditioner; the heat pump in the heating mode draws in cold air and exhausts this air at an even colder temperature. The heat captured is transferred via a refrigerant liquid to the indoor air. The heat pump has the potential to cut in half electric requirements for home or hot water heating.

hepatitis   a disease caused by a virus, in which the liver becomes inflamed. Epidemics of hepatitis have been traced to contaminated water supplies.

herbicide   chemical used to kill weeds.

herbivores   plant-eaters.

highwall   the wall of overburden and coal left behind after contour mining of hillsides.

high-yielding grains   new varieties of corn, wheat and rice developed by agricultural research. These varieties produce much more grain per acre than older varieties. They also require more water, pesticides and fertilizers.

hydrocarbons   a class of air pollutant, principally from the operation of internal combustion engines of motor vehicles; the pollutants contribute to photochemical pollution by the reactions they undergo.

hydroelectric energy   electric energy derived from falling or moving water. The water is commonly stored behind a dam and is released through penstocks to turn a turbine and generate electricity.

hydrogen economy   an energy system which uses hydrogen to store and produce energy. Hydrogen from chemical processing of fossil fuels or from the electolysis of sea water would be used in fuel cells. The cells would generate electricity by combining the hydrogen with oxygen or carbon monoxide.

hydrologic cycle   the cycling of water in the environment, from rainfall to runoff, to evaporation and back again.

indicated and inferred reserves (of oil or gas)   oil or gas known to exist and likely to be recoverable with the application of additional technology (indicated) and oil or gas for which we already have some limited evidence of existence (inferred). (See also proved reserves.)

inversion a weather phenomenon in which cold air lies close to the earth's surface, trapped by a warm air mass above it. This is the inverse of the normal situation in which temperature decreases with increasing distance from earth out into space.

IPM integrated pest management, a combination of techniques which is designed to control pests using a very minimum of chemical sprays.

irrigation scheme to supply water, other than natural rainfall, to farmland.

kwashiorkor a children's disease caused by too little protein.

land application of wastewater a set of three different processes in which wastewater is applied to the land as a means of treatment. The wastewater should previously have undergone primary treatment. The processes are overland flow, spray irrigation and infiltration-percolation.

leaching field a system of underground tiles designed to channel the flow from a septic tank into porous soil.

lead lead is a cumulative poison which can cause brain damage and death; it is present in the air (from the lead in gasoline) and it is in food and in water. It was once in paint as well, and the eating of paint chips by children is a frequent cause of lead poisoning.

lignite a lower form of coal with a little over half the heating value of bituminous or anthracite coal; it constitutes about four percent of the U.S. coal energy resource.

limiting factor whatever nutrient is in shortest supply compared to the amount needed for growth. This factor then limits the growth of plants in a particular environment.

liquefied natural gas (LNG) natural gas made liquid at very low temperatures ($-162\,°C$) in order to transport it economically via special tankers.

liquid metal fast breeder reactor (LMFBR) in this advanced concept, a liquid metal (sodium) circulates through the core of the reactor, removing the heat from the fission of plutonium. The heat in the liquid metal is eventually used to boil water to steam to turn a turbine. The LMFBR fuel is plutonium-239 created by neutrons striking the nucleus of uranium-238. The scheme is designed to use uranium-238 rather than uranium-235 because uranium-238 is so much more abundant.

LMFBR *see* liquid metal fast breeder reactor.

longwall mining imported from Europe, this is a relatively new method of mining coal underground. In this method, the mine roof is allowed to collapse in a controlled way as the mine "room" moves across the coal seam. Greater coal removal and prevention of acid mine drainage are claimed advantages.

loss of coolant accident (LOCA) if the water in the core of a nuclear reactor is lost because of a pipe break, the temperature in the core would rise rapidly as the heat from fission builds up. The fuel rods could melt and radioactive substances would be released through the pipe break if the core is not cooled quickly by the emergency core cooling system.

low-head hydropower in contrast to large dams and their hydropower installations, low-head hydroelectric energy can be generated using water elevations of fifty feet or less. The bulb turbine, a new technology, makes possible relatively efficient capture of energy from low-head dam sites not currently in use for electric generation.

magneto hydrodynamics (MHD) in this concept, the hot gases from the burning of a fossil fuel are seeded with potassium which then ionizes. Electric current is extracted from the hot gases which are then used to boil water to steam for conventional electric generation.

malnourished someone who does not get enough of the various nutrients needed for good health is malnourished.

marine having to do with the oceans or salt waters of the earth.

megawatt a term describing the rate at which electricity can be generated by a power plant. One megawatt is 1000 kilowatts.

methemoglobin a form of hemoglobin in which the iron is oxidized. Methemoglobin is not able to carry oxygen in the blood stream as hemoglobin can.

metric ton (or long ton) 1000 kilograms or about 2,200 pounds, 10% larger than the (short) ton of English measure.

MHD *see* magnetohydrodynamics.

middle distillates that portion of crude oil which is refined to diesel fuel, kerosene (jet fuel), and home heating oil.

multiple use the concept applied to the management of lands in the National Forest. The concept allows timbering, mining, recreation, grazing, watershed protection, fishing, wildlife protection and wilderness as legitimate uses of the same land. The official designation of wilderness, however, most often excludes timbering, mining and grazing.

mutagen   substance that causes an inheritable change in a cell's genetic material.

National Park Service   the managing agency of the National Park System. The National Park Service is within the Department of Interior.

natural gas   derived from chemical and physical processes operating on buried ocean plankton, natural gas consists mainly of methane, a simple hydrocarbon gas. Nitrogen may be present in the gas as well.

NEPA   National Environmental Policy Act. One of the most important parts of this act is that it requires environmental impact statements. These statements are reports based on studies of how a proposed government project will affect the environment.

niche   where an organism lives and how it functions in this environment (i.e., what it eats, who its predators are, what activities it carries out).

NIOSH   National Institute for Occupational Safety and Health.

nitrogen oxides (NO$_x$)   a contributer to photochemical air pollution, nitrogen oxides are produced during high temperature combustion of fossil fuels. Oxygen and nitrogen *from the air* produce the pollutant gas. The oxides of nitrogen (nitric oxide and nitrogen dioxide) are both air pollutants, and nitrogen dioxide has been linked to an increase in respiratory illnesses.

non-point-source water pollution   polluted water arising typically from rural areas and which enters a receiving body from many small widely scattered sources.

nucleic acids   chemical compounds from which important biological materials (such as the hereditary materials DNA and RNA), are made.

ocean thermal energy conversion (OTEC)   a concept now being tested on a pilot scale to use the temperature difference between the warm surface waters of the ocean and the cold deeper waters to produce electrical energy. The warmer waters would boil a working fluid to a "steam" which would turn a turbine. The colder water would condense the vapor for reuse.

oil   derived from chemical and physical processes operating on buried ocean plankton, oil consists mainly of liquid hydrocarbons (compounds of hydrogen and carbon). Nitrogen and sulfur are other elements which may be present.

oil shale   shale rock containing oil; *see* shale oil.

oligotrophic   water which has low concentrations of plant nutrients.

OPEC   Organization of Petroleum Exporting Countries.

organic chemical   one composed mainly of carbon, hydrogen and oxygen.

organic wastes   a class of water pollutants composed of organic substances. When these substances are oxidized by bacteria and other species, oxygen is removed from the water. When high concentrations of organic wastes are present aquatic species may thus be deprived of the oxygen necessary for survival.

OSHA   Occupational Safety and Health Administration.

oxidant   a chemical compound which can oxidize substances that oxygen in the air cannot. Ozone, a prominent photochemical air pollutant is an oxidant, as are nitrogen dioxide, PAN compounds and aldehydes. Levels of photochemical pollution are often reported as oxidant levels.

ozonation   the process of treating water intended for drinking with ozone gas in order to kill microorganisms. Although practiced widely in Europe, most water engineers in the U.S. prefer chlorination because of the simple test for free chlorine.

ozone   a compound made up of three atoms of oxygen (O_3). The oxygen we need to breathe is O_2.

particulate matter   a class of air pollutants consisting of solid particles and liquid droplets of many different chemical types. Examples are fly ash, compounds from photochemical reactions, metal sufates, sulfuric acid droplets, lead oxides. Particles will soil and corrode materials; they may also soil and react chemically with the leaves of plants. Particles are linked firmly to increases in human respiratory illnesses, and some particles types are suspected of causing human cancer.

pathogens   disease producing micro-organisms. These include bacteria, viruses and protozoa. In the context of drinking water, cholera and typhoid are two diseases spread by specific bacterial pathogens.

PCBs   polychlorinated biphenyls; a family of chemicals similar in structure to the pesticide DDT and having a variable number of chlorine atoms attached to a double ring structure.

peak-load pricing   the practice of raising the price of electricity during the hours (or season) of peak demand and lowering the price during times of slack demand. The goal is to level the rate of electric usage through time and decrease the need for bringing into service new electrical generating capacity to meet peak demands.

peat a low heating value fossil fuel derived from wood which decayed while immersed in water.

permafrost frozen layer of soil which underlies the arctic tundra biome.

persistence the length of time a pesticide remains in the soil or on crops after it is applied.

pesticide a substance which will kill pests such as insects or rats.

photochemical air pollution air pollutants such as nitrogen dioxide, ozone, aldehydes, PAN compounds, etc. produced as a consequence of sunlight stimulated reactions involving nitrogen oxides and hydrocarbons. The various compounds have differing effects, but plants are damaged by ozone and PAN compounds. Eye and throat irritation and respiratory illness are common effects of these substances on people.

photovoltaic conversion silicon cells are used to convert sunlight directly to electricity. Although costs of cells are high, attempts to make photovoltaic conversion economical are underway.

phthalates chemical compounds which are esters of phthalic acid and various alcohols for example, diethyl phthalate.

physical–chemical wastewater treatment a relatively new set of methods aimed at replacing conventional primary and secondary wastewater treatment processes. Physical–chemical treatment is designed to remove phosphorus by precipitation and settling and to remove dissolved organics by adsorption on carbon particles.

phytoplankton microscopic, floating plant species.

pica the habit of eating nonfood items; it occurs among about 50 percent of all children independent of social or economic class beginning at about one year of age. Such items as paper, string, dirt, paint chips, etc. may be eaten. Lead poisoning may occur if flakes or chips of leaded paint are eaten by such children.

plankton microscopic plants and animals which float in water, mostly at the mercy of currents and tides.

point-source water pollution those water pollutants which arise in urban areas or from industries and which enter a receiving body from a single pipe.

population (natural) the members of a species living together in a particular locality.

population profile a bar graph showing the number of people in each age group in a population.

power tower a method (in the testing stage) to generate central station electricity using the sun as the energy source. The sun's rays are focussed by banks of mirrors on a tower to provide the heat to boil water to steam. The steam turns a turbine for conventional electric generation.

predator a creature which eats another.

pressurized water reactor (PWR) the water passing through the core of this nuclear reactor is heated by the fission process but does not boil because it is under great pressure. This extremely hot water is used to boil to steam a parallel but separate stream of water in a steam generator. The steam in this second loop is used to turn a turbine for electric generation. *See* boiling water reactor for contrast.

prey a creature who is eaten by another.

primary recovery the quantity of oil that flows out of a well by natural pressure alone; it averages about 20% of the oil in place.

primary treatment of wastewater the first (in sequence) major wastewater treatment process in the typical set of processes at American cities. Primary treatment consists of allowing the organic particles which were in suspension to settle out of the flowing water.

producers organisms who produce organic materials by photosynthesis.

productivity amount of living tissue (plant or animal) produced by a population in a given period of time.

promoter a substance which does not cause cancer itself but which can act to cause another material to be carcinogenic.

proved reserves (of oil or gas) oil or gas *known* to be contained in the portions of fields that have already been drilled and which is profitable to recover. Said another way, it is oil or gas we know we have.

pumped storage at times when spare electric capacity is available, water may be pumped to a high elevation. The water can then be released from its high pool through the turbine pump which raised it in order to generate electricity. Such releases are made when the existing basic generating capacity is insufficient to meet electrical demand.

PWR   *See* pressurized water reactor.

quad   one quadrillion (a million billion) Btu. The annual energy use of an entire nation is often given in quads. The U.S. consumed 77 quads of energy in 1976.

rate of growth   a country's growth rate is the rate of increase (birthrate − deathrate) plus the rates of change due to immigration (people arriving) and emigration (people leaving).

rate of increase   the birthrate minus the death rate of a population is its rate of increase.

reclamation of surface-mined land   in the mining of coal, this step is the replacement of overburden and topsoil and revegetation of the land.

refining (oil)   crude oil before it is used is refined, or separated into less complex mixtures of substances, i.e., gasolines, kerosenes, heating oils, waxes, tars, asphalts.

refuse banks (or "gob" piles)   the wastes from a coal preparation plant which washes and grades coal. The wastes include low grade coal, shale, slate, and coal dust.

residue   the amount of a chemical left in a food by the time it reaches the consumer.

resistance   when a particular germ or pest is no longer killed by a drug or pesticide, that organism is said to have developed resistance.

room and pillar coal mining   in this method of underground coal mining, pillars are left in the mine in a regular pattern in order to support the mine roof while the coal mining is going on.

runoff   water which comes to the earth as rain and runs off the land into lakes, rivers and the oceans.

salinity   salt content.

sand filtration   *see* filtration.

satellite solar power station   a concept in which photovoltaic cells are installed in a satellite orbiting earth. Electricity generated by the cells would be beamed to earth via microwave. The system would be enormously costly if ever undertaken.

scrubbers   a class of devices (differing widely in their chemical process steps) which remove sulfur oxides from the stack gases of power plants that burn coal.

secondary recovery   technique such as gas injection or water injection to place increased pressure on oil in a deep reservoir and thus force more of it to the surface.

secondary treatment of wastewater   the second (in sequence) major wastewater treatment process in the typical set of processes at American cities. Two processes, the trickling filter and the activated sludge plant, are used to remove dissolved organic material from wastewater.

second law of thermodynamics   when energy is changed from one form into another, some energy is always unavailable or lost as heat. Another way of stating this explains the inefficiency of electric power generation, namely, that a natural limit exists on the extent to which heat can be converted to work and hence to electricity. The limit in the conversion of heat to work is about 45%–50% conversion.

sediment   the fine particles of soil washed off land by the erosion of water; these are the particles which become suspended in flowing water and ultimately settle to stream or lake bottoms or which form river deltas.

seed-tree cutting   a logging practice which leaves behind trees to seed the area for regrowth.

selection cutting   cutting only mature timber from an area.

septic tank   a device used in rural areas for partial treatment of wastewater. The device is a concrete or metal tank which detains wastes from a home for up to 3 days, providing settling of solids and partial treatment of dissolved organics.

shale oil   a hydrocarbon liquid derived by retorting (cooking) oil-bearing rock which has been crushed. The hydrocarbon liquid is known as kerogen. The U.S. may have the equivalent of 600 billion barrels of oil in the shale rock of Colorado and neighboring states. The economics of shale oil recovery have prevented previous development of this resource. Potential environmental impacts of shale mining and oil production are thought to be very large.

shelterwood cutting   a three stage cutting plan in which (1) defective trees are first removed, (2) the stand is "opened" by further cutting 10–15 years later, and (3) the mature trees are cut after seedlings are well-established.

siltation   the dropping or settling of sediments to the bottom of a body of water. Water brought to a halt behind a dam will drop much of its load of sediment into the reservoir; the effectiveness of the reservoir for water storage is thus decreased.

solvent refined coal   *see* coal liquefaction.

species   a species is made up of all those organisms who are able to breed successfully (if they are given the opportunity to do so), who share ties of common parentage and who share a common pool of hereditary material.

stratosphere   the layer of the earth's atmosphere directly above the troposhere which is the lowest layer. Scientists are concerned because ozone levels in the stratosphere are declining and carbon dioxide levels in the stratosphere are increasing.

subsidence   as a result of the empty spaces left underground by coal mining, the surface of the land may collapse. Roads may buckle and sewer lines and gas mains may crack as a result of severe subsidence if it occurs in developed areas.

succession   a natural process in which the species found in a given area change conditions to make the area less suitable for themselves and more suitable for other species. This continues until the climax vegetation for the area grows up.

sulfur oxides   a class of air pollutants from the burning of fossil fuels (mainly coal and oil) which contain sulfur. The sulfur is oxidized to sulfur dioxide and sulfur trioxide when the fuel is burned. The oxides of sulfur and the acids they form with water vapor will damage building materials like marble, mortar and metals and they damage plants and the health of people. Respiratory illnesses are increased during times when the levels of sulfur oxides are elevated.

surface mining (or strip mining)   the practice of removing coal by excavation of the surface *without* an underground mine. Area strip mining is practiced on flat lands. Contour strip mining is utilized on hillsides.

sustained yield   the sustained timber yeld in an area is a quantity of timber less than or equal to the natural increase in that area ocurring over the past time period. Increase is measured in volume units, such as cubic feet.

synthetic pesticide   one which was invented in a laboratory and which is not produced by any natural system.

taiga   northern coniferous forest biome, characterized by spruce and other coniferous trees.

tailings   when a substance, for instance, uranium or iron, is mined, it is usually found mixed with various other materials. After the desired substance is extracted from this ore, the remaining waste material along with any other unwanted mining waste is called tailings.

tar sands   a sand coated with bitumen, an oily black hydrocarbon liquid. Extensive tar sand deposits exist in Alberta, Canada, and smaller deposits are in Utah. In the Alberta sands, about two barrels of hydrocarbon liquid can be recovered from three tons of sand; the liquid can be refined to all grades of petroleum.

teratogen   substance that causes a defect during development before birth.

tertiary treatment of wastewater   the third major component of wastewater treatment plants. In tertiary treatment, which consists of a number of processes, nitrogen in its various chemical forms is removed; phosphorous is removed by precipitation and settling; and resistant organics are removed by passage through towers containing activated carbon.

thermal pollution   pollution of the environment with heat.

threatened species   one which is not yet endangered but whose populations are heading in that direction.

tidal power   the capturing of the energy of the tides as electrical energy by the use of dams and a turbine/generator unit. Both the receding and arriving tides can be channelled through the bulb turbine to generate electricity. La Rance power plant in France is the largest tidal power installation in the world.

tolerance   · the amount or residue of a pesticide or drug legally allowed in food.

topsoil   the top few inches of soil, which are rich in organic matter and plant nutrients.

TOSCA   Toxic Substances Control Act.

toxic substance   a material harmful to life.

toxin   a naturally produced poisonous material secreted by certain organisms.

trickling filter (a secondary treatment process)   a device for removing dissolved organics from wastewater. In the device, the wastewater is distributed across a bed of stones on which a microbial slime is growing. The microbes remove the dissolved organics by converting them to simpler substances.

trophic structure   how the various organisms in a community obtain their nourishment.

tundra   arctic biome characterized by a permanently frozen subsoil and low growing plants such as mosses and lichens.

turbidity   a measure of how clear water is, turbidity

depends on the amount of suspended solid materials or organisms in the water.

typhoid an intestinal disease caused by specific bacteria. The disease can be spread by water polluted by human wastes.

ultraviolet light a part of the electromagnetic spectrum, ultraviolet or UV light waves are shorter than visible violet light rays but longer than x rays. These light rays can contribute to the development of skin cancer but are normally absorbed in the upper atmosphere by ozone.

undernourished a person who does not get enough Calories to maintain body weight with normal activity is undernourished.

undeveloped nation a country with little technological development is said to be undeveloped.

undiscovered resources (of oil and gas) oil or gas not yet discovered by drilling, but which on the basis of geological and statistical evidence is expected to be found eventually.

upwelling upwelling is a result of offshore winds that "push" surface waters away from shore and allow nutrient rich bottom waters to rise from the deeper oceans.

U.S. Fish and Wildlife Service the managing agency of the National Wildlife Refuge System. The Service is within the Department of Interior.

U.S. Forest Service the managing agency of the National Forests. The Forest Service is in the Department of Agriculture.

waste stabilization lagoon a large shallow pond into which wastewater is discharged for biological treatment. Solids settle out and dissolved organics are removed by microbes in the pond. The device is used primarily where waste loads are small and land with no conflicting uses is available inexpensively.

watershed the land area which drains into a particular river, lake or reservoir.

windmill or wind turbine a machine whose blades are rotated by the wind and which converts the wind energy into electricity or work (for example, pumping water).

zoning the practice in which local governments designate the allowable uses of a tract or tracts of land as a means to prevent incompatible land uses. Typical zoning classifications are single family residential, multi-family residential, commercial, industrial.

zooplankton microscopic forms of animal life which float about in water, moving mainly in whatever currents there are.

Index